P9-BZJ-899

Inward Bound

Inward Bound

Of Matter and Forces in the Physical World

ABRAHAM PAIS

CLARENDON PRESS · OXFORD
OXFORD UNIVERSITY PRESS · NEW YORK
1986

Oxford University Press, Walton Street, Oxford OX2 6DP
New York Toronto
Delhi Bombay Calcutta Madras Karachi
Kuala Lumpur Singapore Hong Kong Tokyo
Nairobi Dar es Salaam Cape Town
Melbourne Auckland
and associated companies in
Beirut Berlin Ibadan Nicosia

Oxford is a trade mark of Oxford University Press

Published in the United States
by Oxford University Press, New York

British Library Cataloguing in Publication Data
Pais, Abraham
Inward bound: of matter and forces in the physical world.
1. Physics—History—20th century
I. Title
530'.09'034 QC7

ISBN 0-19-851971-0

Library of Congress Cataloging in Publication Data
Pais, Abraham, 1918–
Inward Bound.
Bibliography: p.
Includes Index.
1. Physics. 2. Physics—History. I. Title.
QC21.2.P35 1986 530 85-21587
ISBN 0-19-851971-0

Printed in the United States of America
on acid free paper

10 9 8 7 6 5 4 3 2 1

Preface

This book deals with the times in which man has finally seen what Newton in his *Opticks* foresaw: 'The Attractions of Gravity, Magnetism and Electricity, reach to very sensible distances, and so have been observed by vulgar Eyes, and there may be others which reach to so small distances as hitherto escape Observation'. Between 1895 and 1983, the period spanned in this book, the smallest distances explored have shrunk a hundred millionfold. Along this incompletely traveled road inward man has established markers that later generations will rank among the principal monuments of the twentieth century.

I will endeavor to describe what has been discovered and understood about the constituents of matter, the laws to which they are subject, and the forces that act on them. On these topics Ernest Rutherford, a central figure in this book, once said that the rapidity of advance during the years 1895–1915 has seldom, if ever, been equalled in the history of science. The same is true of the longer period treated here. The advances have by no means come smoothly, however. I will attempt to convey that these have been times of progress and stagnation, of order and chaos, of clarity and confusion, of belief and incredulity, of the conventional and the bizarre; also of revolutionaries and conservatives, of science by individuals and by consortia, of little gadgets and big machines, and of modest funds and big moneys.

Progress has been made possible by the hard labors of many. At the time of writing some ten thousand men and women are devoting their lives to this adventure, now called high energy physics or particle physics. I wish I could, but shall not, name them all. Instead, I have dedicated this book to them.

This book has been long in coming. Fragments, here rewritten, enlarged, and placed in broader context, have appeared through the years: on quantum field theory (in 1948), invariance principles (1963), the status of particle physics (1968), the early history of the electron (1972), the early years of radioactivity (1977), quantum mechanics (1982), and the positron theory (to appear in Heisenberg's collected papers). I have also borrowed from my Einstein biography, *Subtle is the Lord*, in particular in regard to the light-quantum and to relativity, one of the two disciplines on which, since 1926, particle physics is based. The other, quantum mechanics, does not, in fact cannot, receive full treatment in the present enterprise. An essay on the subject is included, however.

The successful interplay between theory and experiment has been a main criterion in the selection of material included in this book, which opens with an introductory chapter that contains an outline of all that comes after. Then follows the main part, a history of the period 1895–1945. The later years are treated as a memoir. Reference to fundamental progress in the recent past needs to be included, even though the perspective is limited, since otherwise I would be telling a joke without giving the punch line. (A word to the wise: no cosmology, no topology, however.)

I have inserted occasional remarks on prominent personalities of earlier generations. (A list is found in Chapter 1.) These comments are meant as vignettes in the spirit of Aubrey's *Brief Lives* rather than as short biographies.

Each chapter has its own set of references. At the end of many chapters I have collected a list of sources that have been particularly helpful to me.

I have many friends to thank for their generous help. Luis Alvarez, Clifford Butler, Rod Cool, Gerson Goldhaber, Leon Lederman, Bill Kirk, Ed Lofgren, Pief Panofsky, Aihud Pevsner, George Rochester, Carlo Rubbia, and Nick Samios have given me counsel on experimental issues. Robley Evans and Lewis Thomas have helped me with Chapter 5. Gloria Lubkin has briefed me on the American Institute of Physics, Laurie Brown and Michiji Konuma on physics in Japan. I am greatly indebted to Howard Georgi, David Gross, Res Jost, and Arthur Wightman for wise comments. Most of all I am beholden to Sam Treiman who has read and discussed with me all of the several stages of the manuscript. His supportive criticisms, encouragement, and friendship have been invaluable. He has seen me through difficult moments. Thank you, all of you.

Finally I should like to express my gratitude to the Alfred P. Sloan Foundation for a grant in support of this project; and to the staff of Oxford University Press for their devoted support.

Abraham Pais
New York
July 1985

To those who built the machines, the beams, and the detectors,
and those who used them,
and those who reflected on their results,
this book is respectfully dedicated.

To Ida.

Contents

1 Purpose and plan 1

(a) From X to Z 1
(b) 1895–1945: a history 7
(c) The postwar years: a memoir 18

Part One. 1895–1945: A History 33

2 New kinds of rays 35

(a) Roentgen: X-rays 35
(b) Becquerel: uranic rays 42
1. January 20–February 24, 1896. 2. February 24–March 2, 1896.
3. The rest of the year

3 From uranic rays to radioactivity 52

(a) The Curies: Becquerel rays 52
(b) Rutherford: α- and β-rays 58

4 The first particle 67

(a) Of Geissler's pump and Geissler's tube and Rühmkorff's coil 67
(b) Faraday, Maxwell, and the atomicity of charge 70
(c) Of Rowland, Zeeman, and Lorentz 74
(d) The discovery of the electron 78
1. The plight of the precursors. 2. Wiechert. 3. Kaufmann.
4. J. J. Thomson
(e) β-rays are electrons 87
(f) Relativistic kinematics 87

5 Interlude: earliest physiological discoveries 93

6 Radioactivity's three early puzzles 103

(a) Introduction 103
(b) The first energy crisis 105
(c) Interlude: atomic energy 115
(d) Metabolous matter 117
(e) Why a half-life? 120
(f) Postscripts: modern times 124
1. 1928: α-decay explained. 2. Non-conservation of radioactivity.
3. The exponential law of radioactive decay

7 Pitfalls of simplicity 129

1. Simplicity as an unnecessary evil. 2. Ripeness of the times. 3. About old pros. 4. About Planck. 5. About Einstein. 6. On the variety of discovery. 7. Simplicity as a necessary evil

8 β-Spectra, 1907–1914 142

(a) Introduction 142
(b) α-decay and the conjecture of β-monochromacy 145
(c) Passage of electrons through matter: the position in 1907 148
(d) The first attack: absorption of β-rays 149
(e) 'A revolutionary conclusion' 152
(f) The second attack: magnetic separation, photographic detection 153
(g) Prenatal nuclear spectroscopy 155
(h) The third attack: magnetic separation, detection by counters 158

9 Atomic structure and spectral lines 163

(a) Introduction 163
(b) Pre-electron preludes 166
1. Spectral analysis. 2. The Balmer formula. 3. Pre-electron models. 4. The Darwinian touch
(c) Early electron models 178
1. In which atoms appear to be composed of electrons by the thousands. 2. A new pitfall: atomic stability and β-decay. 3. The great divide. 4. Pair models. 5. Planetary models. 6. J. J. Thomson, theorist
(d) Ernest Rutherford, theoretical physicist 188
(e) Niels Bohr 193
1. Four roads to the quantum theory. 2. The early papers. 3. Manchester. 4. Spectra, 1903–11. 5. Precursors. 6. Balmer decoded

10 'It was the epoch of belief, it was the epoch of incredulity' 208

1. Reactions to Bohr's theory. 2. Causality. 3. The fine structure constant. Selection rules. 4. Helium. 5. Changing of the guard. 6. A change of pace

11 Nuclear physics' tender age 221

(a) Introduction 221
(b) Niels Bohr on β-radioactivity 223
(c) Isotopes 224
(d) From A to Z 225
(e) A new pitfall: the first model of the nucleus 230
(f) Binding energy 232
(g) Physicists in the First World War 234
(h) Strong interactions: first glimpses 237

12 Quantum mechanics, an essay 244

(a) The status of physics in the spring of 1925 244
1. Two fields. 2. Three forces. 3. Three particles. 4. Two theoretical structures

(b) End of a revolution 250
(c) A chronology 252
(d) Quantum mechanics interpreted 255
(e) Changing of the guard 261

13 First encounters with symmetry and invariance 265

(a) Introduction 265
(b) The exclusion principle 267
(c) Spin 274
(d) Quantum statistics 280
(e) The Dirac equation 286
1. Young Dirac. 2. Relativity without spin; the scalar wave equation. 3. Spin without relativity; the Pauli matrices. 4. The equation

14 Nuclear physics: the age of paradox 296

(a) Quantum mechanics confronts the nucleus 296
(b) In which the proton–electron model of the nucleus runs into trouble 298
1. Nuclear size. 2. Nuclear magnetic moments. 3. Nuclear spin. 4. Nuclear statistics
(c) β-spectra 1914–30, or the life and times of Charles Drummond Ellis 303
(d) New physical laws or new elementary particles? Enter the neutrino 309
1. Yet another pitfall of simplicity: one cure for two ailments. 2. Bohr and the energy law, 1924. 3. Bohr and the energy law, 1929. 4. Pauli's new particle. 5. The years 1931–36, a partial chronology.

15 Quantum fields, or how particles are made and how they disappear 324

(a) The end of the game of marbles 324
(b) Preludes: Goettingen 331
(c) Foundations: Dirac 334
1. Formal introduction of photons. 2. The creation and annihilation of photons. 3. Scattering of photons; virtual states
(d) Quantum fields and quantum statistics 338
(e) Relativistic invariance and gauge invariance: beginnings 341
(f) The positron 346
(g) Appendix 352
Sources: quantum field theory prior to renormalization

16 Battling the infinite 360

(a) Introduction 360
(b) Physics in America: the onset of maturity; the emergence of Oppenheimer 364
(c) A prelude: infinities before the positron theory 370

(d) Quantum electrodynamics in the thirties 374
1. Second-order successes. 2. The Dirac field quantized. 3. Charge conjugation; Furry's theorem. 4. The polarization of the vacuum. 5. The self-energy of the electron, of the vacuum, and of the photon. 6. In which the Maxwell equations become nonlinear. 7. The Pauli–Weisskopf theory
(e) Scientific nostalgia: the search for alternatives 388

17 In which the nucleus acquires a new constituent, loses an old one, reveals new forces with new symmetries, and is explored by new experimental methods 397

(a) Enter the neutron 397
(b) Of the deuteron, of cosmic rays, and of accelerators 402
(c) What is a neutron? 409
(d) Nuclear forces: phenomenological beginnings 413
(e) Fermi's tentativo 417
1. In which Fermi introduces quantized spin-1/2 fields in particle physics. 2. De Broglie on antiparticles. Neutrino theory of light. 3. The tentativo. 4. Further developments in the thirties
(f) How charge independence led to isospin 423
(g) Quantum field theory encounters the nucleus. Mesons 426
1. In which the Fermi theory is taken too seriously and high energy neutrino physics makes a first brief appearance. 2. A meson proposed. Yukawa. 3. A meson discovered. 4. More on pre-war meson theory. The first prediction of a particle on symmetry grounds
(h) Death of Rutherford 436

Part Two. The Postwar Years: A Memoir 445

18 Of quantum electrodynamics' triumphs and limitations and of a new particle's sobering impact 447

(a) Shelter Island and other personal reminiscences 447
(b) Divine laughter: the muon 452
(c) Quantum electrodynamics: the great leap forward 455
1. From the week after Shelter Island to the month after Pocono. 2. About Tomonaga and his group. 3. 1949–84 in a nutshell

19 In which particle physics enters the era of big machines and big detectors and pion physics goes through ups and downs 471

(a) Of new accelerators and physics done by consortium 471
1. The synchrocyclotron (SC). 2. The weak focussing synchrotron. 3. Strong focussing; the alternating gradient synchrotron (AGS).
(b) Home-made pions 479

(c) In which meson field theories fall upon hard times 481
1. Nuclear forces. 2. The proton and neutron magnetic moments. 3. Meson dynamics and renormalizability. 4. Free meson processes
(d) Symmetry saves, up to a point 485
1. Isospin as a free-floating invariance. A new spectroscopy. 2. Antinucleons; nucleon conservation; charge and mass formulae; violation rules. 3. *G*-parity
(e) Of a new spectroscopy and new detectors 490
(f) What use quantum field theory? The years of uncertainty 492
1. The age of diversity. 2. Fermi's assessment in 1951. 3. The semi-classical nucleon. 4. Low energy theorems. 5. The *S*-matrix. Dispersion relations. 6. Regge poles

20 Onset of an era: new forms of matter appear, old symmetries crumble 511

(a) 'Four τ-mesons observed on Kilimanjaro' 511
(b) Early theoretical ideas 517
1. Associated production. 2. The strangeness scheme. 3. The neutral K-particle complex
(c) In which the invariance under reflections in space and conjugation of charge turns out to be violated 523
1. The Dalitz plot thickens. 2. *P*, *T*, and *C* prior to 1956. 3. Spin and statistics; the *CPT* theorem. 4. Toward the universality of weak interactions. Conservation of leptons. 5. Late 1956; *P* and *C* are violated
(d) Being selecta from the exploding weak interaction literature 533
1. The pioneering experiments. 2. *CP* invariance; K°–$\bar{K}^\circ$ revisited. 3. The two-component theory of the neutrino. 4. The universal *V–A* theory. 5. The interplay of weak and strong forces: local action of lepton pairs, conserved vector current, partially conserved axial current
(e) Unfinished story of a near miss: the violations of *CP*- and of time-reversal-invariance 538
(f) Final comments on discrete symmetries 541

21 Essay on modern times: 1960–83 550

(a) In which it is explained why a very rich period is treated with such brevity 550
(b) Higher symmetries take off 552
1. Further extensions of hadron spectroscopy. 2. SU(3). 3. Quarks. 4. SU(6). 5. Color: double SU(3) symmetry. 6. Quarks, the second route: current algebra. 7. Particle physics in the mid-sixties
(c) New tools—new physics 569
1. Reactors—the first neutrino. 2. Neutrino beams—the second neutrino. 3. SLAC—hard scattering. 4. Colliders
(d) Scaling. Partons 576

(e) Conclusion: quantum field theory redux 580
1. Preamble. 2. Non-Abelian gauge theories. 3. Quantum chromodynamics. 4. The electroweak unification. 5. The neutral current; charm. 6. Years of synthesis. The detection of charm, bottom, and, perhaps, top, of a new lepton, and of jets. 7. Years of synthesis. The detection of W and Z

22 Being a conclusion that starts as epilog and ends as prolog 621

APPENDIX A synopsis of this book in the form of a chronology 627

INDEX 639

Inward Bound

1

Purpose and plan

(a) From X to Z

It was the afternoon of 8 November 1895. Wilhelm Conrad Roentgen was all by himself in his laboratory in Würzburg, experimenting once again with cathode rays. His main piece of equipment was a vacuum tube about one meter long in which, as we would now say, electrons were accelerated to an energy in the neighborhood of one hundred thousand electron volts.* The pressure in his tube was about one thousandth of a torr.** On that afternoon Roentgen was quite startled to notice a fluorescence on his detector, a small screen covered with barium platinocyanide, which he was holding in his hand at some distance from the tube. He had discovered what the world would soon know as X-rays, a term he introduced in his first publication on this subject. That paper was signed by him as sole author.

Nearly ninety years later, on 26 January 1983, I left my office at the Rockefeller University for an early lunch, then walked to the New York Hilton on Sixth Avenue where the winter meeting of the American Physical Society was in progress. Carlo Rubbia was to give an invited paper in Section HB. No title for his talk had been announced but word had come from CERN, the European Center for Nuclear Research in Geneva, that the subject was to be the discovery, long anticipated, of the W-boson by the UA1 team (UA = underground area) led by Rubbia. Never mind for the moment what a W-boson precisely is. That will be explained much later. Just accept for now that its discovery would be most important.

I entered the hotel's Sutton Ballroom, joined an audience of several hundred, and listened. First came a report from UA2, another group at CERN. They too were hunting for the W, had promising results, but were not yet ready to commit themselves.† Rubbia spoke next. He began by explaining the experimental arrangement. Antiprotons with a moderate energy of 3.5 GeV (one GeV is a billion (10^9) eV) were collected in the AA, the antiproton accumulator. (Modern physics has its own abundance of acronyms.) This is a roughly square

* One electron volt, or eV, is the energy increment of an electron accelerated by a potential difference of one volt.

** One torr is close to one mm of mercury.

† They were to do so soon afterward.[1]

doughnut-like ring held at a vacuum of about one ten billionth (10^{-10}) of a torr. The pressure had to be low since otherwise the antiprotons would be lost in collisions with gas molecules in the ring. Although this was by no means the lowest pressure then attained at CERN, it sufficed for the purpose because of a brilliant invention by Simon van der Meer, 'our best accelerator man', for keeping antiprotons moving in a disciplined manner. Once every twenty-four hours the AA releases its antiprotons which are then accelerated in two stages, the second of which takes place inside a high vacuum ring, 6 km in circumference, in which these particles reach an energy of 270 GeV. Inside that ring they collided with protons moving in the opposite direction with that same energy. The collisions were analyzed by means of a complex detector, 10 meters high by 5 meters wide, weighing 2000 tons. Rubbia explained next how six out of one billion recorded events had been singled out as bearing the indubitable 'signature' of the W-particle. This production rate, six in a billion, agreed well with theoretical expectations, as did the first crude determination of the W-mass.

Next to me sat an expert in high-energy neutrino physics. As the talk drew to a close we looked at each other and nodded: they had it. Afterwards I talked with Rubbia. He asked me whether I believed it; I said I did. He gave me a preprint of the first UA1 publication[2] on the W, signed by 135 authors from 12 European and 2 American institutions.

Evening had fallen, it was crisp and cold, when I walked back under the bright lights of 57th Street. I kept thinking how the content and style of both experimental and theoretical physics had changed in my lifetime. The discovery of the W surely meant that once again a watershed had been reached. When I came home I went to my desk, cluttered with drafts of chapters for a history of matter. Now I knew the counterpoint with which to start Chapter 1. I sat down and wrote: 'It was the afternoon of 8 November 1895 . . .'

The next time I met Rubbia was the following 9 May. We were in Princeton where that day he was to give a lecture on UA1's further progress. During a long conversation before his talk he told me that the first example had been found of an event that looked just like the signature of a Z-boson. This very heavy neutral particle was the missing ingredient needed, along with positively- and negatively-charged W's, to put on a firm experimental basis the unified field theory of electromagnetic and weak forces that goes by the code name $SU(2)\times U(1)$. I asked Rubbia whether the report he was about to give would be the first in the United States to mention the Z. It is indeed, he replied, we found this event only a few days ago.*

I knew then where to end.

This book will attempt to pull together the strands of the tale spanning nearly a century, from X to Z, of the search for an ever more refined description of

* By June UA1 reported[3] five Z-events, by August UA2 had eight.[4]

matter and forces in the physical world. The questions are old. I could have started by copying from some encyclopaedia the speculations of Empedocles of Akragas about four eternal and unchanging elements—earth, water, air, and fire—and about two fundamental forces, Love or Joy (attractions), and Strife or Hate (repulsions)—terms no more quaint than some in current use. Alternatively, I could have begun with the birth of chemistry and nineteenth-century speculations about the molecular and atomic constitution of compounds and elements. I have chosen, however, to confine myself to the subatomic era with only sporadic backward excursions to earlier times. Roentgen's discovery is a natural starting point for that purpose since it occurred just before the beginning of the explosive period when new laws of nature, quantum theory and relativity, new forms of matter, and evidence (not at all rapidly recognized as such) for new forces made their first appearance. In 1895 Roentgen could not yet know that X-rays may be considered as a stream of particles—photons—with zero mass. He did not know then that cathode rays consist of electrons; those were discovered only two years later. Nor could he have anticipated that within a few months X-rays would be the spur to the discovery of radioactivity. This book concludes with the discovery of the W- and Z-bosons, particles with masses about equal to those of a typical rubidium and ruthenium atom respectively, yet by symmetry arguments intimately related to each other and to the massless photon. The future will decide whether this ending is as natural as the point of departure. At the time of writing it is certain, however, that the journey described in these pages toward higher and higher energies or, which is the same, toward smaller and smaller distances, is by no means over. We continue to be inward bound; the heroic period to be discussed is open-ended. Some believe its closure to be imminent. It could be, however, that writing of its history now is like writing of the French Revolution the week after the Bastille was stormed.

The recitation of facts and dates and a handful of formulae will lend an objective touch to the account to be presented. History is highly subjective, however, since it is created after the fact, and after the date, by the inevitable process of the selection of events deemed relevant by one observer or another. Thus there are as many (overlapping) histories as there are historians. I keep before me two admonitions; one by Macaulay: 'He who is deficient in the art of selection may, by showing nothing but the truth, produce all the effects of the grossest falsehoods. It perpetually happens that one writer tells less truth than another, merely because he tells more truths';[5] and one by Carlyle: 'He who reads the inscrutable Book of Nature as if it were a Merchant's Ledger, is justly suspected of never having seen that Book, but only some school Synopsis thereof; from which, if taken for the real Book, more error than insight is to be derived'.[6]

As with all works of this kind, this author's approach to history will be manifest by and by, as his story unfolds. Nonetheless it seems appropriate at

this point to make a few prefatory remarks on the question of selection. (Further general comments are reserved for Chapter 7.) The emphasis throughout will be on the flow of ideas. Little will be said about the sociological context of the subject, a perspective I find fascinating but to which I do not have much to contribute. As to the community of high energy physicists, in many respects it is not all that different from the rest of humanity. One encounters some noble characters, some bastards, also some who are like poets:[7] 'Strong poets make history . . . by misreading one another so as to clear imaginative spaces for themselves'. Among the distinguishing characteristics of the group, as I see it, are exceptional intellectual fearlessness, and a high incidence of the Shiva complex—unhappiness at not having been born with three eyes and four arms. I shall occasionally indulge in brief comments on specific personalities. These are located as follows:
Balmer (9a), Becquerel (2b1, 2b3), Bohr (9a, 9e2, 9e6, 10.5), William Bragg (8b), van den Broek (11d), the Curies (3a), Dirac (13e1), Elster and Geitel (16b), Fermi (19c3), Goudsmit (13c), Hahn (8d), Lawrence (17b), Meitner (8d), Oppenheimer (16b), Pauli (13b, 14d4, 16a, 16d7), Roentgen (2a), Rowland (4c), Rutherford (1b, 3b, 9d, 17h), J. J. Thomson (4d4, 9c6, 10.1), Tomonaga (18c2), Uhlenbeck (13c), and Yukawa (17g2). Here and otherwise only in the remainder of this first chapter the following notations are used: (4) for Chapter 4; (4d) for Chapter 4, Section (d); (4d1) for Chapter 4, Section (d), Part 1.

I intend to stress how progress leads to confusion leads to progress and on and on without respite. Every one of the many major advances to be discussed created sooner or later, more often sooner, new problems. These confusions, never twice the same, are not to be deplored. Rather, those who participate experience them as a privilege. As Niels Bohr once said to me after an evening of fruitless discussion: 'Tomorrow is going to be wonderful because tonight I do not understand anything'. Since it is my principal purpose to describe how ideas evolved, I shall need to discuss on various occasions how false leads, incorrect improvizations, and dead ends are interspersed between one advance and the next. The omission of such episodes would anaesthetize the story. In that connection I should like to mention an experience after giving a lecture on some of the present material at Fermi National Laboratory. Two younger colleagues came up to me. The first one looked rather glum and said: 'If men as great as these were so often mistaken then what should people like us expect to contribute?' The second one smiled broadly and said: 'So we are not the only fools'. I feel closer to the second response than to the first but stress that I have in mind false leads pursued by a generation as a whole (7) rather than just by one individual or another. It is less astonishing that such divagations happened so often than that, time and again, they were followed by a return to the path of righteousness.

In regard to confusion, the very recent past and the present are of course no exceptions. I strongly believe that a number of current theoretical explorations will turn out to be passing fancies, and I fervently hope that some others will survive. Since it serves no purpose to stretch subjectivity into outright prejudice, I shall increasingly try to use discretion the more I near the present, rather than attempt an up-to-date assessment of developments on all fronts.

A more important reason for augmented selectivity is less a matter of choice than of necessity: the need for finding ways of coping with the outpouring of publications which begins to reach flood level somewhere in the late 1950s (20d). I can well appreciate John Kenneth Galbraith's recent comment[8] regarding his efforts at writing a history of economics: 'As one approaches the present, one is filled with a sense of hopelessness; in a year and possibly even in a month, there is now more economic comment in the supposedly serious literature than survives from the whole of the thousand years commonly denominated as the Middle Ages . . . anyone who claims to be familiar with it all is a confessing liar'.

As to physics, let a few numbers serve to illustrate its growth during the period embraced in this book. In 1899 the American Physical Society was founded by 38 physicists (16b). In 1900* the worldwide number of academic physicists of all ranks was about 1000. Among these the number of senior faculty in theoretical physics was 8 in Germany, 2 in the United States (Gibbs at Yale and Pupin at Columbia[10]), one in Holland (Lorentz), and none in the British Empire. (Maxwell's chair at Cambridge had been in experimental physics (3b, 4b).) The emergence of theoretical physics as a semi-autonomous discipline is a twentieth-century phenomenon. Kirchhoff made fundamental contributions to experimental as well as theoretical physics (9b); so did J. J. Thomson (4d). Maxwell (4b), Boltzmann, Einstein, and Bohr (9a) published experimental results of their own. In the twentieth century Rutherford (9d) and Fermi (9e, 19c) were the leading representatives of this vanishing tradition.

Today the American Physical Society counts over 36 000 members. It is one of the nine member societies of the American Institute of Physics founded in 1931 'for the purpose of coordinating various societies whose interests are primarily in the field of physics and for the purpose of supporting their publications'.[11] The American Physical Society has since become a conglomerate of Divisions, including one of Particles and Fields (the two main topics of this book) which was founded in 1967 and which now has more than 2500 members. The American Institute of Physics currently publishes about 40 journals, 17 of which are exclusively devoted to translations of articles from the Soviet Union, one to articles originally in Chinese. The member societies publish in all some 20 additional journals.

* I found the numbers for 1900 in an interesting account of physics at the turn of the century.[9]

For present purposes the most important American journal is the *Physical Review*. Volume 1 came out in 1894, is 480 pages long and contains 20 articles. In 1950 four volumes appeared with 447 articles on 3167 pages. For fair comparison I have excluded the Letters to the Editor and the abstracts of papers presented at Physical Society meetings, in 1950 all cosily published together with the articles, because there were no letters and abstracts in 1894, while they were published separately by 1980, my third year of comparison. In that year there were two volumes, 29 213 pages, 3403 articles. Since 1970 every volume of the *Physical Review* has appeared in four subvolumes A–D. Nuclear Physics, which gave birth to Particles and Fields (D) only some fifty years ago, is now published separately (C). In 1980 part D alone contained 6651 pages, 793 articles.

The numbers quoted in the preceding two paragraphs refer exclusively to the United States, since only for that region did I know how to make comparisons that span the century. In respect of the most recent twenty-odd years I can add some worldwide figures, however. That period witnessed the creation of central repositories for preprints, presently the common form of rapid communication. One of these, the SLAC preprint library, distributes every week a bulletin listing documents received by title, author(s), and institution. The annual yield,* comprising articles, letters, and reviews, was: 4100 in 1975, 6100 in 1984. One sees that to speak of a flood is no exaggeration. In the course of this century there have been radical changes not only in the content of physical theory and the style of experimentation but also in the ways science is communicated. Physics has become increasingly fragmented by the creation of specialties within specialties (19f1). The phenomenon is universal. There was a time, not long ago, when the existence of two cultures was sufficient cause for complaint. If we only had it so good now.

As the numbers indicate, at the time of writing one would have to digest 17 publications per day every day of the year in order to claim familiarity with the whole body of high energy physics literature. That, fortunately, is not a requirement for being a productive particle physicist. Attend a conference now and then, browse through frequently appearing comprehensive summary reports, read the SLAC list of titles, and, above all, be well locked into the strongly developed modern oral tradition, and one gets quite far. These means of keeping abreast were likewise indispensable to the present enterprise. Yet the double predicament remained of distilling a cohesive account of the postwar era from information that is in flux as to content and in explosion as to volume. How to build the bridge between Roentgen, who did his experiment singlehandedly without yet knowing that electromagnetic radiations cannot vibrate in their direction of motion (2a), and the UA1 experiments

* I owe these numbers to the SLAC librarians.

executed by a multinational consortium in response to predictions of a highly sophisticated relativistic quantum field theory?

The best way that I could devise was to divide this book into two parts. The first: one man's attempt at a history of the period 1895–1945; the second: one man's memoirs of the postwar era. I believe that other members of my own and later generations should be heard before the time will be ripe for an historical assessment of the more recent period. A number of them have already gone on record, others will presumably follow before long. Being one of the above, I consider Part II as a contribution to the pool of information from which a history will have to be discerned in later times.

Inevitably the styles of the two parts are distinct. Part I contains the main thrust of the present work. Among the topics discussed in Part II, several that have reached a fair level of maturity are treated in excellent and easily accessible books and reports which are essentially contemporaneous with the present volume. I shall therefore often content myself with brief indications after which I direct the reader to the literature, especially to such writings as give ample references to original contributions. The number of pages devoted to one or another subject is therefore not necessarily proportional to their relative importance.

As part of the homework for Part II, I have paid particular attention to the published reminiscences of others. I shall not discuss these but note that I do not agree with all opinions expressed therein, nor do these always agree among themselves. In the sources to the appropriate chapters I have recorded references to all reminiscences known to me, in the hope that this may facilitate others' studies.

The remainder of this chapter consists of a non-technical account of the main themes of this book, interspersed with further general comments. Those who prefer to dig in at once may like to turn immediately to Chapter 2. Note: a further guide to the contents of this volume is found in the Appendix.

(b) 1895–1945: A history

The first international post-World War II physics conference, held in Cambridge (England) in July 1946, was devoted to 'Fundamental particles and low temperatures'.[12] Five of its sessions dealt with fundamental particles. In the first, Niels Bohr, Wolfgang Pauli, and Paul Dirac discussed problems and prospects of quantum field theories. Then followed a session on experimental techniques, including a report on plans for a new generation of accelerators aiming at energies that could not be reached by cyclotrons (17b, 19a). The next topic, nuclear forces, included a communication on neutron–proton scattering calculations up to the unheard-of energy of 25 MeV. The final session concerned Heisenberg's *S*-matrix (19f5).

In the following September, the American Physical Society convened in New York a meeting with international participation devoted to cosmic rays, elementary particles, and accelerators. 'Disparate as these three subjects appear to be, the trend of physics is rapidly uniting them' (18a). The ground covered was much the same as at the Cambridge Conference. Among noteworthy items were a report[13] on the detection of one cosmic ray particle with the bizarre mass 990 ± 100 times that of the electron; and a panel discussion on 'Relative advantages of proton and electron accelerators', in which Alvarez, Lawrence, McMillan, and Robert Wilson participated.

Part I of this book begins in 1895, when none of the fundamental particles referred to in those meetings had yet been found, cosmic rays had not yet been discovered, the only accelerator in captivity was a cathode ray tube, and relativity theory and quantum theory were yet to come. It ends in 1945 when physicists had become familiar with the electron, the proton, the neutron, a neutrino, and a meson, when cosmic rays were ardently studied, when cyclotrons had produced particles with energies larger than 10 MeV, and when the first explorations of relativistic quantum field theories had yielded a puzzling mixture of success and failure.

We start with two accidental discoveries made in mid-career by well-established scientists. First, Roentgen stumbled on X-rays (2a). Four months later Becquerel found that uranium spontaneously emits mysterious 'uranic rays' (2b). The brevity of the time interval between these two observations was no accident. Among the many tales of discovery to be encountered in the following, the story of Becquerel starting out to look for X-rays in a sample of a fluorescent uranium salt and ending up discovering radioactivity may well be the most whimsical (2b2).

Continuing his experiments, Becquerel was increasingly struck by the spontaneity of the phenomenon, the apparent absence of an energy source feeding the new radiation. The numerous early speculations on the origin of this 'atomic energy' (the term dates from 1903 (6c)) include doubts about the validity of energy conservation (6b). Also, later in the twentieth century, that principle would be challenged off and on, for a variety of reasons (14d2, 3). The meaning of energy conservation is still a subject of discussion in modern times.[15]

Becquerel's discovery at once attracted a new generation of physicists who had yet to make their mark. By herself, young Marie Curie made the next step, in a paper (1898) that from the point of view of scientific principle I regard as the most important one of her career. She showed (3a) that radioactivity is a property of *individual atoms* of certain species, one of which, she announced, is thorium. This atomic interpretation led her to use radioactivity as a diagnostic tool in the discovery of new elements. That basic idea of radiochemistry led to the celebrated discoveries of polonium and radium, made jointly with her husband Pierre. (Incidentally, from them stems the term 'radioactivity'.)

Meanwhile, a few weeks before the discovery of X-rays, a young New Zealander, by the name of Ernest Rutherford, had arrived at the Cavendish Laboratory in Cambridge, on a research fellowship. He was to end his career as director of that institution, the post held earlier by Maxwell, Rayleigh, and J. J. Thomson. During his first brief but fruitful stay in Cambridge he discovered that there are two distinct kinds of radioactive rays, α-rays and β-rays. Two years later, in 1900, Villard in Paris discovered an even more penetrating type of radiation, γ-rays. The question of the constitution of α- and γ-rays initially caused much confusion (β-rays were much more rapidly identified) and was fully clarified only many years later (3b).

Rutherford's next major discovery came after he had moved to McGill University in Montreal (1898): emanation from thorium, one among the new radium elements he had found, lost half its activity in about a minute. That was the first observation of a new parameter characteristic of a radioactive species: its half life. The discoveries in 1900 by Rutherford of a half life and, later that same year, by Planck of the quantum theory signal the end of the era of classical physics. Neither of them was at once aware how profoundly his work was to change the course of science (7.1, 7.4). It was Einstein who, in 1905, was the first to recognize the revolutionary character of Planck's hypothesis (7.5); and who, in 1917, was the first to grasp that Rutherford's discovery called for no less than a revision of a root concept of classical physics: causality. The fundamental theory of spontaneous decays, developed by Dirac in 1927 (15c, 6f3), was to make clear that it is, and forever will remain, impossible to predict the moment at which a given radioactive atom will decay.

The notion of half life raised a new question: could it be that *all* elements are radioactive but that most of them live too long for such activity to have been noticed? That problem appeared to be definitively settled in the 1920s (6d). The grand unified gauge theories of the 1970s have re-opened the issue, however (22).

During the McGill years Rutherford produced three books,* the first two editions of 'Radioactivity' and the text of his Silliman lectures, all indispensable to the student of the subject's early history. During that period he also revealed his gifts for directing others' research. He cared less for teaching beginners, however. As a high school teacher in New Zealand he had been hopeless (17h). In discussing an offer from Yale he remarked: 'Why should I go there? They act as though the University was made for students'.[16]

I shall return to Rutherford's further career but first pick up another thread: the discovery of the first 'fundamental' particle, the electron. As had been the

* See Chapter 3, Sources.

case for X-rays and radioactivity, this was another novelty not anticipated theoretically.

The electron was first sighted in e/m measurements (e is its charge, m its mass). It began in 1896 when Zeeman succeeded where Faraday had failed, to wit, in showing that spectral lines are split when their source is placed in a magnetic field. Lorentz immediately interpreted the effect to be due to the motion of a charged particle bound within an atom, and showed that the magnitude of the effect fixes that particle's e/m (within a factor two of the modern value of the electron). Next, in 1897, the long-vexing problem of the constitution of cathode rays was resolved by determinations of their e/m, first by Wiechert (4d2), then by Kaufmann (4d3), then by J. J. Thomson (4d4). After Thomson had measured e separately, in 1899, both e and m were now approximately known, and the electron had arrived. It was demonstrated shortly afterward that β-rays and cathode rays were identical (4e). During the following two decades the electron played an important role as an experimental tool for verifying the kinematics of special relativity (4f).

The discovery of the electron provides a splendid example of fundamental advances made possible by new experimental tools.

Zeeman could make his discovery by using Rowland's recent invention of refractive gratings with high optical resolving power (4c).

Advances in cathode ray physics were made possible by improved vacuum technology. In the 1830s Faraday experimented with cathode rays in a vacuum of about 100 torr (4d1). In 1857 Geissler reached 0.1 torr with his much improved version of the mercury vacuum pump (4a). With modifications of that pump vacua of 10^{-6} torr were reached by 1880.*

In the 1830s Faraday had studied cathode rays in an evacuated tube closed by corks at both ends (4d1). Geissler was the first to produce an all-glass tube, the metallic electrodes being inserted by melting the glass at both ends. His technical feat was to match the thermal expansion of glass and metal, so that the electrodes when heated do not crack the glass (4a).

Thomson's measurement of e was the first major application of the newly-invented cloud chamber.

Finally, higher energy became available when Rühmkorff invented a greatly improved version of the induction coil (4a), capable of producing energies of 100 000 eV. His coil was used by many leading physicists of the period, by Faraday in his later years, by Roentgen and Zeeman in making their discoveries, and by others.

Add to the experimental discoveries of X-rays, radioactivity and the electron, the new frontiers opened by Planck's quantum theory, Einstein's light-quantum and his special relativity theory, and one recognizes the extraordinary

* For these numbers see Ref. 17.

variety of discovery during the decade 1895–1905. Some thoughts on the lessons of that period are collected in Chapter 7.

There is yet another facet of the discoveries of that decade. In perusing the early literature on radioactivity I came upon a fascinating joint paper of 1901 by Becquerel and Pierre Curie in which the authors reported on unpleasant personal experiences resulting from exposure to radioactive elements.[18] This led me to wonder: how did one begin to find out about the good and the harmful effects of radioactive radiations? And of X-rays? Now here, now there, I collected information, helped by advice from Robley Evans and Lewis Thomas. The resulting picture is included here (5) as a brief aside that nevertheless does belong to the early history of the new radiations.

Three new themes emerge as we now move further into the twentieth century: first β-spectra, then atomic structure, then early ideas on nuclear structure. That may not be the logical order in which these subjects are taught in physics classes, but then it is one of my aims to show that the evolution of physics does not always follow logical paths.

Let us then begin with β-radioactivity, one of the main subjects of this book which will stay with us till its very end.

It is almost a rule that the first of a new generation of experiments on β-decay are wrong, even when performed by the best of physicists. So it was right at the beginning, in 1906, well before the discovery of the atomic nucleus. In that year it was conjectured, erroneously, that β-rays are monochromatic (8b). This idea was tested experimentally by studying the absorption of β-rays. It was believed at that time, also erroneously, that the β-ray intensity decreases exponentially with distance (8c). This 'law' led to the experimental conclusions (8d) that β-rays were indeed monochromatic and, where not, that the radioactive source consisted of not one but a complex of active elements. In 1909 experiment revealed that the exponential absorption law was incorrect (8e). A new approach was therefore introduced: bend the β-rays magnetically, measure their spectra photographically. Result: β-spectra were now found not to be monochromatic but to consist of discrete lines (8f). In 1912–13 these discrete spectra were studied by various groups. All those engaged were unaware that they were laying the foundations of nuclear spectroscopy (8g).

The change came in 1914 when Chadwick tackled the problem with a third technique: magnetic bending but detection with counters. His conclusions: first, there is a continuous β-spectrum; secondly, superimposed discrete lines do exist but photographic detection vastly overrates their intensity (8h).

The next question was: is the continuous spectrum primary or secondary? The latter view had its proponents who argued for a primary line spectrum washed out in part by secondary effects such as scattering. The debate lasted until 1927 when a difficult and crucial calorimetric experiment revealed a most profound result: the primary spectrum *is* continuous (14c).

I now leave β-decay for a while and turn to questions of atomic and nuclear structure.

The structure of matter is largely determined from the structure of spectra, be they of molecules, atoms, nuclei, or fundamental particles. It is fitting therefore to have a look back at the very beginnings of quantitative spectral analysis, dating from the early 1860s and associated with such names as Kirchhoff, Bunsen, Angström, and Plücker (9b1). Already then a few frequencies in the spectrum of atomic hydrogen were known to an accuracy of one part in 5000. It had also become clear that a spectrum serves as a visiting card for specific types of atoms and hence can serve as a diagnostic for new elements. The search for spectral regularities began not long thereafter, the most memorable result being Balmer's formula (1885) for hydrogen (9b2). Balmer, incidentally, wrote only three physics papers in his life. The first two made him immortal; the third was wrong.

Spectra signified that atoms had to have structure, as Maxwell had been among the first to emphasize (9b1). It is therefore not surprising that the search for atomic models began quite early, long before the discovery of the electron (9b). Many and varied were models containing electrons, invented after that particle had been found (9c).

The great problem with electron models was that they were unstable if the electrons were initially considered at rest. So, Thomson decreed, let them rotate in their positive background. That (he showed) improves stability (9c3). And if there is residual instability could that not explain the expulsion of β-electrons (9c2)? Rotation causes a new form of instability, however, which appeared beyond explanation: rotating electrons lose energy by radiation. To the small band of those engaged in these speculations it all seemed very confusing. Thomson does not appear to have been daunted, however (9c6).

The first important step in the right direction was also made by Thomson. Analyzing experiments such as X-ray absorption he concluded (1906) that the number of electrons inside an atom is very much smaller than initially anticipated (9c6). Then where does the atom's mass come from?

Re-enter Rutherford, the only physicist ever to do his greatest work after receiving his Nobel Prize, in 1908. The previous year he had moved from Montreal to Manchester. It was there that (1909) two of his coworkers found that α-particles impinging on a variety of elements showed a most amazing propensity for 'hard scattering', that is for being deflected by large angles (9d).

We next witness Rutherford engaging in theoretical physics, an activity on which he held lifelong views which, shall we say, were dim. ('How can a fellow sit down at a table and calculate something that would take me, me, six months to measure in a laboratory?'[19]) Be that as it may, in 1911 he announced the reason for hard scattering: it is due to the α-particle's deflection by the 'central charge', the nucleus, in which nearly all the atomic mass is concentrated (9d). In the words of the calypso: the atom is full of empty.

The advent of the nucleus did not cause any immediate sensation. Initially Rutherford himself was reticent about his seminal discovery and barely mentioned it in the new edition (October 1912) of his book on radioactivity. 'It was as if . . . there were a lingering suspicion in Rutherford's mind that [these ideas] were not at that time sufficiently well proven to qualify for inclusion in a student's textbook.'[20] Two years later, however, he spoke forcefully of the nucleus.

In his work on α-scattering Rutherford properly treated the influence of atomic electrons as negligible. The next question was how to link the electrons with the nucleus. That marriage was consecrated by Niels Bohr. His theory of the hydrogen atom (9e6) forged the first link (9e1) between dynamics and the quantum of action.

The Rutherford–Bohr model of the atom established the first solid twentieth-century beachhead in the theory of the structure of matter. The gap between bewildering new phenomena and their interpretation began to close, slowly; accordingly the pace of the present narrative begins to quicken (10.6). Thus I only discuss (9e6) that part of Bohr's 1913 trilogy which deals with the hydrogen atom, with only brief additional remarks on a variety of other items: the early response to Bohr's theory (10.1); the new paradoxes it posed, notably in regard to causality (10.2); the serious difficulties encountered in explaining the next simplest atom, helium (10.4); Sommerfeld's theory of fine structure (10.3); the first introduction of quantum selection rules (10.3). Finally I comment on the change in the cast of characters (10.5).

Meanwhile the young science of nuclear physics was making its first healthy strides. In 1912 Rutherford had recognized α-decay to be a nuclear phenomenon but believed β- and γ-decay to be due to 'instability of the electronic distribution' (9d). In 1913 Bohr noted that the latter, too, were of nuclear origin (11b). Isotopes were recognized (11c). It became clear that atomic number and nuclear charge are two *independent* parameters (11d). The periodic table, dating from about 1870 (11a), came to be vastly better understood, especially as a result of successful interpretation of X-ray spectra (11d). Binding energy was added to the list of nuclear parameters (11f).

After remarks on the role of physicists during the First World War (11g), I turn to the four-part paper of 1919 by Rutherford, Sir Ernest by then (11h). For present purposes its major interest lies in his announcement that his earlier theory of α-nucleus scattering breaks down for light nuclei. That last effect stirred much debate. In 1921 it was stated for the first time that these deviations showed the effect of 'new forces with very great intensity': *the first hints of strong interactions had been discerned* (11h).

Along with all this progress the seeds of serious confusion were sown (1913) when the first picture of nuclear constituents made its appearance: sufficient protons to account for mass, combined with sufficient electrons to account for charge (11e). It was so natural, yet so wrong. Quantum mechanics was to show that this model leads to a series of paradoxes.

Relativity theory and quantum mechanics, the two most excellent achievements in the physics of our century, provide the framework for the entire theory of matter and forces as it has evolved during the last sixty years. In the present synopsis I have already passed without fanfare the years 1905 of the birth of special relativity and 1915 when general relativity reached its present form. Nor shall I now or hereafter return in any detail to these events, the reason being that, as well as I could, I have already dealt with the evolution of relativity in another book[21] that I regard as a companion volume to the present one. I have no such excuse, however, in regard to 1925, the year of birth of quantum mechanics. As the writing of this book proceeded to that period, I found myself therefore in a quandary. I could neither treat that subject with the respect it deserves nor dispense with it in a few mellifluous phrases without thereby throwing the entire enterprise out of balance. I therefore decided upon a chapter 'Quantum mechanics, an essay', in which I touch on some highlights while stressing already in its title that it is not remotely the intent to treat the subject in depth. This essay begins with a survey of the status of physics in the spring of 1925 (12a), continues with comments on the transition from the old quantum theory to the new mechanics (12b) and provides a chronology of main events during 1925–7 (12c), whereafter the origins of the probability interpretation of quantum mechanics are discussed in slightly more detail. Finally, it is noted that 1925 once again brought about a change in cast of principal participants (12e).

The remaining chapters of Part I deal with two main and partially interwoven themes: the evolution of relativistic quantum mechanics and quantum field theory, and further developments in nuclear physics.

Quantum theory before 1925, the old quantum theory, was part craft part art. Old first principles had been found wanting, new ones had not yet been discovered. It is impressive to note how many of the ad hoc rules of those early days, invented to encode experimental regularities, could eventually be incorporated in quantum mechanics. The case of the anomalous Zeeman effect provides an excellent example. As I turn to this topic I am reminded once again of Macaulay's warning to be sparing with truths. Thus from among developments of equal importance I discuss only the anomalous Zeeman effect, the magnetic splittings of spectral lines that do not follow Lorentz' predictions, since that choice provides the natural starting point for the road to the exclusion principle (13b), the discovery of spin (13c), and Dirac's discovery of his equation for the electron (13e), one of the marvels of modern physics.

The evolution from anomalous Zeeman effect to Dirac equation straddles the old and the new quantum theory. The same is true for quantum statistics, the foundations of which were laid between 1924 and 1926, when Dirac established the link between statistics and symmetries of wave functions (13d). The latter property was also the spur to the first introduction in quantum mechanics of group theoretical methods (13a). Later a profound connection was discovered between spin and statistics (20c3).

The stage is now set for an account of the history of relativistic quantum field theory, briefly referred to as field theory in the next few pages. This is a central topic of this book. The first stage of field theory, known as quantum electrodynamics, deals with the extension of quantum rules for mechanical systems to electromagnetic fields. The systematic foundation of quantum electrodynamics begins (1927) with two papers by Dirac (15c).

The technical description of particles and fields now begins to show a distinct and steady increase in mathematical complexity (15a). In Dirac's words (1931):

> The steady progress of physics requires for its theoretical formulations a mathematics that gets continually more advanced.[22]

As can be expected, he had chosen a fitting occasion for that statement. It is found in his paper on magnetic monopoles, a contribution that would be recognized much later as the first application of global topology to physics.* Monopoles are an example of a subject I shall not discuss in spite of its interest, true to my intent of staying away from topics where theory and experiment have not yet met.

At the root of these higher demands on 'mathematical technology' lies the difficult question, by no means resolved to this day: how to satisfy simultaneously the demands of relativity and of quantum mechanics. It took grown men like Heisenberg and Pauli a year of their lives before they found the answer for quantum electrodynamics. In their final proof the newly discovered principle of gauge invariance played an important role (15e).

I return to Dirac's equation for the electron, which was both a grand success and, so it initially seemed, a source of serious trouble: it appeared to have too many solutions. Two years of confusion then followed until Dirac showed (1929) that these extra solutions could be associated with the existence of a particle as yet unseen, the positron, with mass equal to and charge equal and opposite to that of the electron. This prediction and the subsequent discovery of the positron (15f) rank among the great triumphs of modern physics. The relationship between electron and positron is sometimes referred to as the relation between matter and antimatter. That terminology is well and good as long as it is borne in mind that antimatter is as much matter as matter is matter.

As we now move into the 1930s, physics in America begins to take an important share in the further evolution of field theory (16b). During this next phase quantum electrodynamics proved to be both an evident success and an apparent failure.

* See Ref. 23 for a few useful introductory papers that also contain further guidance to the literature.

This is what happened. Neither then nor now can one give exact solutions of the quantum electrodynamic equations. Almost invariably one uses the approximation procedure of solving the equations, step by step, in a perturbative expansion in powers of the dimensionless number $\alpha = e^2/\hbar c$ (e = charge of the electron, c = velocity of light). This approximation procedure looks like a sensible approach since α is small, about 1/137.

Success: to the lowest approximation in α the theoretical predictions are invariably quite close to the corresponding measurements, including such 'new' processes as the creation and annihilation of electron–positron pairs (16d1).
Failure: the contributions to these same effects to higher 'approximations' in α are invariably found to be infinitely large. Small numbers, powers of α, appear multiplied by integrals that are infinite. Hence the predicament of the thirties: how can a theory be essentially correct when treated approximately, yet fail when treated more rigorously?

Some believed that the cure would come by first modifying the classical theory in such a way that this revised version would behave better when subjected to the rules of quantum mechanics (16e). That was not to be the way. The germs of the next advance, the procedure called renormalization (the term dates from 1936 (16d4)) were vaguely discerned already in the 1930s. In Part II we shall see what renormalization is and what it did in the 1940s: to show that quantum electrodynamics is much healthier than was believed earlier, in fact that it is the best field theory we have to date.

The atomic nucleus, when last heard of, was supposed to comprise protons and electrons. Two decades went by before this model was ultimately abandoned. That period divides into three phases. The first of these (1913–25) is pre-quantum mechanical. Without any backing of theory to speak of, there developed a minor industry of comparing nuclear data with models built of subunits, α-particles, and others, themselves considered to be electron–proton composites. One of Rutherford's coworkers delivered himself of the following opinion[24] about those efforts: 'The subject has offered a vast field for what the Germans call *Arithmetische Spielereien* [playing games with arithmetic] which serve rather to entertain the players than to advance knowledge. On this delicate point it is easy to say too much'.

Serious theory began when quantum mechanics confronted the nucleus. This second phase is marked by some successes (14a) but above all by the emergence of a series of paradoxes, every one of them consequences of the presumption that electrons are nuclear constituents (14b). As these structure problems were causing increasing puzzlement, another and, it seemed, quite independent mystery was beginning to draw attention: why are β-spectra continuous?

Two new particles were needed for the resolution of these many difficulties. The neutrino, emitted in β-decay along with the electron, causes the latter to have a continuous spectrum. The neutron solves the structure problems. The third phase of the proton–electron model concerns the years 1932–4, immediately following the discovery of the neutron, during which we witness the gradual expulsion of the electron as nuclear constituent.

It is imperative for an understanding of the years 1929–34 to be aware that at that time one was looking for a common cure for both the structure and the β-problems (14d1). It began with Bohr who read in this complex of mysteries signals for a breakdown of the laws of quantum mechanics, in particular of energy conservation (14d3), an idea he finally abjured in 1936 (14d5). That was the second time he had questioned that conservation law (14d2). Then came Pauli's neutrino hypothesis (which I found first mentioned in a letter by Heisenberg (14d4)). This daring idea, too, was initially conceived as a common cure: the nucleus was briefly believed to have three constituents, protons, electrons, and neutrinos (14d4).

Once again we enter the thirties, the decade not only of travail in field theory (as noted earlier) but also of a great flowering of nuclear physics. At the Cavendish, Cockcroft and Walton observed the first instance of a nuclear process produced by a beam of accelerated particles (protons) (17b). At Berkeley, Lawrence went into action as the builder of the first cyclotrons, and as teacher of a generation that would lead major laboratories after World War II (17b). Cosmic rays, known for two decades (17b), were used by Anderson and his group in discovering the positron and the first meson (17g3). The discoveries of the deuteron (17b), the heavy isotope of hydrogen, of the positron and of the neutron (17a) were made within a period of two months. Unstable isotopes that decay with positron emission were discovered (17a).

Now to the third phase of the proton–electron model. Why did one not at once get rid of electrons as nuclear constituents and simply say: nuclei are made of protons and neutrons? Because it was initially believed by many that the neutron itself was a proton–electron composite (17d). Why this tenacity? Because one still did not see how in β-processes electrons could come out of nuclei without being in there in the first place. It was Fermi's great contribution to apply field theory to β-decay (17e3) and to assert that electrons and neutrinos are created in the very act of β-disintegration. Only after that had sunk in did it finally become clear that neutrons are neither less nor more elementary than protons.

Fermi's paper on β-decay, a major document full of novelty, provided the basis for further elaboration of β-decay theory (17e4); and for the first calculations of high energy neutrino scattering (17g1); all this in the 1930s. Evidently theorists had grown comfortable with the neutrino even though there was as yet no direct evidence for its existence—a situation unique in the history of particle physics. A Royal Society discussion[25] held in 1937 conveys the attitude then current:

'There can be no two opinions about the practical utility of the neutrino hypothesis ... but ... until clear experimental evidence for the existence of the neutrino could be obtained ... the neutrino must remain purely hypothetical ... The detailed theory ascribes to [the neutrino] ... a free path in matter so great that failure to detect any evidence of interaction between neutrino and matter is no evidence against its existence'.

It would not be until 1956 that intense neutrino fluxes generated by a nuclear reactor made it possible to observe neutrinos interacting with matter away from their points of creation (21c1).

The Fermi theory of 1934 also contains the first clue to the existence of the W-boson (21e1).

Fermi's theory is one of three developments in the thirties that set the stage for the postwar era. The other two are: the first appearance of 'isospin', an attribute that is manifest only at nuclear and subnuclear levels (17d, 17f); and the first attempts, initiated by Yukawa (17g2), to associate nuclear forces with fields. By analogy with the electrodynamics of charged particles as well as photons, these new theories deal with nuclear forces and mesons. A world war had to intervene before it became clear that the first meson predicted was not the first meson observed.

Here the history ends. Next, the memoir.

(c) The postwar years: a memoir

> **Memoir** . . . 3. A record of events, not purporting to be a complete history, but treating of such matters as come within the personal knowledge of the writer, or are obtained from certain particular sources of information.
>
> *Oxford English Dictionary*

Right after war's end the physics of particles and fields took a sharp turn rather than continuing on earlier trodden paths. There is an almost abrupt sense of novelty, in regard to instrumentation, new styles of cooperative experimental venture, discoveries of new forms of matter, and evolution of new theoretical methods.

During the war several experimentalists had been pondering how to reach energies well beyond the range of 10–20 MeV that had been attained with the cyclotrons of the late thirties (19a). It was known already then[26] that cyclotrons cannot serve that purpose because of complications caused by the influence of relativity on the motion of high velocity particles. These effects were rendered harmless in the design of a new generation of accelerators aimed at achieving higher energies, either for electrons or for protons (and deuterons and α-particles). Several of these were operational before the forties were over: betatrons, synchrotrons, weak focussing synchrocyclotrons (19a). By 1950 beams of protons with energies of several hundred MeV were available at a number of laboratories. In 1953 the first synchrotron (the Cosmotron), producing protons in the GeV region, was ready for experimentation. Thus during the first postwar decade, energies had been driven up by a factor of over a hundred.

All that cost money, a lot of money. That demanded tapping new financial resources. The first of these was the United States military establishment, which had developed a benevolent attitude toward the scientific community because of its role in developing radar and the atomic bomb. Luis Alvarez told me: 'Right after the war we had a blank check from the military because we had been so successful. Had it been otherwise we would have been villains. As it was we never had to worry about money'. The first military contributions to pure physics were authorized over the signature of General Groves, head of the Manhattan Project (19a). Meanwhile it had become evident that further progress in particle physics, and other areas as well, demanded moneys that no single university could provide. As a result consortia were founded for the purpose of jointly administering new facilities, funding to be provided by a government, or governments. The activities of the first of these groups led to the establishment of Brookhaven National Laboratory in 1946. Others followed. In Western Europe, CERN began its official existence in 1954, governed by a multinational council. The Dubna Laboratory in the Soviet Union was another of the early international ventures. Thus came into being a number of centers that by the late fifties housed accelerators with energies ranging up to 10 GeV (19a).

In regard to developments in the United States, it has often struck me as a curious coincidence that experience gained during the war served to prepare scientists, as it were, for dealing with high-cost projects demanding joint efforts at just about the time these factors became indispensable to the continuation of pure research in several areas. Also before the war science could have profited from more funds, to be sure, yet the natural scientific scope of the enterprise at that time made it possible to keep research moving in single university context. That was out of the question after 1945. Once, years later, when I discussed these questions with I. I. Rabi, he said to me: 'You people were fortunate to be the grandchildren of the atomic bomb and the children of Sputnik'.

I shall turn shortly to the important advances which by 1953 had resulted from experiments at accelerators but comment first on the fundamental insights gained from cosmic ray studies during that same period.

In 1911 an experiment performed in a meadow (now a parking lot) in the city of Vienna had given intimations of rays coming from the skies (17b). These early hints were confirmed when detection equipment was sent aloft attached to balloons, first manned, then unmanned. Detailed study of properties of these cosmic rays (so named by Millikan in 1925) had to await further evolution of detection techniques. The first cosmic ray observations made with cloud chambers date from the late twenties. The discovery of cosmic ray particles with individual energies at least in the billion eV range was made in the early thirties (17b). The first picture of a cosmic ray positron track was recognized in 1931 by putting a cloud chamber in a magnetic field (15f). The further addition of counter-controlled chamber expansion devices, introduced shortly afterward, led to the discovery of cosmic ray showers (17b).

The detection of a meson had been the last major prewar discovery in cosmic rays (17g3). Also from that period dates the first evidence that this meson had properties which were hard to reconcile with those of the particle postulated by Yukawa: its lifetime appeared to be much longer and its scattering much weaker than the theory predicted (18b). Such was the state of affairs when the Second World War broke out. It was only in 1946 that it became known how bad the situation really was: the negatively-charged cosmic ray meson's absorption by atomic nuclei is ten to twelve orders of magnitude weaker than theoretically expected (18b).

Within a year another cosmic ray experiment brought clarification. The cosmic ray meson, since called muon (μ), is not identical with Yukawa's meson, since called pion (π). Rather, the unstable μ is a weakly-interacting decay product of the even less stable but strongly-interacting π, a state of affairs anticipated by several theorists (18b).

The discovery of the muon in 1947 marks the beginning of a new and as yet unfinished chapter in the story of the structure of matter.

Recall the harvest of particles up to that year: electron, proton, neutron, one meson, photon, one conjectured neutrino. Quite a collection but, as noted before, there was still hope that these fundamental particles might be sufficient for understanding the structure of all matter and the nature of all forces. Protons and neutrons form nuclei held together by the force field of 'the' meson. Electrons and nuclei form atoms held together by the electromagnetic field whose quanta are the photons. β-decay was describable in terms of electron–neutrino emission. One did not know then, nor does one now, why, so to speak, these particles are there, why their masses stand in the ratios they do, why forces have the strengths they do. Nevertheless one could still entertain the hope that those six particles and those three forces, along with gravitation, would tell the whole story. The arrival of the μ was a clear harbinger, however, of greater complexity. What else was the muon good for other than being the pion's favorite decay product? Even now, when we know much more about the properties of muons, have grouped them with electrons and neutrinos in a particle family called leptons (18a), and have recognized a new law, lepton conservation (20c4), the muon continues to remind us how little we understand of nature's particles. For that reason physicists have long since dropped the epithet 'fundamental', and prefer to speak of particles *tout court*.

I interrupt the account of cosmic ray physics with a comment on a theoretical development to which the muon gave rise. Soon after its discovery it was found that both its decay and its absorption can be described by an interaction quite similar to the one introduced by Fermi for β-decay. In particular it turned out that these three interactions, collectively called weak interactions (this collection would be added to in later years), have a common strength, a property since known as the universality of weak interactions (20c4).

The universality concept, dating from about 1948, contains the second clue to the existence of the W-boson (21e1).

Now back to the cosmic rays. The year 1947 was not over before the existence of still further new particles was announced, discovered once again in cosmic radiation (20a). It may truly be said that they came out of the blue, there had not been the slightest indication for the new layer of matter they presaged. These objects, initially called V-particles, were rare, slow in coming, and did not all at once cause widespread excitement. That changed in the 1950s. Cosmic ray data, still small in number, revealed (20a) that V-particles come in a variety of forms: a new brand of mesons heavier than the π, called K-mesons, and 'hyperons', particles heavier than the neutron and (hence) the proton, the latter two having meanwhile been given (18a) the collective name nucleon. Later the initial definition of hyperons in terms of mass range only turned out to be inadequate, since we now know that there are mesons heavier than the nucleon, so that 'meson' can be added to the list of oxymorons with which physics abounds. More specifically a hyperon is a half integer spin particle heavier than the nucleon. Hyperons and nucleons are now collectively called baryons (20a).

The announcement in 1953 (20a) that V-particles had been detected in an accelerator experiment marked the beginning of the end of a heroic period in cosmic ray physics. A few years later data on hyperons and K-mesons began to pour out of the accelerator laboratories at a rate with which the relatively very low cosmic ray intensities could not compete (20a). Cosmic ray physics continues to be fit and healthy, however, and is bound, sooner or later, to produce further fundamental novelty. Cosmic rays, like the accelerators mentioned so far, are fixed target devices: a beam hits a target at rest. Arrangements of this kind compete unfavorably with colliders (to be discussed below) in producing high *effective* energies. Nevertheless even colliders are far from competing with the highest cosmic ray energies, which extend beyond 10^{11} GeV. 'Are there important experiments which are, and will remain, unique to cosmic ray investigations?'[27] The particle fluxes at these ultra-high energies are exceedingly low, yet the cosmic ray physicists and astrophysicists may well have the last laugh.

Let us now turn to the physics produced by accelerators.

Bringing a new accelerator to life almost invariably demands refinement, modification, if not complete overhaul of detection techniques. An early example is the first observation of artificially made charged pions. These had certainly been produced in Berkeley before they were discovered in cosmic rays. There was one small detail, however: no one could detect them at first (19b). Their observation shortly afterward was followed by the discovery of the neutral pion (19b), the second example of a particle to be found at an accelerator. The electron, discovered half a century earlier in cathode ray tubes, had of course been the first.

Pion physics was now on its way. Towards the end of 1951 the first clear evidence had been obtained for the existence of a resonance in π-nucleon scattering (19d1). A further resonance was found two years later; others rapidly followed. Thus began the new and still evolving branch of particle physics known as baryon spectroscopy (19e), followed, beginning in 1961, by meson spectroscopy (19e).

The detection of the first meson resonances was the result of an important new development in detection technique: the invention of the bubble chamber. The first version of this device (1953) was a two-cubic-centimeter chamber filled with superheated fluid. By 1975 a 35 000 litre-chamber was completed. Since the late fifties, these and other huge detectors have once again changed the ways experiments are done (19e). Not only the building of accelerators, but also the execution of a single experiment had to be handled by consortia. Data analysis had to be computerized. Data storage in libraries of tapes had to be seen to. A leader of an experiment had now to add managerial skills to scientific excellence.

After this sketch of some of the experimental developments let us see what theorists had been up to meanwhile.

First the good news, the great leap forward in quantum electrodynamics.

It began with the Shelter Island conference (June 1947). At this unexpectedly fruitful meeting (to use Oppenheimer's words) we heard that a new generation of refined spectroscopic experiments in hydrogen and other materials gave the first evidence that corrections to the leading order predictions of quantum electrodynamics had been observed (18a).

It had been known since the 1930s (as noted) that such corrections were infinite, however. Now, with the prospect of new confrontations with experiment, this old problem was revisited with far greater care than had been exercised earlier. These further explorations showed that, without modification of basic principles, all infinities can be absorbed into a redefinition of a *finite* number of constants. This procedure is termed renormalization. In the case of electrodynamics the constants are (up to a fine point) the electron's charge e and mass m, both of which are infinite according to the theory, but of course finite in real life. Renormalization is essentially a trade-off: wherever e and m appear, declare them to have their experimental values. Even then there remain corrections to scatterings and spectral line positions, but those, it turns out, are now all finite and provide tests for the validity of quantum electrodynamics to a hitherto inaccessible level of precision!

I hasten to add that this renormalization program is technically delicate and demanding. It evolved in stages. First it was applied to the lowest relevant order in α. This rapidly led to the demonstration that, quantitatively, the new program works very well (18c). In 1949 the validity of the program had been proved to all orders in α (18c3).

Evaluations of radiative corrections to ever higher order in α have continued. The most advanced of these (the order is α^4), truly a *tour de force*, was completed in the early 1980s. It gives the magnetic moment of the electron to nine significant figures and agrees with the ever more precise experiments. That is the highest level of concordance reached in particle physics. No literature explosion here. We owe these outstanding achievements to a small band of superb experimental and theoretical craftsmen.

Meanwhile, back in the late 1940s, meson theory was in deep trouble.

This theory contains a dimensionless constant which is the analog of α in electrodynamics. Estimates from nuclear forces showed this nuclear coupling constant to be huge, about 15 (19c1), whence the name 'strong interactions'. Thus the expansion in powers of α had no analog in meson theory—and alternative methods did not exist. Application of renormalization to the magnetic moments of proton and neutron gave absurd results (19c2). Attempts at fitting the rapidly increasing body of data on pion processes were failing badly (19c3). The situation was much worse (19f1) than for quantum electrodynamics in the thirties when at least there had been a leading approximation that worked well. All these difficulties were to remain unresolved for a long time. A conversation* between Dirac and Feynman during the twelfth Solvay conference (1961) conveys the mood of the period.

F. I am Feynman.
D. I am Dirac. (Silence.)
F. It must be wonderful to be the discoverer of that equation.
D. That was a long time ago. (Pause.)
D. What are you working on?
F. Mesons.
D. Are you trying to discover an equation for them?
F. It is very hard.
D. One must try.

The resulting lack of consensus on the future of field theory set in motion a marked diversity of efforts at avoiding stagnation (19f). Some believed that the theory simply did not apply to strong interactions and went on to explore alternative avenues. Others turned to a critical re-examination of the foundations of field theory; that led to the creation of a nearly autonomous branch of mathematical physics, axiomatic field theory. Still others focussed on those limited consequences of the theory that remain valid even for a large coupling

* I was sitting at the same table and, as is not my custom, jotted it down right afterward.

constant. In that last category falls an increased reliance on symmetry arguments (19d). None of these efforts produced new physical insights at the fundamental level of particle dynamics, yet many results of lasting importance were obtained during that period.

Evidently the failures of meson field theory, a distinct setback, did not result in unproductive gloom. In fact the fifties were a wonderfully active decade, not just because of the explorations of new theoretical alternatives but also because nature had meanwhile been kind enough to provide new sources of entertainment.

The great question, What business do V-particles have being there? is still unanswered. There were more mundane issues, however. The production mechanisms of V-particles are strong interactions of a new kind. Their decays are caused by weak interactions, also of a new kind. Soon it became clear that an orderly synthesis of the new with the old particles demanded the existence of new kinds of quantum numbers with associated selection rules that are obeyed by strong and electromagnetic but not weak interactions. This led to the proposal of associated production according to which, loosely speaking, the new particles are produced in pairs (20b1). That mechanism found a more refined and highly successful expression through the introduction of a new quantum number, strangeness (20b2). It was realized early that new quantum numbers of this kind indicate the need for an enlargement of isospin to a bigger symmetry, resulting in a hierarchy of interactions, strong, electromagnetic, weak, with progressively less symmetry (20b1). That idea found its implementation in the 1960s (see below).

The strangeness scheme demanded the existence of two neutral K-particles, denoted by K^0 and $\bar{K}^0$, with distinct production, scattering, and absorption properties. Only one neutral K was known at the time (1954) when the decay properties of K^0 and $\bar{K}^0$ came under close scrutiny. The results (20b3) were novel. If, as is common, one defines an unstable particle as having a unique lifetime, then neither K^0 nor $\bar{K}^0$ are particles but, each of them, mixtures of particles—as was verified experimentally a few years later. This situation, unique thus far, makes the neutral K-complex into a veritable laboratory for the study not just of properties of specific particles but also of far more general aspects such as the superposition principle of quantum mechanics. Studies of the latter kind continue.[28] In the words of one veteran of experimental neutral K physics: 'There is scarcely a physical system that contains so many of the elements of modern physics'.[29]

The advent of new particles had made physicists become increasingly accustomed to selection rules, conservation laws, and symmetries that were, from the day they were introduced, meant to apply to some but not to all interactions. Then, in 1956, something different came to pass. Symmetries, earlier believed to hold for all interactions, turned out not to be obeyed by weak interactions. It was the most startling development in all the physics of the 1950s.

At issue were the symmetries with respect to spatial reflections (P), reversal of time (T), and the interchange of matter and antimatter, also known as conjugation of charge (C). By then each of these principles had its own distinct and distinguished history (20c2). As 1956 arrived it was known that the validity under the combined operation CPT could be derived from very general principles of relativistic quantum field theory (20c3). At that time weak decays of K-particles gave some indication, however, that possibly, only possibly, something was amiss with P all by itself (20c1), in other words that parity, the quantum number associated with P, was not conserved.

Pursuing this clue, Lee and Yang noticed that invariance under P, C, and T had never been tested in weak interactions and proposed ways for doing so (20c5). Half a year later a series of brilliant experiments showed that P and C are violated in a variety of weak interaction processes: β-decay, μ-decay, π-decay (20d1).* There appeared to be nothing wrong with T, however.

As a consequence of these revelations the Fermi theory reached a higher level of precision that found its expression in the so-called 'V−A theory' (20d4). The dynamics of weak processes now had reached a stage somewhat reminiscent of electromagnetism at the time when the laws of Coulomb and of Biot and Savart had just been formulated. Also the understanding of the interplay between strong and weak forces improved considerably (20d5).

The V−A theory of 1957 contains the third clue to the existence of the W-boson (21e1).

From 1957 to 1964 it was believed that the new lessons could be subsumed under an elegant new rule: symmetry not with respect to C, P, and T separately but to 'combined inversion' CP; and to T. Experimental indications for these principles looked promising. In 1964 came the next blow. CP is also violated and so, it transpired soon after, is T (20e). Numerically these new violations are very small and very puzzling. As Fitch, one of the discoverers of these effects, said so well[29] in 1980: 'After 16 years the world is still astonished by CP and T noninvariance'. In fact the breakdown of all these symmetries signals clearly that there are two things we need to know much more about: symmetries and weak interactions (20f).

As a fitting conclusion to this part of the story I recount a conversation** with Dirac. 'I asked Dirac why he had not introduced parity in his famous book on quantum mechanics. With characteristic simplicity Dirac answered: "Because I did not believe in it", and he showed me a paper in the *Reviews of*

* It is quite possible though not certain that an experiment performed in 1928 on double scattering of β-particles from a radium source had given valid evidence for parity violation.[29]

** Dates are important here. The conversation took place in 1959 and was recorded in a lecture given in 1963. In those times the violation of P but not of T was known. The lecture was published[30] in 1968.

Modern Physics[31] in which he said so. In that same paper, Dirac expressed doubts about time reversal invariance as well, but, four years ago, I did not pay much attention to that. Unjustifiedly . . . '.

We are now about to enter the 1960s. Following the primitive and illuminating method of marking progress in terms of the highest particle energy attained in a laboratory, let us see how far we have come during the preceding sixty-five years and how far we still have to go in the twenty-odd years remaining.

Roentgen discovered X-rays by accelerating electrons to a tenth of an MeV. Next came the naturally radioactive sources, yielding a few MeV, then the cyclotrons and the synchrotrons. Between 1895 and 1960 the energy had been raised by a factor of one hundred thousand. Another factor of ten thousand was needed before the W and Z could be discovered. Thus on the scale of energy we are a little over half way.

There exist no equally simple criteria characterizing experimental and theoretical advances. In qualitative terms, enormous progress during the sixties and especially the glorious seventies has drastically altered the way we now think about matter and forces in the physical world and has radically changed our perception of the unresolved problems and difficulties encountered in earlier times.

In deciding how to treat the period 1960–83, I was faced most acutely with the earlier mentioned predicament of dealing with an explosive amount of information in flux. On the one hand, some account of the most recent past is evidently called for in order to place earlier developments in a proper context. On the other, instant replay is not the noblest form of history nor even of memoir. I concluded (21a) that at this time and place an essay on the main issues is preferable to an elaborate account and that the introductory comments on that period in this chapter should correspondingly be confined to a brief statement of the main themes, of which there are three.

(1) The 1960s as the decade of new symmetries (21b). New clues from hadron spectroscopy* helped solve the long-standing question (20b1) of what symmetry joins together isospin and strangeness. The answer turned out to be the symmetry known as SU(3) (20b2). In turn, that insight led to the radical hypothesis (21b3) that all hadrons, protons, neutrons, mesons, . . . are, one might say, little atoms composed of new, finer subunits of matter: quarks and antiquarks, particles with fractional electric charge. Quarks had not been seen then as free, isolated objects, nor have they since. In the seventies this absence was greeted as a blessing in disguise.

Elaboration of the quark picture led to further symmetries. One, 'chiral $SU(3)\times SU(3)$', is important for the understanding of the structure of weak

* Since 1962 baryons and strongly interacting mesons are collectively referred to as hadrons.

and electromagnetic currents (21b6). The other, 'static SU(6)', is the most useful tool for systematizing hadron states with mass $\leq$ 2 GeV (21b4). On the whole, the style of the sixties was to admit that quarks came in handy for coding information but should not be taken seriously as physical objects. Originally (1964) it was believed that there are three species of quarks. Now (1985) it is thought that there are at least six (21e6).

Further analysis of static SU(6) led to the conjecture that there exists a new attribute, 'color', non-zero for quarks but always zero, and therefore *hidden*, for the familiar hadrons. Color was associated with yet another symmetry, 'color SU(3)', or $SU(3)_c$, which, it would turn out, is the crucial ingredient for the description of the strong forces. Important bits and pieces such as these were gathered in the sixties (21b7) and put together in the seventies.

(2) Accelerator design during the last quarter-century evolved in three principal directions.

Just about 1960, circular proton accelerators broke into the region beyond 10 GeV as the result of a fundamental advance in magnet design: the strong focussing principle (19a). These new machines have made it possible to extend the accessible energy region to near 1000 GeV (at the time of writing).

The long-known principle of resonance acceleration (17b) spurred major advances in the design and construction of linear accelerators, the most prominent one being the 3 km-long electron accelerator at SLAC in Stanford. The first round of experiments at SLAC led to a discovery (1968) of seminal importance, comparable to the observation of the hard α-nucleus scattering sixty years earlier (9d): high energy inelastic electron–proton (or neutron) scattering is hard; very many more electrons are deflected at large angles than had been anticipated. These most surprising data showed regularities ('scaling') that could be understood if nucleons, when deeply probed, consist of almost free particles, 'partons', that were eventually associated with quarks. The parton model was, one might say, the last missing link in the chain of events that led to the new theory of strong forces.

Experiments with neutrino beams, another major new tool (21c2), have shown that the same parton picture also applies to neutrino–nucleon scattering. That result followed another important discovery also made with these beams: the existence of a second kind of neutrino.

The third new class of accelerators are the colliders. All accelerators mentioned in the preceding are of the fixed target type: a beam hits a target at rest. The fraction of energy wasted in the totally uninstructive process of moving along the center of mass of the system (beam particle plus target particle) increases with energy and reaches, for example, 85 per cent when a 100 GeV proton hits a proton at rest. Colliders, essentially two beams of particles hitting head on, avoid most if not all of this waste. The study of colliders began in the 1950s. The most advanced model at this time, the proton–antiproton collider at CERN, operates at an energy of 270 GeV per beam, which is the equivalent of a fixed target energy of 1.5×10^5 GeV—whence the factor ten thousand mentioned earlier.

This wonderful gain in energy is bought at the heavy price of low event rate. Fixed targets, usually solid foils, provide ever so many more chances for hits by a beam than will occur when two colliding beams, exceedingly dilute gases, encounter each other. Both types of accelerators have their distinct manifest destiny in future developments.

(3) The closing theme of this entire book: the renaissance of quantum field theory, which has brought about an unprecedented high level of synthesis between theory and experiment in the physics of particles and fields.

The theories of strong and weak interactions have been cast in an altogether new frame, the gauge theories. In these theories forces are transmitted by fields with quanta of unit spin. Each such theory has an associated characteristic symmetry. The oldest example is the electromagnetic field, where the quantum is the photon and the symmetry goes by the name U(1). It had long been known that there are infinitely many other gauge theories (described in (21e2)). These had remained no more than elegant structures in search of application, however. Two major technical developments brought them to life: the discovery that the renormalization methods, known from quantum electrodynamics, could be applied here too; and the recognition of novel methods for breaking the associated symmetries (21e4).

The following brand new picture has emerged.

Quantum chromodynamics, the new theory of the strong interactions, is a gauge theory (21e3). Its quanta are eight massless particles, gluons. Its basic units of matter are the quarks. Quarks and their forces behave somewhat like two balls held together by an unbreakable elastic string. The balls appear free when close since, in this condition, the string is slack. That behavior—it is called asymptotic freedom—corresponds to the high energy, short distance, behavior we learned from the parton model just mentioned. The balls cannot escape each other, the string gets tight. That property, called confinement, explains why quarks have not been seen: hadrons are made of quarks that cannot get out. Confinement has not yet been rigorously demonstrated but there are excellent reasons for believing it to be true. Efforts are under way to calculate the hadron mass spectrum and the Yukawa-like forces between the good old hadrons as consequences of the basic quark–gluon forces. Confirmations of other predictions (21e6) have led to the currently common opinion that the theory is in essence correct.

The weak and electromagnetic theories have been unified in another gauge theory (21e4). The symmetry is named $SU(2) \times U(1)$. Its four quanta are: W^+, W^-, and the neutral Z, all massive; and the photon. The theory incorporates all earlier features of the weak processes and predicts new ones, most notably the existence of a new current, known as the neutral current, of no use for lighting the home but just as fundamental as the electromagnetic current.

A consistent description of the neutral current demands the existence of a new species of quarks, 'charmed' quarks. Their discovery (21e5) reaffirmed that the theory was moving in the right direction. More particle discoveries have followed since. There is now evidence for six quark species and an equal number of leptons, e, μ, τ, and their associated neutrinos (21e6).

The discoveries of the W and Z, moments of triumph, mark the capstone of this period and the end of this book.

Not only did the theory predict their existence, but experimental information from the neutral current had made it possible to predict their masses. The discovery of these particles with the right masses brings us to the inwardmost encounter between theory and experiment, at a high point in the history of physics.

A few scattered reflections on these recent years.

While it has of course always been true that theory and experiment prosper by mutual stimulation, there has recently occurred a shift in balance between the two, an increased reliance of experiment on theory in the planning of many—not all—complex and costly major projects. This change is caused by the fact that physical reasoning becomes less intuitive as one advances to ever higher energies. To give but one example, the UA1 and UA2 experiments at CERN were only conceived and approved because theory had not merely predicted the existence of W and Z bosons, but had moreover given cogent reasons for expecting the masses of these particles to lie in the 80–100 GeV range.

Recent theoretical developments have been accompanied once again by a marked increase in complexity of mathematical techniques. The quantum theory of the newly accepted gauge field equations is vastly more intricate than that of any of the fields encountered earlier. The search for the manifold of solutions to these equations, even at the classical level, poses problems not yet fully comprehended at this time. In fact, the theories of strong and weak interactions just sketched are so far not completely developed (22).

Yet, the first steps in these new directions are firm. The physics of particles and fields has taken a recent turn so definitive that already now we can look back with much improved clarity on the earlier parts of the postwar era.

Nucleons and hyperons, π's and K's, and their higher resonances, are composite structures. Meson–baryon interactions, the Yukawa forces, are no longer considered basic but rather as low energy derivative effects of fundamental quark–gluon interactions. It remains a challenge to derive the former from the latter. The approximate symmetries of isospin and its extension to SU(3) will continue their role as guides to the classification of hadron spectra and related parameters. The true strong dynamics resides in the deeply hidden SU(3) color symmetry, however. Fermi's theory, which will forever adequately describe β-decay and related low energy weak processes, is an excellent low energy approximation to a field theory of weak interactions mediated by heavy vector bosons.

A number of new problems attracting widespread attention at the time of writing are enumerated in a brief epilog (22). These include: grand unification, a theory that should tie together SU(3) color and SU(2)×U(1) in a higher synthesis; and the holy grail: the unification of gravity with all other forces in nature, the problem to which Einstein, too early and too narrowly, had devoted so much fruitless effort. It is now conjectured that a new kind of symmetry, tantalizingly tight and elegant, supersymmetry,[32] another achievement of the seventies, will help to incorporate gravity.

And so it goes, on and on, amidst a great deal of excitement.

And also of confusion.

In brief, business is continuing as usual.

References

1. M. Banner *et al.*, *Phys. Lett.* **122B**, 476, 1983.
2. G. Arnison *et al.*, *Phys. Lett.* **122B**, 103, 1983.
3. G. Arnison *et al.*, *Phys. Lett.* **126B**, 398, 1983.
4. P. Bagnaia *et al.*, *Phys. Lett.* **129B**, 130, 1983.
5. T. Macaulay, 'History', repr. in *Essays*, Vol. 1, p. 387, Sheldon, New York 1860.
6. T. Carlyle, 'On history', repr. in *A Carlyle reader*, Ed. G. B. Tennyson, Random House, New York, 1969.
7. H. Bloom, *The anxiety of influence*, p. 5, Oxford University Press 1973.
8. J. K. Galbraith, *N.Y. Times Book Review*, p. 55, October 21, 1984.
9. P. Forman, J. Heilbron, and S. Weart, *Hist. Stud. Phys. Sc.* **5**, 3, 1975.
10. Cf. *Minerva Jahrbuch der gelehrten Welt*, Vol. 10, 1900–1901, Teubner, Strassburg 1901.
11. K. T. Compton, *Physics Today* **5**, 4, Feb. 1952.
12. Held at the Cavendish Laboratory, Cambridge, July 22–7, 1946; Taylor and Francis, London 1947.
13. L. Leprince-Ringuet, *Phys. Rev.* **70**, 791, 1946.
14. Cf. A. Sommerfeld, *Phys. Zeitschr.* **16**, 89, 1915.
15. A. Pais, 'Conservation of Energy', in *Symmetry in physics*, Proc. Sant Felice de Guixols conf. Spain, 1984.
16. H. L. Bronson, in *The collected papers of Rutherford*, Ed. J. Chadwick, Vol. 1, p. 163, Allen and Unwin, London 1962.
17. K. Eichhorn, *Schriftenreihe Deutsches Roentgen Museum*, Nr. 6, 1984.
18. H. Becquerel and P. Curie, *Comptes Rendus* **132**, 1289, 1901.
19. N. F. Mott, *Notes and Rec. Roy. Soc.* **27**, 65, 1972.
20. N. Feather, Ref. 19, p. 50.
21. A. Pais, *Subtle is the Lord*, Oxford University Press 1982.

22. P. A. M. Dirac, *Proc. Roy. Soc. A* **133**, 60, 1931.
23. C. N. Yang, *Ann. N.Y. Ac. Sci.* **294**, 86, 1977; R. Jackiw, *Comm. Nucl. and Part. Phys.* **13**, 141, 1984; J. Preskill, *Ann. Rev. Nucl. Ptcle Sc.* **34**, 461, 1984.
24. E. N. da C. Andrade, *The structure of the atom*, 3rd edn, p. 140, Bell, London 1927.
25. C. D. Ellis, *Proc. Roy. Soc. A* **161**, 447, 1937.
26. H. A. Bethe and M. E. Rose, *Phys. Rev.* **52**, 1254, 1937.
27. W. V. Jones, in *Cosmic rays and particle physics—1978*, p. 41, AIP Conf. Proc. Number 49, Am. Inst. Phys., New York 1979; cf. further T. K. Gaisser and G. B. Yodh, *Ann. Rev. Nucl. Ptcle Sci.* **30**, 475, 1980; A. A. Watson and G. B. Yodh, in *Cosmology and particles* Edition Frontières, Paris 1981.
28. Cf. W. C. Carithers *et al.*, *Phys. Rev.* **D14**, 290, 1976.
29. V. L. Fitch, *Rev. Mod. Phys.* **53**, 367, 1980.
30. A. Pais, 'Invariance principles', in *Spectroscopical and group theoretical methods in physics*, ed. F. Bloch, North Holland, Amsterdam 1968.
31. P. A. M. Dirac, *Rev. Mod. Phys.* **21**, 393, 1949.
32. J. Weiss and J. Bagger, *Supersymmetry and supergravity*, Princeton University Press 1983; P. van Nieuwenhuizen, *Phys. Rep.* **68**, 189, 1981.

Part One

1895–1945: A History

Unconfusion submits
its confusion to proof; it's
not a Herod's oath that cannot change.

Marianne Moore, *The mind
is an enchanting thing*

2

New kinds of rays

Dam: What did you think?
Roentgen: I did not think; I investigated.
Dam: What is it?
Roentgen: I don't know.

(a) Roentgen: X-rays

On the evening of February 19, 1903, William Joseph Hammer, an engineer and close collaborator of Thomas Alva Edison, gave a lecture on 'the much talked of radium' before the New York Academy of Medicine. A brief account of his talk is the first item on radioactivity ever to appear in the *New York Times*.[1] In his lecture, Hammer quoted Pierre Curie as saying that 'he would not like to be in the same room with a kilogram of the substance, which would probably burn the skin off his body and the eyes out of his head'.

In a letter to the *New York Times* of 28 June, 1903, Theodore William Richards, professor of chemistry at Harvard, mentioned the 'highly interesting facts discovered by Roentgen, Becquerel, M. and Mme. Curie, Thomson and Rutherford'.

In a preview of the forthcoming 1904 International Electrical Congress in St. Louis, the *St. Louis Post Dispatch* of 4 October, 1903 carried an article headed: 'Priceless mysterious radium will be exhibited in St. Louis. A grain of the most wonderful and mysterious metal to be shown in St. Louis in 1904'. The text contained these lines: 'Its power will be inconceivable. By means of the metal all the arsenals in the world would be destroyed. It could make war impossible by exhausting all the accumulated explosives in the world .... It is even possible that an instrument might be invented which at the touch of a key would blow up the whole earth and bring about the end of the world'.[2]

1903 was the first year in which radioactivity caught the fancy of the press and the public. Popular books on the subject were not long in coming. The first of these (I believe), entitled 'Radium', written by Hammer,[3] appeared in 1903. In 1904, another such book was published[4] by the physicist Robert John Strutt, the later fourth Baron Rayleigh. In its preface we read that this author too was stimulated by 'the extraordinary properties of radium [which] have excited general interest outside the scientific world'. In discussing the discovery of radioactivity, Strutt remarked: 'It seems a truly extraordinary coincidence that so wonderful a discovery should result from the following up of a series of false clues. And it may well be doubted whether the history of science affords any parallel to it.' Briefly stated (and presently to be discussed at more length) the false clues were these: After the discovery of X-rays it was thought for a short while that these rays might have something to do with phosphorescence. Uranic salts are phosphorescent. Therefore one should (and did) experimentally examine whether the phosphorescence radiations of uranic salts contain X-rays (of course they don't). As will be seen in what follows, these conjectures, so bizarre after the fact, do much to explain the rapidity with which two discoveries followed each other: X-rays on 8 November, 1895; and radioactivity (not as yet so named) on 1 March, 1896. Because of this link, it seems natural to combine in one chapter the work of the respective discoverers: Wilhelm Conrad Roentgen and Antoine Henri Becquerel. This juxtaposition will also facilitate the recognition of striking parallels in the destinies of the two men. Roentgen's traits: '[He] did not belong to the types who bubble with ideas . . . His strength lay in another talent . . . the unremitting critique of the reliability of physical observations and measurements',[5] are not so different from those of Becquerel: 'He had little taste for physical theories . . . His greatest asset was a strong persistent power of afterthought. On those rare occasions when [he] did pursue a hypothesis, this critical power continuously corrected his enthusiasm, and redirected his line of investigation'.[6]

In 1895 we find Roentgen in Würzburg as professor of physics and director of the Physics Institute of the Pleicher Ring, the upper floor of which provided him and his family with commodious living quarters including a well-stocked wine cellar.* Wilhelm and his wife Bertha were devoted to each other and to their niece Berta Ludwig whom they reared and later adopted. (They had no children of their own.) Theirs was a life in which the social graces were not neglected.

Roentgen's main devotion was physics, however. Before he discovered X-rays, at age fifty, he had published 48 papers which had already earned him an excellent name. Indeed, in 1894 he had been unofficially approached for the Directorship of the Physikalisch Technische Reichsanstalt in Berlin, a position of great power and prestige, made vacant by the death of Helmholtz;[9] but Roentgen preferred the life of research and teaching** in Würzburg. To mention but two of the topics to which he had contributed, his earliest papers (1870, 1873) deal with the specific heats of gases. Of this work it has been said: 'Roentgen's first papers, which appeared in the early seventies of the last century and dealt with the determination of the relation of the specific heats of gases, already show that he possessed the essential qualities of the research scientist in the most marked degree—the fine flair in the realization of the problem, the experimental skill in carrying out the examination, the keen

* See Refs. 7 and 8 for a detailed account of Roentgen's younger years.

** In 1895, his weekly teaching consisted of five hours lecturing on experimental physics and ten hours of demonstration lectures in the laboratory.[10]

critical sense which took into account to the greatest possible extent the sources of error and their effect upon the result and thus obtained within the limits of possibility a sharply defined measure of certainty in the final answer'.[11] In 1888, Roentgen, inspired by Maxwell's theory of electric displacement, raised the question: can the motion of an uncharged dielectric placed in a constant homogeneous electric field generate a magnetic field? The answer is yes, of course, since the electric field induces a surface charge distribution in the dielectric. Roentgen was the first to obtain experimental confirmation of this effect by observing deflections in a nearby magnetometer.[12] Lorentz later referred to this induced current as the Roentgen current.[13]

Ludwig Zehnder, Roentgen's assistant at that time, has given a revealing account of Roentgen's ways of working: 'When Roentgen had set himself a problem, he always worked at it quietly and secretly, without allowing anybody a glimpse into his method of working and thinking. Thus I knew nothing of the experiments [on the induced current] with which he was still occupied. But as this Roentgen effect was so weak that to be certain of the movement of the magnetometer required the greatest care, he fetched me one day and told me to look through the telescope. He then said to me that he would make an experiment of which I could see nothing; I was then to tell him whether I could notice anything at the cross-threads of the telescope. I actually did see a very slight movement to the left . . . When the experiment was repeated, I saw a movement of about the same extent to the right. I had to make several such readings without being allowed to know what Roentgen was doing or whether the movement was to be expected towards the right or the left. Roentgen wanted to have his own readings controlled by an unprejudiced observer. It was only after its appearance in the publications of the Berlin Academy that I learned of this discovery of Roentgen's'.[14] This conduct earned him the nickname *der Unzugängliche*, the unapproachable.[15]

This same urge for secrecy also marked events in November 1895.

Roentgen had by then become interested in the experimental results on cathode rays obtained by Heinrich Hertz and the latter's student Philipp Lenard, in particular in Lenard's method of passing these rays (not yet identified as electrons) a short distance outside a vacuum tube (more on that in Chapter 4). Roentgen was in the course of repeating these experiments when, on the afternoon of Friday, 8 November, he returned downstairs to the quiet of his laboratory. What he observed that day is described in the opening lines of his first paper[16] on X-rays, entitled: *Über eine neue Art von Strahlen. Vorläufige Mittheilung* (On a new kind of rays. Preliminary communication):

'If the discharge of a fairly large Rühmkorff induction coil is allowed to pass through a Hittorf vacuum tube . . . and if one covers the tube with a fairly close-fitting mantle of thin black cardboard, one observes in the completely darkened room that a paper screen painted with barium platinocyanide placed near the apparatus glows brightly or becomes fluorescent with each discharge, regardless of whether the coated surface or the other side is turned toward the discharge tube. This fluorescence is still visible at a distance of two meters from the apparatus.

It is easy to prove that the cause of the fluorescence emanates from the discharge apparatus and not from any other point in the conducting circuit.*

In Frau Roentgen's words, the days following that Friday were terrible. Her Willi would come late to dinner, in a state of bad humor. He ate little, did not talk at all, and returned to his laboratory immediately afterward. Nor did his two assistants know what was happening. As Roentgen recalled later, when he first made the discovery of what he was to name X-rays, he was so extraordinarily astonished that he had to convince himself over and over again of their existence. 'To my wife I mentioned merely that I was doing something of which people, when they found out about it, would say, "*Der Roentgen ist wohl verrückt geworden*"' (Roentgen has probably gone crazy).[17] Not until December 28, when he was quite certain and had made a number of observations on X-ray properties did he submit his *Vorläufige Mittheilung* to the Sitzungsberichte of the Physikalisch-medizinische Gesellschaft in Würzburg.[16] He published again on the same subject in March 1896 and in March 1897. Thereafter he never wrote on X-rays again, although he remained scientifically active until a few days before his death in 1923.[5]

The first mention of X-ray photographs is found in Roentgen's very first paper: 'I have photographs . . . of the shadows of the bones of the hand; . . . of a set of weights enclosed in a small box; of a compass in which the needle is entirely enclosed by metal; . . . and so on'. On January 1, 1896, he sent preprints of his preliminary communication to distinguished colleagues in Germany and abroad.** Having done so, he said to his wife: 'Now the fun can start'.[14]

It did.

The worldwide sensation immediately following the publication of Roentgen's work was obviously due to his X-ray photographs. The first daily paper to publish the news was the *Wiener Presse* of 5 January. Two days later the *Frankfurter Zeitung* and the *Standard* of London followed. Thereafter the news spread everywhere. Under the heading 'Hidden Solids Revealed' the *New York Times* published its first article, sensibly written, on 16 January. On 9 February we find in the same paper: 'The wizard of New Jersey [Edison] will try to photograph the skeleton of a human head next week.' (He did not succeed.) In the single year 1896 more than fifty books and pamphlets and more than one thousand scientific or popular scientific articles on the subject were published.[7] On 13 January, Roentgen gave a demonstration before

* Here I use the translation given in Ref. 8 which contains the complete English text of this and the two subsequent Roentgen papers on X-rays.

** Professor A. Romer kindly informed me how this preprint distribution worked. It appears that the practice was limited to Germany, specifically to papers published in the proceedings of local scientific societies. The system was to submit a paper to the secretary of such a society, who endorsed it as accepted. Then one took it to the printer of the proceedings, who set it in the proper format and ran it off immediately as a pamphlet.

Emperor Wilhelm II in Berlin. Otherwise, Roentgen shunned publicity. 'In a few days I was disgusted with the whole thing. I could not recognize my own work in the reports anymore.'[17] He refused interviews, but one enterprising London-based American reporter, Mr. W. J. H. Dam, managed to get through to him. The lines at the head of this chapter are taken from Dam's very interesting interview which one finds reproduced in its entirety in the major Roentgen biographies.[7,8]

Also in the scientific community the response was strong and immediate. Full-length translations of Roentgen's paper appeared in *Nature* (23 January 1896), in *Science* (14 February), and in other learned journals. Among colleagues who wrote letters of congratulation were Kelvin, Poincaré, and Stokes.[18] On 25 January, a young research student named Ernest Rutherford wrote from Cambridge to his fiancée: 'The Professor [J. J. Thomson] has been very busy lately over the new method of photography discovered by Professor Roentgen . . . The Professor, of course, is trying to find out the real cause and nature of the waves, and the great object is to find the theory of matter before anyone else, for nearly every Professor in Europe is now on the warpath'.[19]

On 23 January, Roentgen addressed the Physical–Medical Society at Würzburg on X-rays, the only public lecture he ever gave on this subject. Geheimrat von Köllicker, the famous histologist, was in the audience. After Roentgen's talk 'Köllicker proposes to name the new discovery "Roentgen rays". (Tumultuous acclamation.) Deeply moved, Roentgen expresses his gratitude'.[20] On informal occasions, Roentgen expressed himself well in public, but formal addresses made him uncomfortable in the extreme[21] as is seen, for example, from his reactions on the occasion of his Nobel award in 1901. Upon receiving word of his Prize he wrote to Svante Arrhenius of his intent to come to Stockholm in person. 'For that I will need leave of absence which will probably not be denied me.'[22] He did get his leave and, on 10 December, was present in the Great Hall of the Swedish Academy of Music, where the Swedish Crown Prince handed him his insignia. At the subsequent banquet he delivered an eloquent speech of thanks,[23] but left the next day without giving a Nobel Lecture. He reacted with great shyness to an invitation to return for his Lecture. Zehnder, aware of Roentgen's anxieties, offered to accompany him. Nothing came of it, the Lecture was never given.[23]

In 1905, a plaque commemorating the tenth anniversary of the Roentgen rays was attached to the Physics Institute in Würzburg, at the instigation of a group of most distinguished physicists including Boltzmann, Lorentz, and Planck.

During his later years, Roentgen kept engaged in physics, published little, and tended to avoid meetings with colleagues.[5] The end came in 1923, after a short period of illness during which he kept his own records of his symptoms. In his obituary in the *Annalen der Physik*, Wilhelm Wien wrote of him: 'He knew like no other that Nature plays Her game with us, and that also the experienced and careful researcher is exposed to the danger, time and again, to be caught in one of the traps which have been set around Nature's secrets and which hinder their uncovery.'[5]

In concluding this account of Roentgen and his work on X-rays, I return to his first article on that subject.

There must be few communications marked 'preliminary' which contain such a wealth of meticulously gathered experimental information as does Roentgen's paper. Here is only part of what he reported:

a) He concluded that the effects he had observed were caused by a radiation distinct from cathode rays, in particular because air absorbs his new rays much less than cathode rays.

The cathode ray data used by Roentgen had been obtained in 1894, in 'Lenard's . . . beautiful experiments'. It should be stressed how markedly Roentgen's work was stimulated by Lenard's earlier studies.[24] Lenard had studied cathode rays extracted from the tube in which they were produced; so did Roentgen with his rays. Lenard encased his tube; so did Roentgen. Lenard used phosphorescent materials, including platinocyanides, to follow his rays by visual means; so did Roentgen. Lenard noticed that his rays blacken photographic plates; so did Roentgen. One wonders, therefore, why Lenard himself did not discover X-rays.

The answer, I believe, lies in the kind of encasement with which Lenard had surrounded his tube 'in order to protect the detection space from the light and the electric forces of the discharge'. It consisted of lead and tinned iron. The thickness of the encasement is not recorded, but there is no reason why it should be very thin. In Lenard's experiment the cathode rays escaped from the tube through a very thin aluminum 'window' in its glass wall, and from there entered the detection area through a 0.17-mm-wide aperture in the encasement. It is almost certain, therefore, that this encasement absorbed nearly all the X-rays—in contrast to Roentgen's arrangement in which (as said) the tube was enclosed by 'thin black cardboard'. I do not know of course why Roentgen did not follow Lenard's procedure at this point. But of such details is the kingdom of discovery.

b) Roentgen studied X-ray effects as a function of absorber thickness and density, and, for crystals, as a function of preferred crystallographic directions.

c) He noted also that agents other than barium platinocyanide can be made to fluoresce.

d) He observed the sensitivity to X-rays of photographic dry plates and at once started to use photographic detection methods.

e) From the formation of regular shadows he concluded that the name 'rays' for his phenomenon was justified.

f) He found that even intense magnetic fields do not deflect the rays.

g) When X-rays travel through air, their intensity follows an inverse square law.
h) He looked for but did not find: any appreciable refraction; or regular reflection; or polarization.

Roentgen concluded his paper with a speculation on the nature of X-rays which gives a most interesting perspective on the thinking about electromagnetic radiation prevalent in those days:

> . . . We have known for a long time that there can be in the ether longitudinal vibrations besides the transverse light-vibrations, and, according to the views of different physicists, these vibrations must exist. Their existence, it is true, has not been proved up to the present, and consequently their properties have not been investigated by experiment.
>
> Ought not, therefore, the new rays to be ascribed to longitudinal vibrations of the ether?
>
> I must confess that in the course of the investigation I have become more and more confident of the correctness of this idea, and so, therefore, I permit myself to announce this conjecture, although I am perfectly aware that the explanation given still needs further confirmation.

Roentgen's erroneous conjecture shows that one of the most fundamental qualitative traits of electromagnetism was not clear to him; a light ray vibrates only in directions transverse to its motion. His reference to different physicists indicates that he was not alone in this. Indeed, just two months before he did his work, an article had appeared by George Francis FitzGerald (who in 1899 would be awarded a Royal Medal by the Royal Society 'for his contributions to theoretical physics, especially in the domain of optics and electrodynamics') in which the author analyzed the radiation emitted by a Hertz oscillator and concluded that 'it is evident on the most cursory consideration that these waves must have a longitudinal region'.[25] On 15 January 1896, Boltzmann said in a lecture on X-rays: 'In all elastic bodies, especially in gelatin, the velocity of propagation of longitudinal waves is much larger than for transverse waves. If one supposes the same to be true for the light-aether, then the Roentgen waves would have rather large wavelengths in spite of having a very small vibration period'.[26] Shortly thereafter the *New York Times* quoted Sir Oliver Lodge's opinion that X-rays are longitudinal propagations of the aether.[27] One could no doubt find other similar examples, all of which show that the contents of electromagnetic theory were still far from being understood more than twenty years after the appearance of Maxwell's treatise on electricity and magnetism–aether models still confounded the issue.

Even those who questioned Roentgen's conjecture and correctly pointed out that the absence of refractive effects can be understood if X-rays are a transverse radiation but with wavelengths much shorter than visible light were not always willing to discard the longitudinal option out of hand. Lorentz was noncommittal. In an 1896 lecture he said: 'The phenomena do not in any way forbid seeing in the rays of Roentgen the propagation of a wave motion of a special kind'.[28] Arthur Schuster from Manchester wrote: 'The great argument against the supposition of waves of very small length lies in the absence of refraction; but is this conclusive? . . . It is not my intention to argue in favour of any particular theory, or against Roentgen's suggestion that we have at last found the formerly missed longitudinal wave. I only desire to put those points forward which at first sight seem to go against the supposition of ordinary light vibrations'.[29] Also J. J. Thomson pursued the idea that the waves might be transverse, but found from polarization experiments that the evidence was conflicting.[30] In France, the question was raised by Poincaré. After recalling the similarity between Hertzian and visible rays he asked: what prevents us from thinking that the X-rays form a further extension of this spectrum? Again the lack of refraction made him conclude that such is not the case. By March 1896 the Frenchman Charles Henry interpreted the X-rays correctly as 'ultra-ultra-violet' light and emphasized that refraction causes no problem because of the wavelength dependence of the refractive index.[32] Later that year, J. J. Thomson summarized the situation in a lecture before the Liverpool Science Congress in September 1896: 'Though there is no direct evidence that the Roentgen rays are a kind of light, there are no properties of the rays which are not possessed by some variety of light'.[33] Through the following years the transverse wave interpretation gained in acceptance, but it was not until 1906 that Charles Barkla achieved partial polarization of X-rays[34], and not until 1912 that the issue was fully settled when Max von Laue had the brilliant idea of using crystals as grids to refract the rays, and Walter Friedrich and Paul Knipping performed the first X-ray refraction experiments on crystals.[35]

(b) Becquerel: Uranic rays

1. January 20–February 24, 1896. 'Le 1er mars 1896, Henri Becquerel a découvert la radioactivité.'* Thus, nearly thirty years later, begin the recollections by his son Jean[36] of what happened in Paris one overcast Sunday. In order to appreciate what Henri, then forty-three years of age, had done on that first of March, it will help first to go back two months and see how the news of Roentgen's discovery had reached France.

Le Matin published the X-ray story on 13 January. *L'Illustration* added a detailed description on 25 January. On 27 January, the first French contribution to X-ray physics was communicated by Jean Perrin.[37] It begins as follows: 'J'avouerai d'abord que je n'ai sur la découverte du professeur Roentgen que des renseignements assez vagues, tirés des journaux quotidiens, et que j'ignore encore quelles sont, au juste, ses expériences'.** Then, on 8 February,

* On the first of March, 1896, H. B. discovered radioactivity.

** I should admit first of all that I have only rather vague information, obtained from the daily papers, on Professor Roentgen's discovery, and that I do not know precisely what his experiments are.

L'Eclairage Electrique published a translation of Roentgen's original publication.

Meanwhile, X-rays had been a main topic at the 20 January meeting of the French Académie des Sciences. Two French physicians, Oudin and Barthélemy, had submitted their first X-ray picture of bones of the hand. Poincaré, charged to present this communication, was one of the recipients of the preprints which Roentgen had distributed, and used the occasion to report on the content of Roentgen's paper. In 1903, Becquerel described his own participation in the subsequent discussion as follows: 'I asked my colleague [Poincaré] whether one had determined the place from which these rays were emitted . . . I was told that the origin of the radiation was the luminous spot on the wall which was hit by the cathode rays'.[38] Here Poincaré had in mind the following phrase from the preprint: 'According to experiments especially designed to test the question, it is certain that the spot on the wall [of the vacuum tube] which fluoresces the strongest is to be considered as the main center from which the X-rays radiate in all directions'.[16]

Continuing with his recollections of the 20 January session at the Académie, Becquerel recorded his reaction to Poincaré's response: 'I thought immediately of investigating whether the new emission [X-rays] could not be a manifestation of the vibratory movement which gave rise to the phosphorescence and *whether all phosphorescent bodies could not emit similar rays* [my italics]. I communicated this idea to Monsieur Poincaré and the very next day I began a series of experiments along this line of thought . . .'[38]

On 30 January, 1896, a semi-popular article by Poincaré appeared[39] in which he repeated Becquerel's conjecture, but without acknowledging him: 'Ne peut-on alors se demander si tous les corps dont la fluorescence est suffisament intense n'émettent pas, outre les rayons lumineux, des rayons X de Roentgen *quelle que soit la cause de leur fluorescence*?* [The italics are Poincaré's.] As a result of this article, the idea of a possible connection between luminescence and X-rays is often incorrectly attributed to Poincaré, as witness for example the obituary by W. Wien in praise of Poincaré's significance for physics.**

The Poincaré article attracted immediate and widespread attention. Right away there appeared in the *Comptes Rendus* several statements alleging that new effects has been seen, generated by phosphorescent compounds of zinc

* Can it not be asked then whether all bodies whose fluorescence is sufficiently intense would not emit, apart from the luminous rays, the X-rays of Roentgen, *whatever be the cause of this fluorescence*?

** Wien wrote as follows: 'Eine wichtige Anregung ist von Poincaré ausgegangen, indem er nach der Entdeckung der Röntgenstrahlung auf die Möglichkeit hinweis daß dieses Phänomen mit der Fluorescenz in Zusammenwirkung stehen könnte. Wenn diese Auffassung auch nicht richtig war, so hat sie doch die erste Veranlassung zu den Versuchenvon Becquerel gegeben, die dann später zur Entdeckung des Radiums geführt haben.' (An important suggestion came from Poincaré who, after the discovery of the Roentgen radiation, pointed out the possibility that this phenomenon might be related to fluorescence. Even though this view was not correct, it gave the first stimulus for the experiments of Becquerel which later led to the discovery of radium.)[40]

and of calcium.[41,42,43] Elsewhere, experimental results were published purporting to establish a correlation between the existence of penetrating rays and a variety of luminescent phenomena—including the studies of glow-worms and fire-flies.[44]

Meanwhile, Becquerel was at work. For some weeks he had tested the ability of several phosphorescent bodies to generate X-rays but had found no effects. Yet he did not despair: 'However, in spite of the negative experiences with the other bodies, I had great hopes for the experimentation with the uranium salts the fluorescence of which I had studied on an earlier occasion, following the work of my father . . . among the samples of uranium salts I possessed were very beautiful lamellas of potassium uranyl disulphate which I had prepared some fifteen years earlier . . .* These references to much earlier work call for a digression on that remarkable foursome of lineal descendants, the Becquerel physicists.**

The first of these was Henri's grandfather, the prolific Antoine César. In 1838 a chair of physics was created for him at the Museum of Natural History in Paris which he occupied until his death. His main interest was thermoelectric phenomena. He also studied luminescence, as is seen in Vol. 4 of his seven volume *Traité expérimental de l'électricité et du magnetisme* (1834–40). The first published drawings of phosphorescent spectra appear to be[47] figures in his book *Traité de physique* (1844).

Already as a young man, Antoine César's third son, Alexandre Edmond, was also drawn to the study of luminescence and assisted his father in his experiments. Thus in 1839 a paper on the subject appeared, authored by father, son (then aged nineteen), and Jean Baptiste Biot.[48] Eventually the study of phosphorescent solids became Edmond's main field of endeavor which he practically monopolized during his most productive years.[49] He invented the phosphoroscope with which fluorescence lifetimes as short as 10^{-4} sec can be measured.[50] Among the many substances which he analyzed were compounds containing uranium, an element known since 1789. Indeed he was the first to note that uranic salts are, uranous salts are not phosphorescent.[51] Of particular interest for what follows is Edmond's remark in 1858: 'The bodies which produce the most brilliant [phosphorescent] effects are uranium composites'.[52] (The phosphorescence of uranium compounds was studied in more detail by Stokes[53] who, incidentally, coined the term 'fluorescence'.†)[54] Much of

* Becquerel had lent these lamellas to a colleague. He started to work with them on the day they were returned to him.

** See Refs. 45 and 46 for more details about the Becquerel family.

* In the early literature the terms 'luminescence', 'fluorescence', 'phosphorescence' were used in a fairly arbitrary fashion. It may be helpful to recall the current definitions of those terms.[55] Luminescence: luminous emission which is not purely thermal in origin. Fluorescence: the emitting system is raised to an excited state after which it spontaneously returns to the ground state. (Typical lifetimes are 10^{-8} sec.) Phosphorescence: the system is raised to an excited state. From there it drops into a metastable state where it is trapped until energy is supplied to bring it back to the excited state and thence to the ground state. (Typical lifetimes: $\lesssim 1$ sec.)

Edmond's work is discussed in his book *La Lumière, ses causes et ses effets* (1867). The first of its two volumes is largely concerned with phosphorescence. After this authoritative account no general book on phosphorescence and fluorescence appeared until the next century.[56] Upon his father's death in 1878, Edmond become the next professor at the Museum.

History repeated itself when Antoine Henri, Edmond's second son and our main character, began his scientific career by assisting his father. His early work included studies on infrared spectroscopy and on the absorption of light by crystals. He was the first to observe the phenomenon of rotatory magnetic polarization in gases, a contribution which received favorable comments from Fizeau.[57] His reputation grew rapidly and in 1889 Henri, only in his mid-thirties, was elected to the Académie des Sciences. After the death of his father, two years later, he became the next occupant of the professorship at the Museum. He was of course quite conscious of belonging to a dynastic tradition. His friend Sir William Crookes wrote later:[57] 'Henri Becquerel was acutely interested in the strange fact that three generations of his family had worked successively in the same branch of physical enquiry. This continuity of research held him in a charmed bondage, and irresistibly he was urged to follow the track disclosed by his kinsmen'.

One last time a Becquerel son would come forth to assist his father, to publish jointly with him, to succeed him at the Museum after his death, and to be elected to the Académie des Sciences—Jean, Henri's only descendant. Jean started publishing in 1904. His first love was magneto-optics, a subject on which his grandfather Edmond had already done some work. When he retired in 1948, the chair of physics at the Museum ceased to be occupied by a Becquerel for the first time in 110 years. He spent most of his final years in his villa *Ar Bann* (Breton for 'The Rays') in Brittany. Upon his death in 1953, the lineal Becquerel membership of the Académie des Sciences came to an end. All four physicists rest in Chatillon, in the Loiret.

Of his work on radioactivity, Henri is reported to have said later:[57] 'These discoveries are only the lineal descendants of those of my father and grandfather on phosphorescence, and without them my own discoveries would have been impossible'.

Let us turn next to February 1896, and see how Henri used his uranium compounds in the search for X-rays.

2. 24 February–2 March, 1896. On 24 February, Becquerel submitted his first paper on the subject to the Académie des Sciences. It is entitled 'On radiations emitted in phosphorescence'. The techniques described were photographic and qualitative, typical for the first generation of these investigations. The phosphorescent material used in these investigations were lamellas of potassium uranyl disulfate, $K_2UO_2(SO_4)_22H_2O$, 'which I had prepared some fifteen years earlier'.[58] The gist of the experiment was, in Becquerel's own words: 'One wraps a photographic plate . . . in two sheets of very thick black paper . . . so that the plate does not fog during a day's exposure to sunlight. A plate of the phosphorescent substance is laid above the paper on the outside and the whole is exposed to the sun for several hours. When the photographic plate is subsequently developed, one observes the silhouette of the phosphorescent substance, appearing in black on the negative. If a coin, or a sheet of metal . . . is placed between the phosphorescent material and the paper, then the image of these objects can be seen to appear on the negative'.[59] These procedures are quite reminiscent of Roentgen's arrangement except that the cathode ray tube is replaced by a phosphorescent source pre-excited by sunlight. However, in this first paper there is neither mention of Roentgen nor of X-rays. Becquerel's conclusion is sober and sound: 'The phosphorescent substance in question emits radiations which traverse paper opaque to light.'

On February 26 and 27 there was not much sun in Paris. As a result, Becquerel was back in the Académie the next Monday, 2 March, to report[60] an altogether remarkable discovery: the effect he had observed a week earlier had nothing to do with phosphorescence!

Crookes, who had the opportunity 'on more than one occasion to visit Henri Becquerel's laboratory and to assist at the research of the moment', was there at the time and has given the following description of what had happened. 'He had devised another experiment in which, between the plate and the uranium salt, he interposed a sheet of black paper and a small cross of thin copper. On bringing the apparatus into daylight the sun had gone in, so it was put back in the dark cupboard and there left for another opportunity of insolation. But the sun persistently kept behind clouds for several days, and, tired of waiting (or with the unconscious prevision of genius) Becquerel developed the plate. To his astonishment, instead of a blank, as expected, the plate had darkened as strongly as if the uranium had been previously exposed to sunlight, the image of the copper cross shining out white against the black background. This was the foundation of the long series of experiments which led to the remarkable discoveries which have made '"Becquerel rays" a standard expression in Science.'[57]

This, then, was Becquerel's discovery on Sunday, 1 March, when the cloud deck over Paris was still heavy.[61] There was another eyewitness to the event, Henri's eighteen-year-old son Jean, who later recalled that, 'Henri Becquerel fut stupéfait [H.B. was stupefied] . . . when he found that his silhouette pictures were even more intense than the ones he had obtained the week before'.[62]

Once again, Becquerel stated his own conclusions soberly: the new effect lies 'outside the phenomena which one might expect to observe'. It may be due to 'invisible phosphorescent radiation emitted with a persistence infinitely greater than the persistence of luminous radiation'. His present experiments 'are not contrary to this hypothesis, yet do not prove it'. More experiments were promised.

Much has been written about this accidental discovery. Was it genius? Was it luck? Was it a check on an expected null result? Perhaps it was some of each. Becquerel himself seems to have felt it was destiny:[62] '... Il disait que les travaux qui, depuis une soixantaine d'années s'étaient succédé dans ce même laboratoire formaient une chaine qui devait fatalement, quand l'heure serait propice, aboutir à la radioactivité'.*

The case of the English physicist Silvanus P. Thompson may serve as a counterpoint to Becquerel's experiences. Thompson twice missed his rendezvous with destiny, in the 1870s when he almost discovered electromagnetic waves, and in 1896 when he almost discovered radioactivity. At about the same time as Becquerel, he had observed independently and by virtually identical methods the property of some uranium salts to blacken photographic plates.[63] He too had put various phosphorescent materials on a sheet of aluminum under which a photographic plate was placed.** The arrangement was 'left for several days upon the sill of a window facing south to receive as much sunlight as penetrates in February into a back street in the heart of London'.[65] Among the materials so studied he found that only uranium nitrate and uranium ammonium sulfate produced an effect on the plate. Thompson quickly informed Sir George Stokes who immediately replied, urging Thompson to publish. But only a few days later, Stokes wrote again: 'I fear you have been anticipated ...', referring Thompson to Becquerel's paper of 24 February. Thompson did publish eventually, in July 1896. But he never made the key observation that his correct findings were unconnected with phosphorescence ....

3. The rest of the year. Becquerel's laboratory has been described as follows: 'What struck one as remarkable was the facility with which experimental apparatus was extemporized. Card, gummed paper, glass plates, sealing wax, copper wire, rapidly and almost spontaneously seemed to grow before one's eyes into just the combination suitable to settle the point under investigation. The answer once obtained, the materials were put aside or modified so as to constitute a second interrogation of nature'.[57] This was the environment in which Becquerel went right back to work on the next day, 3 March. He recorded meticulously how he again placed uranium salts on a screened plate,

* He [H.B.] said that the investigations, which during sixty years had followed one another in this same laboratory, formed a chain which, at the propitious hour, were ineluctably to end up with radioactivity.

** Also J. J. Thomson was intrigued by the phosphorescence: 'A very noticeable feature in the bulb producing these Roentgen rays is the phosphorescence of the glass in the bulb. I thought it therefore of interest to try if these rays were generated when the phosphorescence of the glass was produced by other means than the discharge from a negative electrode'. This is stated in a paper dated 27 January in which Thomson reports on an experiment with a ring discharge in an electrodeless bulb.[64] As it should be, he found no evidence for new phenomena.

at 4 p.m. that day. All this was kept in darkness now till 5 March, 4:30 p.m.: he found the same effect as before, but now for a *variety* of uranium compounds.

What intrigued him above all else was the peculiar spontaneity of the phenomenon, the absence of a 'cause excitatrice'. In order to pursue this point further, he mounted another similar experiment on 3 March, but this time kept his arrangement in the dark until 3 May. The result: his new radiation did not weaken perceptibly even over this two months' period. The same remained true as he lengthened the span of time, eventually up to seven years: 'All experiments show qualitatively the quasi-permanence of the phenomenon'.[66]

Another most important property of the new invisible rays was reported by Becquerel on 9 March: they are able to discharge an electroscope.[67] On 23 March he communicated his findings that the effect is independent of the crystalline state of the uranium compounds; and that the effect decreases as the distance between the uranium compounds and the photographic plate increases.[68]

The 9 March paper contained one result which is in error, namely the observation of reflection and refraction properties of the rays. A subsequent experiment appeared to reconfirm this and to show polarization of the rays as well.[69] These results added confusion to paradox. Thus in one of his 1896 Christmas Lectures at the Royal Institution, Thompson said: 'There appears to be no doubt that the uranium rays are a species of extreme ultra-violet light'.[70] Shortly thereafter, Poincaré made the following comparison between X-rays and those found by Becquerel: the former do not reflect and refract (so it seemed at the time); the latter do (or so again it seemed). Moreover, the latter are polarized by tourmaline (not so) and 'cette dernière propriété ne peut appartenir qu'à des ondes transversales. *Les rayons Becquerel sont donc des rayons lumineux*'.[71]*

By May it had become clear to Becquerel that all uranium salts he had examined, whether uranic or uranous, give an effect. Since only the former kind are phosphorescent (as his father had noted) this settled for him once and for all the lack of correlation between the new rays and phosphorescence. He now began to wonder:[72] 'J'ai donc été conduit à penser que l'effet était du à la présence de l'élément uranium dans ces sels, et que le metal donnerait des effets plus intenses que ses composés'.** Are the rays attributable to the element uranium? An experiment with metallic uranium confirmed his guess.

As 1896 draws to a close we see Becquerel continue his studies of the absorption of the rays as a function of material and of thickness. But above all he keeps wondering about the source of energy of what he now calls 'les

* This last property can belong only to transverse waves. The Becquerel rays are therefore lightlike rays.

** I have thus been led to think that the effect was due to the presence of the element uranium in these salts, and that the metal would give more intense effects than its composites.

rayons uraniques',[73] which are ceaselessly emitted by uranium kept in darkness: 'one has not yet been able to recognize wherefrom uranium derives the energy which it emits with such persistence'.

The findings in 1896, a year of culmination in the Becquerel saga, were summarized by Poincaré in terms which sound curiously modern:[71] 'It seemed that the path which was followed by the Becquerels was destined to lead to a dead end. Far from it, one can think today that it will open for us an access to a new world which no one suspected'. Referring both to Roentgen and Becquerel, Poincaré added that other phenomena 'will doubtlessly come . . . and complete a picture of which we barely begin to see the outline'.

In later years Becquerel published numerous brief communications on radioactivity. In 1903 he completed a major review of the subject.[38] Later that year he received half the Nobel Prize for Physics (the other half went to the Curies) 'in recognition of the extraordinary service he has rendered by his discovery of spontaneous radioactivity'. In 1907 he was elected vice-president, in 1908 president of the Académie des Sciences, once again following in the footsteps of his grandfather and his father, both of whom had held these posts in earlier times. Later in 1908 he was elected one of the Perpetual Secretaries of the Académie. Shortly thereafter he unexpectedly died, the victim of a heart attack.

Sources

On Roentgen. The biographies by Glasser[7] and by Nitske[8] are very useful. Nitske's book contains English translations of Roentgen's three papers on X-rays. The sketch by Zehnder,[14] who worked with Roentgen for a number of years, contains revealing personal reminiscences.

On Becquerel. Many details about the Becquerel family are found in Ref. 45. Becquerel's own memoir[38] contains a detailed account of his own work on radioactivity well into 1903, and also a bibliography of his and others' work in the years 1896–1903. English translations of Becquerel's early papers on this subject are found in an anthology by Romer.[74]

The work of Roentgen and Becquerel is also discussed in a recent book by Segré.[75]

References

1. *New York Times*, 20 February, 1903.
2. Quoted in G. Jauncey, *Am. J. Phys.* **14**, 227, 1946.
3. W. J. Hammer, *Radium*, Van Nostrand, New York 1903.
4. R. J. Strutt, *The Becquerel rays and the properties of radium*, Arnold, London 1904.
5. W. Wien, *Ann. der Phys.* **70**, 332 et seq., 1923.

6. A. Romer, in *Dictionary of scientific biography*, Vol. 1, p. 560, Chief Ed., C. Gillispie, Scribner, New York 1970.
7. O. Glasser, *Wilhelm Conrad Roentgen*, 2nd Edn, Springer, Berlin 1959; in English: same title, C. C. Thomas, Springfield, Ill. 1934.
8. W. R. Nitske, *The life of Wilhelm Conrad Roentgen*, Univ. of Arizona Press, Tucson 1971.
9. Ref. 8, p. 80.
10. Ref. 8, p. 84.
11. P. P. Koch, *Zeitschr. techn. Phys.* **4**, 273, 1923.
12. W. C. Roentgen, *Ann. der Phys. u. Chem.* **35**, 264, 1888.
13. H. A. Lorentz, *Enc. d. Math. Wiss.* **5**, part 2, p. 63, Section 17, Teubner, Leipzig 1904; cf. also A. Sommerfeld, *Electrodynamics*, pp. 283–4, Academic Press, New York 1964.
14. L. Zehnder, 'Wilhelm Conrad Roentgen', in *Lebensläufe aus Franken*, Vol. 4, Becker, Würzburg 1930; published in English translation as a pamphlet by Imprimerie James Guichard, Neuchâtel 1930.
15. Ref. 8, p. 212.
16. W. C. Roentgen, *Sitzgber. physik.-med. Ges. Würzburg*, **137**, Dec. 1895.
17. W. C. Roentgen, letter to L. Zehnder, 8 February, 1896, repr. in Ref. 8, p. 100.
18. Ref. 8, p. 103.
19. In N. Feather, *Alembic Club reprint No.* 22, Livingston, Edinburgh 1958.
20. *Fränkisches Volksblatt* Nr. 19, 24 January, 1896.
21. Cf. Ref. 8, p. 162.
22. W. C. Roentgen, letter to S. Arrhenius, 16 November, 1901, repr. in Ref. 23.
23. F. Knutsson, *Acta Radiologica* **8**, 449, 1969.
24. P. Lenard, *Ann. der Phys. u. Chem.* **51**, 225, 1894.
25. G. F. FitzGerald, *Phil. Mag.* **42**, 260, 1896.
26. L. Boltzmann, *Populäre Schriften*, p. 196, Barth, Leipzig 1896.
27. *New York Times*, 25 February, 1896.
28. H. A. Lorentz, *De Gids* **60**, 510, 1896, repr. in *H. A. Lorentz, collected papers*, Vol. 9, p. 149, Nyhoff, The Hague 1939.
29. A. Schuster, *Nature* **53**, 268, 1896.
30. J. J. Thomson, *Nature* **53**, 581, 1896.
31. H. Poincaré, *Rev. Gén. des Sciences* **7**, 52, 1896.
32. C. Henry, *Comptes Rendus* **122**, 787, 1896.
33. J. J. Thomson, in *Sci. Amer.* **75**, 328, 1896.
34. C. G. Barkla, *Proc. Roy. Soc. A* **77**, 247, 1906.
35. W. Friedrich, P. Knipping, and M. von Laue, *Ber. Bayer. Akad. Wiss.* **303**, 1912.
36. J. Becquerel, *La radioactivité et les transformations des éléments*, Libraire Payot, Paris 1924.
37. J. Perrin, *Comptes Rendus* **122**, 186, 1896.
38. H. Becquerel, *Recherches sur une propriété nouvelle de la Matière*, p. 3, Mem. de l'Ac. d. Sci. Paris, Firmin–Didot 1903.
39. H. Poincaré, *Rev. Gén. des Sc.* **7**, 52, 1896.
40. W. Wien, *Acta Mathematica* **38**, 289, 1921.
41. C. Henry, *Comptes Rendus* **122**, 312, 787, 1896.
42. G. H. Niewenglowsky, *Comptes Rendus* **122**, 385, 1896.
43. M. Troost, *Comptes Rendus* **122**, 565, 1896.
44. Cf. L. Badash, *Am. J. Phys.* **33**, 128, 1965, and Ref. 38, pp. 4–7.

45. Lamothe, in *Conférences prononcées à l'occasion du cinquantième anniversaire de la découverte de la radioactivité*, p. 35, Muséum d'Hist. Naturelle, Paris 1946.
46. L. Badash, *Arch. Inst. d'Hist. des Sc.* **18**, 55, 1965.
47. E. N. Harvey, *A history of luminescence*, p. 208, Am. Philos. Soc., Philadelphia 1957.
48. A. C. Becquerel, J. B. Biot, and E. Becquerel, 'Mémoire sur la phosphorescence produite par la lumière électrique', *Arch. du Muséum d'Hist. Naturelle, Paris* **1**, 215, 1839.
49. Ref. 47, p. 359.
50. E. Becquerel, *Comptes Rendus* **46**, 969, 1858.
51. E. Becquerel, ibid. **75**, 296, 1872.
52. Ref. 50, p. 971.
53. G. G. Stokes, *Mathematical and physical papers*, Vol. 3, p. 341; Johnson Repr. Corp., New York 1966.
54. Ref. 53, p. 289, footnote.
55. Cf. D. Curie, *Luminescence in crystals*, Wiley, New York 1963.
56. Ref. 47, p. 221.
57. W. Crookes, *Proc. Roy. Soc. A* **83**, xx, 1910.
58. Ref. 38, p. 8.
59. H. Becquerel, *Comptes Rendus* **122**, 420, 1896.
60. H. Becquerel, ibid. **122**, 501, 1896.
61. L. Badash, *Isis* **57**, 267, 1966.
62. J. Becquerel, in *Conférences* cited in Ref. 45.
63. Cf. L. Badash, Ref. 44.
64. J. J. Thomson, *Proc. Cambr. Phil. Soc.* **9**, 49, 1896.
65. S. P. Thompson, *Phil. Mag.* **42**, 103, 1896.
66. Ref. 38, p. 15.
67. H. Becquerel, *Comptes Rendus* **122**, 559, 1896.
68. H. Becquerel, ibid. **122**, 689, 1896.
69. H. Becquerel, ibid. **122**, 762, 1896.
70. S. P. Thompson, *Light visible and invisible*, p. 281, Macmillan, London 1897.
71. H. Poincaré, *Rev. Scientifique*, **7**, 72, 1897.
72. H. Becquerel, *Comptes Rendus* **122**, 1086, 1896.
73. H. Becquerel, ibid. **123**, 855, 1896.
74. A. Romer, *The discovery of radioactivity and transmutation*, Dover, New York 1964.
75. E. Segré, *From X-rays to quarks*, Freeman, San Francisco 1980.

3

From uranic rays to radioactivity

(a) The Curies: Becquerel rays

Unlike the acclaim accorded to Roentgen's work, the impact of Becquerel's discovery was neither immediate nor widespread. The press, of course, took no notice. Becquerel himself had turned his attention largely to the newly discovered Zeeman effect. In 1897, a few short papers did appear on the subject, including contributions by Kelvin[1] and Silvanus Thompson.[2] More importantly, however, in that year two newcomers were beginning to give serious thought to the uranic rays: Marie Curie and Ernest Rutherford. Their initial studies of the new rays mark the very beginnings of their respective scientific activities. The few topics from their early work selected for discussion in this chapter give but glimpses of what were to become illustrious careers.

The first time George Cecil Jaffe* spent a year at the Curies' laboratory in Paris was in 1905, when he did research under both Pierre and Marie Curie. They impressed him as a most unusual pair: 'There have been and there are, scientific couples who collaborate with great distinction, but there has not been a second union of woman and man who represented, both in their own right, a great scientist. Nor would it be possible to find a more distinguished instance where husband and wife with all their mutual admiration and devotion preserved so completely independence of character, in life as well as in science . . . I was most strongly impressed by their extreme simplicity and modesty together with their extraordinary devotion to their task . . . There was about both of them an unostentatious superiority . . . [Pierre's] disposition made *him* stand aloof when *she* entered upon something like a romantic enterprise: the search for an unknown element'.[3]

Marie Sklodowska, daughter of a Warsaw physics teacher, arrived in Paris in 1891 to study physics at the Sorbonne. She met Pierre Curie in 1894; the next year they married. Pierre's career was already well under way by then. He had done important work on piezo-electricity, together with his brother

* Later professor of physics and theoretical chemistry at Louisiana State University in Baton Rouge.

Paul-Jacques, on symmetries of crystals, and on magnetism. His papers had been well received by distinguished colleagues, notably by Kelvin. His position, director of laboratory work in the School for Industrial Physics and Chemistry, founded by the city of Paris, was a modest one, however. It appears that this was largely due to his aversion to soliciting advancements. As he wrote to Kelvin: 'What an ugly necessity is this of seeking any position whatsoever; I am not accustomed to this form of activity, demoralizing to the highest degree'.[4] After his death, Marie wrote of her husband: 'He sometimes said that he never felt combative, and this was entirely true. One could not enter into a dispute with him because he could not become angry'.[4]

Soon after Marie had completed her first paper, dealing with magnetism of tempered steels, she heard of Becquerel's work, which she discussed with her husband. 'The study of this phenomenon seemed to us very attractive . . . I decided to undertake the study of it . . . In order to go beyond the results reached by Becquerel, it was necessary to employ a precise quantitative method'.[4]

As she thus embarked on what would be her life's task, Marie had just become a mother, after a difficult pregnancy. On 12 September 1897 Irène, the Curies' older daughter, was born. Pierre's mother died a few days thereafter; his father came to live with the young couple and was of great help in taking care of the newborn child. On 12 April 1898, when baby Irène was exactly seven months old, her mother's first paper was presented to the Académie des Sciences. It is a short document[5] which contains three major new points. Before describing these, let me first note what experimental techniques she used and, for that purpose, briefly step back two years.

The study of the conduction of electricity through gases had received much stimulus from the discovery of X-rays. Within months of Roentgen's discovery, their ability to discharge electrified bodies had been discovered almost simultaneously in several laboratories. One of these was the Cavendish, where the subject was pursued by J. J. Thomson together with his student McClelland.[6] In the course of their work they developed a new tool, an early form of the parallel plate ionization chamber. One of two plane electrodes is grounded, a voltage is applied to the other and the leakage of electricity between them is studied as the dielectric between the plates is traversed by X-rays. One of their principal discoveries was the saturation of the leakage as the voltage increases.[7]*

This same method was used by Marie Curie as she set out to study Becquerel's phenomenon with greater precision. From the latter's work she knew the qualitative fact that uranic rays, like X-rays, can discharge electrified bodies.** She decided to investigate this effect quantitatively by observing the

* For the early history of the subject see Thomson's 1896 Princeton Lectures;[8] for a more up-to-date version see Ref. 9.

** See Chapter 2.

rate of leakage obtained if one of the two condensor plates is covered with a fine layer of a uranium compound or some other substance. As had been the case with Becquerel, hers was a table-top experiment. Her condensor plates, 3 cm apart, were each 8 cm in diameter. A modest 100 volt potential difference and a sensitive electrometer were all she needed for her work. Now to her three results.

1) She not only reconfirmed Becquerel's earlier findings for uranium but also discovered a new active substance: thorium. 'Thorium oxide is even more active than metallic uranium'. While she discovered the activity of thorium independently, she was not the first to do so. Unbeknownst to her, the German physicist Gerhard Carl Schmidt from Erlangen had reported the same result somewhat earlier.*

Marie Curie also analyzed 'a large number of metals, salts, oxides, and minerals' which were found to show no activity. Nevertheless, the thorium discovery by itself sufficed to make clear that the new rays were not uniquely associated with uranium and that therefore the appellation 'uranique' had become too narrow. In her next paper[11] a more general name was introduced: 'rayons de Becquerel', Becquerel rays, a term which would be used for a number of years.** In the title of that same paper the expression 'radioactive substance' makes its appearance for the first time.

2) Even more important is the second result announced in her first paper. As we have seen (Chapter 2) Becquerel had concluded that the ability of uranium compounds to emit the new rays is a property of uranium itself. Marie Curie had a more quantitative statement to make on this subject: 'All uranium compounds are active, the more so, in general, the more uranium they contain'.

This experimental conclusion is not as precise as the theoretical statement made later, in 1902, by Rutherford and Soddy:[13] radioactive bodies contain unstable atoms of which a fixed fraction decays per unit time,† which implies, of course, that the radioactivity produced by a substance does not just increase with the amount of substance but, more precisely, is proportional to that amount. It should be recalled, however, that the amount of activity produced is not identical with the amount of activity observed, since radioactive radiations are in general partially absorbed within the finite thickness of the active material. (As will be seen in later chapters, that complication would plague experimentalists for decades.) This circumstance may explain in part why Marie Curie only noted an increase, not a proportionality, with increasing

* Schmidt had communicated his results to the Deutsche Physikalische Gesellschaft in Berlin on 4 February 1898. His paper on the subject[10] was submitted on 24 March. As said, Marie Curie's paper was submitted on 12 April.

** The name 'hyperphosphorescence' had briefly been used in the literature, but Marie Curie realized that this term 'gives a false idea of their [i.e. the rays'] nature'.[12]

† See further, Chapter 6.

amounts of uranium. Nevertheless, only nine months later she jumped to the right conclusion!

In a paper submitted in December 1898, of which she was one of the authors,[14] we find the following reference to her work in April: '*One of us* [*M.C.*] *has shown that radioactivity is an atomic property*' (my italics). It is the first time in history that radioactivity is explicitly linked to individual atoms. In 1920, Soddy, himself co-discoverer of the theory mentioned in the preceding paragraph, properly reminded his readers of 'Madame Curie's theory that radioactivity is an intrinsic property of the atom'.[15]

3) The April 1898 paper clearly shows that Marie Curie's insight into what constituted a 'normal' increase of radioactivity with increasing uranium content was sufficiently quantitative to note that two minerals, pitchblende (rich in uranium oxide) and chalcite (rich in uranyl phosphate), behaved anomalously: '[They are] much more active than uranium itself. This fact is very remarkable and leads one to believe that these minerals contain an element which is much more active than uranium'.[5] With this conjecture she introduced another novelty into physics: *radioactive properties are a diagnostic for the discovery of new substances*, the last of the three major points of her first paper on radioactivity.

Her next assignment was obvious: to verify whether her idea of a new element was indeed true. 'I had a passionate desire to verify this hypothesis as rapidly as possible. And Pierre Curie, keenly interested in the question, abandoned his work on crystals . . . to join me in the search.'[4] He was never again to return to his crystals. Jointly they investigated pitchblende by ordinary chemical methods. In July 1898, they announced that by precipitating with bismuth they had been able to isolate a product about 400 times as active as uranium. 'We believe that [this product] . . . contains a not yet observed metal . . . which we propose to call polonium, after the country of origin of one of us.'[11]

On 17 October, Marie Curie wrote in a private notebook: 'Irène can walk very well, and no longer goes on all fours.'[16]

The analysis of pitchblende continued, Gustave Bémont, a laboratory chief at Pierre's school, aiding the couple in their labors. Meanwhile, the Austrian government had presented them with a gift of 100 kg of residues of pitchblende, material (believed to be without commercial value) obtained from mining operations in Joachimsthal. Separation by chemical methods remained their main procedure. They worked 'in an old, by no means weatherproof, barrack in the yard of [Pierre's] school'.[3] These were no longer table-top operations. As Marie wrote later: 'It was exhausting work to move the containers about, to transfer the liquids, and to stir for hours at time, with an iron bar, the boiling material in the cast iron basin.'[4] On 26 December 1898, they announced[14] that, by precipitating with barium, they had found yet another radioactive substance in pitchblende: radium, a discovery which led Rutherford to remark: 'The spontaneous emission of radiation from this element was so marked that not only was it difficult at first to explain but also, what was more important, still more difficult to explain away'.[17]

On 5 January 1899 Marie wrote in her notebook: 'Irène has fifteen teeth'![16]

Radium was to make radioactivity known to the public at large. The attendant fame, which was to disturb the quiet and concentrated existence to which the Curies were so deeply attached, would 'deafen with little bells the spirit that would think'.[4]

Pierre Curie's discoveries of the experimental laws of piezo-electricity and of what became known as the Curie temperature in ferromagnetism are samples of his outstanding contributions to physics. He also labored mightily alongside his wife on the problems of radioactivity. But, in so far as one can rely on the record of published papers, it is Marie Curie, a driven and probably obsessive personality, who should be remembered as the principal initiator of radiochemistry. That is abundantly clear, it seems to me, from her April 1898 paper, discussed at such length above.

1898 was the heroic year in the Curies' careers. Further important work was to come, yet what followed was to a considerable extent painstaking elaboration of their early discoveries.

As to their later life: in 1900 Pierre was appointed assistant professor at the Sorbonne, Marie teacher at a high school for girls. Continued intense research and teaching duties were so exacting that she had no time until 1903 to complete her Ph.D. thesis, a masterful summary of her work to date. She received her degree on 25 June with the distinction 'très honorable', an elegant understatement. That day was memorable to her for another reason as well: in the evening she met Rutherford for the first time.

In November of that year the Curies were informed that they would share the 1903 Nobel Prize with Becquerel, 'in recognition of the extraordinary services they have rendered by their joint researches on the radiation phenomena discovered by Professor Henri Becquerel.' The Curies' prize was the subject of the following comment by the *New York Times* of 11 December 1903: 'The discoverers of radium have, it is understood, not profited financially from the work as greatly as might have been expected, and their admirers throughout the world will be delighted to hear of this windfall for them'.

The Curies were too unwell and overworked to attend the ceremony in person. When they finally went to Stockholm, in June 1905, only Pierre delivered a Nobel Lecture, as his wife sat and listened. Meanwhile, in 1904 a special chair had been created for him at the Sorbonne.

It appears that Pierre suffered from radiation sickness during the last years of his life, which came to a cruel end on 19 April 1906. He was not yet forty-seven years old. What happened on that day has been described by his wife.

'[As he] was crossing the rue Dauphine, he was struck by a truck coming from the Pont Neuf and fell under its wheels. A concussion of the brain brought instantaneous death. So perished the hope founded on the wonderful being he thus ceased to be. In the study room to which he was never to return, the water buttercups he had brought from the country were still fresh.'[4] In the introduction to his collected works,[18] published in 1908, Marie wrote of the shattered hopes for further research, in the French she mastered so exquisitely: 'Le sort n'a pas voulu qu'il en fût ainsi, et nous sommes constraints de nous incliner devant sa décision incompréhensible'.*

Marie's younger daughter Eve has written of her mother's intense grief after this event and of her determined efforts to avoid speaking of Pierre in later years.[16] Jaffe has left us his impressions of Madame Curie in 1911, when he spent another year at her laboratory as a Carnegie Scholar: 'The stranger saw and admired the power of the intellect. He could not know what was going on behind the air of self-constraint or almost impassiveness which became characteristic of Madame Curie after the death of her husband'.[3]

Shortly after Pierre's death, Marie was named his successor at the Sorbonne. It was the first time in that venerable institution's more than six-hundred-year-long history that a woman was appointed to a professorship. The Paris papers treated this as a major event. On Monday, 5 November 1906, at 1:30 in the afternoon, Marie began her inaugural lecture, continuing a discourse on radioactivity at precisely the point where Pierre had left off in his last lecture.

Marie Curie's life in 1911 was marked by two events, one humiliating, one gratifying.

In the autumn of that year, news items appeared in the French press, quoting private letters, and alleging the existence of an affair between Mme. Curie, widow, and the physicist Paul Langévin, married, a disgusting public treatment of private matters, probably true.[19] In November she received word that she had been awarded the Nobel Prize for chemistry for 1911, 'in recognition of the part she has played in the development of chemistry: by the discovery of the chemical elements radium and polonium; by the determination of the properties of radium and by the isolation of radium in its pure metallic state; and finally, by her research into the compounds of this remarkable element'.[20] In her Nobel Lecture, given on 11 December 1911, she recalled that many of these discoveries were made 'by Pierre Curie in collaboration with me . . . The chemical work aimed at isolating radium in the state of the pure salt . . . was carried out by me, but it is intimately connected with our common work . . . I thus feel that . . . the award of this high distinction to me is motivated by this common work and thus pays homage to the memory of Pierre Curie'.[20]

* Fate has not willed it to be so, and we are forced to bow before its incomprehensible decision.

During the First World War, Marie Curie organized and participated in the work of a number of radiology units for diagnostic and therapeutic purposes. In May 1922 she was named member of the League of Nations' International Committee on Intellectual Cooperation, and took an active part in its work for many years thereafter.

Among the distinctions bestowed on her were numerous honorary doctorates in science, medicine, and law. Meanwhile she continued her research activities. In all, she published about seventy papers, the last one in 1933. Soon thereafter, at age sixty-six, she died in a sanatorium in the Haute Savoye after a brief illness. The death report reads as follows: 'Mme Pierre Curie died at Sancellemoz on 4 July 1934. The disease was an aplastic pernicious anaemia of rapid, feverish development. The bone marrow did not react, probably because it had been injured by a long accumulation of radiations.'[16] She was buried at Sceaux, near her husband.

(b) Rutherford: α- and β-rays

In a report submitted on 27 February 1869, by a Cambridge University committee, it was recommended that 'There shall be established in the University a Professorship to be called the Professorship of Experimental Physics . . . It shall be the principal duty of the Professor to teach and illustrate the laws of Heat, Electricity, and Magnetism'.[21] On 8 March 1871 James Clerk Maxwell was elected as the first occupant of this position, which came to be known as the Cavendish professorship. Ernest Rutherford was not quite two months old when, on 25 October 1871, Maxwell gave his inaugural lecture. 'The opinion seems to have got abroad', Maxwell said in this beautiful talk,[22] 'that in a few years all the great physical constants will have been approximately estimated, and that the only occupation which will then be left to men of science will be to carry on these measurements to another place of decimals . . . But we have no right to think thus of the unsearchable riches of creation, or of the untried fertility of those fresh minds into which these riches will continue to be poured'. He urged his listeners to be devoted to their course: 'A man whose soul is in his work always makes more progress than one whose aim is something not immediately connected by his occupation'.

Maxwell also become the first director of the Cavendish Laboratory, built under his supervision at a cost of £8450, and formally presented to the university on 16 June 1874. The event was duly noted in an editorial in *Nature*: 'The genius for research possessed by Professor Clerk Maxwell and the fact that it is open to all students of the University of Cambridge for researches will, if we mistake not, make this before long a building very noteworthy in English science'. Among the apparatus present in the laboratory, the article noted, were an electro-dynamometer, a quadrant electrometer, three mirror galvanometers, a glass plate electric machine, and an ebonite electric machine.[23]

Upon Maxwell's untimely death in 1879, he was succeeded as director by the third Lord Rayleigh. J. J. Thomson followed Rayleigh in 1884 and was in charge during the period now under discussion.

Thomson was deeply immersed at that time in the problem of the conduction of electricity through gases. He had begun to form an opinion on the mechanism of the conduction induced by X-rays: 'The passage of these rays through a substance . . . seems to be accompanied by a splitting up of its molecules which enable electricity to pass through it by a process resembling that by which a current passes through an electrolyte[24] . . . [We may] regard the conduction as electrolytic, the gas being ionized by the Roentgen rays'.[25] Much remained to be done to put such ideas on a more quantitative basis, however, both theoretically and experimentally. At this stage he turned to Rutherford, who had arrived at the Cavendish half a year earlier, and invited him to work on these problems. Their collaboration started at the beginning of the Easter term, 1896.

Rutherford, born on a farm near the town of Nelson on the South Island in New Zealand, had received his early training on his native island. By 1893 he had obtained his MA at Canterbury College, in the city of Christchurch, where he also began research inspired by Heinrich Hertz's discovery in 1888 of the radio waves predicted by Maxwell's theory. His own efforts toward the development of a magnetic device for the detection of such waves were quite successful. In 1894–5 his first two papers appeared in the *Transactions of the New Zealand Institute*. When in 1895 he won a scholarship offered to New Zealand by the Commissioners of the Exhibition of 1851, he set out for the Cavendish, his detector equipment with him. After arrival in Cambridge he continued work on radio waves and was soon able to receive signals from about 1 km away. Because of this work he made right away a strong impression and one of the younger Cambridge men wrote of him at that time: 'We've got a rabbit here from the Antipodes and he's burrowing mighty deep'.[26]

The timing of Rutherford's arrival in Cambridge could not have been more auspicious. He first set foot in the Cavendish about one month before Roentgen's discovery. A university regulation had just come into effect by which graduates of other universities could be admitted to Cambridge as Research Students, a new category. Rutherford was the first of these.

Of his joint work with Rutherford in 1896 on gas ionization generated by X-rays, Thomson later said, 'Rutherford devised very ingenious methods for measuring various fundamental quantities connected with this subject, and obtained very valuable results which helped to make the subject "metrical" whereas before it had only been descriptive'.[27] This work,[28] dealing with ionization and recombination, has become a classic. No wonder then that Rutherford was drawn to the study of Becquerel rays. His first mention of the ionizing properties of 'uranium rays' is found in a paper completed in February 1898.

Rutherford's first research on this subject was done at the Cavendish. It was published after he had sailed for Canada, in September 1898. For he had meanwhile been appointed to a professorship at McGill University in Montreal, with an annual salary of £500. He was not that keen on leaving Cambridge—but he had had plans to marry ever since leaving New Zealand and was therefore desirous of material improvement. About McGill he wrote to his fiancée: 'I am expected to do a lot of original work and to form a research school to knock the shine out of the Yankees!'[30] In 1900 he went to New Zealand to get married. In 1901 the couples' only child, Eileen Mary, was born; she later married the physicist Ralph Fowler. Her early death, a heavy blow for Rutherford, attached him even more strongly to his four grandchildren of whom he was deeply fond.

Let us now turn to Rutherford's first paper on Becquerel rays,[31] the true beginning of his life's work. In it he established three important facts:

a) The refraction and polarization which Becquerel had believed to have found are not there. (There is some diffuse reflection, but this turned out to be a secondary effect.) Soon thereafter Becquerel concurred with this conclusion.[32]

b) The process of electric discharge transmitted by gases is due to ion formation.

c) The Becquerel rays are inhomogeneous. His absorption experiments 'show that the uranium radiation is complex, and that there are at present at least two distinct types of radiation—one that is very readily absorbed, which will be termed for convenience the α-radiation, and the other of a more penetrative character, which will be termed the β-radiation'.

From the absorption properties of α- and β-rays, we now know that the black wrapper in Becquerel's initial experiment* is thick enough to absorb essentially all the α-radiation. Thus what Becquerel originally saw was the effect of β-rays; and these were not 'uranique' but rather due to the first daughter product, thorium 234.

It took ten years to establish firmly that α-rays consist of particles which are about four times as heavy as hydrogen and which carry two positive units of electric charge, doubly ionized helium. Note that the terrestrial presence of helium was only ascertained in 1895, not long before the discovery of α-rays, when Sir William Ramsay,[33] much to his astonishment,[34] found helium to be present in a uranium-bearing mineral. 'An examination of the physical and chemical properties [of helium] . . . had hardly been completed when Ramsay and Soddy[35] made an examination of the gases liberated from radium and discovered that helium was a product of the transformation of radium.'[36] Because of the presence of helium in several uranium- and thorium-bearing

* See Chapter 2.

minerals, some connection between α-rays and helium now became plausible but was as yet by no means certain.

For some time α-rays were thought to be neutral, since they apparently did not bend in electric and magnetic fields, but in 1903 Rutherford finally showed that they do deviate in strong fields and carry a positive charge.[37] Next, doubts arose whether the e/m for α-particles emitted by polonium on the one hand and by radium on the other were the same. 'Further experimental evidence is required on that important point', Rutherford remarked[38] as late as 1905. Also, the question was raised[39] whether α-particles could initially be neutral, and gain a positive charge 'by throwing off an electron' in subsequent collisions. It became clear soon thereafter that the e/m is unique, and, later in 1905, it was stated with certainty that α-particles are charged 'at the moment of their expulsion from the radium atom'.[40]

It took still longer to determine the actual value of the α-particle's electric charge. In 1905 the situation was as follows:[41] 'Assuming the charge carried by the α-particle to be the same as that carried by the hydrogen atom, the mass of the α-particle is about twice that of the hydrogen atom'. Thus one was closing in on the right e/m.

Rutherford kept after the α-rays. I shall pass by his various experiments and turn directly to the summary 'Nature of the α-particle' found[42] in a 1908 joint paper with Hans Geiger: 'On the general view that the charge e carried by a hydrogen atom [i.e. a proton] is the fundamental unit of electricity . . . the evidence is strongly in favor of the view that that [the charge of the α-particle] = $2e$'. After nearly a decade of labor, Rutherford was finally prepared to state, italicized, what the α-particle really was: 'We may conclude that *an α-particle is a helium atom*, or, to be more precise, *the α-particle, after it has lost its positive charge, is a helium atom*'. In a paper together with Royds,[43] completed in November 1908, he was even more emphatic: 'We can conclude with certainty . . . that the α-particle is a helium atom [sic]'. They had shown that a discharge sent through a volume in which α-particles from radium had been collected produced the characteristic helium spectrum!

In the first-mentioned Rutherford–Geiger paper, the value found for $2e$, the α-particle charge, is 9.3×10^{-10} esu, hence $e = 4.65 \times 10^{-10}$ esu, a fine result (the modern value is $e = 4.803 \times 10^{-10}$ esu). Arthur Steward Eve has recorded an important exchange between himself and Rutherford about this result: 'The value he [R.] found for the unit of electrical charge was a great surprise, for it raised the significant figures [obtained from earlier experiments on the charge of the α-particle] by 36 per cent. When I protested to Rutherford that this result must be wrong, he pointed out that eight years previously Planck had obtained a similar figure to his own. Indeed, that marvellous man Planck had studied the results of radiation distribution from hot bodies . . . and, as a by-product, obtained an excellent value, for that time, of the electronic charge!'[44] I have reviewed elsewhere[45] the steps by which Planck, starting from his radiation law, arrived, in 1901, at the value $e=4.69\times 10^{-10}$ esu (see also Chapter 4) and how Einstein, referring to the Rutherford–Geiger result, called it[46] a 'brilliant confirmation' of Planck's value for e.

'The α-particles were . . . Rutherford's pet—and how he made them work!' one of his contemporaries later recalled.[47] We shall take leave of them here.

The discussion of the nature of β-rays will be left to the next chapter. I conclude with brief remarks on γ-radioactivity. In 1900, Paul Villard, working at the chemical laboratory of the Ecole Normale in Paris, observed with the use of photographic methods that radium was the source of 'radiations très pénétrantes',[48] a result quickly confirmed by Becquerel.[49] Villard had discovered γ-rays. He noted at once that these rays are not deflected by magnetic fields. Two years later Rutherford suggested[50] that γ-rays might be a very hard form of β-rays. This view became less and less tenable as Paschen studied their behavior in very strong magnetic fields and concluded that, if these particles were charged, their mass should be at least 45 times greater than that of the hydrogen atom.[51] Slowly the evidence grew that γ-rays and X-rays were akin, but as late as 1912 Rutherford still wrote with a touch of caution: 'There is at present no definite evidence to believe that X-rays and γ-rays are fundamentally different kinds of radiation'.[52] The matter was finally settled, fourteen years after the first observation of γ-radioactivity, when Rutherford and Andrade observed reflections of γ-rays from crystal surfaces.[52a]

Throughout his life, Rutherford had a loving relationship with his mother. (She died in 1935 at the age of ninety-two, only two years before Rutherford's own death). In 1902, Rutherford wrote to her: 'I have to keep going, as there are always people on my track. I have to publish my present work as rapidly as possible in order to keep in the race. The best sprinters in this road of investigation are Becquerel and the Curies in Paris, who have done a great deal of very important work on the subject of radioactive bodies during the last few years'.[53]

We have seen in this and the previous chapter how the four sprinters laid the foundations of studies in radioactivity, Becquerel by discovering a new radiation emanating from uranium, the Curies by their discoveries of similar effects in thorium, by their isolation of polonium and radium, more generally by pioneering a new field, radiochemistry, and by the realization that radioactivity is an atomic property. Rutherford was the first to observe structure in the new rays by distinguishing between α- and β-radiation.

It is far more true for Rutherford than for the other three pioneers that the work discussed above represents only a small fragment of a much larger oeuvre. The others will appear only sporadically in what follows, Rutherford will reappear prominently. This, in outline, will happen to him up until 1909.

The important role of Rutherford at McGill has been discussed in detail by several authors.[54] His discovery of thorium emanation led to his fundamental new insight of the lifetime concept (see Chapter 6). By 1902, available data on radioelements were sufficient for his formulation, together with Soddy, of the transformation theory of radioactive processes (Chapter 6). In 1903 he was elected to the Royal Society, in 1905 he received its Rumford medal. (From 1925 to 1930 he was to be its president.) He was offered and declined professorships at Yale, Columbia, and Stanford; he wanted to return to England. 'I shall be glad . . . to be nearer the scientific centre as I always feel America as well as Canada is on the periphery of the circle.'[55] In the fall of 1907 he began his next appointment, professor and director of the physics laboratory at the University of Manchester.

In 1908 Rutherford was awarded the Nobel Prize for chemistry 'for his investigations into the disintegration of the elements, and the chemistry of radioactive substances'. His presenter said of his work: 'In a certain way it may be said that the progress of investigation is bringing us back once more to the transmutation theory propounded and upheld by the alchemists of old'.[56] Rutherford, physicist, made a clear concession when he chose 'The chemical nature of the alpha particles from radioactive substances' as the title for his official Nobel Lecture. In his after dinner speech at the Nobel banquet he remarked that he 'had dealt with many different transformations with various time-periods, but the quickest he had met was his own transformation from a physicist to a chemist'.[57] An eye-witness to these festivities later remarked that Rutherford had looked ridiculously young that day.

It has been said that the publicity attendant on this most visible of international awards can be detrimental rather than stimulating to the creativity of its recipients; there is some truth to that. On the other hand it did not prevent men like Emil Fischer, Otto Warburg, and Paul Ehrlich from making major subsequent contributions.[58] No one, however, can compare to Rutherford, who rose to his greatest heights after 1908, most notably because of his discovery of the atomic nucleus, to be discussed in Chapter 9.

Sources

In 1904 publication began of the *Jahrbuch der Radioaktivität und Elektronik*, the earliest forerunner of *Annual Reviews of Nuclear Science*. Its first volume contains the finest bibliography[59] I know of papers on radioactivity from 1896 to 1904.

The Curies. The complete works[18] of Pierre Curie appeared in 1908, those of Marie Curie[60] in 1954. Of particular interest in the latter book are the elegantly written essays on the status of radioactivity at various times. Biographies by Marie Curie of her husband[4] (an autobiographical sketch is attached), by Eve Curie of her mother,[16] and by Robert Reid,[19] also of Marie Curie, are especially important.

Rutherford. In 1903, Rutherford wrote to his mother: 'I have not taken into account Solomon's injunction, "Oh, that mine enemy would write a book".'[61] So much the better for us. Rutherford's *Radioactivity* of 1904 gives a vivid account of the early discoveries and a clear picture of the thinking at that time.[62] Only fifteen months later a second edition had been readied. So rapid was the pace of the developments that this new book,[38] considerably revised, was one and a half times as large as the first edition. In its preface, Rutherford expresses 'some apology due to my readers' for the considerable amount of rewriting which he could not avoid unless he 'were to relinquish his purpose of presenting the subject as it stands at the present moment'. A comparison of the second edition with the first one conveys a keen sense of progress. One derives similar guidance from the third edition, dated seven years later.[52] The last edition, the text by Rutherford, Chadwick, and Ellis[63] was the main manual on radioactivity in the early 1930s. Also interesting are Rutherford's Silliman lectures[36] of 1905. His collected papers, published in three volumes,[64] contain important assessments by others of Rutherford's contributions. These books contain papers published from New Zealand (1894–5) and Cambridge (1896–8), from the years he was professor at McGill (1899–1907) and Manchester (1907–19), and, finally, from his years as Cavendish Professor (1919–37). I refer the reader to a paper by Badash for a detailed list of Rutherford biographies.[65] The best-documented one of these is the book by Eve.[26]

References

1. Lord Kelvin, J. C. Beattie, and M. S. de Smolan, *Nature* **55**, 447, 1897.
2. S. P. Thompson, *Proc. Roy. Soc. A* **61**, 481, 1897.
3. G. Jaffe, *J. Chem. Educ.* **29**, 230, 1952.
4. M. Curie, *Pierre Curie*, Macmillan, New York 1929.
5. S. Curie, *Comptes Rendus* **126**, 1101, 1898.
6. J. J. Thomson, *Proc. Roy. Soc. A* **59**, 274, 1896.
7. J. J. Thomson and J. A. McClelland, *Proc. Cambr. Phil. Soc.* **9**, 126, 1896.
8. J. J. Thomson, *The discharge of electricity through gases*, Scribner's, New York 1898.
9. N. Feather, *Electricity and matter*, Edinburgh Univ. Press 1968.
10. G. C. Schmidt, *Ann. der Phys.* **65**, 141, 1898.
11. P. and S. Curie, *Comptes Rendus* **127**, 175, 1898.
12. M. Curie, *Rev. Gén. des Sc.* **10**, 41, 1899.
13. E. Rutherford and F. Soddy, *Phil. Mag.* **4**, 370, 569, 1902.
14. P. and S. Curie and G. Bémont, *Comptes Rendus* **127**, 1215, 1898.
15. F. Soddy, *The interpretation of radium*, 4th Edn, p. 83, Putnam Sons, New York 1920.
16. Eve Curie, *Madame Curie*, Doubleday, New York 1938.
17. E. Rutherford, *Nature* **134**, 90, 1934.
18. *Oeuvres de Pierre Curie*, Gauthier-Villars, Paris 1908.

19. R. Reid, *Marie Curie*, Dutton, New York 1974.
20. *Nobel Lectures in chemistry, 1901–21*, Elsevier, New York 1966.
21. J. G. Crowther, *The Cavendish Laboratory*, Science History Publ., New York 1974.
22. *The scientific papers of James Clerk Maxwell*, Vol. 2, p. 241, Ed. W. D. Niven, Dover, New York.
23. *Nature* **10**, 139, 1874.
24. J. J. Thomson, *Proc. Roy. Soc. A* **59**, 274, 1896.
25. J. J. Thomson and J. A. McClelland, *Proc. Cambr. Phil. Soc.* **9**, 126, 1896.
26. A. S. Eve, *Rutherford*, p. 14, Cambridge Univ. Press 1939.
27. Ref. 26, p. 42.
28. J. J. Thomson and E. Rutherford, *Phil. Mag.* **42**, 392, 1896.
29. E. Rutherford, *Proc. Cambr. Phil. Soc.* **9**, 401, 1898.
30. E. N. da C. Andrade, *Rutherford and the nature of the atom*, p. 50, Doubleday, New York 1964.
31. E. Rutherford, *Phil. Mag.* **47**, 109, 1899.
32. H. Becquerel, *Comptes Rendus* **128**, 771, 1899.
33. W. Ramsay, *J. Chem. Soc.* **67**, 1107, 1895.
34. R. B. Moore, *J. Franklin Inst.* **186**, 29, 1918.
35. W. Ramsay and F. Soddy, *Proc. Roy. Soc. A* **73**, 346, 190; see also A. Debierne, *Comptes Rendus* **141**, 383, 1905.
36. E. Rutherford, *Radioactive transformations*, p. 18, Yale Univ. Press, New Haven, Conn. 1906.
37. E. Rutherford, *Phil. Mag.* **5**, 177, 1903.
38. E. Rutherford, *Radioactivity*, 2nd Edn, p. 150, Cambridge Univ. Press 1905.
39. See T. J. Trenn, *Hist. St. Phys. Sc.* **6**, 513, 1974, who quotes a letter by J. Larmor to Rutherford in which this question occurs.
40. E. Rutherford, *Phil. Mag.* **6**, 193, 1905.
41. Ref. 38, p. 156.
42. E. Rutherford and H. Geiger, *Proc. Roy. Soc. A* **81**, 162, 1908.
43. E. Rutherford and T. Royds, *Phil. Mag.* **17**, 281, 1909.
44. Ref. 26, p. 176.
45. A. Pais, *Subtle is the Lord*, p. 371, Oxford Univ. Press 1982.
46. Ref. 45, p. 402.
47. Ref. 30, p. 44.
48. P. Villard, *Comptes Rendus* **130**, 1010, 1178, 1900.
49. H. Becquerel, *Comptes Rendus* **130**, 1154, 1900.
50. E. Rutherford, *Phys. Zeitschr.* **3**, 517, 1902.
51. F. Paschen, *Ann. der Phys.* **14**, 164, 389, 1904; *Phys. Zeitschr.* **5**, 563, 1904.
52. E. Rutherford, *Radioactive substances and their radiations*, p. 287, Cambridge Univ. Press 1913.
52a. E. Rutherford and E. N. da C. Andrade, *Phil. Mag.* **27**, 854, 1914.
53. Ref. 26, p. 80.
54. L. Badash, in *Rutherford and physics at the turn of the century*, p. 23, Eds. M. Bunge and W. R. Shea, Science History Publ., New York 1979; J. Heilbron, ibid., p. 42; T. Trenn, ibid., p. 89.
55. E. Rutherford, letter to O. Hahn, January 6, 1907, in Ref. 26, p. 153.
56. *Nobel Lectures in chemistry, 1901–21*, p. 126, Elsevier, New York 1966.
57. Ref. 26, p. 183.
58. Cf. H. Zuckerman, *The scientific élite*, Free Press, New York 1977.

59. M. Iklé, *Jahrb. der Radioakt. und Elektr.* **1**, 413, 1904.
60. *Prace Marii Sklodowskiej-Curie*, Ed. Irène Joliot-Curie, Panstwowe Wydawnictwo Naukowe, Warsaw 1954.
61. Ref. 26, p. 83.
62. E. Rutherford, *Radioactivity*, Cambridge Univ. Press 1904.
63. E. Rutherford, J. Chadwick and C. D. Ellis, *Radiations from radioactive substances*, Cambridge Univ. Press 1930.
64. *The collected papers of Lord Rutherford of Nelson*, Allen and Unwin, London 1962–5.
65. L. Badash, in *Dictionary of scientific biography*, Vol. 12, p. 35, Chief Ed., C. G. Gillispie, Scribner, New York 1975; also D. Wilson, *Rutherford, simple genius*, M.I.T. Press, Cambridge, Mass. 1983.

4

The first particle

(a) Of Geissler's pump and Geissler's tube and Rühmkorff's coil

In the early years following the first observation of the electron, a toast used to be offered at the Cavendish Laboratory annual dinner: 'The electron: may it never be of use to anybody'.[1] That wish has not been fulfilled. The discovery of the electron, the first particle in the modern sense of the word, has brought about profound changes in the world at large. This chapter is devoted to the more provincial but not less interesting questions of how this discovery came about and what it did to the world of physics.

First a synopsis. In Section (b) we will turn to an issue often discussed in the nineteenth century and settled by the electron's discovery: is electric charge a continuous or a discrete attribute? Also mentioned in that section are theoretical estimates of the fundamental unit of charge which predate the observation of free electrons. Sections (c) and (d) deal with the actual discovery, the history of which is largely the history of progress in experimental methods—spectroscopy in the case of the Zeeman effect, vacuum techniques and high energy sources for the study of cathode rays. For nearly a quarter of a century prior to the advent of the electron there had been monumental confusion about the nature of these rays. Were they aether disturbances? Were they molecular torrents? It came as a great surprise that they were neither, but instead an unsuspected new form of matter. In Section (e) the identification of β-rays with electrons will be discussed. The concluding Section (f) deals with early experiments performed with electrons which aimed at determining the energy-mass-velocity relation of free particles. I have dealt at more length with this last subject in a recent discussion of the history of special relativity.[2] I need to repeat here a few of these earlier comments because the question touches so intimately on the role of the electron as a research tool during the early years of the twentieth century.

The electron is but the first of many particles to be discovered as the result of ever improving vacuum techniques. It is therefore most apposite to preface this chapter on the first particle by paying homage to Johann Heinrich Wilhelm Geissler of Bonn, inventor of the Geissler pump and the Geissler vacuum tube, mechanic and glassblower, to whom, in 1865, in Geissler's lifetime,

Johann Christian Poggendorf referred as a 'Glasskünstler', a wizard working with glass.[3]

Geissler invented his mercury pump in the mid-1850s, just about two centuries after Otto von Guericke, then mayor of Magdeburg, had constructed the first vacuum pump—essentially a cylinder with water-saturated washers. That, too, had been a great step. It is well known that in 1657, von Guericke, to the delight of his good citizenry, evacuated a sphere made of two snugly fitting copper hemispheres, which could not be pulled apart by two teams of horses. Poggendorf has reminded us[3] that a first primitive version of a mercury pump was used soon after this event, and has recorded a list of other early efforts for improving vacuum techniques, some realized, some never passing beyond the design stage.*

Geissler's pump consists of a closed bulb connected by a flexible tube to an open reservoir of mercury. A two-way tap can connect the bulb either to the outside air or to the vessel to be evacuated. Mercury makes air-tight contact with the glass walls and penetrates into the bore of the tap, thus driving out all the atmospheric air. Turn the tap to connect with the outside air, then raise the reservoir until all the air has been expelled from the bulb, next turn the tap so as to connect with the vessel, finally lower the reservoir. Repeat the process as often as needed. This device, Poggendorf noted,[3] creates a vacuum of a degree of perfection unmatched by any other pump known at that time.

Geissler's other invention, the vacuum tube, is well appreciated by turning for a moment to the entry[5] in Faraday's diary for January 4, 1838, where one finds a sketch of his experimental arrangement for the study of electrical discharges in gases. (I return to Faraday's results of that year in Section (d).) The central feature is a glass jar containing a fixed electrode. The jar is closed with a cork through which a metal pin, the other electrode, can be moved into a variety of positions—hardly a device for the study of discharges in good vacua. Compare this with Geissler's method: a glass tube is evacuated, then sealed off with glass. Electrodes are made to enter the tube in the sealing process. Thus Geissler's pump produced a vacuum higher than before, his tube a vacuum more stable than before.** It was no longer necessary to continue pumping while an experiment was in progress.

It appears that Geissler's tube was first put to use† early in 1857. Typical pressures at that time were a tenth of a millimeter of mercury. These tubes, provided with simple electrical discharge mechanisms, soon became popular articles in optician's shops, where they were bought for the enjoyment of the pretty colors they produced. To physicists the tubes were a blessing. In 1858

* A more easily accessible and less complete brief history of these developments has been given by Andrade.[4]

** Faraday's friend John Peter Gassiot had attempted to construct similar tubes.

† See an article by Plücker[6] in which also the term 'Geissler's tube' appears for the first time.

Faraday experimented with them. He never published his results of that year, but in his diaries one finds references to 'the Bonn tube', or to 'the German tube of Bonn'. At one point he comments: 'Very beautifully wrought'.[7]

Experimental advances in particle physics are of course due not only to improved vacua but also to higher available energies. We should therefore at this point also pay our respects to Heinrich Daniel Rühmkorff, a native of Hannover, who spent most of his active years in Paris, where in the 1850s he constructed an improved version of the induction coil, an early type of transformer (familiar to car owners) for abruptly generating induction currents. It consists of a cylindrical iron core (preferably made of a bundle of insulated wires) around which a metal wire, the primary coil, is wound. A secondary coil is wound around the primary coil. A direct current, periodically interrupted, is sent through the primary coil. At the beginning and at the end of each interruption a voltage difference is induced between the unconnected ends of the secondary coil. The more turns there are in the secondary wire (typically ~300 000 windings per metre of cylinder) the larger this voltage difference. Long sparks can be made to fly between the end points of the secondary coil, whence the other name for the same device, Rühmkorff's spark inductor. 'Early experimenters obtained the high voltages to operate discharge tubes by wiring large batteries in series and even by using old-fashioned friction machines. By the end of the decade [1850–60], however, they had turned to the battery-powered induction coil. It was convenient, compact, portable, self-operating, easily controlled by a switch in the primary circuit, and the voltage [typically of the order of a few kilovolts] in its secondary circuit was readily adjustable.'[8] Faraday, the father of induction, used it in his later years. 'Leading physicists of Europe are well acquainted with [Rühmkorff's] dingy little bureau in the Rue Champollion, near the University.'[9]

Alternating-current power, step-up transformers, and vacuum tube rectifiers came later. But it was the Rühmkorff coil which served Hertz in 1886–8 in his demonstration of electromagnetic waves and his discovery of the photoelectric effect; Roentgen in 1895 in his discovery of X-rays; Guglielmo Marconi in 1896 in his transmission of telegraph signals without wires; Zeeman in that same year in the discovery of the Zeeman effect; and J. J. Thomson in 1897 in his determination of the e/m ratio of electrons. At least equally decisive for the discoveries of Roentgen and Thomson was their use of simple variants of the tube constructed by Geissler. His inventions, which have been followed by a century of vast improvements in vacuum techniques,[10] were indispensable for the birth of particle physics.

Neither Geissler nor Rühmkorff ever earned an academic degree. In 1864, Rühmkorff received a 50 000 franc prize established by the French emperor for the most important discovery in the application of electricity. In 1868, Geissler was given an honorary degree by the University of Bonn. In 1873, on the occasion of the world fair in Vienna, he received the *Goldene Verdienst-*

kreuz für Kunst und Wissenschaften.* I have held in my hands his early tubes, now in the historical cabinet of the Physics Institute on the Nussallé in Bonn.

(b) Faraday, Maxwell, and the atomicity of charge

In 1733, exactly one hundred years before Faraday formulated his laws of electrolysis, the point of departure for this section, the Frenchman Dufay had written: 'There are two distinct electricities very different from one another; one of which I call *vitreous electricity* and the other *resinous electricity*',[11] and had proclaimed the law that like charges repel, unlike charges attract. As the following list of names and dates shows, the nineteenth-century understanding of electromagnetism was largely made possible by rapid advances in the invention of new instruments and the formulation of phenomenological laws during the period between Dufay and Faraday: the first condensor, the Leiden jar, is invented independently by von Kleist and van Musschenbroek (1745); Benjamin Franklin, inventor of the first lightning rod, introduces the terms 'positive' and 'negative' electricity (1747);[12] Coulomb formulates his inverse square law of electrostatics (1785); Volta invents the first battery, the Voltaic pile (1796); Davy produces the first electric arc (1801); Ampère, inventor of the solenoid, formulates a hypothesis on the relation of electric current and magnetism (1820); Oersted shows that an electric current creates a magnetic field (1820); Biot and Savart discover the force law between an electric current in a straight wire and a magnetic pole (1820); Seebeck discovers thermoelectricity (1822); Schweigger invents the first true galvanometer (1823); Ohm formulates his law of resistance (1827); and Faraday discovers the principle of induction (1831).**

Irrepressible then as now, natural philosophers (they were not yet called scientists) of the eighteenth and early nineteenth century attempted to formulate theories of these phenomena. The first description of the electric current as a stream of discrete electric charges appeared only in the 1840s, however, notably in the work of two Leipzig professors, Fechner[14] and Weber.[15] These and similar efforts are discussed in Whittaker's classic first volume on the history of electricity, where Weber's work is described[16] as 'the first of the electron theories—a name given to any theory which attributes the phenomena of electrodynamics to the agency of moving charges'. By that time, theoretical speculations of this kind were influenced above all by Faraday's researches in 1833 on electrolysis. As Maxwell said sometime later[17] in his treatise on electricity and magnetism: 'Of all electrical phenomena electrolysis appears the most likely to furnish us with a real insight into the true nature of the electric current, because we find currents of ordinary matter and currents of

* Gold cross of merit for arts and sciences.
** For a handy brief history of electricity and magnetism, see for example Ref. 13.

electricity forming essential parts of the same phenomenon'. Let us now turn to Faraday's work.

Pass an electric charge through a solution. Migration of solute particles may result. The process as a whole is called electrolysis, a term introduced by Faraday, who also coined the expressions electrode, anode, cathode, ion,* anion, cation.[18] About these linguistic creations, Faraday wrote (in Gibbon's phrase) with pleasure and contempt: 'These terms being once well defined will, I hope, in their use enable me to avoid much periphrasis and ambiguity of expression. I do not mean to press them into service more frequently than required, for I am fully aware that names are one thing and science another'.[19] To the science of electrolysis Faraday contributed two laws which I state in his own words.

The first law. 'The chemical power of a current of electricity is in direct proportion to the absolute quantity of electricity which passes,'[20] this power being defined as the total mass of the particular element deposited or liberated at an electrode.

The second law. Faraday defines 'electrochemical equivalent' of an element as the amount of mass of that element deposited at an electrode during a fixed time by a fixed amount of external current, taking the mass deposit of hydrogen as the unit. He then states: 'Only single electrochemical equivalents of elementary ions can go to the electrodes, and not multiples . . . the same number which represents the equivalent of a substance A when it is separating from a substance B, will also represent A when separating from a third substance C.'[21]

These are the laws which were to lead to the first determinations of the fundamental unit of electric charge *e*, the charge of the electron. Faraday himself was not yet ready for that step, however, if only because he did not yet appreciate correctly the meaning of electrochemical equivalent: 'A very valuable use of electrochemical equivalents will be to decide, in cases of doubt, what is the true chemical equivalent, or definite proportional, or atomic number of a body . . . I can have no doubt that, assuming hydrogen as 1 . . . the equivalent number of atomic weight of oxygen is 8, of chlorine 36, of bromine 78.4, of lead 103.5, of tin 59, etc., notwithstanding that a very high authority doubles several of these numbers'.[22] Whoever he was, the very high authority was right, Faraday's values for the atomic weights of oxygen, lead, and tin do need to be doubled. This came about because Faraday incorrectly identified electrochemical equivalent with atomic weight, rather than with the ratio of atomic weight and valence of the substance. That is not surprising. The theory

* To Faraday an ion was nothing more precise than one of the two parts into which a molecule breaks up when electrolyzed.

of chemical valence was only developed some twenty years later, in the first instance by Frankland and by Kekulé. In the chapter of his 1873 treatise devoted to electrolysis[23] Maxwell does not repeat Faraday's error but does not refer to valence either. In his Faraday lecture, given in London, in 1881, Helmholtz linked Faraday's second law with valence, thereby giving for the first time the correct interpretation, 'the interpretation Faraday would have given if he had been acquainted with the law of chemical equivalence'.[24] In modern terms, the second law can be stated as follows. The amount of electricity deposited at the anode by a gram atom of monovalent ions is a universal constant, called the Faraday (F), given by

$$F = Ne, \tag{4.1}$$

where N is Avogadro's number, and e is, as said, the charge of the electron.

'Although we know nothing of what an atom is', Faraday wrote, in summarizing his investigations in electrolysis, 'yet we cannot resist forming some idea of a small particle, which represents it to the mind . . . there is an immensity of facts which justify us in believing that the atoms of matter are in some way endowed or associated with electrical powers, to which they owe their most striking qualities, and amongst them their chemical affinity'.[25] This statement might seem to indicate that he was a believer in the reality of atoms, an atomist; it certainly does indicate an early vision of intra-atomic forces. Yet Faraday's position in the matter of atomic reality was not that unambiguous: 'I must confess I am jealous of the term *atom*; for though it is very easy to talk of atoms, it is very difficult to form a clear idea of their nature, especially when compound bodies are under consideration'.[26] That is the true Faraday, exquisite experimentalist, who would only accept what he was forced to believe on experimental grounds.

Maxwell's position on these issues provides another warning against the oversimplified belief that nineteenth-century scientists belonged to either of two camps, atomists or anti-atomists. Of course Maxwell believed in the reality of atoms. How else could he have worked with such devotion (and success) on the theory of gases? More than that, he was convinced that an atom was not just a tiny rigid body but had to have some structure. 'The spectroscope tells us that some molecules* can execute a great many different kinds of vibrations. They must therefore be systems of a very considerable degree of complexity, having far more than six variables [the number characteristic for a rigid body] . . .', he said in a lecture given[27] in 1875. But he also believed that the structured atom was unbreakable: 'Though in the course of ages catastrophes have occurred and may yet occur in the heavens, though ancient systems may be dissolved and new systems evolved out of their ruins, the molecules [i.e. atoms!] out of which these systems [the earth and the whole

* By 'molecules', Maxwell meant our 'atoms'—terminology was not uniform in those years.

solar system] are built—the foundation stones of the material universe—remain unbroken and unworn. They continue to this day as they were created—perfect in number and measure and weight . . .' [28]

In regard to electrolysis, Maxwell's prejudice—natural for the nineteenth century to be sure, but a prejudice nevertheless—of atoms unbroken and unworn had to lead him into a quandary. On the one hand, he was willing to concede that electrolysis indicated the existence of 'the most natural unit of electricity'.[23] On the other, he demanded a *dynamical* understanding of the *universality* of this unit. That, however, is possible only if one understands that an ion is a broken atom (or an atom with an atomic fragment tagged on). He therefore *had* to express grave reservations concerning the atomicity of electricity: 'If we . . . assume that the molecules of the ions within the electrolyte are actually charged with certain definite quantities of electricity, positive and negative, so that the electrolytic current is simply a current of convection, we find that this tempting hypothesis leads us into very difficult ground . . . the electrification of a molecule . . . though easily spoken of, is not so easily conceived'.[23]

Maxwell looked for ways out by studying models in which a pair of molecules touch at one point and then have the rest of their surface charged with electricity due to an electromotive force of contact. But then why, he asks, does this give the same charge to chlorine when it touches zinc as when it touches copper? Only after one has captured the flavor of this struggle can one grasp the significance of Maxwell's next statement: 'Suppose however that we leap over this difficulty by simply asserting the fact of the constant value of the molecular charge and that we call this constant of molecular charge, for convenience of description, one molecule of electricity. This phrase, gross as it is, [is] out of harmony with the rest of this treatise . . .' and he adds: 'It is extremely improbable that when we come to understand the true nature of electrolysis we shall retain in any form the theory of molecular charges . . .'.[23] Thus Maxwell favored a provisional phenomenological over a realistic view of the quantum of charge.

Yet the realistic view gained advocates, the most influential one being Helmholtz, who said in his 1881 Faraday Lecture: 'The most startling result of Faraday's law is perhaps this. If we accept the hypothesis that the elementary substances are composed of atoms, we cannot avoid concluding that electricity also, positive as well as negative, is divided into definite elementary portions, which behave like atoms of electricity'[24]—a statement which explains why in subsequent years the quantity e was occasionally referred to in the German literature as 'das Helmholtzsche Elementarquantum'.[29]

Even before Helmholtz's memorable address, the Irish physicist George Johnstone Stoney, 'veritably a prophet as of olden time',[30] had reported to the 1874 meeting of the British Association for the Advancement of Science an estimate of e, the first of its kind, based on Eq. (4.1). (His paper on this subject[31] was not published until 1881.) The experimental value of F was well determined by then. As I have described elsewhere,[32] there also existed fairly sensible estimates at that time (including one by Stoney[33]) of the value of N. For example, Loschmidt had found[34] that $N \approx 0.5 \times 10^{23}$, Maxwell[28] that $N \approx 4 \times 10^{23}$. (The present best value is $N = 6.02 \times 10^{23}$.) Stoney obtained $e \simeq 3 \times 10^{-11}$ esu, too small by a factor ~ 20, yet not all that bad for a first and very early try.* In 1891 he baptized the fundamental unit of charge, giving it the name 'electron'.[35] Thus this term was coined prior to the discovery of the quantum of electricity *and* matter which now goes by that name.

I cannot conclude the discussion of electrolysis without mentioning what Planck did in 1901 with Eq. (4.1). The year before, he had given his law for blackbody radiation which contains three fundamental constants: the velocity of light, Planck's constant, and the Boltzmann constant k. In 1901 Planck found[36] from experimental data on blackbody radiation that $k = 1.34 \times 10^{-16}$ erg . K^{-1}. From this and the value of the gas constant R he determined N from $R = Nk$. He inserted his result for N into Eq. (4.1) and obtained $e = 4.69 \times 10^{-10}$ esu. Comparing this with the present best value (4.80×10^{-10} esu) we must admire not only Planck's ingenuity but also the quality of the early experiments on blackbody radiation.

In his 1901 paper Planck referred to J. J. Thomson who had meanwhile discovered the electron. Let us then turn to the years 1896 and 1897 and see how that discovery was made.

(c) Of Rowland, Zeeman, and Lorentz

'From 1862 on, Faraday's health steadily declined and his circle of activities grew smaller and smaller . . . From 1865 to 1867 . . . he had only occasional and brief moments of clarity—for the most part he merely sat staring into space.'[37] On 25 August 1867, the life of one of the greatest of experimentalists came to a peaceful end. He had lived to be almost 76 years old.

In his *Encyclopaedia Britannica* article on Faraday, Maxwell recalled that 'In 1862 he [F.] made the relation between electromagnetism and light the subject of his very last experimental work. He endeavoured, but in vain, to detect any change in the lines of the spectrum of a flame when the flame was acted on by a powerful magnet'.[38] This investigation is the subject of the final entry, dated 12 March 1862, in the seven-volume diary of Faraday. Its concluding sentence reads: 'Not the slightest effect on the polarized or unpolarized ray was observed'.[39] It is just possible that Maxwell may have thought of Faraday when, in an address given in 1870, he said: 'Here, and in the starry heavens, there are innumerable multitudes of little bodies of exactly the same

* See Ref. 29 for other early estimates of e.

mass, and vibrating in exactly the same time . . . no power in nature can now alter in the least either the mass or the period in any of them'.[40] In 1875, Tait reported that he had pursued Faraday's last quest 'rather more than twenty years ago . . . I have since that time tried it again and again . . . Hitherto it has led to no result'.[41]

Early in 1896, a young Dutch 'privaatdocent'* at the University of Leiden tried his hand at the problem: Pieter Zeeman, who had studied with Lorentz and Kamerlingh Onnes, and then been Lorentz' assistant. At first, Zeeman had no luck either.[42] Disappointed, he went to Strasbourg for a brief stay, working on the propagation of light through fluids. Upon his return he learned that the Leiden laboratory had acquired a Rowland grating so that higher resolving power of light frequencies than before was now available. Some time during the second half of August he therefore decided to have another try at the possible influence of magnetism on light. He had his first positive results before the month was over.[42] Before discussing these, I should first comment on the role of Rowland.

Henry Augustus Rowland, the first professor of physics at the Johns Hopkins, the first president of the American Physical Society, personally well acquainted with Maxwell, Helmholtz, and Edison, is best known for the concave spectral gratings which bear his name. He was a man endowed with great energy and high powers of concentration. 'I can remember cases', a long-term associate once said, 'when he appeared as if drugged from mere inability to recall his mind from the pursuit of all absorbing problems.'[43] In 1882 he began work on the construction of a machine—he was to call it the 'dividing engine'—which would produce fine equidistant gratings. In his first paper on the subject he stated the main technical obstacle: 'One of the problems to be solved in making a machine is to make a perfect screw, and this, mechanics of all countries have sought to do for over a hundred years and have failed'.[44] For a description of how Rowland solved the problem of the 'process [which] will produce a screw suitable for ruling gratings for optical purposes', the interested reader is referred to his article '*Screw*' in the ninth and many subsequent editions of the *Encyclopaedia Britannica*. Rowland installed his dividing engine in the sub-basement of the Physical Laboratory of the Johns Hopkins, in order to minimize temperature fluctuations as well as vibrations from street traffic. Ruling large diffraction gratings directly on concave mirrors, he was eventually able to emplace 400–800 lines per millimeter with an equidistance accuracy better than 1/4000th of a millimeter. His gratings, which he sold at cost to various institutions, including the Leiden laboratory, were instrumental in bringing about major advances in physics and in astrophysics.

* This is a non-salaried, non-faculty university teaching position, the only remuneration being a small fee paid by each course attendant.

Let us now return to Zeeman. In his first paper on the new effect, dated 31 October 1896, Zeeman recalled how his attention to this question had first been drawn by reading Maxwell's article on Faraday mentioned above. He then continued: 'If a Faraday thought about [this] possibility . . . perhaps it might be yet worthwhile to try the experiment again with the excellent auxiliaries of spectroscopy of the present time',[45] referring here to the 'Rowland grating with a radius of 10 ft and with 14 938 lines per inch',[46] now at his disposal. His other major piece of equipment was an electromagnet of the Rühmkorff type which produced a magnetic field ~10 kilogauss. His first experiment was elementary. A Bunsen burner is placed between the poles of the electromagnet, and a sodium spectrum is obtained by holding in the flame a piece of asbestos drenched in a kitchen salt solution. The two sodium D-lines are seen as narrow, sharply defined lines when the electromagnet is turned off, the lines are broadened when it is turned on, the broadening being ~1/40 of the line separation.

Zeeman was not yet convinced, however, that the effect was caused by the magnetic field. Could the broadening be due to temperature or density fluctuations in the flame? He therefore performed a more sophisticated experiment. A sodium arc spectrum is generated in a porcelain tube which itself is water-cooled. The tube was placed between the poles of the electromagnet. The same broadening was found as before. Zeeman now became a believer and published.

In his next paper (28 November) Zeeman raised the question of whether his effect could be explained by Lorentz' theory of electromagnetism. 'Professor Lorentz, aan wien ik deze beschouwing mededeelde heeft dadelijk de vriendelijkheid gehad mij aan te geven op welke wijze de beweging van een ion in een magnetisch veld volgens zijne theorie wordt bepaald.'*

Lorentz was ready for the assignment. In 1892 he had published his first paper[47] on the atomistic interpretation of the Maxwell equations in terms of charges and currents carried by fundamental particles. In 1892 he called these particles simply 'charged particles'; in 1895 he called them 'ions' (whence Zeeman's use of that term); only in 1899 did he call them 'electrons'. Also in the 1895 paper, Lorentz introduced the new assumption[48] that an 'ion' with charge e and velocity $\vec{v}$ is subject to a force $\vec{K}$ given by

$$\vec{K} = e(\vec{E} + \vec{v} \times \vec{H}/c), \tag{4.2}$$

where $\vec{E}$ and $\vec{H}$ are the electric and magnetic field, respectively, at the position of the ion (considered as a point particle). Lorentz called $\vec{K}$ the *elektrische Kraft*, we call it the Lorentz force. I see the introduction of this force as the most important contribution to theoretical physics in the 1890s. The link

* Prof. L. to whom I communicated these considerations at once kindly informed me of the manner in which, according to his theory, the motion of an ion in a magnetic field is determined.

between electric and magnetic action implied by Eq. (4.2) was clarified in 1905 by Einstein in his first paper on the special theory of relativity.[49]

Zeeman's effect was one of the first applications of the new force to a dynamical problem. Lorentz considered the case of an 'ion' bound by harmonic forces within an atom and subject in addition to the force $\vec{K}$:

$$m\frac{\mathrm{d}^2\vec{x}}{\mathrm{d}t^2}=-k^2\vec{x}+\vec{K} \tag{4.3}$$

for the case of zero electric field. If also $H=0$, we have a motion with period $2\pi\sqrt{m}/k$. For nonzero H, we get two vibrations in the plane $\perp H$ with frequencies $\omega \pm \Delta\nu$, where $\Delta\nu$ is approximately given by

$$\Delta\nu=\frac{e}{4\pi mc}|\vec{H}|. \tag{4.4}$$

In Zeeman's first experiments only a 'Zeeman-broadening' was observed (as we have noted), not a Zeeman splitting. That is sufficient, however, to make an estimate of e/m. That is what Zeeman indeed did with the result[46] that $e/m \sim 10^7$ emu/g, a splendid first result: the modern value is 1.76×10^7 emu/g (or 5.27×10^{10} esu/g).

Important aside. What would Zeeman and Lorentz have thought had Zeeman at once been able to obtain the correct pattern of Zeeman splitting? Lorentz' classical calculation is at best applicable only to the 'normal Zeeman effect', in which spin plays no role, and does not apply to the pair of sodium D-lines which exhibit the spin-dependent so-called anomalous Zeeman effect.* Their fully split Zeeman pattern consists of ten lines.

I return to Zeeman's second paper[46] of 1896. Strikingly absent in this article is any expression of surprise at this large value of e/m. Why is there no mention of the bizarre value for e if m were of atomic order of magnitude? Why did Lorentz not comment on a surprisingly small value for m if e were of the order of magnitude estimated by Stoney, a result with which he undoubtedly must have been familiar? In any event, Zeeman discovered the Zeeman effect, Lorentz showed him the way to an estimate of e/m, neither discovered the electron.

As to Zeeman's subsequent work, in 1897 he reported for the first time a splitting of lines: using the blue line of cadmium ($\lambda = 480\ \mu$m) he found, as Lorentz had taught him to expect, that this line breaks up into a doublet or a triplet according to whether the light is emitted in a direction parallel or perpendicular to the lines of magnetic force.[50] Furthermore, he detected polarization effects which enabled him to deduce that e is negative. Also, now

* This effect is discussed in Chapter 13, Section (b).

he at least remarked that 'it is very probable' that his 'ions' differ from the electrolytical ions![51]

In 1902 Zeeman and Lorentz shared the Nobel Prize 'in recognition of the extraordinary service they rendered by their researches into the influence of magnetism upon radiation phenomena'.

Rowland died in 1901, not quite fifty-three years old. During his last years he knew he had diabetes, still a terminal illness at that time. His efforts to provide for the material security of his wife and children may explain in part why he proposed himself for the Nobel Prize (as the Nobel archives in Stockholm show). On 26 October 1901 his closest friends gathered at the dividing engine vault, and witnessed how 'the ashes of our friend, enclosed in a bronze casket, were, in accordance with his own desire, entombed in a niche in the stone wall of the vault, and the opening of the niche was sealed'.[52] The ashes were presumably removed when, some time after 1916, the old physics building on the intersection of West Monument Street and Linden Avenue in Baltimore was torn down.

(d) The discovery of the electron

J. J. Thomson discovered the electron. Numerous are the books and articles in which one finds it said that he did so in 1897. I cannot quite agree. It is true that in that year Thomson made a good determination of e/m for cathode rays, an indispensable step toward the identification of the electron, but he was not the only one to do so. Simultaneously Walter Kaufmann had obtained the same result. It is also true that in 1897 Thomson, less restrained than Zeeman, Lorentz, and (as we shall see) Kaufmann, correctly conjectured that the large value for e/m he had measured indicated the existence of a new particle with a very small mass on the atomic scale. However, he was not the first to make that guess. Earlier in that year, Emil Wiechert had done likewise, on sound experimental grounds, even before Thomson and Kaufmann had reported their respective results. Nevertheless, it is true that Thomson should be considered the sole discoverer of the first particle, since he was the first to measure not only e/m but also (within 50 per cent of the correct answer) the value of e, thereby eliminating all conjectural elements—but that was in 1899. In what follows I shall discuss, in this order, the independent contributions of Wiechert, Kaufmann, and Thomson. First, however, I need to deal with the preceding debate about the nature of cathode rays, which split the physics community into two camps, very nearly one German, one British, neither side emerging victoriously. Let us go back to these earlier times.

1. The plight of the precursors. The earliest reports of electrical discharges through rarefied gases date from the 18th century. 'It was a most delightful spectacle, when the room was darkened, to see the electricity in its passage',

Benjamin Franklin's friend, the Englishman William Watson, wrote of his observations.[53] These experiments* were, of course, qualitative and remained so until about the time of Faraday's death. Faraday himself also studied these phenomena. In 1838 he found[56] that the glowing column of rarefied gas in his tube is separated from the cathode by a dark space, since known as the Faraday dark space. His views on the future of the subject were prophetic: 'The results connected with the different conditions of positive and negative discharge will have a far greater influence on the philosophy of electrical science than we at present imagine, especially if, as I believe, they depend on the peculiarity and the degree of polarized condition which the molecules of the dielectrics concerned acquire'.[57] Discharge phenomena not only led to the discovery of the electron, but also became a rich source of information on such subjects as atomic and molecular excitation and ionization, recombination, avalanche, space-charge, and plasma effects. We have also learned that the conditions which determine the nature of the discharge are quite complex. They involve the constitution and pressure of the residual gas in the tube, the applied potential difference between the electrodes and the constitution and shape of the latter, as well as the material and shape of the tube itself. Although these complexities have little to do with the present subject, they are mentioned nevertheless for one all-important reason: only if secondary effects are suppressed can one observe cathode rays as simple straight-line particle beams. This demands sufficiently low pressure. For example, for a cathode tube about 30 cm long and 2.5 cm wide, with a potential difference of, say, 1000 volts between the electrodes, one needs pressures $\sim$0.01 mm mercury. Which brings us back to Geissler's pump and Geissler's tube.

During the 1850s and 1860s, physics at the University of Bonn was taught by Professor Julius Plücker, surely best remembered for his contributions to analytic and projective geometry. Remarkably, he developed a taste for experiment. Even so it is doubtful that he would have entered this story were it not for his good fortune to have Geissler nearby, whose vacuum tubes enabled him to study discharge effects in gases. Plücker's results are recorded in a lengthy series of charming, qualitative, and not very lucid papers,[58] which include such interesting details as the increase of the region of glow with decreasing pressure; and changes in shape of that region in the presence of magnetic fields. More important are the observations by Plücker's student Hittorf who, in 1869, was the first to state that 'Glimmstrahlen' (glow rays) are emitted by the cathode and from there follow rectilinear paths.[59] Similar observations were made by Eugen Goldstein who in 1876 introduced the name 'cathode rays'.[60]

Speculations on the constitution of these rays had meanwhile begun. The electrical engineer Cromwell Fleetwood Varley was probably the first to suggest

* See Refs. 54 and 55 for the literature on this period.

that they are corpuscular in nature, 'attenuated particles of matter, projected from the negative pole by electricity',[61] and that their behavior in a magnetic field indicates that they are negatively charged, a view shared by Goldstein.[60]

In 1879, William Crookes also declared himself a champion of the corpuscular view. Having available an even better vacuum, he performed a series of elegant experiments in which he inserted such objects in his vacuum tube as a Maltese cross to demonstrate the existence of sharp shadows cast by the rays; a thin foil which, he showed, gets heated by the rays; and a tiny vaned paddle wheel which served to demonstrate that the rays transmit momentum. Crookes' description of the rays is graphic; he spoke of 'radiant matter', of 'molecular torrents', and of a fourth aggregate state of matter: 'In studying this fourth state of matter we seem at last to have within our grasp and obedient to our control the little indivisible particles which with good warrant are supposed to constitute the physical basis of the Universe'.[62] This sounds better than it actually was—for to Crookes the little indivisible particles were gas molecules, present in his tubes because of residual gas pressure. These molecules were supposed to get negatively charged upon impact with the cathode.

With this interpretation by Crookes the debate about the constitution of cathode rays started in all seriousness. Among early opposition to Crookes' views was the strenuous dissent by Goldstein who had generated cathode rays in a rather long tube of known residual pressure.[63] Given this pressure and assuming a molecular constitution of the rays, he estimated the mean free path of the presumed molecules and found it to be so short that it was impossible for the rays to follow the rectilinear paths which everyone then agreed they did. Hence, Goldstein concluded, cathode ray effects were more probably electromagnetic ray phenomena.

This may seem to be an utterly bizarre idea. It should be remembered, however, that in those days it was often argued that certain effects ascribed to cathode rays might actually be secondary in origin. As an example, consider the 1883 experiments by Hertz in which he found (incorrectly) that external electrostatic fields did not seem to act on the cathode rays, nor that a magnetized needle placed outside a cathode ray tube moved when cathode ray emission was initiated. He did not deny that electric currents were flowing inside the tube but took the view that such currents were probably a secondary phenomenon. 'It does not seem improbable that [cathode rays] have no closer connection to electricity than light which is emitted by an electric bulb . . . cathode rays are electrically indifferent . . . light is the phenomenon most closely akin to them.'[64] It is quite possible (as FitzGerald noted long ago[65]) that many of Hertz's observations were obscured by ionization and space-charge effects in the residual gas in the vacuum system. At any rate his overall conclusions were too sweeping.

As a general illustration of the situation in the early 1880s it is very instructive to read reminiscences, written in 1908 by Arthur Schuster, himself active in cathode ray experimentation in the period under discussion. (He was Rutherford's predecessor as professor of physics at Manchester.) Starting with the work of Crookes he wrote:[66] 'Crookes adopted the corpuscular view . . . and by means of accumulated evidence of most varied nature seemed to many of us to prove his case. Nevertheless a good deal of apathy was shewn, even in this country [England], with regard to the theoretical significance of the experiment, while in Germany the opposition to the corpuscular view was almost universal . . . the view that a current of electricity was only a flow of aether appealed generally to the scientific world and was held almost universally. The most absurd consequences were sometimes drawn from this view. . . . Adopting the view that a current of electricity simply means a flow of aether, it was tempting to attribute the [cathode ray] effects observed under reduced pressures to secondary effects, accompanying longitudinal [!] or other vibrations set up by the discharge. Attempts were made altogether to disconnect the luminous effects observed from the primary effects of the discharge . . .'.

The foregoing should suffice to convey the level of dialog in the 1880s.* Let us next move on to the 90s and the years of denouement.

In 1891 Hertz had found that metal leaf transmits cathode rays. This stimulated his student Philipp Lenard to construct vacuum tubes with a small aperture, over which he placed a thin foil, the 'Lenard window'. In this way he established that the rays can pass through a window thickness opaque to visible light. His studies of the transmission of the rays through outside air or through high vacuum led him to conclude that the rays are not molecular, and that they do bend in external electric fields.[68] Next J. J. Thomson showed in 1894 that the ray velocity is considerably smaller than the velocity of light,[69] and then, in 1895, Perrin, by placing a metal cylinder inside a vacuum tube in order to collect the charge carried away by cathode rays, obtained direct proof that this charge is negative.[70]**

There now follows a significant pause in the development—most physicists concerned with these matters turned their attention to X-rays—until the crucial year 1897. The situation at that time was summarized by Wilhelm Wien in the following way†: 'As is well known there are two opposing views on the nature of the cathode rays. The earlier one, especially adopted by the English physicists, considers the rays as negatively-charged particles. According to the second one, more representative of the German physicists, especially Goldstein, Wiedemann, Hertz and Lenard, the cathode rays are processes

* For a survey of other interesting experiments in that period see Ref. 67.

** Earlier observations on the negative charge of the current emanating from the cathode were based more indirectly on deflection properties.

† Wien wrote these lines shortly after the work by Wiechert and by Thomson which we are about to discuss.

('Vorgänge') in the aether. Especially due to the research of the latter two [physicists], showing the permeability of metallic foils by the rays, the second theory had gained general recognition, since this fact seemed difficult to reconcile with the assumption of charged particles . . .'.[71] Thus the debate continued right down to the wire.*

2. Wiechert. The nature of cathode rays was not the only major bone of contention as the year 1897 began. Equally confusing at that time (see Chapter 2) was the question: are X-rays transverse or longitudinal aether vibrations? Wiechert, since 1890 Privatdozent at the University of Königsberg in East Prussia, was among the first to have the right opinion about both points in dispute. 'Roentgen rays are most probably light rays of the ordinary transverse kind, but with shorter wave length', he had argued[72] in April 1896, only a few months after Roentgen's discovery. In the course of a lecture with demonstrations before Königsberg's Physical Economical Society, on 7 January 1897, he stated his conclusion about cathode rays[73] to which his experiments had led him: 'It showed that we are not dealing with the atoms known from chemistry, because the mass of the moving particles turned out to be 2000–4000 times smaller than the one of hydrogen atoms, the lightest of the known chemical atoms'.** It is the first time ever that a subatomic particle is mentioned in print and sensible bounds for its mass are given. However, it must be stressed that these conclusions depended on his assumption about the charge. 'Als Ladung ist 1 Elektron angenommen' (the charge is assumed to be one electron) he stated, using Stoney's terminology.

Wiechert demonstrated his experimental arrangement to his audience. His vacuum tube was placed in a magnetic field with magnitude H and oriented perpendicularly to the direction of ray propagation. Thus the rays followed a path with radius of curvature r, such that $mv = eHr/c$, where v is the ray velocity. This equation gives an upper bound for m/e if one could show that v is larger than some velocity v_0. Let V be the potential between the electrodes. Then, Wiechert reasoned, the kinetic energy of the rays† is at most equal to eV: $mv^2 \leq 2eV$. Hence

$$\frac{v_0 c}{Hr} \leq \frac{e}{m} \leq \frac{2Vc^2}{H^2 r^2}$$

* Perhaps the last public airing before the resolution came was held at the British Association meeting in Liverpool, September 1896, where Lenard's presentation came under heavy fire from Stokes and FitzGerald.

** 'Sie ergab, dass wir es nicht mit den von der Chemie her bekannten Atome zu thun haben, denn die Masse der bewegten Teilchen zeigte sich 2000–4000 mal kleiner als die der Wasserstofatome, also der leichtesten der bekannten chemischen Atome.'

† In all experiments discussed in this section $V \sim 10$ keV, so that none of the conclusions needed to be revised later because of relativistic effects.

in which V, H, and r are experimentally known. The remaining question was to find a sensitive value* for v_0. To this end Wiechert compared the transit time of the cathode rays with the period of a Hertzian oscillator, obtaining a value for v_0 of the order of one tenth of the velocity of light. For given e his bounds on m resulted.

Later in 1897, Wiechert moved to Goettingen. In 1898 he made[75] an actual determination of cathode ray velocities—but that was after J. J. Thomson had done the same. In that same year he was appointed to the newly created chair in geophysics (another of his early interests) at the University of Goettingen, where he later founded Germany's first institute for geophysics. He did not lose his interest in electrodynamic phenomena: his work on the 'Liénard-Wiechert potentials' dates from 1900. However, his greatest renown stems from his work on the internal structure of the earth, begun in 1897. Perhaps this change of focus explains in part** why he is not as well remembered for his major contribution to the interpretation of cathode rays as, I think, he should be.

3. Kaufmann. Walter Kaufmann's paper, completed in April 1897, on the determination of e/m for cathode rays[76] dates from the time he was an assistant in the Physics Institute at the University of Berlin. His particular interest concerned the question of how the motion of the rays in electric and magnetic fields depends on the pressure (in the range 0.03–0.07 mm) and the constitution of the residual gas in a vacuum tube. Like Wiechert, he used the equation $mv^2=2eV$. In addition he used the following relation. Let the rays move in the x-direction of a right-handed coordinate system, in a homogeneous magnetic field H in the y-direction. This leads to a deviation z of the rays in the z-direction, given by (see Eq. (4.3) with $k=0$)

$$m\frac{d^2z}{dt^2}=\frac{e}{c}Hv. \tag{4.5}$$

It follows that z is given by

$$z=H\frac{x_0^2}{2c}\sqrt{\frac{e}{2mV}} \tag{4.6}$$

where x_0 is the path length traversed by the rays in the x-direction.

Now Kaufmann's experiment showed that z was proportional to $V^{-1/2}$, independently of whether the residual gas in his tube was air, hydrogen, or

* Already in 1890, Schuster had attempted to find a lower bound for e/m by similar arguments. His lower bound for v was far too weak to draw any useful conclusions, however.[74]

** For example, in 1911 Wiechert was nominated for corresponding membership in the geophysics section of the Prussian Academy. The letter of nomination does not contain any reference to the cathode ray work.[75a]

carbon dioxide. This implies that e/m is a *constant*, the same for all his gases. That greatly puzzled him: 'This assumption [of constant e/m] is physically hard to interpret; for if one makes the most plausible assumption that the moving particles are ions [in the electrolytic sense] then e/m should have a different value for each gas.' Furthermore there was, as he perceived it, a second difficulty. Assuming e/m to be a constant, his measurements gave him about 10^7 emu/g for the value of e/m, 'while for a hydrogen ion [e/m] equals only 10^4'. Thus 'I believe to be justified in concluding that the hypothesis of cathode rays as emitted particles is by itself inadequate for a satisfactory explanation of the regularities I have observed'.

To define the 'birth of an era' is perhaps best left for parlor games. Let me write of the birth of particle physics nevertheless, define it to take place in April 1897, and appoint Kaufmann and Thomson as keepers at the gate. Their respective experimental arrangements (the latter to be discussed next) are of comparable quality, their experimental results equally good. Kaufmann's observation that certain properties of cathode rays are independent of the nature of the gas they traverse is, we would say, a clear indication of the universality of the constitution of these rays. The value for e/m he obtained is a good one. Had he added one conjectural line to his paper, something like 'if we assume e to be the fundamental unit of charge identified in electrolysis, then cathode rays must be considered to be a new form of matter', he would have shared equal honors with Thomson for advances made in 1897. Perhaps the thought never struck him, perhaps it did but was rejected as too wild. Perhaps also the Berlin environment was not conducive to uttering speculations of this kind, as is evidenced by a recollection of Jaffe (whom we already met in Chapter 2) about the year 1897: 'I heard John Zeleny say that he was in Berlin at that time, working in the laboratory of Warburg. When the discovery of the electron was announced, nobody in Berlin would believe in it'.[77] It may not have been known at that time what went through Kaufmann's mind; it certainly is not known now.

4. J. J. Thomson. In 1903, after taking his Ph.D. with Ostwald, Jaffe went from Leipzig to Cambridge with a recommendation from Boltzmann, to work with Thomson at the Cavendish. 'I must confess that my first impression of the laboratory itself was a little disappointing. Coming from a modern institution I found the famous school lodged in a somewhat old-fashioned building with bare brick interior walls and not quite corresponding to the standards of neatness to which I had been accustomed . . . It is superfluous to say that I was disappointed only with regard to the external appearance of the laboratory. With respect to the internal scientific life and activity the "Cavendish" would have been up to any expectations.'[77] Jaffe's early impressions of Thomson: 'Mathematics and physics, physics and mathematics, and just as much sport as was required to keep the body in good shape: that was his formula.

It was this intentness of purpose which brought J. J. Thomson to the chair of Maxwell and Lord Rayleigh at the age of 28, which raised him finally to the presidency of the Royal Society and to the mastership of Trinity College in Cambridge . . . When I entered the Cavendish Laboratory . . . students from all over the world looked to work with him . . . Though the master's suggestions were, of course, most anxiously sought and respected, it is no exaggeration to add that we were all rather afraid he might touch some of our apparatus. I hope I will not be misunderstood. I do not mean to detract anything from J. J. Thomson's fame of being one of the foremost experimental, as well as theoretical, investigators of his time. I report this little detail . . . to encourage the younger ones of my readers who are aware of not being quite as handy as the handiest'. The third Lord Rayleigh later reminisced as follows about his successor to the Cavendish Professorship: 'My doubt was whether Thomson should be professor of *experimental* physics. He had done very little experimenting at that time, though enough to show that he could do it. But he has shown since that it was right to appoint him'.[78] Indeed, though it may have been Thomson's main ambition to be a theorist (more about that in Chapter 9) he is remembered most widely for his experimental discoveries of the electron and of the first stable isotopes.*

During a lecture at the Royal Institution on April 30, 1897, the forty-year-old Thomson announced preliminary results for e/m of cathode rays.[80] He began by discussing Lenard's experiments on the absorption of cathode rays in various substances, from which he concluded that 'on the hypothesis that the cathode rays are charged particles moving with high velocities [it follows] that the size of the carriers must be small compared with the dimensions of ordinary atoms or molecules. The assumption of a state of matter more finely subdivided than the atom is a somewhat startling one . . .'. He went on to state his own experimental findings: 'It is interesting to notice that the value of e/m which we have found from the cathode rays is of the same order as the value 10^7 deduced by Zeeman from the experiments on the effect of a magnetic field on the period of the sodium light'. Thomson later recalled that 'I was told long afterwards by a distinguished colleague who had been present at my lecture that he thought I had been "pulling their legs".'[81]

On 7 August 1897, Thomson submitted his memoir on the corpuscular properties of cathode rays to the *Philosophical Magazine*. His first determination of e/m yielded a value 770 times that of hydrogen. He observed that 'the smallness of m/e may be due to the smallness of m or the largeness of e or to a combination of both'. He went on to argue in favor of the smallness of m. '. . . On this view we have in the cathode rays matter in a new state, a state in which the subdivision of matter is carried very much further than in the ordinary gaseous state: a state in which all matter . . . is of one and the same

* For biographies of Thomson see especially Refs. 78 and 79.

kind; this matter being the substance from which all chemical elements are built up.'[82]

Thomson arrived at these conclusions by means of his excellent new method of the crossed electric and magnetic fields: via a hole pierced in the anode, the cathode rays (moving in the x-direction) arrive in another evacuated part of the tube, where they are subjected to a homogeneous magnetic field H in the y-direction and a homogeneous electric field E in the z-direction. Both fields deflect the rays in the z-direction. From Eq. (4.3) (with $k=0$ of course) and the expression (4.2) for the Lorentz force one sees that E can be varied relative to H in such a way that the net ray deflection vanishes, in which case

$$eE=\frac{evH}{c} \quad \text{or} \quad v=\frac{cE}{H}. \tag{4.7}$$

(I leave aside a second method he used for finding v.) Next, turn off the magnetic field. The electric field now deflects the rays by an amount d during the time $t=l/v$ it takes the rays to move a distance l in the x-direction; d and l are measured, v is known, while

$$d=\frac{1}{2}\left(\frac{eE}{m}\right)\left(\frac{l}{v}\right)^2 \tag{4.8}$$

from which one finds e/m.

As I see it, Thomson's finest hour as an experimentalist came in 1899 when he applied the methods just described to photoelectrically produced particles and concluded—he was the first to do so!*—that these particles were electrons: 'The value of m/e in the case of ultraviolet light . . . is the same as for cathode rays'.[84] In the same paper, he announced his experimental results for the value of e, obtained by a method recently discovered by his student C. T. R. Wilson who had found that charged particles can form nuclei around which supersaturated water vapor can condense. Thomson's measurement of e is one of the earliest applications of this cloud chamber technique. He determined the number of charged particles by droplet counting, and their overall charge by electrometric methods, arriving at $e\simeq 6.8\times 10^{-10}$ esu, a very respectable result in view of the novelty of the method. And that is why Thomson is the discoverer of the electron.

When Thomson addressed a joint meeting of British and French scientists in Dover in 1899 most doubts had been resolved. He quoted a mass of 3×10^{-26} g for the electron, the right order of magnitude. Maxwell's difficulty in understanding the universality of the molecule of electricity had been surmounted, the atom had been split. 'Electrification essentially involves the splitting up of the atom, a part of the mass of the atom getting free and becoming detached from the original atom.'[85]

* For the history of the photoelectric effect see Ref. 83.

(e) β-rays are electrons

Zeeman–Lorentz ions, cathode rays, photoelectric ejectamenta—all had been recognized to be electrons as the year 1899 drew to a close. Also on its way by then was the remaining process of identification: β-rays are electrons as well. All went rapidly in that regard; few controversies and no surprises arose. I can therefore be very brief.*

In 1899 the magnetic deflection of β-rays was studied in four laboratories. One claimed that there was no effect,[87] the other three did find a deflection of the same sign as for cathode rays.[88] In 1900 the Curies measured the electric charge of the rays and found it to be negative.[89] Early e/m determinations in that year, including one by Becquerel, gave results not far from the value for cathode rays.[90,91] Shortly thereafter, Walter Kaufmann began a series of detailed experiments with β-rays from a radium source placed in electric and magnetic fields. In 1902, he was ready to state that 'for small velocities, the computed value of the mass of the electrons which generate Becquerel rays . . . fits within observational errors with the value found in cathode rays'.[92] From that time on it was considered settled that β-rays are electrons.

At the time of these early experiments there existed no distinction as yet between electrons of nuclear origin and peripheral atomic electrons—the nucleus had not yet been discovered. Occasional further investigations in later years on the identity of these two kinds of electrons have brought no new surprises. The equality of charge for these two categories has been carefully studied.[93] A mild revival of interest[94] in the precise value of e/m for nuclear electrons during the 1930s led to further precision measurements for this quantity.[95] The most compelling case for the identity of the two types of electrons comes from an ingenious experiment by Gertrude and Maurice Goldhaber.[96] They pointed out that if a nuclear electron were, in whatever way, distinct from an atomic electron, then the former, upon capture in an atom, could drop to a low-lying atomic state without prohibition by the exclusion principle. As a result one could expect characteristic X-ray emission to accompany this capture. No such rays were found—and therewith the matter rests.

(f) Relativistic kinematics

Particle physics is unthinkable without the kinematic rules dictated by the special theory of relativity. In particular, the energy (E)-momentum (p)-velocity (v) relations for a freely moving stable particle with rest mass m:

$$E = \frac{mc^2}{\sqrt{1 - v^2/c^2}} \tag{4.9}$$

* For further details see Ref. 86.

first given by Einstein[97] in 1905, and

$$p=\frac{mv}{\sqrt{1-v^2/c^2}} \tag{4.10}$$

first given by Planck[98] in 1906, are indispensable tools for experimentalists and theorists alike. So is in particular the zero-velocity relation

$$E \to mc^2 \quad \text{as} \quad v \to 0. \tag{4.11}$$

These equations took some time to sink in. (In 1912, Rutherford did not yet have straight[99] the meaning of Eq. (4.9).) There was also a time when these relations served as a test for the validity of special relativity. I conclude this chapter with a few comments on these tests, done with electrons.* These are the earliest high-energy experiments of this century, performed with the fastest particles then known: β-rays from a radium source.

Recall first that the strength of Eqs. (4.9–11) lies in their generality, their independence of dynamical details, in particular their independence of the origin and nature of the mass m. For specific forms of energy the relation (4.11) had been known well before 1905. Already in 1881, J. J. Thomson had noted the energy–mass equivalence for the case of an electrically-charged body.[101] Shortly thereafter, the first theoretical E-m-v relations appeared, based on a specific model of a charged particle: its shape shall be a rigid little sphere, whatever its velocity. This was the model studied in great detail by Max Abraham, theorist in Goettingen.[102] I shall not reproduce here his complicated formulae[103] for E and p as functions of m and v. Suffice it to say that they did, of course, not at all look like Eqs. (4.9) and (4.10), since a rigid sphere is not a relativistic object.

Meanwhile, Kaufmann had moved from Berlin to Goettingen, where he and Abraham became friends. In 1901 Kaufmann published his first paper on β-rays: 'The magnetic and electric deflectability of Becquerel rays and the apparent mass of the electron'.[104] From the outset he had two goals in mind, first to determine e/m for β-rays, secondly to find experimentally the relation between E, m, and v which, as said, had been discussed in the literature since the 1880s. It was at once clear to him that his radium source was well adapted for a simultaneous attack on both problems. He had in fact noted that—for reasons he could of course not fathom!—β-rays emanating from his source exhibit a wide range of velocities. There were slow rays suitable for measuring e/m. There were also very fast rays, in the region $v/c=0.7$–0.98, which were ideal for determining the E-m-v relation. By manipulating electric and magnetic fields he was able to select one velocity v, do his measurements, then move on to another v-value. In 1902 he announced two results in one paper.[92] The

* More details than given here are found in Ref. 2 and in Miller's detailed study of the history of special relativity.[100]

first, already noted in the previous section, was that his e/m agreed within the error with the cathode ray value. The second was that the E-m-v relation 'can be accurately represented by the Abraham formula', a conclusion he reconfirmed in 1903 after further experiments.[105]

The next year, Lorentz unveiled his E-m-v relation, close to but not identical[2] with Eq. (4.9). Then came Einstein. Then, one last time, came Kaufmann again. The new theories had stimulated him to redo his experiments. Once again he found agreement with the Abraham model: 'The measurements are incompatible with the Lorentz–Einstein postulate'.[106] Einstein took note but was unmoved: 'These theories [Abraham's and others] should be ascribed a rather small probability because their basic postulates concerning the mass of the moving electron are not made plausible by theoretical systems which encompass wider complexes of phenomena,' he wrote in 1907.[107]

It took years of further efforts to establish the experimental validity of Eq. (4.9).[2,100] When in 1912 Wilhelm Wien wrote to Stockholm,[108] proposing Lorentz and Einstein for the Nobel Prize, he did not yet consider the issue as settled: 'Concerning the new experiments on cathode rays and beta rays, I would not consider them to have decisive power of proof. The experiments are very subtle . . .'. All doubts were not removed until about 1916.

Slowly, inexorably, special relativity continued to gain ground. In 1905 Einstein had made it very clear that his postulates apply to all of physics. The early verifiable consequences of the theory were essentially all in the domain of electromagnetism, however, for the simple reason that initially there were no other known phenomena with which relativity could be confronted. The verification of the kinetic relations (4.9) and (4.10) were among the early applications of the theory to new areas. Then came Sommerfeld's theory of the structure of atomic spectra (1916), then Debye's and Compton's application to the kinematics of the Compton effect (1923).[109]

The energy–mass relation (4.11) took even longer (until the 1930s) to manifest its enormous importance. About that relation Pauli wrote in 1921: 'Perhaps the law of the inertia of energy will be tested at some future time by observations on the stability of nuclei.'[110] I shall return to that question in Chapter 11.

References

1. E. N. da C. Andrade, *Rutherford and the nature of the atom*, p. 48, Doubleday, New York 1964.
2. A. Pais, *Subtle is the Lord*, Chapter 7, Section (e), Oxford Univ. Press 1982.
3. J. C. Poggendorf, *Ann. der Phys. u. Chem.* **125**, 151, 1865.
4. E. N. da C. Andrade, *Vacuum* **9**, 41, 1959.

5. *Faraday's Diary*, Vol. 3, p. 234, Ed. T. Martin, Bell, London 1933.
6. J. Plücker, *Ann. der Phys. und Chem.* **103**, 88, 1858; Engl. transl. in *Phil. Mag.* **16**, 119, 1858.
7. Ref. 5, Vol. 7, pp. 436, 437, 455.
8. G. Shiers, *Sci. Am.* **224**, May 1971, p. 80.
9. *Nature* **17**, 169, 1877.
10. Cf. e.g. G. L. Weissler and R. W. Carlson, *Vacuum physics and technology*, Academic Press, New York 1979.
11. C. F. de C. Dufay, *Phil. Trans. Roy. Soc.* **38**, 258, 1734, repr. in W. F. Magie, *A source book in physics*, pp. 398–400, Harvard Univ. Press, Cambridge, Mass. 1965.
12. B. Franklin, letter to P. Collinson, May 25, 1747; see *Phil. Trans. Roy. Soc.* **45**, 98, 1750, repr. in W. F. Magie, Ref. 11, pp. 401–2.
13. H. W. Meyer, *A history of electricity and magnetism*, M.I.T. Press, Cambridge, Mass. 1971.
14. G. T. Fechner, *Ann. der Phys. und Chem.* **64**, 337, 1845.
15. W. Weber, *Ann. der Phys. und Chem.* **73**, 193, 1848.
16. E. Whittaker, *A history of the theories of aether and electricity*, Vol. 1, p. 203, Nelson, London 1958.
17. J. C. Maxwell, *A treatise on electricity and magnetism*, p. 307, Clarendon Press, Oxford 1873.
18. M. Faraday, *Experimental researches in electricity*, §§ 662–4, 824, Quaritch, London 1839.
19. Ref. 18, § 666.
20. Ref. 18, § 783.
21. Ref. 18, §§ 830, 835.
22. Ref. 18, § 851.
23. Ref. 17, Part 2, Chapter 4.
24. Cf. *Selected writings by Hermann von Helmholtz*, p. 409, Ed. R. Kahl, Wesleyan University Press, Middletown, Conn. 1971.
25. Ref. 18, § 852.
26. Ref. 18, § 869.
27. J. C. Maxwell, *The scientific papers of J. C. Maxwell*, Ed. W. P. Niven, Vol. 2, p. 418, Dover, New York.
28. Ref. 27, Vol. 2, p. 361.
29. Cf. e.g. F. Richarz, *Ann. der Phys. und Chem.* **288**, 385, 1894.
30. F. Trouton, *Nature* **87**, 50, 1911.
31. G. J. Stoney, *Phil. Mag.* **11**, 381, 1881.
32. Ref. 2, pp. 83–5.
33. G. J. Stoney, *Phil. Mag.* **36**, 132, 1868.
34. J. Loschmidt, *Wiener Ber.* **52**, 395, 1866.
35. G. J. Stoney, *Trans. Roy. Dublin Soc.* **4**, 563, 1888–92; *Phil. Mag.* **38**, 418, 1894.
36. M. Planck, *Ann. der Phys.* **4**, 564, 1901.
37. L. P. Williams, *Michael Faraday*, p. 499, Basic Books, New York 1965.
38. Ref. 27, Vol. 2, p. 790.
39. Ref. 5, Vol. 7, p. 465.
40. Ref. 27, Vol. 2, p. 225.
41. J. G. Tait, *Proc. Roy. Soc. Edinb.* **9**, 118, 1875.
42. H. Kamerlingh Onnes, *Physica* **1**, 241, 1921.

43. T. C. Mendenhall, in *The Physical papers of H. A. Rowland*, p. 16, Johns Hopkins Univ. Press; Baltimore 1902.
44. H. A. Rowland, *Phil. Mag.* **13**, 469, 1882.
45. P. Zeeman, *Versl. Kon. Ak. v. Wetensch. Amsterdam* **5**, 181, 1896, transl. in *Phil. Mag.* **43**, 226, 1897. An appendix to the English translation contains references to the earlier work by Tait and by the Belgian physicist Fievez.
46. P. Zeeman, *Versl. Kon. Ak. v. Wetensch. Amsterdam* **5**, 242, 1896, transl. in *Phil. Mag.* **43**, 226, 1897; cf. also W. F. Magie, Ref. 11, p. 384.
47. H. A. Lorentz, *Arch. Néerl.* **25**, 363, 1892, repr. in H. A. Lorentz, *Collected Papers*, Vol. 2, p. 164, Nyhoff, The Hague 1936.
48. H. A. Lorentz, *Versuch einer Theorie der electrischen und optischen Erscheinungen in bewegten Körpern*, Section 12, Brill, Leiden 1895; *Coll. papers*, Vol. 5, p. 1.
49. See Ref. 2, pp. 145–6.
50. P. Zeeman, *Versl. Kon. Ak. v. Wetensch. Amsterdam* **6**, 13, 99, 1897, transl. in *Phil. Mag.* **44**, 55, 255, 1897.
51. P. Zeeman, *Phil. Mag.* **44**, 60, 1897.
52. J. A. Brashier, *Autobiography*, p. 79, Riverside Press, New York 1924.
53. W. Watson, *Phil. Trans. Roy. Soc.* **45**, 93, 1748; **47**, 362, 1752.
54. Ref. 16, p. 349.
55. J. Müller and C. Pouillet, *Lehrbuch der Physik*, Vol. 4, Bk. 5, p. 1003, Vieweg, Braunschweig 1914.
56. Ref. 18, §§ 1526–43; Ref. 5, Vol. 3, pp. 239–59.
57. M. Faraday, *Phil. Trans. Roy. Soc.* **128**, 125, 1838.
58. J. Plücker, *Ann. der Phys. und Chem.* **103**, 88, 151, 1857; **104**, 113, 622, 1858; **105**, 67, 1858; **107**, 77, 497, 638, 1859.
59. J. W. Hittorf, *Ann. der Phys. und Chem.* **136**, 1, 197, 1869.
60. E. Goldstein, *Monatsber. Ak. der Wiss. Berlin*, *1876*, p. 279.
61. C. F. Varley, *Proc. Roy. Soc.* **19**, 236, 1871.
62. W. Crookes, *Chem. News* **40**, 127, 1879.
63. Cf. G. H. Wiedemann and E. Goldstein, *Phil. Mag.* **10**, 234, 1880.
64. H. Hertz, *Ann. der Phys. und Chem.* **19**, 782, 1883.
65. G. FitzGerald, *Nature* **55**, 6, 1896.
66. A. Schuster, *The progress of physics during 33 years (1875–1908),* Arno Press, New York 1975.
67. Ref. 16, pp. 353–7.
68. P. Lenard, *Ann. der Phys. und Chem.* **51**, 225, 1894; **52**, 23, 1894.
69. J. J. Thomson, *Phil. Mag.* **38**, 358, 1894.
70. J. Perrin, *Comptes Rendus* **121**, 1130, 1895.
71. W. Wien, *Verh. Phys. Ges. Berlin* **16**, 165, 1897.
72. E. Wiechert, *Schriften der Phys.–Ökonomischen Ges. zu Königsberg* (*Abh.*) **37**, 1, 1896; idem. *Sitzungsber.* **37**, 1896, p. 29.
73. E. Wiechert, ibid. **38**, 3, 1897.
74. A. Schuster, *Proc. Roy. Soc.* **47**, 526, 1890.
75. E. Wiechert, *Ann. der Phys. und Chem.* **69**, 736, 1899.
75a. C. Kirsten and H. C. Körber, *Physiker über Physiker*, Vol. 1, p. 197, Akademie Verlag, Berlin 1975.
76. W. Kaufmann, *Ann. der Phys. und Chem.* **61**, 544, 1897.
77. G. Jaffe, *J. Chem. Educ.* **29**, 230, 1952.
78. Lord Rayleigh, *The life of Sir J. J. Thomson*, p. 20, Cambridge Univ. Press 1942.

79. G. P. Thomson, *J. J. Thomson*, Nelson, London 1964.
80. J. J. Thomson, *The Royal Institution Library of Science*, Vol. 5, p. 36, Eds. W. Bragg and G. Porter, Elsevier, Amsterdam 1970.
81. J. J. Thomson, *Recollections and reflections*, p. 341, Bell & Sons, London 1936.
82. J. J. Thomson, *Phil. Mag.* **44**, 311, 1897.
83. Ref. 2, pp. 379–82.
84. J. J. Thomson, *Phil. Mag.* **48**, 547, 1899.
85. Ref. 84, p. 565.
86. M. Malley, *Am. J. Phys.* **39**, 1454, 1971.
87. J. Elster and H. Geitel, *Verh. Deutsch. Phys. Ges.* **1**, 136, 1899.
88. F. Giesel, *Ann. der Phys. und Chem.* **69**, 834, 1899; S. Meyer and E. von Schweidler, *Phys. Zeitschr.* **1**, 90, 113, 1899; H. Becquerel, *Comptes Rendus* **129**, 996, 1205, 1899.
89. M. and P. Curie, ibid. **130**, 647, 1900.
90. H. Becquerel, ibid. **130**, 809, 1900.
91. E. Dorn, ibid. **130**, 1129, 1900.
92. W. Kaufmann, *Phys. Zeitschr.* **4**, 54, 1902.
93. R. Ladenburg and Y. Beers, *Phys. Rev.* **58**, 757, 1940; Y. Beers, *Phys. Rev.* **63**, 77, 1943.
94. Cf. H. R. Crane, *Rev. Mod. Phys.* **20**, 278, 1948.
95. C. T. Zahn and A. H. Spees, *Phys. Rev.* **53**, 357, 1938.
96. M. Goldhaber and G. Scharff-Goldhaber, *Phys. Rev.* **73**, 1472, 1948; cf. further W. Davies and M. Grace, *Proc. Phys. Soc. London* A **64**, 846, 1951.
97. A. Einstein, *Ann. der Phys.* **17**, 639, 1905.
98. M. Planck, *Verh. Deutsch. Phys. Ges.* **4**, 136, 1906.
99. E. Rutherford, *Phil. Mag.* **24**, 893, 1912.
100. A. I. Miller, *Albert Einstein's special theory of relativity*, pp. 325–52, Addison-Wesley, Reading, Mass. 1981.
101. J. J. Thomson, *Phil. Mag.* **11**, 229, 1881.
102. M. Abraham, *Phys. Zeitschr.* **4**, 57, 1902; *Ann. der Phys.* **10**, 105, 1903.
103. Cf. Ref. 2, Eqs. (7.29) and (7.30).
104. W. Kaufmann, *Goett. Nachr.* 1901, p. 143.
105. W. Kaufmann, ibid. 1903, p. 90.
106. W. Kaufmann, *Ann. der Phys.* **19**, 487, 1906.
107. A. Einstein, *Jahrb. Rad. Elektr.* **4**, 411, 1907.
108. W. Wien, letter to the Royal Swedish Academy of Sciences, early January 1912; cf. also Ref. 2, p. 506.
109. Cf. Ref. 2, p. 413.
110. W. Pauli, *Encyklopädie der mathematischen Wissenschaften*, Vol. 5, Part 2, p. 539, Teubner, Leipzig 1921.

5

Interlude: earliest physiological discoveries

> Good applied science in medicine, as in physics, requires a high degree of certainty about the basic facts at hand, and especially about their meaning, and we have not yet reached this point for most of medicine.
>
> Lewis Thomas, *The Medusa and the Snail*, 1979.

In January 1981 the Comptroller General of the General Accounting Office caused to be made public his Report[1] to the Congress of the United States entitled *Problems in Assessing the Cancer Risks of Low-Level Ionizing Radiation Exposure*. Drafted with the assistance of expert consultants, it contains an up-to-date review including a digest of the BEIR-III Report* of 1980, the most recent in a series of documents[2] prepared by distinguished panels of members of the National Academy of Sciences. It is a reflection on the problem's extreme complexity, not only politically but also scientifically, that this last panel was unable to reach a consensus. The uncertainties are clearly expressed in the main conclusions of the GAO Report.[3]

> Scientists are still trying to understand exactly how ionizing radiation causes cancer, and to determine how many cancers are caused by a given amount of radiation . . . Scientists cannot yet characterize what cancer fundamentally is, much less describe precisely what role radiation plays.

It was not for lack of effort. To date, the worldwide literature on the subject is estimated to exceed 80 000 articles. Since 1898 the U.S. Government alone has spent over 2 billion dollars in supporting research in this area, about 80 million per annum in recent times. This modern figure is not particularly impressive, though, if it is compared with the total U.S. expenditure on health care (hospital charges, fees, drugs, insurance, facilities, research, etc.) which is estimated[4] to have run to $140 billion in 1978. It is highly dubious, however, whether more financial support could have made it unnecessary to inform the U.S. Congress[5] in 1981 that 'Low level radiation exposure is a relative term that is difficult to define precisely. There is no consensus on a precise definition. Whether a particular dose and dose rate of radiation is considered "low" may

* BEIR = Biological Effects of Ionizing Radiation.

depend on the person making the judgment, the source and type of radiation, the parts of the body irradiated and other factors'.

Neither the lack of consensus on safe exposure limits nor our ignorance of the nature of cancer should obscure the extraordinary advances, on technical as well as fundamental biological levels, in our knowledge of the influence of ionizing rays on biological organisms achieved since the days of Roentgen and Becquerel. This is perhaps best made clear by considering the impact on the world of medicine of the discoveries recounted in the previous chapters. That impact is the subject of the present chapter. I have included this single excursion beyond the realm of physical science in order to stress that both the sense of novelty generated by the discoveries described earlier and the haphazard nature in which they were made have counterparts in the corresponding applications to physiology and medicine.

To set the stage, it may be well to recall first a few fundamentals which were *not* known in that period. Foremost among these is the insight that the crucial property of radiations as they affect biological systems is their ionizing power. This ignorance is not at all surprising. When Roentgen discovered X-rays in 1895, ions were objects familiar from electrolysis, but the realization that they are broken-up atoms came only four years later.* Moreover, at least until 1910 most physicists' conceptions about the absorption of radiation by matter were incorrect.** An understanding of what happens as X-rays traverse matter is necessary but, of course, not sufficient for an understanding of the havoc they can create as they traverse a living cell. In that last respect our present knowledge, however incomplete, postdates the period I am about to discuss. In this connection it is important to note (a point stressed by Serwer[6]) that during the years preceding the First World War the physics and the clinical radiological communities largely went their separate ways. Thus it took a considerable time before information gathered at the frontiers of physics was assimilated by the medical profession.

The involvement of physicists in clinical radiology increased notably because of the exigencies of the First World War.† Among those who left their early mark, Walter Friedrich, co-discoverer of X-ray diffraction,†† and his collaborators deserve particular mention. It was Friedrich who, in 1918, first introduced an absolute unit of radiation dosage which came to be known as the roentgen: the amount of radiation which by ionization in one cubic centimeter of air (at room temperature and pressure) transfers one electrostatic unit of charge in a saturation current.[7] How slowly quantitative methods penetrated is seen, for example, from the fact that the roentgen was adopted internationally only in 1928, when an International Commission on X-ray units convened—for the first time.[8] The currently used unit, the rem, was

* See Chapter 4, Section (d), Part 4. ** See e.g. Chapter 8, Sections (c) and (d).
† See Chapter 11, Section (g). †† See Chapter 2, Section (a).

introduced much later.* Since 1960 the recommended maximum exposure (whole body) per annum has been set in the U.S. at 12 rems for occupational exposure and at 0.5 rems for any member of the public, a considerable decrease from the permissible levels recommended earlier.[10]

One last example of a basic fact unknown to the pioneers: The mutagenic properties of the new radiations which first became known**[11] in 1927 as a result of Hermann Muller's researches. Experiments could be so arranged that nearly 100 per cent of irradiated fruit flies exhibited mutations. 'A possibility had been created for the first time of influencing the hereditary mass itself artificially.'[12] The need for protection against radiation now became even more urgent. Mutations are on the whole bad things.† How many man-made hits by radiation are tolerable for the hit generation, and for their offspring?

While the debate on this question continues, let us turn to the not so good old days. In his essay 'Medical Lessons from History', Lewis Thomas considers[4] the early history of medicine to be an unrelievedly deplorable story and goes on to remark: 'Virtually anything that could be thought up for the treatment of disease was tried out at one time or another, and, once tried, lasted decades or even centuries before being given up. It was, in retrospect, the most frivolous and irresponsible kind of human experimentation, based on nothing but trial and error, and usually resulting in precisely that sequence'. Drop the 'centuries' and these comments are a fitting summary of what follows.

Roentgen's photograph of the bones of his wife's hand, published as part of his first paper[13] on X-rays, led to an outpouring of articles recording similar pictures. Among my favorites is a publication in a learned journal[14] describing a photograph of 'l'aileron d'un faisan tué à la chasse' (the pinion of a pheasant killed during a hunt). Within a few weeks after Roentgen's discovery, announced on 28 December 1895, its implications for medical diagnostics had occurred to many. This is how the then demonstrator of morbid anatomy at the University of Pennsylvania put it[15] in the weekly *Medical News* of 15 February 1896: 'Roentgen's weird and wonderful discovery is destined to enrich medicine with possibly the most valuable diagnostic process which recent years have witnessed . . . the surgical imagination can pleasurably lose itself in devising endless applications of this wonderful process . . . It is stated that stone in the kidney has already been determined . . . A new means of

* A rad = 100 ergs per gram is the unit of energy absorbed from radiation by a mass of matter. Exposure to one roentgen results in an absorbed dose of about 1 rad for X-, γ-, and β-rays. The absorbed dose in rads multiplied by a rather subjective 'quality factor' Q is known as the dose-equivalent expressed in rems. For X-, γ-, and β-rays, Q is assigned the value unity, so that for those radiations 1 rad corresponds to 1 rem. Q is chosen to be about 10 for neutrons or protons, 20 for α-particles and heavier nuclei.[9]

** It had been suspected earlier that X-rays caused lesions in chromosomes.

† It is believed, however, that genes contain built-in mechanisms for repair of mutational damage.

distinguishing luxations from fractures . . . Obstetricians will readily perceive the immense value of the ability to *see* the fetus in utero after ossification of its bones has occurred . . . The easy application of Roentgen's method of taking a picture on a sensitized plate renders its use at once possible in hospitals . . . The entire cost of such an apparatus need not exceed $50'. In the same paper the name 'skiagraph' (meaning shadow picture) was proposed for an X-ray photograph, a term found frequently in the literature during subsequent years.

The author of this article did not exaggerate. Clinical applications grew rapidly especially after it was realized (very early) that injections with opaque substances enhance skiagraphic contrast. Already in 1896 several hospitals and private practitioners had their X-ray installations.

The speed with which X-ray techniques were adopted is also illustrated by their early use in military campaigns. 'X-ray machines [saw] service in the British army's 1897 Sudan expedition, in the Graeco–Turkish war of that same year, in the Spanish–American war and in the Boer war.'[16] (For the use of X-ray units in the First World War, see Chapter 11, Section (g).)

Other applications came from a quite different direction: already in 1896 X-ray photographs were accepted as evidence in court, first in France and England.[17] Later in that year, the learned judge presiding over a Denver court said,[18] upon admitting the introduction of such a photograph: 'Let the courts throw open the door to all well-considered scientific discoveries. Modern science has made it possible to look beneath the tissues of the human body, and has aided surgery in telling of the hidden mysteries. We believe it to be our duty in this case to be the first, if you please, to so consider it, in admitting in evidence a process known and acknowledged as a determinate science'.*

Also in 1896 the idea gained ground, again independently in several countries, that the study of X-irradiation of microorganisms might lead to medically important results. An editorial in *Medical News* of 22 February 1896 makes clear why that was a plausible thought for its time.[20] 'It is a well-known fact that sunlight possesses a decidedly germicidal effect, and that if the prismatic rays be passed through a culture-tube containing a fluid medium, the various bacilli therein will exhibit a selective action. And it has occurred to more than one worker along the lines of bacteriological investigation that it was quite within the realm of possibility that the Roentgen ray would be capable of killing bacteria within the human system.' Thus, to mention but one example, a few weeks later officers of the New York City board of health announced plans for studying the effect of X-rays on diphtheria bacilli.[21] These earliest studies of the influence on bacterial cultures showed little or no impact.[22]

Direct X-irradiation of human beings also began in that year. Once again, the idea was plausible. Earlier in the decade the Dane Niels Ryberg Finsen

* In 1903, the *Brooklyn Medical Journal* published a review of these early court cases in the United States.[19]

had shown that *lupus vulgaris* (a tuberculosis of the skin) could be successfully treated by exposure to ultraviolet light. In fact, for such purposes a special Light Institute under his direction was opened in April 1896 in Copenhagen. (In 1903 Finsen received the Nobel Prize in medicine for this work.) Dermatologists now set out to treat skin afflictions with the new rays.

Whatever medical successes were thus obtained were almost at once overshadowed by evidence of danger.

Reports of undesired effects induced by X-rays began to appear already in 1896, in many parts of the world.[22] There were cases where irradiation of a skin area caused loss of hair, or erythema (inflammation) or, more seriously, dermatitis (skin burns). Words of caution began to appear in the medical literature, but also such articles[23] as 'A protest against the fear of dermatitis originating from exposure to the Roentgen rays'. Elihu Thomson, a distinguished electrical engineer, became curious about whether rays which do not affect the senses could indeed have such strong impact on human tissue. In the course of a 'Topical Discussion on the Roentgen Ray', held at the American Institute of Engineers in New York City on 16 December 1896, he reported[24] on an experiment he had performed on himself. For several days he had exposed a finger to an X-ray tube, 30 minutes a day, at about 2 cm distance. 'Five, six, seven, eight days passed and nothing happened, and I felt that people had been mistaken about the effect of the rays. But on the ninth day the finger began to redden; on the twelfth day there was a blister, and a very sore blister. On the thirteenth or fourteenth day . . . the blister had included all the skin down to the part not exposed and had gone around the finger almost to the other side. The whole of the epidermis came away and left an ulcer without any possibility of recovering its own epidermis except from the edges. [Recently] the sore actually closed . . . Nature does not appear to have found out how to make a good skin over that finger . . . For a time it was a very angry looking affair.'

Thomson's observation about the week's delay between cause and effect leads us to an important reason why X-rays caused much sickness and death in the pioneering days. It could not be known at that time that it may take not just weeks but years before pathology caused by radiation fully develops. This, I believe, goes a long way toward explaining numerous instances of carelessness in spite of early danger signals.

The accident rate was further enhanced because X-ray apparatus often fell into the wrong hands, since it was neither very expensive nor difficult to acquire, and also because of the public's intense interest in seeing the mysterious pictures produced. There is the case of a faculty member of Columbia University who in 1896 suffered severe skin damage from having given X-ray demonstrations over a period of weeks at Bloomingdale's, already then one of New York City's fine department stores.[22] A Chicago doctor narrates[25] the case of a man '[who] died in 1908 as the result of roentgen injuries, most of which were received while exhibiting his hands and arms at a church fair during the winter of 1907 . . . I personally cautioned him but he laughed at my warnings.'

Tales of accidents multiplied. In 1897 a U.S. Army doctor placed on record[25] 'a case of X-ray burn which has no parallel in medical literature up to date, either in the length of exposure (11 and 5/6 hours) or in the extent (15 in. × 8 in.) and intensity of the resulting lesion . . . The most distressing feature of the case is the intense pain, which nothing but morphine will control . . . The slowness with which this apparently superficial denudation heals is almost incredible'. Also in 1897 the first evidence was reported of deep tissue traumatism from X-ray exposure.[27] The *American X-ray Journal* of that year contains the results of perhaps the first nationwide survey. For each of 69 cases of injury, the apparatus used, number and duration of exposures, and distance from the part exposed are indicated. Such clinical data are obviously useful, yet one is struck by the absence of reference to supplementing laboratory experiments in the early literature. When these were begun, in about the years 1902–3, new dangers were revealed. X-ray exposure can cause sterility, changes in the blood and blood-forming organs, and cancer.[28]

Meanwhile the numerous cases of X-ray injury had begun to cause public outcry. Patients were suing for damages, sometimes successfully, thereby adding to the medical concern for safety measures. In 1903 lead-impregnated glass and rubber were introduced[29] as shielding devices. A Philadelphia doctor recommended[30] that the patient be placed behind a screen of lead 'closely fitted to the part to be treated, leaving the diseased area and a portion of the surrounding normal tissue uncovered'. In addition medical organizations began to stress the need for the introduction of legal measures restricting the use of X-ray equipment to qualified persons only.[31]

In the early years no group paid more heavily in injury and death than the medical profession itself.[32] In his book[25] (published in 1936) on the medical victims in the United States, Percy Brown, at one time the historian of the American Roentgen Society, remarks that this phenomenon is not easy to explain. As contributing factors he mentions the excitement over new discoveries which allowed precaution to lapse, the height of lay interest, and the insistence by the producers of the equipment on the ease of its manipulation. Brown's case histories are grim. There was Dr. Elizabeth Ascheim of San Francisco who 'in spite of warning would not protect herself and exposed herself repeatedly not only during experiments and routine treatments but also for the benefit of patients to prove the operation was painless'. She repeatedly refused medical treatment of her badly ulcerated hands. She died three months after an arm had to be amputated at the shoulder. There was Walter Dodd, Roentgenologist to the Massachusetts General Hospital who 'whether reckless or innocently unaware of the risks he was running certainly did not spare himself'. He died of pulmonary carcinomatous metastasis after fifty operations under ether. There were numerous others.

Radium was to claim its victims too, but the story of its early physiological impact is incomparably less harrowing than that for X-rays. I know of no lethal injuries caused by exposure to radioactive material in the pre-World War I years. The main reason for this difference is of course that radium was a rare commodity and therefore very expensive once commercially available ($100 per milligram until the early twenties). The high price inhibited research, but also kept quacks away. Not only scientific but also economic reasons expedited the introduction of the unit of radioactive strength, the curie,* which was internationally adopted in 1912, long before its X-ray counterpart the roentgen.

Experiments on bacteria and on animals started early. Already in 1901 it was found that populations of *micrococcus prodigiosus* were effectively killed in about three hours when exposed to radium.[33] A review[34] in 1904 reports: 'In very young animals . . . a bit of radium in a sealed glass tube placed against the spine caused death. Larvae and embryos were modified in growth and monstrosities occurred more often. In non-fertilized eggs atypical formation arose'. In 1904 Pierre Curie and coworkers[35] put mice and hamsters inside bottles containing radium emanation. Death followed within four to nine hours, the more rapidly the higher the emanation concentration.**

Meanwhile the first evidence had been obtained for the influence of radium on the human body. Wishing to verify the first report[40] (dating from 1900) of skin infection caused by radium, Friedrich Giesel, the first in Germany to produce radium salts commercially, attached a radium–barium bromide capsule to his arm for a two-hour period. 'After 2–3 weeks a strong infection began followed by loss of skin.'[41] Becquerel had a similar experience.[42] For the purpose of a lecture demonstration, Pierre Curie had lent him a glass phial containing radium-enriched material which he, Becquerel, had put in a waist pocket for a total duration of six hours. Ten days later this exposure caused a red spot to develop on the skin, followed by skin peeling. Upon being informed of this mishap, Pierre Curie decided to check this action on himself, and attached a weak radium source to his arm for a ten-hour exposure. After

* An amount of radioactive substance which generates 3.7×10^{10} disintegrations per second is said to have a strength of one curie.

** Among the more bizarre experiments are those performed in 1919 by the distinguished Dutch physiologist Zwaardemaker who claimed that the natural radioactivity of potassium (discovered in 1907) was responsible for its physiological action. He reached this conclusion from the study of the beating of an isolated frog's heart when perfused with solutions containing potassium chloride.[36] These results attracted considerable attention (Ehrenfest mentioned them in a letter to Bohr[37]) but soon came in for criticism.[38] The matter was finally settled in 1939 when Goudsmit calculated that in Zwaardemaker's experiment only about one K-atom per hour disintegrated in the frog's heart.[39]

confirming his friend's experience, the two men prepared a joint report which they submitted to the Section Physiologie of the Académie des Sciences.[43] In this article Pierre also mentioned some of his past experiences in the handling of radium: an often very painful hardening of the fingers, inflammations which could last weeks . . .

Radium therapy (called 'curiethérapie' in France) began during the years 1903–6. From various countries a number of successful treatments of lupus vulgaris and of superficial cancers were reported.[44]

The bad years for radium came in the late nineteen-twenties (radium was more plentiful by then) when numerous reports appeared of persons severely injured or killed by internally deposited radioactive substances. Most notorious were the 'radium poisonings' at a luminous dial painting plant in East Orange, where workers would lick small paint brushes on which radium was deposited, in order to get a fine brush point.*

In addition there were terrible consequences resulting from the applications of intravenous injections and of tonics, both with radioactive contents, medications mentioned until 1932 by the American Medical Association in their list of 'New and Unofficial Remedies'. Robley Evans, initiator of the medical physics of radioactivity in the United States, has recalled some instances.[46] One concerns a wealthy Pittsburgh bachelor. 'He took a [commercially available] radioactive tonic [containing] a microcurie of radium and a microcurie of mesothorium in about a half ounce of distilled water . . . four of these a day . . . He took a lot of this and so did several of his girl friends . . . and [in 1932] he died a miserable death.' The other concerns a Chicago physician 'who injected more than a thousand patients, the normal regime being 10 microcuries intravenously once a week for a year! That's 500 microcuries or half a millicurie, a very large amount of radium to be administered intravenously! That physician destroyed his records and left town. There were a number of physicians doing that'. Compare these amounts with the current value of the permissible body burden:** 0.1 microcurie for radium, the number first proposed[47] by Evans in 1941, adopted[48] in the U.S. soon thereafter, and now internationally accepted.

Events of the Second World War have transformed the dangers of radiation into a major issue of public policy. They have also made obvious the need for dialog between medicine and physics, currently proceeding at excellent levels. Furthermore, they have led to the creation of facilities for artificially producing radioactive isotopes which by now have almost completely replaced radium for therapeutic purposes.

* Extensive jaw necrosis and severe anemia were among the early symptoms; bone damage and malignancy came later. Follow-up case histories of thirty patients (including fourteen luminous dial painters) who had ingested radioactive material are found in Ref. 45.

** This is the maximum permissible amount of radium the human body is supposed to carry at any given time, an amount which varies for different radioactive materials.

New radiations have been brought to bear on the cure of disease, such as fast neutrons (since 1936) and fast electrons (since 1949). Therapeutic irradiation with external radioactive sources continues. In addition encapsulated radioactive material is inserted interstitially in body tissue or in natural body cavities. There has also been great innovation in diagnostic tools: X-ray methods have been vastly refined by the so-called subtraction methods and by CT and PET scanning. Alternative diagnostic techniques make use of ultrasound and nuclear magnetic resonance imaging.* Much remains to be learned, but there has also been much progress. Evans recently gave me the total number** of medical injuries caused by internal deposits of the highly radioactive and toxic element plutonium since its discovery forty years ago: zero.

Sources

The most important writing on early physiological developments I know is the Ph.D. thesis by Serwer.[6] References to some early medical literature on the subject are found in the Roentgen biography by Glasser,[17] but for more detailed information the *Index Medicus* is indispensable. Percy Brown's account of the early days[25] is very illuminating. The modern vantage point is described in the instructive reports quoted in Refs. 1 and 2.

References

1. U.S. G.A.O. report EMD-81-1, January 2, 1981.
2. *The effects on populations of exposure to low levels of ionizing radiation*, National Academy of Sciences, Washington D.C., 1980.
3. Ref. 1, Vol. 1, pp. 2, 4.
4. L. Thomas, *The medusa and the snail*, Viking Press, New York 1979.
5. Ref. 1, Vol. 2, Chapter 2.
6. D. P. Serwer, 'The rise of radiation protection: science, medicine and technology in society, 1896–1935', Ph.D. Thesis, Princeton University; Xerox University Microfilms, 77-14, 242, Ann Arbor, Mich. 1976.
7. B. Krönig and W. Friedrich, *Physikalische und Biòlogische Grundlagen der Strahlentherapie*, Vol. 3, Urban and Schwarzenberg, Berlin 1918.
8. Ref. 6, p. 223.
9. International Commission on Radiation Protection Publication No. 26, Pergamon Press, Oxford 1977.
10. Ref. 1, Vol. 2, Chapter 3.
11. Ref. 6, p. 117.

* For highly informative brief accounts of the new look in therapy and in diagnostics see Refs. 49 and 50 respectively.

** Private conversations, October 1983. This number excludes of course all consequences of the drop of Fat Man on Nagasaki.

12. T. Caspersson, in *Nobel lectures, physiology or medicine, 1942–1962*, p. 151, Elsevier, New York 1964.
13. W. C. Roentgen, *Sitzgsber. physik.-med. Ges. Würzburg* **137**, December 1895.
14. A. Londe, *Comptes Rendus* **122**, 311, 1896.
15. H. W. Cattell, *The Medical News* (*N.Y.*) **68**, 169, 1896.
16. Ref. 6, p. 126.
17. O. Glasser, *Wilhelm Conrad Roentgen*, Chapter 13, C. C. Thomas, Springfield, Ill. 1934.
18. *Elect. Engineer* **22**, 655, 1896.
19. W. W. Goodrich, *Brooklyn Med. J.* **17**, 515, 1903.
20. *Medical News* (*N.Y.*) **68**, 210, 1896.
21. *New York Times*, March 15, 1896.
22. Cf. Ref. 17, Chapter 14.
23. G. H. Stover, *Western Med. and Surg. Gazette* **1**, 270, 1897.
24. E. Thomson, *Am. Inst. Electr. Eng.* **13**, 418, 1897.
25. P. Brown, *American martyrs to science through the Roentgen rays*, p. 25, C. C. Thomas, Springfield, Ill. 1936.
26. W. B. Banister, *Nat. Med. Rev.* **7**, 127, 1897.
27. *Brit. Med. J.*, 31 July 1897, p. 272, repr. in *Health Phys.* **38**, 885, 1980.
28. Cf. Ref. 6, p. 58.
29. Cf. Ref. 6, p. 66.
30. M. Gramm, *Hahnemann Monthly* **38**, 641, 1903.
31. Cf. e.g. *Bull. de l'Ac. de médicine* **55**, 76, 1906.
32. Ref. 6, p. 60.
33. E. Aschkinass and W. Caspari, *Arch. f. die Ges. Physiol.* **86**, 603, 1901.
34. M. V. Ball, *Med. Fortnightly* **25**, 149, 1904.
35. P. Curie, C. Bouchard, and V. Balthazard, *Comptes Rendus* **138**, 1385, 1904.
36. H. Zwaardemaker, *Pflüger's Arch. f. die Ges. Physiol.* **173**, 30, 1919.
37. P. Ehrenfest, letter to Niels Bohr, June 4, 1919.
38. S. G. Zondek, *Biochem. Zeitschr.* **121**, 76, 1921.
39. S. Goudsmit, *Science* **29**, 615, 1939.
40. Walkhoff, *Photogr. Rundschau* **14**, 28, 1900.
41. F. Giesel, *Ber. der deutsch. Chem. Ges.* **3**, 3569, 1900.
42. H. Becquerel, *Recherches sur une propriété nouvelle de la Matière*, p. 263, Firmin Didot, Paris 1903.
43. H. Becquerel and P. Curie, *Comptes Rendus* **132**, 1289, 1901.
44. Cf. e.g. J. Macintyre, *British Med. J.* **2**, 199, 1903; M. Metzenbaum, *International Clinics* (*Philadelphia*) **4**, 21, 1905.
45. J. C. Aub, R. D. Evans, L. H. Hempelmann, and H. S. Martland, *Medicine* **31**, 221, 1952.
46. C. Weiner, interview with R. D. Evans, May 7, 1974, transcript in Archives of the American Institute of Physics, New York.
47. Cf. R. D. Evans, *Health Phys.* **41**, 437, 1981.
48. 'Safe handling of radioactive luminous compound', *Nat. Bureau of Standards Handbook*, 27 May 1941, later designated NCRP report 5.
49. J. S. Laughlin, *Physics Today* **36**, 26, July 1983.
50. P. R. Moran, R. J. Nickles, and J. A. Zagzebski, *Physics Today* **36**, 36, July 1983.

6

Radioactivity's three early puzzles

(a) Introduction

Radioactivity was discovered in 1896, the atomic nucleus in 1911. Thus even the simplest qualitative statement—radioactivity is a nuclear phenomenon—could not be made until fifteen years after radioactivity was first observed. The connection between nuclear binding energy and nuclear stability was not made until 1920. Thus some twenty-five years would pass before one could understand why some, and only some, elements are radioactive. The concept of decay probability was not properly formulated until 1927. Until that time, it had to remain a mystery why radioactive substances have a characteristic lifetime. Clearly, then, radioactive phenomena had to be a cause of considerable bafflement during the early decades following their first detection. This chapter is devoted to the three principal puzzles of that period.

The first one was: what is the source of the energy that continues to be released by radioactive materials? Already in the year of discovery, Becquerel himself had been quite surprised at the persistence of the energy produced by the 'uranic rays'. From 1898 on, physicists began to pose such questions as: could it be that energy is not conserved in these processes? Does the source of energy reside outside the atom or inside?

The second puzzle was: what is the significance of the characteristic half life for such transformations? (The first determination of a lifetime for radioactive decay dates from the year 1900.) If in a given radioactive transformation all parent atoms are identical, and if the same is true for all daughter products, then why does one radioactive parent atom live longer than another, and what decides when a specific parent atom disintegrates?

The third puzzle: is it really true that some atomic species are radioactive, others not? Or are perhaps all atoms radioactive but many of them with extremely long lifetimes?

Before turning to the discussion of early thoughts on these questions, I should stress that these problems did not hold center stage throughout the period under discussion, a period so rich in other developments. Rather, they were principally the concern of a fairly modest-sized but elite club of experimental radioactivists. In those days, theoretical physicists did not play any role of consequence in the development of this subject, both because they were not particularly needed for its descriptive aspects and because the deeper questions were too difficult for their time. It is true that distinguished theorists (especially those belonging to an older generation) would on occasion express views on these issues from which we gain revealing insights into the climate of thought of the times. But these comments were not to be of lasting significance—with the most notable exception of two contributions by Einstein. One of these I mention right here, the other will be found in Section (e) of this chapter.

In the second of his 1905 papers on relativity Einstein stated that 'If a body gives off the energy L in the form of radiation, its mass diminishes by L/c^2 . . . The mass of a body is a measure of its energy . . . It is not impossible that with bodies whose energy content is variable to a high degree (e.g. with radium salts) the theory may be successfully put to the test'.[1]

With the help of Einstein's discovery of the mass–energy equivalence some of the questions related to the origins of the radioactive energy release could have been answered, at least in principle. However, as a matter of historical fact this did not come to pass in the period under discussion. There appear to be three reasons for this: (1) The precepts of relativity were assimilated rather slowly. (2) The level of accuracy of mass measurements was not adequate during this period. Thus in his 1921 review of relativity theory, Pauli noted that '*perhaps* the theorem of the equivalence of mass and energy can be checked *at some future date* by observations on the stability of nuclei' (my italics).[2] (3) The lifetime question had to remain entirely unresolved until the advent of quantum mechanics, when it became possible for the first time to understand the mechanisms of radioactive decay. Prior to that it was inevitable that the origin of radioactive energy had to remain hazy as well, even though, after the fact, much can be explained about the energy release by the simple application of conservation laws, independently from quantum mechanical arguments.

Yet, well before the proofs were there, a correct consensus began to emerge with regard to the energy puzzle. If around 1910 those who labored and thought seriously about this question had been polled, there is no doubt that a majority would have expressed the belief that energy is conserved, and that the energy source resides in the atomic interior. Had they further been asked about the explanation of the lifetime puzzle, however, then the wisest would have readily admitted that this was a question beyond their horizon. That problem remained unresolved until the summer of 1928, when (see Section (f)) it was noted that '. . . It has hitherto been necessary to postulate some special arbitrary "instability" of the nucleus . . . but . . . disintegration is a natural consequence of the laws of quantum mechanics, without any special hypothesis . . .'.[3]

That was good enough for α- but not for β-radioactivity. In fact, in just about the same year, 1928, new paradoxes emerged regarding the energy loss in β-decay. But that is a subject which must wait until Chapter 14.

(b) The first energy crisis

Between 1898 and the early 1930s it happened three times that the discoveries of new natural phenomena were so unsettling as to make prominent physicists waver in their faith in the universal validity of the law of conservation of energy. The first of these crises, referred to above, concerned radioactivity. Thirty years later it was radioactivity again (more specifically β-decay) which caused temporary doubt in some quarters about energy conservation. In between, agonizing attempts to reconcile quantum effects with classical reasoning led likewise, and again briefly, to suggestions that energy conservation might not hold strictly.* It is the first of these three instances which shall concern us here. As a prelude to this subject, let us look briefly at the status of the conservation law towards the end of the nineteenth century.

In 1775, the Paris Academy of Sciences (still the Académie Royale, at that time) formally announced a significant decision: 'The Academy has resolved, this year, to examine no longer any solutions to problems on the following subjects: the duplication of the cube; the trisection of the angle; the quadrature of the circle, or any machine claiming to be a perpetuum mobile'.[5] The resolution was expatiated upon in a motivation with many a curious turn of phrase; for us, this simple categorical statement is of interest: 'The construction of a perpetual motion machine is absolutely impossible'. Evidently the illustrious Academicians had grown tired of finding the inevitable flaws in papers submitted on that subject.

We now call the machine excommunicated by the Academy a 'perpetuum mobile of the first kind'. The growing insight that such a device, which spontaneously creates energy, cannot be made was one of the main contributing factors to the formulation, more than fifty years later, of the universal energy principle, a major achievement of nineteenth-century physics. In so far as purely mechanical systems are concerned, the law of conservation of energy has much older roots.[6,7] Several of science's most illustrious names are associated with these early developments in mechanics. But the principle in its broader sense emerged only when the need arose to express quantitatively the convertibility of diverse forms of energy (mechanical, electrical, magnetic, chemical, physiological, etc.) into each other. The period of discovery of the macroscopic energy law (the first law of thermodynamics) in its generality,

* This suggestion is the so-called Bohr–Kramers–Slater proposal which I have discussed in detail elsewhere.[4] See also Chapter 14, Section (d), Part 2.

that is, applied to any form or several forms of energy, is approximately 1830–50.

No single year can be associated with this discovery because it was made not by one person but by many, working most often without initial awareness of each other's activities. A list of pioneers on the subject[8] contains no less than twleve names: Sadi Carnot, Colding, Faraday, Grove, Helmholtz, Hirn, Holtzmann, Joule, Liebig, Mayer, Mohr, Séguin. Four of these (Carnot, Hirn, Holtzmann, Séguin) became involved because of their interest in the effectiveness of steam engines, while two others (Helmholtz, Mayer) were initially intrigued by physiological questions. Given this large list of dramatis personae priority disputes were inevitable: 'most intense battles took place about the priority of [these] ideas, during which execrable personal accusations and repugnant national chauvinism came into the open.'[9] These controversies (of which there exist several detailed accounts[8,9,10]) will not be discussed here.

The curious case of Sadi Carnot should be mentioned, however. He is of course justly famous as the discoverer of the second law of thermodynamics (for reversible systems). In actual fact he also discovered the first law. In the early 1820s he stated in his diaries that wherever there is destruction of mechanical work (*puissance motrice*) there is generation of heat (*production de chaleur*) and concluded: 'one can therefore pose the general thesis that mechanical work is an invariable quantity in nature, that properly speaking it is never produced nor destroyed'. In addition he gave an estimate (somewhat low, but not that bad) of the mechanical equivalent of heat. But he never published! Long after his death, in 1878, this material was handed over to the French Academy by his surviving younger brother.[11] As Max Planck put it: 'He [S.C.] has unquestionably the merit of having given the first evaluation of the mechanical equivalent of heat'.[12] As Mach put it: 'Since for practical reasons one cannot name the law [of the equivalence of heat and mechanical work] for all the people who took part in its discovery and its justification, it is advisable to associate [the law] with the names of those who in both respects must be accorded the priority of publication'.[13] For this reason Mach speaks of the Mayer–Joule principle for the case that only mechanical work and heat are considered, and of the energy conservation law when all forms of energy are included. It is understandable that the first law is often referred to as *le principe de Carnot* in the French literature. Since the same appellation is also used for the second law, the reader of such papers is advised to find out from the context what the issue is.

Mach also observed[13] that the strongest emphasis on the universality of the conservation of energy stems from Robert Mayer and Hermann Helmholtz. Already the title of Helmholtz' important essay on the subject:* *Über die*

* Both this essay and the first paper by Robert Mayer on the subject share the distinction of having been rejected for publication by *Poggendorf's Annalen*, the later *Annalen der Physik*. Helmholtz' essay is most easily accessible in Ref. 16.

Erhaltung der Kraft: *eine physikalische Abhandlung* ('On the conservation of force: a physical memoir'), is of considerable interest. What is here called force is what we now call energy. Current terminology in this respect is itself of nineteenth-century origin. The first one to use the term 'energy' in its modern technical meaning was Thomas Young: 'The term energy may be applied with great propriety to the product of the mass or weight of the body into the square of the number expressing its velocity . . .'.[14] This quantity Mv^2 is the *vis viva* of Leibniz. A factor of 1/2 is still lacking before we arrive at our familiar kinetic energy. This factor seems to have been supplied first by de Coriolis.[15]

Nor should one fail to notice Helmholtz's emphasis on his subject as a treatise in physics. As he strongly urges, we are not dealing here with an axiomatic statement or a philosophical tenet, nor with a tautology (all such views were expressed at one time or another) but with a physical hypothesis which needs verification in each instance.* The key to doing this is to find the equivalent of each energy form (via direct or indirect processes) in terms of mechanical work. Moreover, such an equivalence has an unambiguous meaning only after it is realized that the change in energy of a system from an initial to a final state is independent of the way in which the transition between these states takes place.

These, briefly, are the lines along which the conservation of energy came to be clearly understood as a physical principle of universal validity, as the nineteenth century drew to a close. In two respects there have been fundamental developments of the subject in the twentieth century. First, a unification has taken place of the two basic laws: conservation of energy and conservation of matter. Unlike the former law, the latter one, a product of the eighteenth century, is associated with the name of one single scientist: Antoine Laurent Lavoisier, a man to whom a grateful nation expressed its debt by putting him under the guillotine, an event of which Laplace has said: 'It took them only an instant to cut off that head, and a hundred years may not produce another one like it.' Secondly, the energy law appears in its thermodynamic context as a macroscopic principle and on a different footing from the conservation of momentum and angular momentum. The modern association between conservation laws and invariance principles emphasizes the microscopic foundations of all three laws, treats them very much on a common level, and frees the conditions for their validity to a larger extent than before from dynamical details.

As a fitting conclusion to this brief backward glance at the nineteenth century, and at the same time as an epigraph to what comes next, I offer a comment by Planck, found on the opening page of his 1887 prize essay: '. . . If

* 'The principle is presented and should be understood as a hypothesis within physics totally divorced from philosophical considerations.'[16]

today a quite new natural phenomenon were to be discovered, one would be able to obtain at once from [the energy conservation principle] a law for this new effect, while otherwise there does not exist any other axiom which could be extended with the same confidence to all processes in nature'.[10]

To the best of our present knowledge, Planck was right. Yet in later years the paradoxes posed by several new discoveries were initially so grave as to cause a temporary lack of confidence in the energy principle. Let us now turn to the first of these events.*

Becquerel's surprise at the persistence with which his uranic rays kept pouring out energy was mentioned before. In 1910, Marie Curie reminisced as follows about those early days: 'The constancy of the uranic radiation caused profound astonishment to those physicists who were the first to be interested in the discovery of H. Becquerel. This constancy appears in fact to be surprising; the radiation does not seem to vary spontaneously with time . . .'.[19] In order to appreciate this statement fully, three facts should be borne in mind: (1) The radiation emitted by uranium when unseparated from its daughter products does indeed represent, to a very high degree, a steady state of affairs. (2) It took two years from Becquerel's initial discovery until the first parent–daughter separation was effected. (3) It took another two years until it was firmly established that radioactivity does diminish with time.

Speculation on the origin of radioactive energy started with Marie Curie's very first paper on radioactivity (the one in which she announced her discovery of the activity of thorium (1898)). There, cautiously, she suggests the possibility that the energy might be due to an outside source: 'One might imagine that all of space is constantly traversed by rays similar to Roentgen rays, only much more penetrating and being able to be absorbed only by certain elements with large atomic weight, such as uranium and thorium'.[20] Also Becquerel made an analogy with an externally induced process, phosphorescence: 'It would not be contrary to what we know about phosphorescence to suppose that these [U and Th] substances have a relatively considerable energy reserve, which they can emit for years, as radiation, without noticeable weakening'.[21] However, he also stated that this analogy had its limitations. Phosphorescent phenomena exhibit a finite lifetime (as Becquerel and his father well knew), and they can be affected by external agents. Neither of these properties seemed to apply to

* It may be noted that in 1882 Helmholtz expressed uncertainty about the applicability of the thermodynamic principles to 'the fine structures of the organic living tissue'.[17] Likewise Louis-Georges Gouy, one of the pioneers in refined experiments on Brownian motion, wondered in 1888 whether 'le principe de Carnot . . . serait seulement exact pour les mécanismes grossiers . . . et cesserait d'être applicable . . . [pour] des dimensions comparable a 1 micron', ('. . . whether the principle of Carnot would be exact only for large scale mechanisms . . . and would cease to be applicable . . . [for] dimensions of the order of one micron'.[18] However, these comments are in the nature of asides and were not raised as central issues. For what follows, it may be of interest to observe that Marie Curie was aware of these remarks by Helmholtz and by Gouy.

radioactivity: '. . . However, it has not been possible to induce any appreciable variation in the intensity of this emission'.[21]

In that same year, 1898, Marie Curie discovered polonium, for which the liberated energy per unit weight of separated material was even larger than for uranium and thorium. Thus the question of the origin of this energy became an even more burning one and she returned to it, listing a number of possible answers.[22] Here we find the first mention that one might have to face a contradiction with the law of conservation of energy. Furthermore she emphasized that the assumption of an external source would be nothing but an evasion of energy nonconservation—unless the nature of the external source were determined: 'Any exception to Carnot's principle [first law!] can be evaded by the intervention of an unknown energy which comes to us from space. To adopt such an explanation or to put in doubt the generality of the Carnot principle are in fact two points of view which to us amount to one and the same as long as the nature of the energy here invoked stays entirely "dans le domaine de l'arbitraire"'. She also pointed out that the interior of the atom could be the energy source: 'The radiation [may be] an emission of matter accompanied by a loss of weight of the radioactive substances'.[22]

Not only the first but also the second law of thermodynamics was sometimes questioned as a result of this energy puzzle. For example, in his 1898 inaugural address as President of the British Association, the brilliant and erratic Sir William Crookes speculated, somewhere in between dissertations on food shortages and psychical research, whether one can 'mentally modify Maxwell's demons' in such a way that radioactive substances release energy drawn from the air surrounding the active material.[23]

These various speculations set in motion a set of experiments designed to locate a possible outside source of radioactive energy. In an attempt to see whether the sun could be the cause, the Curies looked for diurnal variations in the activity of uranium. They found no effect.[24] Among others who addressed the same question, particular mention should be made of the team of Elster and Geitel.

Julius Elster and Hans Geitel had been high school friends. They both became teachers at the Gymnasium* in Wolfenbüttel near Braunschweig. When Elster married and had a house built, Geitel moved in with the young couple and together the two friends built a laboratory in the new home. Here they started their research (often financed from their own pockets) which was to make them internationally renowned.** They experimented on photoelectric effects, on spectroscopy, on the conduction of electricity through gases, and especially on atmospheric electricity. These last experiments led to their

* It is not evocative enough to translate Gymnasium simply as high school. Let us say it is an academic high school preparatory to going to the university.

** There were others as well who did their most creative work while teaching in a high school, Weierstrass for example.

classic work on the radioactivity of the atmosphere, research about which Rutherford spoke with great respect. Simultaneously with Crookes they discovered the scintillations of zinc sulfide screens by X-rays.[25] In 1905, 1907, 1908, 1910, and 1911 they were nominated for the Nobel Prize.*

In later years the two men loved to relate[26] their experiences in Berlin, where the Prussian Minister of Education tried to convince them to accept a joint offer as university professors at a first-rate institute. They listened modestly but did not react. The minister believed that 'die kleinen Oberlehrer der Provinz' were probably too awed and suggested that they take a few hours to think it over. They did so, came back and said no thank you. They had decided that the transition to the academic world would inhibit their independent research. They were grateful for the honor but preferred to stay in Wolfenbüttel.

The two were inseparable. I cannot resist mentioning an anecdote related by Andrade. 'In their time there was a man who much resembled Geitel in appearance. A stranger meeting him said, "Good morning, Herr Elster . . .", to which he replied, ". . . First, I am not Elster but Geitel, and secondly I am not Geitel".'[27] Almost their complete oeuvre consists of joint publications. 'We shall doubtless search in vain for a similar instance of private scientific partnership throughout a lifelong friendship. Each ascribed to the other the credit for a discovery published jointly.'[28]

Their work on the origins of radioactive processes is contained in two papers. In the first one[29] they begin with the observation that if Crookes were right and the radioactive energy is taken from the surrounding air,[23] then the activity should decrease when the source is placed in a vacuum. They found no such effect. Next they turned to the conjecture of the Curies that the energy may be supplied by an X-ray-like radiation which is all-pervasive in the atmosphere and reasoned that, if this were so, there should be a decrease in activity if the source were placed deep underground. So they requested and obtained permission to do an experiment in the Clausthal mines in the Harz mountains, under 300 m of rock. They found no effect. They admit that perhaps the rock layer may not be all that good an absorber. Nevertheless they conclude, as early as 1898, that 'the hypothesis of the excitation of Becquerel rays by radiation pre-existing in space appears improbable to the highest degree (im höchsten Grade unwahrscheinlich)'. In their second paper[30] they report on attempts to increase the radioactive emissions by exposing a source to cathode rays, or to sunlight. They find no effect and conclude 'man wird vielmehr aus dem Atome des betreffenden Elementes selber die Lichtquellen ableiten müssen'.** It is because of this statement that the house at Rosenwall 114 in Wolfenbüttel

* I am indebted to Professor B. Nagel for permitting me to examine the files of the Nobel Committee for Physics.

** Rather, one will have to derive the source of light [sic] from the atom itself of the element concerned.

where Elster and Geitel lived, now carries a memorial plaque on which the two men are honored as the *Entdecker der Atomenergie*, discoverers of atomic energy.

It is important to stress at this point that the fascinating puzzles discussed in this chapter were never any hindrance to progress in those days. If anything, the contrary is true. The field of radioactivity was young when these questions arose, the tasks were enormous. While these problems were given much thought by the Curies, that never inhibited them from continuing their superb research. The problems were a stimulus to men like Elster and Geitel, as we have just seen. Others chose to state them as unresolved issues and then move on to other pursuits. Such was largely the attitude of the English school. Rutherford, for example, simply noted in his 1899 memoir that 'the cause and origin of the radiation continuously emitted by uranium and its salts still remain a mystery'.[31] J. J. Thomson always took the attitude that the atom itself was the energy source—'. . . the [radioactive] changes we are considering are changes in the configuration of the atom . . .'.[32]

I referred earlier to the article by Marie Curie in which she listed possible options for the explanation of the energy release. It was written before, but published after, the discovery of radium by her, Pierre Curie, and Bémont. This last development once again brought the issue to the fore. The radium radiation was even more intense than the polonium! The question of nonconservation of energy came up once again: 'On réalise ainsi une source de lumière, a vrai dire très faible, mais qui fonctionne sans source d'énergie. Il y a là une contradiction tout au moins apparente avec le principe de Carnot'.[33]*

Nonconservation of energy was never a widely held explanation of these effects. In 1902, the Curies again gave a list of possible interpretations, on which this possibility no longer appears.[34] Yet, in that same year, a visitor to England recalled that he 'had been dining seated between Lord Kelvin and Professor Becquerel, . . . Lord Kelvin had turned to him and said that the discovery of Becquerel radiations had placed the first question mark against the principle of conservation of energy which had been placed against it since the principle was enunciated'.[35]

It should also be stressed that such options as nonconservation of energy or external sources were not proposed lightheartedly. The idea that the atom itself is the source was not easily swallowed at that time, since it meant giving up the concept of an atom as an immutable entity. By 1900, the debate over the reality of atoms was well past its peak; but at that time the question was not universally regarded as settled. The Curies were proponents of the existence of real atoms, as their writings make abundantly clear. But to accept the atom itself as the source of the energy could only mean one thing to them:

* 'Thus one realizes a source of light [sic], quite weak to be sure, but which functions without a source of energy. There is here a contradiction, or so it seems, with the principle of Carnot' [the first law!].

transmutation. And they could not simply accept this, since to them at that time it seemed in conflict with the principles of chemistry as then known—which indeed it was. In 1900 Marie Curie summed up the dilemma in the following way: 'Uranium exhibits no appreciable change of state, no visible chemical transformation, it remains, or so it seems, identical with itself, the source of energy which it emits remains undetectable—and therein lies the profound interest of the phenomenon. There is perhaps a disagreement with the fundamental laws of science which until now have been considered as general . . . The materialistic theory of radioactivity* is very attractive. It does explain the phenomena of radioactivity. However, if we adopt this theory, we have to decide to admit that radioactive matter is not in an ordinary chemical state; according to it, the atoms do not constitute a stable state, since particles smaller than the atom are emitted. The atoms, *indivisible from the chemical point of view* [my italics], are here divisible, and the sub-atoms are in motion . . . The materialist theory of radioactivity leads us . . . quite far. If we refuse to admit its consequences, our embarassment will not lessen. If radioactive matter does not modify itself, then we find ourselves again in the presence of the question: from where comes the radioactive energy? And if the source of energy cannot be found we are in conflict with Carnot's principle, a principle fundamental to thermodynamics . . . . We are then forced to admit that Carnot's principle [second law!] is not absolutely general [and] . . . that the radioactive substances are able to transform heat from the ambient environment into work. This hypothesis undermines the accepted ideas in physics as seriously as the hypothesis of the transformation of the elements does in chemistry, and one sees that the question cannot easily be resolved'. (Cette hypothèse porte une atteinte aussi grave aux idées admises en physique que l'hypothèse de la transformation des éléments aux principes de la chimie, et on voit que la question n'est pas facile a résoudre.)[36]

The transformation theory of Rutherford and Soddy, proposed in 1902, provided the break with the past which was clearly needed in order to answer Marie Curie's question. In this 'great theory of radioactivity which these young men sprung on the learned, timid, rather unbelieving, and, as yet, unquantized world of physics of 1902 and 1903',[37] they unabashedly put forward the idea that some atomic species are subject to spontaneous transmutation.[38] Forty years later, a witness to the events characterized the mood of the times as follows:[39] 'It must be difficult if not impossible for the young physicist or chemist of the present day to realize how extremely bold it was and how unacceptable to the atomists of the time . . . this is a point which must be stressed, for the young generation is more likely to be familiar with the ordered simplicity of the radioactive series as we know them than with the chaotic state which preceded the transformation theory'.**

* According to which radioactive atoms expel subatomic particles.

** The events surrounding the enunciation of the theory have been described in more detail by Badash.[40]

The main tenet of the transformation theory* is: radioactive bodies contain unstable atoms of which a fixed fraction decay per unit time. The rest of the decayed atom is a new radio-element which decays again, and so forth, till finally a stable element is reached.

As Rutherford himself emphasized some time later (see below), there is no explicit reference in this theory to the energy mechanism. Nevertheless the successes of the transformation theory led Rutherford to express the following opinion: 'This [transformation] theory is found to account in a satisfactory way for all the known facts of radioactivity and a mass of disconnected facts into one homogeneous whole. On this view, the continuous emission of energy from the active bodies is derived from the internal energy inherent in the atom, and does not in any way contradict the law of conservation of energy'.[42]

And so the energy debate might have quietened down were it not that, in March 1903, new fuel was added to it by the discovery that radioactive energy release surpassed in magnitude anything that had been known until then from chemical reactions. In that year Pierre Curie and Laborde measured the amount of energy released within a Bunsen ice calorimeter by a known quantity of radium.[43] They found that 1 g of radium can heat $\sim$1.3 g of water from the melting point to the boiling point in 1 hour!

These results caused a tremendous stir. It was in fact the Curie–Laborde paper which was largely responsible for the worldwide arousal of interest in radium described in the opening lines of Chapter 2. Regarding the origins of this energy, the authors themselves referred once again to a possible outside energy source: 'This release of heat can also be explained by supposing that radium utilizes an exterior energy of unknown nature'. In a discussion of the new discovery, Kelvin spoke of THE mystery of radium (his capitals) and continued: 'It seems to me, therefore, absolutely certain that if emission of heat can go on month after month . . . energy must be supplied from without . . . I venture to suggest that somehow etherial waves may supply energy to radium . . .'.[44] In a lecture on 'The present crisis of mathematical physics' given in 1904, Poincaré also brought up the energy conservation question: '. . . These principles on which we have built everything, are they about to crumble away in turn? . . . When I speak thus, you no doubt think of radium, that grand revolutionist of the present time . . . . At least, the principle of the conservation of energy still remained with us, and this seemed more solid. Shall I recall to you how it was in its turn thrown into discredit? This [activity of radium] was itself a strain on the principles . . . But these quantities of [radioactive] energy were too slight to be measured; at least that was the belief, and we were not much troubled. The scene changed when

* Why did Rutherford and Soddy not use the term 'transmutation' but rather the more neutral one, 'transformation'? The following exchange took place while they were at work on the separation of thorium X. *Soddy*: 'Rutherford, this is transmutation . . .' *Rutherford*: 'For Mike's sake, Soddy, don't call it transmutation. They'll have our heads off as alchemists'.[41]

Curie bethought himself to put radium in a calorimeter; it was then seen that the quantity of heat created incessantly was very notable . . .'.[45]

From the physics point of view, these results on energy release became even more remarkable when it was found from additional experiments that seventy-five per cent of this effect was due to a daughter product of radium, the radium emanation (radon, Rn^{222}), although the amount of emanation present was actually extremely small. In fact the energy released by radon[46,47] is more than a million times greater than the heat evolved by the same volume of hydrogen and oxygen when they explode to form water. In 1905 Soddy wrote of these discoveries: 'It is probably the most far reaching and revolutionary fact that has yet transpired in the study of radioactive substances. This enormous evolution of energy which accompanies the production of helium from the radium emanation establishes beyond question the new and fundamental character of radioactive change'.[48]

Soddy also pointed out that the magnitude of the energy production made it even more difficult to imagine it to be due to an external source: 'It has been suggested . . . that all space is traversed by undiscovered radiations to which ordinary matter is completely transparent, but to which radioactive substances are opaque. On this view, the energy traversing a cubic centimeter of space must be at least 60 000 calories per hour [in order to explain the heating effects due to radium]. The total quantity in the universe must therefore be so great that the hypothesis involves far greater difficulties than the effects it is designed to explain'.[48]

Still the external source idea would not quite die.

In 1906, Sagnac raised a new possibility.[49] He asked: could gravitational energy be the external source? Might it be that the Newtonian attraction is universal for nonradioactive bodies but that it has a 'valeur spéciale' for radium? This led him to do a torsion balance experiment in which he compared the oscillations of equal weights of barium and radium. He found no effect.*

In 1911, Rutherford again referred to the energy issue but expressed himself more cautiously than he had done earlier.[42] This time he observed that the transformation theory leaves open the question of the inside versus the outside source, since all results of the transformation theory remain true for either hypothesis.[52]

As late as 1919, Jean Perrin came forth with a new 'ultra X-ray' mechanism for explaining radioactivity as an externally induced effect.[53] This time the new radiation was supposed to come not from outer space but 'from under our feet, from the fiery center of the planet'. The scheme is discussed in detail regarding its astronomical and cosmological implications.

* A related negative result was obtained by Thomson[50] who determined the swinging time of a pendulum to which a bob of RaBr was attached. His work was differently motivated, however. Much later, Rutherford and Compton also obtained a negative result when looking for the influence of gravitation on radioactivity.[51]

The matter was still being argued in the year in which quantum mechanics was born.[54] This is not to say that the external source idea was in any way a serious issue at that late time. It does bring home the fact, however, that only in the quantum-mechanical era could the definitive proof of the existence of internal mechanisms for energy generation be given, which settled the question for the ages.

The negative outcome of these various searches for an external energy source was a positive contributing factor to a major insight which emerged early in the twentieth century: physical and chemical actions do not affect radioactive phenomena. In 1903, Rutherford and Soddy elevated this to a new principle, the 'conservation of radioactivity': 'Radioactivity, according to our present knowledge, must be regarded as a process which lies wholly outside the sphere of known controllable forces, and cannot be created, altered or destroyed'.[55]

There is a vast body of experiments which bear on this question. Temperature independence was established by many. Pierre Curie went to London, to repeat with Dewar his radium heat production experiment, this time at liquid air temperatures. Marie Curie later went to Leiden to do the same at liquid hydrogen temperatures, with Kamerlingh Onnes. Rutherford stuck 4 mg of radium bromide inside a steel-enclosed cordite bomb, exploded the device, and concluded that there was no change in radioactivity at temperatures ~2500°C. Independence of pressure, of concentration, of the presence of strong magnetic fields, and of irradiations of various kinds was established. Detailed discussions of these results are beyond the compass of this section. For more information the interested reader is referred to older textbooks.[56]

Yet strictly speaking, there is no such thing as the conservation of radioactivity. This became clear after the Second World War. I shall come back to this in Section (f). Even so, the 'conservation of radioactivity' served an excellent purpose in its time. In particular it helped Bohr to locate the atomic nucleus as the seat of all radioactive phenomena (see Chapter 11). As this 'principle' illustrates so well, at an early stage it is often better to know 'the truth' than 'the whole truth'.

(c) Interlude: atomic energy

Speculations on the possible good and the possible evil of the atom go back to the founding fathers of radioactivity.

Here is Becquerel, in an early lecture: 'Today the [radioactive] phenomena are of transcendent interest, but in them almost infinitesimal amounts of energy are utilized. Whether ultimately science will have so far advanced as to permit of the practical utilization of the abundant store of energy locked up in every atom of matter is a problem which only the future can answer. Remember, at the dawn of electricity this was looked on as a mere toy, suitable only to amuse children by attracting bits of paper with a stick of rubbed sealing wax'.[57]

Here is Pierre Curie, also in a lecture: '... It can even be thought that radium could become very dangerous in criminal hands, and here the question can be raised whether mankind benefits from the secrets of Nature ...'.[58]

And here is Soddy in an early popular book: 'If we pause but for a moment to reflect what energy means for the present, we may gain some faint notion as to what the question of transmutation may mean for the future to a fuelless world, once more dependent on a hand-to-mouth method of subsistence. It may still be centuries before this occurs, but neither the application of the discoveries of science nor even their achievement is to be compared with the struggle in winning them'.[59]

Also the expression 'atomic energy' entered the language at the very beginning of the twentieth century.

The term was first used in 1903, by Rutherford and Soddy,[55] *not* just for the energy released by a radioactive element, but much more generally for the energy locked up in *any* atom: 'All these considerations point to the conclusion that the energy latent in the atom must be enormous compared with that rendered free in ordinary chemical change. Now the radio-elements differ in no way from the other elements in their chemical and physical behavior. On the one hand they resemble chemically their inactive prototypes in the periodic table very closely, and on the other they possess no common chemical characteristic which could be associated with their radioactivity. Hence there is no reason to assume that this enormous store of energy is possessed by the radio-elements alone. It seems probable that *atomic energy* [my italics] in general is of a similar high order of magnitude, although the absence of change prevents its existence being manifested'. This, truly, is the physics of the twentieth century.

The fact that today 'atomic energy' is an expression firmly anchored in our everyday language has nothing to do, however, with the marvelous lines above. Rather, the present common usage of the term derives in the first instance from a report released by the President of the United States on the evening of Saturday, August 11, 1945. The title of this report came eventually to be 'Atomic Energy for Military Purposes. The Official Report on the Development of the Atomic Bomb under the Auspices of the United States Government, 1940–45'. It is now generally known as the Smyth Report.

I became aware of the way in which the expression 'atomic energy' was reinvented when, one day in 1976, I went to see my friend Henry DeWolf Smyth. On that occasion we had the following conversation, the essence of which I report here with his permission.

A.P. When you were writing your report, were you aware that the term 'atomic energy' dates back to the beginning of this century?

H.D.S. No, I was not.

A.P. I have a second question. In the days of Rutherford and Soddy, the nucleus was not yet discovered. Therefore, the expression 'atomic energy' was, so to speak, the only natural one they could use. Now I have been puzzled, ever since your report came out: why in fact did you not speak of 'nuclear energy', 'nuclear bomb', etc?

H.D.S. Your question comes at a very opportune moment, since I have just published an article on the history of the Smyth Report. You will be glad to know that in my original draft I did use the word 'nuclear' instead of 'atomic'. After the writing was done there followed a period of consultation with [Major General Leslie R.] Groves. In turn, Groves must have discussed my draft with his advisers [James B.] Conant and [Richard C.] Tolman, and possibly with others as well. In a subsequent discussion, we decided that the word 'nuclear' was either totally unfamiliar to the public or had primarily a biological flavor, whereas 'atomic' has a definite association with chemistry and physics. Since it became clear that the report was aimed at a wider audience than nuclear physicists, we decided that 'atomic' was less likely to frighten off readers than 'nuclear'. So I accepted the change after a somewhat painful suppression of my purist principles.

With these words he gave me a copy of his article.[60] I gratefully accepted this gift from a man my respect for whom I have expressed elsewhere.[61]

(d) Metabolous matter

In a report to the British Chemical Society on progress in radioactivity during 1904, Soddy included a list of the eighteen radioactive substances known by then.[62] The first and (with a change of units) third column in the table below are taken from his report. The second column gives the modern symbols for the substances in question. Today's half-life values are given in the fourth column. For the very long lived U^{238} and Th^{232}, the old values are off by an order of magnitude. The other old results are much better.

Elsewhere in his report,[62] Soddy made a most revealing comment about his table: 'In the case of only one [substance], namely, radium, is there direct evidence of a material character of the specific elementary nature of the substance. In the other cases, such evidence is lacking, for all exist in such minute quantity that, apart from the radioactivity, there is no other evidence of their existence'. It is evident from this remark that the existence of a distinct lifetime was not yet (or at least not yet generally) accepted as a signature for a distinct atomic species.

In the same report, Soddy also explained the grounds for his (and others') conviction that the elemental character of radium had been firmly established. 'The well-defined material properties of radium, and especially the spectroscopic evidence, show that radium is as distinct and definite a type of elementary matter as any of the older elements, and furnish a sufficient answer to the suggestion that radium is a compound, using that term in its ordinary chemical significance* ... [The French experimental spectroscopist] Demarçay classed radium as an element giving an exceedingly delicate spectrum reaction, one part of radium being capable of detection by the spectroscope when mixed with 10 000 parts of barium.' Already several years earlier, Marie Curie had emphasized how important these spark spectrum results[64] had been to her and her husband: 'Monsieur Demarçay, whose great competence in spectroscopy is well known, has been willing to occupy himself with the spectral study of our substances. We cannot thank him enough for doing so. He has rendered us the immense service of giving us a certitude, based on a well-proven scientific method, at a time in which we were still in doubt about the value of our own research methods'.[36] Regarding the 'well-defined material properties' of which Soddy wrote, it should be recalled that Marie Curie had by then been able to put the atomic weight of radium close to 225—not a bad result for a Ph.D. thesis completed in 1903!

Substance (1904 Nomenclature!)	Modern symbol	Half-life (1904)	Half-life (1983)
Actinium emanation	Rn^{219}	4 s	3.96 s
Thorium emanation	Rn^{220}	1 m	55.6 s
Actinium B	Bi^{211}	1.5 m	2.14 m
Radium A	Po^{218}	3 m	3.05 m
Radium B	Pb^{214}	21.4 m	26.8 m
Radium C	Bi^{214}	27.6 m	19.8 m
Actinium A	Pb^{211}	41.4 m	36.1 m
Thorium B	Bi^{212}	54.5 m	60.6 m
Thorium A	Pb^{212}	11 h	10.64 h
Radium emanation	Rn^{222}	4 d	3.82 d
Thorium X	Ra^{224}	4 d	3.64 d
Uranium X	Th^{234}	22 d	24.10 d
Radium E	Bi^{210}	—	5.0 d
Radium D	Pb^{210}	—	22.3 y
Actinium	Ac^{227}	—	18.72 d
Radium	Ra^{226}	793 y	1600 y
Uranium	U^{238}	3.5×10^{8} y	4.47×10^{9} y
Thorium	Th^{232}	1.4×10^{9} y	1.4×10^{10} y

As is generally known by now, and as the above table reminds us, the riddle of radioactive substances cannot be unraveled without knowledge of isotopes and isobars, concepts unknown in the early years of this century when

* The debate whether or not radium is a chemical compound continued into 1906. In an article in *The Times*, Kelvin suggested that radium might be a molecular compound of lead and five helium atoms.[63] During the summer of 1906, a lively discussion of this issue took place in the columns of *The Times* and of *Nature*.

(I cannot remind the reader too often) not even the atomic nucleus had been discovered. Physicists of that period were therefore prudent and wise in speaking of 'substances' rather than 'elements'. In fact, Rutherford and Soddy went so far as to coin a new name: 'It seems advisable to possess a special name for those now numerous atom fragments, or new atoms, which result from the original atom after the ray has been expelled, and which remain in existence only a limited time, continually undergoing further change . . . We would therefore suggest the term metabolon for this purpose'.[55]*

The question, how are metabolons related to elements? had a counterpart: how are elements related to metabolons? That is, could 'ordinary' atoms be radioactive but with a very long life? It was of course clear to the experts that impurities would obscure the search for an answer. As was noted in 1907: 'Small quantities of radium appear to be ubiquitous. The air, the oceans and springs, and all soils can be shown to be contaminated with minute quantities of radium and its products: we cannot be sure that the ordinary metals which show a minute activity do not also contain the same universal impurity'.[65] Nevertheless the search went on. In her doctoral thesis (1903), Marie Curie has a section entitled 'Is atomic radioactivity a general phenomenon?', in which she reported on the analysis of substances other than uranium and thorium compounds: 'I undertook this research with the idea that it was scarcely probable that radioactivity, considered as an atomic property, should belong to a certain kind of matter to the exclusion of all other'.[66] It is to her credit that of all the materials she examined (including a dozen rare earths) she found no evidence for radioactivity. Others were either less fortunate or less skillful or both; claims made for new radioactive elements[67] such as berzelium and carolinium were not substantiated. I do not believe that Rutherford made special searches, but he was certainly intrigued by the problem. In 1904 he wrote: 'According to the modern views of the constitution of the atom, it is not so much a matter of surprise that some atoms disintegrate as that the atoms are so permanent as they appear to be';[68] and in 1906: 'Matter may be undergoing slow atomic transformation of a character similar to radium which would be difficult to detect by our present methods'.[69]

In 1907, these ruminations about metabolous matter led to an important positive result: the discovery by Norman Robert Campbell and Alexander Wood that potassium and rubidium are radioactive. Campbell's motivation is clear from a paper he wrote a year earlier: 'Considerable evidence [?] has accumulated in favour of the view that all elements are truly radioactive'.[70] His subsequent papers with Wood[71] have stood the test of time: the rare isotope K^{40} decays with a half-life of 1.28×10^9 yr; Rb^{87} has a half-life of 5×10^{10} yr. (These lifetimes are so long because we are dealing with third forbidden β-transitions.)

* From μεταβολή=change.

As a final word about radioactivity in the first decade, I should like to note a remark by Rutherford, a man not generally given to speculation: 'If elements heavier than uranium exist, it is probable that they will be radioactive'.[72]

During the second decade the advances in understanding radioactivity were spectacular. The nucleus was discovered. The significance of its three global parameters: nuclear charge, nuclear mass number, and nuclear binding energy became clear. The nuclear origins of radioactive processes became evident. Isotopes were introduced. The radioactive displacement laws were discovered. Leaving these topics for later chapters, I conclude this section with a last comment on the stability of ordinary matter.

During the years 1913–20, nuclear masses, more specifically masses of isotopes, were not yet well known. As far as I know, it was only in 1920 when, for the first time, Wilhelm Lenz properly formulated stability against radioactive disintegration in terms of an energy condition based on Einstein's mass–energy equivalence, taking into account all at once isotopic masses, binding energy, and kinetic energy of decay products.[73] Prior to that there was still room for speculation. So it could happen that, in the years 1915–21, a number of authors attempted to develop a disintegration theory of the evolution of the elements. The idea was that all elements in nature occur because of the α-decay of four prime elements, these decays being exceedingly slow. For example, it was conjectured that, starting from Th^{232} one could in this way generate all elements with mass number $4n$ down to the α-particle itself.

In the first edition of his book on isotopes (where one will also find detailed references to these speculations) Aston did not mince words: '[This theory is] unlikely and misleading . . . It starts with at least four elements as complicated as any likely to exist, which does not advance the inquiry very much . . . [This idea] has never been worthy of serious consideration'.[74] It would not be long before an avalanche of data on isotope masses would become available. Starting with the definite proof of the isotopes of neon in November 1919, mass spectrographs began to pour out results, especially in the next five years.[75]

After 1921 the conjecture that perhaps all elements might be radioactive was not heard of again until the appearance of grand unified gauge theories in the 1970s, according to which it is almost inevitable that protons, hence all atoms, are unstable. On that subject see Chapter 22, which the reader will have time to reach without fear of prior disintegration.

(e) Why a half-life?

The first determination of a half-life for a radioactive decay was made by Rutherford.[76] In a study of the properties of thorium emanation (Rn^{220}) he found that the intensity of the radiations given out by his sample fell off with time in a geometric progression. Thus he was the first to note that if $N(t)$ is the number of active atoms at time t, then the decrease of N with t is well described* by

$$dN/dt = -\lambda N \quad \text{or} \quad N(t) = N(o)\,e^{-\lambda t} \tag{6.1}$$

He called λ the radioactive constant, and 'it has been shown that $e^{-\lambda t} = 1/2$ when $t = 60$ sec', a quite respectable half-life determination (the modern value is about 56 sec). It is of course no accident that this first discovery concerned an element of medium short life. Much longer half-lives (such as the one for radium, ~1600 yr) were also well established within the next few years, with the help of the theory of radioactive equilibrium between parent and daughter substances.[77] Equation (6.1) and its generalization to sequential decays was the first of two contributions by Rutherford to theoretical physics, an activity which he did not always hold in the highest esteem. (His second contribution was his theoretical discovery of the central nucleus from the results of scattering experiments.)

Today, even though we may not always be able to compute λ theoretically for any given radioactive decay, the meaning of λ is certainly quite clear. However different the respective mechanisms for α-, β-, and γ-decay are, in each case λ is a quantum mechanical transition probability per unit time. Thus radioactivity represents one instance among many of a situation in which physicists of earlier days were unwittingly dealing with quantum effects.

At the turn of the century there already existed a body of knowledge on unstable systems of atomic dimensions. For example, much work had been done at that time on luminescent phenomena. This had made the lifetime concept familiar. It is true that insuperable problems arose for those who attempted to find mechanisms for these and similar processes; consider for instance Boltzmann's struggles with molecular dissociation.[78] However, if these various unstable systems were not amenable to theoretical treatment, they did not appear to pose any manifest paradoxes, principally since one could (so it was thought) ascribe the instability to *external* causes. For example, for a process like luminescence one always had the excuse that some but not all the irradiated matter had been excited. Note also that, in the early days, lifetimes for ordinary light emission by atoms were too short to be observable. Radioactivity, on the other hand, created problems unique for their time. It seemed (in fact, it was true) that radioactive decays were contrary to the classical concepts of cause and effect. During the first two decades of this century, physicists had no reason to suspect that these paradoxes were not by any means typical for radioactivity only.

Jeans has given a graphic description of the situation: 'Interesting . . . questions arise when we discuss which atoms will disintegrate first, and which will live longest without disintegration. [Suppose that] 500 million atoms are due

* In the absence of any replenishing mechanism.

to disintegrate in the next second. What, we may inquire, determines which particular atoms will fill the quota? . . . It seemed to remove causality from a large part of our picture of the physical world. If we are told the position and the speed of motion of every one [of a set of radium atoms], we might expect that Laplace's super-mathematician would be able to predict the future of every atom. And so he would if their motion conformed to the classical mechanics. But the new laws merely tell him that one of his atoms is destined to disintegrate today, another tomorrow, and so on. No amount of calculation will tell him which atoms will do this . . .'.[79]

Nevertheless, there were those who, early on, began to think of dynamical models which would incorporate radioactivity. One of these model-builders was J. J. Thomson. Some details of his work on models will be discussed in Chapter 9. Here it suffices to state that he attempted to describe radioactivity in terms of classical mechanical pictures. Thus we can well understand the objections which Lord Kelvin wrote to Thomson: 'What would be the difference, between radium atoms in a piece of radium bromide, of the atoms which are nearly ripe for explosion, and those which have the prospect of several thousand years of stable diminishing motions before explosion?'[80] Rutherford also saw this weak point of the classical model: '. . . All atoms formed at the same time should last for a definite interval. This, however, is contrary to the observed law of transformation, in which the atoms have a life embracing all values from zero to infinity'.[81]

In despair, one might of course reply to the question: how is it possible that one species of identical atoms is made up out of particles some of which live longer than others? by saying that 'the different atoms of a radioactive substance are not in every respect identical,[82] a possibility mentioned by Thomson in an address in 1909. However he never came back to this.

Those who wisely left aside the question: 'Why does a radioactive atom change?' and focussed on the more modest problem: 'How does it change?' were able to make some further progress on a more descriptive level, however. They focussed on the essential content of Eq. (6.1), which is probabilistic: the probability that a given unstable atom decays is the same for all atoms (in a sample of a given species) and is independent of its age, but does depend on the specific element under consideration. 'If the destroying angel selected out of all those alive in the world a fixed proportion to die every minute, independently of their age, . . . and chose purely at random . . . then our expectation of life would be that of the radioactive atoms.'[83] With the help of this probability Ansatz, it is of course possible to refine Eq. (6.1) (which can hold strictly only in the limit of very large N, since it ignores the discreteness of N) for the case of finite samples, to predict the average number of events in any finite time interval (for given λ), and to study the fluctuations of that number (the 'Schweidler fluctuations'*) as well as fluctuations for other variables such as heat production, ionization, spatial distributions, etc.[85]

Beyond that, there was not much more that could be done. The lifetime paradox simply did not lend itself to the statement of new hypotheses subject to test. The problem was so difficult that it was hard even to get a wrong idea about it.

In a review of alternatives by Debierne in his Ph.D. thesis[86] an old acquaintance briefly returns: the exterior source of radioactive decay. He observes that exponential decays occur in several chemical processes, such as monomolecular irreversible reactions (dissociations etc.). He notes that in such instances thermal disorder plays a role. This leads him to ask whether the radioactive decay processes could be due to some exterior action which, however, cannot respond to temperature variations. He concludes that such a mechanism is hard to conceive. Pursuing the thermal disorder analogy, he speculates that each unstable atom contains an extremely complex system in which high-velocity particles create a state of 'internal disorder'. Several others likewise tried to associate the decay properties with fluctuations in highly complex internal motions.[87]

Considerations of this kind were the subject of an address by Marie Curie to the second Solvay Conference in 1913. In the subsequent discussion, Rutherford expressed his interest in the ideas of Debierne and summarized his own view as follows: 'The law of radioactive transformation, which is universal for all radioactive substances, seems only to be explicable as a consequence of accidental disturbances ('troubles fortuits') in the nucleus, in conformity with probability laws. But, in the present state of knowledge, it does not seem possible to form a clear idea as to the very constitution of the atomic nucleus, nor of the causes which lead to its disintegration.'[88]

In 1916, Einstein became the first to realize that Eq. (6.1) can only be understood in the context of quantum theory.

In that year he introduced the so-called A- and B-coefficients referring to spontaneous and induced radiative quantum (that is, photon) transitions in atomic systems.[89] In the course of this pioneering work on the quantum theory of radiation, Einstein observed that his A-coefficient for spontaneous emission had the same quantum origins as the λ-coefficient in Eq. (6.1): '*It speaks in favor of* [*my*] *theory that the statistical law* [*equivalent to Eq.* (*6.1*)] *assumed for* [*spontaneous*] *emission is nothing but the Rutherford law of radiative decay*'[90] (my italics). There the matter rested until late 1926, when Dirac, then in Copenhagen, gave the quantum theoretical derivation of the spontaneous emission coefficient in his paper which founded quantum electrodynamics.[91]

* After Egon Ritter von Schweidler who was the first to draw attention to such fluctuation phenomena.[84]

(f) Postscripts: modern times

1. 1928: α-decay explained. It was realized by George Gamow[92] in Goettingen, in August 1928 and independently by Ronald W. Gurney and Edward U. Condon in Princeton, one month later,[93] that α-decay results as a consequence of quantum mechanical tunneling through a potential barrier. Moreover, all these authors had a further significant advance to report: the first explanation of the Geiger–Nuttall relation, known phenomenologically[94] since 1912, which establishes a connection between the lifetime of an α-emitter and the range of the produced α-particles.

In the letter to *Nature* by Gurney and Condon, we hear for the last time the echoes of a confusing past: 'It has hitherto been necessary to postulate some special arbitrary "instability" of the nucleus; but in the following note it is pointed out that disintegration is a natural consequence of the laws of quantum mechanics without any special hypothesis . . . Much has been written about the explosive violence with which the α-particle is hurled from its place in the nucleus. But from the process pictured above, one would rather say that the particle slips away almost unnoticed'.

2. Non-conservation of radioactivity. If the principle of conservation of radioactivity (mentioned toward the end of Section (b)) were strictly valid, then it would follow that the decay constant λ is independent of all chemical and physical changes, for any radioactive process.

The first suggestions that this cannot be universally true were published independently by Segré[95] and Daudel.[96] The process they chose to discuss, K-capture in Be^7, was not known to the founding fathers, to be sure. They noted that the chemical environment should affect the electron capture rate, especially in light nuclei, since chemical changes imply changes in the electron density at the position of the nucleus. During the next decade, effects of $\sim$0.1 per cent were indeed established experimentally.[97] In 1951 it was noted that similar considerations also apply to decays involving internal conversion, and an effect of $\sim$0.3 per cent was observed by comparing different chemical embeddings for a cleverly chosen technetium-99 isomer.[98] In 1965, a niobium-90 isomer turned out to be even better, yielding an effect one order of magnitude larger.[99]

These and other manifestations of 'nonconservation of radioactivity' have become a lively subject of research in recent times.[100] That is of course also due to the discovery of a quite different influence of the environment on radioactive decay: the Mössbauer effect, where a nuclear decay, even though it originates in a single nucleus, is properly described only by treating the decay as a collective quantum-mechanical property of the entire crystal in which that nucleus resides.[101]

3. The exponential law of radioactive decay. The question of how well Eq. (6.1) describes the temporal behavior of radioactive substances is quite an old one. Its early version was: Is λ a constant independent of time? Through the years this was verified experimentally, for long periods of time by Rutherford,[102] and later for 'very young' sources of radium emanation,[103] eventually down to 10^{-5} sec after their creation.[104] A more recent study of manganese-56 showed no deviations of the exponential law during the first 34 half-lives.[105]

Quantum-mechanical arguments show nevertheless that Eq. (6.1) is not mathematically rigorous. Deviations[106] occur for times both very small and very large compared with λ^{-1}. Asymptotically for large times the exponential behavior turns into a power behavior. Experimental situations in which such deviations play a noticeable role have not been found to date.

References

1. A. Einstein, *Ann. der Phys.* **18**, 639, 1905.
2. W. Pauli, 'Relativitätstheorie', in *Enzyklopädie der Math. Wiss.*, Vol. 5, Part 2, Teubner, Leipzig 1921.
3. R. W. Gurney and E. U. Condon, *Nature* **122**, 439, 1928.
4. A. Pais, *Subtle is the Lord*, Chapter 22, Oxford Univ. Press 1982.
5. *Historie de l'Académie Royale des Sciences, Année 1775*, pp. 61–6, Imprimerie Royale, Paris 1778.
6. E. N. Hiebert, *Historical roots of the principle of conservation of energy*, Univ. of Wisconsin Press, Madison, Wis. 1962.
7. Y. Elkana, *The discovery of the conservation of energy*, Harvard Univ. Press, Cambridge, Mass. 1974.
8. T. Kuhn, in *Critical problems in the history of science*, p. 321, Ed. M. Clagett, Univ. of Wisconsin Press, Madison, Wis. 1962.
9. E. Mach, *Die Principien der Wärmelehre*, Barth, Leipzig 1896.
10. M. Planck, *Das Princip der Erhaltung der Energie*, Teubner, Leipzig 1887.
11. H. Carnot, *Comptes Rendus*, **87**, 967, 1878.
12. Cf. Ref. 10, p. 16.
13. Cf. Ref. 9, p. 241.
14. T. Young, *A course of lectures on natural philosophy and the mechanical acts*, Vol. 1, lecture 8, p. 75, Taylor and Walton, London 1807.
15. G. G. de Coriolis, *Du calcul de l'effet des machines, ou considérations sur l'emploi des moteurs et sur leur évaluation pour servir d'introduction à l'étude speciale des machines*, Carilian-Goeury, Paris 1829.
16. See R. Kahl, *Selected writings of Hermann von Helmholtz*, Wesleyan University Press, Watertown, Conn. 1971.
17. H. Helmholtz, *Wissenschaftliche Abhandlungen*, Vol. 2, p. 972, Barth, Leipzig 1883.
18. L.-G. Gouy, *J. Phys (Paris)* **7**, 561, 1888.
19. M. Curie, *Traité de radioactivité*, Vol. 1, Ch. 3, Gauthier-Villars, Paris 1910.
20. S. Curie, *Comptes Rendus*, **126**, 1101, 1898.
21. H. Becquerel, ibid. **128**, 771, 1899.
22. M. Curie, *Rev. Gén. Sci. Pures et Appl.* **10**, 41, 1899.

23. W. Crookes, *Nature* **58**, 438, 1898; cf. also *Comptes Rendus* **128**, 176, 1899.
24. Ref. 19, Vol. 1, p. 129.
25. J. Elster and H. Geitel, *Phys. Zeitschr.* **4**, 439, 1903.
26. R. Pohl, *Naturw.* **12**, 685, 1924.
27. E. N. da C. Andrade, *Rutherford and the nature of the atom*, p. 40, Doubleday, New York 1964.
28. R. W. Lawson, *Nature* **113**, 432, 1924.
29. J. Elster and H. Geitel, *Ann. der Phys.* **66**, 735, 1898.
30. J. Elster and H. Geitel, ibid. **69**, 83, 1903.
31. E. Rutherford, *Phil. Mag.* **47**, 109, 1899.
32. J. J. Thomson, *Nature* **67**, 601, 1903.
33. P. and S. Curie and G. Bémont, *Comptes Rendus* **127**, 1215, 1898.
34. P. and M. Curie, ibid. **134**, 85, 1902.
35. W. J. Hammer, *Radium*, p. 18, Van Nostrand, New York 1903.
36. M. Curie, *Rev. Sci.* **14**, 65, 1900.
37. A. S. Russell, *Proc. Phys. Soc. London* **64**, 217, 1951.
38. E. Rutherford and F. Soddy, *Phil. Mag.* **4**, 370, 569, 1902.
39. H. R. Robinson, *Proc. Phys. Soc. London* **55**, 161, 1943.
40. L. Badash, *Sci. Am.* **215**, No. 2, 289, 1966.
41. M. Howorth, *The life of Frederick Soddy*, p. 83, New World, London 1958.
42. E. Rutherford, *Radioactivity*, pp. 2–4, Cambridge Univ. Press 1904.
43. P. Curie and A. Laborde, *Comptes Rendus* **136**, 673, 1903.
44. Kelvin, *Phil. Mag.* **7**, 220, 1904.
45. H. Poincaré, *The foundations of science*, Chapter 8, Science Press, New York 1913.
46. E. Rutherford and H. T. Barnes, *Phil. Mag.* **1**, 202, 1904.
47. W. Ramsey and F. Soddy, *Proc. Roy. Soc. A* **13**, 346, 1904.
48. F. Soddy, *Radioactivity and atomic theory*, p. 84, repr. Ed. T. J. Trenn, Wiley, New York 1975.
49. G. Sagnac, *J. Phys. (Paris)* **15**, 455, 1906.
50. J. J. Thomson, in *Transactions of the International Electrical Congress*, Vol. 1, p. 234, J. B. Lyons, Albany, N.Y. 1905.
51. E. Rutherford and A. H. Compton, *Nature* **104**, 412, 1919.
52. E. Rutherford, *Radioactive transformations*, p. 267, Yale Univ. Press, New Haven, Conn. 1911.
53. J. Perrin, *Ann. Phys. (Paris)* **11**, 5, 1919; *Revue du Mois*, Feb. 10, 1920, p. 113.
54. E. Briner, *Comptes Rendus* **180**, 1586, 1925.
55. E. Rutherford and F. Soddy, *Phil. Mag.* **5**, 576, 1903.
56. Cf. e.g. S. Meyer and E. Schweidler, *Radioaktivität*, Teubner, Leipzig 1927.
57. Quoted in W. Crookes, *Proc. Roy. Soc. London A* **83**, xx, 1910.
58. P. Curie, in *Nobel Lectures in Physics*, 1901–22, p. 78, Elsevier, New York 1967.
59. F. Soddy, *Matter and energy*, Holt, New York 1912.
60. H. D. Smyth, *The Smyth Report*, *The Princeton University Library Chronicle*, **37**, 173, 1976.
61. A. Pais, in *Oppenheimer*, Scribner, New York 1969.
62. F. Soddy, *Ann. Progr. Rep. to the Chem. Soc. for 1904* **1**, 244, 1905; repr. in T. J. Trenn, *Radioactivity and atomic theory*, Wiley, New York 1975.
63. Kelvin, *The Times*, London, 9 August 1906.
64. E. Demarçay, *Comptes Rendus* **127**, 1218, 1898.
65. N. R. Campbell, *Modern electrical theory*, p. 213, Cambridge Univ. Press 1907.

66. M. Curie, *Recherches sur les substances radioactives*, 1903. English transl. *Radioactive substances*, A. del Vecchio, transl. Philos. Library, New York 1961.
67. E. Rutherford, *Radioactivity*, 2nd Edn, p. 29, Cambridge Univ. Press 1905.
68. E. Rutherford, *Radioactivity*, p. 340, Cambridge Univ. Press 1904.
69. E. Rutherford, *Radioactive transformations*, p. 276. Constable, London 1906.
70. N. R. Campbell, *Proc. Cambr. Phil. Soc.* **13**, 282, 1906.
71. N. R. Campbell and A. Wood, *Proc. Cambr. Phil. Soc.* **14**, 15, 211, 1907.
72. E. Rutherford, Ref. 67, p. 30.
73. W. Lenz, *Naturw.* **8**, 181, 1920.
74. F. W. Aston, *Isotopes*, pp. 116–17, Edward Arnold, London 1922.
75. F. W. Aston, *Mass spectra and isotopes*, pp. 5–7, Edward Arnold, London 1942.
76. E. Rutherford, *Phil. Mag.* **49**, 1, 1900.
77. Cf. e.g. E. Rutherford, J. Chadwick, and C. D. Ellis, *Radiations from radioactive substances*, Cambridge Univ. Press 1930.
78. L. Boltzmann, *Lectures on gas theory*, Part 2, Chapter 6, Univ. of California Press, Berkeley 1964.
79. J. Jeans, *Physics and philosophy*, pp. 149–50, Cambridge Univ. Press 1943.
80. Lord Rayleigh, *The life of Sir J. J. Thomson*, p. 141, Dawsons, London 1969.
81. E. Rutherford, *Radioactive transformations*, p. 267, Yale Univ. Press, New Haven, Conn. 1911.
82. Ref. 80, p. 142.
83. F. Soddy, *The interpretation of radium*, p. 114, Putnam, New York 1920.
84. E. R. von Schweidler, *Premier Congrès de Radiologie*, Liège 1905.
85. Cf. S. Meyer and E. R. von Schweidler, *Radioaktivität*, Teubner, Leipzig 1927; R. Fürth, *Schwankungserscheinungen in der Physik*, Chapter 6, Viewig, Braunschweig 1920.
86. A. Debierne, *Recherches sur les phénomènes de radioactivité*, Gauthier-Villars, Paris 1914.
87. F. A. Lindeman, *Phil. Mag.* **30**, 560, 1915; H. Th. Wolff, *Phys. Zeitschr.* **21**, 175, 1920.
88. E. Rutherford, in *La Structure de la matière*, p. 67, Eds. R. Goldschmidt, M. de Broglie, and F. Lindeman, Gauthier-Villars, Paris 1921.
89. Cf. Ref. 4, pp. 405–7.
90. A. Einstein, *Verh. Deutsch. Phys. Ges.* **18**, 318, 1916.
91. P. A. M. Dirac, *Proc. Roy. Soc. A* **114**, 243, 1927.
92. G. Gamow, *Zeitschr. f. Phys.* **51**, 204, 1928.
93. R. W. Gurney and E. U. Condon, *Nature* **122**, 439, 1928; *Phys. Rev.* **33**, 127, 1929.
94. H. Geiger and J. M. Nuttall, *Phil. Mag.* **23**, 439, 1912.
95. E. Segré, *Phys. Rev.* **71**, 247, 1947.
96. R. Daudel, *Rev. Sci.* **85**, 162, 1947.
97. Cf. R. Bouchez *et al.*, *J. Phys. Radium*, **17**, 363, 1956, also for earlier experimental literature.
98. K. T. Bainbridge, M. Goldhaber, and E. Wilson, *Phys. Rev.* **84**, 1260, 1951; **90**, 430, 1953.
99. J. A. Cooper, J. M. Hollander, and J. O. Rasmussen, *Phys. Rev. Lett.* **15**, 680, 1965.
100. G. T. Emery, *Ann. Rev. Nucl. Sci.* **22**, 165, 1972.
101. Cf. H. Frauenfelder, *The Mössbauer effect*, Benjamin, New York 1962.
102. E. Rutherford, *Wiener Ber.* **120**, 303, 1911.
103. H. H. Poole, *Phil. Mag.* **27**, 714, 1914.

104. F. Joliot, *Comptes Rendus*, **191**, 132, 1930.
105. R. G. Winter, *Phys. Rev.* **126**, 1152, 1962.
106. See e.g. E. Hellund, *Phys. Rev.* **89**, 919, 1953; G. Höhler, *Zeitschr. f. Phys.* **152**, 546, 1958; M. Lévy, *Nuovo Cim.* **14**, 612, 1959; P. Matthews and A. Salam, *Phys. Rev.* **115**, 1079, 1959; J. Petzold, *Zeitschr. f. Phys.* **155**, 422, 1959; **157**, 122, 1959; J. Schwinger, *Ann. of Phys.* **9**, 169, 1960.

7

Pitfalls of simplicity

> No hammer in the Horologe of Time peals through the universe when there is a change from Era to Era. Men understand not what is among their hands.
>
> Thomas Carlyle, *On History*.

1. Simplicity as an unnecessary evil. In 1901, Yale College at New Haven established, by bequest from Augustus Ely Silliman, an annual lectureship designed to illustrate the presence and providence, the wisdom and goodness of God, as manifested in the natural and moral world. J. J. Thomson was the first lecturer; he was followed by Charles Scott Sherrington, the neurophysiologist. In March 1905, Rutherford gave the third set of Silliman lectures. He had chosen 'Radioactive transformations' as his topic, and began the first of his talks as follows:
'The last decade has been a very fruitful period in physical science, and discoveries of the most striking interest and importance have followed one another in rapid succession . . . The march of discovery has been so rapid that it has been difficult even for those directly engaged in the investigations to grasp at once the full significance of the facts that have been brought to light . . . The rapidity of this advance has seldom, if ever, been equalled in the history of science'.[1]

The text of Rutherford's lectures (a handsome, nearly three hundred-page volume) makes clear which main facts he had in mind: X-rays, cathode rays, the Zeeman effect, α-, β-, and γ-radioactivity, the reality as well as the destructibility of atoms, in particular the radioactive families ordered by his and Soddy's transformation theory, and Kaufmann's results on the variation of the mass of β-particles with their velocity. There is no mention, however, of the puzzle posed by Rutherford's own introduction of a characteristic lifetime for each radioactive substance. Nor did he touch upon Planck's discovery of the quantum theory in 1900. He could, of course, not refer to Einstein's article on the light-quantum hypothesis, because that paper was completed on the seventeenth of the very month he was lecturing in New Haven. Nor could he include Einstein's special theory of relativity among the advances of the decade he was reviewing, since that work was completed another three months later.

So much has happened in the more than three-quarters of a century since Rutherford held forth at Yale. So many fruitful theoretical concepts have since been introduced, so many new forms of matter have been discovered experimentally, so many fundamental technological advances have since been made. Yet, it seems to me, Rutherford's remark about the rapidity of advances during the decade 1895–1905 being unparallelled in the history of science remains true to this day, especially since one must include the contributions of Planck and Einstein (about which more anon). Combining the events of that period in experimental and theoretical physics, one is struck not only by the speed but, perhaps even more, by the breadth of the developments.

In his autobiographical notes, Einstein recalled his own perplexities of those days: 'All my attempts . . . to adapt the theoretical foundation of physics to this [new type of] knowledge failed completely. It was as if the ground had been pulled out from under one, with no firm foundation to be seen anywhere, upon which one could have built'.[2] In writing these lines he referred quite specifically to his early recognition of the mysteries revealed by Planck's discovery of the quantum theory. Is it not true that Einstein's vivid phrases apply generally to the entire collection of discoveries and puzzles of that era rather than to Planck's work only? It is not. In fact, such facile and sweeping generalizations only serve to create serious pitfalls of simplicity. I do believe that one can discern general themes in the history of discovery in science and I shall even venture to mention some, but mainly for the purpose of emphasizing variety over uniformity. After decades spent in the midst of the fray I am more than ever convinced, however, that a search for all-embracing principles of discovery makes about as much sense as looking for the crystal structure of muddied waters.

Nevertheless there are those who seek simplicity where there is none. Consider, as but one instance, this declaration about Roentgen: 'For men like . . . Roentgen . . . the emergence of X-rays necessarily violated one paradigm as it created another'.[3] This dictum illustrates not only the pitfall of equating surprise (which Roentgen indeed experienced) with shock (which he did not) but also what bizarre conclusions can be drawn by ignoring the original literature. Indeed, as we saw in Chapter 2, in his very first paper on the subject, Roentgen presented a speculation on the nature of X-rays (longitudinal waves in the aether) which, in his view, fitted perfectly well into the scheme of existing concepts. Never mind that Roentgen's conjecture was wrong; the point is that no one would have been more surprised than Roentgen had he been told in 1895 that this work violated any of his and many others' cherished ideas about physical theory. The emergence of X-rays violated no established pattern, then or at any later time.

The preceding is meant as an exemplification of my very strong reservations concerning the usefulness, let alone the necessity, of a search for general patterns or laws of history, specifically the history of discovery. I am in good company. Let me cite Mommsen: 'History distinguishes itself from its sister disciplines by its inability to bring its elements into a proper theoretical exposition'.[4] The art of history, in my view, is to steer amidst overly tight schema, overly bare fact, overly loose anecdote. As will have emerged from the preceding chapters, I am endeavoring to set my course by adopting the style of an epic (hardly a new device), hoping to serve the reader in deciding whether, and if at all in what way, this will provide a mirror for her or his own reflections. In this spirit, the few general comments on the preceding chapters set forth next are meant as a guide rather than a corset.

2. Ripeness of the times. It may fairly be said that 1897 was the year in which physicists were ready to discover the electron. As was observed in Chapter 4, within the space of a few months, good measurements of its charge-to-mass ratio were made independently in Königsberg, Berlin, and Cambridge. My belief that, regardless of the presence of any specific individual, the time was evidently ripe for finding the electron does not of course diminish my respect for the contributions of Wiechert, Kaufmann, and expecially J. J. Thomson.

Why was it that, of these three men, Thomson carried away the laurels? Because (see Chapter 4) it was only he who measured the charge of the electron. Why was he the only one of the triumvirate to do so? Because it was in his laboratory that C. T. R. Wilson developed the cloud chamber. To reiterate, the fundamental discoveries in physics described in earlier chapters are very strongly linked to advances in instrumentation. It is hard to risk an opinion as to when cathode rays, X-rays, the photoelectric effect, or the Zeeman effect would have been found without men like Geissler, Rühmkorff, and Rowland. Once again, without disrespect to Roentgen, the time for X-rays was at hand as soon as techniques became available for letting cathode rays out of vacuum tubes. Nor (as I have described elsewhere[5]) could the quantum of action have been discovered in 1900 without the major advances in experimental infra-red spectroscopy made during the closing years of the nineteenth century. There is more to be said on that last subject, and I shall say it shortly.

What about Becquerel's discovery of the uranic rays? In his Silliman lectures, Rutherford had something to say about that question. 'The discovery of the radioactive property of uranium* might have been made accidentally a century ago [i.e. in the early 1800s] for all that was required was the exposure of a uranium compound on the charged plate of a gold-leaf electroscope . . . the discharging property of [uranium] could not fail to have been noted if it had been placed near a charged electroscope. It would not have been difficult to deduce that the uranium gave out a type of radiation capable of passing through

* Uranium was discovered in 1789.

metals opaque to ordinary light. The advance would probably have ended there, for the knowledge at that time of the connection between electricity and matter was far too meagre for an isolated property of this kind to have attracted much attention . . . If the discovery had been made even a decade earlier [than 1896], the advance must necessarily have been much slower and more cautious.'[6]

As usual, Rutherford had a point worth enlarging on. As we have seen (Chapter 2), Becquerel's discovery of uranic rays in 1896 was as accidental and as qualitative as Rutherford's fictitious physicists' might have been much earlier. Also, 'the advance would probably have ended there', and indeed did, briefly, until 1898, when Marie Curie and Rutherford took matters in hand, swiftly transforming the isolated qualitative uranic phenomenon into a new field of quantitative research (Chapter 3). Their principal tool, a simple version of an ionization chamber, had been developed only a few years earlier, mainly for the purpose of studying X-rays. Rutherford had this new instrumentation in mind when speaking of how slow and cautious advances would have been 'had the discovery [of radioactivity] been made even a decade earlier'. This technological link between X-rays and radioactivity does not make for as tall a tale as the link between Roentgen and Becquerel (Chapter 2), but is obviously at least as important.

In turn (as Rutherford did not fail to note[7]) advances in radioactivity are connected with Kaufmann's work on the E–m–v relation (Chapter 6, Section (f)), of importance for experimental support for the special theory of relativity. At that time, high energy (a few MeV) β-rays constituted the only available particle source for this type of experiment. Clearly these various links are of help in getting some feeling for the reasons why events followed one another with such rapidity during the period 1895–1905, though one will of course never be able to explain in its fullness the density of discovery.

Times were not yet ripe, however, for a great impact of theoretical ideas on experimental developments.* Nevertheless one should neither forget the influence of Abraham's electron model, sensible for its time though incorrect, on Kaufmann's measurements, nor Lorentz's interpretation, correct though only qualitatively so (Chapter 4, Section (c)) of the Zeeman effect. It should be especially remembered how innovative for their time were the applications of the Maxwell equations and the Lorentz force to the design of experiments on the deflection of charged particles.

3. About old pros. In the minds of many, including myself, discovery in physics tends to be associated with work done by people in their twenties. I was struck, therefore, by the number of middle-aged men who played dominant roles in

* The strong feedback on experiments of Planck's and Einstein's ideas came later.

creating the transitions described in previous chapters. At the time of their decisive discoveries, Roentgen was fifty, Becquerel almost forty-two, Thomson forty-one, and Planck forty-two.*

On further reflection I realized that the ages of the three experimentalists among this foursome are perhaps not that surprising. These were not young Turks out to set the world on fire, but rather seasoned pros, systematically extending and refining work done by their experimentalist predecessors. Revolution was not on their minds, it was even alien to them. Their discoveries, made in the course of continuous progress, represent unforeseen discontinuities in the advances of science. Consummate professionalism was brought to bear, times were ripe. I am not tempted to weigh the relative importance of these two factors.

4. About Planck. After receiving his Ph.D. in 1879, Planck had published some forty papers, mainly on the theory of heat, by the time he discovered the quantum theory. Thus he too was an old pro by then, but there all resemblance with Roentgen, Becquerel, and Thomson ends. Nor would it do justice to Planck to consider him merely as the right man at the right time. It *is* true that Planck was there at the right time: the late 1890s, and at the right place: Berlin, where crucial measurements of the spectral distribution of blackbody radiation were under way, the results of which enabled him, in October 1900, to make a stunningly successful guess at the explicit expression for the spectral distribution as a function of frequency and temperature.[8] However, that was not the end, but rather the beginning of his heroic period. I have described elsewhere[9] how he made his guess at what now is called Planck's radiation law, and also how, not content with this surmise, he set out to try and justify this law from first principles. Two months later he had done so,[10] by means of a new assumption, the quantum hypothesis. At that time, the new postulate served one and only one purpose: to derive—if one may call it that—his radiation law which, Planck was informed by his experimental colleagues, agreed exceedingly well with experiment.

In his autobiographical sketch (written when he was about ninety) Planck has described[11] how at first he attempted, in vain of course, to incorporate his quantum postulate into the scheme of classical physics: 'I tried immediately to weld the elementary quantum of action h somehow into the framework of classical theory. But in the face of all such attempts, this constant showed itself to be obdurate . . . My futile attempts to fit the elementary quantum of action into the classical theory continued for a number of years, and they cost me a great deal of effort'.

* Note also that Lenard, Zeeman, Wiechert, and Marie Curie were in their early thirties when they did their work discussed earlier.

There is no need to discuss here in detail how Planck formulated his quantum postulate, nor what he thought about it at that time or in later years.* The only point about his December 1900 paper on the quantum theory I would like to make is that it serves as a most outstanding example in the history of all science of the precept that to make a discovery is not necessarily the same as to understand a discovery. Not only Planck but also other physicists were initially at a loss as to what the proper context of the new postulate really was. For example, there is an interesting recollection[14] by Debye about a discussion of Planck's work (soon after its publication) which took place in Sommerfeld's colloquium: 'We did not know whether the quanta were fundamentally new or not'.

Earlier in this chapter I mentioned how, in his Silliman lectures, Rutherford kept silent about the implications of the lifetime concept, another typical quantum phenomenon (as we now know). I regard this as a related example of a discoverer who did not at once grasp the enormity of what he had wrought.

5. On Einstein. Were I asked to designate just one single discovery in twentieth-century physics as revolutionary I would unhesitatingly nominate Planck's of December 1900. As noted, it was not at once clear either to Planck himself or to other physicists that his contribution demanded no less than a revision of the tenets of classical physics. Evidently one cannot use terms like 'scientific revolution', 'seminal developments', etc., without raising the question: as seen by whom?

I count it among Einstein's great achievements that he may well have been the first to realize that the advent of the quantum theory represented a crisis in science. 'It was as if the ground was pulled from under one.' Thus (as mentioned earlier) he expressed his state of mind shortly after Planck's paper had appeared. As said, I consider it a pitfall of simplicity to apply Einstein's phrase too broadly. This is not to imply that the discoveries described in the previous six chapters pale by comparison with Planck's work. In fact no physicist would think of comparing in importance or of ranking the discoveries of the quantum theory, of radioactivity, of the electron . . . Nevertheless the discovery of the quantum theory occupies a special position in that it signalled not only that new concepts had come but also that old first principles had to go, or, better, had to be revised.

Einstein later wrote of Planck's so-called derivation of the radiation law: 'The imperfections of [this derivation] . . . remained at first hidden, which latter fact was most fortunate for the development of physics'.[15]** The same

* As was already stated in the Preface, the plan of this book does not in fact include an account of the history of the quantum theory. For more details on Planck see especially a paper by Martin Klein,[12] and also Ref. 9. For the history of events leading up to Planck's discovery see Ref. 13.

** Here Einstein had in mind that, had Planck been systematic, he would have ended up with the Rayleigh–Einstein–Jeans law, the approximation to Planck's law for $h\nu/kT \ll 1$ (ν = frequency, k = Boltzmann's constant, T = temperature).[16]

can be said of his own discovery of the light-quantum, another milestone in the decade 1895–1905. As has been known since 1924, light-quanta obey the quantum statistics known as Bose–Einstein statistics.[17] Yet Einstein managed to introduce light-quanta by means of an argument based on classical, that is Boltzmann, statistics. Briefly, his reasoning went like this. Consider the energy density $\rho\, d\nu$ of blackbody radiation per unit volume in the frequency interval between ν and $\nu + d\nu$. If $h\nu/kI \gg 1$, then, in modern notation,

$$\rho(\nu, T) = \frac{8\pi h\nu^3}{c^3}\, e^{-h\nu/kT} \tag{7.1}$$

the so-called Wien law. The exponential looks like a typical Boltzmann factor, familiar from classical physics. It was precisely this parallel which struck Einstein. Comparing Eq. (7.1) with the energy distribution of a *classical* ideal gas of pointlike particles (as a function of volume and temperature) he concluded:[18] in the region of validity of the Wien law, blackbody radiation with frequency ν behaves like mutually independent energy quanta with energy $h\nu$.

Thus the photon concept was born by applying a classical argument, Boltzmann statistics, to a quantum law, Eq. (7.1). The imperfection, which at first remained hidden, was the absence of the proof (the need for which was only perceived twenty years after Einstein's work) that, for $h\nu/kT \gg 1$, the physical consequences of Bose–Einstein and Boltzmann statistics coincide.[17] I find this a particularly illuminating example of how the best of minds are forced to operate when, to an extent not at all obvious, the ground is pulled away from under them. Attempts to understand such times, some of physics' finest hours, when logic fails, when art and craft combine, are enough to drive many a philosopher to distraction.

The history of the prediction and discovery of the photon ranks among the most fascinating chapters of twentieth-century science. Never, to this day, has the suggestion of a new particle caused more and longer-lasting confusion, the principal reason being that one was not confronted with never yet seen phenomena but, instead, with old phenomena in a new guise: free electromagnetic radiation, thought to be well understood as waves, was supposed to behave, in some circumstances, as if it consists of particles. Inevitably, this, the first manifestation of particle-wave duality, had to throw the world of classical physics into disarray. In addition, one cannot associate the prediction of the photon with one single theoretical paper. In 1905 Einstein wrote of light as energy quanta. Only in 1916 did he write of light as momentum quanta, again abstracting this property from the analysis of blackbody radiation. The birth of the photon as a true particle, endowed with well-defined energy and momentum, was therefore not only a painful but also a protracted process. Still another seven years passed until the experimental discovery of the photon as an individual object, capable of exchanging energy and momentum with another particle (the electron) in accordance with the energy–momentum conservation laws. (Here I refer to the Compton effect, first observed in 1923). My only reason for not treating the photon more extensively in this book is that I have already done so previously.[19]

I conclude this brief survey of a decade of discovery with a comment on the special theory of relativity. That advance may have pulled the ground away from under some, but not from under Einstein. Note first that the need for precise operational definitions of distance and duration had been well appreciated before 1905. So too for the concept of simultaneity. The notion of relativity had also long been recognized, though only in the area of mechanics. This is the so-called Galilean relativity (a term introduced only later[20]). Einstein's theory revised the meaning of all these concepts, with new experimental implications. From his very first paper on this subject, the theory was put on a firm, logically complete, axiomatic basis—quite unlike the quantum theory which remained a logically muddled subject for the first twenty-five years of its existence. If, in the early years of this century, some physicists could not accept the quantum theory, that was because to them it seemed (as indeed it was) to be too complicated a hodgepodge of *ad hoc* rules. If, at that same time, some physicists could not accept special relativity, that was, largely I believe, because for them the theory was too simple to grasp.

6. On the variety of discovery. As said, the speed and breadth of discoveries in physics during the period we are now about to leave was extraordinary. Experimentally there were the X-rays, radioactivity, the Zeeman effect, the electron. Theoretically there were the quantum theory, special relativity, and the realization that electromagnetic radiation can behave like a beam of particles. After the fact these theoretical developments served to define the notion of 'classical physics': that part of physics in which actions are large on the scale set by Planck's constant, velocities small on the scale set by the light velocity. All that lies beyond is of twentieth-century vintage.

I remark one last time on my failure to see what is gained by strong emphasis on common traits in discovery. What is shared tends to be obvious; individuality is the key. I mention a few obvious general features. There is the influence of advances in instrumentation on advances in fundamental physics. There is the ripeness of the times. There is the strikingly frequent initial lack of understanding of the import of a given discovery (Einstein and special relativity being a notable exception). 'No hammer in the Horologe of Time peals through the universe when there is a change from Era to Era.' Thus there was Maxwell's prejudice about unbroken and unworn atoms, and his opinion that the discreteness of electric charge was a passing fancy;[21] Roentgen's belief that his rays were longitudinal excitations;[22] Becquerel's conjecture that X-rays are associated with luminescence;[23] the debate on whether cathode rays are molecular torrents or aetherial disturbances;[24] doubts about energy conservation stemming from the mysterious radioactive energy release;[25] Planck's attempts to incorporate the quantum into the classical theory; and Rutherford's silence about the meaning of lifetime. Responses such as these are inevitable. 'Men understand not what is among their hands.' Such initial prejudices are also harmless as long as they are given up in the face of better evidence. If there is a moral to all this, it is that revolutions in science (whatever that *precisely* means) are rarely brought about by revolutionary souls. Like the rest of humanity, physicists tend to cling tenaciously to what they know or think they know, and give up traditional thinking only under extreme duress.

Another characteristic of that tumultuous decade is that experimental advances took place on one front, those in theory on an almost separate front. Thus Planck's theory addressed blackbody radiation and (during those early years) nothing else; relativity gave an underpinning to Maxwell's electromagnetic theory, yet, in spite of its recognized universal nature, could not at once be brought to bear on other kinds of observations. On the other hand, radioactivity represented a class of phenomena in search of a theory.

In less tumultuous times one often encounters discoveries of a quite different nature—how great is the variety of discovery!—in which theory and experiment, though out of phase, have both reached mature levels. I mention but two famous examples. First, superconductivity. Experiment preceded theory, until there came a time when this remarkable phenomenon could be understood with the use of theories long familiar by then: quantum mechanics, Coulomb forces, and the exclusion principle. Secondly, phase transitions. The time of discovery that water vapor can be made to condense is lost in antiquity. The quantitative dependence of condensation on pressure and temperature is of course part of much more recent knowledge. There is no doubt that we possess the right theory for calculating the condensation point. However, no one has done so yet. The problem is mathematically still too difficult. As is said in physics' finest clerihew:

> Sir James Dewar
> Is cleverer than you are
> None of you asses
> Can condense gases

Lastly there are intermediary periods during which theory and experiment confront each other while both are still in primitive stages. This is the subject of my final comment, which also serves as a general introduction to subsequent chapters.

7. Simplicity as a necessary evil. In the course of a recent discussion with an experimentalist colleague about a subtle weak-interaction experiment he planned to perform, the question came up of how long this effort would take. Such actuarial questions are inevitable these days, not only because of the high technical and logistic demands of modern experimentation, especially in high-energy physics, but also because an experimental proposal must first be scrutinized by a committee of peers and, even when accepted, there remain the dictates of beam availability for its execution. As we went on to reflect on these factors which influence the pace of discovery, one of his younger colleagues who was present wondered out loud about the days of yore when physics was done by small groups, if not by individuals, when the space needed for an experiment was still table-top-sized, and when the time needed could often be counted in weeks if not less. This led one of us to remind him that it took more than thirty years from the discovery of β-radioactivity to the postulation of the neutrino, and nearly another twenty-five years from its postulation to its detection. He was astonished and asked: What did people do in between?

The answer to this question depends on a complex of factors, ranging from problems in hard science (instrumentation, experimentation, as well as theory) to problems of hard cash. Since the speed of development of twentieth-century physics in general, and of particle physics in particular, is such that fifty years must be considered a very very long interval, the question is approached more easily by first dividing this period into smaller parts. Let us then consider the following mini-chronology of main events marking the first half-century of studies in β-radioactivity.

1898: Rutherford notes the distinction between α- and β-rays.
About 1900: It is firmly established that β-rays are electrons, as defined in terms of cathode ray constituents.
1914: Discovery of the continuous character of the β-spectrum.
1930: The neutrino is postulated to explain the continuous β-spectrum.
1934: A new force, the weak interaction, is introduced to describe β-decay.
1939: The first reliable shapes of β-spectra are obtained.
1956: The neutrino is detected.

The one initial question now decomposes into many. Why did it take sixteen years between the discovery of β-radioactivity and the first observation that the β-spectrum is continuous? Why did it take another sixteen years from this last discovery to the insight that some new postulate was needed to explain this spectral property? Why did it take altogether two generations before decent β-spectra became available? Why was the neutrino not found until the nineteen-fifties?

One by one, these issues will be dealt with in later chapters. At this point I only note that they are linked by the sequence:

incomplete experimental guidance → pitfalls of simplicity → paradox
→ progress → incomplete experimental guidance → . . .

The first link of this chain, which may or may not ever end, refers to information received at some fixed time from the laboratory. Since not even the best experimentalist can overcome the inevitable limitations set by his equipment, he is bound to give us incomplete information (and of course knows he does so). On this basis, the theorist devises a clever scheme for interpreting these results, basing his reasoning on the simplest assumptions he can think of that are compatible with the data. All goes well for some time until either the repetition of an earlier experiment with improved accuracy or experiments of a new kind reveal that his simplest assumptions were in fact a pitfall. The experimentalist is now very happy because he has refuted the theorist's constructs. The theorist is now very happy* because the new experimental information has given him a paradox to resolve. He now has an improved basis for trying again with a revised set of simplest assumptions. Progress eventually results. But even the new experimental basis is bound to be incomplete. The sequence: pitfall, paradox, progress, repeats itself. And so it continues.

Pitfalls of simplicity of this kind are neither avoidable nor a mark of stupidity. They should rather be considered a necessary evil. It is the scientist's task to seek for simplicity at the fuzzy frontiers of knowledge. He is privileged in that his latitude for choice of simplicity, necessarily an ill-defined concept, is held in tight check by experiment. In that respect his labors differ from those of many others engaged in intellectual pursuits.

The search for simplicity can be illuminated by past successes and past failures. In the final analysis it is an intuitive process which cannot be taught. At times, what is simple to one is simple to many, at other times the choice is a lonely one. One individual's simple ideas may triumph at one time, and fail at another, as happened most notably to Einstein. The pursuit of simplicity is an adventurous undertaking, as was expressed so well by Sigmund Freud in a letter to Wilhelm Fliess: 'You often estimate me too highly. For I am not really a man of science, not an observer, not an experimenter, and not a thinker. I am nothing but by temperament a conquistador—an adventurer, if you want to translate the word—with the curiosity, the boldness, and the tenacity that belongs to that type of being. Such people are apt to be treasured if they succeed, if they have really discovered something; otherwise they are thrown aside. And that is not altogether unjust'.[26] Those lines are especially fitting since they were written in 1900, in the middle of the decade to which this chapter has been mainly devoted.

* It would perhaps be better to say that he *should* be very happy. Experimental disproof of hypotheses has sometimes led to despair, even to suicide.

The pursuit continues. There will be later times when some older particle physicists will sit together and pontificate about events past to them, present and future to us. A younger colleague will listen and ask in astonishment: What did people do in between?

The next chapter deals with the first of the series of questions asked previously: Why did it take sixteen years between the discovery of β-radioactivity and the first observation of the continuous β-spectrum? There one will find striking examples of the scheme pitfall–paradox–progress. The other questions will be dealt with later. The answer to my last question, why was the neutrino not found until the 1950s, has everything to do with instrumentation and nothing to do with paradoxes. Thus my scheme provides yet another example of facile and sweeping generalizations, which only serve to create serious pitfalls of simplicity.

References

1. E. Rutherford, *Radioactive transformations*, pp. 1 and 16, Constable, London 1906.
2. A. Einstein, 'Autobiographical notes', in *Albert Einstein: philosopher scientist*, p. 45, Ed. P. A. Schilpp, Tudor, New York 1949.
3. T. Kuhn, *The structure of scientific revolutions*, 2nd edn, p. 93, Univ. of Chicago Press 1970.
4. Th. Mommsen, 'Reden und Aufsätze', p. 10, Weidmann, Berlin 1905; transl. in *The varieties of history*, p. 192, Ed. F. Stern, Vintage Books, New York 1973.
5. A. Pais, *Subtle is the Lord*, Chapter 19, Section (a), Oxford Univ. Press 1982.
6. Ref. 1, p. 17.
7. Ref. 1, p. 19.
8. M. Planck, *Verh. Deutsch. Phys. Ges.* **2**, 202, 1900.
9. Ref. 5, pp. 364–72.
10. M. Planck, *Verh. Deutsch. Phys. Ges.* **2**, 237, 1900.
11. M. Planck, *Scientific autobiography and other papers*, Philosophical Library, New York 1949.
12. M. Klein, in *History of twentieth century physics*, Academic Press, New York 1977.
13. H. Kangro, *History of Planck's radiation law*, Taylor and Francis, London 1976.
14. U. Benz, *Arnold Sommerfeld*, p. 74. Wissenschaftliche Verlagsgesellschaft, Stuttgart 1975.
15. Ref. 2, p. 39.
16. Cf. Ref. 5, Chapter 19, Section (b).
17. Cf. Ref. 5, Chapter 23.
18. Ref. 5, Chapter 19, Section (c).
19. Ref. 5, Chapters 19 and 21.
20. P. Frank, *Sitz. Ber. Akad. Wiss. Wien IIa*, **118**, 373, 1909, esp. p. 382.
21. Chapter 4, Section (b).
22. Chapter 2, Section (a).

23. Chapter 2, Section (b).
24. Chapter 4, Section (d).
25. Chapter 6, Section (b).
26. S. Freud, letter to W. Fliess, February 1, 1900, quoted in E. Jones, *The life and work of Sigmund Freud*, Vol. 1, p. 348, Basic Books, New York 1953.

8

β-Spectra, 1907–1914

(a) Introduction

The bulk of the preceding account of radioactivity concerned problems and progress of a nature so general that there was hardly any need to distinguish between α-, β-, and γ-radiations beyond their constituent properties. Indeed, recall the main topics discussed up to this point: (1) uranium emits a new kind of radiation, the uranic rays; (2) certain other elements emit similar radioactive radiations; (3) these emissions are atomic properties; (4) the radiations are generally inhomogeneous and consist of three types: α-, β- and γ-radiations, distinguished in the first instance by their varying degrees of absorptivity; (5) radioactive processes occur spontaneously. Each such process has a characteristic lifetime; (6) even the most qualitative description of radioactive phenomena demands that atoms are not the ultimate building blocks of matter. This view becomes inevitable also because of the discovery of the electron; (7) the transformation theory brings a first phenomenological order in the complexity of radioactive processes; (8) even so, the origins of the radioactive energy release and the significance of the lifetime remain obscure. As said, these are mainly generic problems, depending little on specific properties of the three classes of decays.

Once again, radioactivity will be the subject of this chapter, but with a marked shift in emphasis. Very little will be heard hereafter about α- and γ-radioactivity. On the other hand, and starting with this chapter, β-decay will become a major issue. In order to explain why this narrowing of focus is dictated by the aims of this book, let us begin by listing the three radioactive decay types and the forces or (which amounts to the same) interactions dominantly responsible for their occurrence.

Process	*Dominant force*
α-decay	strong
β-decay	weak
γ-decay	electromagnetic

The story of the strong interactions is yet to come. Let it suffice for now to state that these forces are responsible for holding atomic nuclei together, and, in the case of α-decay, for tearing certain nuclei apart. Qualitatively, α-

radioactivity provides an important but very complex source of information about strong interactions; and the same is true for γ-decay in relation to electromagnetism.

By sharpest contrast, until 1947—the year in which μ-meson decay was discovered—β-decay was the *only* manifestation of a specific type of force rather than one among many. Indeed, β-decay served to reveal that there *is* a novel force, the weak interaction. Because of this unique position, the early history of β-radioactivity has bizarre touches. Conjectures about the nature of this process often led to pitfalls; analogies with better-known phenomena were doomed to failure.

The struggles and confusion out of which grew the recognition that the β-process occupies a very special position are of course very much a part of the historical developments addressed in this book. On the other hand, the story of the α- and γ-decays will not be pursued further, since it is not a primary source for our understanding of fundamental forces. It is true that these processes continue to be important for our understanding of nuclear structure. But that is not to be one of our subsequent topics.

This chapter, the first of several devoted specifically to β-radioactivity, deals with the earliest experimental studies of β-spectra. The period covered is 1907–14, when neither the neutron, nor the proton, nor the (anti) neutrino had been postulated or observed, and when, therefore, there was no remote hint that β-decay is due to the fundamental process in which a neutron (bound or free) disintegrates according to

$$\text{neutron} \rightarrow \text{proton} + \text{electron} + \text{antineutrino}.$$

In those years of the early pioneers, no one could possibly have had any inkling that the energy of the emitted electron is not unique but covers a range, essentially because it has to share the liberated energy with the neutrino, more precisely because neutron decay is a three-body process.

The plan of this chapter is as follows:

Section (b) The discovery (1904) that α-rays from a pure α-emitter are monochromatic* leads to a pitfall: it is conjectured that, similarly, the β-rays from a pure β-emitter have a monochromatic primary energy. Brief comments are made on the discoveries of α-particle scattering (1906) and straggling (1912).

Section (c) Since at first it was believed that the absorption of electrons in metal foils would be a good tool to verify this conjecture, it is necessary to digress and describe the ideas about this absorption process prevalent in 1907. Here we encounter a second pitfall: it was thought at that time that the absorption of monoenergetic electrons as a function of foil thickness satisfies a simple exponential law.

* It was discovered much later that α-rays in general have an energy fine structure. This important detail is of no concern to our account.

Section (d) This law is promoted to a diagnostic for the conjecture mentioned in Section (b). The first β-spectrum measurements are described. It appears at first as if the monochromatic conjecture is confirmed!
Section (e) We return to the question of the absorption law. In 1909 it begins to become evident that this law is far more complicated than was thought earlier. It is realized that the simple absorption tests for the structure of β-spectra are meaningless.
Section (f) A second phase in the study of β-spectra is initiated in 1910. β-rays are magnetically separated and the resulting velocity spectrum is analyzed by photographic methods. At first it seems once again that the β-spectrum is monochromatic. Further investigations by the same method finally show that the idea of a unique electron energy must be abandoned.
Section (g) During the years 1912–13 experiments appear to indicate that the β-spectrum is a discrete line spectrum. At this stage the foundations are laid, unwittingly, for nuclear spectroscopy.
Section (h) Magnetic deflection of the β-rays but detection by counters finally reveals that the β-spectrum is continuous (1914).

At this stage the present chapter concludes. I can now answer one of the questions posed at the end of Chapter 7: why did it take sixteen years between the discovery of β-radioactivity and the first observation that the β-spectrum is continuous? Answer: not only because electrons were quite novel objects of investigation, but also because there was no qualitative guidance from other phenomena regarding what to expect. In addition, secondary effects strongly obscured the nature of the primary β-spectrum.

Before turning to the main topic, I remark briefly on a related but much simpler problem. As was seen in Chapter 3, Marie Curie had observed in 1898 that radioactivity is an atomic property. Applied to β-decay this means that a fixed number of electrons is produced in a given β-radioactive process. Question: *how many*?

The informed reader knows the answer: one. The earliest work bearing on this question dates[1,2,3] from the years 1903–5. However, the problem was still rather open in 1912, when Moseley went to work on it. His plan[4] was to determine the amount of electric charge carried by β-rays produced by a known weight of radioactive material: 'It is . . . of importance to know the average number of [β-] particles emitted by an atom. If this proves to be an integer, the process of disintegration is probably much simpler than appears at first sight'. The way Moseley phrased his conclusions indicates the prevailing uncertainties about secondary processes: 'It will perhaps be best to accept provisionally the value 1.10 particles . . . both for an atom of RaB and of RaC, deferring however any explanation of the departure from a whole number until more is known about the initial absorption of heterogeneous β-radiation'. Moseley's result, a number near one, was an important contributing factor to the formulation, in 1913, of the radioactive displacement laws according to which a single electron is emitted in the fundamental β-process. Experimentally, the question was still under investigation[5] in 1925. Finally it should be noted that the cloud chamber as a tool for counting the number of electrons per β-disintegration was first used in 1926, by a Japanese group in Tokyo.[6]

(b) α-decay and the conjecture of β-monochromacy

William Henry Bragg, a late developer, received his education at Cambridge University. The bright young student was going along the King's Parade one morning, when he ran into J. J. Thomson, his teacher and tennis partner. As the two walked along together, the conversation turned to Horace Lamb, already then a renowned applied mathematician, who had recently resigned his position as the first professor in mathematics and physics at the University of Adelaide, in order to return to England. Thomson suggested that Bragg apply for the vacant post; Bragg did. A few days later he was interviewed in London by Lamb and by the Agent General of South Australia. That same evening he was informed that the position was his. A month thereafter, the twenty-four-year-old professor sailed for Australia. He had as yet no published work to his credit.[7]

Until his early forties, Bragg devoted himself mainly to teaching, to the affairs of the Australasian Society for the Advancement of Science and, generally, to the good life of Australia. (He helped to lay out the first golf course in Southern Australia.) The few papers he wrote during that period were not particularly memorable.[8] Then abruptly, his life changed. In Bragg's own words, this is what happened next.[7] 'For seventeen years I worked steadily in Adelaide. Then came [a] crisis. It had never entered my head that I should do any research work. I was to give the presidential address to Section A of the Australasian Association for the Advancement of Science . . . in Dunedin, New Zealand, in January 1904. I thought that I could make an interesting address if I spoke of the recently discovered electron and of the phenomenon of radioactivity. While reading up the subject, I came upon some results described by Mme Curie which seemed to me capable of only one interpretation, and that an interpretation which had not yet been suggested. It was known that when the radium atom broke up into two parts, one large and one small, the latter . . . was driven into the surrounding air . . . Mme Curie described experiments which implied that all the alpha particles thus expelled went about the same distance. This interested me greatly. All ordinary radiations fade away gradually with distance; the alpha particles seemed to behave like bullets fired into a block of wood. But, if this were so, the particle must travel in a straight line through the air, as the bullet does through the block. Now, some hundreds of thousands of air atoms would necessarily be met with on its journey. How did it get past? . . . There was only one answer to the problem. The particle must go *through* the air atoms that it met. There must be a moment when two atoms, the alpha particle and the atom being crossed, *occupied the same space*. This was contrary to all the teachings that I knew. Still, it seemed to be right . . . When I got back [from Dunedin] to Adelaide I was given funds for the purchase of a small quantity [of radium] . . . and so I was able to try my special experiment, and all went well.'

In Bragg's special experiment,[9] performed together with Richard Daniel Kleeman, his assistant, use was made of a pencil of α-rays emanating from a *thin* layer of radium salt.* The rays, confined to a narrow stream by suitable lead stops, pass into a shallow ionization vessel. The distance between this vessel and the source is varied, and the α-particle range in air is defined as the distance at which ionization ceases to be registered. (The range in other gases was determined by the same method.[11]) In 1905 they summarized their findings as follows:[12] 'Each α-particle possesses . . . a definite range in a given medium the length of which depends on the initial velocity of the particle and the nature of the medium. Moreover, the α-particles of radium which is in radioactive equilibrium can be divided into four groups, each group being produced by one of the first four** radioactive changes in which α-particles are emitted. All the particles of any one group have the same range and the same initial velocity'.

This simple and fundamental property was to become a powerful diagnostic for the detection of new α-emitters: if a new α-particle range is observed a new α-emitter has been found. As an example, in 1907, this reasoning provided the most succinct argument for the identification[13] of a new radioactive substance, ionium (Th^{230}), by Bertram Borden Boltwood from Yale, the first leading authority on radiochemistry in the United States.

For our present purposes it is not necessary to discuss in detail certain revisions that were found necessary in the analysis of α-range experiments.[14] The conclusions were correct in their essentials and their great interest was described[15] by Soddy in 1905: '. . . The Daltonian conception that the atoms of the same element are all exactly alike applies even to the velocity with which they expel radiant particles on disintegration. This is probably the most severe experimental test to which this conception has ever been subjected, for it might be imagined hardly possible for any two systems, so extraordinarily complex as the heavy atoms are known to be, to be so absolutely alike that exactly the same velocity should be impressed in each case on the fragments during their explosive disruption'.

It should be remembered that no one had as yet observed the phenomenon of α-particle scattering at the time Bragg did these experiments. In fact, it

* Bragg was aware of the fact that his source should not be 'so thick that the α-rays from the bottom of it are unable to reach the air above: Such a thickness is of the order of 0.002 cm.'[10]

** In 1904 it was known that if one de-emanates radium (Ra^{226}), itself an α-emitter, then within a matter of a few days near equilibrium is established with three other α-emitters: radium emanation (Rn^{222}), radium A (Po^{218}) and radium C (Bi^{214}).

appeared to Bragg[16] that his results confirmed his view, held already before he started these investigations, that 'the [α-] radiation must be absolutely rectilinear ... and that there must be no scattered α-particles, as in the case of β-radiation'. The discovery of α-scattering dates from 1906 when Rutherford observed[17] an 'undoubted slight scattering or deflexion of the α-particle in passing through matter'. Thereafter it became gradually clear that Bragg's correct conclusions depended on the happy circumstance that for his purposes α-scattering could be neglected to a very good approximation.

In 1904, Bragg knew that his α-particles lost their energy by ionization, but was not yet aware of the fact that the energy lost by charged particles traversing a fixed path length with given initial energy is subject to statistical fluctuations. Consequently there are statistical fluctuations in their range. These effects are known as straggling, a concept not introduced[18] until 1912. By way of transition to our principal subject, β-rays, it is instructive to quote here a comment on the subject of straggling made in 1928.[19] 'In the case of α-rays straggling is small and only detected by refined measurements. It was therefore possible to carry the study of α-rays a long way without taking it into account. With β-rays, however, the situation is entirely different ... the study of β-rays has been held up by our ignorance of the true form of this large scattering so fundamental to the whole problem of their behavior.'

In 1906, when Otto Hahn and Lise Meitner began their investigations of the primary β-spectrum, they could scarcely have anticipated the complexities of the problem they hoped to resolve. The thought which occurred to them was sensible enough for its day: if it is true that α-rays emerge with a unique velocity, then the same might perhaps also hold for electrons coming from a pure β-emitter. What to do in order to check this?

They were well aware that they could not simply adapt the Bragg–Kleeman method for α-rays to their study of β-rays, since, unlike the former, electrons do not have a clearly marked range in a gaseous medium. In 1904, Bragg had written: 'If ... a jet of electrons be projected into the air, some will go far without serious encounter with the electrons of the air molecules; some will be deflected at an early date from their original directions. The general effect will be that of a stream whose borders become ill-defined, which weakens as it goes, and is surrounded by a haze of scattered electrons. At a certain distance from the source all definition is gone, and the force of the stream is spent'.[20]

Thus Hahn and Meitner had to seek another criterion for testing their conjecture. This, they believed, could be provided by the absorption law for monoenergetic electrons in metal foils. In order to follow their line of thought it is necessary to digress. We need to inquire into the state of knowledge in 1907 concerning the passage of electrons through matter.

Before doing so, let us take leave of Bragg. His rise to scientific eminence began with his work on the range of α-particles. He was elected to the Royal

Society in 1907 (and was later to be its President). The next year he was appointed Cavendish professor of physics at the University of Leeds. It was there that he did his best-known work, on X-ray crystallography, for which he shared the 1915 Nobel Prize with his son.

(c) Passage of electrons through matter: the position in 1907

'Although the interaction of particulate radiation with matter has been one of the most studied of "atomic" phenomena, the theory having been developed and fairly reliable measurements made as early as 1912, there has never been a sufficiently accurate theory nor precise enough measurements to enable the worker who uses such radiation to apply these results to a general case where such interactions might obscure other and possibly more interesting effects.' Such was the state of affairs in 1951 when these lines were written.[21] An example par excellence of a 'possibly more interesting effect' is the shape of the β-spectrum, which was not at all understood until the 1930s (and even then only partially) just because some aspects of the passage of electrons through matter had not yet been fully grasped.

It will therefore not come as a great surprise that the state of experimental and theoretical knowledge in 1907 about the transition of a beam of electrons through matter was embryonic. There exists a good historical account of the history of the subject during the first decade of the twentieth century,[16] as well as good technical reviews dealing with the great efforts spent on these problems during the next fifty years.[22] A discussion of these very interesting topics goes beyond the limited purpose of this digression which (to repeat) is to make clear how the first attack on the β-spectrum was planned.

First a brief reminder. At typical β-ray energies, the most important deflection mechanism of electrons is the Coulomb scattering with atomic nuclei. The physicists of 1906 could not know this, there was no nucleus as yet. They did know that at these energies ionization is the main cause of energy loss, but had as yet no good theory for that process. There were first glimpses of secondary phenomena involving electrons. For example, Thomson[23] and Rutherford[24] had made the first observations of δ-rays (knock-on electrons) resulting from the passage of α-particles through matter. There had also been several studies of the absorption properties of electrons in solids, our main topic of concern. By about 1907 the opinion was widely held that absorption of electrons came about by a 'sudden death' mechanism.

From experiments conducted since the turn of the century, the view had begun to gain ground that the electron intensity falls off exponentially as a function of absorber thickness.[16] A series of experiments performed by Heinrich Willy Schmidt from Giessen appeared to add further evidence for such exponential behavior. Schmidt had studied the absorption of β-radiations from RaB (Pb^{214}) and RaC (Bi^{214}) in aluminum foils with thicknesses varying from 0 to 4 mm. He claimed to be able to fit his data to an absorption curve consisting of a superposition of a few exponentials (three for RaB, two for RaC). In 1906 he summarized his results as follows: 'We have seen that the β-rays from radium are absorbed according to a pure exponential law within certain filter thicknesses. Should this not be taken to mean that there exists a certain group [of rays] with a constant absorption coefficient among the totality of β-radiations? Indeed, could we not go one step further and interpret the total action of β-rays in terms of a few β-ray groups [each] with constant absorption coefficient?'[25] Thus Schmidt proposed that the intensity I of a monoenergetic beam of electrons traversing a thickness d satisfies

$$I(d) = I(0)\,e^{-\nu d}, \tag{8.1}$$

where the value of the 'absorption coefficient' ν depends on the nature of the absorbing material (but not on d), and *where ν is in general different for different energies*. As to the meaning of this exponential behavior, he had these comments to make: 'Since ν . . . takes on values independent of the filter thickness used . . . one has to assume that, within the errors, the penetrating power [of electrons] and therefore their velocity is not influenced by the thickness of the layer it has already passed. It may seem somewhat improbable that β-particles would not change their velocity when they traverse matter. In particular it seems hard to explain how it is possible for the rays to produce an ionization energy without loss of velocity. However one can assume (and this does not at all seem far-fetched) that the same percentage of the available particles is always fully stopped by the same [absorber] thickness, while the others go on with unchanged velocity.'[25]

To illustrate this sudden death mechanism, Schmidt appealed to a picture in which electrons behave like small bullets shot through a rigid wide-mesh screen. Those that hit the mesh are 'eliminated from the beam', the others continue with unchanged velocity, the result being an exponential fall-off with target thickness.

Eq. (8.1) was widely accepted in the years 1906–7. In a review of the absorption problem, Soddy noted in 1908 that this view 'has gained some ground during the year'.[26] As will be discussed in Section (e), things changed drastically in 1909. But first we must see what happened to Hahn and Meitner when they applied the incorrect assumption (8.1) to the incorrect conjecture that pure β-emitters generate monochromatic electrons.

As for Schmidt, he went to Manchester in 1907, to work with Rutherford, who thought well of him.[27] He was on his way to becoming one of Germany's leading authorities on β-rays, but died young, aged thirty-eight.

(d) The first attack: absorption of β-rays

Otto Hahn was a close contemporary of Einstein. He was five days older; and his first important paper,[28] announcing the discovery of a new radio-element, radiothorium (Th^{228}), was read before the Royal Society the day before Einstein completed *his* first important paper, the one on the light-quantum.

Hahn studied organic chemistry at the University of Marburg. After receiving his Ph.D. in 1901, he reported as a one-year volunteer to the 81st Infantry Regiment in Frankfurt am Main, his native city. Upon returning to civilian life he was for two years an assistant at the university laboratory for organic chemistry in Marburg. He had no plans for a scientific career, however, but rather intended to go into industry. Mastery of the English language would be very useful for that purpose. Thus in 1904, financed by his parents, he set out for London, where he was to work for some time at the Chemical Institute, University College, headed by William Ramsay. When the latter asked him whether he wanted to work with radium, Hahn replied that he did not know anything about radium. Nevertheless he began research in radiochemistry. His discovery of radiothorium, which followed soon, was initially received with scepticism. One expert wrote that Hahn's new element was probably 'a mixture of ThX [Ra^{224}] and stupidity'.[29] Hahn insisted and proved to be right. This discovery marked the end of his aspirations to a position in industry, and the beginning of a brilliant career in radiochemistry,[30,31,32] which was fostered further by a stay with Rutherford in Montreal. In the autumn of 1906, Hahn left Canada for Berlin, where he had obtained a junior position in Emil Fischer's Chemistry Institute.

In the beginning he felt somewhat isolated, and wrote to Rutherford: 'in our country there are so few people who work on radioactivity that I have to make clear the simplest fact . . . If they hear something about radium they always seem sceptical.'[33] Rutherford replied in an encouraging vein: 'I am glad you are getting things in shape. I trust you will be able to convince the "savants" that there is some method in our madness'.[34] A week later Hahn mentioned to Rutherford some work on radioactivity done by 'a young lady from Vienna'.[35] In early 1907 things were getting better: 'Radioactivity is getting more and more in fashion in Germany . . . '[36] At the end of that year Hahn wrote about the arrival of the young lady from Vienna: 'Lise Meitner has come to Berlin to work with Planck theoretically and comes to my place 2 hours every day'.[37] They began their joint work in October 1907; their close scientific association was to last for three decades.

Lise Meitner was born and raised in Vienna, where she studied physics at the old physics laboratory in the Turkenstrasse, an institution with which such names as Loschmidt, Stefan, and Boltzmann were associated. Initially her main interest was in theory. She was much influenced by Boltzmann and recalled, in later years, how well he could convey to his audience the beauty of theoretical physics.[38] It was Stefan Meyer, Boltzmann's assistant, who suggested she work on radioactivity. Her first research on the subject, done in Vienna, (and which had caught Hahn's eye) concerned the absorption of α- and β-rays in metal foils.[39] At that time, however, theory remained her main interest. Boltzmann's suicide in September 1906 changed the course of her life. She met Planck briefly when, after Boltzmann's death, he visited Vienna in response to an invitation to become Boltzmann's successor. This encounter made her decide to go to Berlin for some time. to study under Planck.[40]

Those were the paths which brought together two of the main characters of this chapter, Hahn and Meitner. Their stage: the old carpenter shop on the main floor of Fischer's Institute, Hahn's quarters of research in radioactivity. (In accordance with existing rules, Meitner was not permitted access to the laboratory space of male students on the upper floors.)* Their plan: to see whether they could prove for β-rays what Bragg and Kleeman had shown for α-rays. 'Our conviction that every element that emitted beta rays would only emit beta rays of uniform velocity and could thus be identified was, in the last analysis, based on a comparison with alpha rays, which I [i.e. Hahn] had studied with Rutherford in Montreal. . . . In 1904 W. H. Bragg and R. O. Kleeman published in the *Philosophical Magazine* a beautiful piece of work in which they proved that the alpha rays emitted by radioactive elements had a range which was characteristic for each element. . . . Both Lise Meitner and I thought that these researches on alpha rays could be applied to beta radiation.'[41]

Hahn and Meitner took their initial cue from Schmidt's work and adopted as a working hypothesis: the absorption of monoenergetic electrons satisfies Eq. (8.1.). They knew of objections by Bragg[42] against such a simple picture. However, they argued, the radioactive source he had used might perhaps be too complex a mixture of substances to settle the issue. And so they started a series of absorption experiments. In their first joint paper they remarked that β-rays of thorium did seem to yield an exponential behavior characteristic (they believed) for a single discrete primary energy: 'We conjecture ("vermuten") from our results that pure β-ray emitters only emit β-rays of one kind [i.e. one velocity], similar to what happens with α-rays. The deviation from this supposition, which doubtless occurs for mesothorium, can perhaps be explained in terms of a complex nature of thorium 2'.[43] And in a sequel: 'Our assumption, made earlier, that pure [radioactive] products emit unique β-rays and that their absorption in aluminum follows an exponential law has been shown to be valid as a working hypothesis also for actinium'.[44] However, beginning with radium things became complicated (we are now in September 1909): 'On the grounds of our hypothesis that complex [i.e. non-unique velocity] rays correspond to complex substances . . . one must conclude to a complex nature of radium'.[45]

At about that same time, their diagnostic, the exponential law, began to fall apart.

* This ban was lifted in 1909 when women were admitted to academic studies in Germany.[40]

(e) 'A revolutionary conclusion'

Soddy's comment, mentioned earlier, that the idea of an exponential absorption law for electrons was gaining ground is found in his report[26] on progress in radioactivity during 1907. In his next review, covering the years 1908–9 he was much less sanguine: 'The exact meaning to be attached to [β-ray] "absorption" in passage through matter is, in spite of numerous researches, still in a highly controversial state ... relatively very thin screens—0.01 cm of aluminum, or 0.002 cm of gold— completely scatter electrons in all directions .... This easy scattering is one of the many difficulties which make any exact theory of the absorption of β-rays difficult to establish'.[46] His change of mind had come about because of results obtained by William Wilson[47] from Manchester who had been set to work on the absorption problem by Rutherford. Indeed, Wilson's main conclusion (June 1909) overturned all previous views on this subject: an electron beam *guaranteed* to be monoenergetic does not remotely follow exponential behavior in absorption.

Already the opening lines of Wilson's article are memorable: 'The present work was undertaken with a view to establishing, if possible, the connection between the absorption and velocity of β-rays. So far no actual experiments have been performed on this subject ... '. The exponential law Eq. (8.1) had in fact largely been guesswork.

Wilson's method, classic by now, is described in any textbook on β-ray spectrography.[48] β-particles pass through a slit into a space where the action of a homogeneous magnetic field perpendicular to their velocity bends the rays into circular orbits. After traversing a semicircle they hit a screen with another slit in it. Those rays which can go through the slit have (approximately) homogeneous velocities. Their absorption is measured by covering the slit with metal foil of variable thickness; the ionization current due to the rays which traverse the foil is measured with an electroscope. The magnetic field can be adjusted so as to select in turn a variety of homogeneous velocities.

For *each* of a set of velocities ranging from 2.1 to 2.9×10^{10} cm/sec, Wilson found that the variation of absorption with filter thickness was not exponential but approximately linear. The next year, 1910, it became evident that this linear behavior had only a limited range of validity.[49] One general conclusion, first drawn in 1909 by Wilson, has stood up, however: when β-rays traverse matter they do not suffer 'sudden death' but lose velocity in some gradual and complicated manner.

It is not necessary for our purposes to pursue the history of the passage of electrons through matter in further detail. Rather we should return to our main subject, the primary β-spectrum, and see what happened after it was realized that the strategy: exponential absorption → monochromatic spectrum described in Section (d) had to be abandoned.

Wilson himself was the first to stress the need for revision. His paper of 1909 contains a section entitled 'Explanation of the exponential law found by various observers for the absorption of rays from radioactive substances'. He accepts the earlier experiments but draws a new conclusion from them: '... It is preferable to try to explain why various observers have found that the [β-] rays ... are absorbed according to an exponential law with the thickness of the matter traversed. *The fact that homogeneous rays are not absorbed according to an exponential law suggests that the rays from these substances are heterogeneous*'.[47]

Soddy referred to the phrase I italicized as a 'revolutionary conclusion', adding that further experiments were needed in order to make these results more definite.[46] One may wonder why Wilson did not go on to prove that the primary β-spectrum is continuous (I shall comment further on this in Section (h)). He probably did not take that step because he was too involved in absorption questions.

Rutherford must have had these developments in mind when, in October 1909, he wrote to Hahn: '*Don't* necessarily commit yourself to the view that each line corresponds to a distinct product ...' (his italics).[50] Hahn's reply shows that he was by no means swayed, however: 'It seems that there may be in some cases very soft β-rays of a uniform type in addition. But those Miss Meitner and myself could not detect in our electroscope. We therefore believe to have a proof that radioelements emit single and definite types of β-rays and very likely only one type for single products'.[51]

The letter by Hahn is dated 13 May 1910. Of course, he knew very well by then that the original Hahn–Meitner working hypothesis of a pure exponential absorption law for homogeneous velocities was incorrect. Nevertheless he wrote the above lines with renewed confidence because, meanwhile, he and his friend Otto von Baeyer from the Physics Institute of the University of Berlin had finished a first paper on an alternative method for the study of β-spectra which appeared to bear out the earlier conjecture of a unique primary β-ray energy.

(f) The second attack: magnetic separation, photographic detection

The von Baeyer–Hahn paper,[52] submitted in May 1910, begins by addressing a question which may meanwhile have occurred to the attentive reader of Chapter 4, Section (f). How can one insist on a discrete β-spectrum in view of the fact that Kaufmann, experimenting years earlier with the β-rays from a radium source, had found that these rays have a broad velocity spectrum which he in fact had used for the determination of the energy–velocity relation of electrons? That proves nothing about the primary spectrum of a *pure* β-emitter, Hahn and von Baeyer argued, because, first, Kaufmann's source actually consisted of several β-active elements in equilibrium, and, secondly, 'it is known that β-rays generate strong secondary rays when penetrating matter . . . Hardly anything is known about the velocity of these secondary rays'. That was the truth, but not the whole truth.

Hahn and von Baeyer, using pure β-emitters, followed Wilson's technique, described above, for the magnetic separation of velocities. Their detection method was a different one, however. Electrons of all velocities, after following semicircular trajectories (a different one for each velocity) are incident on a photographic plate. Then the blackening of the plate gives a record of the initial velocity spectrum. Their conclusion: 'The present investigation shows that, in the decay of radioactive substances, not only α-rays but also β-rays leave the radioactive atom with a velocity characteristic for the species in question. This lends new support to the hypothesis of Hahn and Meitner . . .'.

Thus the conjecture that a pure β-emitter generates monochromatic electrons was still alive in 1910. It died in April 1911, when Hahn and von Baeyer, together with Lise Meitner, submitted a sequel[53] to the previous paper. Their methods were the same as before. This time, however, they were forced to admit that the *effective* β-spectrum of a pure substance is *inhomogeneous*. Nevertheless, they still held open the possibility that this effective inhomogeneity is a secondary modification of initially monochromatic β-emission: 'The inhomogeneity of fast β-rays can have its origin in the fact that the rays were initially emitted by the radioactive substance with unequal velocities . . . It is more plausible to look for a secondary cause which render inhomogeneous the emitted homogeneous rays . . . the exponential law [in absorption] cannot be a criterion for the homogeneity of the rays, as was supposed by Hahn and Meitner in contrast to [the view of] other investigators'.

Hahn and Meitner's persistence is impressive, their admission of earlier mistakes forthright. The same frankness strikes me in reading Hahn's scientific autobiography: 'Our earlier opinions were beyond salvage now. It was impossible to assume a separate substance for each beta line. Our original explanation of the exponential absorption had been wrong because we had assumed that we actually had measured the absorption. What we had principally measured was the dispersion and the greater the distance of our preparations from the bottom of the electroscope, the more dispersion we had obtained. In increasing the distance, we had dispersed the weakest beta rays so much that they failed to register, and we had also slowed down the fastest beta rays, so that the average velocity of the non-homogeneous beta rays remained fairly constant over a short length of time . . . Though our opinion about the "absorption law" had been wrong, the work had considerably improved our techniques. We had learned how to produce different substances in thin layers and also how to handle them, especially those with short half-lives.'[54]

In 1911, it had therefore become clear that the issue of the β-spectrum was very complicated. In May of that year Rutherford wrote to Boltwood: 'You will have seen that Hahn is finding that the β-ray problem is more complex than first appeared. Geiger and Kovarik have been investigating the matter in another way and find that, while there is some order in the β-rays, there are some notable exceptions. The whole problem is very peculiar . . .'.[55] Indeed, in preparing a new edition of his book on radioactivity,[56] Rutherford had difficulties in giving a succinct overview of β-spectra. In October 1911 he wrote to Hahn: 'I am at present trying to write up the subject of beta rays for my new edition, and I find it the most difficult task in the book, as there has to be a good deal of compression to bring it within reasonable compass'.[57]

(g) Prenatal nuclear spectroscopy

Having beheld the disengagement from the presupposition that β-spectra are monochromatic, the reader may well expect the next item of business to be the discovery of the continuous β-spectrum. Not yet, dear reader, not quite yet. We should first pay attention to the years 1912–13, during which several experiments seemed to indicate that β-spectra consist not of one discrete line but of a set of such lines.

The first intimations of line spectra are found in the 1910 paper[52] by von Baeyer and Hahn: 'In all cases examined a clearly discontinuous spectrum was obtained'. During the next few years, the study of β-spectra by means of the magnetic separation/photographic detection method continued in several laboratories. All research groups reported complicated discrete spectra. Von Baeyer, Hahn, and Meitner extended their earlier investigations.[58,59] Important contributions were also made by Jean Danysz in Paris.[60] Rutherford, too, went to work on the problem. Together with Robinson, he studied the spectra of RaB and RaC and found 16 lines for RaB, 48 lines for RaC, the lines falling into seven classes of intensity.[61] But there were also difficulties in fitting all results by means of line spectra. Already in 1911, Hahn had written to Rutherford: 'RaE is the worst of all. We can only obtain a fairly broad band. We formerly thought that it was as narrow as the other bands, but that is not true. It looks as if secondary or such effects had a maximum influence on rays of a medium velocity like RaE'.[62] A few months later Hahn wrote him again: 'The trouble with the soft rays is very great and we do not feel sure that we can overpass [sic] the difficulties to obtain good lines'.[63] In a general discussion of the subject, Rutherford did note that continuous bands are sometimes observed. However, he tentatively concluded that, 'The continuous β-ray spectrum observed for uranium X [Th^{234}] and radium E [Bi^{210}] may be ultimately resolved in a number of lines'.[64] And so it came to pass that general physics textbooks of the period describe β-radioactivity as a discrete phenomenon and display elaborate tables of discrete spectral lines.[65]

Do these discontinuous spectra have anything to do with the real world? In order to answer this question I interrupt the historical account. It is indeed true that discrete lines are very often superimposed over the fundamental continuous β-spectrum. In general terms this comes about because the β-decay of a nucleus may lead either to the ground state of another nucleus or to one of its many excited states. Two examples: (1) in the β-decay of radium E: $Bi^{210} \rightarrow Po^{210}$, nearly 100 per cent of the decays go directly to the Po ground state; (2) in the case of radium A: $Pb^{214} \rightarrow Bi^{214}$, only 6 per cent of the β-decays go directly to the Bi ground state.[66]

Transitions from an excited nuclear state to the corresponding nuclear ground state are important sources for the appearance of discrete electron energies. The two main mechanisms are[67]: (1) internal conversion electrons. A transition excited state → ground state + photon, followed by photon + electron bound in the same atom → free electron with a discrete energy equal to the photon energy plus the (negative) binding energy of the ejected electron; (2) Auger electrons. An electronic level emptied by internal conversion is reoccupied by an electron from a higher level. The discrete energy thus liberated is transferred to still another electron in an outer level, which moves to the continuum in a radiationless transition.*

It is not necessary at this point to discuss some less important additional causes which generate discrete electron energies. Suffice it to say that the resulting line spectrum can be quite complicated,[66] and that these discrete lines give most valuable information on nuclear energy levels.

Let us now return to our history. None of the mechanisms just described were known in the days when Hahn, Meitner, von Baeyer, Danysz, Rutherford, and others found the first evidence for discrete lines. It is now clear that '... numerous erroneous disintegration schemes have been published as a result of the use of spectrographs using photographic recording'.[68] Be that as it may, what happened was that, in the quest for the fundamental β-spectrum, the physicists just mentioned unwittingly became the pioneers of nuclear spectroscopy!

The origins of the discrete electron energy spectrum did not begin to be understood until about 1920. What were some of the thoughts about this subject around 1912?

In October of that year, well after the discovery of the nucleus, the new edition of Rutherford's book[56] was ready for press. It came out in 1913. In this volume Rutherford expressed his belief in two distinct causes for radioactive instability, '... the instability of the central mass** and the instability of the electronic distribution. The former type of instability leads to the expulsion of an α-particle, the latter to the appearance of β- and γ-rays ... part of the surplus energy of a ring of [peripheral] electrons is released in the form of a high speed β-particle and part in the form of γ-rays, the division between the two forms of energy depending on factors which are not at present understood.'[69]

* The relative importance of these two mechanisms for creating discrete electron lines depends on the species (A, Z).
** That is, the nucleus, which had meanwhile been discovered, see the next chapter.

It should also be noted that, in 1912, Rutherford still considered it possible that the primary β-spectrum might in fact be monochromatic:[70] the electron was supposed to convert part of its energy E_0 into discrete γ-ray energies of which, he supposed there were two, E_1 and E_2. Next he assumed that the observed β-spectrum is not the manifestation of what happens in a single atom but rather that it is a statistical effect due to action by a large number of atoms. The statistical aspect arose, he thought, from the probabilities that p γ-rays of kind E_1 and q γ-rays of kind E_2 are produced when the initial energy E_0 of the electron is partially converted as it passes through many atoms. Thus he arrived at an effective electron spectrum consisting of a set of lines with respective energies $E_0-(pE_1+qE_2)$, p, q, integer, and found fair fits with the observed line spectrum for suitably chosen E_0, E_1, E_2, these last numbers being characteristic for specific radioactive materials. In regard to the case of RaE, 'where the γ-rays are very weak, it seems probable that the β-rays originate near the surface of the atom and consequently do not traverse the regions where penetrating γ-rays can be set up'.

Labyrinthine speculations such as these partially explain why all those excellent physicists who worked with the photographic detection method completely missed the continuous β-spectrum. It is indeed evident that, around 1912, prevailing prejudice still strongly favored a discrete spectrum possibly due to a monoenergetic primary energy. This may well have inhibited researchers from reading carefully between the lines. A more objective and important factor was, however, the ignorance at that time about the relation between the blackening of a photographic plate and the intensity of irradiation. This blackening depends on the energy and intensity of the radiation, on exposure time, and on many details of the development procedure. It will be recalled that systematic studies of this problem in places like the Kodak laboratories did not start until the early 1920s.[71] Note also that the photographic plate as a tool for intensity measurements of atomic spectral lines came into its own only in the mid-1920s, largely because of the work done by the Utrecht school.[72] In the years 1910–13 the photographic plate was therefore still an extremely primitive tool for intensity measurements.

As we shall see in the next section, April 1914 brought the first observation of the continuous spectrum, as reported by Chadwick, who had abandoned photographic detection methods. My search for comments in the subsequent literature on the failure to observe the continuous spectrum by photographic methods has yielded little. Later in 1914, Soddy wrote: 'Probably the photographic plate exaggerates the relative importance of the line as compared with the continuous spectrum'.[73] In the well-known textbook (published in 1930) by Rutherford, Chadwick, and Ellis, the only comment on the subject is: 'Chadwick showed that the prominence of these groups [of lines] was due chiefly to the ease with which the eye neglects background on a plate'.[74]

Indeed, Chadwick not only discovered the continuous spectrum by counter detection but (in the same paper) also gave a prescription for the photographic production of fake line spectra.

It would be improper to write of Hahn and Meitner's trials and errors without adding a brief word about their achievements. Jointly they discovered the so-called C″ β-emitters: actinium C″ (Tl^{207}), thorium C″ (Tl^{208}), and radium C″ (Tl^{210})[75]; and the long-lived isotope of protactinium (Pa^{231}). Hahn discovered mesothorium 1 and 2 (Ra^{228} and Ac^{228}). He was also the first to detect recoil in radioactive transformations; and to discover nuclear isomerism. Lise Meitner's distinguished career was not made any easier by the facts that she was a woman, and that she was Jewish. The best account of her life and work was written by her nephew, my late friend Otto Frisch.[40]* Many of her later papers deal with β- and γ-ray spectroscopy. She was among the first to verify the Klein–Nishina formula experimentally.[76]

In 1938, after collaborations which spanned three decades, the joint work of Hahn and Meitner came to a conclusion with an article, together with Fritz Strassman, on transuranic elements. Then Hitler and his friends intervened. Shortly thereafter Hahn, with Strassman, and Meitner, with Frisch, did their best-known work. Hahn and Strassmann, in Berlin, produced incontrovertible evidence for the formation of elements like barium and lanthanum when uranium is bombarded by neutrons. Meitner, in Stockholm, and Frisch, in Copenhagen (they composed their joint paper by telephone) followed with a paper** in which they interpreted these results as nuclear fission, a phrase they coined. It may be noted finally that neither Hahn nor Meitner participated in the nuclear weapons projects of the Second World War.

(h) The third attack: magnetic separation, detection by counters

James (later Sir James) Chadwick studied at Manchester. 'Chadwick's decision to proceed with physics was much influenced by Rutherford's presence even though he gave relatively few lectures to the physics students—with his robust personality he was much in demand for lecturing to the engineers who were notoriously difficult to control.'† In 1913 Chadwick was awarded an Exhibition of 1851 Senior Research Studentship. Under its terms he was obliged to carry out research in some place other than Manchester. He chose to go to Berlin, to work with Geiger at the Physikalisch Technische Reichsanstalt.

In a letter which Chadwick sent from there to Rutherford we get the first intimation that a major turning point in the history of particle physics was near. The letter, dated January 14, 1914, contains these lines: 'We [Geiger and Chadwick] wanted to count the β-particles in the various spectrum lines

* See also Ref. 38.
** For a compact useful documentary book on the discovery of fission see Ref. 77.
† For details on Chadwick's career see Ref. 78.

of RaB+C and then to do the scattering of the strongest swift groups. *I get photographs very quickly easily, but with the counter I can't even find the ghost of a line. There is probably some silly mistake somewhere*'.[79]

The lines which I have italicized contain two key statements. First, Chadwick had been able to obtain line spectra by photographic detection, like everyone else. Secondly, far from having made a silly mistake, he had found that these lines are vastly exaggerated in relative importance. Chadwick gave full details in a paper[80] submitted in April 1914.

On its opening page we find the first expression of concern about the meaning of β-ray line intensities as seen on the photographic plate: 'Since . . . the photographic action of β-rays with variable velocity is unknown, one does not obtain in this way reliable information about the intensity of the individual groups of rays. It is . . . also . . . hard to decide whether or not a continuous spectrum is superposed over the line spectrum'. Chadwick then described another detection method. Just as in the Wilson experiment described above,[47] electrons of fixed velocity are bent 180°, then pass through a slit. Thereafter, their intensity is measured by the discharge they cause in an electric potential maintained between a metal plate and a very clean needle with a sharp point.* Different velocities are selected by varying the magnetic field strength. Their source was a mixture of radium B (Pb^{214}) and radium C (Bi^{214}). The result: a continuous spectrum on which are superposed four, and only four, lines at the lower energy portion. The position of these lines coincided with some of the strong lines found by others in previous investigations. In addition, he made some tests which convinced him that the continuous energy distribution was not due to secondary scattering effects. Compare these findings with the earlier results of Rutherford and Robinson[61] for the same source: more than sixty discrete lines, no continuous spectrum!

In the same paper Chadwick also gave prescriptions for the production of fake lines: 'The difference [with the photographic method] could be explained by the circumstance that the photographic plate is extraordinarily sensitive for small changes in the intensity of the radiation'. For example, he irradiated a photographic plate for 100 minutes, then put a sheet of lead with a narrow slit on top of the plate and re-exposed for an additional 5 minutes. The resulting picture turned out to depend on the details of the development of the plate: already normal development generates a sharp line. Very slow development, terminated at a convenient time, made it even possible to obtain a nearly black line against a clear background!

And so he drew his final conclusion: the β-rays of RaB and RaC consist of a continuous spectrum. There is an additional line spectrum. With a few exceptions the lines have only a very small intensity.

* Alternatively a small ionization vessel of more conventional shape was also used. Wilson[47] could have done the same experiment if he had simply omitted all absorbing foils over the slit!

This was Chadwick's first major contribution. What happened to him next is recorded in a letter to Rutherford: 'I was in the middle of the experiments on the scattering of β-rays when the war broke out . . . Radioactivity is naturally not in a flowering condition here . . .'. The letter[81] was written on September 14, 1915. The sender's address: Barack 10, Engländerlager, Ruhleben. Chadwick had been interned in the stables of a racecourse near Spandau. Yet, under very primitive conditions, he kept working on physics. On March 31, 1917 he wrote[82] to Rutherford from the camp: 'I believe I wrote you before that we had a small lab. A space was granted for scientific work . . . you will see that I am not getting rusty for want of work . . .'.

While interned, Chadwick was able to receive such journals as *Nature*. He also was allowed to maintain some communication with Geiger. Otto Frisch told me the following story about physics in Ruhleben. An important piece of equipment secured by Chadwick was a Bunsen burner. Since he had no foot-operated bellows (a common tool at that time) for increasing the airflow in the burner, he asked Charles Drummond Ellis, a British army officer and fellow prisoner, to kneel on the floor and blow air through a tube into the burner. At one point, Chadwick, by force of habit, put his foot on Ellis' back, bore down, and shouted: 'Blow, Charlie, blow'.

Chadwick stayed in the camp for the duration of the war. Later in this book we shall catch up with him again, this time at the Cavendish.

References

1. E. Dorn, *Phys. Zeitschr.* **4**, 507, 1903.
2. W. Wien, *Phys. Zeitschr.* **4**, 624, 1903.
3. E. Rutherford, *Phil. Mag.* **10**, 193, 1905.
4. H. G. J. Moseley, *Proc. Roy. Soc. A* **87**, 230, 1912.
5. R. W. Gurney, *Proc. Roy. Soc. A* **109**, 540, 1925.
6. S. Kinoshita, S. Kikuchi, and Y. Hagimoto, *Jap. J. of Phys.* **4**, 49, 1926; S. Kikuchi, ibid., 143.
7. G. M. Caroe, *William Henry Bragg*, Cambridge Univ. Press 1978.
8. Cf. E. N. da C. Andrade, *Obit. notices Fell. Roy. Soc.*, **4**, 277, 1942–4; L. Bragg and G. M. Caroe, *Notes and Rec. Roy. Soc.* **17**, 169, 1962.
9. W. H. Bragg and R. Kleeman, *Phil. Mag.* **8**, 726, 1904.
10. See Ref. 9 and W. H. Bragg, *Phil. Mag.* **11**, 754, 1906.
11. W. H. Bragg, *Phil. Mag.* **11**, 617, 1906.
12. W. H. Bragg and R. Kleeman, *Phil. Mag.* **10**, 318, 1905.
13. B. B. Boltwood, *Am. J. Sci.* **24**, 370, 1907; **25**, 365, 493, 1908.
14. W. H. Bragg, *Phil. Mag.* **10**, 600, 1905.
15. F. Soddy, *Radioactivity and atomic theory*, p. 80, Ed. T. J. Trenn, Wiley, New York 1975.
16. J. Heilbron, *Arch. Hist. Ex. Sci.* **4**, 247, 1967, see esp. p. 255.
17. E. Rutherford, *Phil. Mag.* **11**, 166, 1906; **12**, 134, 1906.
18. C. G. Darwin, *Phil. Mag.* **23**, 901, 1912.

19. P. White and G. Millington, *Proc. Roy. Soc. A* **120**, 701, 1928.
20. W. H. Bragg, *Phil. Mag.* **8**, 719, 1904.
21. J. J. L. Chen and S. D. Warshaw, *Phys. Rev.* **84**, 355, 1951.
22. See e.g. W. T. Scott, *Rev. Mod. Phys.* **35**, 231, 1963, also for detailed references to the literature.
23. J. J. Thomson, *Proc. Cambr. Phil. Soc.* **13**, 49, 1904.
24. E. Rutherford, *Phil. Mag.* **10**, 193, 1905.
25. H. W. Schmidt, *Phys. Z.* **7**, 764, 1906; **8**, 361, 1907; *Jahrb. der Rad. u. Elek.* **5**, 451, 1908.
26. Ref. 15, p. 154.
27. A. S. Eve, *Rutherford*, p. 196, Cambridge Univ. Press 1939.
28. O. Hahn, *Proc. Roy. Soc. A* **76**, 115, 1905.
29. L. Badash, *Rutherford and Boltwood*, p. 81, Yale Univ. Press, New Haven, Conn. 1969.
30. O. Hahn, *A scientific autobiography*, transl. W. Ley, Scribner, New York 1968.
31. E. H. Berninger, *Otto Hahn*, Rowohlt, Hamburg 1974.
32. R. Spencer, *Biogr. Mem. Fell. Roy. Soc.* **16**, 279, 1970.
33. O. Hahn, letter to E. Rutherford, August 21, 1906, Cambridge University Manuscript Collection, on microfilm at the Niels Bohr Library, American Inst. of Phys., New York.
34. E. Rutherford, letter to O. Hahn, September 25, 1906, see further Ref. 33.
35. O. Hahn, letter to E. Rutherford, October 3, 1906, see further Ref. 33.
36. O. Hahn, letter to E. Rutherford, February 10, 1907, see further Ref. 33.
37. O. Hahn, letter to E. Rutherford, December 15, 1907, see further Ref. 33.
38. B. Karlik, *Almanach der Österreichischen Ak. der Wiss.* **119**, 345, 1969.
39. L. Meitner, *Phys. Zeitschr.* **7**, 588 (1906); **8**, 489, 1907.
40. For more details on Meitner's life and career, see O. R. Frisch, *Biogr. mem. Fell. Roy. Soc.* **16**, 405, 1970.
41. Ref. 30, pp. 54–5.
42. W. H. Bragg, *Trans. Roy. Soc. South Australia* **31**, 79, 1907.
43. O. Hahn and L. Meitner, *Phys. Zeitschr.* **9**, 321, 1908.
44. O. Hahn and L. Meitner, *Phys. Zeitschr.* **9**, 697, 1908.
45. O. Hahn and L. Meitner, *Phys. Zeitschr.* **10**, 741, 1909.
46. Ref. 15, p. 192.
47. W. Wilson, *Proc. Roy. Soc. A* **82**, 612, 1909; **84**, 141, 1910.
48. E.g. K. Siegbahn, *Alpha-, beta- and gamma-ray spectroscopy*, p. 87, North Holland, Amsterdam 1965.
49. J. A. Gray and W. Wilson, *Phil. Mag.* **20**, 870, 1910; J. A. Crowther, *Proc. Cambr. Phil. Soc.* **15**, 442, 1910.
50. E. Rutherford, letter to O. Hahn, October 24, 1909, see further Ref. 33.
51. O. Hahn, letter to E. Rutherford, May 13, 1910, see further Ref. 33.
52. O von Baeyer and O. Hahn, *Phys. Zeitschr.* **11**, 488, 1910.
53. O von Baeyer, O. Hahn, and L. Meitner, *Phys. Zeitschr.* **12**, 273, 1911.
54. Ref. 30, p. 57.
55. Ref. 29, p. 249.
56. E. Rutherford, *Radioactive substances and their radiations*, Cambridge Univ. Press 1913.
57. A. S. Eve, *Rutherford*, p. 207, Cambridge Univ. Press 1939.
58. O. Hahn and L. Meitner, *Phys. Zeitschr.* **12**, 378, 911).
59. O. von Baeyer, O. Hahn, and L. Meitner, *Phys. Zeitschr.* **13**, 264, 1912.

60. J. Danysz, *Comptes Rendus* **153**, 339, 1066, 1911; *Le radium* **9**, 1, 1912; *J. de Phys.* **3**, 949, 1913.
61. E. Rutherford and H. Robinson, *Phil. Mag.* **26**, 717, 1913.
62. O. Hahn, letter to E. Rutherford, January 11, 1911, see further Ref. 33.
63. O. Hahn, letter to E. Rutherford, May 8, 1911, see further Ref. 33.
64. Ref. 56, p. 256.
65. J. Müller and C. Pouillet, *Lehrbuch der Physik*, 10th Edn, Vol. 4, pp. 1272–4, Vieweg, Braunschweig 1914.
66. See e.g. M. Lederer, J. Hollander, and I. Perlman, *Table of isotopes*, Wiley, New York 1968.
67. Cf. Ref. 48, Chapters 16 and 25.
68. Ref. 48, p. 468.
69. Ref. 56, p. 622.
70. E. Rutherford, *Phil. Mag.* **6**, 453, 1912.
71. L. Silberstein, *Phil. Mag.* **44**, 956, 1922; **45**, 1062, 1923.
72. Cf. L. S. Ornstein, W. J. H. Moll, and H. C. Burger, *Objektive Spektralphotometrie*, Vieweg, Braunschweig 1932; see also Ref. 48, p. 409.
73. Ref. 15, p. 375; see also p. 376.
74. E. Rutherford, J. Chadwick, and C. D. Ellis, *Radiations from radioactive substances*, p. 399, Cambridge Univ. Press 1930.
75. Ref. 30, p. 57.
76. L. Meitner and H. H. Hupfeld, *Phys. Zeitschr.* **31**, 947, 1930; *Naturw.* **18**, 534, 1930.
77. H. H. Graetzer and D. L. Anderson, *The discovery of nuclear fission*, Van Nostrand, New York 1971.
78. H. Massey and N. Feather, *Biogr. Mem. Fell. Roy. Soc.* **22**, 11, 1976.
79. J. Chadwick, letter to E. Rutherford, January 14, 1914; see further Ref. 33.
80. J. Chadwick, *Verh. d. Deutsch. Phys. Ges.* **16**, 383, 1914.
81. J. Chadwick, letter to E. Rutherford, September 14, 1915; see further Ref. 33.
82. J. Chadwick, letter to E. Rutherford, March 31, 1917; see further Ref. 33.

9

Atomic structure and spectral lines

> Do not all fix'd Bodies, when heated beyond a certain degree, emit Light and shine; and is not this Emission perform'd by the vibrating motions of their parts?
>
> Newton, *Opticks*, Query 8

(a) Introduction

On 25 October 1957, Robert Oppenheimer and I took an early train to Washington D.C. We were on our way to the Great Hall of the National Academy of Sciences, where, that afternoon, President Eisenhower was to present the first Atoms for Peace Award to Niels Bohr. It was a festive event. James Killian, president of M.I.T., read the citation: 'You have given man the basis for a greater understanding of matter and energy. You have made contributions to the practical uses of this knowledge. You have exerted great moral force in behalf of the utilization of atomic energy for peaceful purposes. At your Institute at Copenhagen, which has served as an intellectual and spiritual center for scientists, you have given scholars from all parts of the world an opportunity to extend man's knowledge of nuclear phenomena'. The President spoke next, then Bohr responded. Afterwards I had a chance to congratulate Bohr and to tell him how much I looked forward to his forthcoming visit to the Institute for Advanced Study, of which I was then a faculty member.

In December 1957, Bohr arrived in Princeton for a visit of several months. It was to be the last of his stays at the Institute,* and also the last period during which I had the opportunity of long discussions with him about physics. I recall an evening when we talked till quite late about a paper on Rutherford which Bohr had to present that coming autumn before the Physical Society in London.** As he had done so often before, Bohr spoke to me of his

* Bohr was a visiting member of the Institute during the following periods: the academic year 1938–9, February–June 1948, February–May 1950, September–December 1954, December 1957–February 1958.

** This Rutherford Memorial Lecture was given in November 1958 but was not published[1,2] until 1961.

veneration for Rutherford. 'To me he had almost been as a second father.'[1,2] (The Bohrs named one of their sons after him.) It was only natural that our discussion would turn to Bohr's paper of 1913 on the hydrogen atom. I said to him how that work reminded me of the current need for understanding the mass spectrum of the new particles which had been discovered during the past ten years (a need still undiminished) and wondered out loud how, before 1913, he must have been puzzled for years about the significance of the Balmer formula for the spectral frequencies of hydrogen. (This formula is found later in this chapter, see Eq. (9.1).) Oh no, Bohr said, he had not been aware of Balmer until shortly before his own work on the hydrogen atom. In fact, he added, everything fell into place for him as soon as he heard of that formula.

I was amazed.* While not very familiar at that time with history and dates and such things, I was aware that Balmer's formula had been known quite some time before 1913. Now, much later, I am somewhat puzzled by Bohr's statement. The formula was proposed in 1885, long before Bohr did his work. By 1913, it was not at all an esoteric proposal known to just a few cognoscenti. In 1897, Lord Rayleigh wrote of 'the remarkable law of Balmer'.[5] In 1901, Sutherland wrote from Melbourne: 'The great heartening to theoretical spectroscopic students undoubtedly came from Balmer's discovery of his formula for hydrogen . . .'.[6] Balmer's work was discussed[7] during the major international physics congress in Paris, in 1900; in meetings of learned academies in Paris[8] and Vienna;[9] in papers from the United States.[10,11] Walter Ritz, of the famous combination principle of spectral lines, frequently mentioned the formula in his publications.[12] Arthur Schuster recorded it in his article 'Spectroscopy' for the eleventh edition of the Encyclopaedia Brittannica, which came out in 1910. It is hard to believe that a result so well known would never have come Bohr's way prior to 1912.** It is more probable that he may have heard of it early but without registering it as relevant for his own thinking, and then have forgotten all about it—a far from uncommon phenomenon.

My telling this tale of Bohr and Balmer is perhaps of some mild anecdotal interest in that it marks a crucial turning point in one man's life. It is of far greater importance, however, as a reminder of how much Bohr owed to nineteenth-century physics, when, in 1913, by reading spectra, he laid 'the basis for a greater understanding of matter', being the first to demonstrate unequivocally that the hydrogen atom contains one and only one electron. At that time he drew not only on Balmer's excellent phenomenology, but also, indirectly, on those nineteenth-century experiments which had confirmed Balmer's formula to considerable accuracy. Thus the present chapter, while steering towards twentieth-century developments, is unlike all previous ones

* I found out only later that Bohr had expressed himself similarly to others as well.[3,4]
** This opinion is also expressed by Heilbron and Kuhn.[13]

and all those that follow in that its content is rooted in detailed knowledge of a much earlier vintage. The amount of information on spectra available in 1900 was indeed quite substantial. In that year, Heinrich Kayser completed an 800-page book, the first volume of his 'Handbuch der Spectroscopie' (which, of course, contains the Balmer formula). By the time Bohr published his work, five more volumes had been made ready, the whole series being over 5000 pages long.[14]

It would be out of place to devote much space to these earlier developments, but ill-advised to ignore them completely. Where to begin? Awareness of spectra predates recorded history—primitive man must have worshiped the rainbow. Aristotle produced a theory for this phenomenon, elaborated in learned discourses during medieval times.[15] Descartes, later Newton, grasped the true origins of the rainbow effect. Though fascinating, none of that is pertinent to our purposes. We come closer when we recall that Newton was the founder of spectral analysis, and, as the epigraph to this chapter shows, that he was wondering about the dynamical origin of spectra. However, his equipment did not have sufficient resolution to detect spectral lines. The earliest observations of such lines, first in absorption, then in emission, dates from the early years of the nineteenth century. So do the discoveries that the visible spectrum extends into the infrared and the ultraviolet. We are now getting closer to the appropriate starting point for a brief look backward at spectroscopy (to be given in Section (b)): the mid to late 1850s, when spectral frequencies were first measured with some accuracy.

At that very time, spectra showed the way to an enormous extension of man's grasp of the universe. In 1859, Kirchhoff found that there is sodium in the sun,[16] a discovery which marks the beginning of a new branch of astrophysics: stellar spectroscopy.* While this development does not bear directly on our main subject, it serves all the same to illustrate how abrupt and, it is no exaggeration to say, how dramatic the advances in the physics of spectra were in those years. Thus, as late as 1835, the influential French philosopher Auguste Comte had written[17] in the nineteenth lesson of his 'Cours de la Philosophie Positive': 'On the subject of stars, all investigations which are not ultimately reducible to simple visual observations are ... necessarily denied to us. While we can conceive of the possibility of determining their shapes, their sizes, and their motions, we shall never be able by any means to study their chemical composition or their mineralogical structure ... Our knowledge concerning their gaseous envelopes is necessarily limited to their existence, size ... and refractive power, we shall not at all be able to determine their chemical composition or even their density ... I regard any notion concerning the true mean temperature of the various stars as forever

* It is the second oldest branch, predated only by photometry, which became a quantitative science in the eighteenth century, mainly through the contributions of Pierre Bouguer and Johann Heinrich Lambert.

denied to us.' Spectral analysis is but one branch of science which has taught philosophers caution.

This chapter is laid out in four stages. The first (Section (b)) concerns developments prior to the discovery of the electron and discusses the following topics: the early years of quantitative spectral analysis; Balmer's and other's spectral formulae; early atomic models; and speculations on the evolution of matter. The second (Section (c)) deals with the period from the discovery of the electron to the discovery of the nucleus, and covers the early electron models with their thousands of electrons per atom, criteria for model building during that period, and J. J. Thomson's plum pudding model. Section (d) deals with Rutherford's discovery of the nucleus, Section (e) with Bohr's realization that it takes the quantum of action to stabilize the atom.

A word on the title of this chapter. It is identical with the title of Sommerfeld's masterwork, often referred to in earlier years as the bible of spectroscopists.[18] In Sommerfeld's own words, his book is devoted to 'the language of spectra . . . a true atomic music of the spheres'.[19] For many years it was the only comprehensive text available to those desirous of learning the techniques and results of atomic physics. It is also of considerable interest in the history of the subject. Laying side by side the first four editions* of *Atombau und Spektrallinien*, one gains vivid insights into the development of a subject in rapid flux during the years just prior to quantum mechanics.

This chapter and Sommerfeld's book have only the title in common. For Sommerfeld, Bohr's quantum theory of the atom is the point of departure, here it is a point of culmination. I could not possibly do better than advise the reader interested in the further evolution of the Bohr theory to enjoy one or, preferably, several editions of Sommerfeld's wonderful opus. This will also serve to remind him, if at all necessary, that the brevity with which Bohr's work will be discussed below is disproportionate to its profound significance and subsequent influence.

(b) Pre-electron preludes

1. Spectral analysis. 'To my knowledge, I was the first who, in 1853, observed the spectrum of hydrogen.' So wrote Anders Jonas Ångström[20] in 1872 about his earlier experiments[21] in which he had studied spark spectra of various gases enclosed in a glass tube. These first observations had revealed three emission lines, one in the red, one in the blue-green, one in the violet; a fourth line, also in the violet, was found soon thereafter. These are the first four

* These appeared in 1919, 1921, 1922, and 1924, respectively. To this day four more editions have appeared. The first edition of a 'Wellenmechanischer Ergänzungsband' (the subtitle of Volume 2) came out in 1929. Three further editions of this second volume have seen the light.

lines of what came to be known as the Balmer series (a name which, for ease, I shall use from now on).

Old acquaintances reappear when we look for the earliest quantitative measurements of Balmer frequencies: Geissler's tube, Rühmkorff's coil, Plücker at the helm. In 1859 Plücker studied the discharge through a Geissler tube filled with hydrogen.[22] In the following table, the first column gives the names, still in use, introduced by Plücker for the first three Balmer lines. The second column gives his results for the respective wavelengths, the third Ångström's results[23] of 1868 which were used by Balmer in guessing his formula (more about that shortly), the fourth gives the modern values.* It needs to be remembered that, more than a century ago, one had spectral results good to one part in ten thousand.

	Plücker	Ångström	Modern
H_α	6533	6562.1	6562.8
H_β	4843	4860.7	4861.3
H_γ	4339	4340.1	4340.5

Already the year before, Plücker had speculated[25] that spectra are unambiguous visiting cards for the gases which emit them: 'Undoubtedly the spectrum determines the constitution of the gas or vapor in the [Geissler] tube . . . we encounter here, to coin a phrase, a kind of microchemistry'. The definitive proof of this statement is associated first and foremost with two names: Gustav Kirchhoff, the grandfather of the quantum theory, and Robert Bunsen, his chemist colleague, both from Heidelberg.

It began with Kirchhoff's paper[16] of October 1859 in which he came upon an 'unerwartete Aufschluss' (unexpected explanation): some of the dark Fraunhofer lines in the solar spectrum are due to sodium. He arrived at this conclusion by interposing a flame containing kitchen salt between the solar spectrum and his detector. 'If the sunlight is sufficiently damped, then two luminous lines appeared at the position of two dark [solar] D-lines; if the solar intensity surpassed a certain amount, then the two dark D-lines appeared much more pronounced . . . the dark D-lines lead one to conclude that there is sodium in the solar atmosphere'. Six weeks later[26] he gave the theoretical interpretation of his observations on the D-lines. The effect follows from what is now known as Kirchhoff's law of blackbody radiation, according to which (I state it somewhat loosely) the ratio of emissive to absorptive power of a body in thermal equilibrium with radiation is a universal function of frequency ν and temperature T. 'It is a highly important task to find this function', Kirchhoff wrote. How true. The function in question is proportional to the

* Wavelengths are expressed in Ångström (10^{-10} m) units. The modern values are taken from Ref. 24. In current terminology, the respective transitions are $n^2D \rightarrow Z2P$, $n = 3, 4, 5$. All fine and hyperfine structure effects are suppressed.

spectral density $\rho(\nu, T)$ of blackbody radiation (see Chapter 7). Forty years later Planck would decode $\rho(\nu, T)$, thereby founding the quantum theory. Incidentally, Planck succeeded Kirchhoff as professor of physics in Berlin (where the latter had moved in 1875).

Next came Kirchhoff and Bunsen's collaboration in which they founded spectral analysis. 'They had the great advantage over their predecessors that they could use Bunsen's non-luminous burner, which freed them of the extraordinarily disturbing spectrum of a luminous flame.'[27] The strength of their work lies primarily in the systematic way in which they eliminated spurious influences. They heated small amounts of various metals and salts in flames of varying constitution and of varying temperatures. They compared flame spectra with spark spectra, using (what else?) a Rühmkorff coil. The main conclusions of these 'comprehensive and time consuming investigations' can be summarized in the following three statements.[28]

(1) 'In spectrum analysis ... the colored lines appear unaffected by ... external influences and unchanged by the intervention of other materials. The positions occupied [by the lines] in the spectrum determine a chemical property of a similar unchangeable and fundamental nature as the atomic weight ... and they can be determined with an almost astronomical accuracy. What gives the spectrum-analytic method a quite special significance is the circumstance that it extends in an almost unlimited way the limits imposed up till now on the chemical characterization of matter.'

This statement is of course true in essence but not in detail. Already in 1861, the authors qualified it: 'It is conceivable that a chemical compound always shows other lines than the elements of which it consists'.[29] It became clear in 1862 that compounds show one spectrum, while the spectra of the composing elements appear when dissociation takes place upon raising the temperature.[30]

These results on compounds were on the whole readily accepted. However, a related paper[31], by Plücker and Hittorf in 1865 caused much commotion. It is entitled 'On the spectra of ignited gases and vapours with especial regard to the different spectra of the same elementary gaseous substance'. Their principal (italicized) finding: 'There is a certain number of elementary substances which, when differently heated, furnish two kinds of spectra of a quite different character, not having any line or any band in common'.

It took many years and much labor before it became clear that the principal issue was: what are elementary substances? Vacuum techniques had done much for the study of spectra of pure substances in isolation. Nevertheless, there remained problems of unanticipated impurities caused, for example, by sputtering electrodes. There were also more fundamental problems of principle. Oxygen, for example, is an 'elementary gaseous substance' in the sense of Plücker and Hittorf. But spectroscopically it could produce, depending on circumstances, lines and bands corresponding to atomic, diatomic, triatomic, and ionic spectra. The unraveling of these problems will not be discussed here.

(2) Kirchhoff and Bunsen's second statement: 'Spectrum analysis might be no less important for the discovery of elements that have not yet been found'. They themselves were the first to show the fruitfulness of this idea, by announcing the discovery of caesium in their first joint paper,[28] and of rubidium in the second one.[29] Before the nineteenth century was over, spectral methods had played crucial roles in the identification of ten more elements: thallium, indium, gallium, scandium, germanium, and the five stable noble gases, splendid examples of how great advances in physics are possible in the absence of first principles. In all those years, there was not a clue to the structure of the atom and, therefore, to the origin of spectra. Yet the insight that an atomic spectrum, even when incompletely known, is a unique label for an element was sufficient for making major progress.

(3) Their third statement, perhaps the most stirring one: '[Spectrum analysis] opens ... the chemical exploration of a domain which up till now has been completely closed ... It is plausible that [this technique] is also applicable to the solar atmosphere and the brighter fixed stars'.

Already the 1860s showed the extraordinary fertility of this idea. Among the pioneers of stellar spectroscopy we find Lewis Morris Rutherfurd, the gentleman physicist from Manhattan's Lower East Side,* and William Huggins, the gentleman physicist from Tulse Hill, England, who in 1864 discovered spectroscopically that a number of nebulae are luminous gas glouds.[32] Also the first comet spectra were discovered in that decade. Maxwell foresaw how much stellar spectroscopy could teach when he wrote,[33] in 1875: 'The discovery of particular lines in a celestial spectrum which do not coincide with any line in a terrestrial spectrum does not much weaken the general argument, but rather indicates that a general substance exists in the heavenly body not yet detected by chemists on earth, or that the temperature of the heavenly body is such that some substance, indecomposable by our methods, is there split up into components unknown to us in their separate state'. That, however, does not cover all that the heavenly spectra would reveal. Take the case of helium. Analyzing his data on solar prominences, Joseph Norman Lockyer, the noted astrophysicist and founder of the Journal *Nature*, observed[34] a mysterious yellow line, which he named D_3. Soon he became convinced that this line belonged to a new element which he called helium. The matter remained in suspense, however, until William Ramsay's confirmation, in 1895, from terrestrial data. Take, on the other hand, the case of nebulium. Already in 1864, Huggins had found lines in the spectrum of nebulae 'indicating the presence of nitrogen, hydrogen, and a substance unknown'.[32] This last material was assumed to be a new element, nebulium. For more than sixty years references to this otherwise elusive new substance are found in the literature. For example, Rutherford mentioned it in his 1921 Bakerian lecture.[35] The

* Then the site of the Stuyvesant Estate. The independently wealthy Rutherfurd had married into the Stuyvesant family and had his own observatory constructed on their estate, near what is now Second Avenue and 11th Street.

nebulium lines were finally identified at Caltech, in 1927, by young Ira Bowen, later director of the Mount Wilson and Palomar Observatories. They are due to transitions from metastable states of oxygen and nitrogen.[36] These lines had never been seen under terrestrial conditions, where pressures are such that these states more readily lose energy through collisions of the second kind or collisions with walls.* Not even Maxwell could have divined all the lovely things one may see in huge masses of gas at very low pressures, and not hemmed in by walls.

As a final comment on the early spectroscopic experiments I stress the utter simplicity of the tools employed. All Kirchhoff and Bunsen used was a Bunsen burner, a platinum wire with a ringlet at the end for holding the material to be examined, a sulfur dioxide prism, and a few small telescopes, mirrors, and scales. Those were the days prior to Rowland's gratings, which appeared around 1890, and so much improved the observational accuracy, as we saw from Zeeman's experiences (Chapter 4).

The great strength of Kirchhoff and Bunsen's work lies less in the novelty or surprises of the idea that spectra uniquely label elements than in the firmness of their proofs. (Recall for example Plücker's pertinent remark of 1858 about microchemistry, mentioned earlier.) It is therefore not surprising that the Kirchhoff–Bunsen papers were subject not only to acclaim but also to widespread priority dispute.[37] In 1862, Kirchhoff responded to these allegations with dignity and firmness in a paper, undoubtedly the first, on the history of spectral analysis.[38] He quoted William Herschel's words of 1827: 'The colours ... communicated by the different bases to the flame afford, in many cases, a ready and neat way of detecting extremely minute quantities of them'; William Talbot's of 1834: 'I hesitate not to say that optical analysis can distinguish the minutest portions of ... substances from each other with as much certainty, if not more, than any other known method'; and cited related early statements by others as well.

He insisted, nevertheless, and I think rightly so, that not until the work by Bunsen and himself had spectral analysis been given a firm basis—an appropriate comment with which to conclude this ever so brief sketch on spectra in the nineteenth century.

2. The Balmer formula. This is (in modern notation**) the formula for the wave numbers ν of the discrete atomic hydrogen spectrum which Balmer found[39]

$$\nu = R\left(\frac{1}{n^2} - \frac{1}{m^2}\right) \qquad n = 1, 2, 3, \ldots, m > n \text{ and integer.} \qquad (9.1)$$

* Another hypothesized stellar element, coronium, turned out to be very highly ionized iron.

** Balmer's own version is $\lambda(=\nu^{-1}) = hm^2/(m^2 - n^2)$ cm, where $h = n^2/R = 4/R$. He found that $h = 364.56 \times 10^{-7}$ cm.

This is the value of R (the Rydberg–Ritz constant) in the formula which Balmer found

$$R = 109\,721\ \text{cm}^{-1}. \tag{9.2}$$

This (to my knowledge) is the best value obtained in a dye laser experiment,[40] a century later, for the constant R, first calculated by Balmer for the formula which Balmer found:

$$R = 109\,737.315\,21\,(11)\ \text{cm}^{-1} \tag{9.3}$$

showing that Balmer was correct to typical late 1880s accuracy: about one part in ten thousand.

Searches for patterns or formulae for spectral lines began in the late 1860s.[41,42] Among the pioneers looking for relations between frequencies we encounter Stoney once again who, in those years, seems to turn up wherever physics reaches its frontiers.* Stoney must in fact be considered as a direct precursor of Balmer, for the following reason. I mentioned earlier Ångström's 1868 values for the wavelengths of the first three Balmer lines. Actually he had also measured[23] the fourth line, H_δ:

$$H_\delta\text{: } 4101.2\ \text{Å}.$$

In 1871, Stoney noted[43] the following ratios of wavelengths

$$H_\alpha : H_\beta : H_\delta = \frac{1}{20} : \frac{1}{27} : \frac{1}{32} \tag{9.4}$$

which represent, in his words, 'the 32nd, 27th, and 20th harmonics of a fundamental vibration'. These are, on the dot, the ratios which follow from Balmer's formula (9.1) with $n = 2$, $m = 3, 4, 6$!

As to other harmonics, 'the 19th, 21st, 22nd etc. harmonics are not found . . . possibly the missing harmonics will be found when [the positions of other lines] shall have been sufficiently accurately mapped down'. Possibly also, Stoney speculated, 'there may be several distinct motions in each molecule'. (As was often the case in those days, molecule here means atom.) He made one subsequent attempt at searching for harmonics in other spectra,[44] then left the subject alone. His observation also led to a futile search for harmonic ratios by various other physicists.[42]

Why did Stoney not include H_γ in his considerations? He does not quite say,[44a] but one can guess! It may be that, if for no other reason, this omission made him miss the Balmer formula.

It is clear from Balmer's celebrated first communication of 1885 on the hydrogen spectrum[39] that he, too, had reflected on the question of harmonics. 'Hydrogen, the atomic weight of which is by far the smallest of all substances

* See Chapter 4.

known to date ... seems more qualified than any other body to open new vistas in the investigation about the nature and properties of matter. In particular, the wavelengths of the first four hydrogen lines excite and arrest attention ... One believed it possible to interpret the vibrations of the individual spectral lines of a material as overtones of, so to say, one specific keynote. However, all attempts to find such a keynote for hydrogen, for example, have not turned out satisfactorily ... Nevertheless the idea suggested itself that there should be a simple formula.'* Then Balmer goes on to tell what experimental information he had used to obtain his formula: nothing more, nothing less than Ångström's data about the four mentioned lines. This is also confirmed by a story which Eduard Hagenbach-Bischoff, professor of physics at the University of Basel and a friend of Balmer, once told to his son.[47]

Two comments. In his first paper,[39] Balmer is so sure of himself that he compliments Ångström: '[My results are] a brilliant testimony to the great conscientiousness and care with which Ångström must have proceeded'. Secondly, Balmer predicted an infinity of lines, not only those for what we call the Balmer series, $n = 2$, but also those for all other n! It is a piece of ***chutzpah*** only matched by the many and less successful ways with which various particle physicists are wont to predict new particles in the second half of the twentieth century. There is nothing in Balmer's earlier œuvre that would presage such exuberance, for the simple reason that, though nearly sixty years old, he had never before published any research paper in physics.

Balmer, born in 1825, studied mathematics in Karlsruhe and Berlin. In 1849 he received his Ph.D. in Basel, on a thesis 'On cycloids'. Some time later he obtained a position, kept till retirement, as teacher at a girl's school in Basel. In 1865 he presented his Habilitationsschrift** at the university there. His subject: 'The prophet Ezekiel's vision of the temple, clearly portrayed and architecturally explained', a piece of biblical geometry (undoubtedly *Ezekiel* 40–3). He was Privatdozent until 1890 and died in Basel in 1898. As said, the paper just mentioned was his first on a physics subject. Also in 1885, he wrote an article on 'Health, a word to the healthy and the sick'. He was 'neither an inspired mathematician nor a subtle experimentalist, [but rather] an architect ... [To him] the whole world, nature and art, was a grand unified harmony, and it was his aim in life to grasp these harmonic relations numerically.'[47]

In his first note,[39] Balmer related how he went to tell his first results to Hagenbach, who informed him that actually more lines were known, due to work by Huggins and by Hermann Vogel from Potsdam (the discoverer of spectroscopic binary stars, another recent novelty). In his second note,[46] Balmer

* Unfortunately, these and other colorful phrases are missing from the paper[45] submitted to the *Annalen der Physik* later in January in which Balmer summarized the content of his first[39] and second[46] communications to the Naturforschende Gesellschaft in Basel.
** An original paper required for obtaining the position of Privatdozent.

compared these data with his formula, for $n = 2$, $m = 5$–16, and found 'Übereinstimmung die im höchsten Grade überraschen muss' (agreement which must surprise to the highest degree).* In his third and last physics paper,[49] written at the age of seventy-two, a year before his death, he attempted to find spectral formulae for elements other than hydrogen. Those results have not survived.

Not only Balmer, but also numerous other physicists were in hot and vain pursuit of such more general formulae, for line as well as band spectra. A number of these were presented and discussed by Rydberg at the 1900 Paris Conference.[7] Such attempts continued until 1913, when Bohr interpreted the Balmer formula. We have in fact a precise record of all spectral formulae in captivity in April 1913, when Bohr's paper was submitted. One month later, Konen completed a textbook on spectroscopy, in which he recorded all spectral formulae which had been confronted with data.[50] There are no less than twelve entries on his list. These efforts are nearly all forgotten now. Two points should be mentioned however.

(*a*) In 1889, Rydberg submitted to the Royal Swedish Academy of Science a memoir[51] which contains the spectral formula

$$\nu = \nu_0 - \frac{R}{(m+\mu)^2} \tag{9.4}$$

meant to describe all series of all atomic line spectra. Here m runs through the positive integers, R is supposed to be universal for all series (whence the name Rydberg constant), $0 < \mu < 1$. ν_0 and μ are adjusted for each series separately. It appears[52] that Rydberg first presented his formula at a meeting in Lund in 1888. The Balmer formula is a special case of Eq. (9.4). At the Paris Conference, Rydberg stated that he had found his formula already in 1885, unaware of Balmer's work.[53] During the Rydberg centennial conference, Pauli said: 'I think one has to admit that Rydberg's speculations were sometimes rather wild, but on the other hand they were always controlled again by his study of the empirical material'.[54]

(*b*) Subsequently, Ritz produced a formula more general than Rydberg's, which I shall not reproduce. Suffice it to say that it again contains the universal R (whence the name Rydberg–Ritz constant), that of course it again contains Eq. (9.1) as a special case, and that it again has the form of the difference between two terms. This difference structure led Ritz to his combination principle: 'By additive or subtractive combination, be it of the series formulae, be it of the constants in these formulae, one obtains formulae which permit the complete calculation of certain newly discovered lines from those known earlier'.[55] This formulation, too much tied to spectral formulae, is not felicitous. Nevertheless, the principle is of profound significance. We now state it more readily as follows: The wave number of any spectral line can be represented

* Hagenbach also published a short note on this subject.[48]

as the difference between two terms, such that each of these terms represents an atomic energy level. Ritz's principle was absolutely crucial to Bohr in his formulation of the quantum theory of spectra.

3. Pre-electron models. 'When the theory of which we have the first instalment in Clausius and Maxwell's work, is complete, we are brought face to face with a superlatively grand question, What is the inner mechanism of the atom? In the answer to this question we must find the explanation not only of the atomic elasticity, by which the atom is a chronometric vibrator according to Stokes' discovery but of chemical affinity and of the differences of quality of different chemical elements, at present a mere mystery in science.' Thus W. Thomson (Kelvin) in his inaugural presidential address of August 1871 before the British Association.[56] There was much to look forward to at that time. The periodic table of elements, which encodes regularities in the family of atomic species, had been discovered quite recently. Spectroscopy was in rapid ascendant. The work on gases by Clausius and Maxwell had added to the conviction that atoms are real. Boltzmann had begun his struggles with thermodynamic probability and was to enunciate his H-theorem one year later. Thomas Young, Loschmidt, Stoney, and Kelvin himself, had already made theoretical estimates of the sizes of molecules, and Maxwell and van der Waals were soon to do the same. All these masters took the reality of atoms as their point of departure, not overly worried by numerous voices, some influential as well, which, speaking in opposition to such realism, cautioned that atoms were at best nothing more than convenient coding and counting devices.*

The question which Thomson asked in his address evidently went beyond the reality of atoms. He was raising the ancient Epicurean proposition of their structure, giving moreover his motivation for doing so. The atom, he went on to declare, is 'a piece of matter with shape, motion and laws of action, intelligible subjects of scientific investigation'.[56] Already in 1867 he had taken issue with 'the monstrous assumption of infinitely strong and infinitely rigid pieces of matter, the existence of which is asserted as a probable hypothesis by some of the greatest modern chemists in their rashly worded introductory statements'.[60]

For one reason or another, the issue of atomic structure had been cropping up time and again, ever since, early in the nineteenth century, chemistry began to develop into a science. In 1815, William Prout claimed to have shown that the specific gravities of atomic species can be expressed as integral multiples of a fundamental unit,[61] and had gone on to surmise that this fundamental unit may be identified with the specific gravity of hydrogen.[62] From this speculation, controversial though it became in later years, it is but a small step to imagine that where there is a subunit there must be structure. For example, as organic chemistry began to flourish and homologous series of

* For more details on atomic reality in the nineteenth century, see Refs. 57, 58, 59.

compounds differing in composition by a number of CH_2 groups were discovered, the supposition was raised that the elements themselves were, in some sense, a homologous series as well.[63] More generally, as the number of known elements increased, so did the justified unease, also reflected in Kelvin's words, that elements could not all be that elementary.

However, long before the discovery of the electron made the compositeness of atoms explicit, the clearest signals that structure was called for came from spectra. Already in 1852, Stokes, in the important memoir[64] 'On the change of refrangibility of light' (to which Kelvin referred) had written: 'In all probability ... the molecular vibrations by which ... light is produced are not vibrations in which the molecules move among one another, but vibrations among the constituent parts of the molecules themselves, performed by virtue of the internal forces which hold the parts of the molecules together'. Please note once again that, at that time, the term 'molecule' often meant what we call 'atom'. Kelvin certainly thought (as the above quotation shows) that Stokes meant to include 'atoms'. So, I believe, did Stoney in 1868: 'How wonderfully regular the internal motions of the molecules are, and at the same time how complex, appears to be revealed to us by the fixity of the rays of the spectrum in each gas and by their number ... '.[65] Maxwell thought of atoms when, in 1875, he laid down three 'conditions which must be satisfied by an atom ... permanence in magnitude, capability of internal motions or vibration, and a sufficient amount of possible characteristics to account for the difference between atoms of different kinds'.[33] (His reference to 'permanence in magnitude' reflects once again his belief in unbreakable atoms, discussed in Chapter 4, Section (b).)

In the same year he stated these desiderata, Maxwell, most thoroughly versed in kinetic theory, reasoned[66] that atomic model building would be a very difficult task: 'The spectroscope tells us that some molecules can execute a great many different kinds of vibrations. They must therefore be systems of a very considerable degree of complexity, having far more than six variables [the number characteristic for a rigid body] ... every additional variable increases the specific heat ... every additional degree of complexity which we attribute to the molecule can only increase the difficulty of reconciling the observed with the calculated value of the specific heat. I have now put before you what I consider the greatest difficulty yet encountered by the molecular theory'. Recall that, according to the classical equipartition theorem, an internal atomic degree of freedom corresponding (for example) to harmonic binding contributes an amount k (the Boltzmann constant) to the specific heat (at constant volume). Maxwell was saying that, thermodynamically speaking, there was not enough room for such additional variables. The resolution of this paradox belongs to the twentieth century, quantum theory being a necessary prerequisite.*

* I have discussed elsewhere at more length the nineteenth-century problems concerning specific heats.[67]

Nineteenth-century physicists did not just argue that the atom had to be structured. They also endeavored to divine what that structure had to be. Among their early attempts, one encounters quaint mechanical contraptions, largely made in Britain, consisting of strings, pulleys, springs, etc.[68,69] perhaps best characterized as atomic versions of Searle's celebrated *Punch* cartoons.* Then there were what may be called acoustical models, exemplified by Stoney's work mentioned earlier. These models, growing in sophistication as it became clear that simple harmonic ratios for spectral frequencies would not do, reached such levels of complexity that, as Maxwell put it, 'we have no right to expect any definite numerical relations among the wavelengths of the bright lines of a gas'.[33] Out of these considerations grew an 'inverse problem': what vibrates in such a manner as to provide a model for the actual frequencies in spectra? The study of this question (by Ritz, for example) did not illuminate the physics question asked, but led to significant contributions to the theory of partial differential equations.

The last and most discussed of the pre-electron atom models was William Thomson's vortex atom, of which Maxwell (though not uncritical of it) said: '[It] satisfies more of the conditions than any atom hitherto imagined'.[33]

The inspiration in this case came from hydrodynamics. In 1858, Helmholtz had published a fundamental paper[71] on *Wirbelbewegung*, vortex motion: 'On the Integrals of the Hydrodynamic Equations which express Vortex Motion'. An English translation, written by Peter Guthrie Tait from Glasgow, was published some time later.[72] In mid-January 1867, Thomson visited Tait in Glasgow, who showed him several ingenious experiments with vortex rings of smoke in air designed to illustrate some of Helmholtz's points. 'With the lightning rapidity of thought and the dominant passion for physical interpretation characteristic of his mind, Thomson, by a flash of inspiration, perceived in the smoke ring a dynamical model of the atom of matter.'[73] In a letter dated January 22, 1867, he wrote to Helmholtz: 'A vortex ring would be as permanent as the solid hard atoms by Lucretius'.[74] In February, he submitted his first paper, a rather lyrical document, on the vortex atom.[60]

What had struck Thomson was a theorem in Helmholtz's memoir: in an ideal fluid, each vortex line remains continually composed of the same elements of fluid. A vortex tube can never end within a fluid, but is either ring-shaped, or else it reaches the boundary of the fluid.** This property of the vortex ring, Thomson asserted in his paper, 'diminished by one the number of assumptions required to explain the properties of matter, on the hypothesis that all bodies

* Sample: in 1897, Stokes suggested[70] that radioactivity could be viewed as due to anharmonicities of an uranium atom which might be compared with a 'flexible string with a weight at the end'.
** A fluid is called ideal when thermal conductivity and viscosity are negligible. A vortex line is a curve everywhere tangential the local axis of rotation of fluid elements. A vortex tube is a portion of the fluid bounded by vortex lines drawn through every point of an infinitesimal closed curve.

are composed of vortex atoms in a perfect homogeneous liquid', the one assumption no longer necessary being the ad hoc introduction of forces which hold the structured atom together. He went on to propose that an atom is a closed vortex tube, or a set of such tubes, in the infinite, homogeneous, incompressible, frictionless aether. This, he believed, would account for three facts: (1) atomic stability. Here Thomson appealed to Helmholtz's proof of the indestructibility of vortex rings; (2) atomic variety, 'infinitely perennial specific quality . . . Diagrams and wire models . . . were shown to the [Royal] Society [of Edinburgh], to illustrate knotted and knitted vortex atoms, the endless variety of which is infinitely more than sufficient to explain the varieties and allotropies of known simple bodies and their mutual affinities'. For example, he conjectured, 'the sodium atom may . . . very probably consist of two approximately equal vortex rings passing through one another like two links of a chain', the doubling being related to the yellow sodium doublet; (3) possibly, the explanation of spectra. After referring to Kirchhoff and Bunsen, Thomson noted: 'The vortex atom has perfectly definite fundamental modes of vibration. . . . The discovery of these fundamental modes forms an intensely interesting problem in pure mathematics'. He also speculated on ways to determine the size of his fundamental rings.

Off and on, Thomson continued to work on his vortex atom for the next twenty years.* Some followed his progress with interest. 'Kirchhoff, a man of cold temperament, could be roused to enthusiasm when speaking about it. It is a beautiful theory he once told [Schuster[77]] because it excludes everything else.' Others addressed these problems themselves. The subject selected by the Examiners for the Adams Prize for 1882 was 'A general investigation of the action upon each other of two closed vortices in a perfect incompressible fluid'. The prize was won by J. J. Thomson, whose essay[78] concludes with a brief 'Sketch of a chemical theory'. For various reasons, W. Thomson himself began to doubt his model by that time. Some years later he abandoned it altogether.[79] His model did more for hydrodynamics than for atomic theory. Modern textbooks[80] continue to refer to Kelvin's law of conservation of circulation, according to which the velocity circulation round a closed fluid contour is constant in time.

4. The Darwinian touch. Like the decade 1895–1905, so the years 1858–64 were a vintage period of discovery in science. Clausius wrote of the motion we call heat, Maxwell discovered the equilibrium molecular velocity distribution of gases, Le Verrier found that an anomaly in the precession of Mercury could not be explained by known perturbations, Kirchhoff found sodium in the sun, Darwin published his *Origin of Species*, Kirchhoff and Bunsen founded spectral analysis, Huggins realized that nebulae are gaseous clouds, and

* See the Kelvin bibliography in Ref. 73, and also an essay on the vortex atom.[76]

Maxwell wrote down his electromagnetic equations. Man's sense of place in the universe was evolving, his grasp of it expanding.

In subsequent decades, a phenomenon occurs which I find curious and unsurprising. Darwinian imagery enters into writings about physics, particularly in regard to whether evolution of matter is an ongoing process in stars. This strikes me as interesting but I have no wise comments to offer. Instead, I give some examples.

Maxwell, in 1875: 'It has been found possible to frame a theory of the distribution of organisms into species by means of generation, variation, and discriminative destruction. But a theory of evolution of this kind cannot be applied to the case of molecules, for the individual molecules are neither born nor die. . . . The constitution of an atom is such as to render it, so far as we can judge, independent of all the dangers arising from the struggle for existence'.[33] As has been noted repeatedly, to Maxwell atoms were unbroken and unworn.

Lockyer proposed a theory of 'chemical' evolution in stars (the details of which are of no present interest) about which he wrote, in 1881: 'I think it derives its whole force from the fact that along many lines it seems parallel with the evolutionary processes in the different kingdoms of nature. . . . All these evolutionary processes, obtained from different regions of thought, have such a oneness about them'.[81]

Crookes, in 1888, considered 'existing elements not as primordial but as the gradual outcome of a process of development, possibly even a "struggle for existence". Bodies not in harmony with the present general conditions have disappeared, or perhaps never existed'.[82]

(c) Early electron models

1. In which atoms appear to be composed of electrons by the thousands. Physicists can as little be held away from the search for fundamental principles, as fortunately, they can from tinkering with models. Whatever blocks they have, models they must build. At the turn of the century they had only one species of block: electrons. Accordingly they set out to build atoms from electrons only. 'They', of course, refers to only that handful of theorists actively involved at that time with the structure of matter. 'It is perhaps not unfair to say that for the average physicist of the time, speculations about atomic structure were something like speculations about life on Mars—very interesting for those who like that kind of thing, but without much hope of support from convincing scientific evidence and without much bearing on scientific thought and development.'[82a]

Polyelectron models of atoms originated with J. J. Thomson. In his paper[83] of 1897 on the determination of e/m for cathode rays (see Chapter 4, Section (d)) there is, at once, reference to 'the hypothesis . . . enunciated by Prout [according to which] the atoms of the different elements were hydrogen atoms; in this precise form the hypothesis is not tenable, but if we substitute for hydrogen some unknown primordial substance X, there is nothing known which is inconsistent with this hypothesis'. He left no doubt as to what X had to be: '... These primordial atoms, which we shall for brevity call corpuscles' (the way Thomson continued to refer to electrons for some years). Two years later[84] he wrote: 'I regard the atom as containing a large number of ... corpuscles'.

Gradually, these qualitative initial statements became a little bit more precise. In 1900, FitzGerald[85] wrote of 'a very interesting suggestion that all matter is built up of electrons. That an atom of hydrogen, for example, consists of some 500 electrons, one of oxygen of some 8000, and so forth'. (Evidently, the electron's mass was not yet well known.) In 1902, Rutherford[86] referred to 'the view [which] has been put forward that all matter is composed of electrons. On such a view an atom of hydrogen for example is a very complicated structure consisting possibly of a thousand or more electrons'. In 1903, Thomson, first Silliman lecturer, said: 'The atom of hydrogen contains about a thousand electrons'.[87] In 1904 Strutt wrote likewise.[88]

What agent neutralizes the huge charge within an atom due to all these electrons? And should one not consider that that agent may contribute to the atom's mass?

In 1899, Thomson[84] chose to be very vague on the issue of charge: 'When [corpuscles] are assembled in a neutral atom the negative effect is balanced by something [sic] which causes the space through which the corpuscles are spread to act as if it had a charge of positive electricity equal in amount to the sum of the negative charges on corpuscles'. Even more obscure is his letter[89] to Oliver Lodge, written in 1904: 'I have ... always tried to keep the physical conception of the positive electricity in the background because I have always had hopes (not yet realized) of being able to do without positive electrification as a separate entity, and to replace it by some property of the corpuscles [?]'.

Thomson clearly was not predisposed to associating the charge-compensating agent with a new form of matter, and, therefore with an additional contribution to the atom's mass. This same view was expressed more explicitly by Arthur Kimball from Amherst in his report[90] to the Congress of Arts and Science, held in 1904, on the occasion of the Universal Exposition in St. Louis: 'The whole mass of the atom is supposed to be due to the negative electrons ... which it contains. As to the *positive* charge, although it determines the apparent *size* of the atom, it appears to make no contribution to its mass'. So, without good reason, it went on, at least until 1907. That is the last year for which I have found a reference to the purely electronic origin of the atom's mass, in a textbook on electricity[91] written by a Cavendish man: 'In the present state of our knowledge no certain statement can be attained as to the whole number of electrons in any atom, but the conclusion that it is such that the mass of the atom is the sum of the masses contained in it is so attractive that it seems desirable to accept it provisionally in the absence of any conclusive evidence to the contrary'.

However, as I shall discuss shortly, in 1907 the art of atom building had progressed beyond mere speculation about the origins of atomic mass. By that time Thomson *had* committed himself to a model of the positive charge and *had* concluded that atoms do not contain thousands of electrons. In order to appreciate these developments, it is best to comment first on the criteria, then current, for model building.

2. A new pitfall: atomic stability and β-decay. 'We have evidence that some of the elements have existed for many thousands, nay, millions of years; we have, indeed, no direct evidence of any change at all in the atom. I think however that some phenomena of radioactivity . . . afford, I will not say a proof of, but a very strong presumption in favor of some secular changes taking place in the atom.' Thus, in his Silliman lectures,[92] Thomson mentioned atomic stability and radioactive instability in the same breath. It was of course quite natural that he should do so. Cathode rays are ejected from atoms, β-rays are ejected from atoms. By 1900 it was well established that both are electrons (Chapter 4). Should one not attempt, therefore (I paraphrase Thomson's subsequent comments in his lectures at Yale) to find a common dynamical mechanism for explaining atomic stability and β-decay? So, following the dictates of what then appeared to be simplicity, ignorant of the distinctions between atomic and nuclear physics, Thomson raised an issue which was sensible for its day, but which in fact was a pitfall. In his eighties, Kelvin was one among several who liked to speculate upon this question.[93]

What about α-decay? In 1904, Rutherford thought[94] that here one witnessed the 'expulsion . . . of a connected group of electrons'.

3. The great divide. Progress in science depends vitally on a backlog of experimental data in need of interpretation. The whole purpose of this chapter up to this point can be encapsulated in one phrase: to demonstrate that in all of the twentieth century (to date) the experimental backlog in physics was never greater than during its opening years.

For a century, chemists had amassed data which called for physical interpretation, most particularly the regularities in the periodic table of elements. Half a century of spectroscopy had yielded a wealth of results about which, as a matter of principle, nothing more was known than that they revealed that something was moving inside the atom; the Zeeman effect indicated that this 'something' was a universal atomic constituent. It was known that all atoms could break up by ionization, emitting electrons in the process. It was known that some species of atoms could break up in more violent ways, the radioactive disintegrations, but not whether or not this property was common to all species (Chapter 6, Section (d)).

Even without an inkling about what any and all of this meant, the experimentalists of the first decade of the twentieth century were eminently capable of asking questions which, operationally, were entirely sensible; their rapid progress continued. As that decade draws to a close we are very near the great divide: the discovery of the nucleus and closely related to that, finding a primitive but essentially correct form of the quantum dynamics of the atom.

In retrospect (it was not at all that clear at the time) the extraordinary advances in theoretical physics during the years 1900–10 were dominated by the contributions of Planck and Einstein. However, there was almost no advance in regard to the list of atomic problems just mentioned, in spite of a fair amount of effort. The individual atom remained a mystery. I shall next give a brief sketch of the thinking on the atom during that decade, beginning with the criteria for model building.

Almost from the start it was clear that electrons within the atom had to be in motion, for two reasons. First, this seemed indispensable for the understanding of spectra. Secondly, according to a theorem by the Reverend Samuel Earnshaw, M.A. Cambridge University 1831, a system of particles interacting via forces varying as the inverse square of the distance cannot be in stable static equilibrium.[95]* Thus the only hope for a stable atom was for its electrons to be in motion. That, however, raised problems of a new kind. Since these electrons are confined to a finite volume, their motion is not uniform. Hence, by the laws of classical physics, the atom is unstable because of energy loss by electromagnetic radiation. This was stressed by Joseph Larmor[97] within two months of the appearance of Thomson's very first paper[83] on the electron: 'In motion with uniform velocity there is no loss [of energy]; during uniformly accelerated motion the rate of loss is constant'. Nevertheless he hoped for a way out: 'It would ... appear that when the steady orbital motions in a molecule are so constituted that the vector sum of the accelerations of all its ions or electrons is constantly small, there will be no radiation, or very little, from it, and therefore this steady motion will be permanent'. This vector constraint does indeed suppress radiation, but of course not sufficiently for the purpose.

Shortly thereafter[98] Larmor made his statement significantly more precise: 'It is here implied that the electrons are contained by the attraction of an electron of opposite sign at the centre of the ring: as otherwise their mutual repulsions and the centrifugal forces would produce their dispersion'. Thus, as early as 1900, one vaguely sees a hint of truth emerge, but only a hint. There are additional desiderata. The electrons' motions should be such as to

* All I know about Earnshaw stems from an article by Bill Scott.[96]

explain why the spectra emitted by atoms consist of a system of sharp lines. Jeans[99] stressed this criterion as early as 1901, but was of course not the only one unable to cope with it. Lastly, there was a need to interpret the patterns set by the periodic table.

All in all, it was an awesome assignment. Larmor[100] said it well, in 1900: 'The problem . . . involved is not to assign a structure so minutely definite that it will include the whole complex of chemical actions, but rather to ascertain how much must be postulated in order to co-relate the main features of those universal agencies', the atoms. It is not surprising that the model builders of that time would often close an eye on one or another of the numerous criteria.

Let us next see, briefly, what concrete ideas they had.

4. Pair models. In 1901, Jeans made a proposal[99] which 'should not be judged as an attempt to attain the ultimate truth', but rather as 'perhaps [giving] something of the foreshadowing of the real truth'. He suggested that the atom is electroneutral because it contains, in addition to electrons, another kind of particle of equal mass and opposite charge. However, 'the predominance of the negative ion [electron] in most material phenomena and in the emission of light (as evidenced by the Zeeman effect) seems to suggest that positive and negative ions differ in something more than mere sign.' Jeans thought that perhaps the pairs of negative and positive electrons were spatially oriented in such a way that the positives would always point inward and so could not readily be dislodged, but realized that this is not easy to achieve, since to each oriented set of pairs there should correspond an equally probable configuration in which the role of the pair members is interchanged. The difference between these two configurations, he conjectured, 'can only arise from a difference in initial conditions'. Jeans made an unsuccessful attempt to explain spectra in terms of his model.

Also Lenard proposed[101] that each electron is paired with a 'positive elementary quantum', which, however, has a 'grosz erscheinende Masse' (apparently large mass).[102] He called such a pair a 'dynamid'. His paper deserves to be remembered as the first in which it is stated that *the atom is almost completely empty*. 'For example . . . the space occupied by a cubic meter of solid platinum is empty . . . in the same sense . . . as the skies.' He drew this conclusion from the ease with which cathode rays traverse large numbers of atoms. Neither Jeans nor Lenard had anything to say about chemical regularities.

5. Planetary models. In the 1920s, Carl Runge, reminiscing about the days, forty years earlier, when he and Kayser did their spectroscopic researches, recalled how he had gone to ask Helmholtz's opinion about certain spectral formulae which he and Kayser had devised. Helmholtz 'sasz eine Weile im

Gedanken und dann hörte ich ihn sagen: Hm, ja, die Planeten, wie ist das doch? Die Planeten—ach nein, dasz geht wohl nicht'.[103]*

Thus, images of the atom somewhat analogous to a solar system emerged long before Rutherford discovered the nucleus. I give a few more examples.

Perrin[104] in 1901: 'Each atom might consist ... of one or more positive suns ... and small negative planets ... If the atom is quite heavy, the corpuscle farthest from the centre—the Neptune of the system—will be poorly held by the electron attraction ... The slightest cause will detach it; the formation of cathode rays [electrons] will become so easy that [such] matter will appear spontaneously radioactive', a good example of the mixed stability/instability metaphors characteristic for that period.

Beginning in 1903, Nagaoka,[105] professor of physics at Tokyo University, proposed a Saturnian atom in which the electrons move in one or more rings around a central body. Spectra are supposed to be due to various perturbations in the motion of the rings. This work has the distinction of having been quoted by Poincaré[106] and Rutherford.[107] It was soon realized, however, that the comparison with Saturn fails. Nagaoka's atom is seriously unstable.**

Finally, there is the vision[108a] of Fernando Re, pupil of Becquerel, concerning the birth of atoms: 'It seems natural to suppose that the particles which constitute the atom were free at one time, and that they constituted a nebula of extreme tenuity; that thereupon they united themselves around condensation centers, giving birth to infinitely small suns which, by a further process of contraction, have taken on stable and definitive forms, which would be the atoms of the elements that we know and which we could compare to small burnt-out suns. The bigger suns, which are not burnt out, would constitute the atoms of the radioactive bodies'.

The period 1897–1913 consists of two distinct parts. In the first, it was believed that the number of electrons in the atom is large. In the second, it was realized that this number is of the order of the atomic number. This change was wrought by Thomson, in 1906.

6. J. J. Thomson, theorist. Alfred Marshall Mayer was a fine self-taught physicist and a character. He never received an earned academic degree. 'The first lecture upon Physics which he ever heard was one he gave himself when Assistant Professor of that science in the University of Maryland at the age of twenty-one.'[109] He was accomplished in research,† personally acquainted with Rayleigh, Regnault, and Tyndall, a friend of Joseph Henry, founder of the physics department at Stevens Institute, Hoboken (which in his time had

* H. sat and thought for a while, and then I heard him say: Hm, yes, the planets, how is that again? The planets—but no, that would not work.

** For more details see Ref. 108.

† See Ref. 109 for a list of his publications.

one of the best-endowed laboratories), inventor of a new minnow-casting rod, member of the National Academy of Sciences and other learned academies. He delighted in devising 'simple, entertaining and inexpensive experiments for the use of students of every age'.[110] Among these, his investigations recorded in two papers,[111] both entitled 'Floating Magnets', are of particular interest to us. In these articles, Mayer described 'a system of experiments which illustrate the action of atomic forces, and the atomic arrangements in molecules, in so pleasing a manner that I think these experiments should be known to those interested in the study and teaching of physics'. Mayer's experiments were described by Thomson[87] in his Silliman lectures: 'A number of little magnets are floated in a vessel of water. The magnets are steel needles magnetized to equal strengths and are floated by being thrust through small disks of cork. The magnets are placed so that the positive poles are either all above or all below the surface of the water. These positive poles, like the corpuscles, repel each other with forces varying inversely as the distance* between them. The attractive force is provided by a negative pole (if the little magnets have their positive poles above the water) suspended some distance above the surface of the water. This pole will exert on the positive poles of the little floating magnets an attractive force the component of which, parallel to the surface of the water, will be radial, directed to the center 0, the projection of the negative pole on the surface of the water, and, if the negative pole is some distance above the surface, the component of the force to 0 will be very approximately proportional to the distance from 0'.

Mayer had found experimentally that the most stable configurations of n floating magnets have quite remarkable 'shell' properties: up to $n=5$ they arrange themselves in a single regular polygon. But for $n=6$ one magnet moves to the centre while the other five remain on the polygon. From $n=15$ a third shell begins to develop: $15=9+5+1$, and so on.

Mayer's results, published in 1878, drew the immediate comment from W. Thomson (Kelvin) that they are 'of vital importance in the theory of the vortex atom'.[112] J. J. Thomson devoted a section[113] to the subject in his Adams Prize essay of 1882, expressing the hope that this idea would teach him something about multi-vortex atoms. In 1897, he at once returned to the same topic in his paper[83] on the nature of cathode rays: 'A study of the forms taken by these magnets seems to be suggestive in relation to the periodic law'. As we shall see, for years thereafter Mayer's magnets would inspire Thomson to erect a theory of atomic structure.

By the time he measured the e/m of cathode rays, Thomson, past forty, had published some sixty papers, about half of these on theoretical subjects, and four books, all of a theoretical nature.** I have little doubt that his prime ambition was to be a theorist. As his son describes him: 'J. J. spent a good

* He meant: as the square of the distance.

** Thomson's papers were never published in collected form. For a bibliography, see Ref. 114, for an addendum thereto see Ref. 115. (The list is still incomplete.)

part of most days in the armchair of Maxwell, doing mathematics . . . In the period with which we are now concerned, two major fields . . . occupied J. J.'s attention. One was virtually to rewrite physics . . . in terms of the newly discovered electron . . . the other was to get beyond Maxwell'.[116] It is evident that Maxwell was Thomson's demon. Qua style, the two men were utterly different: '[J. J.] usually preferred to dwell on what a theory *would* explain than on what it would not . . . He was so fertile in suggesting ways of getting out of a conclusion he did not want to accept, that the idea of a crucial test was apt to fade away in talking matters over with him . . . J. J. was not inclined to be dogmatic about his atomic theories, and indeed he was quite prepared to change them sometimes without making it altogether clear that he had wiped the slate clean, and that what he had written before must now be considered canceled'.[117] Niels Bohr later said of Thomson's model building: 'Things needed not to be very correct, and if it resembled it a little then it was so'.[118]

Six years passed after Thomson remarked on the magnets and the periodic table, twenty-five papers were written, nearly all experimental, some quite important, until in 1903 he returned to the structure of the atom. The year before, Kelvin, in an otherwise forgettable paper, had proposed that the positive charge be homogeneously distributed over the atomic volume, supposed to be spherical.[119] Thomson now adopted this picture too.[120] The positive charge density ρ equals $3Q/4\pi r^3$ where r is the 'atomic radius' and Q the total positive charge. An electron inside the sphere at a radius a from its centre experiences a restoring force $(4\pi\rho a^3 e/3)/a^2 = eQa/r^3$—the same linear (in a) force law as for Mayer's magnets. Aware of Earnshaw's theorem,[121] Thomson let his electrons rotate with a common uniform angular velocity ω. For the case that the number n of electrons is small, he made a few calculations for three-dimensional electron configurations. Otherwise he considered the simpler case that all electrons move in a plane, distributed with equal angular intervals over one or more rings.

For this model (sometimes called the plum pudding model) Thomson considered (1903) the issue of radiation loss raised earlier[97] by Larmor. I paraphrase his reasoning. Forgetting about radiation to begin with, n electrons move in a ring, all with uniform velocity $v = \omega a$. Radiation disturbs this uniform motion. Question: how does the radiation per electron compare with the radiation of a single electron (i.e. $n = 1$) which initially was moving with the same velocity v in the same orbit? There is considerable radiation suppression, for example, for $n = 6$ and $v/c = 1/10$ $(1/100)$ by a factor $\sim 10^{-7}$ (10^{-17}). Here lay the hope—unfulfilled—for a sufficiently stable atom.

In his next paper[122] (1904) Thomson tried to make a virtue of this radiative energy loss. He had noted that certain electron configurations are stable only if v is larger than a critical value. 'In consequence of the radiation from the moving corpuscles, their velocities will slowly—very slowly—diminish; when, after a long interval, the velocity reaches the critical velocity, there will be what is equivalent to an explosion of the corpuscles . . . The kinetic energy gained in this way might be sufficient to carry the system out of the atom, and we should have, as in the case of radium, a part of the atom shot off.'

It is as true to Thomson's style that he would write about radium as he would not, then or later, refer in any detail to atomic spectra.

However, in his 1904 paper he did mention a result which, it seemed to him, was suggestive of the regularities in the periodic table. His calculations showed that, even in the absence of radiation effects, the steady motion of $n > 6$ corpuscles in a single ring is unstable however rapid their rotation may be. Very much like Mayer's magnet configurations, for $n > 6$ a single ring of electrons is less stable than a set of concentric rings. He noted in particular that, if the outer ring contains 20 electrons, then all configurations with 39, 40, . . . , 47 additional electrons distributed over inner rings are stable (always barring radiation, of course) and went on to suggest that these sets of 59, 60, . . . , 67 particles might be akin to a sequence of nine elements beginning with an inert gas and ending with the next inert gas. It is typical for Thomson that he would draw analogies between elements and systems of some sixty corpuscles, and that, in the same paper, he would 'suppose that the mass of an atom is the sum of the masses of the corpuscles it contains'.

In 1905 Thomson lectured on these ideas at the Royal Institution and must have delighted his audience with the demonstrations he gave of Mayer's magnet configurations.[121] To recapitulate, up to this point all he had were three hazy ideas: one about radiation loss, one about radioactivity, and one about the periodic table.

1906 was a very important year in Thomson's life. In June, a paper by him appeared which contains great discoveries, perhaps his greatest as a theoretician.[123] He reported that 'the number of corpuscles in an atom . . . is of the same order as the atomic weight of the substance'. In particular for hydrogen this number 'cannot differ much from unity'. Moreover, 'the mass of the carriers of positive electricity cannot be small compared to nm, the mass of the carriers of negative electricity'. In December of that year he received the Nobel Prize 'in recognition of the great merits of his theoretical and experimental investigations on the conduction of electricity by gases'.[124] A week thereafter he celebrated his fiftieth birthday.

Thomson arrived at his conclusion about the number n of electrons per atom by three independent methods. First, he derived a formula for the refractive index of monatomic gases,* and compared his answer (by a slightly

* Assuming that the electrons can be treated as free, Thomson found for the dispersion of light with frequency ν:

$$\frac{\mu^2-1}{\mu^2+2}=\frac{NQ(Me+mQ)}{\rho(Me+mQ)-Mm\nu^2}.$$

μ is the refractive index, N the number of atoms per unit volume, m the electron mass. M, ρ, and Q are the mass, charge density, and total charge of the positive sphere. ρ was eliminated in favor of μ for zero frequency. The number n enters in the formula because $Q=-ne$.

indirect argument) with experimental data for hydrogen. Secondly, he considered the scattering of X-rays by gases as due to the scattering off the intra-atomic electrons, supposed to be free particles. To this problem he applied a result he had derived elsewhere[125]: the fractional energy loss of an X-ray beam per unit path length of scattering substance is given by $\sigma N n$, where N is again the number of atoms per unit volume, and where σ is given by

$$\sigma = \frac{8\pi}{3}\left(\frac{e^2}{mc^2}\right)^2. \tag{9.5}$$

Again he compared this answer with available data.* Finally he discussed the absorption of β-rays in matter, assumed to be due to single scattering off intra-atomic electrons treated as fixed centers, calculated the absorption coefficient, and compared it with Rutherford's data for absorption of rays with $v/c \simeq 0.5$ in copper and silver. His overall conclusion from the three methods: the number of electrons per atom lies somewhere between 0.2 and 2 times the atomic number.

In many respects these are primitive calculations. The possible influence of the positive sphere was ignored. Electrons were treated as free. In the third method, the motion of atomic electrons was neglected, so was multiple scattering. Furthermore, it was assumed that β-rays follow an exponential absorption law which, as discussed at length in the previous chapter, is far from true. Thomson himself attempted to improve his calculations soon thereafter.** Be all that as it may, his paper of 1906 must be considered as a fundamental advance, indeed as the first paper of substance on the physics of atomic structure. In 1909, young Max Born called Thomson's work 'an excerpt for piano of the great symphony of the radiant atom'.[127] It is also fitting to note that Thomson's Eq. (9.5) has survived in quantum field theory, including the renormalization program, as the zero-energy limit of the Compton effect. σ will forever be known as the Thomson cross-section.

It would have enhanced the stature of Thomson's 1906 paper even more had he stated another evident conclusion: if the number of electrons in a hydrogen atom indeed lies near unity, then there is a crisis in regard to radiative stability of atoms. But that was not his style.

1906 was the year of Thomson's last hurrah as a theorist. In 1913 he returned to the atom in a paper[128] which is not good. Evidently dissatisfied with the stability problem, he proposes a modification of Coulomb's law at small distances and attempts to derive Planck's constant from classical dynamics. It is not necessary to go into details except for the following:

there is no word about the hydrogen atom;

* These data had very recently been obtained by Barkla; it is possible that they may have been the crucial stimulus for Thomson's 1906 paper.

** The reader is urged to consult a paper by Heilbron[126] for more details on Thomson's calculations of 1906 and on his subsequent efforts at refinement.

there is no word about Rutherford;
there is no word about Niels Bohr.

Now Thomson read his paper before the September 1913 meeting of the British Association (and enlarged on it at the Solvay Congress the month thereafter[129]). Yet two years before, Rutherford had discovered the nucleus, while, earlier in 1913, Bohr had cracked the hydrogen atom!

I leave Thomson at this point, come back to him briefly in Chapter 10 and turn to Rutherford.

(d) Ernest Rutherford, theoretical physicist

On 14 December 1910, Rutherford wrote to his friend Boltwood: 'I think I can devise an atom much superior to J.J.'s, for the explanation of and stoppage of α- and β-particles, and at the same time I think if will fit in extraordinarily well with the experimental numbers'.[130]

During the Montreal period Rutherford had been the first to observe[131] the scattering of α-particles in matter (see Chapter 3). That happened in 1906. The year thereafter he left Canada for England, arriving in Manchester in June 1907. In October he assumed his duties as Langsworthy Professor of Physics at the Victoria University in that city. One of his first acts was to compose a list of 'Researches possible', one entry of which reads 'Scattering of α-rays'.[132] It was one of several topics he took up in collaboration with Hans Geiger who had been in Manchester since 1906 as assistant to Schuster, Rutherford's predecessor.

On 18 June 1908, Rutherford communicated two papers to the Royal Society. The first, a joint article with Geiger[133] in which it was shown that the charge of an α-particle equals in magnitude twice that of an electron, has been described earlier (Chapter 3). The second was a paper by Geiger alone, a preliminary note on α-particle scattering.[134] His source was a well-defined pencil of α-particles from a few mg of $RaBr_2$, the scatterer a thin foil either of gold or of aluminum; the α's were detected by counting scintillations. Scattering was observed in both materials, but more so in gold (for equivalent foil thickness). Geiger also concluded: 'Some of the α-particles . . . were deflected through quite an appreciable angle . . . A fuller investigation will also enable us to treat the matter from a theoretical point of view'.

About half a year later, Rutherford went to Stockholm to receive his chemistry Nobel Prize. At a dinner there in his honor, Mittag-Leffler addressed him with these words: 'Mr. Rutherford knows how to work with mathematics . . . he knows how to plan and carry out experiments . . . many future discoveries are doubtless in store for him'.[135] It is the perfect epigraph for what happened in the course of the next two years when Rutherford's scientific career reached its acme.

It began one day, early in 1909, when Rutherford stepped into Geiger's room. Also present there was a young helper of Geiger, a twenty-year-old undergraduate, later described as 'genial, puckish, never boring, prone to infectious laughter, full of fun and excitement in science'. His name was Ernest Marsden.* What happened next has been recalled by Marsden himself: 'One day Rutherford came into the room where we were counting . . . α-particles . . . turned to me and said: "See if you can get some effect of α-particles directly reflected from a metal surface". I do not think he expected any such result, but it was one of those "hunches" that perhaps some effect might be observed . . . To my surprise, I was able to observe the effect looked for . . . I remember well reporting the result to Rutherford a week after, when I met him on the steps leading to his private room'.[137]

These findings were reported in a paper by Geiger and Marsden[138] submitted in May 1909. The α-particle source was radium emanation (Rn^{222}). Detection was again by scintillations. Their α-beam was not too well collimated. The major conclusion: 'Of the incident α-particles about 1 in 8000 was reflected', or, as we would say, scattered by more than 90°. The paper also contains preliminary information concerning the total number of α's scattered as a function of the metal chosen for the scattering foil.

In one of the last lectures Rutherford ever gave (and of which a record exists) he described his reaction to the back-scattering effect: 'It was quite the most incredible event that has ever happened to me in my life. It was almost as incredible as if you fired a 15-inch shell at a piece of tissue paper and it came back and hit you'.[139]**

From the point of view of atom models then prevalent it was indeed a staggering result. Imagine a big α-particle traveling with a velocity $\sim$10 000 km per second hitting a bunch of tiny electrons and a jelly of positive charge and coming right back at you! Already in 1906, when Rutherford had observed α-scattering angles $\sim 2°$ in a mica sheet 0.003 cm thick he had remarked that this 'would require over that distance an average transverse electric field of about 100 million volts per cm'.[137] But now, what to do with angles $>90°$?

Referring to his results with Marsden, Geiger wrote, early in 1910: 'It does not appear profitable at present to discuss the assumption which might be made to account [for the large back-scattering]'.[140] It is plausible that Geiger knew by then that the master was brooding. All remained quiet until December, when Rutherford wrote the letter to Boltwood from which I quoted earlier.[130] Geiger's own recollection of that time, late 1910 or early 1911, goes like this:

* For an account of the life and work of Sir Ernest see Ref. 136.
** Samuel Devons and Maurice Goldhaber who as young men heard Rutherford lecture at the Cavendish both told me that they remember Rutherford recalling his surprise in similar language. We cannot know how early this picturesque comparison struck Rutherford for the first time.

'One day [R.] came into my room, obviously in the best of moods, and told me that now he knew what the atom looked like and what the strong scatterings signified'.[141] On 7 March 1911, Rutherford presented his principal result to the Manchester Literary and Philosophical Society.[142] The definitive paper came out in the May issue of *Philosophical Magazine*.[143]

Rutherford felt compelled to do theoretical physics if, and only if, he otherwise could not interpret data of his own or from his laboratory. More broad-based theoretical issues tended to be alien to him, I think, especially if they were speculative in character. Thus he was distinctly reserved in his early response to Bohr's quantum theory of the atom (more about that presently). Then there is the story (recorded by a witness, the time was 1910) of Rutherford twitting Willy Wien about relativity. After expounding some points of that theory Wien said to Rutherford that no Anglo-Saxon could understand that. Whereupon Rutherford laughingly replied no, they have too much sense.[144] Good anecdotes like good caricatures display facets of truth by hiding the rounded portrait. Even so, Rutherford's occasional sarcasms about sophisticated theorists and theories are amply documented.

I never was fortunate enough to meet Rutherford (having just left high school when he died) but have heard stories about him from Bohr, from some of his students, and have also read many of them. The portrait which emerges is of a figure humane, no-nonsense (of a self-important official: 'He is like a Euclidean point: he has position without magnitude'[145]) and rugged. *A propos* of nothing I recount one more Rutherford story. On a certain occasion, Rutherford told a group of his students about an exchange with a bishop during a formal luncheon. The bishop had asked him how many people lived in his native South Island. When Rutherford told him: about 250 000, the bishop noted with amazement that that was only about the population of Stoke-on-Trent. 'So I said to him', continued Rutherford, and then he paused and looked kindly at [his group of students]. 'I hope there are none of you here from Stoke-on-Trent. So I said to him: "Maybe the population is only about that of Stoke-on-Trent. But let me tell you, Sir, that every single man in the South Island of New Zealand could eat up the whole population of Stoke-on-Trent, every day, before breakfast, and still be hungry" '.[145]

Let us turn to Rutherford's work as a theorist. He made his first important contribution in 1902, when he and Soddy developed the formalism for sequential radioactive decays, known as the transformation theory (Chapter 6). Early in 1909, when he realized that he did not know enough probability theory for handling small samples of data, the Nobel laureate became a regular student in a course on that subject by Horace Lamb.[146] His most outstanding theoretical contribution is of course his paper[143] of 1911 on the scattering of α- as well as β-particles. I shall not discuss his comments on β-scattering beyond noting that it was still in an immature state at that time,* much like β-absorption discussed in Chapter 8.

Rutherford's model for an N-electron atom has 'a charge $\pm Ne$ at its centre . . . for convenience the sign will be assumed positive'. He noted that his final answer for α-scattering does not depend on this sign ambiguity. (It appears that he briefly toyed with the odd idea of a negative central charge because he thought that a negative core could more easily explain the absorption of β-rays.[148]) The (positive) central charge is surrounded by a sphere in which the negative charge $-Ne$ due to the electrons is uniformly distributed. This last assumption is actually irrelevant for his theory of α-scattering, since he approximates this phenomenon by the action of a pointlike central charge (we would say the bare nucleus) on a pointlike α-particle, neglecting the recoil of the central charge.

I do not know what took Rutherford more time: his brilliant break with the model of Thomson (to whom he made courteous reference in his paper) or the actual evaluation of the cross section (a term he did not yet use) $\sigma(\vartheta)$ for the scattering of an α-particle over an angle ϑ. This calculation, now an undergraduate exercise in the Newtonian mechanics of hyperbolic orbits in a $1/r$-potential, was not all that difficult even then; but Rutherford cannot be called an accomplished mathematician. More important, however, than the reason why he took so much time is the fact that he did take the time. The facts were there for all to see, but no one else bothered, as best I know.

In modernized notation Rutherford's result can be written as

$$\sigma(\vartheta)=\frac{(NeQ)^2}{4m^2v^4\sin^4\vartheta/2}, \tag{9.6}$$

where v, m, Q are the velocity, mass, and charge of the α-particle respectively.

The 'Rutherford scattering cross-section' Eq. (9.6) obviously contains far more information than the data which inspired its derivation. Thus, after verifying that his theory is in qualitative agreement with the work of Geiger and Marsden[138] on large-angle scattering ('the scattering observed is about that to be expected on the theory') and on the atomic number dependence, and with Geiger's results[140] on the average α-scattering angle, Rutherford concluded his paper with the remark that further discussion will be reserved for such times 'when the main deductions of the theory have been tested experimentally'. Geiger remembers that 'It may have been on the same day [the day that Rutherford came into his room, see above] that I began to test the relation between the number of particles and the scattering angle predicted

* For a discussion of Rutherford's work on β-scattering see a paper by Heilbron[147] which also contains remarks on an early draft of the 1911 paper.

by Rutherford'.[141] These experiments, performed again together with Marsden, were completed late in 1912, with satisfactory results.*

Let us briefly depart from the historical account and step into the present.

The earliest successes of Rutherford's model were not only a mark of his great ingenuity, but also of his good luck. All data at that time were obtained from scattering low energy ($\approx$5 MeV) α-particles on targets with high nuclear charge. Therefore the penetration of the nuclear Coulomb barrier was negligible so that the deviations from the Rutherford formula due to strong interactions still lay hidden, which was just as well. Yet the α-particles were fast enough to justify the neglect of scattering off atomic electrons. Moreover, for a Coulomb potential the classical and quantum mechanical scattering cross sections are identical, at least in the nonrelativistic limit—which applied of course to the Geiger–Marsden data. Finally, one shudders to think what Rutherford and coworkers would have concluded had they shot α-particles into a helium-filled vessel!

Many refinements of Rutherford's formula have been studied since those early days: screening of the nuclear Coulomb field by atomic electrons, scattering contributions from the atomic electrons themselves, spin and relativistic effects, the influence of finite nuclear size, solid state effects—and the influence of strong interactions. Only the last of these modifications will be returned to later on.

No hammer in the Horologe of Time pealed through the universe when the nuclear age in science arrived.

I find the lack of immediate response to Rutherford's 1911 article not in the least surprising. Fundamental though the advance was, his model emphasized one and only one feature of a full-fledged atom: the localization of the positive charge. In so far as α-scattering is concerned, electrons played no role at all in his paper. Then what about the conundrums of the other model-builders, what about radiative stability of electron orbits, spectra, the periodic table? Rutherford, obviously aware of these problems, had the sagacity to bypass them. As he wrote in his paper: 'The question of the stability of the atom proposed need not be considered at this stage, for this will obviously depend on the minute structure of the atom and on the motion of the constituent charged parts'.[143]

Since, as said, Rutherford preferred to wait and see[152] whether Eq. (9.6) was true in detail, it is not remarkable that he remained silent on his model while attending the Solvay Conference of 1911. In his book on radioactivity, completed in October 1912, he used the term 'nucleus' for the first time, but only once: 'The atom must contain a highly charged nucleus'.[153] The new

* The results were first published in the *Wiener Berichte*,[149] shortly thereafter also in *Philosophical Magazine*.[150] Different nuances in these two articles were noted in an interesting paper by Trenn.[151]

Geiger–Marsden results[150] which had just come out are mentioned, but only in passing.[154] Andrade, who arrived at Manchester in 1913, has written: 'Rutherford does not appear to have considered his discovery as the epoch making event that it turned out to be'.[82a]—at least not at once. Rutherford did briefly discuss the nuclear atom[155] as a discussant during the second Solvay Congress (October 1913). The first time he spoke forcefully on the subject in a physics meeting was in March 1914, during a Royal Society discussion on the structure of the atom.[156]*

In Chapter 8, I mentioned Rutherford's views on radioactivity as of 1912. To repeat: 'The instability of the atom may be conveniently considered to be due to two causes . . . the instability of the central mass and the instability of the electronic distribution. The former . . . leads to the expulsion of an α-particle, the latter to the appearance of β- and γ-rays'.[157] Thus in the year 1912 α-decay was for the first time correctly diagnosed as a nuclear process. Another year went by before it was understood that β-decay was a nuclear process as well. That contribution was made by Niels Bohr, whose work is the final subject of this chapter.

(e) Niels Bohr

> One thought [spectra are] marvelous, but it is not possible to make progress there. Just as if you have the wing of a butterfly then certainly it is very regular with the colors and so on, but nobody thought that one could get the basis of biology from the coloring of the wing of a butterfly.
>
> Niels Bohr

1. Four roads to the quantum theory. Like the conquest of territory by multi-pronged attack, so the entry into the quantum domain proceeded by distinct roads. First there was Planck, who in 1900 started it all with his blackbody radiation law. Five years later, Einstein introduced light-quanta, an idea so alien to Planck that he resisted it for a long time. One year thereafter, Einstein ushered in the quantum theory of the solid state in order to explain long-standing specific heat anomalies. Common to these three roads is their origin in statistical mechanics. New dynamics was implied, of course, but was not yet explicit. Einstein came close, however, when in 1906, he wrote: 'We must consider the following theorem to be the basis of Planck's radiation theory: the energy of a [linear material oscillator] can take on only those values that are integral multiples of $h\nu$; in emission and absorption the energy of [this oscillator] changes by jumps which are integer multiples of $h\nu$'.[158]

The first strides on the fourth, the dynamical road, were made by Bohr.

* For early responses to the nuclear atom by others, see Refs. 82a and 147.

2. The early papers. Niels Henrik David Bohr was born in Copenhagen, in 1885. His father was a distinguished physiologist, his mother hailed from a cultured, wealthy Jewish banker's family. Niels' family ties were always close, as I had occasion to notice in 1946, when I saw him together with his brother Harald, a renowned mathematician, and Hanna Adler ('Moster Hanna'), his mother's surviving sister, then nearly ninety. (I first met Niels Bohr in January 1946, when I came to Copenhagen on a postdoctoral fellowship, and have written elsewhere of my personal recollections of my acquaintance with him.[159])

Bohr entered the University of Copenhagen in 1903. In 1906 he won a gold medal of the Royal Danish Academy of Science for a theoretical and experimental investigation of ripples on vibrating liquid jets, work which led to two papers in Royal Society journals. In May 1911, his thesis 'Studies on the Electron Theory of Metals' earned him the Ph.D. degree. This very thorough work is based on Lorentz' electron theory. By then Bohr was already aware of the limitations of the classical description. Specifically he mentioned two difficulties in his thesis, first the strange behavior of specific heats,* secondly 'it must be assumed that the Maxwell–Lorentz equations are not strictly satisfied' because these cannot explain the high-frequency behavior of black-body radiation. Of J. J. Thomson's and Jeans' attempts to give a classical interpretation to this behavior he argued: 'This . . . does not seem to be correct'. To these difficulties Bohr added one he himself had discovered: 'It does not seem possible, at the present stage of development of the electron theory, to explain the magnetic properties of bodies from this theory'. I have not found in the thesis any reference to the question of spectra.**

My first stay in Copenhagen lasted from January to August, 1946. Thereafter I went to Princeton but on the way spent a few weeks in my native Holland. During that visit I had occasion to call on the physicist Adriaan Fokker, a contemporary of Bohr, whom I told of my recent experiences in Denmark. This led Fokker to recall his own contact with Bohr around 1913. I found his stories so interesting that, much against my custom, I made some notes after my return from this visit. Part of my jottings have to do with Bohr's first experiences in England. In October 1911 he had gone to Cambridge, hoping to work with J. J. Thomson. His knowledge of English was rather limited at that time. I now turn to my notes. 'Visit with Fokker in Haarlem. According to F., Bohr's first meeting with J. J. went about as follows. B. entered, opened J. J.'s book *Conduction of Electricity through Gases* on a certain page, pointed to a formula concerning the diamagnetism of conduction electrons, and politely said: "This is wrong". Some of the subsequent encounters went similarly,

* Cf. Section (b), part 3 of this chapter.

** For an English translation of the thesis see Ref. 160.

until J. J. preferred to make a detour rather than meet Bohr'. Rosenfeld later told me a rather similar story. Shortly before his death, Bohr said: 'I considered Cambridge as the center of physics, and Thomson as a most wonderful man. It was a disappointment to learn that Thomson was not interested to learn that his calculations were not correct. That was also my fault. I had no great knowledge of English, and therefore I did not know how to express myself . . . The whole thing was very interesting in Cambridge, but it was absolutely useless'.[161] The lack of contact was especially disappointing to Bohr because, then and later, he looked up to J. J. as a great man. He kept busy in Cambridge, however, attending lectures, writing a short paper on the electron theory of metals, and reading the *Pickwick Papers* (a book he always remained fond of) in order to improve his English.

3. Manchester. 'The first time I had the great experience of seeing and listening to Rutherford was [in Cambridge] in the autumn of 1911', Bohr has recalled.[162] Rutherford's address included reference to his new atom model, published in May of that year. 'Thomson was absolutely against it'.[162] Bohr's rapidly conceived plan to go to work in Manchester took a while to mature. He arrived there in mid-March 1912 for a stay till late July. During those months his attempt at learning experimental techniques in radioactivity was only an aside, his theoretical work profound. References to detailed analyses of this work are found in the 'Sources' at the end of this chapter. I confine myself to two conclusions which he reached during that period.

Using Rutherford's nuclear model, Charles Galton Darwin (grandson of the biologist), also in Manchester, had just completed a theoretical discussion[163] of the energy loss of α-particles in matter, a process almost entirely due to collisions with electrons and therefore of particular interest since it complements α-scattering which is almost entirely due to the nucleus. He treated the n atomic electrons as free within a sphere of radius r and derived a velocity-range relation which depends on n and r. Using available data, Darwin found r-values greatly at variance with estimates from kinetic gas theory. In particular 'in the case of hydrogen it seems possible that the formula for r does not hold on account of there being only very few electrons in the atom. If it is regarded as holding, then $n = 1$ almost exactly'.

Bohr, realizing that Darwin's neglect of electron binding was the source of the trouble, especially for large impact parameters, published a paper in which he took this binding into account.[164] His work, a distinct advance, was in need of improvement too, to which he himself contributed in later years. For our purposes one of his conclusions is most striking.

'If we adopt Rutherford's conception of the constitution of atoms, we see that the experiments on absorption of α-rays *very strongly suggest* that a hydrogen atom contains only one electron outside the positively charged nucleus.'

I have italicized three words in order to stress that, less than a year before Bohr derived the Balmer formula, he was very nearly convinced, but not yet completely certain, that the hydrogen atom is a one-electron system.

The methods Bohr used in his thesis were quite helpful for his paper on α-absorption, which constitutes the bridge between his early work on metals and his concern, never wavering from then on, with the atom's internal structure, the problem on which he started to work intensely while still in Manchester. These early efforts have not been published, but are preserved in a memorandum entitled 'On the constitution of atoms and molecules' which he prepared for showing to Rutherford.* This document, containing several germs of his 1913 theory of atomic and molecular structure, begins with a last adieu to calculations in the Thomson style, as Bohr discovers new instabilities (*disregarding* radiative losses) of a ring of electrons, no longer moving in Thomson's positive sphere, but in the field of Rutherford's nucleus.**

Then comes the turning point.

Nothing in this model, Bohr notes, determines the radius of the atom or the orbital frequencies of its electrons. Therefore, he states in the memorandum, 'We shall introduce . . . a hypothesis, from which we can determine the quantities in question. This hypothesis is: that there for any stable ring (any ring occurring in the natural atoms) will be a definite ratio between the kinetic energy of an electron and the time of rotation. This hypothesis, for which there will be given no attempt at a mechanical foundation (*as it seems hopeless*), is chosen as the only one which seems to offer a possibility of an explanation of the whole group of experimental results which gather about and seems to confirm conceptions of the mechanisms of the radiation as the ones proposed by Planck and Einstein'.

The words italicized by me are the first clear statement that (in present language) classical physics does not suffice for explaining the atom. Bohr was on his way, but one crucial ingredient was not yet there: nowhere in this document is there any reference to spectra, let alone to the Balmer formula. That came after he returned home.

Bohr left Manchester for Copenhagen in late July. There, on August 1, 1912, he married Margrethe Nørlund. It was one of the rare matches which may be called blessed, the partners staying young even as they matured. Bohr's ties to his children were strong as well. I count the times I spent with Bohr and his family among the best in my life.

4. Spectra, 1903–11. In 1913, Bohr made clear that spectra are due to transitions between stationary states (a term he coined on that occasion) either of

* The memorandum is reproduced in Refs. 4 and 165.
** This calculation is not free of errors.[13]

neutral or ionized molecules (band spectra), or of neutral or ionized atoms (line spectra), an individual excited atom in general contributing to one single line. What ideas about the origins of spectra were in vogue just prior to that time? The following is a sample which probably could be extended. I shall not discuss the experimental arguments which were adduced for each view.

Lenard (1903) raised the question whether 'at all times in every atom there exist as many vibratory systems as there are series in its spectrum, in other words: whether every excited atom simultaneously emits all series of its spectrum'.[166] He concluded that this is not the case.

J. J. Thomson (1906): '[Line spectra] would be due, not to the vibrations of corpuscles inside the atom, but of corpuscles vibrating in the field of force outside the atom'.[167]

Stark (1907) opines that band spectra are due to excitations of neutral bodies, line spectra to excitations of ionized atoms.[168]

Wien (1909) criticizes Stark and states that (specifically for monatomic mercury vapor) neutral atoms can produce line spectra.[169]

Stark (1911) maintains his previous position.[170]

So it was when Bohr came along. In his words, in those early days spectra were as interesting and incomprehensible as the colors on a butterfly's wing.[171]

5. Precursors. When Bohr laid the foundations for the dynamics of the atom by establishing a link between its structure and Planck's constant h, he was not the first to propose that this link was the key. Much later he remarked: 'It was in the air to try to use the Planck ideas in connection with such things'.[171] As a final preliminary to Bohr's contribution, let us briefly consider his precursors.

Planck's constant was first introduced into an attempted theory of atomic structure in 1910, by the Australian physicist Arthur Erich Haas[172] (who was to end his days as professor of physics at Notre Dame). His model of the hydrogen atom is a single electron (charge $-e$, mass m) describing a periodic orbit with frequency f on the surface of a sphere with positive charge e and radius a. Electric attraction balances centrifugal repulsion:

$$\frac{e^2}{a^2} = ma(2\pi f)^2. \tag{9.7}$$

The total energy E of the electron is the sum of its kinetic energy E_{kin} and its potential energy E_{pot},

$$E_{\text{kin}} = \tfrac{1}{2}m(2\pi f a)^2 = \frac{e^2}{2a} \tag{9.8}$$

$$E_{\text{pot}} = -\frac{e^2}{a}. \tag{9.9}$$

Next, Haas introduces h by imposing the constraint

$$|E_{\text{pot}}| = hf. \tag{9.10}$$

His justification for this relation is not terribly interesting. It is otherwise with the consequence of Eqs. (9.7), (9.9) and (9.10):

$$a = \frac{h^2}{4\pi^2 e^2 m} \tag{9.11}$$

the correct expression for what is now called the Bohr radius of the hydrogen atom!

How can this be? Where are the quantum numbers of the Bohr theory? The answer is (as will shortly be clear) that Eq. (9.10) is indeed a correct consequence of Bohr's theory, but for the *ground state* of hydrogen *only*; for a general stationary state, Eq. (9.10) must be replaced by

$$|E_{\text{pot}}| = nhf, \qquad n = 1, 2, 3, \ldots. \tag{9.12}$$

Haas also realized that the Rydberg constant R is related to e, m, and h. However, his expression for R is incorrect by a numerical factor (which escaped his attention because he substituted an incorrect experimental value for e). Was there any mention of the radiative instability of his model? Or of spectra? Not a word. Like so many others, Haas simply could not cope with all desiderata at once.

Then there was the case of John William Nicholson,* a mathematical physicist (at that time) in Cambridge, later a Fellow of the Royal Society. Now hear this. On the basis of the Larmor–Thomson argument for zero vector sum of electron accelerations in an atomic ring (Section (c)) Nicholson argued in 1911 that a one-electron atom cannot exist. The simplest and lightest atoms, he proposed,[174] are, in this order, coronium (which is actually, as noted,** highly ionized iron), hydrogen, and nebulium (metastable nitrogen), with 2, 3, 4 electrons respectively. (The atomic weight of coronium was supposed to be about half that of hydrogen. Helium is considered to be a composite.) In a series of papers Nicholson associated spectral lines with various modes of vibration of electrons around their equilibrium orbits in the field of a central charge. His calculations appeared to fit numerous frequency ratios of lines. One of his papers, which appeared in June 1912, contains the following memorable phrases.[175]† '*Angular momentum*. It is possible to hold another view of Planck's theory, which may be briefly pointed out. Since the variable part of the energy of an atomic system of the present form is proportional to $mna^2\omega^2$, the ratio of energy to frequency is proportional to $mna^2\omega$, or $mnav$,

* For the life and work of Nicholson see Ref. 173.

** See Section (b).

† In the following lines n denotes the number of electrons, v the linear orbital velocity. A few trivial typos in the original text have been corrected.

which is the total angular momentum of the electrons round the nucleus. If, therefore, the constant h of Planck has, as Sommerfeld* has suggested, an atomic significance, it may mean that the angular momentum of an atom can only rise or fall by discrete amounts when electrons leave or return. It is readily seen that this view presents less difficulty to the mind than the more usual interpretation, which is believed to involve an atomic constitution of energy itself'.

For some of his atoms, Nicholson went on to calculate the angular momentum (in their ground states, we would say), obtaining integral multiples of $h/2\pi$, the multiples being 18 for hydrogen and 22 for nebulium.

In that bizarre way quantized angular momentum entered physics.** A Dutch proverb applies well to the precursors: They could hear the clock ringing, but could not find the clapper.

It is now time to return to Bohr.

6. Balmer decoded. The connected steps of Bohr's progress toward the quantum theory of the atom are:

The doctor's thesis → absorption of α-particles →
memorandum to Rutherford → the great trilogy of 1913,

the last expression, often used, being exalted and justified. The preceding account of Bohr's work up to the summer of 1912 demonstrates what was mentioned at the beginning of this chapter: His unawareness of the Balmer formula, which, it appears,[179] was brought to his attention early in 1913 by a Danish friend and colleague. It is indeed true (as Bohr had said) that this at once made everything clear to him: the first part[180] of his *On the Constitution of Atoms and Molecules* carries the date April 5, 1913, part two was ready in June, part three in August.

What was Bohr like in those days? According to the mathematician Richard Courant: 'Somewhat introvert, saintly, extremely friendly, yet shy'.[181] According to the physicist Rud Nielsen: 'He was an incessant worker and seemed always in a hurry. Serenity and pipe smoking came later . . . very friendly'.[182]

At the beginning of his first paper, Bohr refers, for the first time I believe, to the fact that, according to the then standard theory, 'the electron will no longer describe stationary orbits', but will fall inward to the nucleus due to energy loss by radiation. Then he plunges into the quantum theory. His first postulate: an atom has a state of lowest energy (he calls it the permanent state, we the ground state) which, *by assumption*, does not radiate, one of the most audacious hypotheses ever introduced in physics. It implied that the efforts of men like Larmor and J. J. Thomson to partially stabilize orbits against

* This is in reference to Sommerfeld's contribution to the Solvay Conference of 1911 in which, incidentally, reference is made to the work of Haas.[176]

** Cf. also Refs. 177 and 178.

radiative energy loss were irrelevant. More importantly, it implied that the electromagnetic theory of Maxwell and Lorentz reaches limits of validity on the atomic scale. Bohr's second postulate: higher 'stationary states' of an atom will turn into lower ones such that the energy difference E is emitted in the form of a light-quantum with frequency f given by $E = hf$.

True to my plan of not treating the quantum theory in this book, I shall not discuss the trilogy in detail. Rather, I shall deal only, and even that cursorily, with Bohr's analysis of the hydrogen atom.

'General evidence indicates that an atom of hydrogen consists simply of a single electron rotating round a positive nucleus of charge e', Bohr says in his first paper.[180] To this system he applies the equilibrium condition Eq. (9.7). Then he turns to the quantum condition. Recall his hypothesis in the memorandum to Rutherford: E_{kin} is proportional to f. Already in that document he tried to determine the proportionality factor.* This time, without ado, he introduces the assumption that

$$E_{\text{kin}} = \frac{n}{2} hf, \qquad n = 1, 2, 3, \ldots \tag{9.13}$$

(which implies Eq. (9.12)). That is not the quantum condition which we learned at school, which is

$$M = n \cdot \frac{h}{2\pi} \tag{9.14}$$

where M is the orbital angular momentum. However, the two are equivalent, as follows at once, for circular orbits,** from Eqs. (9.7), (9.8) and

$$M = 2\pi f a^2 m \tag{9.15}$$

This derivation of Eq. (9.14) is found in Bohr's paper, along with the comment: 'The possible importance of angular momentum in the discussion of atomic systems in relation to Planck's theory is emphasized by Nicholson'.

A brief digression on the influence of the precursors on Bohr's thinking. Bohr's first paper contains a reference to Haas, though Bohr said repeatedly in later years that he was unaware of Haas when he did his own work. Nicholson is also referred to in the paper, and, in addition, in Bohr's correspondence during the preceding months.[4] Later Bohr said that Nicholson's work was nonsense.[3] That is true but not relevant. I do not disagree with McCormmack's observation[183] that Nicholson's ideas on angular momentum could possibly have influenced Bohr at that time.†

* Cf. especially Rosenfeld's discussion of this point.[4]

** In his paper, Bohr confined himself mainly, but not exclusively, to circular orbits.

† Note also that, in 1913, Bohr gave no less than three distinct justifications (analyzed in Refs. 4 and 13) for his relation Eq. (9.13), two in the first paper, one shortly thereafter.[184]

The Balmer formula follows at once.* From Eqs. (9.7), (9.14) and (9.15):

$$a_n = \frac{n^2h^2}{4\pi^2e^2m} \tag{9.16}$$

which includes Eq. (9.11) for $n = 1$. From Eqs. (9.8), (9.9) and (9.16)

$$E_n = -\frac{2\pi^2e^4m}{n^2h^2}. \tag{9.17}$$

From Bohr's second postulate, the energy released in a transition from a state $n = a$ to a lower state $n = b$ is given by $E_a - E_b = hf_{ab}$, where

$$f_{ab} = R'\left(\frac{1}{b^2} - \frac{1}{a^2}\right) \tag{9.18}$$

$$R' = \frac{2\pi^2e^4m}{h^3}\ \text{sec}^{-1}.$$

This derivation of the Balmer formula is the first triumph of quantum dynamics.

R' is related to the Rydberg constant R defined in Eq. (9.1) by $R' = cR$, $c = 3 \times 10^{10}$ cm/sec. Balmer's best phenomenological fit to Eq. (9.2) corresponds to $R' = 3.29 \times 10^{15}\ \text{sec}^{-1}$, about 6 per cent too small but 'inside the experimental errors in the constants entering in the expression for the theoretical value'.[180]

Colleagues certainly did not accept at once all Bohr said. But from that time on, there was never any doubt that a hydrogen atom consists of a nucleus and of one electron.

The rich harvest on the constitution of atoms and molecules contained in other parts of the trilogy as well as in Bohr's subsequent papers will not be discussed here, except for a final remark—on He^+, singly ionized helium.

Very briefly, the following came to pass. For the case of an electron moving in the field of a nucleus with charge Ze, the corresponding Rydberg constant $R'(Z)$ equals Z^2R'. An experimental objection arose: it was found that for He^+

$$\frac{R'(\text{He}^+)}{R'(\text{H})} = 4.0016(\textit{experimental}) \tag{9.19}$$

instead of the naive guess: 4. Bohr at once rose to the occasion: he had neglected the mass of the electron relative to the mass of the nucleus. Thus, in addition to the factor Z^2, one had to modify Eq. (9.17) by substituting the reduced

* I short-cut Bohr's own derivation somewhat, and attach subscripts n to some quantities defined previously.

mass $mM/(M+m)$ for m, where M is the appropriate nuclear mass. Doing so Bohr found,[185] using $M(\text{He}^+)=4M(\text{H})$, $M(\text{H})=1835m$:

$$\frac{R'(\text{He}^+)}{R'(\text{H})}=4.00163 \tag{9.20}$$

'in exact agreement with the experimental value', and independent of the actual values of e and h. Immediately and graciously, the opposition conceded.[186] The response to the Bohr theory will be found in the next chapter.

Sources

As always, I worked from the original papers but in addition have made extensive use of the voluminous literature on the subjects treated in this chapter. The following books and articles were particularly helpful to me.
Nineteenth Century Spectroscopy, Spectral Formulae. By far the most important are Kayser's historical chapters in Volume 1 (Chapter 1) and Volume 2 (Chapter 8) of his *Handbuch*.[14] Also useful are Rydberg's report to the Paris Conference[7] and a monograph by McGucken.[42]
On Kelvin: The biography by S. P. Thompson.[73]
On J. J. Thomson: The biography by the fourth Lord Rayleigh.[89] The biography by J. J.'s son[116] adds personal touches.
On early atom models through J. J. Thomson: An article by Heilbron,[68] reproduced in a book by the same author.[69]
On Rutherford: Volume 2 of his collected works;[141] the biographies by Feather,[132] Andrade,[139] and Eve.[144] On α-scattering: an article by Heilbron,[147] also reproduced in his book.[69] Particularly handy is *Rutherford at Manchester*,[2] a book which contains reprints of the Geiger–Marsden papers,[138,149] of Rutherford's paper on the nucleus,[143] and of part one of the Bohr trilogy.[180]
On Haas: A biography by Hermann.[187]
On Nicholson: An essay by McCormmack.[183]
On Bohr: The first two volumes of his collected works.[160,165] On α-absorption: a monograph by Hoyer.[188] On the memorandum to Rutherford and the trilogy: articles by Rosenfeld;[4] and by Heilbron and Kuhn.[13] Especially interesting is the transcript of the interviews of Bohr by Kuhn, made available to me by the Niels Bohr Library, American Institute of Physics, New York.[3]

Finally, I mention the historiographical review by Heilbron,[189] containing bibliography up to 1968 of papers of historical interest dealing with the quantum theory from 1900 to the early 1930s.

References

1. N. Bohr, *Proc. Phys. Soc. London* **78**, 1083, 1961.
2. N. Bohr, in *Rutherford at Manchester*, Ed. J. B. Birks, p. 114, Benjamin, New York 1963.

3. Oral history interview of N. Bohr by T. S. Kuhn, 1962, in Archives of the History of Quantum Physics, Niels Bohr Library, American Institute of Physics, New York.
4. L. Rosenfeld, Introduction to *On the constitution of atoms and molecules*, Benjamin, New York 1963.
5. Rayleigh, *Phil. Mag.* **44**, 356, 1897.
6. W. Sutherland, *Phil. Mag.* **2**, 245, 1901.
7. J. Rydberg, in *Rapports présentés au Congrès International de Physique*, Vol. 2, p. 200, Eds. Ch. Guillaume and L. Poincaré, Gauthier-Villars, Paris 1900.
8. C. Fabry and H. Buisson, *Comptes Rendus* **154**, 1500, 1912.
9. K. F. Herzfeld, *Sitz. Ber. Akad. Wiss. Wien* **121**, 593, 1912.
10. E. A. Partridge, *J. Frankl. Inst.* **149**, 193, 1900.
11. R. W. Wood, *Phil. Mag.* **18**, 530, 1909.
12. W. Ritz, *Comptes Rendus* **145**, 178, 1907; *Ann. der Phys.* **25**, 660, 1908; *Astrophys. J.* **28**, 237, 1908.
13. J. L. Heilbron and T. S. Kuhn, *Hist. St. Phys. Sci.* **1**, 211, 1969, repr. in J. L. Heilbron, *Historical studies in the theory of atomic structure*, Arno Press, New York 1981.
14. H. G. J. Kayser, *Handbuch der Spectroscopie*, 6 volumes, Hirzel, Leipzig 1900–12. (Additional parts appeared later.)
15. R. C. Dales, *The scientific achievement of the middle ages*, Chapter 5, Univ. of Pennsylvania Press, Philadelphia 1978.
16. G. Kirchhoff, *Ber. der Berliner Akad. 1859*, p. 662; transl. *Phil. Mag.* **19**, 193, 1860.
17. A. Comte, *Cours de la philosophie positive*, Vol. 2, pp. 2, 4, 5, Bachelier, Paris 1835, repr. by Editions Anthropos, Paris 1968.
18. U. Benz, *Arnold Sommerfeld*, p. 111, Wissensch. Verlag MBH, Stuttgart 1975.
19. A. Sommerfeld, *Atombau und Spektrallinien*, Preface to the first edition, Vieweg, Braunschweig 1919.
20. A. Ångström, *Ann. der Phys. und Chem.* **144**, 300, 1872.
21. A. Ångström, *Kongl. Svensk. Vet. Ak. Handl. 1852*, p. 229; transl. *Ann. der Phys. und Chem.* **94**, 141, 1855; *Phil. Mag.* **9**, 327, 1855.
22. J. Plücker, *Ann. der Phys. und Chem.* **107**, 497, 638, 1859.
23. A. Ångström, *Recherches sur le spectre solaire*, p. 31, Uppsala Press 1868.
24. S. Bashkin and J. O. Stoner, *Atomic energy levels and Grotrian diagrams*, Vol. 1, p. 2, Elsevier, New York 1975.
25. J. Plücker, *Ann. der Phys. und Chem.* **104**, 113, 1858; *Phil. Mag.* **16**, 408, 1858.
26. G. Kirchhoff, *Ber. der Berliner Akad. 1859*, p. 783; *Ann. der Phys. und Chem.* **109**, 275, 1859.
27. Ref. 14, Vol. 1, p. 85.
28. G. Kirchhoff and R. Bunsen, *Ann. der Phys. und Chem.* **110**, 160, 1860; transl. *Phil. Mag.* **20**, 89, 1860.
29. G. Kirchhoff and R. Bunsen, *Ann. der Phys. und Chem.* **113**, 337, 1861; transl. *Phil. Mag.* **22**, 329, 448, 1861.
30. Cf. A. Mitscherlich, *Ann. der Phys. und Chem.* **116**, 499, 1862; H. E. Roscoe and J. Clifton, *Chem. News* **5**, 233, 1862.
31. J. Plücker and W. Hittorf, *Phil. Trans. Roy. Soc.* **155**, 1, 1865.
32. W. Huggins, *Proc. Roy. Soc. A* **13**, 492, 1864.
33. J. C. Maxwell, '*Atom*', *Enc. Britannica*, 9th edn. 1875; repr. in *Collected Works*, Vol. 2, p. 445, Dover, New York.

34. J. N. Lockyer, *Phil. Trans. Roy. Soc.* **159**, 425, 1869.
35. E. Rutherford, *Proc. Roy. Soc. A* **97**, 374, 1920; see esp. p. 395.
36. I. S. Bowen, *Nature* **120**, 473, 1927; *Astrophys. J.* **67**, 1, 1928.
37. Ref. 14, Vol. 1, pp. 91–8.
38. G. Kirchhoff, *Ann. der Phys. und Chem.* **118**, 94, 1862; transl. *Phil. Mag.* **25**, 250, 1863.
39. J. Balmer, *Verh. Naturf. Ges. Basel* **7**, 548, 1885.
40. S. R. Amin, C. D. Caldwell, and W. Lichten, *Phys. Rev. Lett.* **47**, 1234, 1981.
41. Ref. 14, Vol. 1, pp. 123–7.
42. W. McGucken, *Nineteenth century spectroscopy*, Ch. 3, Johns Hopkins Press, Baltimore 1969.
43. G. J. Stoney, *Phil. Mag.* **41**, 291, 1871.
44. G. J. Stoney and J. E. Reynolds, *Phil. Mag.* **42**, 41, 1871.
44a. G. J. Stoney, *Nature* **21**, 508, 1880.
45. J. Balmer, *Ann. der Phys. und Chem.* **25**, 80, 1885.
46. J. Balmer, *Verh. Naturf. Ges. Basel* **7**, 750, 1885.
47. A. Hagenbach, *Naturw.* **9**, 451, 1921. Several dates in this paper are incorrect.
48. E. Hagenbach, *Verh. Naturf. Ges. Basel* **8**, 242, 1886.
49. J. Balmer, *Verh. Naturf. Ges. Basel* **11**, 448, 1897, also *Ann. der. Phys.* **60**, 380, 1897; transl. *Astrophys. J.* **5**, 199, 1897.
50. H. Konen, *Das Leuchten der Gase und Dämpfe*, pp. 71 ff. Vieweg, Braunschweig 1913.
51. J. R. Rydberg, *Kongl. Svenska Vet. Handl.* **23**, Nr. 11, 1890; summarized in J. R. Rydberg, *Phil. Mag.* **29**, 331, 1890.
52. A. von Oettingen, German transl. of Ref. 51, p. xiii, Ostwalds Klassiker Nr. 196, Akademische Verlagsges., 1922.
53. Ref. 7, p. 203.
54. W. Pauli, *Lunds Universitets Årsskr.* **50**, part 2, 1954.
55. W. Ritz, *Gesammelte Werke*, p. 162, Gauthier-Villars, Paris 1911.
56. W. Thomson, *Popular lectures and addresses*, Vol. 2, p. 164, Macmillan, London 1894.
57. A. Pais, *Subtle is the Lord*, Chapters 5, 20, Oxford Univ. Press 1982.
58. W. H. Brock, *The atomic debates*, Leicester Univ. Press 1967.
59. M. J. Nye, *Molecular reality*, Elsevier, New York 1972.
60. W. Thomson, *Phil. Mag.* **34**, 15, 1867; *Proc. Roy. Soc. Edinburgh* **6**, 94, 1867.
61. W. Prout, *Ann. of Philosophy* **6**, 321, 1815.
62. W. Prout, *Ann. of Philosophy* **7**, 111, 1816.
63. W. V. Farrar, *British J. Hist. of Sci.* **2**, 297, 1964.
64. G. G. Stokes, *Mathematical and physical papers*, Vol. 3, p. 267, Cambridge Univ. Press 1901.
65. G. J. Stoney, *Phil. Mag.* **36**, 132, 1868,
66. J. C. Maxwell, *Scientific papers*, Vol. 2, p. 418.
67. Ref. 57, Chapter 20.
68. J. L. Heilbron, in *History of twentieth century physics*, p. 40, Academic Press, New York 1977.
69. J. L. Heilbron, *Historical studies in the theory of atomic structure*, Arno Press, New York 1981.
70. G. G. Stokes, *Mem. Proc. Manchester Lit. Phil. Soc.* **41**, 1, 1897; cf. also L. Badash, *Am. J. Phys.* **33**, 128, 1965.

71. H. von Helmholtz, *Crelle's Journal* **55**, 25, 1858.
72. P. G. Tait, *Phil. Mag.* **33**, 485, 1867.
73. S. P. Thompson, *The life of Lord Kelvin*, Vol. 1, p. 512, Chelsea Publ. Co., New York 1976.
74. Ref. 73, Vol. 1, p. 514.
75. Ref. 73, Vol. 2, p. 1223.
76. R. H. Silliman, *Isis* **54**, 461, 1963.
77. A. Schuster, *The progress of physics*, p. 34, Arno Press, New York 1975.
78. J. J. Thomson, *A treatise on the motion of vortex rings*, Macmillan, London, 1883; repr. by Dawsons of Pall Mall, London 1968; also *Proc. Roy. Soc. A* **33**, 145, 1881; *Phil. Mag.* **13**, 493, 1881.
79. Ref. 73, Vol. 2, pp. 1046–9.
80. Cf. L. D. Landau and E. M. Lifshitz, *Fluid mechanics*, p. 15, Addison-Wesley, Reading, Mass. 1959.
81. R. N. Lockyer, *Nature* **24**, 39, 1881.
82. W. Crookes, *J. Chem. Soc.* **53**, 487, 1888.
82a. E. N. da C. Andrade, *Proc. Roy. Soc. A* **244**, 437, 1958.
83. J. J. Thomson, *Phil. Mag.* **44**, 293, 1897, esp. p. 311.
84. J. J. Thomson, *Phil. Mag.* **48**, 547, 1899, esp. p. 565.
85. G. F. FitzGerald, *Nature* **62**, 524, 1900.
86. E. Rutherford, *Trans. Roy. Soc. Canada* **8**, 79, 1902.
87. J. J. Thomson, *Electricity and matter*, p. 114, Scribner, New York 1904.
88. R. J. Strutt, *The Becquerel rays and the properties of radium*, p. 183, Edward Arnold, London 1904.
89. Rayleigh, *The life of Sir J. J. Thomson*, p. 140, Cambridge Univ. Press 1942.
90. A. L. Kimball, in *Congress of arts and science*, Ed. H. J. Rogers, Vol. 4, p. 69, Houghton Mifflin, New York 1906.
91. N. R. Campbell, *Modern electrical theory*, p. 251, Cambridge Univ. Press 1907.
92. J. J. Thomson, Ref. 87, p. 108.
93. Kelvin, *Phil. Mag.* **8**, 528, 1904; **10**, 695, 1905.
94. E. Rutherford, *Radioactivity*, p. 342, Cambridge Univ. Press 1904.
95. S. Earnshaw, *Trans. Cambr. Phil. Soc.* **7**, 97, 1842.
96. W. T. Scott, *Am. J. of Phys.* **27**, 418, 1959.
97. J. Larmor, *Phil. Mag.* **44**, 503, 1897.
98. J. Larmor, *Aether and matter*, p. 27 and Chapter 14, Cambridge Univ. Press 1900.
99. J. Jeans, *Phil. Mag.* **2**, 421, 1901, Secs. 2–8.
100. Ref. 98, p. 193.
101. P. Lenard, *Ann. der Phys.* **12**, 714, 1903.
102. Ref. 101, p. 743, footnote 1.
103. I. Runge, *Carl Runge*, p. 197, Vandenhoeck and Ruprecht, Goettingen 1949.
104. J. Perrin, *Rev. Scientifique* **15**, 447, 1901.
105. H. Nagaoka, *Proc. Tokyo Math. Phys. Soc.* **2**, 92, 129, 240, 1903–6.
106. H. Poincaré, see p. 317 in 'The value of science'; repr. in *The foundations of science*, Sci. Press, New York 1913.
107. E. Rutherford, *Phil. Mag.* **21**, 669, 1911.
108. Ref. 69, pp. 52, 53; also E. Yagi, *Jap. St. Hist. Sci.* 1964, No. 3, p. 29.
108a. F. Re, *Comptes Rendus* **136**, 1393, 1903.
109. A. M. Mayer and R. S. Woodward, *Biogr. Mem. Nat. Acad. Sci.* **8**, 243, 1919.
110. A. M. Mayer, *Sound*, Appleton, New York 1882.

111. A. M. Mayer, *Am. J. Sci.* **15**, 276, 477; **16**, 247, 1878; repr. in *Nature* **17**, 487; **18**, 258, 1878.
112. W. Thomson, *Nature* **18**, 13, 1878.
113. Ref. 78, Section 54.
114. Rayleigh, *Obit. notices Fell. Roy. Soc.* **3**, 587, 1941.
115. Ref. 89, p. 292.
116. G. P. Thomson, *J. J. Thomson and the Cavendish Laboratory in his day*, p. 115, Nelson, London 1964.
117. Ref. 89, pp. 136, 141, 142, 151.
118. Ref. 3, interview on November 1, 1962.
119. Kelvin, *Phil. Mag.* **6**, 257, 1902.
120. J. J. Thomson, *Phil. Mag.* **6**, 673, 1903; cf. also J. J. Thomson, *Proc. Cambr. Phil. Soc.* **13**, 49, 1904.
121. J. J. Thomson, *The Royal Institution Library of Science*, Eds. W. L. Bragg and G. Porter, Vol. 6, p. 165, Elsevier, New York 1970.
122. J. J. Thomson, *Phil. Mag.* **7**, 237, 1904.
123. J. J. Thomson, *Phil. Mag.* **11**, 769, 1906.
124. *Nobel Lectures in Physics 1901–22*, p. 139, Elsevier, New York 1967.
125. Cf. J. J. Thomson, *Conduction of electricity through gases*, 2nd. edn, p. 325, Cambridge Univ. Press, 1906.
126. J. Heilbron, *Arch. Hist. Ex. Sci.* **4**, 247, 1968; repr. in Ref. 69.
127. M. Born, *Phys. Zeitschr.* **10**, 1031, 1909.
128. J. J. Thomson, *Phil. Mag.* **26**, 792, 1913.
129. J. J. Thomson, in *La structure de la matière*, p. 1, Gauthier-Villars, Paris 1921.
130. L. Badash, *Rutherford and Boltwood*, p. 235, Yale Univ. Press, New Haven, Conn. 1969.
131. E. Rutherford, *Phil. Mag.* **12**, 143, 1906.
132. N. Feather, *Lord Rutherford*, p. 117, Blackie, Glasgow 1940.
133. E. Rutherford and H. Geiger, *Proc. Roy. Soc. A* **81**, 162, 1908.
134. H. Geiger, *Proc. Roy. Soc. A* **81**, 174, 1908.
135. Ref. 132, p. 129.
136. C. A. Fleming, *Biogr. Mem. Fell. Roy. Soc.* **17**, 463, 1971.
137. E. Marsden, in *Rutherford at Manchester*, p. 8, Ed. J. B. Birks, Benjamin, New York 1963.
138. H. Geiger and E. Marsden, *Proc. Roy. Soc. A* **82**, 495, 1909.
139. E. N. da C. Andrade, *Rutherford and the nature of the atom*, p. 111, Doubleday, New York 1964.
140. H. Geiger, *Proc. Roy. Soc. A* **83**, 492, 1910.
141. H. Geiger, in *Collected papers of Rutherford*, Vol. 2, p. 295, Interscience, New York 1963.
142. Ref. 141, p. 212.
143. E. Rutherford, *Phil. Mag.* **21**, 669, 1911.
144. A. S. Eve, *Rutherford*, p. 193, Cambridge Univ. Press 1939.
145. A. S. Russell, *Proc. Phys. Soc. London* **64**, 217, 1951.
146. Ref. 132, p. 129.
147. J. Heilbron, *Arch. Hist. Ex. Sci.* **4**, 247, 1967; see also Ref. 69.
148. Ref. 143, Section 7, and Ref. 144, p. 195.
149. H. Geiger and E. Marsden, *Wiener Ber.* **121**, 2361, 1912.
150. H. Geiger and E. Marsden, *Phil. Mag.* **25**, 604, 1913.

151. T. Trenn, *Isis* **65**, 74, 1974.
152. Ref. 132, Chapter 5.
153. E. Rutherford, *Radioactive substances and their radiations*, p. 184, Cambridge Univ. Press 1913.
154. Ref. 153, p. 619.
155. E. Rutherford, in *La structure de la matière*, p. 53, Gauthier-Villars, Paris 1921.
156. E. Rutherford, *Proc. Roy. Soc. A* **90**, 1914, insert following p. 462.
157. Ref. 153, p. 622.
158. A. Einstein, *Ann. der Phys.* **20**, 199, 1906.
159. A. Pais, in *Niels Bohr*, p. 215, Ed. S. Rozental, Wiley, New York 1967.
160. *Niels Bohr, collected works*, Vol. 1, p. 291, especially pp. 300, 379, 395; Ed. J. R. Nielsen, North Holland, Amsterdam 1972.
161. Ref. 3, interviews on November 1 and 11, 1962.
162. Ref. 3, interview on November 1, 1962.
163. C. G. Darwin, *Phil. Mag.* **23**, 901, 1912.
164. N. Bohr, *Phil. Mag.* **25**, 10, 1913.
165. *Niels Bohr, collected works*, Vol. 2, p. 135, Ed. U. Hoyer, North Holland, Amsterdam, 1981.
166. P. Lenard, *Ann. der Physik*, **11**, 636, 1903; cf. also ibid. **17**, 197, 1905.
167. Ref. 123, p. 774; cf. also P. V. Bevau, *Proc. Roy. Soc. A* **84**, 209, 1911.
168. J. Stark, *Jahrb. der Radioakt. und Elektr.* **4**, 231, 1907, esp. p. 244.
169. W. Wien, *Ann. der Phys.* **30**, 349, 1909, cf. also F. Horton, *Phil. Mag.* **22**, 214, 1911.
170. J. Stark, *Prinzipieen der Atomdynamik*, Vol. 2, Secs. 19, 25, Hirzel, Leipzig 1911.
171. Ref. 3, interview on November 7, 1962.
172. A. E. Haas, *Wiener Berichte IIa* **119**, 119, 1910; *Jahrb. der Radioakt. und Elektr.* **7**, 261, 1910; *Phys. Zeitschr.* **11**, 537, 1910.
173. W. Wilson, *Biogr. mem. Fell. Roy. Soc.* **2**, 209, 1956.
174. J. W. Nicholson, *Phil. Mag.* **22**, 864, 1911.
175. J. W. Nicholson, *Monthly Not. Roy. Astr. Soc.* **72**, 677, 1912.
176. A. Sommerfeld, in *Théorie de rayonnement et les quanta*, p. 362, Eds. P. Langevin and M. de Broglie, Gauthier-Villars, Paris 1912.
177. P. Langevin, ref. 176, p. 403.
178. M. Abraham and R. Gans, *Phys. Zeitschr.* **12**, 952, 1911.
179. Ref. 4, p. xl.
180. N. Bohr, *Phil. Mag.* **26**, 1, 1913.
181. R. Courant, Ref. 159, p. 301.
182. J. R. Nielsen, *Phys. Today* **16**, October 1963, p. 22.
183. R. McCormmack, *Arch. Hist. Ex. Sci.* **3**, 160, 1966.
184. N. Bohr, *Fysisk Tidsskr.* **12**, 97, 1914, English transl. in Ref. 165, p. 281.
185. N. Bohr, *Nature* **92**, 231, 1913.
186. A. Fowler, *Nature* **92**, 232, 1913.
187. A. Hermann, *Arthur Erich Haas, der erste Quantumansatz für das Atom*, Battenberg, Stuttgart 1965.
188. U. Hoyer, *Die Geschichte der Bohrschen Atomtheorie*, Weinheim, Stuttgart 1974.
189. J. Heilbron, *Hist. of Sci.* **7**, 90, 1968.

10

'It was the epoch of belief, it was the epoch of incredulity.'

1. Reactions to Bohr's Theory. First I list without comment a number of responses dating from years immediately following the appearance of Bohr's papers in 1913. Thereupon I add some remarks of my own.

From my jottings about my visit to Fokker.[1] 'In 1913–14 Fokker studied with Einstein in Zürich* where he gave the first colloquium on Bohr's theory of the hydrogen atom. Einstein, von Laue, and Stern in the audience. Einstein did not react immediately but kept a pensive silence. In 1914 F. spent six weeks with Rutherford in Manchester where he met Bohr** who asked everyone: "Do you believe it?"'

A recollection Otto Stern told me when I visited him in Berkeley in 1960. It was not long after the publication of Bohr's papers that Stern and von Laue went for a walk up the Uetliberg, a small mountain just outside Zürich. On the top they sat down and talked about physics, in particular about the new atom model. There and then they made the 'Uetli Schwur'†: If that crazy model of Bohr turned out to be right, then they would leave physics. It did and they didn't.

Hevesy in a letter to Bohr, September 1913.[3] 'I spoke this afternoon to Einstein . . . I asked him about his view of your theorie . . . He told me, it is a very interesting one if it is right and so on and he had very similar ideas many years ago but had no pluck to develop it.' When Hevesy explained Bohr's work on helium,[4] Einstein commented: 'This is an *enormous achiewement*' (H.'s orthography).

Robert John Strutt on a discussion with his father, the third Lord Rayleigh, then past seventy.[5] 'I asked him if he had seen Bohr's paper on the hydrogen atom which had then recently appeared. He replied, "Yes, I have looked at it, but I saw it was no use to me. I do not say that discoveries may not be

* See Ref. 2.

** From October 1914 until July 1916, Bohr was Schuster reader in mathematical physics at the University of Manchester.

† 'Pledge of the Uetli', a pun on the *Ruetli Schwur*, according to legend the event that marked the federation of some Swiss cantons.

made in that sort of way. I think very likely they may be. But it does not suit me". I have quoted these words exactly as they were spoken.'

Early September 1913. Bohr receives a complimentary letter from Sommerfeld who clearly is much interested in the new theory.[6]

From a report in *Nature*[7] of a debate on 'Radiation', held on September 12, 1913, at the British Association meeting in Birmingham: 'Mr Jeans . . . quoted the recent work of Dr. Bohr who has arrived at a convincing and brilliant explanation of the laws of spectral series . . . Dr. Bohr . . . gave a short explanation of his atom . . . Prof. Lorentz intervened to ask how the Bohr atom was mechanically accounted for. Dr. Bohr acknowledged that this part of his theory was not complete, but the quantum theory being accepted, some sort of scheme of the kind suggested was necessary'. This was Bohr's first report on his theory before an international forum.

From *The Times*, 13 September, page 6: 'Full dress debate on Radiation . . . a pitched battle between the adherents of Young and Fresnel, Maxwell and Hertz on the one hand, and the revolutionary followers of Planck, Einstein and Nernst on the other . . . Sir J. J. Thomson with his commanding voice and genial presence taking, as usual, an independent view of his own . . . keen delight among the audience . . . when Prof. Lorentz remarked that Sir Joseph Thomson's model of an atom* was ingenious—it could not be anything else—but the point was whether it represented the truth, a point which seemed to occur to the audience at the same time as to the laughing Cambridge professor'.

The Times, September 13, page 10, contains rerference to Dr. Bohr and his theory—the first press report ever (outside his native country) of young Bohr's scientific activities.

The precursors.[9] Nicholson in a letter to *Nature* in which he comments sympathetically on Bohr's work: 'It is not difficult to obtain fair mechanical [i.e. classical] models of atoms the angular momentum of which can only have a discrete set of values'.[10] Haas in a letter to Bohr: 'I have eagerly studied your highly important papers'.[11]

At the second Solvay Conference (October 1913), J. J. Thomson, rapporteur, and Rutherford, discussant, both refer to work of Bohr, but not to his theory of the atom.[12]

Soddy on Bohr's theory in his Annual Progress Report to the Chemical Society for 1913:' . . . In striking accord with experimental determinations'.[13]

Rutherford, in February 1914: 'The theories of Bohr are of great interest and importance to all physicists as the first definite attempt to construct simple atoms and molecules and explain their spectra'.[14] In March 1914: 'While it is

* This is his model of 1913, referred to earlier,[8] in which Thomson proposed a modification of Coulomb's law.

too early to say whether the theories of Bohr are valid, his contributions . . . are of great importance and interest'.[15] In August 1914: 'N. Bohr has faced the difficulties [of atomic structure] by bringing in the idea of the quantum. At all events there is something going on which is inexplicable by the older mechanics'.[16]

Studies of Thomson-type models continue. In 1914 a new model with modified Coulomb potential is proposed: 'It does not seem that we are yet under compulsion to forsake the laws of ordinary dynamics in connection with atomic phenomena'.[17] In 1915 an attempt is made to marry the Thomson atom to the quantum theory.[18]

Then the building of alternative models comes to an end.

Except for Thomson.

In a lecture given in 1914 Thomson does mention Bohr.[19] In 1923 he gives five lectures at the Franklin Institute in which he discusses, without using or mentioning the quantum theory or the names of Rutherford and Bohr, a classical atom model with a nucleus and with a modified Coulomb law between the nucleus and the electrons.[20] In 1928 he expresses the opinion that electron diffraction phenomena signify that the electron has a substructure.[21] Ultimately, at age eighty, he wrote: 'At the end of 1913 Niels Bohr published the first of a series of researches on spectra, which it is not too much to say have in some departments of spectroscopy changed chaos into order, and which were, I think, the most valuable contributions which quantum theory has ever made to physical science'.[22]*

Bohr could hardly be called a reconteur, but there were a few stories he was fond of telling and we never grew tired of listening to even if we had heard them before. One of these was about a visitor to the Bohr country home in Tisvilde who noticed a horseshoe hanging over the entrance door. Puzzled, he turned to his host and asked him if he really believed that this brings luck. 'Of course not', Bohr replied, 'but I am told it works even if you don't believe in it'.[25]

The old quantum theory, developed between 1900 and 1925, was much like that horseshoe above the door.

In 1900, Planck did not at once realize that his blackbody radiation law was in conflict with the theoretical concepts of his day, the classical theory, but it did not take him long to find out.[26] The foundations of his law were to remain a mystery until 1926, yet Planck knew all along that he had to be right. The superb agreement between his law and the experimental data simply could not be fortuitous. In 1905, when Einstein proposed the light-quantum, he

* For other reactions to Bohr's trilogy see Ref. 23. For a review of the first decade following Bohr's 1913 papers see Ref. 24.

understood at once that his hypothesis flouted classical physics. Initially there was little hard quantitative experimental evidence in support of his view, yet he never doubted that he was on the right track.[27] Like Einstein in 1905, so Bohr in 1913 knew at once that his theory was in conflict with classical theory. As mentioned earlier, already in 1912 he had written about his quantum hypothesis: 'No attempt will be given at a mechanical [i.e. classical] foundation as it seems hopeless'.[28] Like Planck's in 1900, so Bohr's work of 1913 found immediate experimental support. The foundations might be obscure but it could not be accidental that his expression for the Rydberg constant in terms of the charge and mass of the electron and of Planck's constant agreed so well with experiment and that his answer for the helium/hydrogen ratio was correct to five significant figures.[29] His theory appeared to work even if you didn't believe in it because of inconsistent assumptions. Thus one can well understand Bohr's buttonholing everyone in 1914. 'Do you believe it?'

Nevertheless Bohr's ideas were rapidly taken seriously, even if at times with justified reservations, as in the case of Rutherford. It is only natural that the building of alternative models would not come to an abrupt end, or that some, like Nicholson, did not at once catch the main point. It is natural as well that older physicists, Rayleigh for example, would not go along with the new style of doing physics. Expressions of reservation were not confined to older generations, however. Stern was 25, von Laue 34, when in 1913 the two of them walked up the Uetliberg. Nor did approbation come from youngsters only. Sommerfeld was 45 when he complimented Bohr.

The decade preceding the beginnings of quantum mechanics (1925) was more one of truth than of proof. There was promise, yet deepening crisis. It was a decade resembling in some respects the one of which Charles Dickens wrote: 'It was the epoch of belief, it was the epoch of incredulity'.

J. J. Thomson obviously represents a special case. Who can refrain from admiring his contributions to experimental physics, the measurements of the electron's mass and charge, the work on X-rays and on the conduction of electricity through gases?[30] Who can fail to respect his theoretical analysis of vortex rings and his discussion in 1906 of the number of electrons in an atom?[31] His manifest obstinacy should be judged, it seems to me, with Lorentz' lenience. By his achievements he ranks among the imposing founders of twentieth-century physics.

Often in his later years Bohr used to reminisce about earlier times. On such occasions he would nearly always hark back[32] to the early days of quantum mechanics, rarely to 1913 and the years immediately following. From my own experiences in talking with him I would say that to Bohr himself his struggles with complementarity were ultimately of greater importance than his work on the hydrogen atom, on the old quantum theory.

2. Causality. The first to raise the issue of causality in regard to Bohr's theory was Rutherford, who did so even before Bohr's paper on the hydrogen atom[33] had appeared. On March 6, 1913, Bohr had sent this paper to Rutherford, requesting that it be submitted to *Philsosophical Magazine*.[34] In his reply[35] (March 20), Rutherford remarked: 'There appears to me one grave difficulty in your hypothesis which I have no doubt you fully realize, namely, how does an electron decide with what frequency it is going to vibrate and when it passes from one stationary state to another? It seems to me that you would have to assume that the electron knows beforehand where it is going to stop'.

I find it as worthy of note that Rutherford would make this well-taken point as that he would fail to make the same observation in regard to his own work. It was remarked before[36] that Rutherford's introduction in 1900 of the lifetime concept provided the very first instance of a conflict with classical causality; it was further pointed out[37] that he remained curiously silent on this issue in subsequent years.

Rutherford's letter bears of course on Bohr's relation $E_m - E_n = hf_{mn}$. The situation was more serious, however, than Rutherford thought or Bohr knew. In 1917 Einstein pointed out that the momentum of a photon emitted in an atomic transition has a magnitude hf_{mn}/c. How, Einstein asked, does the photon know in which *direction* to move?[38] Momentum considerations such as these play no role in the Bohr theory of 1913. In fact, Bohr was among the last to accept the photon as a real particle.[39]

Where Bohr, Einstein, Rutherford, and some select others differed is far less important, however, than where they agreed: long before quantum mechanics came along they knew that causality had become a grave issue. What quantum mechanics did subsequently was to assert that classical causality was irrevocably gone.

3. The fine structure constant. Selection rules. It came to pass, soon after 1913 when quantum dynamics began, that the first steps were made toward a union which has not been fully consummated to this day, and which is of immense importance for all that is to follow: relativity met quantum theory. The *apropos* was that the Balmer formula (9.1) does not explain an important detail of the hydrogen spectrum. Every line predicted by that formula is in actuality a narrowly-split set of lines. In 1914–15 Sommerfeld made a first attempt at explaining this phenomenon, called fine structure. In the course of doing so he had occasion[40] to introduce a 'neue Abkürzung', a new abbreviation, the symbol 'α':

$$\alpha = \frac{2\pi e^2}{hc}. \tag{10.1}$$

This dimensionless quantity has been known ever since as the fine structure constant. It is one of the most fundamental numbers in all of physics. More about that later.

I introduce the subject of fine structure with a mini-calendar of events.

1892. Michelson, that master of high-precision measurement, observes that, for the hydrogen lines* H_α and H_γ, 'the visibility curve is practically the same as that for a double source'.[41] What he had found was that the intensity of each of these two lines as a function of frequency shows a double peak. Along with his other distinctions, Michelson must therefore (as best I know) be considered as the discoverer of fine structure.

March 1914. Commenting on these and subsequent results, Bohr writes: 'The lines of the ordinary hydrogen spectra . . . appear as close doublets . . . it seems probable that the lines are not true doublets but are due to an effect of the electric discharge'.[42]

September 1914. From a paper reporting on experiments done at Imperial College, London: 'There is a general consensus of opinion that [H_α] and [H_β] are double, but considerable disagreement as to the precise separations of the components. Balmer's formula has been found inexact'.[43]

January 1915. Bohr now takes the results seriously: 'Assuming that the orbit of the electron is circular but replacing the expressions for the energy and momentum of the electron by those deduced on the theory of relativity', he finds corrections of order α^2 to the Balmer formula.[44] The order is correct but the corrections are incomplete.

Winter 1914–15. Sommerfeld computes relativistic orbits for hydrogen-like atoms.[45] Paschen, aware of these studies, carefully investigates fine structures, especially for He^+, singly ionized helium.

January 6, 1916. Sommerfeld announces his fine structure formula,[40] citing results to be published by Paschen in support of his answer.

February 1916. Einstein to Sommerfeld: 'A revelation!'[46]

March 1916. Bohr to Sommerfeld: 'I do not believe ever to have read anything with more joy than your beautiful work'.[47]

September 1916. Paschen publishes his work,[48] acknowledging Sommerfeld's 'indefatigable efforts'.

By taking into account the influence of relativity theory, Sommerfeld had shown that the orbits of an electron in the field of a nuclear charge Ze are approximate ellipses which exhibit a perihelion precession. His generalization of the Balmer formula can be written as

$$E_{n,k} = -R'\left[\frac{Z^2}{n^2}+\frac{\alpha^2 Z^4}{n^3}\left(\frac{1}{k}-\frac{3}{4n}\right)\right]+O(\alpha^4)$$
$$n=1,2,3,\cdots;\qquad k=1,2,\cdots,n. \qquad (10.2)$$

* The symbols for the hydrogen lines were defined in Chapter 9, Section (b), part 1.

R' is the Rydberg constant (Eq. (9.18)). The first term in the square brackets gives the Balmer formula, the second the main corrections. k is related to the orbital angular momentum quantum number l by

$$k = l + 1. \tag{10.3}$$

Thus fine structure occurs because relativistic effects lift the degeneracy in l characteristic for the non-relativistic theory of states in a $1/r$-potential.

According to Bohr all transitions from any n to any smaller n correspond to a possible spectral line. If all changes in k were also allowed, then (to give an example) the fine structure of the line $\lambda = 4686$ Å, studied in detail by Paschen, which corresponds to an $n = 4 \rightarrow n = 3$ transition, should consist of 12 components. Most of these were eventually seen by Paschen,[49] but only if an external field was present. If the latter is absent then, Paschen found, the number of components drops considerably.

There now followed a period of trial and error* until Bohr[51] and, independently, Adalbert Rubinowicz,[52] a collaborator of Sommerfeld, were able to explain, almost, Paschen's results. With the help of semi-classical arguments, they showed (or, better, proposed) that the difference Δl between initial and final l-value is restricted to satisfy the selection rule**

$$\Delta l = \pm 1 \tag{10.4}$$

reducing the number of components in the He^+ line from 12 to 5. This was a distinct improvement, but not perfection. It transpired that experimentally there were not 5 but 6 components: there was clear evidence for the transition $(n, k) = (4,1) \rightarrow (3,1)$—the 4686 Å line with the frequency precisely as predicted by Sommerfeld's Eq. (10.2), but strictly forbidden by the rule Eq. (10.4)![53]

In 1925 two young Dutch physicists from Leiden, George Uhlenbeck and Samuel Goudsmit, believed they saw a way out of this problem: Keep Eq. (10.2) but replace Eq. (10.3) by

$$k = j + \tfrac{1}{2} \tag{10.5}$$

which implies that j is half-integer. Further, replace Eq. (10.4) by†

$$\Delta j = 0, \pm 1. \tag{10.6}$$

* For some time there was a 'first Sommerfeld theory', and a 'second Sommerfeld theory', cf. a review[50] written in 1925.

** Only electric dipole transitions are considered.

† Eq. (10.6) allows for an 8-component He^+ line two of which turn out to be weak, however. The additional restriction $j = 0 \rightarrow 0$ forbidden is of later date.

When I asked Uhlenbeck what on earth possessed them in doing this at that time, he replied that they were just guessing, and reminded me that playing with half-integer values for quantum numbers had been done earlier, for the Zeeman effect.[54]* He also told me that Ehrenfest looked dubious when they told him about this idea but suggested that they write a little note—which they did. It was their first joint paper, a little known article written in Dutch, completed a few months before they came upon the electron spin.[55] Uhlenbeck calls it the August paper.**

Then came October 1925, when Uhlenbeck and Goudsmit announced their discovery of spin. Now they realized[57] that their rules (10.5) and (10.6) could be interpreted: j is the total (that is, orbital plus spin) angular momentum quantum number. Furthermore, a calculation in which they treated spin semi-classically convinced them that Eq. (10.2) indeed remains unchanged.[58] The real and ultimate derivation of the fine structure of energy levels had to await the Dirac equation. It was given[59] in 1928.

For some years it appeared that the quantity $1/\alpha$ might be an exact integer, 137. This number caused speculation, sleepless nights, mystic visions,[60] and the publication of arguably the best physics joke ever to slip by an editor of a first-rate physics journal.[61] All that has subsided: $1/\alpha \simeq 137.036$.

4. Helium. For a proper perspective it must be recalled that the Bohr–Sommerfeld theory, while by and large successful for hydrogen, was a disaster for neutral helium. This was first pointed out by Nicholson in 1914: 'If we take the same premises [as Bohr] and try to get . . . the ordinary helium spectrum . . . the attempt fails altogether'.[62] So it remained all during the days of the old quantum theory. A new generation of young scientists including Kramers[63] and Heisenberg[64] made heroic efforts, to no avail. Sommerfeld in 1923: 'All attempts made hitherto to solve the problem of the neutral helium atom have proved to be unsuccessful'.[65] Helium was only understood after quantum mechanical methods could be brought to bear, including important applications of spin and of the exclusion principle.†

I hereby leave the old quantum theory.

5. Changing of the guard. Kelvin's work mentioned in the preceding,[67] though interesting, does of course not remotely do justice to his great importance as a physicist. He published his first paper in 1841, aged seventeen. His last work, *On the formation of concrete matter from atomic origins*, occupied his attention to the last few days of his life. It deals with 'dynamical antecedents

* The points raised in this and the next paragraph are discussed in more detail in Chapter 13.
** Shortly afterward the same ideas were proposed independently by Slater.[56]
† For more on helium see Mehra and Rechenberg's history of the quantum theory.[66]

of Earth, Moon and Sun', and was published posthumously.[68] He died peacefully in December 1907. Larmor, Rayleigh, and J. J. Thomson were among the multitude attending the funeral services in Westminster Abbey, where he was interred in the Science Corner, in the north aisle of the nave, next to Newton, Herschel, and Darwin.

In 1913, Thomson, at age 56, made his last important contribution to science: the first separation of stable isotopes (Ne^{20} and Ne^{22}).* In 1919 he relinquished his position at the Cavendish to become Master of Trinity College, a position he held until his death in 1940. He outlived Rutherford, his junior, by fourteen years. He too was interred in Science Corner (as were the ashes of Rutherford before him).

Not much has been heard in this book about Lorentz since the discussion of the Zeeman effect in Chapter 4. During the years immediately following that event he devoted himself almost exclusively to his electron theory, work which made him the most important precursor to the special theory of relativity, as both Poincaré and Einstein repeatedly acknowledged.[70] He did follow the developments in quantum physics but, on the whole, only from the sidelines. In his lectures at Columbia University (1906) he referred to Thomson's atom model[71] but never got actively involved with atomic structure or spectra. During the final decade of his life, his leadership in the planning for the partial reclamation of the Zuiderzee brought him national acclaim.

Lorentz died on 4 February 1928, two days after a paper by Dirac containing the relativistic quantum theory of the Zeeman effect had been received by the Royal Society. On 9 February, the day of his funeral, all telegraph service in the Netherlands was closed from noon to 12:03 by order of the Postmaster General. The burning streetlamps along the funeral route through Haarlem were draped in crepe. Ehrenfest, Rutherford, Langevin, and Einstein spoke at the grave.[72]

Rutherford was knighted in 1914. His last seminal contribution to physics was also the last paper he published from Manchester: the first instance of induced nuclear transmutation (of nitrogen).[73] It is as well to remember that this was a one-man effort. The only assistance Rutherford had was from his laboratory steward, William Kay, whom he acknowledged 'for his invaluable help in counting scintillations'. In 1919 he succeeded Thomson at the Cavendish. In 1931 he was created Baron Rutherford of Nelson. His Armorial Bearings may be blazoned as follows: Per saltire arched gules and or, two inescutcheons voided of the first within each a martlet sable; the crest is described as upon a rock a kiwi proper and for the Supporters on the dexter side a figure representing Hermes Trismegistus, and on the sinister side a Maori holding in the exterior hand a club all proper.**

* This interpretation of his measurements was not at once clear to him.[69]
** I thank the Norfolk Herald for preparing this blazon for me.

Bohr was appointed professor of physics at the University of Copenhagen in April 1916. A few months later his first collaborator arrived, young Hans Kramers from Holland. Two years later they were joined by Oskar Klein. In January 1921, Bohr moved into his own 'Universitetets Institut for teoretisk Fysik'. For the next four decades he was to be the scientific inspiration, and, in certain respects, a father to all physicists who flocked to Copenhagen from all over the world.* Of his style it has been well written: 'The tentative character of all scientific advance was always in his mind, from the day he first proposed his hydrogen atom, stressing that it was merely a model beyond his grasp. He was sure that every advance must be bought by sacrificing some previous "certainty" and he was forever prepared for the next sacrifice'.[74]

We shall hear more of Rutherford and Bohr later on.

6. A change of pace. From an address by Rutherford to the Roentgen Society, April 1918: 'The two decades, 1895 to 1915, will always be recognized as a period of remarkable scientific activity which has no counterpart in the history of Physical Science . . . While it has been to some extent a time of speculation, yet it has not been as a whole a time of rash speculation, for the main ideas which governed the advance have been shown to have a solid foundation in fact'.[75]

Apart from a few brief but important points to be raised at the beginning of the next chapter, my account of these two decades has now come to an end.

Never in this century was the gap between bewildering experimental facts and at least some semblance of their coherent interpretation wider than in those twenty years. To be sure, the time of puzzling, even startling, new information received from the laboratories was not by any means over by then; it probably never will be. Nevertheless it seems fair to say that at about the time of Rutherford's speech the gap began to close, slowly. Correspondingly, the pace of the present story begins to accelerate. Let us see where we stand around 1915.

Quantum physics was not *in* a crisis. Quantum physics *was* a crisis. Wigner has told me that as a result of listening to colloquia held in Berlin in the early twenties he had begun to believe that man might never grasp the meaning of the quantum theory. He also mentioned to me that Dirac ('my famous brother-in-law') told him much later of his own similar opinions in those early years. Yet after Bohr and Sommerfeld there could be no doubt that the quantum was the key to the structure of the atom. That in itself was an advance compared to the rudderless drifting of the early years of the century, described in the previous chapter.

† For the story of the first ten years of Bohr's Institute see Ref. 25.

By 1915 the all-important distinction between atomic and nuclear phenomena had become manifest. Nuclear physics could make its very tentative beginnings, although, as the next chapter will show, a serious pitfall of simplicity still lay ahead.

Experimental techniques were fairly advanced in the area of atomic and molecular spectroscopy, less so for radioactivity and for the scattering and absorption of electrons, α- and γ-rays, yet in those domains improvement was quite noticeable as well. For example, radiochemistry (a subject not discussed in what follows) was flourishing.

It is also important to recall that in 1915 Einstein completed his fundamental equations of gravitation. That theory never (at least not to date) went through crises in any way comparable with the happenings in quantum theory.

During the early 1920s, the beginning of the final decade in which physics at the frontiers was quintessentially European, a new generation was preparing at four main schools for what was to come: Bohr's in Copenhagen, Born's in Goettingen, Rutherford's in Cambridge, and Sommerfeld's in Munich.

The years 1925–7 marked the beginning of a new era. Non-relativistic quantum mechanics was discovered, its main principles were understood. In Chapter 12 I give a mini-chronology of the main events. Explosive theoretical activity in atomic and molecular physics resulted. The logic of physics and of chemistry became interwoven. The theory of scattering and absorption of particles in matter was given its foundation. New avenues opened for the theory of matter in bulk, such as the theory of metals.

We, however, shall leave most of this marvelous physics aside and move inward, to shorter distances, to higher energies, to the nucleus. To begin with we must return to the year 1913.

References

1. Chapter 9, Section (e), part 2.
2. A. Pais, *Subtle is the Lord*, pp. 236, 487, Oxford Univ. Press 1982.
3. G. von Hevesy, letter to Niels Bohr, September 23, 1913, repr. in *Niels Bohr, collected works*, Vol. 2, p. 532, Ed. U. Hoyer, North Holland, Amsterdam 1981.
4. Chapter 9, Section (e), part 6.
5. R. J. Strutt, *Life of John William Strutt, third baron Rayleigh*, p. 357, Univ. of Wisconsin Press, Madison, Wis. 1968.
6. A. Sommerfeld, letter to N. Bohr, September 3, 1913, repr. in Ref. 3, p. 123.
7. *Nature* **92**, 304, 1913.
8. Chapter 9, Section (c), part 6.
9. Chapter 9, Section (e), part 5.
10. J. W. Nicholson, *Nature* **92**, 199, 1913.
11. A. E. Haas, letter to N. Bohr, January 6, 1914, repr. in Ref. 3, p. 127.
12. *La structure de la matière*, pp. 20, 50, Eds. R. Goldschmidt, M. de Broglie, and F. Lindemann, Gauthier-Villars, Paris 1921.

13. Cf. T. J. Trenn, *Radioactivity and atomic theory*, p. 341, Taylor and Francis, London 1975.
14. E. Rutherford, *Phil. Mag.* **27**, 488, 1914.
15. E. Rutherford, *Proc. Roy. Soc. A* **90**, 1914, insert after p. 462.
16. E. Rutherford, *Nature* **94**, 350, 1914.
17. W. Peddie, *Phil. Mag.* **27**, 257, 1914.
18. A. W. Conway, *Phil. Mag.* **26**, 1010, 1913.
19. J. J. Thomson, *Atomic theory*, Clarendon Press, Oxford 1914.
20. J. J. Thomson, *The electron in chemistry*, Lippincott, Philadelphia 1923.
21. J. J. Thomson, *Beyond the electron*, Cambridge Univ. Press 1928.
22. J. J. Thomson, *Recollections and reflections*, p. 425, Bell, London 1936.
23. Ref. 3, p. 122.
24. *Naturw.* **11**, pp. 535 ff., 1923.
25. P. Robertson, *The early years, the Niels Bohr Institute* 1921–30, Universitetetsforlaget, Copenhagen 1979.
26. Chapter 7, part 4.
27. Chapter 7, part 5.
28. Chapter 9, Section (e), part 3.
29. Chapter 9, Section (e), part 6.
30. Chapter 9, Section (c); Chapter 4, Section (d).
31. Chapter 9, Section (c), part 6.
32. A. Pais, in *Niels Bohr*, p. 215, Ed. S. Rozental, Wiley, New York 1967.
33. N. Bohr, *Phil. Mag.* **26**, 1, 1913.
34. N. Bohr, letter to E. Rutherford, March 6, 1913, repr. in Ref. 3, p. 111.
35. E. Rutherford, letter to N. Bohr, March 20, 1913, repr. in Ref. 3, p. 112.
36. Chapter 6, Section (e).
37. Chapter 7, Part 1.
38. Ref. 2, Chapter 21, Section (d).
39. Ref. 2, Chapter 22.
40. A. Sommerfeld, *Sitz. Ber. Bayer. Akad. Wiss.* 1915, p. 459.
41. A. A. Michelson, *Phil. Mag.* **34**, 280, 1892.
42. N. Bohr, *Phil. Mag.* **27**, 506, 1914, footnote on p. 521.
43. W. E. Curtis, *Proc. Roy. Soc. A* **90**, 605, 1914, esp. pp. 614, 620.
44. N. Bohr, *Phil. Mag.* **29**, 332, 1915.
45. U. Benz, *Arnold Sommerfeld*, p. 87, Wissenschaftliche Verlagsges., Stuttgart 1975.
46. *Einstein/Sommerfeld Briefwechsel*, Ed. A. Hermann, Schwabe, Stuttgart 1968.
47. Ref. 3, p. 603.
48. F. Paschen, *Ann. der Phys.* **50**, 901, 1916.
49. Cf. A. Sommerfeld, *Atombau und Spektrallinien*, 4th edn, p. 436, Vieweg, Braunschweig 1924.
50. L. Janicki and E. Lau, *Zeitschr. f. Phys.* **35**, 1, 1925.
51. N. Bohr, *Dansk. Vid. Selsk. Skrifter* **4**, 1, 1918; see esp. p. 34 and the footnote on p. 60; repr. in *Niels Bohr, collected works*, Vol. 3, p. 67, Ed. J. R. Nielsen, North Holland, Amsterdam 1976.
52. A. Rubinowicz, *Naturw.* **19**, 441, 465, 1918; repr. in *A. Rubinowicz, selected papers*, p. 31, Polish Scientific Publishers, Warsaw 1975.
53. G. E. Uhlenbeck and S. Goudsmit, *Physica* **5**, 266, 1925; W. Pauli, *Handb. der*
54. A. Landé, *Zeitschr. f. Phys.* **5**, 231, 1921; **11**, 353, 1922.
55. G. E. Uhlenbeck and S. Goudsmit, Ref. 53.

56. J. Slater, *Proc. Nat. Acad. Sci.* **11**, 732, 1925.
57. G. E. Uhlenbeck and S. Goudsmit, *Nature* **117**, 264, 1926.
58. G. E. Uhlenbeck and S. Goudsmit, *Physica* **6**, 273, 1926.
59. W. Gordon, *Zeitschr. f. Phys.* **48**, 11, 1982; C. G. Darwin, *Proc. Roy. Soc. A*, **118**, 654, 1928.
60. A. S. Eddington, *Relativity theory of protons and electrons*, Macmillan, New York 1936.
61. G. Beck, H. A. Bethe, and W. Riezler, *Naturw.* **19**, 39, 1931.
62. J. W. Nicholson, in 'Discussion on the structure of the atom', *Proc. Roy. Soc. A* **90**, 1914, insert following p. 462.
63. H. A. Kramers, *Zeitschr. f. Phys.* **13**, 312, 1923.
64. M. Born and W. Heisenberg, *Zeitschr. f. Phys.* **16**, 229, 1923.
65. A. Sommerfeld, *Rev. Sci. Instr.* **7**, 509, 1923.
66. J. Mehra and H. Rechenberg, *The historical development of quantum theory*, Vol. 1, Chapter 4, Section 2; Vol. 3, Chapter 5, Section 6, Springer, New York 1982.
67. Chapter 9, Section (b), part 3.
68. Kelvin, *Phil. Mag.* **15**, 397, 1908.
69. J. J. Thomson, in *Royal Institution Library of Science*, Vol. 7, p. 296, Eds. W. Bragg and G. Porter, Elsevier, New York 1970.
70. Ref. 2, Chapters 7 and 8.
71. H. A. Lorentz, *The theory of electrons*, Chapter 3, Teubner, Leipzig 1909.
72. G. de Haas-Lorentz, *H. A. Lorentz*, North Holland, Amsterdam 1957; also *Haarlem's Dagblad*, February 9, 1928.
73. E. Rutherford, *Phil. Mag.* **37**, 581, 1919.
74. *New York Times*, November 19, 1962.
75. E. Rutherford, *J. Roentgen Soc.* **14**, 81, 1918.

11

Nuclear physics' tender age

(a) Introduction

The concept of atomic weight A of an element is as old as the science of chemistry founded by Dalton. The first experimental information on A is not much younger. One appreciates the quality of the early data by considering the A-values used in 1819 by Dulong and Petit[1] in formulating their rule[2] about specific heats of solids. Their numbers are (for comparison modern A-values are added in parentheses): nine good ones: bismuth 213 (209), lead 207 (207), gold 199 (197), silver 108 (108), zinc 64 (65), copper 64 (63), nickel 59 (59), iron 54 (56), sulfur 32 (32); two poor ones: platinum 178 (195), cobalt 39 (59); and one clearly resting on a misunderstanding about valences, tellurium 64 (128).* Dulong and Petit chose the A of oxygen as their unit. I multiplied their numbers by 16, using hydrogen as a unit. For the purposes of this chapter, it makes no difference whether one chooses $A(\text{oxygen}) = 16$ (the standard adopted in 1898 by the German Chemical Society) or $A(\text{hydrogen}) = 1$.

As time went by, the number of known elements increased, A-data improved, flaws of the tellurium type lingered on. By 1832 it was known[3] that $A \simeq 35.5$ for chlorine, an illustration of the disarray concerning Prout's hypothesis that all A's should be integer multiples of the atomic weight of hydrogen.** Prout himself remarked at that time: 'I see no reason why bodies still lower in the scale than hydrogen . . . may not exist, of which other bodies may be multiples, without being actually multiples of the intermediate hydrogen'.[4] In 1869, Mendeléev had A-values for seventy-two elements at his disposal when he stated the rule that elements arranged according to the value of their atomic weights present a clear periodicity of properties.†

Mendeléev's first version[6] of the periodic table bears only a moderate resemblance to the modern one. Tellurium is assigned $A = 128$, but with a question mark, thorium is put down for $A = 118$ (true value 232), also with a question mark. His uranium value: $A = 116$ was deduced from the incorrect

* For the confusion occasionally created by the role of chemical valence see Chapter 4, Section (b).
** Cf. Chapter 9, Section (b), part 3.
† Mendeléev was most generous in acknowledging the contributions of others, especially of Lothar Meyer, to the founding of the periodic system.[5]

assumption that uranium is trivalent. As late as 1880, Mendeléev remarked: 'We cannot give uranium any determinate place in the system'.[7] Shortly thereafter he bit the bullet: '$U = 120$ ... did not correspond to the periodic law. I therefore proposed to double its atomic weight'[8] (true value: 238), a first example of his bold use of the table for making new predictions. In addition he deduced the existence of three gaps in his table from considerations of chemical valence, his principal criterion for periodicity. By 1886 the corresponding elements (gallium, scandium, germanium) had all been discovered and lingering doubts about the usefulness of his ideas largely quelled.

It does not diminish Mendeléev's achievements if we note that actually there were not 3 but 20 gaps in his table (if we stop at uranium).* Among the elements which he could not possibly predict in 1869 was the entire zero-valence group, the noble gases. In the standard arrangement, elements with the same valence occupy the same column in the table. It is obvious that one can only observe a gap in a column if at least one element in that column is known. When the periodic table was first written down, the entire noble gas column was dormant.

At the beginning of the twentieth century, when the noble gases had been found, the periodic table began as follows:

Element	H	He	Li	Be	B	C...
N	1	2	3	4	5	6...
A	1	4	7	9	11	12...

where the serial number N simply denotes the place of the element in a linear array ordered by increasing A. Although the number N had no physical meaning yet at that time, it was nevertheless conjectured that it had some importance. For example, attempts were made**—purely by playing with numbers—to find a 'law' which expresses A as a function of N, a hazardous enterprise if it were only because one had no grasp yet of possible gaps in the table. As to these gaps, speculation was recurrent that there might be elements with $A = 2$ and 3, nebulium and coronium being favorite candidates.

There were also peculiar irregularities. To repeat, the two-dimensional table of elements was obtained by first ordering them by increasing A, then splicing this array so that elements with common valence appear in a common column, the noble gases, the alkalis, etc. It became clear however, that these two steps were not unexceptionally compatible. In particular, by weight nickel should precede cobalt, but by valence cobalt should precede nickel. In 1908, Charles Barkla devoted a special paper to that subject.[10] Other such irregular pairs were discovered, argon and potassium, tellurium and iodine.

* This follows from the criterion, to be discussed in Section (d), that to each value 1–92 of the nuclear charge there belongs an element.
** See for example Ref. 9, also for further references.

By 1911, the importance of the periodic table was beyond dispute. But what did the table mean? Where were its gaps? How many elements were there?

This chapter is devoted to the earliest period, 1911–25, in the development of a new branch of physics inaugurated by Rutherford's interpretation of α-particle scattering: nuclear physics. It was a period of major advances and of a major pitfall. Section (b) deals with Bohr's realization that β-decay is a nuclear process, Section (c) with the discovery of isotopes. Section (d), in which the story of the periodic table is continued, concludes with van den Broek's conjecture and Moseley's experimental proof of the relation

$$N = Z \tag{11.1}$$

where Z is the nuclear charge in units $-e$, e being the charge of the electron. Section (e) contains a new example of simplicity as a necessary evil: the hypothesis that nuclei are built up from hydrogen nuclei and electrons. In Section (f) the first faltering steps in the understanding of nuclear binding energy are discussed. Section (g) speaks of physicists and the First World War. Finally, in Section (h) an account is given of the first intimations that new forces are at play inside the nucleus. I leave for later only one topic bearing on nuclear physics in the period at hand. In Chapter 14 I shall discuss what happened to β-spectra during those years.

(b) Niels Bohr on β-radioactivity

In the course of discussions at the first Solvay Conference, held in October 1911, Marie Curie observed that thermal, optical, elastic, magnetic, and other phenomena all appear to depend on the peripheral structure of the atom, then she continued: 'Radioactive phenomena form a world apart, without any connection with the preceding [phenomena]. It seems therefore that radioactive phenomena originate from a deeper region of the atom, a region inaccessible to our means of influence and probably also to our means of observation, except at the moment of atomic explosions'.[11]

Rutherford's discovery[12] of the nucleus had been announced the preceding May. I do not know whether Marie Curie did not follow up on her wise remark with a comment on the nucleus because she was unaware of this work, or did not believe the results, or failed to see the connection. Nor do I know why Rutherford, who was in the audience, refrained from drawing Curie's attention to the nucleus. I do know (and have mentioned earlier) what were Rutherford's own views in 1911 and 1912: α-decay is due to nuclear instability, β-decay to instability of the peripheral electron distribution.[13]

It was Bohr who finally set the matter straight. The second of his 1913 papers[14] on the constitution of atoms and molecules contains a section entitled 'Radioactive Phenomena', in which he states: 'A necessary consequence of

Rutherford's theory of the structure of atoms is that α-particles have their origin in the nucleus. On the present theory it seems also necessary that the nucleus is the seat of the expulsion of the high speed β-particles.'

His arguments in support of this picture are twofold. First, he knew enough by then about orders of magnitude of orbital (peripheral) electron energies to see that the energy release in β-decay simply could not fit with a peripheral origin of that process. His second reason had to do with isotopes (a term he did not yet use, a subject to which I shall return in the next section): there appear to be substances, he noted, 'which are different only in radioactive properties and atomic weights but identical in all other physical and chemical aspects . . . the only difference being the mass and the internal constitution of the nucleus'. Now then, he reasoned, there are known instances in which two such apparently identical elements emit β-rays with different velocities. He also knew that 'all other physical and chemical aspects' are governed by the peripheral electrons. Therefore, he concluded, β-rays must have their origin in the nucleus. Thus it was finally settled in the autumn of 1913 that all radioactive processes are nuclear.

Shortly thereafter, in October 1913, Marie Curie ventured similar ideas[15] during the second Solvay Conference, but without mentioning Bohr. Rutherford, once again in the audience, did not draw Curie's attention to Bohr's work either . . . .

(c) Isotopes

The term 'isotope' was coined[16] in 1913, but already early in 1911, even before the discovery of the nucleus, the underlying physical concept was formulated correctly in its essentials. There is of course nothing startling about that sequence of events. The statement: 'There exist elements with identical physical and chemical properties except for atomic weight', which does not contain reference to the nucleus, is in need of refinement but will obviously do for the diagnosis of the phenomenon.

By 1910 it was believed, correctly, that there exist several groups of chemically identical elements with different A. Just one example: the five substances* thorium, radiothorium, radioactinium, ionium, and uranium-X. Soddy must be credited with the discovery of isotopes because of what he wrote[18] early in 1911: 'The conclusion is scarcely to be resisted that we have in these examples no mere chemical analogues, but chemical identities. . . . The recognition that elements of different atomic weight may possess identical chemical properties seems destined to have its most important application in the region of inactive elements** . . . Chemical homogeneity is no longer a guarantee

* These were eventually identified as thorium isotopes with $A = 232$, 228, 227, 230, and 234 respectively. For a review of these early developments in radiochemistry see an account[17] by Soddy in 1923.

** The first inactive isotopes, Ne^{20} and Ne^{22}, were discovered in 1913.

that any supposed element is not a mixture of several of different atomic weights, or that any atomic weight is not merely a mean number. The constancy of atomic weight, whatever the source of the material, is not a complete proof of homogeneity'. All that is physics at its best. So is Soddy's concluding remark: 'The absence of simple numerical relationships between the atomic weights becomes a matter of course rather than for surprise.'

Thus, in 1911, it began to dawn that the nearly one-century-old and often maligned rule of Prout might not be so bad after all if one bears in mind that the quantity A for a given element is the weighted mean of distinct isotopic masses.

By 1913 it was clear[19] that the occurrence of isotopes was the rule rather than the exception: 'Not a single one of the radio-elements, known at the commencement of the year, has a peculiar chemical nature unshared by others'. Forty distinct radioactive substances had been distributed over only ten slots in the periodic table.

In 1917 Soddy, lecturing before the Royal Institution, reported[20] on attempts to find spectral distinctions among isotopes: 'No certain differences have been found and it may be concluded that the spectra of isotopes are identical'. It was realized soon thereafter that this is not rigorously true.[21] One of the important isotope effects in spectra concerning hyperfine structure was first found in 1927 for the case of neon.[22]

The order thus created in radiochemistry was further enhanced by the so-called displacement laws according to which an α-emission sends the initial element to another two columns to the left in the periodic table, while a β-emission corresponds to a move one column to the right.* This simple regularity was initially either incompletely or incorrectly treated by all its originators.** The reader may wonder how this can be since these laws are obvious if one knows that the position of an element in the periodic table is dictated by its nuclear charge. That simple fact was not known, however, when the displacement laws were first formulated. Let us see next how the role of nuclear charge was discovered.

(d) From A to Z

The Rutherford scattering cross-section $\sigma(\vartheta)$ given by Eq. (9.6) is proportional to the square of the nuclear charge Z. (In deference to Rutherford I used the symbol N for Z in Eq. (9.6).) In the same paper[12] in which Rutherford derived his scattering formula he also checked the Z-dependence of $\sigma(\vartheta)$ against

* Here one imagines the periodic table wound around a cylinder with axis in the 'vertical' direction of the table.

** Soddy stated it correctly, but for α-decay only.[23] Russell[24] and Hevesy[25] discussed both α- and β-decay but claimed that in α- (β-) decay one moves two (one) steps either to the right or to the left as the case may be. Fajans[26] stated the laws correctly but believed that they implied a non-nuclear origin of radioactive processes.

Geiger–Marsden data[27] for scattering on a variety of targets ranging from aluminum to lead, and found the agreement between theory and experiment to be reasonably good '. . . *assuming* [my italics] that the central charge [i.e. Z] is proportional to the atomic weight A.'[28] An assumption was indeed called for. In 1911 Rutherford knew many things, but not what Z was. From subsequent improved measurements by Geiger and Marsden[29] he concluded in February 1914 that Z is given by[30]

$$Z \simeq \frac{A}{2}. \tag{11.2}$$

One need not of course be a Rutherford to realize that this relation does not work so well for hydrogen. All Rutherford meant was that Eq. (11.2) appeared to be a good first guess for a sizable part of the periodic table—which indeed it was.

Remarkably, Barkla had reached the same conclusion in 1911, but on entirely different grounds. Reanalyzing data for Roentgen scattering[31] he found: 'The theory of scattering as given by Sir J. J. Thomson* leads to the conclusion that the number of scattering electrons per atom is about half the atomic weight in the case of light atoms' ($A \leq 32$). Barkla's and Rutherford's respective conclusions are of course identical for neutral atoms.

A third atomic regularity involving A, also dating from 1911, has to do with X-ray spectra, a novelty at that time. During the years just preceding, Barkla and coworkers had made the important discovery that X-rays incident on various elements generate discrete secondary X-rays characteristic for these respective elements but essentially independent of the incident radiation, as long as it is sufficiently hard (that is, energetic), and of the density, state of aggregation, and chemical composition of the target.[32] For the reader's convenience I note a few facts about X-ray spectra which became clear later. The electronic structure of atoms consists of a set of shells, the innermost being called the K-shell, the next one the L-shell (notations introduced by Barkla[33]), then the M-shell . . . If a sufficiently hard X-ray knocks an electron out of the K-shell, the empty position can be reoccupied by an electron from the L, M, . . . shell dropping to the K-shell accompanied by discrete X-ray emission. These are the so-called K-lines; K_α corresponds to L→K, K_β to M→K . . . . Of these K_α has the lowest energy (frequency), K_β is next . . . . Qua intensity K_α is stronger than K_β . . . . The M→L lines are L_α, etc. All this is well-explained in standard textbooks.

Now in 1911 it occurred to Richard Whiddington from Cambridge to ask how the frequency of a given line, say K_α, depends on the choice of target. To this end he directed a stream of cathode rays of known velocity v onto

* It was explained in Chapter 9, Section (c), Part 6, how this scattering yields information on the number of electrons per atom. Barkla's reference to Thomson concerns Eq. (9.5).

an anti-cathode where it produces X-rays. These rays are made to hit a secondary target of varying constitution, producing K, L, . . . lines. His result[34] (in present language): for elements from Al to Se, K_α is produced if and only if v exceeds a critical velocity v_c given by

$$v_c \simeq A \times 10^8 \text{ cm/sec.} \tag{11.3}$$

This result, known for some years as Whiddington's law, was important to Bohr's thinking in 1913 about the innermost electron orbits in atoms.[14]

Thus α-scattering, X-ray scattering, and X-ray spectra all led to a focus on A as the ordering parameter of the periodic system. But then, all this changed because of fundamental new experimental discoveries in X-ray spectroscopy and of theoretical speculations by a rather improbable character whom I introduce next.

Antonius Johannes van den Broek, a Dutchman, studied law in Leiden and at the Sorbonne, and econometrics in Vienna and Berlin. He earned his living as a lawyer specializing in real estate transactions.* In addition, he was an amateur theoretical physicist intensely interested in the numerical regularities embodied in the periodic table and in Prout's rule, the subjects of all his published physics papers (some twenty of them). In his first paper (1907) he proposed the approximate rule that A increases by 2 from element to element, concluding that there should be 120 elements.[36] Following a comment made in passing by Mendeléev he next proposed a cubic rather than a plane periodic table.[37] Shortly thereafter (July 1911) he published a twenty-line letter to *Nature*[38] entitled 'The number of possible elements and Mendeléev's "cubic" system', in which he referred to the very recent papers by Rutherford on the nucleus[12] and by Barkla on X-ray scattering.[31] In his opinion these results, specifically Eq. (11.2), confirmed not only his earlier idea of 120 elements (since $A \simeq 240$ for uranium) but also his cubic table. Then comes his striking conclusion: 'If this cubic periodic system should prove to be correct, then the number of possible elements is equal to the number of possible permanent charges of each sign per atom, or to each possible permanent charge (of both signs) per atom belongs a possible element'.**

Thus, based on an incorrect version of the periodic table, and on the incorrect relation (11.2) between Z and A, did the primacy of Z as the ordering number of the periodic table enter physics for the first time.†

In his next paper[39] (January 1913) he replaced his cubic system by a planar one, far more complicated than the modern version. The main result of this paper became known as van den Broek's rule: 'The serial number of every element in the sequence ordered by increasing atomic weight equals half the

* The most detailed account of van den Broek's life and activities in physics is found in Ref. 35.

** The charge of both signs refers of course to the electrons and the nucleus.

† van den Broek never properly incorporated the notion of isotopes. He also had difficulties fitting β-emitters into his scheme.

atomic weight and therefore the intra-atomic charge'. Bohr quoted this result in the second paper of the 1913 trilogy. The rule is a large step in the right direction, yet far from perfect. In the first place it continues to adhere to Eq. (11.2). Secondly it passes over the nickel–cobalt and other irregularities mentioned earlier.

Then, in a letter to *Nature*[40] of November 27, 1913, van den Broek took the liberating step of breaking the alleged link between A and Z: 'The hypothesis [about the serial number of elements being equal to Z] holds good for Mendeléev's table but the nuclear charge is not equal to half the atomic weight'.

His change of mind had come about because of improved results by Geiger and Marsden, just published,[29] which indicated, van den Broek noted, that the relation (11.2) does not fit very well to the data for copper and gold.* Thus van den Broek deserves the credit for having introduced Z as an independent nuclear parameter along with A. (In the next section I shall have occasion to refer to another interesting comment made in this paper.)

Reactions were immediate and positive. In the next issue of *Nature*, Soddy pointed out[41] that van den Broek's conclusion made obvious the meaning of the displacement laws: Z changes by -2 $(+1)$ for α- (β-) decay. One week later Rutherford wrote,[42] also in *Nature*: 'The original suggestion of van den Broek that the charge on the nucleus is equal to the atomic number [i.e. the serial number in the periodic table] and not to half the atomic weight seems to me very promising'. As to van den Broek, he remarked next[43] that Whiddington's law Eq. (11.3) gives improved fits if A is replaced by $2Z$. His later publications, increasingly speculative, are of no interest and he vanished from the scene as abruptly as he had entered it.

I now turn to the point of culmination of this section, Moseley's precision measurements of the wavelengths of K_α X-rays. Moseley began this work in 1913 in Manchester where since 1910 he had been in various junior capacities including 'teaching idiots elements'.** In 1914 he explained[46] what had motived his X-ray work: 'My work was undertaken for the express purpose of testing van den Broek's hypothesis, which Bohr has incorporated as a fundamental part of his theory'.† Bohr, who visited Manchester in July 1913 (just after his paper referred to by Moseley had been completed), recalled[47] discussions with Moseley at that time: 'I got to know Moseley really partly in this discussion about whether the nickel and cobalt should be in the order of their atomic weights. Moseley . . . asked what I thought about that. And I said "There can be no doubt about it. It has to go according to the atomic

* Geiger and Marsden were less exercised: 'The deviations . . . are nearly within the experimental error.'

** For details of Moseley's life and work see especially Heilbron's biography,[44] and also Ref. 45.

† Here Moseley referred to the imperfect version of the hypothesis which van den Broek had given in January 1913.[40]

number [i.e. *not A*]". And then he said, "We will try and see". And that was the beginning of his experiments . . . . And then he did it at tremendous speed'.

Moseley's experimental arrangement was ingenious and delightful. A little train, each car carrying a different target, is made to move inside a one-metre-long cathode ray tube (evacuated with the help of a pump borrowed from Balliol College, Oxford). Each target in turn, upon being hit by cathode rays, produces characteristic discrete X-rays. These are let out of the tube through a thin window, then fall on the cleavage face of a crystal which serves as a diffraction grating for determining their wavelengths. In two papers[48] Moseley recorded the frequencies ν_α of K_α-lines of 21 elements.* Inspired by van den Broek, Barkla, and Bohr he analyzed his data in the following way. Number the elements by Z, let $\nu(Z)$ be the wavenumber of K_α for element Z, and R the Rydberg constant for hydrogen. His experiment revealed a beautiful regularity: the quantity $\{4\nu(Z)/3R\}^{1/2}$ increases by one as $Z \to Z+1$, and is therefore of the form $Z-\sigma$, where σ is a constant. This led him to write a Balmer-type formula:

$$\nu(Z) = R(Z-\sigma)^2\left(\frac{1}{1^2}-\frac{1}{2^2}\right). \tag{11.4}$$

If it is true, he argued next, that $Z=20$ for calcium (it is) then his data for $\nu(20)$ told him that

$$\sigma = 1. \tag{11.5}$$

Inverting Eqs. (11.4) and (11.5) from a record of observation to a diagnostic, Moseley was now ready (as it has been said of him) to call the roll of the elements. Ending with uranium there had to be 92 of them. Seven were missing. All these have since been discovered, some by X-ray analysis, some by techniques of much later vintage.** Rutherford put Moseley's work on a par with the discovery of the periodic table.[50] In the opinion[50] of the French chemist Georges Urbain: 'Le travail de Moseley substituait à la classification un peu romantique de Mendeléev une précision tout scientifique'.†

As to the meaning of Eqs. (11.4) and (11.5): they account in correct first approximation for the frequency of X-rays accompanying a transition from the L-shell ($n=2$) to the K-shell ($n=1$). There are sizable corrections to Eq. (11.4) due to fine structure terms (which vary as Z^4, see Eq. (10.2)). The appearance of the σ-term is due to the screening of the nuclear charge. Moseley realized this, and, splendid physicist that he was, puzzled (in vain) about the

* In addition he reported many more results on other K-lines and on L-lines.

** These seven elements are, with their Z-value and year of discovery: protactinium (91, 1917), hafnium (72, 1923), rhenium (75, 1925), technetium (43, 1937), francium (87, 1939), astatine (85, 1940), and prometheum (61, 1945).[49]

† The work of Moseley substituted a quite scientific precision for the rather romantic classification of Mendeleev.

reason for $\sigma=1$. Bohr (to whom he wrote about this[51]) was of no help because of his incorrect belief at that time that for $Z>7$ the K-shell contains 4 or more electrons.[14] Actually it always contains two electrons. Thus if one of these is knocked out, the subsequent transition $L\rightarrow K$ occurs in the field of the nucleus screened by one unit of charge: $\sigma=1$.

(e) A new pitfall: the first model of the nucleus

Early in 1914 it was known that the nucleus is the seat of all radioactive processes and that a nuclear species is specified by the numbers A and Z. It was believed that the interaction between α-particles and nuclei is purely electromagnetic. Darwin had concluded[52] from the known data: 'No force proportional to some power of the distance other than the inverse square can give the dependence of [the Rutherford scattering cross section] on [the initial velocity]'. Assuming this to be true and then calculating the distance of closest α-particle–nucleus approach, Rutherford had realized[12,29] already in 1911 that the nuclear radius is small: $r\lesssim 3\times 10^{-12}$ cm. By the same argument Darwin noted[52] in 1914 that 'the radii of hydrogen and helium are certainly less than 10^{-13} cm'. It was obvious that the nucleus of hydrogen is particularly important. 'The hydrogen nucleus is the *positive electron*', Rutherford wrote[30] early in 1914 (his italics). This nucleus was often called the H-particle in those days, the name 'proton' came later. To summarize: early in 1914 one knew some basic facts about radioactivity, about A and Z and r, and about the basic character of the H-particle.

What are nuclei made of?

In the same paper[40] in which he introduced Z as an independent parameter, van den Broek was the first to propose that α-particles and electrons are nuclear constituents. The contribution of the electrons to the nuclear mass is quite small. The electrons mainly serve, he remarked, to compensate the charge to the correct value. Of course, not all nuclei can consist of α-particles and electrons only, since in that case the only possible mass numbers A would approximately be multiples of 4. Obviously the H-particle could help out. Indeed, in February 1914 Rutherford conjectured[30] about the structure of the α-particle itself: 'It is to be anticipated that the helium atom [i.e. the α-particle] contains four positive electrons [H-particles] and two negative', symbolically $\mathrm{He}=4\mathrm{H}+2\mathrm{e}$; generally for a given isotopic species X with mass number A and nuclear charge Z (treating A as an integer):

$$\mathrm{X}=A\mathrm{H}+(A-Z)\mathrm{e}. \tag{11.6}$$

In a Royal Society discussion on 19 March 1914, Rutherford commented further[53] on nuclear structure: 'The general evidence indicates that the primary β-particles arise from a disturbance of the nucleus. The latter must consequently be considered as a very complex structure consisting of positive particles and electrons but it is premature (and would serve no useful purpose) to discuss at the present time the possible structure of the nucleus itself'.

Thus Rutherford, though always cautious and averse to speculation, blithely assumed that electrons are nuclear constituents. Actually he would not have conceived of this as an assumption. Was it not self-evident? Did one not see electrons come out of certain nuclei, in β-processes? To Rutherford, as to all physicists of that time, it was equally sensible to speak of electrons as building blocks of nuclei as it was to speak of a house built of bricks, or of a necklace made of pearls.

In actual fact, the H-particle–electron picture of the nucleus is another example of simplicity as a necessary evil. It was a model as inevitable as it was wrong. It is *not* true that electrons are building blocks of nuclei, Eq. (11.6) is incorrect. Then how can one understand that electrons do emerge from nuclei in β-decay? Almost exactly twenty years after Rutherford spoke in the Royal Society, Fermi found the answer to that question, using the tools of quantum field theory. That will be a subject for later chapters, as will be the evidence, accumulating well before Fermi's work, that the H-particle–electron model was in serious trouble.

I return to the years immediately following 1914. In view of the importance of H-particles it was natural to search for 'spontaneous H-decay'. In 1915 experiments showed 'a strong suspicion that H-particles are emitted from radioactive atoms'.[54] It passed. At about that time we also witness the beginning of a search for substructure within the nucleus. Already in 1914 Rutherford had written: 'The helium nucleus is a very stable configuration which survives the intense disturbances resulting in its expulsion with high velocity from the radioactive atom, and is one of the units of which possibly the great majority of the atoms are composed'.[30] New subunits were proposed: the α'-particle (4H+4e), the μ-particle (2H+2e), and others, mainly for reasons of numerology. In 1921 Rutherford wrote to Boltwood: 'It is exceedingly easy to write about these matters but exceedingly difficult to get experimental evidence to form a correct decision'.[55] For some time Rutherford himself was led astray by apparent evidence for another subunit, X_3 (3H+e). With some imagination one can see certain of these papers as forerunners of the α-particle model, even the shell model.* Nevertheless I do not believe it is unfair to say that none of this work has left any mark on physics.

What forces hold the nucleus together?

Rutherford in 1914: 'The nucleus, though of minute dimensions, is in itself a very complex system consisting of positively and negatively charged bodies bound closely together by *intense electrical* forces', (my italics).[57] What else could he say? At that time there simply were no other than electromagnetic**

* For detailed references to these articles see an informative paper by Stuewer.[56]

** In 1915 it was suggested[58] that magnetic dipole interactions might contribute importantly to intra-nuclear forces.

and gravitational forces, the latter being manifestly negligible for the problem at hand. Even so, one notes an element of wonder in another of Rutherford's statements[30] that same year: '[The nuclear electrons] are packed together with positive nuclei and must be held in equilibrium by forces of a different order of magnitude from those which bind the external electrons'. As we shall see in Section (h) of this chapter, it did not take very long before the first evidence for strong nuclear forces of a new kind made its appearance.

What about the nucleus and the quantum theory?

In the first edition of *Atombau und Spektrallinien* Sommerfeld ventured the opinion[59] that 'nuclear constitution is governed by the same quantum laws as the [periphery] of atoms'. At that time there was not much one could do to verify that idea. If the atomic spectrum of helium would not yield to the quantum theory, who would dare tackle the α-particle?

(f) Binding energy

In 1905, when Einstein derived[60] for the first time the equation

$$E = mc^2 \tag{11.7}$$

he remarked at once: 'It is not out of the question that one can devise a test of the theory for bodies the energy content of which is variable to a high degree (as for example for radium salts)'. By 1907 he had reached the conviction,[61] however, that it was 'of course out of the question' to reach the experimental precision necessary for his test;* in 1910 he remarked[62] that 'for the moment there is no hope whatsoever' for the experimental verification of Eq. (11.7).

While Einstein was the first to propose a check for Eq. (11.7) in terms of loss of weight in radioactive decays, Planck was the first to draw attention[63] to another test: a bound system should weigh less than the sum of its constituents. Until 1932 the answer remained Eq. (11.6). As long as one had molecular binding energy for a mole of water. The effect was very small ($\sim 10^{-8}$ g) but the idea was new and excellent.

In 1913 Langevin applied[64] Planck's remark to the nucleus: 'It seems to me that the inertial mass of internal [i.e. binding] energy is made evident by the existence of certain deviations from the law of Prout.' Unfortunately he did not consider the influence of isotope mixing and therefore overrated binding energy effects.

In order to give a precise meaning to the magnitude of the nuclear binding energy B for a nuclear species X one must of course know what are X's constituents. Until 1932 the answer remained Eq. (11.6). As long as one had

* Radium loses weight at a rate of 1 part per hundred thousand per year.

nothing better one was inevitably stuck with the following incorrect mass formula for nuclei (in obvious notation)

$$m_X = Am_H + (A-Z)m_e - \frac{B}{c^2}. \qquad (11.8)$$

Intra-nuclear forces were supposed to be electromagnetic. So therefore was the dynamical origin of nuclear binding B. Rutherford in 1914: 'As Lorentz has pointed out, the electrical mass of a system of charged particles, if close together, will depend not only on the number of these particles, but on the way their fields interact. For the dimensions of the positive and negative electrons considered, the packing must be very close in order to produce an appreciable alteration of the mass due to this cause. This may, for example, be the explanation of the fact that the helium atom has not quite four times the mass of the hydrogen atom'.[30] Quantitative estimates showed that an assumed electromagnetic origin of nuclear binding would give much too small an effect, however. This was emphasized by Wilhelm Lenz (a coworker of Sommerfeld) who was the first to concentrate[65] on Eq. (11.8) which, whatever its other flaws, is independent of the dynamical origins of B.

So, inconclusively, the best of physicists kept stumbling along. As late as 1921 Pauli remarked: 'Perhaps the theorem of the equivalence of mass and energy can be checked *at some future date* [my italics] by observations on the stability of nuclei'.[66] Meanwhile mass spectrographs had begun to pour out good data on isotope masses. In 1927 Aston gave a list of thirty of these.[67] Binding energies are recorded in terms of the packing fraction, the ratio B/A on the scale $A = 16$ for oxygen, so that by definition the packing fraction for oxygen equals zero. It is one of the least transparent ways of representing the data I know of, also because oxygen has three stable isotopes.

The book by Rutherford, Chadwick, and Ellis, which appeared in 1930, and which for some years was the most influential nuclear physics text, contains the statement that the α-particle has a composition 4H+2e and a binding energy of 27 MeV.

The year 1932 brought clarity.

After the discovery of the neutron was published in February, it soon became clear that the nuclear constituents are protons and neutrons (see Chapter 17). The discovery of deuterium, the hydrogen isotope with mass number two, was announced that same month. Perhaps the last reference to the old H-particle–electron model is found in a paper[68] by Ken Bainbridge (October 1932) entitled 'The isotopic weight of H^2': 'On the assumption that the nucleus is composed of two protons and one electron, the energy binding is approximately 2×10^6 electron volts. If the H^2 nucleus is made up of one proton and one Chadwick neutron of mass 1.0067 then the binding energy of these two particles is 9.7×10^5 electron volts', (the correct value is $B \simeq$ 2.15 MeV). This last value was found from possibly the first application of the correct equation for B which, to this day, replaces the old Eq. (11.8):

$$m_X = Zm_p + (A - Z)m_n - \frac{B}{c^2} \tag{11.9}$$

where m_p, m_n are the proton and neutron mass respectively.

In June, Cockcroft and Walton's paper on the first nuclear transformation produced by artificially accelerated particles appeared.[69] Their result:

$$Li^7 + \text{proton} \rightarrow 2\alpha + (14.3 \pm 2.7)\ \text{MeV} \tag{11.10}$$

verified Eq. (11.7) within the errors, the masses of all particles appearing in the reaction being well known. Eq. (11.10) is the oldest example of a new category of tests for $E = mc^2$.

In 1937 a value for the velocity of light, correct to within less than one half of one per cent, was obtained from nuclear reactions in which all relevant masses and kinetic energies were known.[70] Thus the 1930s brought the correct model of the nucleus and advances in experimental techniques necessary for numerous verifications of $E = mc^2$ with the help of nuclear phenomena.

Then, of course there was the A-bomb . . .

(g) Physicists in the First World War

> Goethe ridicules the 'fatherland' talk of the Germans, Chateaubriand and Taine that of the French, and Shakespeare and Shaw that of the English.
>
> G. F. Nicolai, *The Biology of War*[71]

I return to the year 1914.

The war which broke out in August was variously known as the Great War, the war to end all wars, the World War, until there came a time in which it had to be remembered as the First World War. It was a man-willed slaughter the likes of which had never been witnessed before, the number of killed, wounded, and missing exceeding thirty-seven million. I cannot quite escape a sense of disproportion in focussing next on the fate of a handful of physicists. Yet a few comments are in order.

Probably the most notorious document defending certain acts of that war was a German manifesto,[72] published in October 1914, known as *Aufruf an die Kulturwelt* (appeal to the civilized world), also as *The manifesto of the 93*, for the number of its signatories, also as *Es ist nicht wahr* (it is not true), for the opening words of each of its six sections. After a preamble: 'We as representatives of German science and art protest before the entire civilized world the lies and slander . . . ' etc., six points are categorically denied: that Germany had provoked the war; that it had wantonly violated the neutrality of Belgium; that it had encroached upon the life and property of a single

Belgian citizen without grimmest necessity; that German troops had brutally destroyed the city of Louvain;* that the German command disregarded the rights of nations; finally: *Es ist nicht wahr* that the battle against our so-called militarism is not a battle against our culture, as our enemies hypocritically pretend. Without German militarism, German culture would have been wiped off the surface of the earth long ago. The former arose out of and for protection of the latter in our land which has been exposed to raids for centuries. The German army and the German people are as one . . . ' etc. Among the signatories we find names encountered in the foregoing: Emil Fischer, Lenard, Ostwald, Planck, Roentgen, W. Wien; also Nernst, the chemist Fritz Haber, and the mathematician Felix Klein. Subsequently more than 3000 German professors signed a document endorsing the manifesto.[74] It is fitting to note that the official French translation of the manifesto was more inflammatory than the original text.[75]

One can only conjecture what convictions, reservations, or obligations affected the individual signers, or what pressures were exerted on them.** Regardless, their participation impoverishes our memories of them. The bitter anger in the world of science caused by their words took years to subside.[77]

Within days after the publication of this manifesto, Georg Friedrich Nicolai, professor of physiology at the University of Berlin, composed a 'Manifesto to Europeans', which contains these lines: 'Anyone who cares in the least for a common world culture is now doubly committed to fight for the maintenance of the principles for which it must stand. Yet, those from whom such sentiments might have been expected—primarily scientists and artists—have so far responded, almost to a man, as though they had relinquished any further desire for the continuance of international relations. They have spoken in a hostile spirit, and they have failed to speak out for peace . . . . All those who truly cherish the culture of Europe [should] join forces'.[71] Nicolai circulated this document among colleagues. Only three were prepared to sign. One of these was Einstein, who had given his inaugural address before the Prussian Academy of Sciences in Berlin one month before the outbreak of hostilities.† The signers dropped plans to publish since the impact would be too small. Professor Nicolai, a volunteer army doctor with officer's rank, was demoted to hospital orderly.[79]

There were those who served behind the lines.

Fritz Haber was in charge of chemical warfare projects at the Kaiser Wilhelm Institute in Dahlem near Berlin. Nernst participated in this effort, as did James

* 'The brutal behavior of the German armies in Belgium was undeniable . . . . Belgian resistance infuriated them . . . . They took hostages and executed them when they found opposition . . . . In the belief that sniping had occurred and that Louvain was full of *franc-tireurs*, the Germans shot a large number of citizens and set the town on fire. The famous old library of the university was entirely destroyed.'[73]

** Already during the war Planck distanced himself from the manifesto.[76]

† For further details on Einstein's reactions to the war see Ref. 78.

Franck, Otto Hahn, and Gustav Hertz.[80] An explosion at Haber's laboratory took the life of the distinguished physical chemist Otto Sackur.[81]

The British scientific war effort was coordinated by a Board of Invention and Research of which J. J. Thomson was a member and to which W. Bragg, Crookes, Lodge, Rutherford, Stoney, and R. J. Strutt were consultants.[82] Perhaps their main problem was defense against submarine attack. 'Rutherford was . . . entrusted with the task of making the preliminary report on possible methods . . . . As a result of this appointment, within a short time the laboratory at Manchester became the centre of research on underwater attack, and a large tank was installed on the ground floor of the building.'[83] Shortly after the United States entered the war, Rutherford participated in a scientific mission to America for discussions on problems of common concern. On that occasion he conferred with Millikan who had joined the army as head of the U.S. Signal Corps' science and research division with the rank of major, later of lieutenant-colonel.[84] Also Michelson, graduate of Annapolis, was commissioned, as lieutenant-commander in the Naval Coast Reserve; he later joined the Bureau of Ordnance.[85]

Marie Curie arranged for the equipment of some twenty automobiles with Roentgen apparatus in order that soldiers could be operated on near the battlefield. She also oversaw the installation of some two hundred radiological rooms in various hospitals. She trained others as X-ray diagnosticians and often acted as one herself.[86] Lise Meitner volunteered as an X-ray nurse with the Austrian army. 'It was a harrowing time for her, working up to twenty hours a day with inadequate equipment and coping with the large numbers of Polish soldiers with every kind of injury, without knowing their language.'[87] Great Britain and the United States made considerable exertions in readying X-ray installations. The United States Army, for example, which before the war had only five mobile X-ray units mounted on four-mule escort wagons, managed to send more than seven hundred units to the European theatre, many of them automobile-mounted.[88] (During the war, radium played no significant role from the medical point of view, but was in demand for other reasons such as the application of self-luminescent paints to gunsights and engine dials.[88])

There were those who did battle.

Geiger and Marsden were at the Western front, in opposing camps. Andrade served as artillery officer;* so did Schroedinger. Reinganum, known for his work on the equation of state and the theory of conductivity, fell at le Ménil in the Vosges.[89] Hasenöhrl, successor to Boltzmann as professor of theoretical physics at the University of Vienna, was killed at the Isonzo.[90] Schwarzschild (of the Schwarzschild radius) died of illness contracted at the Eastern front.[91] Planck's younger son was taken prisoner during the battle of the Marne, his

* Chadwick and Ellis were interned in Germany, see Chapter 8, Section (h).

elder son fell at Verdun.[92] Moseley, second lieutenant Royal Engineers, was killed by a bullet in the head at Suvla Bay when in the act of telephoning an order to his division while the Turks were attacking two hundred meters away.[50] He left his earthly goods 'to the Royal Society of London to be applied to the furtherance of experimental research in Pathology, Physics, Physiology, Chemistry, or other branches of science but not in pure mathematics, astronomy, or any branch of science which aims merely at describing, cataloguing, or systematizing.'[44]

In his obituary of Moseley,[93] Rutherford remarked: 'His services would have been far more useful to his country in one of the numerous fields of scientific inquiry rendered necessary by the war than by the exposure to the chances of a Turkish bullet', an issue which will be debated as long as the folly of resolving conflict by war endures.

(h) Strong interactions: first glimpses

In 1911, Rutherford the theorist had deduced the existence of the nucleus. His principal assumptions had been that the scattering of α-particles off atoms is mainly due to the $1/r^2$ Coulomb force between a pointlike α-particle and a pointlike nucleus. In 1919, Rutherford the experimentalist reported that these assumptions do not always hold true, that the law of force between α's and nuclei is more complicated than he originally had thought.

In February 1916 Bohr wrote[94] from Manchester to Fokker: 'Here things have changed very much on account of the war . . . you will understand that there are not many left to do research work . . . . You have of course heard the horrible story of Moseley's death . . . he was considered to be probably the most promising young physicist in England'.

Bohr was no longer in Manchester when, in September 1917, Rutherford began a series of experiments 'carried out at very irregular intervals, as the pressure of routine and war work permitted'.[95] The results were published in 1919 in a four-part article entitled 'Collision of α-particles with light atoms'. His best known discovery in these papers is the first example[96] of an induced nuclear reaction.* Another finding reported in this work is not as widely remembered, though it is at least as important: the scattering of α-particles on hydrogen does not always obey the Rutherford scattering law.[95] It was inevitable that this new phase of α-particle physics would once again start in the Manchester laboratory. α-particle sources were available in other places as well but not the expertise of the Rutherford school in handling them.

Had Geiger and Marsden not been at their respective fronts, they might well have made this next step. The new experiment was not all that different

* The reaction is $\alpha + N^{14} \rightarrow \text{proton} + O^{17}$. Rutherford did not yet directly verify the production of the oxygen isotope.

from what these men had done earlier. In particular, Rutherford's source, a brass disk coated with RaC (Bi^{214}), once again produced α-particles in the 5 MeV range. The main distinction between the Geiger–Marsden experiment[29] of 1913 and the new one was that high Z targets were used in the former, while Rutherford used the lowest possible Z: hydrogen. Since this minimizes the Coulomb repulsion (I next briefly anticipate the ultimate interpretation) the intrinsic nuclear forces have the best chance to stand out. As is known from years of subsequent experimentation, 5 MeV is plenty of energy for penetrating the α-hydrogen Coulomb barrier and thus for detecting nuclear force effects.

Back to 1919. Rutherford placed his source inside a vessel filled with hydrogen at atmospheric pressure. By covering the source with gold or aluminum foil of known stopping power, carefully treated against hydrogen occlusion, he could vary the range (i.e. the velocity) of the α-particles between 7 and 3 cm air equivalent. His detection method consisted of counting scintillations produced by the scattered H-particles (protons).

For a comparison of his experimental results with his theoretical picture of pointlike particles interacting via a $1/r^2$-potential, Rutherford could of course not use his cross-section (9.6) since that expression refers to the case of no target recoil (quantities $O(m_\alpha/m_{\text{nucleus}})$ were neglected). In the present case the H-particle recoils strongly. The corresponding theory had meanwhile been given by Darwin[52] who had shown that the number $\mathrm{d}n(\phi)$ of H-particles (supposed to be initially at rest) scattered over an angle between 0 and ϕ by a single α-particle in its passage through a thickness $\mathrm{d}x$ of H-gas is given by

$$\frac{\mathrm{d}n(\phi)}{\mathrm{d}x}=\frac{100\pi Ne^4}{m^2v^4}\cdot tg^2\phi. \tag{11.11}$$

Here N is the number of H atoms per cc; v, m, and $2e$ are the initial velocity, mass, and charge of the α-particle respectively; the H-mass has been put equal to $m/4$.

Let us see what Rutherford found on comparing his data with Eq. (11.11).

(a) For α-particles with range 7 cm, corresponding to the full 5 MeV energy, 'the number of swift H-atoms produced ... is 30 times greater than the theoretical number', not some small correction but a major new effect.

(b) 'For α-particles of range less than 4 cm of air, the distribution and absorption of H-atoms are in fair accord with the simple theory.' The 'simple theory' from which Eq. (11.11) is derived implies that $n(\phi)$ rises rapidly for decreasing v, so rapidly in fact that, even for scattering on hydrogen, the Coulomb forces dominate if v is sufficiently small.

(c) It was therefore possible to characterize the new effect by a range r, as follows. Use the Coulomb picture for sufficiently small v, where it is valid, to calculate the distance of closest approach between α and H. This distance decreases with increasing v. According to (a) and (b) there evidently exists a critical v at which the Coulomb picture ceases to hold.* Define r as the distance of closest approach for this critical v. From his experiments, Rutherford found that

$$r \simeq 3.5 \times 10^{-13}\,\text{cm}, \tag{11.12}$$

the distance characteristic for whatever caused this new effect.

What did Rutherford have to say about this cause?

'We have every reason to believe that the α-particle has a complex structure, consisting probably of four hydrogen nuclei and two negative electrons ... it appears significant that [r] is about the same as the accepted value of the diameter of the negative electron, viz. 3.6×10^{-13} cm .... It is of course possible to suppose that the actual law of force ... does not follow the inverse square for very small distances ... [However] it seems simpler to suppose that the rapid alteration ... of the force close to the nucleus is due rather to a deformation of its structure and of its constituent parts.' In other words, Rutherford thought it more prudent to continue assuming that all forces involved are electromagnetic, and that the new effect might be ascribed to deviations from point structure of the colliding particles. That is why in the quoted lines he cited the near equality of the values for r and for the classical electron radius** e^2/mc^2 (modern value $\simeq 2.8 \times 10^{-13}$ cm). After all, wasn't the electron 'probably' a constituent of the α-particle?

In a subsequent paper[97] Rutherford reported deviations from the 'simple theory' for the scattering of α's off N and O. He also assigned a young Ph.D. the task to pursue these matters further.[98] Right thereafter he left for his new post at the Cavendish. 'He brought with him a considerable amount of apparatus from Manchester ... he brought the large quantity of radium lent him by the Academy of Sciences of Vienna in 1908, he brought with him one of his research students, James Chadwick [now released from detention in Germany] .... Finally, Rutherford brought to Cambridge, what he was never long without, a list of twenty or thirty "projected researches". '[99]

One of the projects, continued study of α-hydrogen scattering, was tackled by Chadwick and Bieler, using much improved techniques.[100] They confirmed Rutherford's findings but were more emphatic in their conclusions: 'No system of four hydrogen nuclei and two electrons united by inverse square law forces could give a field of force of such intensity over so long an extent. We must conclude either that the α-particle is not made up of four H-nuclei and two electrons, or that the law of force is not the inverse square in the immediate neighborhood of an electric charge. It is simpler to choose the latter alternative

* This is of course an over-simplified picture; there is no abrupt change of régime. However, the argument is good enough for order of magnitude estimates.

** For a long time it was thought, incorrectly as we shall see later, that this radius followed from an assumed purely electromagnetic origin of the electron's mass. Consider the electron as a little sphere with radius a. Then the assumption implies that $e^2/a \simeq mc^2$.

particularly as other experimental as well as theoretical considerations point in this direction'. Their mention of theoretical considerations may refer to calculations which had just been published[101] by Darwin, who, stimulated by Rutherford's experiments, had examined how the simple theory is modified if hydrogen remains a point particle but the α-particle is taken to be either a charged hard sphere, or a charged hard disk, or a 'square nucleus' with an H-particle at each corner and two electrons in the center, models chosen simply because one could calculate with them. None of these models, Darwin concluded, were particularly convincing.

In any event, Chadwick and Bieler's final conclusion[100] avoids all reference to a possible electromagnetic cause for the deviations from the simple theory: 'The present experiments do not seem to throw any light on the nature of the law of variation of the forces at the seat of an electric charge, but merely show that the forces are of very great intensity . . . . It is our task to find some field of force which will reproduce these effects.'

I consider this statement, made in 1921, as marking the birth of the strong interactions.

Phenomenological attempts at modifying the $1/r^2$ Coulomb force were not long in coming. Terms were added behaving as inverse third[102] or fourth[103] or fifth[104] powers of the distance. Theoretical speculations on models of complex nuclei fit to describe the growing number of observed nuclear transmutations induced by α-bombardment also date from this period.[105] Experimental work at the Cavendish continued, important new results kept coming. However, for a number of years an entirely new and different part of physics took center-stage: quantum mechanics.

References

1. A. T. Petit and P. L. Dulong, *Ann. de Chimie et Phys.* **10**, 395, 1819.
2. A. Pais, *Subtle is the Lord*, Chapter 20, Oxford Univ. Press 1982.
3. E. Turner, *Phil. Mag.* **1**, 109, 1832.
4. D. M. Knight, *Classical scientific papers, chemistry*, p. 62, Elsevier, New York 1970.
5. D. Mendeléev, *The principles of chemistry*, Vol. 2, p. 16, footnote, Longmans Green, London 1891.
6. D. Mendeléev, *Zeitschr. f. Chem.* **12**, 405, 1869, repr. in Ref. 4, p. 273.
7. D. Mendeléev, *Chem. News* **41**, 39, 1880, repr. in Ref. 4, p. 291.
8. Ref. 5, Vol. 2, p. 25.
9. J. H. Vincent, *Phil. Mag.* **4**, 103, 1902.
10. C. G. Barkla, *Phil. Mag.* **14**, 408, 1908.
11. M. Curie, in *Théorie du rayonnement et les quanta*, p. 385, Eds. P. Langevin and M. de Broglie, Gauthier-Villars, Paris 1912.
12. E. Rutherford, *Phil. Mag.* **21**, 669, 1911.
13. E. Rutherford, *Radioactive substances and their radiations*, p. 622, Cambridge Univ. Press 1913.

14. N. Bohr, *Phil. Mag.* **26**, 476, 1913, repr. in *Niels Bohr, collected works*, Vol. 2, p. 188, Ed. U. Hoyer, North Holland, Amsterdam 1981.
15. M. Curie, in *La structure de la matière*, p. 56, Gauthier-Villars, Paris 1921.
16. F. Soddy, *Nature* **92**, 400, 1913.
17. F. Soddy, in *Royal Institution Library of Science* **8**, 449, 1923; Eds. W. Bragg and G. Porter, Elsevier, New York 1970.
18. F. Soddy, *Ann. Report to the London Chem. Soc. for 1910*, repr. in T. J. Trenn, *Radioactivity and atomic theory*, esp. pp. 251–2, Taylor and Francis, London 1975.
19. F. Soddy, Ref. 18, pp. 332, 333.
20. F. Soddy, Ref. 17, **8**, 123, 1917.
21. Cf. F. W. Aston, *Isotopes*, Chapter 10, Edward Arnold, London 1921.
22. G. Hansen, *Naturw.* **15**, 163, 1927.
23. F. Soddy, *The chemistry of radioelements*, p. 30, Longmans Green, London 1911.
24. A. S. Russell, *Chem. News* **107**, 49, 1913.
25. G. von Hevesy, *Phys. Zeitschr.* **14**, 49, 1913.
26. K. Fajans, *Phys. Zeitschr.* **14**, 131, 136, 1913; *Verh. Deutsch. Phys. Ges.* **15**, 240, 1913; cf. also D. U. Anders, *J. Chem. Educ.* **41**, 522, 1964.
27. H. Geiger and E. Marsden, *Proc. Roy. Soc. A* **82**, 495, 1909.
28. Ref. 12, Section 6.
29. H. Geiger and E. Marsden, *Phil. Mag.* **25**, 604, 1913.
30. E. Rutherford, *Phil. Mag.* **27**, 488, 1914.
31. C. G. Barkla, *Phil. Mag.* **21**, 648, 1911.
32. C. G. Barkla and C. A. Sadler, *Phil. Mag.* **14**, 408, 1907; **17**, 739, 1909; C. G. Barkla, *Proc. Cambr. Phil. Soc.* **10**, 257, 1909; C. G. Barkla and J. Nicol, *Nature* **84**, 139, 1910.
33. C. G. Barkla, *Phil. Mag.* **22**, 396, 1911.
34. R. Whiddington, *Proc. Roy. Soc. A* **85**, 323, 1911.
35. H. A. M. Snelders, *Nederl. Tydsschr. v. Natuurk.* **40**, 241, 1974; *Janus* **61**, 59, 1974; cf. also T. Hirosige, *Jap. St. Hist. Sc.* **10**, 143, 1971.
36. A. J. van den Broek, *Ann. der Phys.* **23**, 199, 1907.
37. A. J. van den Broek, *Phys. Zeitschr.* **12**, 490, 1911.
38. A. J. van den Broek, *Nature* **87**, 78, 1911.
39. A. J. van den Broek, *Phys. Zeitschr.* **14**, 32, 1913.
40. A. J. van den Broek, *Nature* **92**, 372, 1913.
41. F. Soddy, *Nature* **92**, 399, 1913.
42. E. Rutherford, *Nature* **92**, 423, 1913.
43. A. J. van den Broek, *Nature* **92**, 476, 1913.
44. J. L. Heilbron, *H. G. J. Moseley*, Univ. California Press 1974.
45. B. Jaffé, *Moseley*, Doubleday, New York 1971.
46. H. G. J. Moseley, *Nature* **92**, 554, 1913.
47. T. Kuhn, interview with Niels Bohr, November 1, 1962, Niels Bohr Library of the American Institute of Physics, New York.
48. H. G. J. Moseley, *Phil. Mag.* **26**, 1024, 1913; **27**, 703, 1914.
49. M. E. Weeks and H. M. Leicester, *The discovery of the elements*, 7th edn, Mack Printing Co., Easton, Pa. 1968.
50. E. Rutherford, *Proc. Roy. Soc. A* **93**, xxii, 1917.
51. Ref. 14, p. 544.
52. C. G. Darwin, *Phil. Mag.* **27**, 499, 1914.
53. E. Rutherford, *Proc. Roy. Soc. A* **90**, 1914, insert after p. 462.

54. E. Marsden and W. C. Lantsberry, *Phil. Mag.* **30**, 240, 1915.
55. E. Rutherford, letter to B. Boltwood, February 28, 1921, repr. in L. Badash, *Rutherford and Boltwood*, p. 343, Yale Univ. Press, New Haven, Conn. 1969.
56. R. H. Stuewer, *The nuclear electron hypothesis*, Univ. of Minnesota preprint, to be published.
57. E. Rutherford, *Scientia* **16**, 337, 1914.
58. W. Duane, *Phys. Rev.* **5**, 335, 1915.
59. A. Sommerfeld, *Atombau und Spektrallinien*, p. 540, Vieweg, Braunschweig 1919.
60. A. Einstein, *Ann. der Phys.* **18**, 639, 1905.
61. A. Einstein, *Jahrb. der Rad. und Elektr.* **4**, 411, 1907.
62. A. Einstein, *Arch. Sci. Phys. Nat.* **29**, 5, 125, 1910, see esp. p. 144.
63. M. Planck, *Verh. Deutsch. Phys. Ges.* **4**, 136, 1906; *Ann. der Phys.* **26**, 1, 1908; cf. Rayleigh, *Nature* **56**, 58, 1902.
64. P. Langevin, *J. de Phys.* (Paris) **3**, 553, 1913.
65. W. Lenz, *Naturw.* **8**, 181, 1920.
66. W. Pauli, *Theory of relativity*, p. 123, transl. G. Field, Pergamon, Oxford 1958.
67. F. W. Aston, *Proc. Roy. Soc. A* **115**, 487, 1927.
68. K. T. Bainbridge, *Phys. Rev.* **42**, 1, 1932.
69. J. Cockcroft and E. Walton, *Proc. Roy. Soc. A* **137**, 229, 1932.
70. W. Braunbeck, *Zeitschr. f. Phys.* **107**, 1, 1937.
71. G. F. Nicolai, *The biology of war*, p. 303, The Century Co., New York 1918.
72. Frankf. Ztg. October 4, 1914, Engl. trans. in Ref. 71.
73. F. Gilbert, *The end of the European era, 1890 to the present*, 2nd edn, p. 137, Norton, New York 1979.
74. K. Schwabe, *Wissenschaft und Kriegsmoral*, p. 23, Musterschmidt, Goettingen 1969.
75. L. Dimier, *L'appèl des intellectuels allemands*, Nouvelle Librairie Nationale, Paris 1915.
76. Ref. 74, p. 195.
77. Cf. e.g. D. J. Kevles, *The physicists*, Chapter 10, Random House, New York 1979.
78. Ref. 2, p. 242.
79. O. Nathan and H. Norden, *Einstein on peace*, Chapter 1, Schocken, New York 1968.
80. E. H. Beininger, *Otto Hahn*, p. 51, Rowohlt, Hamburg 1974.
81. W. Herz, *Phys. Zeitschr.* **16**, 113, 1915.
82. Rayleigh, *The life of Sir J. J. Thomson*, Chapter 10, Cambridge Univ. Press 1942.
83. N. Feather, *Lord Rutherford*, Chapter 5, Blackie, Glasgow, 1940.
84. R. H. Kargon, *The rise of Robert Millikan*, p. 87, Cornell Univ. Press, Ithaca 1982.
85. D. M. Livingston, *The master of light*, p. 284, Scribner, New York 1973.
86. M. Curie, *La radiologie et la guerre*, F. Alcan, Paris 1921.
87. O. R. Frisch, *Biogr. Mem. Fell. Roy. Soc.* **16**, 405, 1970.
88. D. P. Serwer, 'The rise of radiation protection', Ph.D. thesis, Princeton Univ. 1977.
89. E. Marx, *Phys. Zeitschr.* **16**, 1, 1915.
90. S. Meyer, *Phys. Zeitschr.* **16**, 431, 1915.
91. C. Runge, *Phys. Zeitschr.* **17**, 545, 1916.
92. A. Hermann, *Max Planck*, Rowohlt, Hamburg 1973.
93. E. Rutherford, *Nature* **96**, 331, 1915.
94. N. Bohr, letter to A. Fokker, February 14, 1916, repr. in Ref. 14, p. 501.
95. E. Rutherford, *Phil. Mag.* **37**, 537, 1919; correction in *Phil. Mag.* **41**, 307, 1921.
96. E. Rutherford, *Phil. Mag.* **37**, 581, 1919.

97. E. Rutherford, *Phil. Mag.* **37**, 571, 1919.
98. L. B. Loeb, *Phil. Mag.* **38**, 533, 1919.
99. Ref. 83, p. 159.
100. J. Chadwick and E. S. Bieler, *Phil. Mag.* **42**, 923, 1921.
101. C. G. Darwin, *Phil. Mag.* **41**, 486, 1921.
102. E. S. Bieler, *Proc. Roy. Soc. A* **105**, 434, 1924.
103. E. S. Bieler, *Proc. Cambr. Phil. Soc.* **21**, 686, 1923.
104. P. Debye and W. Hardmeier, *Phys. Zeitschr.* **27**, 196, 1926.
105. E. Rutherford and J. Chadwick, *Phil. Mag.* **42**, 809, 1921; **44**, 417, 1922; **50**, 889, 1925; H. Petterson, *Proc. Phys. Soc.* **36**, 194, 1923.

12

Quantum mechanics, an essay

(a) The status of physics in the spring of 1925

In the spring of 1925 physicists had in hand: two fundamental fields, three basic forces, three fundamental particles, and two logically unconnected theoretical structures which appeared to be in distinct conflict with each other, yet each of which already then seemed destined to play a role in the ultimate description of physical reality.

1. Two fields. The purpose of Maxwell's memoir *A dynamical theory of the electromagnetic field* is, the author informs us, 'to explain the [electromagnetic] action between distant bodies without assuming the existence of forces capable of acting directly at sensible distances. The theory I propose may therefore be called a theory of the Electromagnetic Field'.[1] The rejection of forces acting instantaneously at a distance in favor of forces being transmitted from point to neighboring point by continuous fields meant the end of a purely mechanical picture of the physical world. Of this innovation Einstein later wrote: 'Since Maxwell's time, Physical Reality has been thought of as represented by continuous fields . . . not capable of any mechanical interpretation. This change in the conception of Reality is the most profound and the most fruitful that physics has experienced since the time of Newton'.[2]

No consequence of the new theory was more profound than the unification of light with electromagnetism: light is composed of electromagnetic waves. No part of the new theory appeared to be better understood than the wave nature of light as it travels through empty space. Consequently no part of the old quantum theory appeared more dubious than Einstein's idea that under certain circumstances light behaves as if it has particle structure.

The electromagnetic theory proposed by Maxwell in 1864 is not what we today call the Maxwell theory. It remained for the original version of the theory to be freed from mechanical remnants, in particular from Maxwell's belief that electromagnetic forces are transmitted through the aether. This medium is, in his words, 'a material substance or body, which is certainly the largest, and probably the most uniform, body of which we have any knowledge'.[3] Einstein's special theory of relativity, twenty years old in the spring of 1925, demonstrated the need to dispense with the aether altogether.[4] Thus

Maxwell had important successors. He also had important precursors. For a guide to the evolution of the electromagnetic field concept I refer the reader to a bibliography compiled by Scott.[5]

Maxwell was the first to inquire whether gravitation could perhaps also be described by some field which, as for electromagnetism, possesses its own dynamical degrees of freedom. In the same memoir of 1864 he was 'naturally led to inquire whether the attraction of gravitation, which follows the same law of the distance [as Coulomb's law] is not also traceable to the action of a surrounding medium'.[6] Fifty years of trial and error went by[7] until, about a decade prior to the spring of 1925, Einstein's general relativity provided the answer. The high point of the classical era had now been reached. In 1915 two kinds of forces were known, each associated with a multi-component field.

2. Three forces. By 1925 a third force had been identified: whatever holds the atomic nucleus together cannot be due to electromagnetism and gravitation alone (see the preceding chapter). Nothing I know indicates that this novelty attracted the attention of any but those few concerned with α-particle physics. Another decade was to pass before the third force was also associated with a field.

3. Three particles. In the spring of 1925 the electron and the proton were well-established objects, but the existence of the third particle, the light-quantum, had only just been experimentally confirmed. Because of its intimate links with the confusions surrounding quantum physics, the evolution of the light-quantum into a full-fledged particle was a far more tortuous, complex, and controversial process than the steps which led to the discovery of the electron and the proton. Elsewhere I have dealt at length with the light-quantum story.[8] Here I mention only those few points which are of particular relevance in the present context.

When initially proposed[9] by Einstein in 1905, the light-quantum was not yet a particle[10] but only a parcel of energy E related to the light-frequency ν by

$$E = h\nu. \tag{12.1}$$

In the course of subsequent theoretical developments it became clear, but only gradually,[10] that the light-quantum is also to be endowed with a momentum p:

$$\vec{p} = h\vec{k} \tag{12.2}$$

where $\vec{k}$ is the wave vector corresponding to ν, $|\vec{k}| = \nu/c$. Accordingly E and p are related by

$$E = c|\vec{p}| \tag{12.3}$$

which is the special case of the relation Eq. (4.11) between the energy, momentum, and mass of a particle in which the rest mass equals zero.*

The relations (12.1) and (12.2) had both been theoretically abstracted from the properties of electromagnetic radiation in or near thermal equilibrium long before experiment demonstrated that the light-quantum is a particle in the same sense that an electron or a proton is a particle. The crucial experiments were performed by Compton and coworkers in the years 1923–5. They showed[12] that the scattering of a light-quantum on an electron at rest obeys the conservation laws—as should be true for the scattering of any two particles—

$$h\vec{k} = \vec{p} + h\vec{k}' \tag{12.4}$$

$$hc|\vec{k}| + mc^2 = hc|\vec{k}'| + (c^2p^2 + m^2c^4)^{1/2} \tag{12.5}$$

where m is the electron's mass, $\vec{p}$ the final electron momentum, and $\vec{\mathrm{k}}$, $\vec{\mathrm{k}}'$ are the light-quantum wave vectors before and after scattering respectively. Of particular interest are those parts of the experiments[12] in which the scattering was observed by means of a cloud chamber. This technique made it possible *for the first time* to check the validity of the energy–momentum conservation laws as applied to individual events of scattering between fundamental particles.

The relatively late date of this discovery explains why it took so long, until 1926, before the light-quantum received its modern particle name: the photon.[13]

4. Two theoretical structures. As the spring of 1925 arrived, the old quantum theory entered its final season. It is debatable whether theory is the right appellation for the set of rules designed to bring some preliminary order in a set of phenomena which defied the logic of classical theory. Earlier in the century the postulates of classical physics, the collective domain of Newtonian mechanics, thermodynamics, statistical mechanics, and Maxwell–Lorentz electrodynamics, had been found to be incomplete. First the special, then the general theory of relativity had led to modifications of the classical postulates. These revisions were profound yet, one might say, not disturbing (except to a small minority). It was at once clear that classical physics emerged unscathed by these innovations as long as one defined, with greater precision than before, its domain of validity: the classical descripton remains true in essence as long as velocities of material objects are negligible compared to the velocity of light, and as long as their weights are not excessive, the sun (for example) belonging to the non-excessive category.

In that same springtime it was still utterly unclear, however, how the classical theory and the quantum hypothesis could be reconciled. Ever since

* For experimental upper limits on the mass of the light-quantum see Ref. 11.

1913, when Bohr had advanced his theory of the hydrogen atom, nearly all activity in quantum physics had focussed on problems concerning the spectra and structure of atoms and molecules. The period from 1913 to 1925 was one of improvization, sometimes faltering, other times brilliant, the efforts centering mainly on Copenhagen, Munich, and, beginning somewhat later, on Goettingen. An expert elementary text[14] on the quantum theory written in 1923 describes those years like this: 'It has been necessary in many respects to grope in the dark, guided in part by the experimental results and in part by various assumptions, often very arbitrary'. A wealth of spectral data, primitive theoretical tools, and courage served to fashion patterns among spectral frequencies, intensities, polarizations, and selection rules. I must pass by this gripping tale* except for a brief mention of one concept which bears directly on the relations between classical and quantum physics: the correspondence principle.

Consider as an example Eq. (9.1):

$$\nu = R\left(\frac{1}{m^2} - \frac{1}{n^2}\right) = \frac{R(n-m)(n+m)}{m^2 n^2}$$

for the frequency of light emitted as an electron in the hydrogen atom jumps from orbit n to orbit m. As this formula illustrates, the classical connection between the frequency of emitted light and the initial frequency of orbital motion is given up in quantum theory; ν depends not only on the initial but also on the final orbit. If, however, $n - m = 1$ and $n \gg 1$ then, approximately,

$$\nu \simeq \frac{2R}{n^3},$$

an expression which depends only on the initial orbit. It is therefore plausible to pose a correspondence: for neighboring orbits and for high quantum numbers, identify the quantum frequency with the known classical expression for ν. This correspondence has predictive power; it determines R. This reasoning is one of several used by Bohr to find the Rydberg constant in terms of e, m, and h.

It takes artistry to make practical use of the correspondence principle: 'It is difficult to explain in what [the principle] consists, because it cannot be expressed in exact quantitative laws, and it is, on this account, also difficult to apply. In Bohr's hands it has been extraordinarily fruitful in the most varied fields; while other more definite and more easily applicable rules of guidance have indeed given important results in individual cases, they have shown their limitations by failing in other cases'.[14] The correspondence principle, then, was the only link between the old and the new. But what was

* In the next chapter we shall catch some glimpses of this way of doing quantum physics, however.

the new? What new first principles would justify the quantum rules introduced in *ad hoc* fashion?

During those years of exploring the atom, electrons were supposed to describe planetary orbits around the nucleus; they were particles. Their wave aspects were yet to be discovered. Even before 1925, the quantum aspects of electromagnetic radiation posed less central but no less baffling problems which lay entirely beyond the reach of the correspondence principle. Evidence for the wave nature of light had long been abundant, yet, ever since the discovery of the Compton effect there was also incontrovertible evidence for particle-like behavior of light in certain experiments. But a particle is not a wave and a wave is not a particle.* How can light behave now as waves, now as particles?

Einstein was the first to have an inkling of the answer. In 1909 he had analyzed the energy fluctuations in a subvolume of a cavity filled with thermal electromagnetic radiation and had found** that these fluctuations are the sum of two terms, one which would be present all by itself if radiation were a pure wave phenomenon, the other, again all by itself, if radiation consisted of 'pointlike moving quanta'.[17] Thus, according to his fluctuation formula, the particle and the wave aspects appeared to occur side by side, as it were. As early as 1909 he therefore concluded:[17] 'It is my opinion that the next phase of theoretical physics will bring us a theory of light that can be interpreted as a kind of fusion of the wave and the emission theory'.†

Thus Einstein may be considered as the godfather of complementarity, a notion which caused him the greatest discontent after its formal introduction by Bohr in 1927.

Between 1909 and 1925 nothing happened to give substance to Einstein's hopes for fusion. In 1924, Einstein had written[18] 'There are therefore now two theories of light, both indispensable and—as one must admit today despite twenty years of tremendous effort on the part of theoretical physicists—without any logical connection'.

Beginning in the summer of 1925 the fusion was on its way, not only for light but also for matter.

The way I was first exposed to quantum mechanics, not long after 1935 when I began my university studies, was no different from the way I learned, say, thermodynamics. There were courses on the subject and there were books, some more helpful for an understanding of the principles, some better for learning how to solve problems. I learned some experimental facts about

* The last-ditch effort by Bohr, Kramers, and Slater to evade this issue was abandoned[15] early in 1925.

** For more details see e.g. Ref. 16.

† By 'emission theory' Einstein meant the Newtonian conception of light as a stream of tiny particles.

electrons behaving as particles in collision processes, as waves in diffraction effects. I was awed by the success of the Schroedinger equation for the hydrogen atom and found the introduction of quantum mechanical probabilities via the continuity equation a most plausible step. Nor did I experience any difficulty in accepting Heisenberg's uncertainty relations, served up with the help of a classical picture of the dispersion of wave packets combined with $E = h\nu$ and $p = hk$. Soon I was happily making quantum mechanical exercises. I had no sense whatever at that time of the stir and struggle which, only ten years earlier, had accompanied the introduction of the new mechanics. I knew a few dates but those seemed to belong to antiquity.

In 1946 I went to Copenhagen and for a brief period became Niels Bohr's close collaborator. Of that experience I have written: 'I must admit that in the early stages of the collaboration I did not follow Bohr's line of thinking a good deal of the time . . . I failed to see the relevance of such remarks as that Schroedinger was completely shocked in 1926 when he was told of the probability interpretation of quantum mechanics, or a reference to some objection by Einstein in 1928, which apparently had no bearing whatever on the subject at hand. But it did not take long before the fog started to lift . . . Bohr would relive the struggles which it took before the content of quantum mechanics was understood and accepted . . . Through steady exposure to Bohr's "daily struggle" and his ever repeated emphasis on "the epistemological lesson which quantum mechanics has taught us", to use a favorite phrase of his, my understanding deepened not only of the history of physics but of physics itself'.[19]

In the course of time, other physicists who were active during the years of discovery of quantum mechanics also told me occasionally of their reactions. Uhlenbeck said to me that it was as if within the span of a few years his life had changed. Wigner told me of his astonishment upon reading Born and Jordan's paper which explained that Heisenberg had unwittingly introduced matrix methods and of his sense that now there seemed to be hope after all for a rationale of quantum theory. Several members of the 1925 generation have told me that Heisenberg's paper which marks the beginning of the new era took a while, but not long, to sink in. Even now this paper, one of the most admirable contributions to physics, is hard to read without knowledge of its subsequent elaboration.

Since there is no particle physics without quantum mechanics, I decided to include a chapter, the present one, on the innovations brought about by non-relativistic quantum mechanics. The title of this chapter was chosen to stress from the outset that it does not pretend to be a systematic historical account. Rather, its aim is to convey some of the flavor of those times of great and rapid metamorphosis, with particular emphasis on the emergence of the new concept of quantum mechanical probability (see Section (d)). Section (b) contains general remarks on the change of style brought about by quantum mechanics. Section (c) contains a succinct chronology of events beginning with de Broglie's hypothesis of matter waves (1923) and ending with Bohr's formulation of complementarity (1927). For perspective I have included dates referring to the exclusion principle, spin, and statistics, although these topics actually belong to the next chapter. The final section (e) contains further comments on the demarcation, marked by the arrival of quantum mechanics, between older and younger generations.*

(b) End of a revolution

The introduction of probability in the sense of quantum mechanics—that is, probability as an inherent feature of fundamental physical law—may well be the most drastic scientific change yet effected in the 20th century. At the same time, this advent marks the end rather than the beginning of a 'scientific revolution', a term often used but rarely defined.

In the political sphere, revolution is a rather clear concept. One system is swept away, to be replaced by another with a distinct new design. It is otherwise in science, where revolution, like love, means different things to different people. Newspapermen and physicists have perceptions of scientific revolution which need not coincide Nor would individual members of these or other professions necessarily agree on what a scientific revolution consists of. For example, *The Times* of 7 November 1919 headed its first article on the recently discovered bending of light: 'Revolution in science . . . Newtonian ideas overthrown'. Einstein, on the other hand in a lecture given in 1921, deprecated[21] the idea that relativity was revolutionary and stressed that his theory was the natural completion of the work of Faraday, Maxwell, and Lorentz. I happen to share Einstein's judgement, while other physicists will quite reasonably object that the abandonment of absolute simultaneity and of absolute space are revolutionary steps.

However, all of us would agree, I would think, that the *Times* statement 'Newtonian ideas overthrown', being unqualified, tends to create the incorrect impression of a past being entirely swept away. That is not how science progresses. The scientist knows that it is in his enlightened self-interest to protect the past as much as is feasible, whether he be a Lavoisier breaking with phlogiston, an Einstein breaking with aether, or a Max Born breaking with classical causality.

These tensions between the progressive and the conservative are never more in evidence than during a revolutionary period in science, by which I mean a period during which (*i*) it becomes clear that some parts of past science have to go, and (*ii*) it is not yet clear which parts of the older edifice are to

* Part of the material of this chapter has appeared earlier in an article written on the occasion of the centenary of Max Born's birth.[20]

be reintegrated in a wider new frame. Such periods are initiated either by experimental observations that do not fit into accepted pictures or by theoretical contributions that make successful contact with the real world, at the price of one or more assumptions which are in violation of the established corpus of theoretical physics.

The era of the old quantum theory, the years from 1900 to 1925, constitutes the most protracted revolutionary period in modern science. Six theoretical papers appeared during that time which are revolutionary in the above sense: Planck's on the discovery of the quantum theory (1900); Einstein's on the light-quantum (1905); Bohr's on the hydrogen atom (1913); Bose's on what came to be called quantum statistics (1924); Heisenberg's on what came to be known as matrix mechanics (1925); and Schroedinger's on wave mechanics (1926). If these papers have one thing in common it is that they contain at least one theoretical step which (whether the respective authors knew it then or not) could not be justified at the time of writing.

The end of this revolutionary period (I consider only non-relativistic quantum mechanics) is not marked by a single date, nor was it brought about by a single person, but rather by four: Heisenberg, Schroedinger, Born, and Bohr. The end phase begins in 1925 with the abstract of Heisenberg's extraordinary first paper[22] on quantum mechanics, which reads: 'In this paper it will be attempted to secure foundations for a quantum theoretical mechanics which is exclusively based on relations between quantities which in principle are observable'. With these words Heisenberg states specific desiderata for a new axiomatics. His paper is the correct first step in the new direction. The end phase continues in 1926 with Schroedinger's papers on wave mechanics and Born's remarks on probability and causality, and comes to a conclusion in 1927 with Heisenberg's derivation of the uncertainty relations and Bohr's formulation of complementarity. At that stage the basic ingredients had been provided which, in the course of time, were to allow for a consistent theoretical foundation of quantum mechanics, including a judgment of the way the new theory contains the old, the classical, theory as a limiting case.

For good reasons the period 1925–7 would become known in Goettingen as the years of Knabenphysik: boy physics. Indeed, consider the respective ages of the architects of the new dynamics in July 1925 when Heisenberg submitted his first paper on quantum mechanics. At that time Heisenberg (Ph.D. with Sommerfeld on problems in turbulence) was 23; Jordan (Ph.D. with Born on the old quantum theory of radiation) was 22; Pauli (Ph.D. with Sommerfeld on the old quantum theory of the hydrogen molecule ion) was 25; Dirac (Ph.D. entitled 'Quantum Mechanics', the first Ph.D. in history awarded (in 1926) for topics in quantum mechanics) was 22. Schroedinger (Ph.D. 1910 with Hasenöhrl on the conduction of electricity on the surface of insulators exposed to humid air) does not fit easily into this simplistic scheme: he was thirty-seven at that time. It does not seem out of place, however, to note here the remark once made to me by Hermann Weyl that Schroedinger did his great work during a late erotic outburst in his life. Nor should it be forgotten that Schroedinger was the only one among the creators of the new mechanics who never found peace with what he had wrought.

Born, Bohr, and Sommerfeld, men of riper age, made major contributions to quantum mechanics, yet in the first instance they must be regarded as the foremost teachers of the period, each with his distinctive style. Born was in his middle forties when he did his work on the statistical interpretation of quantum mechanics. By that time he was already a renowned physicist and teacher, had published more than a hundred research papers, and had written six books. Likewise, Bohr was already a stellar figure in his forties when he gave the complementarity interpretation of quantum mechanics. Sommerfeld, in his late fifties when quantum mechanics arrived, founded the quantum mechanical theory of metals. He was past sixty when he wrote one of the best early textbooks on wave mechanics, the supplement[23] to his *Atombau und Spektrallinien*.

(c) A chronology*

De Broglie, 10 September 1923. 'After long reflection in solitude and meditation, I suddenly had the idea, during the year 1923, that the discovery made by Einstein in 1905 should be generalized by extending it to all material particles and notably to electrons.' Thus Louis de Broglie's recollections[24] of the way in which he became the first to associate wave-like behavior with matter. His train of thought is daring and evident. In 1905 Einstein had associated particles (light-quanta) with electromagnetic wave phenomena. De Broglie suggests that the association particle ↔ wave shall be universal. On September 10, 1923 he proposes that $E = h\nu$ hold not only for photons but also for a 'fictitious associated wave' assigned to electrons.[25] On September 24 he notes that accordingly one might anticipate diffraction phenomena for electrons.[26] On 25 November, 1924, he defends his Ph.D. thesis,[24] an extended version of these two articles.** Einstein says of this work: 'I believe it is a first feeble ray of light on this worst of our physics enigmas'.[27]

Bose, 2 July 1924. Satyendra Nath Bose introduces a new coarse-grained statistical counting procedure which leads to Planck's radiation law.[31] (See Chapter 13.)

Einstein, 10 July 1924, extends Bose's procedure to a gas of free material particles.[32] His study of fluctuations around equilibrium of this quantum gas

* Dates either refer to the day of receipt of an article by a journal or the day of presentation before a learned gathering.

** For a detailed analysis of de Broglie's thesis see Ref. 28. For the links between de Broglie, Einstein's work on the quantum gas, and Schroedinger's work on wave mechanics see Refs. 29, 30.

leads him (8 January 1925) to associate waves with gas particles by an argument independent of de Broglie's.[33] (See Chapter 13.)

Pauli, 16 January 1925, enunciates his exclusion principle.[34] (See Chapter 13.)

Heisenberg, 25 July 1925. In this seminal paper,[22] Heisenberg bids a not too fond goodbye to the old quantum theory and, in the same phrase, goes his own way: 'It is better . . . to admit that the partial agreement of the quantum rules [of the old theory] with experiment is more or less accidental, and to try to develop a quantum theoretical mechanics, analogous to the classical mechanics in which only relations between observable quantities appear'. Here are but a few of his main points.* Consider the classical one-dimensional equation of motion (a dot means differentiation with respect to time):

$$m\ddot{x}+f(x)=0 \tag{12.4}$$

with the classical energy integral

$$W=\tfrac{1}{2}m\dot{x}^2+\int_0^x f(x)\,\mathrm{d}x. \tag{12.5}$$

Heisenberg introduces a new mechanics** by associating with $x(t)$ the 'ensemble of quantities' (Gesamtheit der Gröszen)

$$x(t)\rightarrow x_{nm}\,\mathrm{e}^{i\omega_{nm}(t)}. \tag{12.6}$$

Since Eq. (12.4) will contain terms non-linear in x for all cases but the harmonic oscillator, Heisenberg is led to ask: What ensemble must be associated with x^2? And is it indeed true that, also in the new mechanics, W (always nonlinear in x) remains a conserved quantity, that is, time independent? He attacks these questions with the Ansatz

$$x^2(t)\rightarrow (x^2)_{mn}\,\mathrm{e}^{i\omega_{mn}t} \tag{12.7}$$

and by making 'the simplest and most natural assumption'

$$(x^2)_{nm}=\sum_s x_{ns}x_{sm}. \tag{12.8}$$

He treats 'in a similar manner' quantities $[x(t)]^m$ and remarks that 'whereas in classical theory $x(t)y(t)$ is always equal to $y(t)x(t)$, this is not necessarily the case in quantum theory'. As a first example he considers the harmonic oscillator: $f(x)=m\omega^2x$. By detailed calculations he shows that W is not only conserved, it is quantized: W can only take on the values W_n given by

$$W_n=(n+\tfrac{1}{2})\frac{h\omega}{2\pi}. \tag{12.9}$$

* For an English translation of this paper as well as for revealing letters by Heisenberg written at that time see Ref. 35.

** I do not follow Heisenberg's notation in detail.

Furthermore the ω_{mn} are constrained by a Bohr-like condition, they are non-zero only if $m = n-1$, in which case

$$\omega_{n,n-1} = \frac{2\pi}{h}(W_n - W_{n-1}) = \omega. \tag{12.10}$$

In addition, he verifies explicitly that for $f(x) = m\omega^2 x + \lambda x^p$, $p = 2$ or 3, W is once again conserved, at least to $O(\lambda^2)$. He concludes by remarking that his method has been 'only very superficially employed'. That is true only in so far as the examples treated are among the simplest.

Born and Jordan, 27 September 1925, note[36] that Heisenberg's rule Eq. (12.8) is 'none other than the well-known mathematical rule for matrix multiplication'. First derivation of

$$pq - qp = \frac{h}{2\pi i} \tag{12.11}$$

where p and q are matrices and p represents the momentum conjugate to q.

Uhlenbeck and Goudsmit, 17 October 1925, announce their discovery of the spin of the electron.[37] (See Chapter 13.)

Dirac, 7 November 1925, gives an independent derivation of Eq. (12.11). Introduction of the 'commutator'

$$[p, q] \equiv pq - qp. \tag{12.12}$$

Derivation of

$$ih\dot{x} = 2\pi[x, H] \tag{12.13}$$

where H is the Hamiltonian. Generalizations of Eq. (12.11) to systems with many degrees of freedom.

Born, Heisenberg, and Jordan, 16 November 1925. First comprehensive treatment of the foundations of matrix mechanics.[38] Introduction of canonical transformations, perturbation theory, treatment of degenerate systems, introduction of the commutation relations

$$[M_x, M_y] = \frac{h}{2\pi i} M_z, \text{ cycl.} \tag{12.14}$$

for the components of angular momentum $\vec{M}$ of a many-particle system.

Pauli, 17 January 1926. Derivation of the discrete spectrum of the hydrogen atom by matrix methods.[39]

Schroedinger, 27 January 1926. First of a series of papers all entitled 'Quantisierung als Eigenwertproblem'. Derivation of the hydrogen spectrum, including its continuous part.[40] Even before it was published, Schroedinger spoke about this work at a colloquium in Zürich. Felix Bloch, then a young student, remembered[41] of this event: 'I was still too green to really appreciate the significance of this talk, but from the general reaction of the audience I realized that something rather important had happened'.

Fermi, 7 February 1926. The first paper[42] on 'Fermi–Dirac statistics'. (See Chapter 13.)

Born, 25 June 1926. First paper on the probability interpretation of quantum mechanics.[43]

Dirac, 26 August 1926. Derivation of Planck's radiation law from first principles. Independent presentation of 'Fermi–Dirac statistics'.[44]

Davisson and Germer, 3 March 1927. First detection of electron diffraction by a crystal.[45]

Heisenberg, 23 March 1927, presents his uncertainty relations.[46]

Bohr, 16 September 1927, states the principle of complementarity for the first time.[47]

(d) Quantum mechanics interpreted

> This problem of getting the interpretation proved to be rather more difficult than just working out the equations.
>
> P. A. M. Dirac[48]

If the early readers of Heisenberg's first paper on quantum mechanics had one thing in common with its author it was an inadequate grasp of what was happening. The mathematics was unfamiliar, the physics opaque. In September, Einstein wrote to Ehrenfest about Heisenberg's paper: 'In Goettingen they believe in it (I don't)'.[49] At about that same time, Bohr considered the work of Heisenberg to be 'a step probably of fundamental importance' but noted that 'it has not yet been possible to apply [the] theory to questions of atomic structure'.[50] Whatever reservations Bohr initially may have had were dispelled by early November[51] when word reached him[52] that Pauli had done for matrix mechanics what he himself had done for the old quantum theory: derive the Balmer formula for the discrete spectrum of hydrogen.[39]

Nor did Schroedinger's discovery of wave mechanics bring immediate salvation in regard to new first principles, but at least physicists were on the whole more comfortable manipulating partial differential equations, as required by wave mechanics, than matrices. Uhlenbeck told me: 'The Schroedinger theory came as a great relief. Now we did not any longer have to learn the strange mathematics of matrices'. Rabi told me how he looked through Born's book *Atommechanik* for a nice problem to solve by Schroedinger's method, found the symmetric top, went to Kronig, and said: 'Let's do it'. They did.[53] Wigner told me: 'People began making calculations but it was rather foggy'.

Indeed, until the spring of 1926 quantum mechanics, whether in its matrix or its wave formulation, was high mathematical technology of a new kind, manifestly important because of the answers it produced, but without clearly stated underlying physical principles. Schroedinger was the first, I believe, to propose such principles in the context of quantum mechanics, in a note completed not later than May, which came out on 9 July.[54] He suggested that waves are the only reality; particles are only derivative things. In support of this monistic view he considered a suitable superposition of linear harmonic oscillator wave functions and showed (his italics): 'Our wave group holds *permanently together*, does *not* expand over an ever greater domain in the course of time', adding that 'it can be anticipated with certainty' that the same will be true for the electron as it moves in high orbits in the hydrogen atom. Thus he hoped that wave mechanics would turn out to be a branch of classical physics—a new branch, to be sure, yet as classical as the theory of vibrating strings or drums or balls.

Schroedinger's calculation was right; his anticipation was not. The case of the oscillator is very special: wave packets do almost always disperse. Being a captive of the classical dream, Schroedinger missed a second chance at interpreting his theory correctly. On 21 June 1926, his paper[55] on the nonrelativistic time-dependent wave equation was received. It contains in particular the one-particle Schroedinger equation (I slightly modify his notations)

$$\frac{ih}{2\pi}\frac{\partial\psi}{\partial t}=\left(-\frac{h^2}{2m}\Delta+V\right)\psi \tag{12.15}$$

(where ψ is the wave function, t is time, Δ is the Laplace operator, and V is a potential), and its conjugate as well as the corresponding continuity equation,

$$\frac{\partial\rho}{\partial t}-\operatorname{div}\vec{j}=0 \tag{12.16}$$

$$\rho=\psi^*\psi \tag{12.17}$$

$$\vec{j}=\frac{ih}{4\pi m}(\psi^*\vec{\nabla}\psi-\vec{\nabla}\psi^*\cdot\psi). \tag{12.18}$$

Eq. (12.16), Schroedinger believed, had to be related to the conservation of electric charge.

The break with the past came in a paper by Born received 4 days later, on 25 June 1926. In order to make his decisive new step, 'It is necessary [Born wrote half a year later[56]] to drop completely the physical pictures of Schroedinger which aim at a revitalization of the classical continuum theory, to retain only the formalism and to fill that with new physical content'.

In his June paper,[43] entitled 'Quantum mechanics of collision phenomena', Born considers (among other things) the elastic scattering of a steady beam of particles with mass m and velocity v in the z-direction by a static potential which falls off faster than $1/r$ at large distances. In modern language, the stationary wave function describing the scattering behaves asymptotically as

$$e^{ikz}+f(\theta,\phi)\frac{e^{ikr}}{r},\qquad k=\frac{2\pi mv}{h}. \tag{12.19}$$

The number of particles scattered into the element of solid angle $d\omega = \sin\theta d\theta d\phi$ is given by $N|f(\theta,\phi)|^2 d\omega$, where N is the number of particles in the incident beam crossing unit area per unit time. In order to revert to Born's notation, replace $f(\theta,\phi)$ by Φ_{mn}, where 'n' denotes the initial-state plane wave in the z-direction and 'm' the asymptotic final state in which the wave moves in the (θ,ϕ)-direction. Then, Born declares, 'Φ_{mn} determines the probability for the scattering of the electron from the z-direction into the direction $[\theta,\phi]$'.

At best, this statement is vague. Born added a footnote in proof to his evidently hastily written paper: 'A more precise consideration shows that the probability is proportional to the square of Φ_{mn}'. He should have said 'absolute square'. But he clearly had got the point, and so that great novelty, the correct transition probability concept, entered physics by way of a footnote.

I will return shortly to the significant fact that Born originally associated probability with Φ_{mn} rather than with $|\Phi_{mn}|^2$. As I learned from recent private discussions, Dirac had the same idea at that time. So did Wigner, who told me that some sort of probability interpretation was then on the minds of several people and that he, too, had thought of identifying Φ_{mn} or $|\Phi_{mn}|$ with a probability. When Born's paper came out and $|\Phi_{mn}|^2$ turned out to be the relevant quantity, 'I was at first taken aback but soon realized that Born was right', Wigner said.

If Born's paper lacked formal precision, causality was brought sharply into focus as the central issue: 'One obtains the answer to the question, *not* "what is the state after the collision" but "how probable is a given effect of the collision". Here the whole problem of determinism arises. From the point of view of our quantum mechanics there exists no quantity which in an individual case causally determines the effect of a collision . . I myself tend to give up determinism in the atomic world'. However, he was not yet quite clear about the distinction between the new probability in the quantum mechanical sense and the old probability as it appears in classical statistical mechanics: 'It does not seem out of the question that the intimate connection which here appears between mechanics and statistics may demand a revision of the thermodynamic-statistical principles'.

One month after the June paper, Born completed a sequel with the same title.[57] His formalism is firm now and he makes a major new point. He considers a normalized stationary wave function ψ referring to a system with discrete, nondegenerate eigenstates ψ_n and notes that in the expansion

$$\psi=\sum c_n\psi_n \tag{12.20}$$

$|c_n|^2$ is the probability for the system to be in the state n. In June he had discussed probabilities of transition, a concept that, at least phenomenologically, had been part of physics since 1916, when Einstein had introduced his A- and B-coefficients in the theory of radiative transitions—and at once had begun to worry about causality.[58] Now Born introduced the probability of a state. That had never been done before. He also expressed beautifully the essence of wave mechanics: 'The motion of particles follows probability laws but the probability itself propagates according to the law of causality'.

During the summer of 1926, Born's insights into the physical principles of quantum mechanics developed rapidly. On August 10 he read a paper before the meeting of the British Association at Oxford[59] in which he clearly distinguished between the 'new' and the 'old' probabilities in physics: 'The classical theory introduces the microscopic coordinates which determine the individual processes only to eliminate them because of ignorance by averaging over their values; whereas the new theory gets the same results without introducing them at all ... We free forces of their classical duty of determining directly the motion of particles and allow them instead to determine the probability of states. Whereas before it was our purpose to make these two definitions of force equivalent, this problem has now no longer, strictly speaking, any sense'.

The history of science is full of gentle irony. In teaching quantum mechanics, most of us arrive at Eq. (12.16), note that something is conserved, and identify that something with a probability. But Schroedinger, who discovered that equation, did not make the connection and never liked quantum probability, while Born introduced probability without using Eq. (12.16).

In December 1926, the probability for a many-particle system with coordinates $q_1, \ldots, q_f$ was introduced for the first time: '$|\psi(q_1, \ldots, q_f)|^2 \cdot \mathrm{d}q_1 \ldots \mathrm{d}q_f$ is the probability that, in the relevant quantum state of the system, the coordinates simultaneously lie in the relevant volume element of configuration space'. The paper is by Pauli and deals with gas degeneracy and paramagnetism. The remark was inspired by Born's work and is found—once again—in a footnote.[60]

What made Born make his step?

In 1954 Born was awarded the Nobel Prize 'for his fundamental research, especially for his statistical interpretation of the wave function'. In his acceptance speech Born, then in his seventies, ascribed his inspiration for the statistical interpretation to 'an idea of Einstein's [who] had tried to make the duality of particles—light-quanta or photons—and waves comprehensible by interpreting the square of the optical wave amplitudes as probability density for the occurrence of photons. This concept could at once be carried over to the Ψ-function: $|\Psi|^2$ ought to represent the probability density for electrons'.[61] Similar statements are frequently found in Born's writings in his late years. On the face of it, this appears to be a perfectly natural explanation. Had Einstein not stated that light of low intensity behaves as if it consisted of energy packets $h\nu$? And is the intensity of light not a function quadratic in the electromagnetic fields? In spite of the fact that I must here dissent from the originator's own words, I do not believe that these contributions by Einstein were Born's guide in 1926.*

My own attempts at reconstructing Born's thinking (necessarily a dubious enterprise) are exclusively based on his two papers on collision phenomena and on a letter he wrote to Einstein, also in 1926. Recall that Born initially thought, however briefly, that Ψ rather than $|\Psi|^2$ was a measure of the probability. I find this impossible to understand if it were true that, at that time, he had been stimulated by Einstein's brilliant discussions of the fluctuations of quadratic quantities (in terms of fields) referring to radiation. Nevertheless, it is true that Born's inspiration came from Einstein: not Einstein's statistical papers bearing on light, but his never published speculations during the early 1920s on the dynamics of light-quanta and wave fields. Born states this explicitly in his second paper: 'I start from a remark by Einstein on the relation between [a] wave field and light-quanta; he said approximately that the waves are only there to show the way to the corpuscular light-quanta, and talked in this sense of a "ghost field" [Gespensterfeld] [which] determines the *probability* [my italics] for a light-quantum . . . to take a definite path'.[57]

It is hardly surprising that Einstein was concerned so early with these issues. In 1909 he had been the first to write about particle–wave duality. In 1916 he had been the first to relate the existence of transition probabilities (for spontaneous emission of light) to quantum theoretical origins—though how this relation was to be formally established he did not of course yet know. Little concrete is known about his ideas of a ghost field or guiding field (Führungsfeld). The best description we have is from Wigner,[64] who knew Einstein personally in the 1920s: '[Einstein's] picture has a great similarity with the present picture of quantum mechanics. Yet Einstein, though in a way he was fond of it, never published it. He realized that it is in conflict with the conservation principles . . . This Einstein never could accept and hence never took his idea of the guiding field quite seriously . . . The problem was solved as we know, by Schroedinger's theory'.**

Born was even more explicit about his source of inspiration in a letter to Einstein[65] written in November 1926 (for reasons not clear to me this letter is not found in the published Born–Einstein correspondence): 'About me it can be told that physicswise I am entirely satisfied since my idea to look upon Schroedinger's wave field as a "Gespensterfeld" in your sense proves better

* Nor do I believe that Born was guided by the Bohr–Kramers–Slater theory, proposed in 1924, abandoned in 1925,[62] nor that his ideas were 'formed in the trend of the Einstein–de Broglie dualistic approach'.[63]

** The conflict with the conservation laws arose because Einstein had in mind one guide field per particle. By contrast, the Schroedinger waves are 'guiding fields' in the configuration space of all particles at once.

all the time. Pauli and Jordan have made beautiful advances in this direction. The probability field does of course not move in ordinary space but in phase- (or rather, in configuration-) space . . . Schroedinger's achievement reduces itself to something purely mathematical; his physics is quite wretched [recht kümmerlich]'.

Thus it seems to me that Born's thinking was conditioned by the following circumstances. He knew and accepted the fertility of Schroedinger's formalism but not Schroedinger's attempt at interpretation: 'He [Schroedinger] believed . . . that he had accomplished a return to classical thinking; he regarded the electron not as a particle but as a density distribution given by the square of his wave function $|\Psi|^2$. He argued that the idea of particles and of quantum jumps be given up altogether; he never faltered in this conviction . . . I, however, was witnessing the fertility of the particle concept every day in [James] Franck's brilliant experiments on atomic and molecular collisions and was convinced that particles could not simply be abolished. A way had to be found for reconciling particles and waves'.[66] His quest for this way led him to reflect on Einstein's idea of a ghost field. It now seems less surprising that his first surmise was to relate probability to the ghost field, not to the '(ghost field)^2.' His next step, from Ψ to $|\Psi|^2$, was entirely his own. We owe to Born the original insight that Ψ itself, unlike the electromagnetic field, has no direct physical reality.

Born may not have realized at once the profundity of his contribution, which helped bring the quantum revolution to an end. Much later he reminisced as follows[67] about 1926: 'We were so accustomed to making statistical considerations, and to shift it one layer deeper seemed to us not very important'.

This frank statement brings to mind once again Carlyle's words: 'Men understand not what is among their hands'.

It is a bit odd—and caused Born some chagrin—that his papers on the probability concept were not always adequately acknowledged in the early days. Heisenberg's own version of the probability interpretation, written in Copenhagen in November 1926, does not mention Born.[68] One finds no reference to Born's work in the two editions of Mott and Massey's book on atomic collisions, nor in Kramer's book on quantum mechanics. In his authoritative *Handbuch der Physik* article of 1933, Pauli refers to this contribution by Born only in passing, in a footnote. Jörgen Kalckar from Copenhagen wrote to me about his recollections of discussions with Bohr on this issue. 'Bohr said that as soon as Schroedinger had demonstrated the equivalence between his wave mechanics and Heisenberg's matrix mechanics, the "interpretation" of the wave function was obvious . . . For this reason, Born's paper was received without surprise in Copenhagen. "We had never dreamt that it could be otherwise", Bohr said.' A similar comment was made by Mott: 'Perhaps the probability interpretation was the most important of all [of Born's contributions to quantum mechanics], but given Schroedinger, de Broglie, and the experimental results, this must have been very quickly apparent to everyone, and in fact when I worked in Copenhagen in 1928 it was already called the "Copenhagen interpretation"—I do not think I ever realized that Born was the first to put it forward'.[69] In response to a query, Casimir, who started his university studies in 1926, wrote to me: 'I learned the Schroedinger equation simultaneously with the interpretation. It is curious that I do not recall that Born was especially referred to. He was of course mentioned as co-creator of matrix mechanics'.

(e) Changing of the guard

Born wrote to Einstein about the ghost field on November 30, 1926. Einstein's reply of 4 December is the oft-quoted[70] letter in which he wrote: 'The theory [quantum mechanics] says a lot but does not really bring us any closer to the secret of the "old one". I, at any rate, am convinced that *He* is not playing at dice'.* Also, the attitudes of the other leaders of the once dominant Berlin school—Planck, von Laue, and Schroedinger—continued to range from scepticism to opposition. In the first week of October 1926, Schroedinger went to Copenhagen, at Bohr's invitation, to discuss the status of the quantum theory. Heisenberg also came. Later Bohr often told others (including me) that Schroedinger reacted on that occasion by saying that he would rather not have published his papers on wave mechanics, had be been able to foresee the consequences. Schroedinger continued to believe that one should dispense with particles. Born continued to refute him. After Schroedinger's death, Born, mourning the loss of his old friend, wrote of their arguments through the years: 'Extremely coarse [saugrob] and tender, sharpest exchange of opinion, never a feeling of being offended'.[71]

After Born's work, Lorentz could no longer grasp the changes wrought by the quantum theory. In the summer of 1927 he wrote to Ehrenfest: 'I care little for the conception of $\Psi\Psi^*$ as a probability . . . In the case of the H-atom, the difficulty in making precise what is meant if one interprets $\Psi\Psi^*$ as a probability manifests itself in that for a given value of E (one of the eigenvalues) there is also a [nonvanishing] probability outside the sphere which electrons with energy E cannot leave'.[72]

In March 1927 Heisenberg published[46] the uncertainty relations. In a given experimental arrangement, let Δx be the latitude within which a coordinate x is determined and, in that same arrangement, Δp the latitude for the conjugate

* For more on Einstein's views see Ref. 70.

momentum. Then, he showed,

$$\Delta p \Delta x \geq \frac{h}{2\pi}. \tag{12.21}$$

Likewise, if in a given arrangement ΔE and Δt are the respective latitudes in energy and time of observation then*

$$\Delta E \Delta t \geq \frac{h}{2\pi}. \tag{12.22}$$

I have often felt that the expression 'uncertainty relation' is unfortunate since it has all too often invoked imagery in popular writings utterly different from what Heisenberg very clearly had in mind, to wit, that the issue is not: what don't I know? but rather: what can't I know? In common language, 'I am uncertain' does not exclude 'I could be certain'. It might therefore have been better had the term 'unknowability relations' been used. Of course one neither can nor should do anything about that now.

Bohr's complementarity principle is an elaboration of the content of the uncertainty relations. In September 1927 he stated[47] the principle as follows: 'The very nature of the quantum theory . . . forces us to regard the space-time coordination and the claim of causality, the union of which characterizes the classical theories, as complementary but exclusive features of the description, symbolizing the idealization of observation and definition, respectively'. Through the years he was to refine his analysis of the foundations of quantum mechanics. His best presentation[74] of his views dates from 1949.

The quantum revolution was over by October 1927, the time of the fifth Solvay Conference, largely devoted to the new mechanics. The printed proceedings of this meeting[75] appeared in 1928. They open with a tribute by Marie Curie to Lorentz, who had presided over the conference in October and who had died shortly thereafter. Next follows a list of the participants, which includes Planck, Einstein, Bohr, Ehrenfest, de Broglie, Born, Schroedinger, and the youngsters, Dirac, Heisenberg, Kramers, and Pauli. Einstein raised his first public objections to quantum mechanics. Planck remained silent. It was a changing of the guard.

Sources

I found particularly helpful the source book by van der Waerden[35] which contains important letters, a reprint of a series of original papers on the subject 'Toward quantum mechanics', as well as reprints of the first six fundamental papers on matrix mechanics, all in English or English translation; and the

* It was shown shortly thereafter[73] that the right-hand sides of Eqs. (12.21) and (12.22) can be sharpened to read $h/4\pi$.

book by Jammer[73] on the conceptual foundations of quantum mechanics. Note that the collected papers on wave mechanics by Schroedinger also exist in English translation.[40]

References

1. J. C. Maxwell, *Collected papers*, Vol. 1, p. 526, Dover, New York 1952.
2. A. Einstein, in *James Clerk Maxwell*, p. 66, Cambridge Univ. Press 1931.
3. J. C. Maxwell, *Encycl. Britannica*, 9th edn. Vol. 8, 1878, repr. in Ref. 1, Vol. 2, p. 763.
4. A. Pais, *Subtle is the Lord*, Chapters 6, 7, Oxford Univ. Press 1982.
5. W. T. Scott, *Am. J. Phys.* **31**, 819, 1963.
6. Ref. 1, p. 570.
7. Ref. 4, Chapter 13, Section (a).
8. Ref. 4, Chapters 19, 21.
9. A. Einstein, *Ann. der Phys.* **17**, 132, 1905; Engl. transl. A. B. Arons and M. B. Peppard, *Am. J. Phys.* **33**, 367, 1965.
10. Ref. 4, Chapter 19.
11. L. Davis, A. S. Goldhaber, and M. M. Nieto, *Phys. Rev. Lett.* **35**, 1402, 1975.
12. A. H. Compton, *Phys. Rev.* **21**, 483, 1923; A. H. Compton and A. W. Simon, *Phys. Rev.* **26**, 889, 1925.
13. G. N. Lewis, *Nature* **118**, 874, 1926.
14. H. A. Kramers and H. Holst, *The atom and the Bohr theory of its structure*, Knopf, New York 1923.
15. Ref. 4, Chapter 22.
16. Ref. 4, Chapter 21, Section (a).
17. A. Einstein, *Phys. Zeitschr.* **10**, 185, 817, 1909.
18. A. Einstein, *Berliner Tageblatt*, April 20, 1924.
19. A. Pais, in *Niels Bohr, his life and work*, p. 215, North Holland, Amsterdam 1967.
20. A. Pais, *Science* **218**, 1193, 1982.
21. *Nature* **107**, 504, 1921.
22. W. Heisenberg, *Zeitschr. f. Phys.* **33**, 879, 1925.
23. A. Sommerfeld, *Atombau und Spektrallinien, wellenmechanischer Ergänzungsband*, Vieweg, Braunschweig 1929.
24. L. de Broglie, preface to his re-edited 1924 Ph.D. thesis, *Recherches sur la théorie des quanta*, p. 4, Masson, Paris 1963.
25. L. de Broglie, *Comptes Rendus* **177**, 507, 1923.
26. L. de Broglie, *Comptes Rendus* **177**, 548, 1923.
27. A. Einstein, letter to H. A. Lorentz, December 16, 1924.
28. F. Kubli, *Arch. Hist. Ex. Sci.* **7**, 26, 1970.
29. M. Klein, *Nat. Phil.* **3**, 1, 1964.
30. Ref. 4, Chapters 23, 24.
31. S. N. Bose, *Zeitschr. f. Phys.* **26**, 178, 1924.
32. A. Einstein, *Sitz. Ber. Preuss. Ak. Wiss.* 1924, p. 61.
33. A. Einstein, *Sitz. Ber. Preuss. Ak. Wiss.* 1925, p. 3.
34. W. Pauli, *Zeitschr. f. Phys.* **31**, 765, 1925.
35. B. L. v.d. Waerden, *Sources of quantum mechanics*, Dover, New York 1968.
36. M. Born and P. Jordan, *Zeitschr. f. Phys.* **34**, 858, 1925; English transl. in Ref. 35.

37. G. Uhlenbeck and S. Goudsmit, *Naturw.* **13**, 953, 1925.
38. M. Born, W. Heisenberg, and P. Jordan, *Zeitschr. f. Phys.* **35**, 557, 1925; Engl. transl. in Ref. 35.
39. W. Pauli, *Zeitschr. f. Phys.* **36**, 336, 1926; Engl. transl. in Ref. 35.
40. E. Schroedinger, *Ann. der Phys.* **79**, 361, 1926; Engl. transl. in *Collected papers on wave mechanics by E. Schroedinger*, transl. J. Shearer and W. Deans, Blackie, Glasgow 1928.
41. F. Bloch, *Physics Today*, December 1976, p. 23.
42. E. Fermi, *Rend. Acc. Lincei* **3**, 145, 1926, repr. in *Enrico Fermi, collected papers*, Vol. 1, p. 181, Univ. Chicago Press, 1962.
43. M. Born, *Zeitschr. f. Phys.* **37**, 863, 1926.
44. P. A. M. Dirac, *Proc. Roy. Soc. A* **112**, 661, 1926.
45. C. J. Davisson and L. H. Germer, *Nature* **119**, 558, 1927.
46. W. Heisenberg, *Zeitschr. f. Phys.* **43**, 127, 1927.
47. N. Bohr, *Nature* **121**, 580, 1928.
48. P. A. M. Dirac, *Hungarian Ac. of Sci. Report KFK*-62, 1977.
49. A. Einstein, letter to P. Ehrenfest, September 20, 1925.
50. N. Bohr, *Nature* **116**, 845, 1925.
51. Ref. 50, footnote 17.
52. *W. Pauli, scientific correspondence*, Vol. 1, pp. 252–4, Springer Verlag, New York 1979.
53. R. de L. Kronig and I. I. Rabi, *Phys. Rev.* **29**, 262, 1927.
54. E. Schroedinger, *Naturw.* **14**, 644, 1926.
55. E. Schroedinger, *Ann. der Phys.* **81**, 109, 1926.
56. M. Born, *Goett. Nachr. 1926*, p. 146.
57. M. Born, *Zeitschr. f. Phys.* **38**, 803, 1926.
58. Ref. 4, Chapter 21, Sections (b) and (d).
59. M. Born, *Nature* **119**, 354, 1927.
60. W. Pauli, *Zeitschr. f. Phys.* **41**, 81, 1927, footnote on p. 83.
61. M. Born, in *Nobel lectures in physics* 1942–62, p. 256, Elsevier, New York 1964.
62. W. Heisenberg, in *Theoretical physics in the twentieth century*, p. 44, Interscience, New York 1960.
63. H. Konno, *Jap. Stud. Hist. Sci.* **17**, 129, 1978.
64. E. Wigner, in *Some strangeness in the proportion*, p. 463, Ed. H. Woolf, Addison-Wesley, Reading, Mass. 1980.
65. M. Born, letter to A. Einstein, November 30, 1926.
66. M. Born, *My life and my views*, p. 55, Scribner, New York 1968.
67. Oral history of Born by T. Kuhn, 1962, Archives of the History of Quantum Physics, Niels Bohr Library, American Institute of Physics, New York.
68. W. Heisenberg, *Zeitschr. f. Phys.* **40**, 501, 1926.
69. N. F. Mott, Introduction to Ref. 66, pp. x–xi.
70. Ref. 4, Chapter 25.
71. M. Born, *Phys. Bl.* **17**, 85, 1961.
72. H. A. Lorentz, letter to P. Ehrenfest, August 29, 1927.
73. M. Jammer, *The conceptual development of quantum mechanics*, pp. 335 ff, McGraw-Hill, New York 1966.
74. N. Bohr, in *Albert Einstein: philosopher–scientist*, Ed. P. Schilpp, p. 199, Tudor, New York 1949.
75. *Electrons et Photons*, Gauthier-Villars, Paris 1928.

13

First encounters with symmetry and invariance

'It seems to me that the deliberate utilization of elementary symmetry properties is bound to correspond more closely to physical intuition than the more computational treatment.'

E. P. Wigner in 1931[1]

(a) Introduction

Nearly a year after Heisenberg had reconsidered the theory of one linear oscillator and so discovered quantum mechanics, he had something interesting to say about two identical oscillators symmetrically coupled to each other.[2] The quantum states of this system, he found, separate into two sets, one symmetric, the other anti-symmetric under exchange of the oscillator coordinates. Assuming further that the oscillators carry electrical charge, he noted that radiative transitions can occur only between states within each set, never between one set and the other. He further conjectured that non-combining sets should likewise exist if the number n of identical particles is larger than two, but had not yet found a proof.[3] He left this problem aside; another question was on his mind. Six weeks later he gave the theory of the helium spectrum, that bane of the old quantum theory.[4] To Pauli he complained that his calculations were 'imprecise and incomplete'.[5] It is true that others were able to refine his answers in later years. Nevertheless, his outstanding paper contains all the basic ingredients used today. The first quantum mechanical application of the Pauli principle is given: two-electron wave functions are antisymmetric for simultaneous exchange of space and spin coordinates. That principle is the subject of the following Section (b).

Meanwhile in Berlin a young Hungarian chemical engineer, Jenö Pál (better known since as Eugene Paul) Wigner, had become interested in the $n>2$ identical particle problem. He rapidly mastered[6] the case $n=3$ (without spin). His methods were rather laborious; for example, he had to solve a (reducible) equation of degree six. It would be pretty awful to go on this way to higher n. So, Wigner told me, he went to consult his friend the mathematician Johnny von Neumann. Johnny thought a few moments then told him that he should read certain papers by Frobenius and by Schur which he promised to bring the next day. As a result Wigner's paper on the case of general *n* (no spin), was ready soon[7] and was submitted in November 1926. It contains an acknowledgment to von Neumann, and also the following phrase: 'There exists a well-developed mathematical theory which one can use here: the theory of transformation groups which are isomorphic with the symmetric group (the group of permutations)'.

Thus did group theory enter quantum mechanics.

Groups, symmetries, invariance, these are many-splendored themes which will recur, now in one manifestation, now in another, through the rest of this book. I shall have numerous occasions to enlarge on their place in physical theory. At this point I confine myself to remarks on the different roles of symmetries in classical as compared to quantum theory. First, there are quantum symmetries which simply have no place in the classical context, a first example being the just-mentioned permutation symmetry of states of *n* identical particles. These symmetries hold the key to the exclusion principle and to quantum statistics. They served to resolve, finally, a twenty-five-year-old puzzle, the meaning of Planck's blackbody radiation law.

There is a second difference, even more profound, which in quantum theory affects alike the roles of 'new' and 'old' groups, an example of the latter being three-dimensional rotations (the Euclidean group). Another discussion with Wigner may serve to illustrate the point. I once asked him whether the transition from three to four identical particles marked his first full awareness of the power of group theory (as it had done for me in a similar problem[8]). He replied that of course that had been important, but that his acquaintance with symmetry arguments had earlier origins. Already in his chemical engineering days he had needed the classical space groups for a paper on the lattice structure of rhombic sulfur.[9] More importantly (he said) already before 1925 he did not believe that a hydrogen atom in its ground state was a plane but rather a sphere, 'something like a shell on which you had to introduce spherical harmonics'. The different role of symmetries in classical theory and in quantum mechanics cannot be illustrated better than by comparing two pictures of this ground state: a circle before 1925, a sphere thereafter.

More generally, the state of a system is given classically by a point in phase space. Symmetry operations act on coordinates and velocities in that space. Quantum mechanically, a state is given by a vector in Hilbert space; symmetry operations are implemented by *linear* (more precisely, unitary or antiunitary) operations in that space. This linearity, the real novelty, is directly related to the superposition principle: in quantum mechanics one can add two states. There is nothing like that in classical physics.* Linearity is also the basic reason why symmetries in quantum mechanics determine, all by themselves,

* For more on the comparison between symmetries in classical and in quantum theory see especially Wigner's Gibbs lecture.[10]

many properties of systems.* Therein lies their power. Where the dynamical equations of a system are sufficiently precise, one is free to choose between direct computation (a method preferred by some[11]) and reasoning as much as possible by symmetry.

In spite of some initial resistance, the novel group theoretical methods rapidly took hold. In fact two books on the subject, still among the best, appeared quite early: one by Weyl (in 1928),[12] one by Wigner (in 1931).[1] Wigner's is still an ideal introductory text. Weyl is much harder going but well worth the effort. The late Giulio Racah told me how he spent a full year studying Weyl's book during the isolation following his move from Florence to Jerusalem. That was all he needed to get started on his subsequent well-known work on complex atoms.

Since early applications of group theory to atomic spectra are still modern and readily available in current books, there is no reason to dwell on them here, important though they are. It is necessary, on the other hand, to discuss a number of other developments during the 1920s, all bearing on symmetry and invariance, all of the utmost significance for the subsequent evolution of atomic, nuclear, as well as particle theory. These topics, to which this chapter is devoted, are: the discoveries of the exclusion principle, of quantum statistics, of spin, and of the Dirac equation. The first three of these represent the closing chapters of the old quantum theory. Pauli discovered his principle by thinking hard upon 'old' rules, those of Landé for the anomalous Zeeman effect, and those of Stoner for the closure of electron shells in atoms (see Section (b)). Likewise, quantum statistics had its beginnings in *ad hoc* rules devised in the pre-quantum mechanical period (Section (d)). The discovery of spin, though occurring between the beginnings of matrix mechanics and wave mechanics, was nevertheless an advance made entirely independent of quantum mechanics (Section (c)). With an account of these discoveries I ring out the old quantum theory, paying homage one last time to seventy-five years of work by the experimental spectroscopists, in the visible and in the infra-red, and to those few theorists who could read so much in these data with so little else to rely on.

Dirac's discovery rings in the new connection between spin, relativistic invariance, and quantum mechanics. In Section (e) I shall discuss only the beginnings of this new phase, leaving in particular the positron till later chapters.

(b) The exclusion principle

In 1918, less than three months after finishing high school and before entering the University of Munich, the eighteen-year-old Wolfgang Pauli submitted

* The separation of the helium spectrum in ortho- and parastates was one of the earliest examples.

his first physics article, on a problem in general relativity. His paper on unified field theory written a year later caused Hermann Weyl to welcome the freshman student as a coworker in this new field.[13] In 1920 he wrote his encyclopaedia article on relativity, much admired by Einstein and others. The year after, he received his Ph.D. summa cum laude with Sommerfeld and became Born's assistant. In September 1922 he went for a year to Bohr in Copenhagen. As we meet him later in this section he is Privatdozent in Hamburg. The time is December 1924. By then he has published fifteen papers ranging from relativity to the old quantum theory—including the Zeeman effect. That December Pauli reaches a creative peak, his highest I would think. In one paper he finds that the electron possesses a fourth quantum number. In his very next publication he states the exclusion principle.*

'During the last quarter of a century the development of physics has been so rich and varied that not without effort can we recall the conditions existing, as regards knowledge of facts and theoretical insight, when Zeeman made the discovery which we now commemorate.' Thus begins Lorentz' contribution to the October 1921 issue, devoted to the twenty-fifth anniversary of the Zeeman effect, of the Dutch journal *Physica*.[15] After recalling the lasting consequences of Zeeman's measurements** Lorentz continued: 'Unfortunately, however, theory could not keep pace with experiment and the joy aroused by his first success was but short-lived. In 1898 Cornu discovered—it was hardly credible at first!—that the line D_1 is decomposed into a quartet . . . . Theory was unable to account . . . for the regularities observed . . . to accompany the anomalous splitting of the lines . . .'.

It will help the understanding of Lorentz' comment to recall his prediction of 1897 for the Zeeman splitting.** Observed in the direction of an external magnetic field, spectral lines should split into doublets displaced in equal amounts and opposite directions from the field-free position, one component being right-, the other left-polarized. Perpendicular to the magnetic field a third, undisplaced, linearly polarized line should be seen. This Lorentz triplet is called the normal Zeeman effect.[16] To explain Cornu's result I employ modern quantum numbers, also to be used as much as possible in what follows later. The D_1-line is the member of the yellow sodium doublet (a doublet in the field-free case, that is) corresponding to the atomic transition $2P_{1/2} \rightarrow 2S_{1/2}$, its partner ($D_2$) being $2P_{3/2} \rightarrow 2S_{1/2}$. Here '2' denotes the value of the principal quantum number (the one occurring in the Balmer formula). P and S correspond respectively to the values 1 and 0 (in units $h/2\pi$) of the orbital angular momentum quantum number l; the corresponding vector is called $\vec{l}$. The subscripts 1/2, 3/2 refer to the respective values of the total angular momentum

* See also Karl von Meyenn's and Heilbron's accounts of Pauli's road to the exclusion principle.[14]
** See Chapter 4, Section (c).

quantum number j; the corresponding vector, called $\vec{j}$, is given by

$$\vec{j} = \vec{l} + \vec{s} \tag{13.1}$$

where $\vec{s}$ is the spin vector. $\vec{m}$, the component of $\vec{j}$ in the direction of observation, has a corresponding quantum number m which can take on the values

$$m = -j, -j+1, \ldots, +j. \tag{13.2}$$

The change Δm of m in a Zeeman transition is constrained by

$$\Delta m = 0, \pm 1. \tag{13.3}$$

Thus the D_1 line splits into $2\times 2 = 4$ components, D_2 into $4\times 2-2=6$ components. Any pattern of splitting other than the classical triplet is commonly called the anomalous Zeeman effect. Spin had not yet been discovered in 1921, so that Lorentz, like his contemporaries, had every reason to be mystified by the early measurements of the French physicist Cornu. (He could have mentioned similar experimental results by others as well.*)

In the early twenties the understanding of the Zeeman effect was marred not only by ignorance about spin; there were experimental complications as well. As we now know, the Zeeman splitting for the sodium D-doublet has its exact analog for all other alkalis as well as for hydrogen. Yet in Pauli's *Handbuch* article on the old quantum theory (the most complete review of that subject) we find the following remarkable passage: 'In a magnetic field, many spectral lines, especially those of hydrogen, show a splitting into a Lorentz triplet which is called the "normal Zeeman effect",'[19] clear evidence that experiments were not good enough at the time. At just about the same time Goudsmit[20] wrote: 'It has not been explained why the Zeeman effect of hydrogen is not similar to that of the alkali atom'. It should be noted that these two articles were completed in the fall of 1925, after the discovery of the spin![21] Also in other respects the hydrogen spectrum looked deceptively 'normal' in the early twenties, as we shall see in the next section.

Remark. The splitting of the D-doublet in ten lines and the selection rule (13.2) hold strictly only for the *linear* Zeeman effect, that is, if the splitting is proportional to the magnitude H of the magnetic field strength. We are in this linear regime if

$$\mu H \ll \delta$$

where

$$\mu = \frac{eh}{4\pi m_0 c} \tag{13.4}$$

* For the literature on the Zeeman effect up to 1913 see Ref. 17. For the history of that period see Mehra and Rechenberg.[18]

is the Bohr magneton (m_0 is the electron mass) and δ is the $2P_{1/2}-2P_{3/2}$ fine structure splitting. The experiments to which Pauli and Goudsmit referred do not satisfy $\mu H \ll \delta$ but rather $\gg \delta$. In that case one has the so-called complete Paschen–Back effect* which has its own distinct selection rules: write

$$m = m_l + m_s \tag{13.5}$$

m_l, m_s are the magnetic quantum numbers corresponding to l and s respectively. In the complete Paschen–Back region l and s decouple and the selection rules become[23]

$$\Delta m_s = 0, \qquad \Delta m_l = 0, \pm 1 \tag{13.6}$$

so that each D-line splits into a triplet! Why were these triplets seen for hydrogen but not for alkalis? Because for hydrogen δ is so much smaller. Yes, the Zeeman effect was confusing business in the early twenties.

Having jumped from the days of Zeeman to modern times and modern notation, I now turn back to earlier days, skip many interesting developments, and move to 1919, the year in which the old quantum theory of the anomalous Zeeman effect began** with a proposal by Tennis van Lohuizen, a pupil of Zeeman. Consider the frequency ν of light emitted, in the absence of a magnetic field, if an electron jumps from energy E_1 to energy E_2: $h\nu = E_1 - E_2$. In a magnetic field let ν be displaced by an amount $\Delta\nu$ (for one of the Zeeman lines). Lohuizen suggested that[25]

$$h\Delta\nu = \Delta E_1 - \Delta E_2 \tag{13.7}$$

where ΔE is the displacement of E due to the field. This relation seems self-evident, but remember: the year was 1919, the anomalous Zeeman effect was mysterious. At any rate, Eq. (13.7) proved to be very helpful for organizing experimental data.

The inevitable next undertaking was to make a good guess for ΔE. The first to do so was Alfred Landé who in 1921 proposed the formula[26]

$$\Delta E = mg\mu H. \tag{13.8}$$

The quantum number m has been defined earlier. The factor g, ever since called the Landé factor, equals one for normal Zeeman splitting.[24] Landé made two new assumptions.

(1) In general $g \neq 1$, but remains independent of m. This implies that the gyromagnetic ratio ρ, the ratio of the magnitudes of the magnetic moment and the angular momentum, no longer has the value $e/2m_0c$ dictated by the

* Named after Paschen and Back who in 1912 first observed the 'normal triplet' in a very strong field. In their paper the term 'anomalous Zeeman effect' appears for the first time.[22]

** The old quantum theory of the normal Zeeman effect had been treated[24] in 1916.

Larmor theorem of classical theory. Instead

$$\rho = \frac{e}{2m_0c} g. \tag{13.9}$$

(2) In order to preserve the reader's sanity I preface Landé's second assumption with the following remark. Well before spin had been discovered one already worked with the two distinct quantum numbers, j (called the inner quantum number in those days) and l. Since we are now back in pre-spin days, Eq. (13.1) did not yet exist. How can there be two angular momenta if there is no spin? No problem: j and l refer to the assembly of electrons inside an atom, not to one electron in some orbit, so there is plenty of room. Up till the time that Landé made his second assumption it was assumed, moreover, that *both* j and l are always integer-valued. Then came Landé's proposal: m shall be half-integer for the alkalis. Also, Eq. (13.2) shall survive so that j is half-integer when m is.

The introduction of half-integer quantum numbers was very new and very audacious. It appears that young Heisenberg independently had the same idea but did not publish because it quite shocked his teacher Sommerfeld.[27] However, shortly after Landé, Heisenberg published a paper, his very first, in which he proposed a model for the half-integer quantum numbers.[28] In the alkalis there is one valence electron and an inner complex of electrons which he calls the 'Atomrumpf', the atomic core. Heisenberg proposed* that the core and the valence electron each have angular momentum 1/2.

For reasons of no interest here this model was soon abandoned, but Landé took up the idea of a core angular momentum quantum number which he called R. R *shall be equal to one half for alkalis.* The corresponding vector is $\vec{R}$. In 1923 he proposed that[30]

$$\vec{j} = \vec{l} + \vec{R} \tag{13.10}$$

where $\vec{l}$ is the orbital angular momentum of the valence electron in the alkali atom, and that his g-factor is given by

$$g = 1 + \frac{j(j+1) + R(R+1) - l(l+1)}{2j(j+1)}. \tag{13.11}$$

The arguments, partly geometrical, partly empirical, by which Landé arrived at this expression are spelled out in Ref. 18. Note that $j = R$ if the orbital angular momentum l vanishes, so that

$$g = 2 \text{ for the core by itself.} \tag{13.12}$$

In his 1925 review[20] Goudsmit justly called this last relation 'completely incomprehensible', but added that, on accepting Landé's assumption, 'one

* For more on Heisenberg and the *Rumpf* model see Ref. 29.

masters completely the extensive and complicated material of the anomalous Zeeman effect'. Landé's formula in fact demonstrates excellently how, in the days of the old quantum theory, gifted physicists were able to make important progress without quite knowing what they were doing. The formula seemed mighty peculiar but it worked. What was going on?

Enter Pauli.

His December 1924 paper on the Zeeman effect[31] begins by recalling the then current view that the properties of the core must be ascribed to 'a magneto-mechanical anomaly of the K-shell by assuming that the quotient of the magnetic moment and the angular momentum of this shell is twice as large [see Eqs. (13.9) and (13.12)] as its classical value'. The two electrons inhabiting the K-shell, the innermost part of the atomic core, were supposed to be the culprits. As Pauli noted further: 'It is often assumed that the angular momentum of the K-shell differs from zero in the noble gas configuration, in contrast with the angular momentum of the higher electron groups (L-, M-, ... shell).' That is of course not true. The point of Pauli's paper was in fact to show by the following semi-classical reasoning that this leads to a conflict with experiment. Write Eq. (13.11) as $g = 1 + x$. x is due to the core, more specifically to the K-shell electrons, since $R = 0$ if the core were to play no role, in which case $j = l$, hence $x = 0$. According to Landé, x depends only on the quantum numbers of the configuration, not on the nuclear charge Z of the atomic species considered. However, said Pauli, that cannot be strictly true. He shows that relativistic effects correct the gyromagnetic ratio (13.9), hence g, by a factor which is the average over an electron orbit of $(1 - v^2/c^2)^{1/2}$, v being the electron's instantaneous velocity. This multiplicative correction turns out to equal $(1 - \alpha^2 Z^2)^{1/2}$ for K-shell electrons (α is the fine structure constant). Pauli goes on to compute this effect for the alkalis. The deviations are 'practically unobservable' for sodium (Z is not big enough) but are large for heavier nuclei, 'up to 18 per cent for the alkali-like thallium lines'. Yet 'in reality the observations agree to an accuracy of about one per cent with Landé's g-values'. There is a paradox which Pauli resolves by the 'most natural' (correct) assignment of zero angular momentum to *all* closed shells.

We have now arrived at one of those marvelous moments in science when the lessons of logic are in conflict with the lessons of the laboratory. The idea of a core angular momentum is nonsense, Pauli had shown. But the Landé formula for g works, experiment had shown! Pauli found the correct way out: we need an R, but since R has nothing to do with the core it must have something to do with the valence electron itself. In Pauli's words, the anomalous Zeeman effect 'according to this point of view is due to a peculiar not classically describable two-valuedness [Zweideutigkeit] of the quantum theoretical properties of the valency electron'.

All was there now to associate Landé's value $R = 1/2$ for alkalis with the spin of the valence electron. I do not fully understand why Pauli himself did not make that step but can at least give one important contributing reason. Four days after his paper on the anomalous Zeeman effect had been received, Pauli wrote[32] to Sommerfeld: 'I have made headway with a few points ... [regarding] . . . the question of the closure of atomic groups in the atom. Your book* has helped me much, in particular your special emphasis on the paper by Stoner'. Stoner's work deflected Pauli from the question: what can the two-valuedness mean? to the question: has the two-valuedness something to do with the closed electron shells in atoms? The answer, he found,[33] can be expressed by a new rule which, one year later, Dirac was to name Pauli's exclusion principle.[34] Skipping many interesting spectroscopical arguments in Pauli's paper I steer directly to the exclusion principle, adapting notations to modern usage.

In October 1924 Stoner had proposed[35] the following rule: 'The number of electrons in each completed shell is equal to double the sum of the inner quantum numbers'. I should explain. Imagine what we would call an independent particle model for the atom: each electron moves around the nucleus independently of all the others, and therefore describes a hydrogen-like orbit. Then we can ascribe to *each* electron the following quantum numbers: a principal quantum number n (which occurs in the Balmer formula), and the quantum number l which (as known from hydrogen) can take the values $l=0, 1, \ldots, n-1$. Imagine further that these independent electrons are in an external magnetic field. A third quantum number appears: m_l (see Eq. (13.5)), and each level (n, l) is split into $2l+1$ levels corresponding to $-l \leq m_l \leq l$. The number N of levels is:

$$n=0:\quad l=0,\qquad N=1$$

$$n=1:\quad l=0, 1,\qquad N=1+3=4$$

$$n=2:\quad l=0, 1, 2,\qquad N=1+3+5=9, \text{ etc.}$$

Stoner's rule says: a shell corresponds to a fixed n and the number of electrons in that shell, if completely filled, equals twice N.

Why twice?

Here Pauli takes over.[33] He proposes 'to pursue as far as possible the working hypothesis [about *Zweideutigkeit*] also for atoms other than alkalis', and goes on to introduce new postulates about two-valuedness: first, it applies to *every* atomic electron; secondly, it is formally expressed by a two-valued new quantum number. Thus he is led to introduce *four* quantum numbers for each electron. Instead of giving Pauli's choice of quantum numbers I shall use the simpler (but equivalent) ones introduced a few months later by Goudsmit.[36] These are: n, l, m_l (in the notation introduced earlier) and another magnetic

* The fourth edition of *Atombau und Spektrallinien* which was completed in October 1924.

quantum number called m_R by Goudsmit and satisfying

$$m_R = \pm 1/2 \tag{13.13}$$

(m_R is of course the same as the spin quantum number m_s in Eq. (13.5), but remember once again: there was no spin yet!) Pauli used (in Goudsmit's notation) $m_l + m_R$ and $m_l + 2m_R$ rather than m_l, m_R. Whatever convention is used, a doubling of states occurs which, Pauli suggests, accounts for the doubling of N in Stoner's rule. Why can a shell not contain *more* than $2N$ electrons? 'In the atom there can never be two or more equivalent electrons for which in strong fields* the values of all quantum numbers [n, l, m_l, m_R] coincide. If there is an electron in the atom for which these quantum numbers have definite values then the state is "occupied".'

Thus Pauli arrived at the exclusion principle via the route

Anomalous Zeeman effect → Landé rule → two-valuedness
↓
Stoner's rule → exclusion principle.

He knew well that this was not the end of the story: 'We cannot give a more precise reason for this rule'. Improved precision in terms of symmetries was to come soon, as we shall see at the end of Section (d).

(c) Spin

Near Düsseldorf, capital of the Duchy of Berg, there lived at one time a family named Üllenbeck or Uhlenbeck. One of its members went into military service under King Frederick the Great of Prussia. He had to flee on account of a duel, eventually joined the military in the Dutch colony of Ceylon, and became the founder of the Dutch branch of the family. Three of his great-great-grandsons saw service with the Dutch East Indian army. During the Atjeh (Achin) wars in Northern Sumatra two of them threw themselves on their sabres to avoid capture by brutal tribes. The third rose to the rank of lieutenant-colonel and married the daughter of a Dutch major-general. They were the parents of George Eugene Uhlenbeck, born in 1900, who received his first schooling at a kindergarten in Padang Panjang on Sumatra.

In 1907 the family moved to Holland and settled in the Hague. Uhlenbeck was first drawn to physics by a high school course. Eager to learn more he would bike to the Koninklyke Bibliotheek (Royal Library) where he absorbed Lorentz' *Lectures on Physics*. After finishing high school in 1918 he could not enter a Dutch university since his school had not offered him Greek and Latin, at that time a prerequisite by law for university study in whatever discipline.

* The reference to strong fields concerns the complete Paschen–Back effect in which, we would say, m_l and m_s decouple and are both good quantum numbers.

(Van der Waals and van 't Hoff, in a similar position at earlier times, had only been able to enter a university by special governmental dispensation.) In September 1918 he therefore entered the Technische Hogeschool (Institute of Technology) in Delft, planning to study chemical engineering. A new law, enacted almost immediately thereafter, dispensed with the Greek and Latin requirements for university training in the sciences. In January 1919 Uhlenbeck left Delft and enrolled in the University of Leiden to study physics and mathematics.

All through his student years Uhlenbeck commuted by train between the family home in the Hague and Leiden. His mother would pack his lunch and give him a 'kwartje' (twenty-five cents) for coffee which he saved until he had enough to buy a second-hand copy of Boltzmann's *Vorlesungen über Gastheorie*. After graduating he attended courses by Ehrenfest and Lorentz and also the celebrated Wednesday evening 'Ehrenfest colloquium' which one could attend by invitation only, but to which one had to go once admitted. Ehrenfest even took attendance. Ehrenfest was by far the most important scientific figure in Uhlenbeck's life. In all the years I have known Uhlenbeck, in Utrecht, in Ann Arbor, in New York, only one picture always stood on his office desk: a small photograph of a warmly smiling Ehrenfest.*

One day Ehrenfest asked in class whether anyone might be interested in a teaching position in Rome. Uhlenbeck raised his hand. So it came to pass that from September 1922 until June 1925 he became the private teacher of mathematics, physics, chemistry, Dutch, German, and Dutch history to the younger son of the Dutch ambassador. He continued his university studies and in 1923 obtained the Leiden degree of 'doctorandus', the equivalent of a master's degree.

While in Rome he attended courses at the university, and met the one year younger Fermi. They became friends and organized a small colloquium. 'Fermi was the born leader and did most of the talking.' Meanwhile Uhlenbeck became deeply involved in cultural history. His first publication (in Dutch) deals with Johannes Heckius, one of the four founders of the *Academia dei Lincei*.[38] It was the Dutchman Uhlenbeck who for the first time introduced Fermi, born and raised a Roman, to the Moses of Michelangelo in the church of San Pietro in Vincoli.

When, in June 1925, Uhlenbeck returned to Holland he was seriously considering giving up physics and becoming a historian. Ehrenfest was sympathetic but suggested that he first find out what was currently happening in physics by working with him for a while and learning from Goudsmit, a graduate student, what was going on in 'Spektralzoologie' (as Pauli used to call the study of spectra). Uhlenbeck followed his advice. All through the summer of 1925—he later called it the Goudsmit summer—Goudsmit would

* For Uhlenbeck's recollections of Ehrenfest see Ref. 37.

come to his home and educate him in spectra. Then, in the middle of September 1925, doctorandus Uhlenbeck and graduate student Goudsmit discovered spin. Gone were Uhlenbeck's aspirations of becoming a historian.

Samuel Abraham Goudsmit (Sem to his friends) was born in 1902 in the Hague, the son of a prosperous wholesale dealer in bathroom fixtures. His mother owned a fashionable hat shop. He got his first taste of physics at the age of eleven when browsing through an elementary physics text, being particularly struck by a passage explaining how spectroscopy had shown that stars are composed of the same elements as the earth. 'Hydrogen in the sun and iron in the Big Dipper made Heaven seem cozy and attainable.'* After finishing high school in one year less than the usual time, he became a physics student in Leiden, where Ehrenfest turned his initial interest into devotion. Soon his bent became evident for intuitive rather than analytical thinking, starting from empirical hunches. Uhlenbeck later said of him: 'Sem was never a conspicuously reflective man, but he had an amazing talent for taking random data and giving them direction. He's a wizard at cryptograms . . .' Rabi said: 'He thinks like a detective. He *is* a detective'.[39] In fact, Goudsmit once took an eight-month course in detective work in which he learned to identify fingerprints, forgeries, and bloodstains. A two-year university course taught him to decipher hieroglyphics. In physics the decoding of spectra became his passion. At age eighteen he completed his first paper, on alkali doublets.[40] Uhlenbeck called it 'A most presumptuous display of self-confidence but . . . highly creditable'.[39]

In August 1925 the two men started their regular meetings in the Hague. George, the more analytic one, better versed in theoretical physics, a greenhorn in physics research, aspiring historian with a paper on Heckius to his credit; Sem, the detective, thoroughly at home with spectra on which he had already published several papers, known in the physics community, part-time assistant to Zeeman in Amsterdam. In almost no time Sem's teachings turned into joint research and publication, their relation into a close and lasting friendship. I know from my own later personal friendship with both even more than from their writings[41,42,43,44] how each of them remained forever beholden to the other for his share in the work during those months. That was no *politesse* but deep appreciation.

Sem knew everything there was to know about half-integer quantum numbers. His note[36] on the simplification of Pauli's work (Eq. (13.13)) had been completed in the preceding May. One of their earliest topics of discussion (George told me) was the *Rumpf* model. Sem explained how it worked for the alkalis. Hydrogen was a different story, however; there one used Sommerfeld's theory of the fine structure (see Eq. (10.2)) which worked very well. Nor was there any problem with the Zeeman effect for hydrogen which

* Many personal details about Goudsmit are found in a *New Yorker* profile.[39]

appeared to be normal (see the preceding section). Thus hydrogen appeared to be in good shape. Yet George was unhappy. ('He knew nothing, he asked all those questions which I had never asked.'[41]) Why two distinct models if the alkalis and hydrogen were so much alike? Why not try the half-integer quantum numbers of the *Rumpf* model on hydrogen as well? This led to their first joint paper,[45] discussed in Chapter 10, Part 3, where we saw how they were the first to introduce half-integer quantum numbers for hydrogen and how their work yielded a bonus: the explanation of the 4686 Å fine structure in He^+, an improvement over the Sommerfeld picture.

Goudsmit has written about what happened next: 'Our luck was that the idea [of spin] arose just at the moment when we were saturated with a thorough knowledge of the structure of atomic spectra, had grasped the meaning of relativistic doublets, and just after we had arrived at the correct interpretation of the hydrogen atom'.[42] One of Sem's next lessons dealt with the exclusion principle including his own contribution: $m_R = \pm 1/2$ for the fourth quantum number. George has written: 'It was then that it occurred to me that since (as I had learned) each quantum number corresponds to a degree of freedom of the electron, the fourth quantum number must mean that the electron had an additional degree of freedom—in other words the electron must be rotating!'[44] Everything fell into place. The electron has a spin s:

$$s = \frac{1}{2}\left(\text{in units } \frac{h}{2\pi}\right). \tag{13.14}$$

Sem's m_R was the associated magnetic quantum number, called m_s from then on. Landé's relation (13.12) was to be reinterpreted:

$$g = 2 \text{ for the electron.} \tag{13.15}$$

Sem asked whether this g-value could be given a physical meaning.[44] Following a hint by Ehrenfest, George found in an old article by Abraham[46] that an electron considered as a rigid sphere with surface charge only does have $g = 2$. All this was written up in a short note[47] which includes the Abraham model, but with a caveat: if that model were the explanation of $g = 2$, then the peripheral rotational velocity should be much larger than the velocity of light.* It is clear that the spirit of the discovery of spin was classical and old-quantum theoretical.

The note was published with Uhlenbeck as first, Goudsmit as second author because (George told me) Ehrenfest suggested that this order would avoid the impression that George was only Sem's student, while Sem himself preferred this because George had first thought of spin.

The note is dated October 17, 1925. One day earlier Ehrenfest had written to Lorentz[48] asking him for an opportunity to have 'his judgment and advice

* This follows from $m_0 vr = h/4\pi$, with r taken equal to the classical electron radius e^2/m_0c^2.

on a *very* witty idea of Uhlenbeck about spectra'. Lorentz listened attentively when George went to see him soon thereafter, and then raised an objection. The spinning electron should have a magnetic energy $\sim\mu^2/r^3$, where $\mu = eh/4\pi m_0 c$ is its magnetic moment, r its radius. Equate this energy to m_0c^2. Then r would be $\sim 10^{-12}$ cm, too big to make sense.* (The weak point in this argument was to be revealed years later by the positron theory.) George, upset, went to Ehrenfest to suggest that the paper be withdrawn. Ehrenfest replied that he had already sent off their note, adding that its authors were young enough to be able to afford a stupidity.[44] Sometime later Lorentz handed Uhlenbeck a sheaf of papers with calculations of spinning electrons orbiting a nucleus. This work was to become the last paper[50] by the grand master of the classical electron theory, presented to the Como conference in September 1927.

No sooner had George and Sem's note appeared when Goudsmit received a letter from Heisenberg[51] congratulating him on his 'mutige Note' (brave note) and inquiring 'wie Sie den Faktor 2 losgeworden sind' (how you have got rid of the factor 2) in the formula for the fine structure splitting in hydrogen, as derived from a semi-classical treatment of spin precession. The young Leideners had not even thought of calculating this splitting. After some struggle they found that Heisenberg was right; the fine structure came out too large by a factor two.** This puzzle was still unresolved when in December 1925 Bohr arrived in Leiden to attend the festivities for the golden jubilee of Lorentz's doctorate. One evening in 1946, the hour was late, Bohr told me in his home in Gamle Carlsberg what happened to him on that trip.

Bohr's train to Leiden made a stop in Hamburg, where he was met by Pauli and Stern who had come to the station to ask him what he thought about spin. Bohr must have said that it was very very interesting (his favorite way of expressing his belief that something was wrong), but he could not see how an electron moving in the electric field of the nucleus could experience the magnetic field necessary for producing fine structure. (As Uhlenbeck said later: 'I must say in retrospect that Sem and I in our euphoria had not really appreciated [this] basic difficulty'.[44]) On his arrival in Leiden, Bohr was met at the train by Ehrenfest and Einstein who asked him what he thought about spin. Bohr must have said that it was very very interesting but what about the magnetic field? Ehrenfest replied that Einstein had resolved that. The electron in its rest frame sees a rotating electric field; hence by elementary relativity it also sees a magnetic field. The net result is an effective spin–orbit coupling. Bohr was at once convinced. When told of the factor two he expressed confidence that this problem would find a natural resolution. He urged Sem

* Essentially the same objection was raised shortly afterward by Rasetti and Fermi who wondered how an atomic nucleus could contain many electrons of such size;[49] see the next chapter.

** This calculation is reproduced in Ref. 52 and also in the review article of August 1926 by Goudsmit and Uhlenbeck.[53]

and George to write a more detailed note on their work. They did; Bohr added an approving comment.[54]

After Leiden Bohr traveled to Goettingen. There he was met at the station by Heisenberg and Jordan who asked what he thought about spin. Bohr replied that it was a great advance and explained about the spin–orbit coupling. Heisenberg remarked that he had heard this remark before but that he could not remember who made it and when. (I return to this point shortly.) On his way home the train stopped at Berlin where Bohr was met at the station by Pauli, who had made the trip from Hamburg for the sole purpose of asking Bohr what he now thought about spin. Bohr said it was a great advance, to which Pauli replied: 'eine neue Kopenhagener Irrlehre' (a new Copenhagen heresy). After his return home Bohr wrote to Ehrenfest that he had become 'a prophet of the electron magnet gospel'.[55]

I conclude the spin story with a number of scattered comments.

(1) In February 1926 the missing factor two was supplied by Llewellyn Thomas and has since been known as the Thomas factor. Thomas noted that earlier calculations of the precession of the electron spin had been performed in the rest frame of the electron, without taking into account the precession of the electron orbit around its normal. Inclusion of this relativistic effect reduces the angular velocity of the electron (as seen by the nucleus) by the needed factor 1/2. Einstein was surprised.[57] Pauli became converted.[58]

(2) Concerning precursors. Already in 1900, FitzGerald had raised the question whether magnetism is due to rotation of electrons.[59] In 1921 Arthur Compton had a similar idea: 'It is the electron rotating about its axis which is responsible for the ferromagnetism . . . . The electron itself, spinning like a tiny gyroscope, is probably the ultimate magnetic particle.'[60] The same proposal was made in 1922 by Kennard[61] who made a calculation identical to Abraham's[46] (of which he was unaware) to show that an electron could have $g=2$. In all these instances electrons were pictured as rotating rigid bodies of finite extent, in Compton's case with quantized angular momentum.

(3) In August 1924 (before the fourth quantum number and the exclusion principle) Pauli proposed[62] an explanation of hyperfine structure: 'The nucleus possesses in general a non-vanishing angular momentum . . . . In the future one might hope to learn something [from this hypothesis] about the structure of nuclei'. I have heard it said that this was the first suggestion of spin. With all appreciation for this nice paper by Pauli, I cannot agree.*

* Pauli himself said later (1946) about this paper: 'Already in 1924 before the electron spin was discovered I proposed to use the assumption of nuclear spin . . . [which] influenced Goudsmit and Uhlenbeck in their claim of an electron spin'.[63] It seems to me that Pauli's use of the term nuclear spin is but one of untold many examples of an adaptation of language to later usage. As to the influence on the discovery of electron spin, Goudsmit has written[64] (1961): 'We were not aware of this paper till five years later'. For a more detailed discussion of the issue see Ref. 65. For the discovery of proton spin see the next section.

(4) Unbeknownst to Goudsmit and Uhlenbeck, the idea of spin had occurred to Kronig in January 1925 when, on a visit to Tübingen, he heard from Landé about the exclusion principle. According to Kronig[66] this led him at once to the idea of spin. 'The same afternoon . . . I succeeded in deriving with it the so-called relativistic doublet formula'—which he found to be off by the factor two mentioned above. When Pauli also arrived in Tübingen shortly thereafter, Kronig explained his ideas to him. Pauli was sceptical, as were Heisenberg and Kramers to whom Kronig talked on a subsequent visit to Copenhagen.[67] As a result Kronig did not publish.* The discussion with Kronig is the one that had escaped Heisenberg's mind when Bohr was in Goettingen. About these events Uhlenbeck has remarked: 'There is no doubt that Ralph Kronig anticipated what certainly was the main part of our ideas'.[44] It is certain that it was because of this episode that no Nobel Prize for spin was ever awarded.

It has happened before and will happen again that destinies will be affected by the choice between submitting to authority and going one's own way. It has happened before and will happen again that debates will arise about priorities of unpublished material. In the case of spin it should be added that others also had the idea without publishing: Urey for the electron,[68] Bose (in 1924) for the photon (see the next section). It is my own belief that in the final analysis only the published record should decide.

(5) On 7 July 1927 both George and Sem received their Ph.D.'s in Leiden. A month thereafter they and their wives set sail for the United States. In September they started their academic careers as physics instructors at the University of Michigan in Ann Arbor.

(d) Quantum statistics

The counting procedures of classical statistical mechanics need modification when quantum effects are important. That insight is as old as the old quantum theory. When in 1900 Planck tried to justify his successful guess at the blackbody radiation law, he could do so only by using a statistical reasoning which—he knew—is at variance with the Boltzmann statistics of classical physics.[69] Strong experimental indications that there is something amiss with the classical ways are even older: deviations of observed specific heats of solids from the predictions of the classical equipartition law date back[70] to the 1870s.

Since 1926 we have known that in quantum mechanics one does not need one, but two, new ways of counting, Bose–Einstein (BE) statistics and Fermi–Dirac (FD) statistics; and that permutation symmetries of many-particle wave functions are the key to their justification. Yet Bose's discovery of a new statistics for photons and Einstein's extension to material gases—including the phenomenon of BE condensation—were made well before anyone had

* I shall mention in the next chapter an objection to spin raised by Kronig in April 1926.

ever heard of a Schroedinger wave function. How can that be? And how was it sorted out which quantum statistics is appropriate for which system of particles?

Let us recall first how Boltzmann counted. Consider N identical (structureless) non-interacting particles whose individual energies can take on only the discrete values $\varepsilon_1, \varepsilon_2, \ldots$. In this model for a classical ideal gas we have

$$N = \sum_i n_i, \qquad E = \sum_i \varepsilon_i n_i, \tag{13.16}$$

where n_i is the number of particles with energy ε_i and E is the total energy. Find the number w of ways in which the particles can be partitioned over the energies ε_i. Since the particles do not interact, the presence of five of them at energy ε_4 does not affect the probability of a sixth to be also at ε_4; the particles are statistically independent. Even though they are identical, one can imagine (Boltzmann argued[71]) marking them at a given time with their individual position coordinates and then following each so-marked particle. Thus the particles are distinguishable and the answer for w is

$$w = \frac{N!}{n_1!\, n_2! \ldots}. \tag{13.17}$$

w, the number of microstates, is the probability for the partition $(n_1, n_2, \ldots)$ up to a normalizing factor. Along with Boltzmann's w, now called a fine-grained probability, we must consider W, the coarse-grained probability, introduced later by Gibbs[72] and defined as follows. Divide the one-particle phase space into cells such that a particle in cell number A has a *mean* energy E_A. Let N_A be the number of particles in cell A:

$$N = \sum_A N_A \tag{13.18}$$

$$E = \sum_A E_A N_A. \tag{13.19}$$

Further let there be g_A energy levels (of the ε_i-type) in cell A. Then *it follows* from Eq. (13.17) that the number W of N-particle states is given by*

$$W = \prod_A \frac{g_A^{N_A}}{N_A!} \quad \text{(Boltzmann)}. \tag{13.20}$$

By a standard argument[73] the entropy in equilibrium, S, is given by

$$S = k \ln W_{\max} \tag{13.21}$$

where k is the Boltzmann constant and $W_{\max}$ is the maximum of W subject

* For a derivation of Eq. (13.20) and also of Eqs. (13.25), (13.26) see for example Ref. 73.

to the constraints

$$\delta N = \sum \delta N_A = 0 \tag{13.22}$$

$$\delta E = \sum E_A \delta N_A = 0. \tag{13.23}$$

Using Eqs. (13.20–23) and

$$\frac{\partial S}{\partial E} = \frac{1}{T} \tag{13.24}$$

one derives the Maxwell–Boltzmann distribution.

Let us change the rules of the game. Retain the g_A levels per cell as well as Eqs. (13.18, 19, 21–24) but drop Boltzmann's distinguishability of particles, the key to the derivation of Eq. (13.17). In fact drop the earlier connection between w and W altogether. Adopt instead either of the following rules of counting:

(1) the particles are indistinguishable. Then, instead of Eq. (13.20)[73]

$$W = \prod_A \frac{(N_A + g_A - 1)!}{N_A!(g_A - 1)!} \quad \text{(BE)}, \tag{13.25}$$

or

(2) the number of particles in each cell is 0 or 1. Then[73]

$$W = \prod_A \frac{g_A!}{N_A!(g_A - N_A)!} \quad \text{(FD)}. \tag{13.26}$$

Eqs. (13.25, 26) correspond to BE and FD statistics respectively. All their thermodynamic consequences follow by combining either of these two expressions with Eqs. (13.21, 24). In particular, $\bar{N}_A$, the equilibrium value of N_A is given by[73]

$$\bar{N}_A = g_A(z^{-1}\, e^{\beta E_A} + \lambda)^{-1}, \qquad \beta = \frac{1}{kT}, \tag{13.27}$$

where z is the fugacity and

$$\lambda = \begin{cases} -1 & \text{BE} \\ \;\;\,0 & \text{Boltzmann} \\ +1 & \text{FD} \end{cases} \tag{13.28}$$

We have arrived at the new statistics by applying *ad hoc* new counting rules. No reference to symmetries of wave functions has been made; in fact Planck's constant has not even appeared yet. $\bar{N}_A$ will come to depend on h only after we specify E_A. This derivation does not perhaps excel in pedagogical virtue. It does facilitate the understanding of the founders' reasonings, however. Let us see what they did, in chronological order of appearance.

Ladislas Natanson from Cracow (1911) was the first to state[74] that distinguishability has to be abandoned in order to arrive at Planck's law: 'The energy units [i.e. light-quanta] cannot be treated as identifiable'.

Bose's derivation of Planck's law (July 1924) is a confused masterpiece.[75] His reasoning is correct but, as he himself once said, he had no idea that it was novel. I have explained elsewhere and shall not repeat here how Bose applied Boltzmann statistics not to particles but to cells filled with variable numbers of particles.[76] Eq. (13.25) does not appear in his work.

In applying Eq. (13.27) (with $\lambda = -1$) to photons one must beware. First, in that case $z = 1$. This constraint is a consequence of the fact that the number of photons is not conserved so that one must drop Eq. (13.22). Bose does not refer explicitly to this non-conservation (nor does Einstein) which is first mentioned, I believe, by Fowler in November 1926.[77] Secondly, one must multiply the right-hand side of Eq. (13.27) by a factor two because of the photon's two helicity states. But in 1924 there was as yet no spin or helicity! Bose did introduce the factor two, noting merely that 'it seems required' to do so.[75] Much later he told a visitor[78] of his thoughts on this point in 1924: 'He [B.] had proposed that this factor came from the fact that the photon carried one unit of angular momentum and that this could only take 2 orientations. "But Einstein wasn't satisfied with that", he said smilingly but sadly. "The old man crossed this portion out from the paper* and wrote to me saying it wasn't necessary at that stage to introduce such a concept. He considered it would suffice to say it came from the two states of polarization of light. But in those days the polarization of a particle didn't make sense, you see", he said with an amused and distant gaze. Amazed, I asked why, once spin was discovered, he had not bothered to write to Einstein claiming his priority. "Surely Einstein would have stood by you", I remonstrated. "How does it matter who proposed it first? It *has* been found, hasn't it?" he said triumphantly. That was Bose'.

Also in July 1924, Einstein[79] applied Bose's methods to an ideal monatomic (and non-relativistic) gas of particles with mass m and found deviations from the classical ideal gas law in the régime (ρ is the gas density)

$$\rho\left(\frac{h^2}{mkT}\right)^{3/2} \ll 1. \tag{13.29}$$

In a sequel[80] he was the first to write down Eq. (13.25). There he mentions an objection by Ehrenfest: 'The quanta and the molecules are not treated as statistically independent'. Einstein agrees and remarks that the new ways of counting 'express indirectly a certain hypothesis on a mutual influence of the molecules which is of a quite mysterious nature'. That influence is, of course,

* Bose had sent an English version of his paper to Einstein who translated it and submitted it to the *Zeitschrift für Physik*.[76]

the particle correlations implied by the restriction to symmetric wave functions—recall, however, that we are still a year away from the discovery of wave mechanics.

The paper also contains the discovery of BE condensation, an effect without application at that time. The HeI–HeII phase transition was only discovered[81] in 1928; its interpretation as a BE condensation came another ten years later.[82]

Fermi's first paper[83] on the new statistics (February 1926), submitted after the discoveries of Heisenberg and Schroedinger, is still in the style of the old quantum theory. He introduces* the Pauli principle as an 'additional rule to the Sommerfeld quantization conditions'. There is no spin in his paper, so that he uses a *three*-quantum number version of the exclusion principle: two gas molecules cannot have the same translational velocity. However primitive, it is the first application of the exclusion principle to systems other than atomic levels. In a sequel[84] Fermi is the first to write down Eq. (13.26) from which, using also Eqs. (13.21–24), he derives the equation of state in the region where Eq. (13.29) holds true. With Fermi's two papers the non-quantum mechanical stage of quantum statistics, the time of mere counting, comes to a conclusion.

The first to make a stab at the relation between the symmetry of N-particle wave functions and statistics was Heisenberg (June 1926). In the paper mentioned at the very beginning of this chapter[2] he noted that the condition of total antisymmetry 'corresponds to the reduction of the statistical weights from $N!$ [see the factors in the denominator of Eq. (13.17)] to the 1 of the Bose–Einstein counting. The formulation given here . . . prescribes the choice of a very definite system from among the $n!$ solutions, namely . . . the one . . . which agrees with Pauli's prohibition [the exclusion principle] . . . Pauli's prohibition and the Einstein statistics have the same origin . . . and are not in conflict with quantum mechanics'. The identification of BE counting with the exclusion principle is of course incorrect, yet Heisenberg makes an important point: the restriction to the one anti-symmetric wave function among all possible $N!$ functions is consistent with quantum mechanics. As long as the Hamiltonian is totally symmetric in the particle variables, the property of anti-symmetry is an integral of the motion.

The matter was straightened out by Dirac (August 1926): 'The solution with symmetric eigenfunctions must be the correct one when applied to light-quanta, since it is known that the Einstein–Bose statistical mechanics leads to Planck's law for blackbody radiation. The solution with antisymmetrical eigenfunctions is . . . the correct one for electrons in an atom.'[85]

* His motivation had to do with the third law of thermodynamics for gases, a subject in which Einstein also was greatly interested at that time. He makes use of a clever model: particles moving independently in a harmonic oscillator potential. This makes it easy to label and count states explicitly.

Dirac had therefore found the quantum analog of Boltzmann's formula Eq. (13.17) for the number of microstates:

$$w = 1 \quad \text{for BE and FD} \tag{13.30}$$

corresponding to the restriction to the single symmetric and anti-symmetric state respectively. The link between fine- and coarse-grained counting was now re-established, as Dirac exemplified by his independent[85] derivation of Eq. (13.26).

Thus by August 1926 photons and electrons had been put in their statistical place. What about other particles?

Einstein (January 1925) suggests applications of the BE equation of state to electrons and to helium.[80]

Fermi (March 1926) does not quite commit himself but uses helium to illustrate the FD equation of state.[84]

Dirac (August 1926): 'The solution with anti-symmetrical eigenfunctions . . . is probably the correct one for gas molecules, since it is known to be the correct one for electrons in an atom and one would expect molecules to resemble electrons more closely than light-quanta'.[85]

Pauli (December 1926) follows Dirac: 'We wish to take the view advocated by Dirac . . .'.[87]

Friedrich Hund (February 1927) concludes from an analysis of the specific heat of hydrogen molecules at low temperatures that an ideal gas of protons would probably satisfy BE statistics.[88] Except for the application of BE statistics to helium, all these suggestions are incorrect.

David Dennison (June 1927), assuming (contrary to Hund) that the time of ortho–para transitions in hydrogen is long compared to the time for observing specific heats so that molecular hydrogen at low temperatures is effectively a mixture of two distinct gases, finds[89] that he can interpret all data 'If the nuclear [i.e. proton] spin is taken equal to that of the electron and only the complete[ly] antisymmetric solution of the Schroedinger wave function [is] allowed'.

Wigner (October 1928) proves a theorem.[90] Consider a system (a molecule, a nucleus) which contains N particles each of which satisfies FD statistics. Then a gas of such systems follows BE (FD) statistics if N is even (odd).*

Chadwick (April 1930) analyzes his data on the scattering of α-particles in helium, taking into account BE effects.[92]

Thus, step by step (not all of them correct), quantum statistics evolved in the early years. At this point I end my account of the subject, leaving the connection between spin and statistics for Chapter 20.

* This result was later rediscovered by Ehrenfest and Oppenheimer.[91]

(e) The Dirac equation

> 'It was found that this equation gave the particle a spin of half a quantum. And also gave it a magnetic moment. It gave just the properties that one needed for an electron. That was really an unexpected bonus for me, completely unexpected.'
>
> P. A. M. Dirac[93]

1. Young Dirac. In the year 1902 the literary world witnessed the death of Zola, the birth of John Steinbeck, and the first publication of *The Hound of the Baskervilles*, *The Immoralist*, *Three Sisters*, and *The Varieties of Religious Experience*. Monet painted 'Waterloo Bridge', and Elgar composed 'Pomp and Circumstance'. Caruso made his first phonograph recording, and the Irish Channel was crossed for the first time by balloon. In the world of science, Heaviside postulated the Heaviside layer, Rutherford and Soddy published their transformation theory of radioactive elements, Einstein started working as a clerk in the patent office in Bern, and, on August 8, Paul Adrien Maurice Dirac was born in Bristol, one of the children of Charles Dirac, a native of Monthey in the Swiss canton of Valais, and Florence Holten. There also was an older brother, whose life ended in suicide, and a younger sister. About his father Dirac has recalled: 'My father made the rule that I should only talk to him in French. He thought it would be good for me to learn French in that way. Since I found that I couldn't express myself in French, it was better for me to stay silent than to talk in English. So I became very silent at that time—that started very early'.*

Dirac received his secondary education at the Merchant Venturers' School in Bristol where his father taught French. 'My father always encouraged me toward mathematics .... He did not appreciate the need for social contacts. The result was that I didn't speak to anybody unless spoken to. I was very much an introvert, and I spent my time thinking about problems in nature'.[94] Throughout his life Dirac maintained a minimal, sparse (not terse), precise and apoetically elegant style of speech and writing. Sample: his comment on the novel *Crime and Punishment*: 'It is nice, but in one of the chapters the author made a mistake. He describes the sun as rising twice on the same day.'[96]

Dirac completed high school at age 16, then went to the University of Bristol, where he graduated in 1921 with a degree in engineering. He stayed on for further study in pure and applied mathematics until, in the autumn of 1923, he enrolled at Cambridge where, nine years later, he would succeed Larmor to the Lucasian Chair of Mathematics once held by Newton.** It was Fowler who in Cambridge introduced Dirac to the old quantum theory and it was

* This description is found in the interview of Dirac by Kuhn,[94] an important source for Dirac's recollections and opinions. See also Ref. 95.

** For an account of Dirac's Cambridge days see Ref. 97.

from him that he first learned of the atom of Rutherford, Bohr, and Sommerfeld.

Dirac first met Bohr in May 1925 when the latter gave a talk in Cambridge on the fundamental problems and difficulties of the quantum theory. Of that occasion Dirac said later: 'People were pretty well spellbound by what Bohr said.... While I was very much impressed by [him], his arguments were mainly of a qualitative nature, and I was not able to really pinpoint the facts behind them. What I wanted was statements which could be expressed in terms of equations, and Bohr's work very seldom provided such statements. I am really not sure how much my later work was influenced by these lectures of Bohr's.... He certainly did not have a direct influence, because he did not stimulate one to think of new equations'.[98]

A few months later Heisenberg's first paper on quantum mechanics came out.* 'I learned about this theory of Heisenberg in September and it was very difficult for me to appreciate it at first. It took two weeks; then I suddenly realized that the noncommutation was actually the most important idea that was introduced by Heisenberg.'[93] The result was Dirac's first paper on quantum mechanics[99] containing the relation $pq - qp = h/2\pi i$, independently derived shortly before by Born and Jordan. The respective authors were unaware of one another's results. Born has described his reaction upon receiving Dirac's paper: 'This was—I remember well—one of the greatest surprises of my scientific life. For the name Dirac was completely unknown to me, the author appeared to be a youngster, yet everything was perfect in its way and admirable'.[100]

In those days Dirac invented several notations which are now part of our language: q-numbers, where 'q stands for quantum or maybe queer'; c-numbers, where 'c stands for classical or maybe commuting'.[98] He has described his work habits in those years: 'Intense thinking about those problems during the week and relaxing on Sunday, going for a walk in the country alone'.[98]

In May 1926 Dirac received his Ph.D. on a thesis entitled 'Quantum Mechanics'. Meanwhile Schroedinger's papers on wave mechanics had appeared, to which Dirac reacted with initial hostility, then with enthusiasm. He quickly applied the theory to systems with many identical particles[85] (see the previous section). Beware of the constant h introduced in this paper which 'is $(2\pi)^{-1}$ times the usual Planck's constant'. This quantity, often used by Dirac, is the now familiar $\hbar$:

$$\hbar = \frac{h}{2\pi} \tag{13.31}$$

where h remains the usual Planck constant.

* For this and other dates the reader may like to consult the chronology in Chapter 12, Section (c).

In September Dirac went to Copenhagen. 'I admired Bohr very much. We had long talks together, long talks in which Bohr did practically all the talking'.[98] It was there that he worked out the theory of canonical transformations in quantum mechanics known since as the transformation theory.[101] 'I think that is the piece of work which has most pleased me of all the works that I've done in my life . . . . The transformation theory [became] my darling'.[93] Also in Copenhagen he completed the first paper[102] on quantum electrodynamics, work to which I return in Chapter 15.

In February 1927 Dirac left Denmark, and next spent time in Goettingen. From there he went to Leiden, and concluded his travels by attending the Solvay conference in Brussels (in October) where he met Einstein for the first time. From discussions with Dirac I know that he admired Einstein. The respect was mutual ('. . . Dirac to whom, in my opinion, we owe the most logically perfect presentation of [quantum mechanics]'[103]). Yet the contact between the two men remained minimal. I do not believe that this is due to Einstein's critical attitude to quantum mechanics, expressed first at that 1927 Solvay conference. Indeed, as time went by, Dirac himself developed reservations not only regarding quantum field theory but also, though less strongly, in relation to ordinary quantum mechanics.[104] I would rather think that it was not in Dirac's personality to seek for father figures.

Dirac has recalled a conversation with Bohr during the 1927 Solvay conference. Bohr: 'What are you working on?' Dirac: 'I'm trying to get a relativistic theory of the electron.' Bohr: 'But Klein has already solved that problem.'[98]

Dirac disagreed.

2. Relativity without spin; the scalar wave equation. Let us see what was known about relativistic wave equations by the time of that Solvay conference.

Already in his first paper on wave mechanics (February 1926) we find that Schroedinger had tried to take a relativistic starting point for his theory.[105] He must have written down the wave equation

$$\left(\Box-\frac{m^2c^2}{\hbar^2}\right)\psi=0, \tag{13.32}$$

$$\Box=\frac{\partial^2}{\partial x^2}+\frac{\partial^2}{\partial y^2}+\frac{\partial^2}{\partial z^2}-\frac{1}{c^2}\frac{\partial^2}{\partial t^2} \tag{13.33}$$

then have introduced electromagnetic interactions, and then have calculated the hydrogen spectrum (as indeed he did in his first paper for Eq. (12.15)). This procedure yields the Balmer formula but not the correct fine structure: the hydrogen D-doublet is too wide by a factor 8/3.

Interruption. Advice to the reader. At this point the reader may like to have at hand a simple but good book on quantum mechanics (like the one by

Schiff[106]) where results merely stated here and in the rest of this section are derived in detail. End of interruption.

Between April and September 1926, Eq. (13.32) was independently stated by at least six authors,[107] Schroedinger and Klein among them. I shall therefore only refer to it by its technical name: the scalar wave equation.

From Eq. (13.32) and its conjugate one derives a continuity equation (12.16). $\vec{j}$ is again given by Eq. (12.18) but now

$$\rho = \frac{i\hbar}{2mc^2}\left(\psi^* \frac{\partial\psi}{\partial t} - \frac{\partial\psi^*}{\partial t}\psi\right). \tag{13.34}$$

This new ρ, unlike its non-relativistic version Eq. (12.17), can of course not be interpreted as a probability density since it is not positive definite.

That is why Dirac disagreed with Bohr.

A positive definite ρ is central to the transformation theory. 'The transformation theory had become my darling. I was not interested in considering any theory which would not fit in with my darling.'[93] Dirac believed at that time that the requirement of positive definite ρ implied that the wave equation had to be linear in $\partial/\partial t$. 'The linearity in $\partial/\partial t$ was absolutely essential for me, I just couldn't face giving up the transformation theory.'[93] Thus, with no thoughts of spin in mind, did Dirac set out to find an alternative relativistic wave equation. In Chapter 16 I shall come back to the scalar equation which is not as bad as Dirac thought in 1927.

3. Spin without relativity; the Pauli matrices. Before turning to the Dirac equation I must introduce the Pauli matrices.

In February 1927 Darwin suggested that 'the electron is to be taken as a wave of two components, like light, not of one like sound.'[108] The two-component idea was sound, the comparison with light was not. In May the right formulation was given by Pauli.[109] One must introduce 'the intrinsic angular momentum [i.e. the spin] of the electron in a fixed direction as a new variable'. Since spin is capable of two values only we must, he argued, find a realization of the angular momentum commutation relations (12.14) in terms of a vector $\vec{s}$ which is represented by 2×2 matrices. His result:

$$\vec{s} = \tfrac{1}{2}\hbar\vec{\sigma} \tag{13.35}$$

where the components of $\vec{\sigma}$, the Pauli matrices, are

$$\sigma_1 = \begin{pmatrix} 0 & 1 \\ 1 & 0 \end{pmatrix}, \quad \sigma_2 = \begin{pmatrix} 0 & -i \\ i & 0 \end{pmatrix}, \quad \sigma_3 = \begin{pmatrix} 1 & 0 \\ 0 & -1 \end{pmatrix}. \tag{13.36}$$

Thus in the 'fixed direction' 3:

$$s_3\begin{pmatrix} \psi_1 \\ \psi_2 \end{pmatrix} = \frac{\hbar}{2}\begin{pmatrix} \psi_1 \\ -\psi_2 \end{pmatrix}, \tag{13.37}$$

that is, ψ_1, ψ_2 are eigenfunctions of spin corresponding to eigenvalues $\hbar/2$, $-\hbar/2$ respectively. The components of $\vec{s}$, a *special* realization of Eq. (12.14), satisfy special additional relations:

$$\sigma_i\sigma_j + \sigma_j\sigma_i = 2\delta_{ij}$$
$$\delta_{ij} = 1\ (i=j),\ 0\ (i \neq j) \tag{13.38}$$

which were pointed out to Pauli by Jordan who also drew his attention to the connection between these matrices and quaternions.[110] Dirac later commented: 'I believe I got these [matrices] independently of Pauli, and possibly Pauli also got them independently from me.'[98]

The two-component wave function satisfies a Schroedinger equation of the general form

$$\sum_\beta H_{\alpha\beta}\psi_\beta = i\hbar\frac{\partial\psi_\alpha}{\partial t} \quad (\alpha, \beta = 1, 2) \qquad \text{or} \qquad H\psi = i\hbar\frac{\partial\psi}{\partial t}. \tag{13.39}$$

In its second form, the one commonly used, it is understood that H is a 2×2 matrix and ψ a 1×2 matrix. I shall not discuss Pauli's H in detail. Suffice it to say that it contains a spin–orbit coupling—treated as a perturbation—with a coefficient inserted by hand ('without new justification') to fit the relativistic Thomas factor. As Pauli emphasized, his theory is non-relativistic and therefore 'provisional and approximate'. Yet what an advance compared to the Goudsmit–Uhlenbeck model with its classical echoes of rotating small charged spheres! Pauli's work is the first step toward what two years earlier he had called 'not classically describable two-valuedness'.[31]

4. The equation. The relativistic wave equation of the electron ranks among the highest achievements of twentieth-century science. Dirac's two papers on that subject[111,112] were presented early in 1928, the second one two days before Lorentz' death.

'A great deal of my work is just playing with equations and seeing what they give', Dirac has said.[113] The game which led him to his goal was his observation that

$$(\vec{\sigma}\vec{p})^2 = \vec{p}^2 \cdot 1$$

(where 1 is the 2×2 unit matrix). 'That was a pretty mathematical result. I was quite excited over it. It seemed that it must be of some importance.'[98] How to generalize it to the sum not of three but of four squares, so that one can play with

$$p_1^2 + p_2^2 + p_3^2 + p_4^2 = -m^2c^2, \quad p_4 = \frac{iE}{c}?$$

'It took me quite a while . . . before I suddenly realized that there was no need to stick to quantities σ . . . with just two rows and columns. Why not go to four rows and columns?'[98] Thus was born the Dirac equation[114]

$$\sum_{\beta}\left[\sum_{\mu}(\gamma_\mu)_{\alpha\beta}\frac{\partial}{\partial x^\mu}+\frac{mc}{\hbar}\delta_{\alpha\beta}\right]\psi_\beta=0,$$
$$x^\mu=\vec{x},\, ict \tag{13.40}$$

or in the now common shorthand

$$\left[\gamma_\mu\frac{\partial}{\partial x^\mu}+\frac{mc}{\hbar}\right]\psi=0 \tag{13.41}$$

linear in $\partial/\partial t$ as its author so fervently had desired.

The rest is standard textbook stuff.[106] Each ψ_α satisfies Eq. (13.32) provided that the constants $(\gamma_\mu)_{\alpha\beta}$ satisfy

$$\gamma_\mu\gamma_\nu+\gamma_\nu\gamma_\mu=2\delta_{\mu\nu}\cdot 1. \tag{13.42}$$

The continuity equation (12.16) becomes, in covariant form,

$$\frac{\partial j_\mu}{\partial x^\mu}=0, \qquad j_\mu=i\bar{\psi}\gamma_\mu\psi \tag{13.43}$$

where

$$\bar{\psi}=\psi^\dagger\gamma_4, \qquad \text{i.e. } \bar{\psi}_\alpha=\psi^\dagger_\beta(\gamma_4)_{\beta\alpha} \tag{13.44}$$

is a one-row matrix ($\psi^\dagger$ is the hermitian conjugate of ψ). In particular

$$j_4=i\rho, \qquad \rho=\psi^\dagger\psi=\psi^*_\alpha\psi_\alpha. \tag{13.45}$$

Dirac had found his positive definite probability density.

Some of the other results Dirac obtained are:

(1) The proof that in a central potential

$$\vec{J}=\vec{x}\times\vec{p}+\tfrac{1}{2}\hbar\vec{\sigma} \tag{13.46}$$

is a constant of the motion: spin is an automatic consequence of the theory.*

(2) The proof of relativistic invariance of Eq. (13.40).

(3) For eigenvalues $E\ll mc^2$, the results of the ordinary Schroedinger theory are recovered.

(4) The equation in the presence of an external electromagnetic field described by a four-vector potential A_μ:

$$\left[\gamma_\mu\left(\frac{\partial}{\partial x^\mu}-\frac{ie}{\hbar c}A_\mu\right)+\frac{mc}{\hbar}\right]\psi=0. \tag{13.47}$$

An approximate treatment of the case of an external Coulomb field automatically yields the Thomas factor! About his use of approximation methods Dirac

* In Eq. (13.46) $\vec{\sigma}$ may be understood to mean $\vec{\sigma}=\begin{pmatrix}\vec{\sigma}_P & 0\\ 0 & \vec{\sigma}_P\end{pmatrix}$ where $\vec{\sigma}_P$ are the 2×2 Pauli matrices defined by Eq. (13.36) and 0 is the 2×2 null matrix.

said later: 'I thought that if I got anywhere near right with an approximation method, I would be very happy about that.... I think I would have been too scared myself to consider it exactly.... It leads to great anxiety as to whether it's going to be correct or not'.[115] The exact treatment of fine structure in the Dirac theory was given later, in 1928.[116]

Physicists of the time realized at once how enormous the harvest was. A new mystery arose, however: a ψ with two components was to be expected since Pauli's work, but what is the physical meaning of *four* components? I shall leave this question for now and return to it in Chapter 15 where the positron will make its appearance.

Two final comments:

(1) Mathematicians too became quickly interested in the new theory. Von Neumann wrote in his paper of March 1928: 'That a quantity with four components is not a four-vector has never yet happened in relativity theory'.[117] In that paper he also gives the five fundamental covariants of the Dirac theory. They are*

$$\bar{\psi}O\psi, \quad O=1\text{: scalar} \tag{13.48}$$

$$O=\gamma_\mu\text{: four-vector} \tag{13.49}$$

$$O=\sigma_{\mu\nu}\text{: tensor;} \quad \sigma_{\mu\nu}=-\frac{i}{2}[\gamma_\mu, \gamma_\nu] \tag{13.50}$$

$$O=\gamma_5\gamma_\mu\text{: pseudo-four-vector} \tag{13.51}$$

$$O=\gamma_5\text{: pseudoscalar} \tag{13.52}$$

$$\gamma_5=\gamma_1\gamma_2\gamma_3\gamma_4. \tag{13.53}$$

(2) The term 'spinor' for Dirac's ψ_α was first introduced by Ehrenfest in a letter to van der Waerden[118] in which he asks whether there exists a spinor analysis similar to the long familiar tensor analysis. Van der Waerden's resulting paper[118] provides a systematic answer.**[119]

References

1. E. P. Wigner, *Gruppentheorie und ihre Anwendung auf die Quantenmechanik*, Preface, Vieweg, Braunschweig 1931.
2. W. Heisenberg, *Zeitschr. f. Phys.* **38**, 411, 1926.
3. Ref. 2, p. 425.
4. W. Heisenberg, *Zeitschr. f. Phys.* **39**, 499, 1926.
5. W. Heisenberg, letter to W. Pauli, July 28, 1926; repr. in *W. Pauli, scientific correspondence*, Vol. 1, p. 337, Springer, New York 1979.

* I have modernized the notation of von Neumann who neither uses the prefix 'pseudo' nor calls the right-hand side of Eq. (13.53) by the name γ_5.

** Here dotted and undotted indices are introduced for the first time.

6. E. P. Wigner, *Zeitschr. f. Phys.* **40**, 492, 1926.
7. E. P. Wigner, *Zeitschr. f. Phys.* **40**, 883, 1926.
8. A. Pais, *Ann. of Phys.* (*N.Y.*) **9**, 548, 1960; **22**, 274, 1963.
9. H. Mark and E. P. Wigner, *Zeitschr. f. phys. Chem.* **111**, 398, 1924.
10. E. P. Wigner, *Bull. Am. Math. Soc.* **74**, 793, 1968.
11. Cf. E. U. Condon and G. H. Shortley, *The theory of atomic spectra*, p. 11, Macmillan, New York, 1935.
12. H. Weyl, *Gruppentheorie und Quantenmechanik*, Hirzel, Leipzig 1928; in English: *The theory of groups and quantum mechanics*, transl. H. P. Robertson, Dover, New York.
13. H. Weyl, letter to W. Pauli, May 10, 1919, repr. in Ref. 5, p. 3.
14. K. von Meyenn, *Phys. Bl.* **36**, 293, 1980; **37**, 13, 1981; J. L. Heilbron, *Hist. St. Phys. Sc.* **13**, 261, 1983.
15. H. A. Lorentz, *Physica* **1**, 228, 1921; Engl. transl. in *H. A. Lorentz, collected papers*, Vol. 7, p. 87, Nyhoff, The Hague 1934.
16. T. van Lohuizen, *Physica* **1**, 288, 1921.
17. P. Zeeman, *Researches in magneto-optics*, pp. 191 ff., Macmillan, London 1913.
18. J. Mehra and H. Rechenberg, *The historical development of quantum theory*, Vol. 1, Chapter 4, Springer, New York, 1982.
19. W. Pauli, *Quantentheorie*, p. 147, in *Handbuch d. Phys.*, Vol. 23, Springer, Berlin 1926. Repr. in *Collected papers by W. Pauli*, Eds. R. Kronig and V. Weisskopf, Vol. 1, p. 417, Wiley, New York 1964.
20. S. Goudsmit, *Physica* **5**, 281, 1925.
21. Cf. e.g. W. Pauli, *Quantentheorie*, p. 223, footnote (see Ref. 19).
22. F. Paschen and E. Back, *Ann. d. Phys.* **39**, 897, 1912.
23. H. E. White, *Introduction to atomic spectra*, p. 168, McGraw-Hill, New York 1934.
24. A. Sommerfeld, *Phys. Zeitschr.* **17**, 491, 1916; P. Debye, ibid., 507.
25. T. van Lohuizen, *Proc. Ak. v. Wet. Amsterdam*, **22**, 190, 1919.
26. A. Landé, *Zeitschr. f. Phys.* **5**, 231, 1921.
27. Ref. 18, Vol. 2, p. 30.
28. W. Heisenberg, *Zeitschr. f. Phys.* **8**, 273, 1921.
29. D. Cassidy, *Hist. St. Phys. Sc.* **10**, 187, 1979.
30. A. Landé, *Zeitschr. f. Phys.* **15**, 189, 1923.
31. W. Pauli, *Zeitschr. f. Phys.* **31**, 373, 1925.
32. W. Pauli, letter to A. Sommerfeld, December 6, 1924, repr. in Ref. 5, Vol. 1, p. 182.
33. W. Pauli, *Zeitschr. f. Phys.* **31**, 765, 1925.
34. P. A. M. Dirac, *Proc. Roy. Soc. A* **112**, 661, 1926.
35. E. C. Stoner, *Phil. Mag.* **48**, 719, 1924.
36. S. Goudsmit, *Zeitschr. f. Phys.* **32**, 794, 1925.
37. G. E. Uhlenbeck, *Am. J. Phys.* **24**, 431, 1956.
38. G. E. Uhlenbeck, *Comm. Dutch Hist. Inst. Rome* **4**, 217, 1924.
39. D. Lang, *New Yorker*, November 7 and 14, 1953.
40. S. Goudsmit, *Arch. Néerl. des Sc. ex. et nat.* **6**, 116, 1922; summarized in *Naturw.* **9**, 995, 1921.
41. S. Goudsmit, *Ned. Tydschr. v. Natuurk.* **37**, 386, 1971 (in Dutch). *English*: *Delta* **15**, summer 1972, p. 77.
42. S. Goudsmit, *Phys. Bl.* **21**, 445, 1965.
43. S. Goudsmit, *Physics Today* **29**, June 1976, p. 40.

44. G. E. Uhlenbeck, *Physics Today* **29**, June 1976, p. 43, 45.
45. S. Goudsmit and G. E. Uhlenbeck, *Physica* **5**, 266, 1925.
46. M. Abraham, *Ann. der Phys.* **10**, 105, 1903, Section 11.
47. G. E. Uhlenbeck and S. Goudsmit, *Naturw.* **13**, 953, 1925.
48. P. Ehrenfest, letter to H. A. Lorentz, October 16, 1925.
49. F. Rasetti and E. Fermi, *Nuovo Cim.* **3**, 226, 1926, repr. in *Enrico Fermi, collected works*, Vol. 1, p. 212, Univ. Chicago Press 1962.
50. H. A. Lorentz, *Collected works*, Vol. 7, p. 179, Nyhoff, The Hague 1934.
51. W. Heisenberg, letter to S. Goudsmit, November 21, 1926, repr. in Ref. 41.
52. L. Pauling and S. Goudsmit, *The structure of line spectra*, p. 58, McGraw-Hill, New York 1930.
53. S. Goudsmit and G. E. Uhlenbeck, *Physica* **6**, 273, 1926.
54. S. Goudsmit and G. E. Uhlenbeck, *Nature* **117**, 264, 1926.
55. N. Bohr, letter to P. Ehrenfest, December 22, 1925.
56. L. H. Thomas, *Nature* **117**, 514, 1926; *Phil. Mag.* **3**, 1, 1927. Cf. also Y. I. Frenkel, *Nature* **117**, 653, 1926; *Zeitschr. f. Phys.* **37**, 243, 1926.
57. A. Pais, *Subtle is the Lord*, p. 143, Oxford Univ. Press 1982.
58. W. Pauli, *Scientific correspondence* (Ref. 5), Vol. 1, pp. 296–312.
59. G. F. FitzGerald, *Nature* **62**, 564, 1900.
60. A. H. Compton, *J. Franklin Inst.* **192**, 145, 1921.
61. E. H. Kennard, *Phys. Rev.* **19**, 420, 1922.
62. W. Pauli, *Naturw.* **12**, 741, 1924.
63. W. Pauli, Nobel lecture 1946, repr. in *Collected papers* (Ref. 19), Vol. 2, p. 1080.
64. S. Goudsmit, *Physics Today* **18**, June 1961, p. 21.
65. L. Belloni, *Am. J. Phys.* **50**, 461, 1982.
66. R. Kronig, in *Physics in the twentieth century*, pp. 20–28, Eds. M. Fierz and V. Weisskopf, Interscience, New York 1960.
67. B. L. van der Waerden, ibid., pp. 210–12.
68. F. R. Bichowski and H. C. Urey, *Proc. Nat. Ac. Sci.* **12**, 801, 1926.
69. M. Planck, *Verh. Deutsch. Phys. Ges.* **2**, 237, 1900.
70. Ref. 57, Chapter 20, Section (a).
71. L. Boltzman, *Vorlesungen über die Principe der Mechanik*, Vol. 1, p. 9, Barth, Leipzig 1897. Repr. by Wissenschaftliche Buchges., Darmstadt 1974.
72. Cf. P. and T. Ehrenfest, *Enc. der Math. Wiss.*, Vol. 4, Part 2, Sec. 23, Teubner, Leipzig 1911. In English: *The conceptual foundations of the statistical approach in mechanics*, transl. M. J. Moravcsik, Cornell Univ. Press, Ithaca 1959.
73. K. Huang, *Statistical mechanics*, pp. 192–7, Wiley, New York 1965.
74. L. Natanson, *Phys. Zeitschr.* **12**, 659, 1911.
75. S. N. Bose, *Zeitschr. f. Phys.* **26**, 178, 1924.
76. Ref. 57, Chapter 23, Section (b).
77. R. H. Fowler, *Proc. Roy. Soc. A* **113**, 432, 1926.
78. P. Ghose, *Physics News, Bull. of the Indian Phys. Ass.* **13**, 130, 1982.
79. A. Einstein, *Sitz. Ber. Preuss. Ak. Wiss. 1924*, p. 261.
80. A. Einstein, *Sitz. Ber. Preuss. Ak. Wiss. 1925*, p. 3.
81. W. H. Keesom, *Helium*, Elsevier, New York 1942.
82. F. London, *Nature* **141**, 643, 1938; *Phys. Rev.* **54**, 1947, 1938.
83. E. Fermi, *Rend. Acc. Lincei* **3**, 145, 1926, repr. in *Collected works* (Ref. 49), Vol. 1, p. 181.

84. E. Fermi, *Zeitschr. f. Phys.* **36**, 902, 1926, *Collected works*, Vol. 1, p. 186.
85. P. A. M. Dirac, *Proc. Roy. Soc. A* **112**, 661, 1926.
86. Cf. P. A. M. Dirac, in *History of twentieth century physics*, p. 133, Academic Press, New York, 1977.
87. W. Pauli, *Zeitschr. f. Phys.* **41**, 81, 1927.
88. F. Hund, *Zeitschr. f. Phys.* **42**, 93, 1927.
89. D. Dennison, *Proc. Roy. Soc. A* **115**, 483, 1927.
90. E. P. Wigner, *Math. und Naturwiss. Anzeiger der Ungar. Ak. der Wiss.* **46**, 576, 1929.
91. P. Ehrenfest and J. R. Oppenheimer, *Phys. Rev.* **37**, 333, 1931.
92. J. Chadwick, *Proc. Roy. Soc. A* **128**, 114, 1930.
93. P. A. M. Dirac, report KFKI-1977-62, Hung. Ac. of Sc.
94. T. Kuhn, interview with Dirac, April 1, 1962, Niels Bohr Library, Am. Inst. of Phys., New York.
95. Ref. 18, Vol. 4, Chapter 1.
96. G. Gamow, *Thirty years that shook physics*, p. 121, Doubleday, New York 1966.
97. R. J. Eden and J. C. Polkinghorne, in *Aspects of quantum theory*, Eds. A. Salam and E. P. Wigner, p. 1, Cambridge Univ. Press 1972.
98. P. A. M. Dirac, Ref. 86, p. 109.
99. P. A. M. Dirac, *Proc. Roy. Soc. A* **109**, 642, 1925.
100. M. Born, *My life*, p. 226, Scribner, New York 1978.
101. P. A. M. Dirac, *Proc. Roy. Soc. A* **113**, 621, 1927.
102. P. A. M. Dirac, *Proc. Roy. Soc. A* **114**, 243, 1927.
103. A. Einstein, in *James Clerk Maxwell*, p. 66, Macmillan, New York 1931.
104. Cf. P. A. M. Dirac, *Proc. Roy. Soc. Edinburgh* **59**, 122, 1939.
105. E. Schroedinger, *Ann. der Phys.* **79**, 361, 1926, esp. p. 372.
106. L. I. Schiff, *Quantum Mechanics*, 2nd edn, McGraw-Hill, New York 1955.
107. O. Klein, *Zeitschr. f. Phys.* **37**, 895, 1926; E. Schroedinger, *Ann. der Phys.* **81**, 109, 1926, Section 6; V. Fock, *Zeitschr. f. Phys.* **38**, 242, 1926; Th. de Donder and H. van den Dungen, *Comptes Rendus* **183**, 22, 1926; J. Kudar, *Ann. der Phys.* **81**, 632, 1926; W. Gordon, *Zeitschr. f. Phys.* **40**, 117, 1926.
108. C. G. Darwin, *Nature* **119**, 282, 1927; also *Proc. Roy. Soc. A* **116**, 227, 1927.
109. W. Pauli, *Zeitschr. f. Phys.* **43**, 601, 1927.
110. Ref. 109, p. 607, footnote 2.
111. P. A. M. Dirac, *Proc. Roy. Soc. A* **117**, 610, 1928.
112. P. A. M. Dirac, *Proc. Roy. Soc. A* **118**, 351, 1928.
113. Ref. 94, interview May 7, 1963.
114. Ref. 111, Eq. (11).
115. Ref. 94, interviews May 7 and 14, 1963.
116. W. Gordon, *Zeitschr. f. Phys.* **48**, 11, 1928; C. G. Darwin, *Proc. Roy. Soc. A* **118**, 654, 1928.
117. J. von Neumann, *Zeitschr. f. Phys.* **48**, 868, 1928.
118. B. van der Waerden, *Goett. Nachr. 1929,* p. 100.
119. Cf. also O. Laporte and G. E. Uhlenbeck, *Phys. Rev.* **37**, 1380, 1552, 1931.

14

Nuclear physics: the age of paradox

(a) Quantum mechanics confronts the nucleus

On March 19, 1914, Rutherford opened a 'Discussion on the structure of the atom' at the Royal Society.[1] After a comparison between Thomson's model and his own picture of the nucleus he singled out the discovery of radioactive isotopes and the work by Moseley as major advances (see Chapter 11).

On February 7, 1929, Rutherford opened a 'Discussion on the structure of atomic nuclei' at the Royal Society.[2] He began by listing the main advances during the intervening fifteen years:

(1) The proof of the isotopic constitution of non-radioactive elements, mainly the result of Aston's work with the mass spectrograph.

(2) His own work on the artificial transmutation of elements.

(3) Progress in γ-ray spectroscopy.

(4) The deviations from Rutherford scattering (cf. Chapter 11, Section (h)). 'Bombarding hydrogen with α-particles the scattering is completely abnormal . . . the hydrogen and helium nucleus appears to be surrounded by a field of force of unknown origin.'* As to nuclear constituents, 'probably in the lighter elements the nucleus is composed of α-particles, protons, and electrons'. In 1929 the proton–electron model (pe-model) of the nucleus was still in force. The term 'proton', coined by Rutherford, which first appeared in print[3] (I believe) in 1920 had gained general acceptance by then.

(5) Finally Rutherford mentioned Gamow's and Gurney and Condon's interpretation of α-decay as the penetration of a potential barrier of finite height by the α-particle.[4]

Thus α- and γ-radioactivity were given prominence, but there was not one word about β-decay, either by Rutherford or by other discussants, even though there was very important news on that score by the time of the Royal Society discussion. Equally notable is the absence of any reference to the increasingly bizarre consequences following from the assumption that electrons are nuclear constituents.

* In 1929 Rutherford still toyed with the idea that the unknown forces might be of magnetic type.

During the year following the meeting Rutherford, together with Chadwick and Ellis, completed his fourth* and final edition of a textbook on radioactivity.[5] It contains the following comment on β-decay: 'We must conclude that in β-ray disintegration the nucleus can break up with emission of an amount of energy that varies within wide limits. No satisfactory interpretation has yet been given of this curious result so much in contrast with the high degree of homogeneity and definiteness shown by the α-ray disintegrations'.[6] This remark helps to explain the silence about β-decay at the Royal Society gathering: the primary β-spectrum had at long last generally been conceded to be continuous but nobody had anything sensible to say about that mystery.

Another year later, in 1931, the textbook on nuclear physics by Gamow appeared, the first of its kind written by a theoretical physicist.[7] On its first page the author defines his model of the nucleus: 'In accordance with the concepts of modern physics we assume that all nuclei are built up of elementary particles—protons and electrons'. Gamow was aware, however, of the difficulties the pe-model seemed to be running into. In the course of preparing his manuscript, 'Gamow had a rubber stamp made with a skull and crossbones with which he marked the beginning and end of all passages dealing with eleectrons'.[8] When the Oxford University Press objected to this symbol, Gamow replied: 'It has never been my intention to scare the poor readers more than the text itself will undoubtedly do'.[8] The skull and bones were replaced by a boldface sleepy *S*.

This chapter deal with two independent topics. Continuing the account of nuclear structure begun in Chapter 11, I treat the pe-model's paradoxes as they came to be recognized beginning in 1926. Then the story of β-spectra, begun in Chapter 8, is taken up again and attempts to interpret their continuous character are reviewed.

We are now entering nuclear physics' third phase. The first, which may be called nuclear physics without nucleus, deals with the discoveries of radioactivity and qualitative properties of its radiations (Chapters 2–4, 6–8). The second phase covers the discovery of the nucleus, the early exploration of its global properties, and the beginning stages of nuclear reactions (Chapter 11). In all these developments theoretical physicists played no role to speak of. Indeed the two main theoretical contributions during those early days, the transformation theory and the analysis of α-particle scattering, are due to the experimentalist Rutherford. In the third phase, beginning in 1926 and ending with the discovery of the neutron in 1932, theoreticians began to play a role of importance. This marked change was caused by the advent of quantum mechanics, as is very clear from the topics treated in Gamow's book:[7] the theory of α-decay, of γ-ray emission and absorption, of inelastic effects in

* For the earlier editions see the section on Sources of Chapter 3.

α-particle bombardment, and others. The puzzles which we are about to discuss in the next section came into focus when quantum mechanics confronted the nucleus. As Gamow wrote[9] in his book: 'The usual ideas of quantum mechanics absolutely fail in describing the behavior of nuclear electrons; it seems that they may not even be treated as individual particles'.

The present chapter is devoted to this third phase, the age of paradox. A variety of difficulties with the pe-model are discussed in the next section. Section (c) deals with experimental work on β-spectra from 1914 until 1929. Bohr's suggestion that energy might not be conserved in β-decay and Pauli's alternative, the neutrino hypothesis, are treated in Section (d). By then we shall be prepared for nuclear physics' new look resulting from the discovery of the neutron and Fermi's theory of β-decay, to be discussed in Chapter 17.

(b) In which the proton–electron model of the nucleus runs into trouble

Since 1913 it had been assumed that a nucleus (A, Z) with mass number A and charge number Z consists of A protons and A–Z electrons:

$$(A, Z) = Ap + (A - Z)e. \tag{14.1}$$

It was seen in Chapter 11, Section (e), that this pe-model was almost inevitable. Passing by the efforts of some physicists and physical chemists who in subsequent years labored mightily[10] to adorn this model with elaborate substructure, let us next examine a variety of arguments which led to a growing discomfort with this picture.

1. Nuclear size. Imagine that an electron is a little sphere with radius a and that its mass is entirely electromagnetic in origin. (Both assumptions were widespread until well into the 1920s). Then, classically, $e^2/a \sim mc^2$ or

$$a \sim \frac{e^2}{mc^2} = 2.8 \times 10^{-13}\ \text{cm}. \tag{14.2}$$

As was well known in the teens, this length is of the same general order of magnitude as nuclear radii. Question: how is it possible to stuff many electrons this size in a box as small as a nucleus? As Andrade phrased it in 1923: 'Coulomb's law cannot account for the stability of a nucleus composed of positive and negative charges which when free would be as large as the nucleus itself'.[11]

In 1926 Rasetti and Fermi remarked that spin exacerbates this problem.[12] A spinning electron, still conceived as a little sphere, acquires a magnetic energy $\sim(e\hbar/2mc)^2/a^3$. Put *this* equal to mc^2 and a becomes larger by a factor $\sim(137)^{2/3}$ than its previous valuc. The spirit of their paper was to note a difficulty for spin rather than for the pe-model. However, as Rasetti recalled:

'We didn't mean that paper very seriously. Fermi, I don't think, did. He said "Well, it's sort of a cute point to make" . . . He was very enthusiastic about the theory of spin'.[13]

Quantum electrodynamics was to show that all these arguments are specious. The electromagnetic mass behaves neither like $1/a$ nor like $1/a^3$ but like $e^2/\hbar c \cdot mc^2 \cdot \ln \hbar/mca$.

More interestingly, applications of qualitative quantum mechanics (especially the uncertainty relations) to electrons confined to a nucleus with a typical radius $\sim 5 \times 10^{-13}$ cm showed that it is not the electron's size but rather its de Broglie wavelength λ which causes problems. How can a β-ray with energy a few MeV wait inside the nucleus to be released if its λ is large compared to the nuclear radius? (Attempts to treat β-decay by barrier penetration, like α-decay, failed of course.[14]) Moreover, the typical (relativistic) kinetic energy of a confined electron is $\gtrsim 40$ MeV, implausibly large compared to the average nuclear binding energy per particle.[15] I do not know who first raised these issues and find it interesting that they are not mentioned at all in several places natural for the purpose, such as Gamow's book[7] or Bohr's address at the nuclear physics conference in Rome (1931) on the complications caused by electrons in the nucleus.[16] (Bohr loved qualitative arguments of the kinds just mentioned.) These two sources indicate that other problems appeared more compelling: those of nuclear spin and statistics, to which I turn next, and the Klein paradox, to be discussed in Chapter 15. That is also in accord with what Wigner told me: the consequences of the uncertainty relations for the nucleus were widely known, but the experimental results about nitrogen molecules made a much stronger impression on him.

2. Nuclear magnetic moments. In 1926 Kronig pointed out[17] a difficulty related to the magnitude $e\hbar/2mc$ (the Bohr magneton) of the electron's magnetic moment. Since there are supposed to be electrons in the nucleus, 'the nucleus too will have a magnetic moment of the order of a Bohr magneton unless the magnetic moments of all the nuclear electrons happen to cancel [the probability for which] seems *a priori* to be very small'. Nuclear magnetic moments had not yet been measured directly at that time. (The first direct measurements of the proton moment dates from 1933.[18]) It was clear, however, that nuclear moments of the order of an electron Bohr magneton would give rise to unacceptably large hyperfine splittings of spectral lines, due to the interaction between the magnetic moments of peripheral electrons and of the nucleus.

Kronig's note makes it quite clear that, in his opinion, the magnetic moment question must be seen as a difficulty for spin rather than for nuclear constitution. All the same he had injected an important new issue into the discussion.

3. Nuclear spin. Two years later Kronig spotted another strange consequence of the pe-model. On a visit to Utrecht, where Ornstein directed an important center for precision measurements of intensities in optical spectra, he learned something very interesting about the rotational band spectra of N_2^+, singly ionized molecular nitrogen.

A few reminders about spectra of diatomic molecules may help to understand Kronig's point. Imagine the two nuclei in the molecule held fixed. Then the spectrum would be purely due to electronic de-excitations (frequencies ν_e). Additional vibrational modes (frequencies ν_v) come into play if the motion of the nuclei in the direction of their joint axis is included. Considering next that this axis also rotates, one has further rotational modes (frequencies ν_r). Since typically $\nu_e \gg \nu_v \gg \nu_r$, the overall spectrum consists of ν_e-lines each of which has a fine structure of ν_v-lines, while each electronic/vibrational line has a finer structure of ν_r-lines.

For many purposes, including our nitrogen problem, these three types of excitations can to a good approximation be treated as mutually independent. In that case one can assign a nuclear orbital angular momentum quantum number J to the rotational levels. In general the change ΔJ in a rotational transition (accompanied by dipole radiation) satisfies

$$\Delta J = 0, \pm 1. \tag{14.3}$$

However, for the lines considered by the Utrecht group, $\Delta J = \pm 1$ only.* Let ν_0 be the frequency for some fixed electronic/vibrational transition. Then the frequencies ν of the associated rotational band are of the form

$$\nu = \nu_0 + \nu_r(J). \tag{14.4}$$

This band consists of two branches: one (the R–branch) with $J \to J-1$ for which $\nu_r(J)$ is positive and (for not too large J) increases monotonically with J, the other (the P-branch) with $J \to J+1$ for which $\nu_r(J)$ is negative and (for not too large J) decreases monotonically with J.

By a general theorem of quantum mechanics, the intensity of a spectral line is proportional to the statistical weight of the final state.** The electronic/vibrational weight, common to all lines in a given rotational band, drops out if we consider only relative intensities. The remaining weight factor $2J+1$ would indicate that the intensities change monotonically as we move away from ν_0 along each branch. In the case of nitrogen (and other molecules as well) one observes, however, that the intensities alternate along each branch. A 'strong' line is followed by a 'weak' line, then again a strong line, etc. This is where two new factors come in: nuclear spin and the circumstance that the two nuclei are identical.

* The transitions in question are all of the type $^2\Sigma \to {}^2\Sigma$, where Σ denotes the eigenvalue zero of the total electron angular momentum in the direction of the nuclear axis, while the superscript '2' means that we deal with doublets of total electron spin. $J \to J$ is forbidden for $\Sigma \to \Sigma$ transitions.[19]

** The subsequent argument is neither affected by the dependence of the intensity on the Boltzmann factor specifying the occupation of the initial state nor by at most very weak dependences of matrix elements on J or I.

Consider the case of two identical nuclei with spin I. The total number $(2I+1)^2$ of spin states (degenerate for our purposes) can be divided into a number $g_s(g_a)$ of states which are nuclear spin symmetric (antisymmetric), where*

$$g_s=(I+1)(2I+1),\ g_a=I(2I+1). \tag{14.5}$$

Now the two-nucleus system by itself satisfies either BE or FD statistics.** As we move along each branch the consecutive final state J-values are . . . , even, odd, even, . . . so that these states are . . . , symmetric, antisymmetric, symmetric, . . . under exchange of the spatial nuclear coordinates. Correspondingly there is an additional weight factor due to nuclear spin which equals . . . , $g_s, g_a, g_s, \ldots$ if the nuclei satisfy BE statistics, and . . . , $g_a, g_s, g_a, \ldots$ for the FD case. Thus it follows from Eq. (14.5) that, *whatever the statistics may be*, nuclear spin introduces a 'strong to weak' alternating intensity ratio R given by

$$R=\frac{I+1}{I}. \tag{14.6}$$

The Utrecht measurements showed[20] that, for N_2^+, R lies close to 2, so that the spin of the nitrogen nucleus equals one.

'One might in the first instance be surprised about this result', Kronig remarked.[21] Indeed. According to the pe-model the N^{14} nucleus consists of 14 protons + 7 electrons, that is, of an odd number of spin-1/2 particles.† (Kronig knew about the proton spin from Dennison's results; see Chapter 13, Section (d).) The resultant nuclear spin should therefore be half-integer. Then how can it be equal to one? 'Probably one is therefore forced to assume that protons and electrons do not retain their identity to the extent they do outside the nucleus.'[21]††

Hyperfine structure, the only other then available source of information about nuclear spin, began to show similar oddities. It was found[23] in 1929 that cadmium has at least one isotope with $I=1/2$. Since in this case $Z=48$ it follows from Eq. (14.1) that all cadmium isotopes contain an even total number of protons and electrons so that they all should have integer I!

4. *Nuclear statistics*. Inelastic scattering of photons by molecules, first detected by Raman in 1928, obeys the energy relation

$$h\nu+E_a=h\nu'+E_b \tag{14.7}$$

* Denote the states by (m_1, m_2) where the m's are the respective magnetic quantum numbers. All $(2I+1)$ states (m_1, m_1) are symmetric. For $m_1\neq m_2$ the sum (difference) of (m_1, m_2) and (m_2, m_1) is symmetric (antisymmetric).

** We are entitled to consider this system separately as long as the total molecular wave function is of the form electronic × vibrational × rotational factor. Small deviations from this decomposition do not qualitatively change the argument.

† The ~0.4 per cent admixture of N^{15} can of course be ignored.

†† The status in 1931 of nuclear spin information from rotational band spectra, Raman spectra, and other sources is reviewed in Ref. 22.

where $\nu(\nu')$ is the initial (final) photon frequency and $E_a(E_b)$ the initial (final) state molecular energy level. The Raman spectrum consists of Stokes lines ($\nu' < \nu$) and anti-Stokes lines ($\nu' > \nu$), the latter appearing only if E_a is an excited state. The hierarchy of electronic/vibrational/rotational lines which we just met occurs also in the Raman effect. Once again we shall be interested in the rotational bands. In the Raman case the selection rules are in general[24]

$$\Delta J = 0, \pm 2. \tag{14.8}$$

In the examples* we are about to meet, $\Delta J = 0$ is forbidden, however.

Two bands of approximately equidistant rotational lines are associated with a given electronic/vibrational line: the Stokes band for which $J \to J+2$, and the anti-Stokes band ($J \to J-2$), monotonically displaced to lower and higher frequencies respectively as J increases. The statistical weights, including nuclear spin effects, follow the rules mentioned previously.

Let us now turn to Rasetti's experimental results[25] of early 1929 for the molecules H_2 and N_2. For hydrogen he detected the line $J = 1 \to 3$, which turned out to be stronger than the already known[26] line $J = 0 \to 2$. This, he noted, is as it should be. Indeed, protons have spin 1/2 so that the ratio R, Eq. (14.6), equals 3 in this case. Moreover since protons obey FD statistics the ratio favors J odd$\to$odd transitions by the argument given previously. Comparing this result with his measurements of many nitrogen lines he found something remarkable: 'N_2 and H_2 which have a similar electronic structure behave in opposite ways as to the relative weight of odd and even rotational states'!

Rasetti was entirely correct in calling this conclusion 'perhaps significant for the properties of nuclei'. Right after his paper had reached Goettingen, Heitler and Herzberg sent a letter to *Naturwissenschaften*[27] with the title: 'Do nitrogen nuclei obey Bose statistics?' Their reasoning was clear: Rasetti's result meant that J even$\to$even transitions are favored in N_2 and therefore that nitrogen nuclei satisfy BE statistics by the reasoning given earlier. 'This fact is extraordinarily surprising.'[27] According to the pe-model the N^{14} nucleus contains an odd number (21) of spin-1/2 particles and should therefore obey FD statistics by Wigner's rule (see Chapter 13, Section (d)) with which Heitler and Herzberg were familiar.** Therefore, they concluded: 'This rule is no longer valid in the nucleus . . . it seems as if the electron in the nucleus loses, along with its spin [reference to Kronig's earlier conclusion], also its right of participation in the statistics of the nucleus'. Shortly thereafter, hyperfine structure measurements indicated that Li^6 also has the 'wrong' statistics.†

* They are $\Sigma \to \Sigma$ transitions.

** Their remark that the N_2^+ spectrum discussed above cannot give information about the statistics of the nitrogen nucleus is not correct. In fact Kronig could have deduced the statistics paradox as well. It is my impression that he did not do so because he did not know Wigner's theorem.

† Cf. the comments in Ref. 28.

Wherever one looked, at nuclear sizes, spins, magnetic moments or statistics, intranuclear electrons caused grave problems. The answer was to come soon: there are no electrons in the nucleus, (A, Z) corresponds to Z protons and $(A-Z)$ neutrons. Therefore the number of spin-1/2 particles in the nitrogen nucleus is not odd (14 protons + 7 electrons) but even (7 protons + 7 neutrons). It cannot be held against the physicists of the late twenties that they could not and would not do better than suppose that electrons lose their identity when entering the nucleus. Only if one had an alternative to present would it be possible to abandon electrons as old-fashioned nuclear building blocks. The first to do so (to my knowledge) was Yakov Gregorievitch Dorfman from Leningrad* who in 1930 did propose the absence of electrons in the nucleus as a way out of the difficulties. (He thought well enough of his ideas to publish them both in the Paris *Comptes Rendus*[29] and in *Zeitschrift für Physik*.[30]) He dared proffer this thought because he was aware of a suggestion by others[31] that β-decay need not imply the existence of real electrons in the nucleus. That suggestion, though wrong, is of some interest since it is the first attempt to link nuclear phenomena with quantum field theory.**

This concludes the recitation of nuclear constitution's perplexities which the discovery of the neutron was to resolve. Even after that event constituent electrons were not banned all at once. That postscript can wait until Chapter 17. We should next catch up with the developments regarding β-spectra.

(c) β-spectra 1914–30, or the life and times of Charles Drummond Ellis

> Some very interesting phenomenon seems to be involved in β-ray disintegration.
>
> Ellis and Wooster in 1925[32]

In a lecture before the Royal Institution given on June 4, 1915, Rutherford gave his audience an update[33] on the status of 'Radiations from Exploding Atoms'. His discourse on β-rays included the following remarks: 'Chadwick has recently shown that the fraction of the rays which give a line spectrum is only a few per cent of the total radiation. The general evidence shows that β-radiation . . . gives a *continuous* spectrum due to β-rays of all possible velocities on which is superimposed a *line* spectrum due to a small number

* Dorfman worked at the Physical Technical Institute at that time. He was an expert on magnetism. Together with Yakov Ilyich Frenkel he gave the first theory of ferromagnetic domain structure. Later he became director of the Physics section at the Institute for the History of Science and Technology of the Soviet Academy of Sciences.

** Here I run slightly ahead of the story of the Dirac equation. The suggestion[31] was that 'a very large number of [negative energy] electrons are latently present, so to say' in the nucleus, that β-decay amounts to an electron transition from a negative to a positive energy state, and that the β-spectrum is continuous because one has so many initial negative energy levels to choose from. Weird!

of β-particles of definite velocities comprising each group . . . It is thus necessary that each atom [of a given nuclear species] does not emit an identical β-radiation'. The italics are Rutherford's.

In Chapter 8, which dealt with the earliest researches on β-spectra and concluded with an account of Chadwick's work, we saw how photographic detection methods failed to draw attention to the continuous spectrum, how in 1914 Chadwick made his discovery by switching to counter techniques for the observation of magnetically separated β-rays, and how his preliminary tests indicated that the continuous nature of the spectrum cannot be attributed to secondary scattering effects. It was also noted (Chapter 8, Section (g)) that already in the years 1912–13 a wealth of data was collected concerning the discrete lines in the effective β-spectrum, well before it was known that these can largely be ascribed to Auger transitions and internal conversion.

I now continue beyond 1914, the year at which I left off in Chapter 8. Except for one experiment which confirmed[34] Chadwick's continuous spectrum, there is nothing memorable to report about the next seven years. There was a war going on. The first item of business thereafter is Charles Drummond Ellis' recognition in 1921 of the internal conversion mechanism.

When last heard of, Ellis was detained in the Ruhleben camp (Chapter 8, Section (h)) where he met Chadwick, who instilled in him an interest in physics so strong that he gave up his plans for a career as an artillery officer, deciding to become a physicist instead. In 1919 he secured a place at Trinity College and became a student of Rutherford when the latter arrived in Cambridge shortly thereafter. 'From the moment he arrived at the Cavendish I was one of Rutherford's men.'[35] Ellis rapidly developed into a highly skilled experimentalist with a fine taste in his choice of problems. During the most productive period of the Cavendish (where he stayed until 1936) he became its leading authority in γ- and especially β-radioactivity. He and his coworkers are primarily responsible for those experimental advances during the 1920s which were to lead to the neutrino hypothesis.*

I now turn to Ellis' paper[36] of 1921, his very first and a pretty one. It takes its cue from a 1914 study[37] by Rutherford and coworkers of electrons produced when lead and gold are bombarded by γ-rays from a radium -B source. These authors had found that 'the groups of rays [that is, the discrete energies of some emitted electrons] . . . are due to the conversion into β-rays of the more penetrating γ-rays in radium B'. They had further noted that the electrons emitted from gold have a 2 per cent higher energy than those coming from lead. This last remark was Ellis' starting point. 'This fact receives a simple explanation if it be assumed that the energy of the emitted electron is equal to some energy characteristic only of the γ-ray, minus the energy necessary to remove the electron from the atom.' This is a refined statement of the

* For many details on Ellis' life and work see Ref. 35.

Einstein photoelectric effect relation

$$E = E_\gamma - P \qquad (14.9)$$

where E_γ and E are the respective energies of the incoming γ-ray and the emitted electron. P is the so-called work function. Ellis went a step further by identifying P with some ionization energy, the magnitude of which depends on the atomic level from which the electron is ejected.

Ellis' radioactive source was a small narrow tube filled with Ra-emanation which deposits RaB and RaC on the walls. Three strong γ-ray lines of RaB (with energies in the 200–300 kilovolt region) eject electrons from foils of tungsten, platinum, lead, or uranium wrapped in succession around the tube. After these photoelectrons are semicircularly bent in a magnetic field (see Chapter 8) they hit a photographic plate. The position of the electron images on the plate determines their respective energies. Ellis now had a set of E-values for use in Eq. (14.9). Next he assumed that these electrons were emitted from the K-shell of the atomic species at hand. The corresponding P-values had been known for some years. If his assumptions were correct, then adding E to P should give the *same* three E_γ-values for all choices of targets. This, he found, was indeed the case to 1 per cent or better. He had now established three *primary* γ-energies. Note also that this experiment proved the validity of the photoelectric equation at much higher energies than had been explored so far.

Thereupon Ellis turned his attention specifically to the lead target. Rutherford already knew[37] that the electron lines produced from lead have energies identical with those produced by the radium-B source itself. 'This result was repeated in the present work.' Ellis saw at once the reason for this identity: radium B *is* lead, the isotope Pb^{214}. Therefore, he argued, the discrete lines of the RaB spectrum are due to an inner photoelectric effect, that is, the ejection of electrons from the K-shell of a RaB atom when they are hit by γ-rays coming from that atom's *own* nucleus. That is how internal conversion was discovered.

Ellis' second paper[38] (1922) contains another novelty: the first sketch of a nuclear energy level diagram, the first attempt to seek 'support [for] the view that quantum dynamics apply to the nucleus, and that part at least of the structure of the nucleus can be expressed in terms of stationary states'. His elaboration of this picture in a series of papers[39] published in 1924 need not concern us here. In fact I now leave the subject of γ-ray spectroscopy for good with an invitation to the reader to have a look at recent handbooks on the subject. He will find, to quote but one example, that in the β-decay of RaC, $Bi^{214} \rightarrow Po^{214}$, the final polonium nucleus can be in any of twenty states between which fifty-one distinct γ-ray transitions are recorded.

While the boys from the Cavendish were merrily steaming along, a note of discord was heard from the Kaiser Wilhelm Institut für Chemie in Berlin–Dahlem.

About a year after Ellis' first paper[36] Lise Meitner came forth[40] with a completely different and complicated scheme for β-decays based on the following assumptions: (1) All β-rays start out with a unique* energy E inside the nucleus (we are still in the days of the pe-model of course); (2) they may escape with this full energy; (3) or they may convert the portion $E-w$ of their energy into a single γ-ray; (4) the γ-ray will give secondary β-rays due to internal conversion in the K, L, M, . . . shell. I shall not discuss the experimental evidence Meitner claimed[40] for her model—which is of course a variant of the old ideas of Hahn and herself of a monochromatic primary β-ray energy (Chapter 8, Section (b)). In her paper she refers[43] to Ellis' 'other point of view' but makes no mention of Chadwick's continuous spectrum.

Chadwick and Ellis were not amused.

First, Ellis repeated some of Meitner's measurements, found[44] different results, and concluded: 'There would appear therefore to be no evidence in favor of [Meitner's] theory, but on the contrary very direct evidence against it', adding: 'A most serious objection is that this theory gives no possibility of explaining the general [i.e. the continuous] spectrum of β-rays, in fact it appears to deny its existence'. Elsewhere[45] he came to the root of the problem: 'To all appearances the simple analogy between α- and β-decay cannot be maintained'. Shortly thereafter Meitner conceded that some of Ellis' objections were to the point.[46] From his side Ellis acknowledged[32] some time later that he had been wrong on another issue between himself and Meitner.**

Next, Chadwick and Ellis jointly returned[48] to the continuous spectrum which 'on Meitner's view . . . is presumably held to be a fortuitous occurrence due perhaps to the scattering of the homogeneous groups, whereas on our view the continuous spectrum consists of the actual disintegration electrons'. In their paper they report on new experiments which show that 'The continuous spectrum has a real existence which is not dependent on the spectrum . . . Any explanation of it as due to secondary causes is untenable'.

The main point of their work is this. In 1914 Chadwick had used[49] RaB and RaC generated inside a thin tube filled with radium emanation. This time they deposited RaB and RaC on a brass plate, thereby increasing on purpose the possibility that the electrons get scattered back. If the continuous spectrum were a secondary effect of the line spectrum, then the lines in the old experiment should stand out much more against the continuous electron background than in the new one, in which (if the continuum were due to scattering) the lines would be depleted, the continuum enhanced. No such effect was found. 'Any theory of the β-ray disintegration must take this [result] into account.' It is most fortunate that Meitner forced the Cavendish group to pursue the con-

* In a sequel[41] she considered the possibility that there may not be one but two initial energies. For attempts to justify the Meitner model see Ref. 42.

** Meitner had shown that β-decays precede γ-decays if both kinds of radiations are emitted from a given nucleus.[47] For some time Ellis had maintained the opposite.

tinuous spectrum further. She herself next took up more detailed studies of β-ray line spectra, with important results.[50]

The debate about the primary β-spectrum was nearing its end now but was not quite over yet. After the discovery of the Compton effect in 1923, Meitner suggested[51] that here perhaps was the origin of the continuous spectrum: 'Surely the assumption is justified that the γ-rays which emerge from the nucleus will be scattered by electrons in the same atom'. Marie Curie, too, wrote[52] of 'a spectrum the appearance of which would suggest an interpretation by means of the Compton effect'. Both authors remarked that problems would remain even if this effect were important: 'The situation remains unexplained for those special substances which have no γ-rays, like RaE'.[51] In actual fact this so-called internal Compton effect is small, the attenuation length of γ-rays being large compared to typical β-decay source thicknesses.*

In 1924 it was suggested[54] that 'a conceivable explanation of the heterogeneity of the [β]-rays might be found in the assumption that the atoms do not all possess the same probability of decay'. However, an experiment in which the RaE β-spectrum was separated into a high energy and a low energy part followed by lifetime measurements for these separate samples gave the same lifetime in both cases, within the experimental error.[54]

In the fall of 1925 Ellis and William Alfred Wooster reviewed[32] the various proposals for coping with the continuous spectrum and concluded: 'There seems to be no doubt that [this spectrum] exists and that the explanation of its occurrence is not to be sought in any ordinary effect'. They also announced the beginnings of a difficult experiment which, they hoped, would further clarify the situation. This was their idea: imagine one could measure calorimetrically the total energy released per individual β-decay. If the continuous spectrum has an 'ordinary' origin, that is, if there were a unique primary β-energy which is redistributed by 'ordinary' processes (involving electromagnetic radiation for example) then the energy registered in the calorimeter should be the total primary energy, to be identified, they sensibly suggested, with the upper limit of the β-spectrum. Suppose on the other hand that nothing but the electron itself heats up the calorimeter so that the registered energy per decay would be the average over the spectrum of electron energies. That would be 'extraordinary'.

It was.

After two years of hard work Ellis and Wooster announced[55] the result of their experiment. Their source was RaE, chosen because it has practically no line spectrum, so that all complications of internal conversion are avoided. The decays involved are

$$\mathrm{RaE(Bi^{210})} \overset{\beta}{\rightarrow} \mathrm{Po^{210}} \overset{\alpha}{\rightarrow} \text{stable } \mathrm{Pb^{206}} \tag{14.10}$$

* For modern aspects of the internal Compton effect see Ref. 53.

RaE has a half-life ~5 days. Its spectrum has a mean energy ≃0.39 MeV and an upper limit ≃1 MeV. Polonium (half-life ~14 days) emits monochromatic α's with a known energy (~5.2 MeV). They determined the ratio x of the energy release per polonium decay to the average energy per RaE decay. Consistency demands that x be independent of the time t of observation. An $x(t)$ curve was obtained spanning a period of 26 days. After that time only ~3 per cent of the initial RaE remains and the heating effect (with a known extrapolation) is due to polonium only.* Technically, the source was a platinum wire 1 mm in diameter on which RaE is deposited. The wire just fits in the cylindrical hole of a lead calorimeter 13 mm long and 3.5 mm in external diameter. The temperature effects, measured by thermocouples, were $\sim 10^{-3}$ °C. Over the 26 days x remained constant to within 2 per cent. The energy detected by the calorimeter corresponded to 0.34 MeV with an error of about 10 per cent—close to the average spectral energy of RaE, far from the upper limit of the spectrum.

This momentous result ruled out for good any attempt to explain the β-spectrum by 'ordinary' means. The authors stated their main conclusion in these words:[55] 'We may safely generalize this result for radium E to all β-ray bodies and the long controversy about the origin of the continuous spectrum of β-rays appears to be settled'.

A few final remarks.

Regarding the interpretation of their results, Ellis and Wooster made only some hazy comments having to do with then current ideas of Rutherford (no longer of interest) about nuclear structure.

The news from the Cavendish came as 'a great shock' to Meitner.[56] Good physicist that she was, she and Wilhelm Orthmann repeated the experiment. The resulting paper[57] was completed in December 1929. It handsomely acknowledges Ellis and Wooster, reports agreement with their results, and remains silent about what it all may mean.

Ellis continued working on β-radioactivity. Another of his important results will be found in the next section. In 1936 he left the Cavendish 'on warm and friendly terms with their autocratic and sometimes unpredictable leader',[35] to take up the position of Wheatstone Professor at King's College London. (The obituary Ellis wrote[58] about his erstwhile teacher, who died a year later, ranks among the most perceptive writings about Rutherford.) His scientific productivity came essentially to an end. He was knighted in 1946 in recognition of

* Actually they measured the ratio y of the total heating effect of RaE to that of polonium, given by

$$y = \frac{(\lambda_P - \lambda_E)\, e^{-\lambda_E t}}{x\lambda_P(e^{-\lambda_E t} - e^{-\lambda_P t})} \tag{14.11}$$

where λ_P (λ_E) is the decay constant Po (RaE).

his war work as scientific adviser to the British Army Council. After the war he did not return to the academic world but held various scientific advisory posts. Among his later concerns was the influence of smoking on health. Toward the end of his life he destroyed all his letters and papers. In 1981 he entered a nursing home. (His wife had died before him.) He died two weeks later.

(d) New physical laws or new elementary particles? Enter the neutrino

> As soon as we inquire . . . into the constitution of even the simplest nuclei the present formulation of quantum mechanics fails entirely.
>
> N. Bohr, 8 May 1932[59]

> There could exist in the nucleus electrically neutral particles . . .[60]
>
> W. Pauli, 4 December 1930[60]

1. Yet another pitfall of simplicity: one cure for two ailments. Ellis and Wooster's remark (just quoted) on the long controversy about β-spectra being settled was correct only up to a point. It is true that they had silenced for good all claims about discrete primary β-energies. However, neither they nor Meitner and Orthmann had had anything of substance to say about the interpretation of their results. It would be quite interesting to know how rapidly this puzzle became the subject of ardent discussion in the corridors of physics. I regret to have no certain answer to this question. It is my impression, however, that the seriousness of the situation was not immediately widely appreciated. Ellis and Wooster's article[55] appeared in December 1927. Yet in all the literature of 1928 I found only one reference[61] to their paper, due to George Paget Thomson (to whose remarks I shall come back). Pauli's complaint[62] in February 1929, in a letter to Oskar Klein, that Bohr was serving him up all kinds of ideas about β-decay 'by appealing to the Cambridge authorities but without reference to the literature' would indicate that even then he had not yet seen Ellis and Wooster's paper. Pauli's letter also contains the first of the sarcasms found in his correspondence of the next few years regarding Bohr's ideas about β-decay: 'With his considerations about a violation of the energy law Bohr is on a *completely wrong* track' (his italics).

Thus, beginning in 1929, did the meaning of the continuous spectrum become a subject of controversy with Bohr and Pauli as protagonists. The respective causes they championed can be simply stated. From 1929 until not later than 1936 Bohr advocated non-conservation of energy. In 1930 Pauli proposed that in a β-process an as yet undetected particle, now called the neutrino, carries off an amount of energy dictated by the condition of overall energy conservation.

It is imperative to bear in mind that Bohr and Pauli did not just seek an interpretation of β-decay but were, each in his own way, striving for a common cure for two ailments: the continuous spectrum *and* the paradoxes of nuclear spins, magnetic moments, and statistics. Bohr believed that new physical laws were called for, Pauli that the neutrino alone could save all. Neither anticipated that the two ailments demanded two distinct cures: the neutron and the neutrino.

It is curious that even after the discovery of the neutron (1932) everything did not at once fall into place. Leaving that final phase in the unraveling of the nuclear paradoxes until Chapter 17, I now turn first to Bohr's thinking, then to the neutrino as initially conceived by Pauli, and conclude this chapter with Bohr's concession that energy non-conservation was not the right idea.

It will add perspective to recall that in 1924, only a few years before the events to be related, the energy law had been put in doubt for quite different reasons. Also then Bohr was at the centre of debate but, at one time or another, doubts about energy conservation had been expressed by others as well, Einstein, Nernst, and Sommerfeld among them. I digress briefly in order to explain.

2. *Bohr and the energy law, 1924.* Let us return for a moment to those times before quantum mechanics when (as noted earlier*) the particle–wave duality of electromagnetic radiation was already known, a difficulty so profound that it was suggested from various sides to evade light-quanta altogether by assuming that energy is not conserved in individual processes. It was hoped that this sacrifice might reconcile the existence of matter (oscillators,** atoms) in discrete, quantized, energy states with transitions between these states being caused by emission or absorption of continuous (unquantized) electromagnetic radiation—an impossibility if energy were conserved.†

The first (I believe) to hold this view, though very briefly, was Einstein. In 1910 he wrote to Laub: 'At present I have high hopes for solving the radiation problem, and that without light-quanta. I am enormously curious how it will work out. One must renounce the energy principle in its present form'.[66] Three days later he wrote again: it does not work.[67] At the first Solvay conference (1911) he mentioned energy non-conservation as an option to be rejected.[68] The idea did not die, however. In 1916 Nernst suggested (again with quantum theory in mind) that energy conservation is only valid in some statistical sense.[69] In 1919 C. G. Darwin wrote in an unpublished†† manuscript: 'It is impossible to believe that if the science of the present time had not been saturated with the idea of conservation of energy, these [quantum]

* See Chapter 12, Section (a), Part 4.
** Einstein knew already in 1906 that the material harmonic oscillator is quantized.[63,64]
† This idea is discussed in somewhat more detail in Ref. 65.
†† I thank Professor J. M. Sanchez Ron for drawing my attention to this manuscript.

complications would have been avoided by saying that there is no *exact* conservation in such cases'.[70] In 1922 Sommerfeld remarked that the 'mildest cure' for reconciling the wave theory of light with quantum phenomena would be to relinquish energy conservation.[71]

The final and extreme example of this position is the proposal made in 1924 by Bohr, Kramers, and Slater[72] (BKS) that energy conservation as well as causality are only valid statistically in quantum transitions.[65] In October 1924 Pauli informed Bohr[73] that Einstein was much against these ideas, and added his own opinion: 'One cannot prove anything logically and also the available data are not sufficient to decide for or against your view'. This last comment serves to remind us that *until early 1925 there did not exist any experimental proof of energy and momentum conservation in individual physical processes between elementary particles*. Rutherford and coworkers had of course used these laws in analyzing α-particle scattering data. Compton had done the same for the Compton effect. However, experiments of this kind, done with counters of some kind or another, could not refute the possibility that these laws might only hold statistically. Then, early in 1925, a study of Compton scattering in a cloud chamber demonstrated that energy and momentum conservation do hold true in individual events.[74] That and the arrival of quantum mechanics a few months later caused doubts about energy conservation to evaporate.

This interlude is meant as a reminder that in the year 1929, when the debate about energy conservation in β-decay began, experimental information about the validity of the energy law in individual elementary particle processes was limited and of very recent vintage. It may also be well to remember that in the twentieth century β-decay has not been the only cause for doubting conservation of energy. (Recall further the earlier confusion about the origins of radioactive energy, discussed in Chapter 6*.)

3. Bohr and the energy law, 1929. The first time Bohr publicly proposed energy non-conservation in β-decay was in his Faraday lecture[76] on 8 May 1930. We do know, however, that this idea had been on his mind since early 1929. I already mentioned Pauli's February letter[62] in which, incidentally, Pauli gave his own opinion: 'I am rather certain . . . that γ-rays must be the cause of the continuous spectrum'. A few weeks later, Pauli to Bohr: 'Do you intend to mistreat the poor energy law further?'[77] October 1929, Bohr to Mott: 'I am preparing an account on statistics and conservation in quantum mechanics in which I also hope to give convincing arguments for the view that the problem of β-ray expulsion lies outside the reach of the classical conservation principles of energy and momentum'.[78] November, Rutherford to Bohr: 'I have heard that you are on the warpath and wanting to upset the Conservation of Energy both microscopically and macroscopically. I will wait and see before

* The twentieth-century history of the energy law is discussed at more length in Ref. 75.

expressing an opinion but I always feel "there are more things in Heaven and Earth than are dreamt of in our philosophy"'.[79] Meanwhile, in July, Meitner had lectured in Zürich on the calorimetric experiments,[80] causing Pauli to write to Bohr that he now was '*almost* convinced that the β-spectrum *cannot* be explained by secondary processes'.[81]

Further information about Bohr's thoughts is found in a manuscript he prepared in the summer of 1929 but never published.[82] It begins with a critique of recent remarks by G. P. Thomson.[61,83] In a discussion of 'free electrons and their waves' Thomson had concluded* that quantum mechanics implies that 'an electron's speed can never be exactly known . . . the electron *has* no definite velocity . . . the work of Ellis and Wooster on radium E seems to show that there is an uncertainty of this kind'. Bohr stressed[82] that conservation laws are not contradictory but complementary to the space–time description.[84] 'The quantum laws in free space offer no basis for a violation of the conservation principles.' This statement brings us to the root of Bohr's own line of thought. If energy is not conserved then, since the quantum mechanical laws do not provide for this contingency, new physical laws are needed. These, Bohr believed, might be necessary when lengths of the order of the classical electron radius a (see Eq. (14.2)) come into play. He had in mind, one might say, a new sort of complementarity in which a was to play a role as fundamental as Planck's constant in quantum mechanics.** He further conjectured that non-conservation might help explain energy production in the interior of stars via inverse β-processes. It goes without saying that Bohr knew how high the stakes were. 'The loss of the unerring guidance which the conservation principles have hitherto offered in the development of atomic theory would of course be a very disquieting prospect.'[82]

Bohr sent a copy of his manuscript to Pauli[86] who did not like it. 'I must say that your paper has given me *little* satisfaction . . . There is the disagreeable introduction of the electron's diameter . . . I do *not* exactly mean that this is unpermissible but it is a risky business . . . Let the stars radiate in peace!'[87]

Another facet of Bohr's ideas is gleaned from this correspondence with Dirac in November–December 1929. At that time, Klein (then in Copenhagen) had just derived[88] an apparently disquieting consequence of the Dirac equation: a slow electron can pass through a steep potential wall higher than $2mc^2$ and emerge with a negative energy. This correct conclusion, the so-called Klein paradox, was to become harmless in the positron theory. (See the next chapter.) Bohr, impressed by Klein's result, wrote to Dirac inquiring whether 'the difficulties in relativistic quantum mechanics might perhaps be connected with the apparently fundamental difficulties as regards conservation of energy in

* These conclusions resulted from inappropriate manipulations with phase velocities and group velocities.

** The statement repeatedly found in the literature[85] that Bohr's ideas of 1929 are related to his earlier speculations (of 1924) about statistically valid energy conservation is without foundation.

β-ray disintegration and the interior of stars'.[89] Dirac replied: 'I should prefer to keep rigorous conservation of energy at all costs'.[90] (I come back later to his reaction to the problems with the Dirac equation.) In a next letter Bohr tells of his admittedly vague ideas on the limitations of quantum mechanics. 'We can . . . never determine the position of an electron with an accuracy comparable with [its radius, e^2/mc^2].'[91] Dirac to Bohr: 'Although I believe that quantum mechanics has its limitations . . . I cannot see any reason for thinking that quantum mechanics has already reached the limit of its development'.[92] Bohr to Dirac: 'I am . . . inclined to take the present difficulties as an indication that we have not yet obtained the proper expression for the correspondence with the classical electrodynamics'.[93] On that inconclusive note this phase of their correspondence ends.

It should be evident by now that Bohr's visions of new physical laws sprang in fact from three rather than two ailments, all diagnosed when electrons made their appearance in small-distance physics: as nuclear constituents, in β-decay, and in the Klein paradox. Who, in 1929, could have had the wild idea that it would take three new particles, the neutron, the neutrino, and the positron to cure all these diseases?

We are now in a position to appreciate Bohr's comments on nuclear constitution made in his Faraday lecture.[76] He remarked that 'quantum mechanics fails essentially' to explain how four protons and two electrons are held together in the α-particle, the radius of which 'is of the same order as the classical electron diameter'. He refers to Kronig[17] when noting that 'the idea of spin is found not to be applicable to intra-nuclear electrons', and emphasizes the 'remarkable "passivity" of the intra-nuclear electrons in the determination of the statistics'. His final comment concerns β-decay: 'At the present stage of atomic theory we have no argument, either empirical or theoretical, for upholding the energy principle in the case of β-ray disintegrations, and are even led to complications and difficulties in trying to do so . . . Still, just as the account of those aspects of atomic constitution essential for the explanation of the ordinary physical and chemical properties of matter implies a renunciation of the classical idea of causality, the features of atomic stability, still deeper-lying, responsible for the existence and the properties of atomic nuclei, may force us to renounce the very idea of energy balance. I shall not enter further into such speculations and their possible bearing on the much debated question of the source of stellar energy. I have touched upon them here mainly to emphasize that in atomic theory, notwithstanding all the recent progress, we must still be prepared for new surprises'.

Seven months after Bohr's lecture it was Pauli's turn.

4. *Pauli's new particle*. When I first met Pauli, early in 1946 at Niels Bohr's home in Copenhagen, he was kind enough to invite me for dinner the next evening at Krog's fish restaurant. In the course of that meal I witnessed for

the first time his chassidic mode, a gentle rhythmic rocking to and fro of the upper torso. Something was on his mind. He began to talk of his difficulties in finding a physics problem to work on next, adding 'Perhaps that is because I know too much'. Silence; more rocking. Then: 'Do you know much?' I laughed and said, no, I did not know much. Another silence while Pauli seriously considered my reply, then: 'No, perhaps you don't know much, perhaps you don't know much'. A moment later: 'Ich weiss mehr, I know more'. That was said in the Pauli style, without aggression, merely an expression of a statement of fact.

In subsequent years I saw Pauli frequently, in Zürich and in Princeton, until shortly before his death in 1958. We also corresponded. Among my fond memories are the times when a small party, including Pauli and myself, went wood-chopping on the grounds behind the Institute for Advanced Study. I can still see Pauli, wearing mud boots, a grey sweatshirt, and a basque beret, holding a crowbar as if it were a bishop's staff, rocking gently as he watched others saw down a tree.

In our frequent exchanges the early times of the neutrino were never mentioned, we were too engaged in the physics of the day. Nor did I ever discuss that period with Bohr, despite many opportunities. All I learned about Pauli's hypothesis is the result of reading in later years.

In a letter to Max Delbrück written two months before Pauli died, he referred to the neutrino as 'that foolish [närrisch] child of the crisis of my life (1930–1)—which also further behaved foolishly'.[94]* While it is difficult, and often impossible, to grasp cause and effect in human endeavor, most particularly in regard to creativity, I tend to regard Pauli's association between his time of personal turmoil and the moment at which he stated his new postulate as highly significant. Revolutionary steps were out of line with his general character. Indeed, he once said of himself, also late in life: 'When I was young I believed myself to be a revolutionary . . . [but] I was a classicist, not a revolutionary'.[95] From personal knowledge I would agree with this self-assessment. In any event, there is a striking confluence of dates. On 26 November 1930, Pauli's brief and unhappy first marriage ended in divorce.** Our first information about his new hypothesis dates from that same week.

In his letter to Delbrück[94] Pauli recalled: 'The history of that foolish child . . . begins with those vehement discussions about the continuous β-spectrum between [Meitner] and Ellis which at once awakened my interest'. It seems plausible that he would first confide about his new ideas in his friend

* 'Dieses närrische Kind meiner Lebenskrise (1930–1)—das sich auch weiter recht närrisch aufgeführt hat.'

** I owe this information to Hermann and von Meyenn.[96] Not long before, Pauli was shaken by the death of his beloved mother. That was also the brief period during which Pauli drank too much, and during which he was in Jungian analysis. His second marriage, in April 1934 to Franca Bertram, was successful and lasted for the rest of his life.

Heisenberg, who had demonstrated so well that daring thoughts can at times be highly successful. At any rate the earliest reference I know to the new particle is Heisenberg's mention of 'your neutrons' in a letter to Pauli[97] dated 1 December. More details are found in Pauli's letter (its main part follows) of 4 December to a gathering of experts on radioactivity in Tübingen.[60]

Dear radioactive ladies and gentlemen,
I have come upon a desperate way out regarding the 'wrong' statistics of the N- and the Li 6-nuclei, as well as to the continuous β-spectrum, in order to save the 'alternation law' of statistics* and the energy law. To wit, the possibility that there could exist in the nucleus electrically neutral particles, which I shall call neutrons, which have spin 1/2 and satisfy the exclusion principle and which are further distinct from light-quanta in that they do not move with light velocity. The mass of the neutrons should be of the same order of magnitude as the electron mass and in any case not larger than 0.01 times the proton mass.—The continuous β-spectrum would then become understandable from the assumption that in β-decay a neutron is emitted along with the electron, in such a way that the sum of the energies of the neutron and the electron is constant.

There is the further question, which forces act on the neutron? On wave mechanical grounds . . . the most probable model for the neutron seems to me to be that the neutron at rest is a magnetic dipole with a certain moment μ. Experiments seem to demand that the ionizing action of such a neutron cannot be bigger than that of a γ-ray, and so μ may not be larger than $e \times 10^{-13}$ cm.

For the time being I dare not publish anything about this idea and address myself confidentially first to you, dear radioactive ones, with the question how it would be with the experimental proof of such a neutron, if it were to have a penetrating power equal to or about ten times larger than a γ-ray.

I admit that my way out may not seem very probable *a priori* since one would probably have seen the neutrons a long time ago if they exist. But only he who dares wins, and the seriousness of the situation concerning the continuous β-spectrum is illuminated by my honored predecessor, Mr. Debye, who recently said to me in Brussels: 'Oh, it is best not to think about this at all, as with new taxes'. One must therefore discuss seriously every road to salvation.—Thus, dear radioactive ones, examine and judge.—Unfortunately I cannot appear personally in Tübingen since a ball** which takes place in Zürich the night of the sixth to the seventh of December makes my presence here indispensable . . . Your most humble servant, W. Pauli'.

Some thirty years after this letter was written one begins to find theoretical papers in the literature containing predictions of many, sometimes infinitely many, new particles all at once. But this was 1930 when only three particles, the electron, the proton, and the photon were known. Only one of these (the photon) had been predicted on theoretical grounds. Pauli's unwillingness to rush into print is therefore neither a sign of coyness nor of undue reticence. Nor should one consider his use of 'a desperate way out' (an expression he repeated a week later[98]) as overdramatic. As Wigner once told me, his first

* An even (odd) number of spin-1/2 particles has integer (half-integer) spin and satisfies BE (FD) statistics, Wigner's theorem.

** The Italian student ball in the Hotel Baur au Lac.

reaction upon hearing of Pauli's postulate was that this was crazy—but courageous.

Let us look at Pauli's letter more closely. His use of 'neutron' (I shall call it the Pauli neutron for a while) is natural, 'our' neutron was not there yet. It is evident that (as said) Pauli looked upon his particle as the answer to the problems of nuclear constitution as well as of β-decay. He was unable to ban electrons as constituents, however. The Pauli neutron weighed little; protons still had to supply the nuclear mass, electrons still had to compensate the charge. In its original conception his picture of the nucleus is therefore a 'three species model'. The Pauli neutron had to bind to nuclear matter—as was still supposed to be the case for the electron as well—whence his need for a magnetic moment. Writing to Klein, Pauli was quite explicit[98] about this: 'Now it matters importantly which forces act on the neutron since they could not stay in the nucleus if there were no such forces or if these were too weak'. In that letter he proposes the following Dirac-type equation for the neutron as 'almost the only possible model'. ($F_{\mu\nu}$ is the electromagnetic field; I have changed the notation a bit.)

$$\left(\gamma_\mu \frac{\partial}{\partial x^\mu} + i\mu\gamma^\mu\gamma^\nu F_{\mu\nu} + \frac{mc}{\hbar}\right)\psi = 0. \tag{14.12}$$

However, Pauli notes, this equation may lead to trouble: if μ were of the order of an electron magneton then cloud chamber pictures should 'teem with neutrons . . . therefore I myself do not quite believe in neutrons'[98]—another reason for his reticence, his main one I believe. In fact, Pauli adds: 'I have been thinking about a failure of the energy law in case the neutron idea might turn out to be wrong'. Not that he stopped brooding about his own proposal. A month later he wrote to Klein: 'I do not believe that the existence of neutrons is definite, but do think that this hypothesis should be seriously checked or definitely disproved . . . I would be very happy if the momentum balance of individual β-processes could be explored in a direct way. The difficulty lies of course in the determination of the momentum of the recoil atom . . . that would definitely clear up the existence of neutrons. But how to proceed in practice?'[99]

5. *The years 1931–36, a partial chronology*. At the beginning of 1931 the issue: new laws versus a new particle was sharply drawn. A brief account of how the neutrino gained favor over the violation of the energy law concludes this chapter. The format will be a chronology which leads into 1936. Major developments in the intervening years such as the discovery of 'the' neutron, Heisenberg's treatment of the nucleus, and Fermi's theory of β-decay, are not included however; these topics will be left for Chapter 17.

June 1931. On June 16 Pauli spoke in Pasadena on 'Problems of hyperfine structure' at a symposium on 'The present status of the problem of nuclear structure' organized jointly by the American Physical Society and the American Association for the Advancement of Science.[100] On that occasion, Pauli has recalled, 'I reported publicly for the first time on my idea of very penetrating neutral particles . . . it seemed to me to be still quite uncertain, however, and I did not have my lecture printed'.[60] The next day Pauli made (as we Americans say) the New York Times,[101] for the first time: 'A new inhabitant of the heart of the atom was introduced to the world of physics today when Dr W. Pauli of the Institute of Technology in Zürich, Switzerland, postulated the existence of particles or entities which he christened "neutrons"'.*

On his way back from California, Pauli stopped at Ann Arbor where he lectured at the physics summer school, along with Sommerfeld, Kramers, and Oppenheimer. His subject was 'Problems in nuclear physics'. He also gave a colloquium on his new particle. Uhlenbeck told me that there was little discussion afterward. 'I was very impressed but also found it all very strange.'

October 1931. A nuclear physics conference in Rome. At that meeting Goudsmit, who had also spoken in Pasadena, gave a report[103] on hyperfine structure in which he mentioned Pauli's neutrons, 'which might remove present difficulties in nuclear structure and at the same time in the explanation of the β-spectrum'. Just as he was delivering his talk (Goudsmit told me) Pauli made his appearance at the conference. Pauli later recalled two memorable experiences at the Rome meeting: 'Horribile dictu, I had to shake hands with Mussolini', and 'Fermi asked me to talk about my new idea, but I was still cautious and did *not* speak in public . . . only privately'.[104] Nor did Bohr mention Pauli's proposal in his address on 'Atomic stability and conservation laws', a broad assessment of all the problems (described previously) that were exercising him.[16] Regarding the nucleus, 'the experimental evidence [finds] an immediate explanation on the view that all nuclei are built up of protons and electrons . . . the paradoxes of relativistic quantum mechanics . . . here take a most acute form'. He concluded by alluding to his vision of a future theory. 'Just as we have been forced to renounce the ideal of causality in the atomistic interpretation of the ordinary physical and chemical properties of matter, we may be led to further renunciations in order to account for the stability of the atomic constituents themselves'.

January 1932. Discovery of 'the' neutron.

May 1932. In a letter to Rutherford, Bohr expresses the hope[105] that as a result of future experiments 'it would perhaps be possible to settle this fundamental problem . . . about the possible failure of energy conservation'.

* Pauli's recollection[60] that he did not use this name any more at that time is not correct, see also Ref. 102.

July 1932. Fermi on July 7, in a report[106] on 'the current status of the physics of the atomic nucleus', given at a conference in Paris. About nuclear spin and statistics: 'One must never forget that even the fundamental ideas of quantum mechanics are probably no longer applicable in the study of the interior of the nucleus'. Fermi notes that these problems as well as β-decay could find 'a very simple explanation' in terms of Pauli's particle, still called a neutron. (A note added after the completion of the manuscript mentions 'the' neutron.)
April 1933. Pauli to Blackett[107] on the name 'neutrino': 'The italian name (in contrast to neutron) is made by Fermi'; see also Ref. 108.
May 1933. An important paper by Ellis and Mott.[109] Consider the β-decay of a nucleus P (energy E_P) into a nucleus Q (E_Q). 'We make the new assumption that . . . $E_P - E_Q$ is *equal to the upper limit of the* β-ray spectrum.' Armed with this proposition they consider the branching in the thorium series

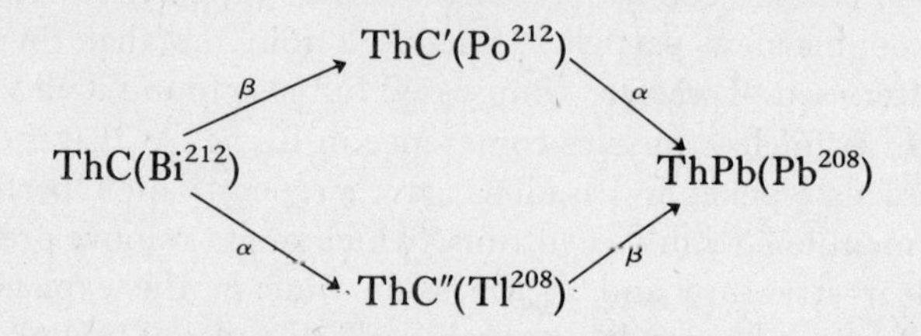

Adding the maximum β-energy to the monochromatic α-energy they find for the ThC–ThPb energy difference: $2.2+8.95=11.15$ MeV via the upper, $6.2+5.0=11.2$ MeV via the lower branch. *Something* involving β-decay energies is conserved![110]* Improved measurements a year later confirm their results.[111]
June 1933. Pauli to Heisenberg: '*Once again* I believe very much in the energy law' (my italics).[113]
July 1933. Pauli to Heisenberg.[114] Stresses that to him the conservation of angular momentum and of statistics are 'almost more important than the conservation laws of energy and momentum'.
October 1933. The seventh Solvay conference.[115] Ellis reports[116] on his work[109] with Mott. Pauli discusses the neutrino. '[Bohr's] hypothesis does not seem satisfactory nor even plausible to me.' He insists on conservation of energy, momentum, angular momentum, and statistics 'in all elementary processes'. He regards it as 'conceivable' that neutrinos have spin 1/2 and obey FD statistics but notes that this has not yet been experimentally verified.[117] There is no longer any mention of the neutrino as a nuclear constituent. Perrin remarks that relativistic kinematics applied to the radium-E spectrum shows 'that the neutrino has *zero intrinsic mass*, like the photon' (his italics).[118]** Bohr

* Also in May: An abortive attempt at a theory of β-decay (based on the Dirac equation) in which energy is not conserved.[112]

** Perrin does not say so, but he must have represented the spectrum by the statistical weight factor—a very bright idea.

comments: 'As long as we have no new experimental data it is wise not to give up the conservation laws, but on the other hand no one knows what surprises may still await us'.[119]

Fermi is present but does not participate in the β-decay discussion. His theory is ready a few months after the conference.

June 1936. Energy nonconservation would no doubt have died a quiet unheralded death had it not been for a last flare-up from an unanticipated direction.

Early in 1936 it was claimed on experimental grounds[120] that there was something wrong after all with the conservation laws as applied to the Compton effect. At this point, Dirac entered the arena.[121] In an article entitled: 'Does conservation of energy hold in atomic processes?' he wrote that these experiments, 'require for their explanation something on the lines of the BKS theory. (See part 2 of this section.) Thus physics is now faced with the prospect of having to make a drastic change in the fundamentals . . .'. He goes on to suggest that energy and momentum are not in general conserved in processes involving large velocities. What should be given up of the existing theory? 'The only important part that we give up is quantum electrodynamics . . . we may give it up without regrets—in fact, on account of its extreme complexity, most physicists will be very glad to see the end of it'.

What underlies these strong statements is above all the despair which was experienced quite widely, though perhaps most keenly (and longer) by no one but Dirac, about the status of quantum field theory. To be sure, the prediction and subsequent discovery of the positron had eliminated many of the problems and paradoxes of the late twenties concerning the Dirac equation—including the Klein paradox. Meanwhile new difficulties had arisen. All these are matters to be discussed at length in later chapters.

Otto Frisch was in Copenhagen in 1936 and has written of how the Compton scattering experiment 'gave rise to agonizing discussions, which made it ever clearer that it could not be reconciled with the advances that physics had made since those early days [of the BKS proposal]'.[122] Fortunately the confusion did not last long. Quite soon a number of further experiments (including one by the original proponent) showed[123] that there was nothing the matter with the conservation laws.*

It was most particularly Dirac's comment which compelled Bohr to take a public stand. He did so in July 1936, in a note[125] entitled 'Conservation laws in quantum theory'. There he refers one last time to the BKS theory, but as a thing of the past: 'The situation was quite different from what it is today'. After commenting on the problems of measurement, he notes that 'the still unsolved difficulties of quantum electrodynamics, emphasized by Dirac in connection with this discussion, can scarcely be attributed to any incompatibility between the foundations of quantum theory and relativity theory'. In 1929

* For later more precise measurements see Ref. 124.

Bohr had challenged quantum mechanics, Dirac had defended it. Now the roles were reversed.

Then, in conclusion, Bohr takes the opportunity to side firmly with the neutrino hypothesis: 'It may be remarked that the grounds for serious doubts as regards the strict validity of the conservation laws in the problem of the emission of β-rays from atomic nuclei are now largely removed by the suggestive agreement between the rapidly increasing experimental evidence regarding β-ray phenomena and the consequences of the neutrino hypothesis of Pauli so remarkably developed in Fermi's theory'.

And so the controversy came to an end. It seemed that in β-decay there was nothing new under the sun as far as conservation principles were concerned.

Twenty years later that view was to change drastically.

Sources

A detailed discussion of the proton–electron model is found in an article by Stuewer.[10] Pauli's recollections about the history of the neutrino are found in Ref. 60. A brief informative article on the neutrino by L. M. Brown is also recommended.[126] Pauli's scientific correspondence[62] is a treasure trove.

References

1. E. Rutherford, *Proc. Roy. Soc. A* **90**, 1914, insert after p. xxiii.
2. E. Rutherford, *Proc. Roy. Soc. A* **123**, 373, 1929.
3. *Nature* **106**, 357, 1920.
4. G. Gamow, *Zeitschr. f. Phys.* **51**, 204, 1928; R. W. Gurney and E. U. Condon, *Nature* **122**, 439, 1928.
5. E. Rutherford, J. Chadwick, and C. D. Ellis, *Radiations from radioactive substances*, Cambridge Univ. Press 1930.
6. Ref. 5, p. 404.
7. G. Gamow, *Constitution of atomic nuclei and radioactivity*, Clarendon Press, Oxford 1931.
8. H. Casimir, *Haphazard reality*, p. 117, Harper and Row, New York 1983; also C. Weiner, interview with G. Gamow, April 25, 1968, Niels Bohr Library, Am. Inst. of Phys. New York.
9. Ref. 7, p. 5.
10. R. H. Stuewer, in *Otto Hahn and the rise of nuclear physics*, Ed. W. R. Shea, p. 19, Reidel, Boston 1983.
11. C. N. da C. Andrade, *Proc. Phys. Soc. London* **36**, 202, 1923.
12. F. Rasetti and E. Fermi, *Nuovo Cim.* **36**, 226, 1926, repr. in *Enrico Fermi collected works*, Vol. 1, p. 212, Univ. of Chicago Press 1962.
13. T. Kuhn, interview with F. Rasetti, April 8, 1963, Niels Bohr Library, American Institute of Physics.
14. R. W. Gurney and E. U. Condon, *Phys. Rev.* **33**, 127, 1929; cf. also J. Kudar, *Zeitschr. f. Phys.* **57**, 259, 1929; **60**, 168, 1930.
15. Cf. H. Bethe and R. Bacher, *Rev. Mod. Phys.* **8**, 82, 1936, Sections 3, 38.
16. N. Bohr, in *Convegno de Fisica Nucleare, October 1931*, p. 119, Reale Accademia d'Italia, Rome 1932.

17. R. de L. Kronig, *Nature* **117**, 550, 1926.
18. R. Frisch and O. Stern, *Zeitschr. f. Phys.* **85**, 4, 1933; I. Esterman and O. Stern, ibid. **85**, 17, 1933; I. I. Rabi, J. M. B. Kellogg, and J. R. Zacharias, *Phys. Rev.* **46**, 157, 1934.
19. G. Herzberg, *Molecular spectra and molecular structure*, 2nd edn, p. 169, van Nostrand, New York, 1950.
20. L. S. Ornstein and W. R. van Wyk, *Zeitschr. f. Phys.* **49**, 315, 1928; also W. R. van Wyk, ibid. **59**, 313, 1930.
21. R. de L. Kronig, *Naturw.* **16**, 335, 1928; also ibid. **18**, 205, 1930; cf. further S. D. Bryden, *Phys. Rev.* **38**, 1989, 1931.
22. R. de L. Kronig and S. Frisch, *Phys. Zeitschr.* **32**, 457, 1931.
23. H. Schüler and H. Benck, *Zeitschr. f. Phys.* **56**, 291, 1929; see also L. Pauling and S. Goudsmit, *The structure of line spectra*, p. 222, footnote, McGraw-Hill, New York 1930.
24. Ref. 19, p. 88.
25. F. Rasetti, *Proc. Nat. Ac. Sci.* **15**, 515, 1929; also *Nature* **123**, 757, 1929; *Phys. Rev.* **34**, 367, 1929; *Zeitschr. f. Phys.* **61**, 598, 1930.
26. J. C. McLennan and J. H. McLeod, *Nature* **123**, 160, 1929.
27. W. Heitler and G. Herzberg, *Naturw.* **17**, 673, 1929.
28. P. Güttinger and W. Pauli, *Zeitschr. f. Phys.* **67**, 743, 1931.
29. J. Dorfman, *Comptes Rendus* **190**, 924, 1930.
30. J. Dorfman, *Zeitschr. f. Phys.* **62**, 90, 1930.
31. V. Ambarzumian and D. Iwanenko, *Comptes Rendus* **190**, 582, 1930.
32. C. D. Ellis and W. A. Wooster, *Proc. Cambr. Phil. Soc.* **22**, 849, 1925.
33. E. Rutherford, *Nature* **95**, 494, 1915.
34. J. Danysz, *Comptes Rendus de la Soc. des. Sc. de Varsovie*, Fasc. 8, 1916.
35. K. Hutchison, J. Gray, and H. Massey, *Obit. notices Fell. Roy. Soc.* **27**, 199, 1981.
36. C. D. Ellis, *Proc. Roy. Soc. A* **99**, 261, 1921.
37. E. Rutherford, H. Robinson, and W. F. Rawlinson, *Proc. Roy. Soc. A* **28**, 281, 1914.
38. C. D. Ellis, *Proc. Roy. Soc. A* **101**, 1, 1922.
39. C. D. Ellis and H. W. B. Skinner, *Proc. Roy. Soc. A* **105**, 60, 165, 185, 1924.
40. L. Meitner, *Zeitschr. f. Phys.* **9**, 131, 1922.
41. L. Meitner, *Zeitschr. f. Phys.* **9**, 145, 1922.
42. A. Smekal, *Zeitschr. f. Phys.* **10**, 275, 1922.
43. Ref. 40, p. 131, footnote.
44. C. D. Ellis, *Proc. Cambr. Phil. Soc.* **21**, 121, 1922.
45. C. D. Ellis, *Zeitschr. f. Phys.* **10**, 303, 1922.
46. L. Meitner, *Zeitschr. f. Phys.* **11**, 35, 1922.
47. L. Meitner, Ref. 40 and *Zeitschr. f. Phys.* **26**, 169, 1924; **34**, 807, 1925.
48. J. Chadwick and C. D. Ellis, *Proc. Cambr. Phil. Soc.* **21**, 274, 1922.
49. J. Chadwick, *Verh. d. Deutsch. Phys. Ges.* **16**, 383, 1914.
50. Cf. O. Hahn and L. Meitner, *Zeitschr. f. Phys.* **26**, 161, 1924; L. Meitner, *Zeitschr. f. Phys.* **26**, 169, 1924.
51. L. Meitner, *Zeitschr. f. Phys.* **19**, 307, 1923.
52. M. Curie, *J. de Phys. et le Rad.* **7**, 97, 1926.
53. B. G. Petterson, in *Alpha-, beta-, and gamma-ray spectroscopy*, Ed. K. Siegbahn, Vol. 2, p. 1579, North Holland, Amsterdam 1965.
54. L. Bastings, *Phil. Mag.* **48**, 1075, 1924.
55. C. D. Ellis and W. A. Wooster, *Proc. Roy. Soc. A* **117**, 109, 1927.

56. O. Frisch, *Biogr. Mem. Fell. Roy. Soc.* **16**, 408, 1970.
57. L. Meitner and W. Orthmann, *Zeitschr. f. Phys.* **60**, 143, 1930.
58. C. D. Ellis, *Proc. Phys. Soc. London* **50**, 463, 1938.
59. N. Bohr, *J. Chem. Soc.* **135**, 349, 1932.
60. W. Pauli, letter to a physicists' gathering at Tübingen, December 4, 1930, repr. in *W. Pauli, collected scientific papers*, Eds. R. Kronig and V. Weisskopf, Vol. 2, p. 1313, Interscience, New York 1964.
61. G. P. Thomson, *Nature* **122**, 279, 1928.
62. W. Pauli, letter to O. Klein, February 18, 1929, repr. in *Wolfgang Pauli, scientific correspondence* (referred to as '*PC*' below), Eds. A. Hermann, K. von Meyenn, and V. Weisskopf, Vol. 1, p. 488, Springer, New York 1979.
63. A. Einstein, *Ann. der Phys.* **20**, 199, 1906.
64. A. Pais, *Subtle is the Lord*, Chapter 19, Section (d), Oxford Univ. Press 1982.
65. Ref. 64, Chapter 22.
66. A. Einstein, letter to J. J. Laub, November 4, 1910.
67. A. Einstein, letter to J. J. Laub, November 7, 1910.
68. A. Einstein, *Proc. first Solvay conference*, pp. 429, 436, Gauthier-Villars, Paris 1912.
69. W. Nernst, *Verh. Deutsch. Phys. Ges.* **18**, 83, 1916.
70. C. G. Darwin, Handwritten manuscript dated July 1919, copy in Niels Bohr Library, Am. Inst. Phys., New York.
71. A. Sommerfeld, *Atombau und Spektrallinien*, 3rd edn, p. 311, Vieweg, Braunschweig 1922.
72. N. Bohr, H. A. Kramers, and J. C. Slater, *Phil. Mag.* **47**, 745, 1925.
73. W. Pauli, letter to N. Bohr, October 2, 1924; *PC*, Vol. 1, p. 163.
74. A. H. Compton and A. W. Simon, *Phys. Rev.* **26**, 889, 1925.
75. A. Pais, 'Conservation of energy', in Proc. conf. on the history of scientific ideas, Sant Felice de Guixols, Spain, September 1983, to be published.
76. N. Bohr, *J. Chem. Soc.* **135**, 349, 1932.
77. W. Pauli, letter to N. Bohr, March 5, 1929, *PC*, Vol. 1, p. 493; cf. also W. Heisenberg and W. Pauli, *Zeitschr. f. Phys.* **56**, 61, 1929.
78. N. Bohr, letter to N. F. Mott, October 1, 1929, copy in Niels Bohr Library, Am. Inst. Phys. New York.
79. E. Rutherford, letter to N. Bohr, November 19, 1929, copy in Niels Bohr library, Am. Inst. Phys. New York.
80. L. Meitner, *Phys. Zeitschr.* **30**, 515, 1929.
81. W. Pauli, letter to N. Bohr, July 17, 1929, *PC*, Vol. 1, p. 512.
82. N. Bohr, *β-ray spectra and energy conservation*, unpublished, dated summer 1929, copy in Niels Bohr Library, Am. Inst. Phys., New York.
83. G. P. Thomson, *Phil. Mag.* **7**, 405, 1929.
84. Cf. N. Bohr, *Nature* **121**, 580, 1928, esp. p. 590.
85. Cf. F. G. Houtermans, *Erg. Ex. Naturw.* **9**, 123, 1930; C. Enz, in *The physicist's conception of nature*, Ed. J. Mehra, p. 785, Reidel, Boston 1973.
86. N. Bohr, letter to W. Pauli, July 1, 1929, *PC*, Vol. 1, p. 507.
87. W. Pauli, letter to N. Bohr, July 17, 1929, *PC*, Vol. 1, p. 512.
88. O. Klein, *Zeitschr. f. Phys.* **53**, 157, 1929.
89. N. Bohr, letter to P. A. M. Dirac, November 24, 1929, copy in Niels Bohr Library, Am. Inst. Phys. New York.
90. P. A. M. Dirac, letter to N. Bohr, November 26, 1929, copy in Niels Bohr Library, Am. Inst. Phys. New York.

91. N. Bohr, letter to P. A. M. Dirac, December 5, 1929, copy in Niels Bohr Library, Am. Inst. Phys., New York.
92. P. A. M. Dirac, letter to N. Bohr, December 9, 1929, copy in Niels Bohr Library, Am. Inst. Phys., New York.
93. N. Bohr, letter to P. A. M. Dirac, December 23, 1929, copy in Niels Bohr Library, Am. Inst. Phys., New York.
94. W. Pauli, letter to M. Delbrück, October 6, 1958, *PC*, in preparation.
95. Statement by W. Pauli to R. Jost (Jost, private communication).
96. A. Hermann and K. von Meyenn, *PC*, vol. 2, Springer, New York 1985.
97. W. Heisenberg, letter to W. Pauli, December 1, 1930, *PC*, Vol. 2, p. 37.
98. W. Pauli, letter to O. Klein, December 12, 1930, *PC*, Vol. 2, p. 43.
99. W. Pauli, letter to O. Klein, January 8, 1931, *PC*, Vol. 2, p. 51; cf. also W. Pauli, *Phys. Zeitschr.* **32**, 664, 1931.
100. W. Pauli, *Phys. Rev.* **38**, 579, 1931.
101. *New York Times*, June 17, 1931.
102. *Science* **74**, 111, 1931.
103. S. Goudsmit, Ref. 16, p. 41.
104. W. Pauli, letter to F. Rasetti, October 6, 1956, copy in Niels Bohr Library, Am. Inst. Phys., New York.
105. N. Bohr, letter to E. Rutherford, May 2, 1932, copy in Niels Bohr Library, Am. Inst. Phys., New York.
106. E. Fermi, in *Comptes rendus du congrès international d'électricité*, Ed. R. de Valbreuze, 1st section, p. 798, Gauthier-Villars, Paris, 1932.
107. W. Pauli, letter to P. M. S. Blackett, April 19, 1933, *PC*, Vol. 2, p. 158.
108. G. Gamow, *Thirty years that shook physics*, p. 141, Doubleday, New York 1966.
109. C. D. Ellis and N. F. Mott, *Proc. Roy. Soc. A* **141**, 502, 1933.
110. Cf. also G. Beck, *Nature* **132**, 967, 1933.
111. W. J. Henderson, *Proc. Roy. Soc. A* **147**, 572, 1934.
112. G. Beck, *Zeitschr. f. Phys.* **83**, 498, 1933; G. Beck and K. Sitte, *Zeitschr. f. Phys.* **86**, 105, 1933.
113. W. Pauli, letter to W. Heisenberg, June 2, 1933, *PC*, Vol. 2, p. 166.
114. W. Pauli, letter to W. Heisenberg, July 14, 1933, *PC*, Vol. 2, p. 184.
115. *Structure et propriétés des noyaux atomiques*, Gauthier-Villars, Paris 1934.
116. Ref. 115, p. 284.
117. Ref. 115, pp. 324–5.
118. Ref. 115, p. 327; also F. Perrin, *Comptes Rendus* **197**, 1625, 1933.
119. Ref. 115, p. 328.
120. R. S. Shankland, *Phys. Rev.* **49**, 8, 1936.
121. P. A. M. Dirac, *Nature* **137**, 298, 1936; cf. also E. J. Williams, ibid., p. 614; R. Peierls, ibid., 904.
122. O. Frisch, in *Niels Bohr*, Ed. S. Rozental, p. 139, North Holland, Amsterdam 1967.
123. H. R. Crane, E. R. Gaerttner, and J. J. Turin, *Phys. Rev.* **50**, 302, 1936; J. C. Jacobsen, *Nature* **138**, 25, 1936; W. Bothe and H. Maier–Leibnitz, *Zeitschr. f. Phys.* **102**, 143, 1936; R. S. Shankland, *Phys. Rev.* **52**, 414, 1937.
124. W. G. Cross and N. Ramsay, *Phys. Rev.* **80**, 929, 1950.
125. N. Bohr, *Nature* **138**, 25, 1936.
126. L. M. Brown, *Physics Today*, September 1978, p. 23.

15

Quantum fields, or how particles are made and how they disappear

(a) The end of the game of marbles

> The properties of elementary processes . . . make it almost seem inevitable to formulate a truly quantized theory of radiation.
>
> A. Einstein in 1917[1]

In Chapter 1 I gave reasons for dividing the present account into a history and a memoir. This book separates into two parts in a different way as well: the years before, and those after, the birth of quantum field theory. With the exception of parts of Chapter 14 all the preceding dealt with the years before, the period up till late 1926, when relativity and quantum mechanics had not yet begun to play their joint role in the theory of particles and fields. That impact began with the arrival of quantum field theory when particle physics acquired, one might say, its own unique language. From then on particle theory becomes much more technical and much more focussed. A new central theme emerges: how good are the predictions of quantum field theory? That is the principal issue of the rest of this book, beginning with the present chapter. As the reader will see, confusion and insight will continue to alternate unabated, but these ups and downs will from here on mainly occur within a tight theoretical framework, the quantum theory of fields.

At a moment which (quantum mechanics tell us) cannot be predicted an excited atom makes a transition to its ground state by emitting a photon. Where was the photon before that time? It was not anywhere; it was created in the act of transition.

At a moment which cannot be predicted a β-radioactive nucleus decays into another nucleus, an electron, and a neutrino. Where were the electron and neutrino before that time? They were not anywhere; they were created in the act of β-disintegration.

An atom absorbs a photon and goes into an excited state. Where is the photon after absorption? It is not anywhere; it is extinct, annihilated.

Is there a theoretical framework for describing how particles are made and how they vanish? There is: quantum field theory. It is a language, a technique, for calculating the probabilities of creation, annihilation, scatterings of all sorts of particles: photons, electrons, positrons, protons, mesons, others, by methods which to date invariably have the character of successive approximations. No rigorous expressions for the probability of any of the above-mentioned processes has ever been obtained. The same is true for the corrections, demanded by quantum field theory, to the positions of energy levels of bound-state systems. There is still a Schroedinger equation for the hydrogen atom, but it is no longer exactly soluble in quantum field theory. In fact, in a sense to be described, the hydrogen atom can no longer be considered to consist of just one proton and one electron. Rather it contains infinitely many particles.

In quantum field theory the postulates of special relativity and of quantum mechanics are taken over unaltered, and brought to a synthesis which perhaps is not yet perfect but which indubitably constitutes a definitive step forward. It is also a theory which so far has not yielded to attempts at unifying the axioms of general relativity with those of quantum mechanics. Is quantum field theory the ultimate framework for understanding the structure of matter and the description of elementary processes? Perhaps, perhaps not.

In its relatively short life, a little more than half a century, this theory, or rather this class of theories, has gone through several upheavals. The beginnings (late twenties, early thirties) were splendid. Theoretical predictions for processes like Compton scattering (the Klein–Nishina formula), the production of electron–positron pairs, and others, agreed very well with experiment as long, however, as calculations stopped short at the first approximation (to be defined in detail later on). 'Corrections' to this approximation proved to be calamitous. The mass and charge of the electron, probabilities for any electromagnetic process one can name, all these quantities turned out to be infinitely large. No wonder therefore that in the 1930s the whole approach was viewed with deep distrust, especially by Dirac, its founder, and believed to be seriously wrong. In the late forties a new method was developed, the renormalization program, for bypassing these infinities and for extracting physical information from the higher corrections. As a result, it came about that one was dealing with a theory which, while incomplete, was far better than had been realized before, capable of producing agreement with experiment to an unparalleled degree of accuracy—but only in so far as electromagnetic processes are concerned.

Meanwhile the first mesons had been discovered and a new branch of the theory began to develop, meson field theory. Hopes were briefly high in the late forties that what worked so well for photons in interaction would do likewise for mesons. For reasons to be explained later, these illusions were rapidly quashed. Once again enthusiasm was followed by despair. Through much of the fifties and sixties many experts considered quantum field theory a lost cause in so far as nuclear processes (meson scattering, nuclear forces, etc.) were concerned.

From that same period date the beginnings of a new endeavor: axiomatic field theory. Its practitioners asked: Do we know what we mean when we speak of quantum field theory? Have we perhaps been too uncritical in our procedures, which basically consist in applying the rules of quantum mechanics for systems with a finite number of degrees of freedom to fields where this number is invariably infinite? They answered no to the first question, yes to the second, and proceeded to introduce and study additional axioms (refined as time went by) which do not contravene the principles of 'ordinary' quantum mechanics (and of special relativity), but which supplement them in order to cope with field problems. During this very period of despair, in some quarters the axiomatic method began to flourish and yield highly non-trivial and potent results—which, however, did not (and still do not) help one bit in calculating those quantities which can be confronted with laboratory measurements. The gap thus created between the axiomatic and the less rigorous and more commonly followed paths has narrowed very considerably in the course of time but has not closed as yet.

In the seventies quantum field theory (of the common variety) once again manifested unexpected new vitality in its application to such subjects as electroweak unification, quantum chromodynamics, and others.

In the preceding few paragraphs I have laid out very briefly the panorama of much of the rest of this book. I now turn to the far more specific main topic of this chapter, the very beginnings of quantum electrodynamics. Since this oldest branch of quantum field theory is the only one to grow out of long familiar classical field theory, and since, in a sense, it has its beginnings in the old quantum theory, I begin with a brief look back at old times.

It is not quite precise but not far off the mark to consider the beginnings of modern chemistry, dating from the early nineteenth century, as the time of transition between the eras of speculation, going back to antiquity, and rigorous scientific inquiry regarding the finer structure of matter. Amidst often intense controversy among chemists and among physicists, fraternities still largely separate, evidence grew slowly but inexorably that atoms had to be taken seriously.* Some of those who had the faith began making estimates of the two parameters which, it seemed, fully specify a given atomic species: atomic weight and atomic size. Thus, with some results that lasted, began the serious game of marbles, in which matter is conceived to consist of tiny spheres which collide, link, or disconnect but which are immutable by themselves, 'the only material things which remain in the precise condition in which they first began to exist', in Maxwell's phrase.[3]

* Ref. 2 contains a brief account of these developments as well as references to more detailed monographs on this subject.

As the century wore on it became increasingly evident that the extrapolation from the marbles of childhood to the tiny toys of physics and chemistry was faulty. Spectroscopy indicated that atoms have movable parts; radioactivity showed that not all atoms remain in the same precise condition; the discoveries of the electron and the ionization mechanism made manifest the existence of smaller material subunits. Yet the shooting of marbles continued, the electron now being the tiniest new plaything, labeled by its weight m, its charge e, and its radius a defined by (c is the velocity of light)

$$a = \frac{e^2}{mc^2}, \tag{15.1}$$

the classical electron radius so beloved by Lorentz. As will be discussed in the next chapter, one of the advances of quantum field theory is to have rendered obsolete all arguments for assigning the finite extent a to the electron.

In some games, success depends on your ability to shoot your opponent's marble out of the way. This process happens most often without breakage (elastic collisions) but occasionally target or projectile or both may fragment (inelastic collisions). Either way, energy, momentum, and angular momentum are conserved. So, to an incredibly good approximation, is mass. All these laws for macroscopic bodies applied equally well to the small particles—until 1905, when the rules of the game were refined by a new kinematics, special relativity. Momentum and angular momentum conservation remained in force, but the recognition that mass times c^2 is a special mode of energy resulted in the merger of the two conservation laws of energy and mass into one. The connection between energy E, momentum p, and mass M of a free particle took on the new form

$$E^2 = c^2p^2 + M^2c^4 \tag{15.2}$$

corresponding to which mass was defined more sharply as the energy/c^2 of a particle observed at rest.

Since special relativity tells us that matter is but one among many forms of energy, it should be possible, under suitable conditions, to convert some matter into some other kind of energy. Einstein realized at once that radioactive energy release might thus be explained. Similarly, sensible speculation that annihilation of matter might result in large release of radiation and thus solve the vexing problem of stellar energy's origin has a long history. All these were phenomenological considerations without dynamical underpinnings, however.

No less momentous for the evolution of particle physics than Einstein's introduction of special relativity in 1905 was his light-quantum proposal of that same year. I have already indicated how strenuously physicists tried to deny that there was any need for discrete parcels of electromagnetic energy and how long it took before the light-quantum was understood to be a particle, the photon, participating in collision processes by the rules of energy, momentum, and angular momentum conservation, like a good old marble.* Recall that the turning point was the Compton effect, the elastic scattering of photons off electrons. The game of marbles came to an end, however, when the question arose of how photons 'are made'; for example, how does an excited atom eject a photon.

That issue could first have been raised in 1913 by Bohr when he gave his theory of the hydrogen atom. Bohr, however, concentrating on the dynamics of stationary atomic states, initially left aside the problems of the detailed mechanism by which photons are produced. (We saw in the previous chapter that he was in fact among the last to seek an alternative to the photon picture.) The first serious attempt to build a bridge between the quantum theory of atomic states and the quantum theory of radiation came almost four years later when Einstein, forever pondering the meaning of Planck's blackbody radiation law, gave his phenomenological discussion of spontaneous and induced radiative transitions.[1,5,6] In the course of that work he raised some questions which form the very point of departure for quantum field theory. Let us see what they were.

Einstein studied the thermal equilibrium of an atomic gas in interaction with electromagnetic radiation. Let $E_m > E_n$ be a pair of atomic energy levels occupied by N_m, N_n atoms respectively. Further let $\rho\, \mathrm{d}\nu$ be the equilibrium energy of the radiation per unit volume and in the frequency interval $\mathrm{d}\nu$. Einstein assumed that the number $\mathrm{d}W$ of atomic transitions per time interval $\mathrm{d}t$ is given by

$$\begin{aligned} \mathrm{d}W_{m\to n} &= N_m(\rho B_{mn} + A_{mn})\,\mathrm{d}t \\ \mathrm{d}W_{n\to m} &= N_n \rho B_{nm}\,\mathrm{d}t. \end{aligned} \tag{15.3}$$

A_{mn}, B_{mn}, B_{nm} are known as the coefficients of spontaneous and induced emission and induced absorption respectively. About the A's and B's Einstein remarked: 'The constants A and B could be computed directly if we were to possess an electrodynamics and mechanics modified in the sense of the quantum hypothesis'[5]—his equations were meant as an Ansatz, not as the first principles of a new theory. How Einstein, without a theory for A and B, nevertheless managed to derive Planck's law with the help of other plausible ingredients need not concern us here,[7] except for one intermediate result: he obtained besides $B_{mn} = B_{nm}$

$$A_{mn} = \frac{8\pi h\nu^3}{c^3} B_{mn}. \tag{15.4}$$

In the course of these investigations Einstein also showed (by analyzing the Brownian motion of the atoms due to the presence of radiation) that, in

* See Chapter 7, Part 5; Chapter 12, Section (a); Chapter 14, Section (d) Part 2, and also Ref. 4.

spontaneous emission, the photon is born not only with an energy $h\nu = E_m - E_n$ but also with a momentum $h\nu/c$ in some direction. These implications rightly troubled him. Neither the moment of birth of the photon nor the direction in which it takes off were dictated by his arguments: '[It is] a weakness of the theory that it leaves time and direction of elementary processes to chance'.[7] New first principles were called for. 'The properties of elementary processes . . . make it almost seem inevitable to formulate a truly quantized theory of radiation.'[1]

It is not a weakness but a strength of quantum mechanics that it ineluctably leaves the time and direction of certain elementary processes to chance. Individual events do not obey the classical principle of causality. That much became gradually clear after Heisenberg and Schroedinger had given us, in Einstein's words, 'a mechanics modified in the sense of the quantum hypothesis'. As Einstein had correctly foreseen, however, that is a necessary but not a sufficient condition for calculating his B- and, especially, his A-coefficients. To that end electrodynamics had to be modified as well.

That last step marks the beginning of quantum field theory.

The earliest reference to quantum electrodynamics dates from the fall of 1925. Already in the second paper ever written on quantum mechanics there is mention of 'matrix electrodynamics'. The first clear statement of the physical idea known as field quantization or second quantization* also dates from 1925. These early contributions from Goettingen are treated in Section (b). It was not until late 1926, however, that Dirac, then in Copenhagen, laid the systematic foundations of quantum field theory in a paper which concludes with the calculations of the A- and B-coefficients from first principles. At first Dirac, the central figure of this chapter, thought that his new theory forbade photon scattering. Second-order perturbation theory did not yet exist! So he invented it and gave the theory of photon scattering in a sequel to his previous article. These two papers are discussed in Section (c). Then follows the connection between second quantization and quantum statistics, Section (d). Section (e) covers the earliest work on the relativistic and gauge invariance of the theory. We salute the principal characters: Pascual Jordan who first saw second quantization, both for BE and FD statistics; Dirac who created its technical language; Jordan, Heisenberg, and Pauli who pioneered the relativistic formulation of the theory; Walter Gordon who first wrote down the gauge invariance principle in its modern form; Hermann Weyl who first saw the connection between gauge invariance and charge conservation; and Heisenberg and Pauli who used both relativistic and gauge invariance in fashioning the theoretical tool which would be used almost exclusively in the nineteen-thirties and which goes by the name of Coulomb gauge.

* I do not much care for this second name since it may give the erroneous impression that the quantum of action is introduced twice in the same theory.

Section (f) deals with a subject not (or rather, not yet at that time) connected with the preceding. It is a continuation of the account, begun in Chapter 13, Section (e), of the Dirac equation of the electron. The topics discussed are the diagnosis of apparent difficulties raised by the negative energy solutions to that equation, Dirac's proposal for resolving these problems by the positron postulate, and his realization (first expressed in a letter to Bohr in December 1929) that the scattering of a photon on an electron is not a two-body but an infinitely-many-body problem. The section ends with Anderson's discovery of the positron. It may well be[7a] that others had seen positrons before him. But then many people had seen apples fall from trees before Newton discovered the law of gravitation.

It may fairly be said that the theoretical approach to the structure of matter began its age of maturity with Dirac's two papers published in early 1927. It is true that the surface had barely been scratched at that time. There were still only three particles: the electron, the proton, and the photon; and only a single quantum field: the electromagnetic field. Furthermore, by present standards the new theoretical framework, as it was developed in the late twenties, looks somewhat primitive. Nevertheless the principal foundations had been laid by then for much that has happened in particle theory to this day.

Many years after these events Heisenberg, himself one of the founding fathers of quantum field theory, was asked what the early days of quantum electrodynamics had been like. He replied: 'You know, it was not like quantum mechanics. In quantum mechanics everything came out much simpler and much better than what I expected. Somehow when you touched it and you had a disagreeable difficulty, at the end you saw "Well, was it that simple?" Here in electrodynamics, it didn't become simple. Well, you could do the theory, but still it never became that simple'.[8]

So it is to this day, and it will never be otherwise. In the beginning it may not have been that easy to assimilate the concepts of special relativity or non-relativistic quantum mechanics. Yet, at a technical level, these theories are like child's play compared with quantum field theory, which has developed into a discipline that makes use of mathematical tools which, even to the practitioner, are oftentimes formidably complex. As time went by, quantum field theory developed, so to speak, a life of its own. Important progress was made in later years, not just because of surprising new experimental discoveries, not just because of the introduction of qualitatively new theoretical ideas, but also because it turned out on several occasions that the content of quantum field theory was much richer than expected—in some ways similar to what has happened in the classical theory of general relativity. New insights emerged not only from new equations but also from a better grasp on handling old equations.

There can be no question but that a history of our penetration into the mysteries of matter must take cognizance of this inner life of quantum field theory. This inevitably raises issues of a highly technical nature. I shall address these, but only in so far as (in my judgment) they bear closely on the increase in our understanding of what particles are, and what are the forces acting on them. Thus what follows is not a full-fledged history of quantum field theory. Yet, I hope, it is more than a cursory essay on that subject. In an endeavor to accommodate those readers who may be less interested in technicalities, I place some of these in an Appendix, Section (g). Furthermore I can legitimately ease my task by often referring to textbooks which are essentially contemporaneous with the present volume. This chapter concludes with a listing of those sources on the history of quantum field theory prior to renormalization which I found particularly helpful.

(b) Preludes: Goettingen

Quantum mechanics was barely two months old when the first intimations of quantum electrodynamics appeared in the literature, to wit in Born and Jordan's paper[9,10] of September 1925, the one in which they also showed that Heisenberg's antics amounted to a matrix mechanics. In the final section of that paper we read: 'Electromagnetic processes in the vacuum can be represented by a superposition of plane waves. We shall consider the electric and magnetic field strengths $\vec{E}$, $\vec{H}$ in such plane waves as *matrices* the elements of which are harmonically vibrating plane waves; for example, for a suitable choice of coordinate system

$$E = E_{nm} \exp 2\pi i \nu_{nm}\left(t - \frac{x}{c}\right) \tag{15.5}$$

. . . The Maxwell equations shall be retained as matrix equations'. This, the first equation ever published on what its authors call 'matrix electrodynamics', is of historical interest. Born and Jordan do not give any hint, however, as to what the E_{nm} and ν_{nm} might be. All they do is note some properties (Appendix, Note 1) of the energy density operator W for a pure radiation field:

$$W = \tfrac{1}{2}(\vec{E}^2 + \vec{H}^2) \tag{15.6}$$

where the fields $\vec{E}$ and $\vec{H}$ may be written as

$$\vec{E} = -\frac{1}{c}\frac{\partial \vec{A}^{tr}}{\partial t} \tag{15.7}$$

$$\vec{H} = \operatorname{curl} \vec{A}^{tr}. \tag{15.8}$$

Here $\vec{A}^{tr}$ is the transverse vector potential, that is

$$\operatorname{div} \vec{A}^{tr} = 0. \tag{15.9}$$

Only one month had gone by when Heisenberg wrote to Pauli: 'Something done by Jordan for our paper is the calculation of the statistical behavior of proper vibrations . . . in the new theory. Jordan claims that one gets the right answer [for energy fluctuations, see below] . . . and believes to see an analogy between our calculations and Bose statistics. I am a bit unhappy about that since I do not know enough statistics to judge if there is something to it'.[11] The paper[12] to which Heisenberg referred was the Born–Heisenberg–Jordan collaboration, completed in November 1925. Taking Heisenberg's words at face value, it would appear that the first vision of second quantization stems from Jordan.*

The statistical problem discussed in the three-man paper had first been raised[14,15] in 1909 by Einstein: what are the energy fluctuations in a subvolume of a cavity filled with electromagnetic radiation in thermal equilibrium? In deriving a formula (Appendix, Note 2) for these fluctuations Einstein had used Planck's radiation law as an input. Now, in 1925, the issue was: can one do without this input and derive the same result from first quantum mechanical principles?

For an appreciation of the jump which Born, Heisenberg, and Jordan made it is most helpful to begin with their critique of earlier work by Debye. Accordingly, I take the reader back, for the last time in this book, to the old quantum theory.

While reanalyzing Planck's own derivation of his law, Einstein had shown in 1906 that the energy of a material oscillator 'can take on only those values that are integer multiples of $h\nu$; in emission and absorption [this] energy changes by jumps which are multiples of $h\nu$'.[16] This is the oldest example of quantum rules for a simple dynamical system. In 1910, Debye applied Einstein's result to the set of uncoupled oscillators which represent the electromagnetic field vibrations in a cavity; that is, he introduced the prescription[17] that the only allowed energies of such an oscillator with fundamental frequency ν are given by

$$E_n = nh\nu, \qquad n = 0, 1, 2 \ldots \tag{15.10}$$

From this assumption he derived Planck's law in a few lines (Appendix, Note 3). It appeared equally straightforward to use the same method for deriving Einstein's fluctuation formula (Appendix, Note 3). In 1919 this was indeed tried but with disastrous results: the answer could not be recovered this way.[18]

Thus when, in 1925, Born, Heisenberg, and Jordan were able to obtain Einstein's result from quantum mechanics, they rightly considered this as 'particularly encouraging for the further development of the theory'.[12] Two years later, Pauli began his first paper on quantum electrodynamics with the remark that this result 'made it appear certain that essential advances should

* Jordan had worked (less successfully) on the old quantum theory of radiation a year earlier.[13]

be achievable along these lines'.[19] Uhlenbeck told me that in Leiden also this treatment of fluctuations was ranked among the early successes of quantum mechanics.*

Even more important, it seems to me, than the discussion of fluctuations is their remark, made almost in passing, about a reinterpretation of the formalism. 'It seems to us that [Debye's] blend of wave theoretical and light-quantum-like concepts hardly corresponds to the essence of the problem.' Instead they propose a new interpretation of Eq. (15.13) which, with suitable adaptation of the label k, applies of course just as well to the three-dimensional case of radiation oscillators. '*The quantum number* $[n_k]$ *of an oscillator is equal to the number of quanta with the corresponding* $[\nu]$.'**

Let us have a closer look at this simple yet quite extraordinary remark which opens the door to quantum field theory.

Step one: consider one single particle with coordinate q_k, momentum p_k, with energies labelled by n_k. Step two: *second quantization*. Reinterpret n_k *so as to refer not to one but to* n_k *particles, photons. One has jumped from a one-body problem* (*better*: *from a problem with one degree of freedom*) *to an* n_k*-body problem where* n_k *is variable*. To distinct energy levels of the initial oscillator correspond (they propose) distinct numbers of photons. In the new interpretation a transition from one level to another must therefore mean that particles with energy $h\nu$ are either made or else disappear.

* Actually the three men treated a one-dimensional toy model, the energy fluctuations in a segment of a vibrating string with length l and fixed end points, writing the string displacement $u(x, t)$ at point x and time t as

$$u(x,t)=\sum_{k=1}^{\infty} q_k(t)\sin \nu_k x$$
$$\nu_k = k\frac{\pi}{l}, \qquad k=1,2,\ldots, \tag{15.11}$$

which reduces the problem to that of an infinite set of uncoupled oscillators with coordinates $q_k(t)$. In suitable units the total string energy is

$$H=\tfrac{1}{2}\sum_k (p_k^2+\omega^2 q_k^2), \qquad \omega=2\pi\nu \tag{15.12}$$

or, in diagonal form

$$H=\sum_k (n_k+\tfrac{1}{2})h\nu. \tag{15.13}$$

The 1/2 term is the zero-point energy. $p_k = \mathrm{d}q_k/\mathrm{d}t$ is the momentum conjugate to q_k, so that Eq. (12.11) generalizes to

$$[p_i(t), q_j(t)]=\frac{h}{i}\delta_{ij}, \qquad [p_i(t), p_j(t)]=[q_i(t), q_j(t)]=0 \tag{15.14}$$

($\delta_{ij}=1$ for $i=j$; $=0$ otherwise). These commutation relations play a vital role in their derivation of the energy fluctuations which contain expressions quartic in the p's and q's. The Born–Heisenberg–Jordan treatment of fluctuations is not the last word on the subject.[20]

** The italics are mine. I slightly changed their phrase without changing its meaning. The authors refer to 'the number of quanta in the corresponding cell [of size h^3, in phase space]'—a coarse-grained version of what is written above.

Nor did Born, Heisenberg, and Jordan fail to observe that the new interpretation automatically implements BE statistics since, first, the number of photons with energy $h\nu$ is unrestricted and, secondly, by its very nature the formalism prohibits any possibility of giving meaning to 'which photon is which'—Boltzmann's distinguishability criterion is gone, as it should be for BE statistics (Chapter 13, Section (d)).

In order to find out how the mechanism for creation and annihilation of photons actually works one has to consider the coupling of quantized radiation to matter. The first steps on this long road were taken by Dirac.

(c) Foundations: Dirac

> A great deal of my work is just playing with equations and seeing what they give. Second quantization I know came out of playing with equations.
>
> P. A. M. Dirac in 1963[21]

1. Formal introduction of photons. Already in August 1926 Dirac had given[22] a theory of the B-coefficients for induced transitions by treating atoms quantum mechanically, but still considering the Maxwell field as a classical system. This procedure, now known as the semi-classical theory* of radiation processes, does allow for a good approximate but not a rigorous treatment of these induced processes.** However, 'one cannot take spontaneous emission into account without a more elaborate theory'.[22] Early in 1927 Dirac gave[25,26] the foundations for that theory, quantum electrodynamics. At that time his relativistic equation for the electron was still a year off (see Chapter 13, Section (e)). Dirac, aware already then that relativity requirements would pose several thorny problems, chose 'to build up a fairly satisfactory theory . . . on the basis of a kinematics and dynamics which are not strictly relativistic'[25]—a very wise move. I shall confine myself to those main points in his work which are of particular importance for what will come later, and adapt notations accordingly. Also, I shall treat these two papers jointly, leaving aside the evolution from the first to the second one. For an admirable analysis which goes into more detail, I refer to an article by Res Jost.[27]

The dynamical system studied by Dirac consists of an atom interacting with a radiation field. 'We consider the atom to consist of a single electron moving in an electrostatic field.'[28] The associated energy operator (Hamiltonian) H of the whole system is

$$H = H^0 + H^1 + H^{\text{rad}} \tag{15.15}$$

H^0, the Hamiltonian of the atom, includes the electrostatic field.[29] H^{rad}

* For a detailed discussion of this theory see Refs. 23 and 24.

** In modern parlance, the semi-classical method does not give an account of radiative corrections.

depends only on the variables of the radiation field and is given by

$$H^{\mathrm{rad}} = \int W\, \mathrm{d}\vec{x}; \tag{15.16}$$

see Eq. (15.6). H^1, the interaction between the atom and the radiation, is 'taken over from the classical theory . . . [and serves] to obtain the correct results for the action of the radiation and the atom on one another'. Dirac took H^1 to be of the form*

$$H^1 = -\frac{e}{c}\vec{A}^{\mathrm{tr}}(\vec{x}, t)\frac{\mathrm{d}\vec{x}}{\mathrm{d}t} \tag{15.17}$$

where $\vec{x}$ is the coordinate of the electron. Eq. (15.17) is the classical expression for the coupling of the vector potential to the time derivative of the atom's electric dipole moment.

Following Dirac, let us consider first the free field case when only H^{rad} is present. As in Eq. (15.12) H^{rad} can be written as a sum of oscillator terms, a representative of them being of the form

$$H(p, q) = \tfrac{1}{2}(p^2 + \omega^2 q^2); \qquad [p, q] = \frac{\hbar}{i}. \tag{15.18}$$

Define a and its conjugate $a^\dagger$ by

$$a = \frac{\omega q + ip}{\sqrt{2\hbar\omega}}, \qquad a^\dagger = \frac{\omega q - ip}{\sqrt{2\hbar\omega}}. \tag{15.19}$$

Then Eq. (15.18) becomes

$$H = \hbar\omega(a^\dagger a + \tfrac{1}{2}) \tag{15.20}$$

$$[a, a^\dagger] = 1. \tag{15.21}$$

A matrix representation which satifies Eq. (15.21) and diagonalizes Eq. (15.20) is given by[30]

$$a_{n,n+1} = a^\dagger_{n+1,n} = \sqrt{n+1}, \qquad n = 0, 1, 2, \ldots \tag{15.22}$$

which yields

$$H = (n + \tfrac{1}{2})\hbar\omega = (n + \tfrac{1}{2})h\nu; \qquad \omega = 2\pi\nu. \tag{15.23}$$

Textbooks on ordinary quantum mechanics[30] explain how to go from the one oscillator to the infinitely many that make up H^{rad}. I give a transcription table.

$$a, a^\dagger \to \text{the set } a_j(\vec{k}), a_j^\dagger(\vec{k}), \tag{15.24}$$

* See Ref. 25, p. 262. Dirac wrote $\vec{A}^{\mathrm{tr}}$ as $\vec{h}$.

where $j=1$ or 2 denotes the state of polarization of an electromagnetic oscillator, $\vec{k}$ its wave vector,* and

$$\nu = c|\vec{k}|. \tag{15.26}$$

Furthermore**

$$\text{Eq. (15.20)} \rightarrow H^{\text{rad}} = \sum_{j,\vec{k}} h\nu(a_j^\dagger(\vec{k})a_j(\vec{k})+\tfrac{1}{2}) \tag{15.27}$$

$$\text{Eq. (15.21)} \rightarrow [a_j(\vec{k}), a_j^\dagger(\vec{k}')] = \delta_{jj'}\delta_{\vec{k}\vec{k}'}, \tag{15.28}$$
$$[a_j(\vec{k}), a_{j'}(\vec{k}')] = [a_j^\dagger(\vec{k}), a_{j'}^\dagger(\vec{k}')] = 0$$

$$\text{Eq. (15.22)} \rightarrow (a_j(\vec{k}))_{n_j(\vec{k}),n_j(\vec{k})+1} = (a_j^\dagger(\vec{k}))_{n_j(\vec{k})+1,n_j(\vec{k})} \tag{15.29}$$
$$= \sqrt{n_j(\vec{k})+1}, \qquad n_j(\vec{k}) = 0, 1, 2, \ldots$$

$$\text{Eq. (15.23)} \rightarrow H^{\text{rad}} = \sum (n_j(\vec{k})+\tfrac{1}{2})h\nu. \tag{15.30}$$

To these equations apply the Born–Heisenberg–Jordan interpretation: a state labelled by the set of numbers $n_j(\vec{k})$, one such number for each j, $\vec{k}$, describes an assembly of $n_j(\vec{k})$ photons with energy $h\nu$. Dirac stressed the point already made that BE statistics is automatic.

Now add H^0 to H^{rad}. Then we have an atom with quantum states n, m, ... and free photons. In the absence of H^1 the atom in state n stays there, nor do the $n_j(\vec{k})$ change.

2. The creation and annihilation of photons. Dirac's invention for handling the full problem specified by Eq. (15.15) was to treat H^1 as a perturbation. By the general rules of quantum mechanics this leads to energy- and momentum-conserving transitions from the initial state $i=(n, n_j(\vec{k}))$ to the final state $f=(m, n_{j'}(\vec{k}'))$. To lowest order the transition probability is proportional to $|H_{fi}|^2$ where

$$H_{fi} = (H^1)_{(m,n_{j'}(\vec{k}')),(n,n_j(\vec{k}))}. \tag{15.31}$$

What are these matrix elements? The factor $\mathrm{d}\vec{x}/\mathrm{d}t$ in Eq. (15.17) allows for transitions between atomic states. According to Eq. (15.29) the factor $\vec{A}^{\text{tr}}$, which depends linearly[30] on the a's and the $a^\dagger$'s, allows for transitions due to the $a^\dagger$'s in which one of the $n_j(\vec{k})$ is raised by one, while the a's lower that

* Following Dirac we enclose the system in a cube. Periodic boundary conditions are used: each of the three components k_i of $\vec{k}$ takes the same value on two opposite planes of the box $V=l^3$, so that $k_i = 2\pi n_i/l$, $n_i = 0, 1, 2, \ldots$ Note that

$$\frac{1}{V}\int \mathrm{d}\vec{x}\, \mathrm{e}^{i(\vec{k}-\vec{k}')\vec{x}} = \delta_{\vec{k},\vec{k}'} \tag{15.25}$$

where the right-hand side $=1$ for $k_i = k_i'$; $=0$ otherwise.

** Dirac derived these equations via a number of variable transformations which I skipped. The quantities here called $a_j(\vec{k})$ and $a_j^\dagger(\vec{k})$ correspond to Dirac's b_r and b_r^*; see Ref. 25, p. 251.

number by one. But by the interpretation just given this action of $a^\dagger$ and a leads to the creation and annihilation of one photon respectively!

Thus with Dirac's work we see for the first time in full-fledged action the new game, quantum field theory, which replaced the old game of marbles. Consider in particular a state containing n photons (for some fixed j, $\vec{k}$). It follows from Eqs. (15.29) and (15.31) that the probability for absorption ($n \to n-1$, a acts) is proportional to n, while that for emission ($n \to n+1$, $a^\dagger$ acts) is proportional to $n+1$. This '$n+1$' consists of two parts. The 'n' corresponds to the induced processes, proportional to the density of radiation present, as in the B-terms of Eq. (15.3). The '1' accounts for the n-independent spontaneous emission, the A-term in Eq. (15.3). It is now obvious that the two emission mechanisms should be related in a simple manner. It was indeed the first major achievement of quantum electrodynamics that in his first paper on the subject Dirac was able to derive in essence Einstein's relation (15.4) from first principles. Thus did Dirac achieve his aim of 'a more elaborate theory'.

Remark. For ease of presentation I have embellished Dirac's own presentation in one respect. In the original work the polarizations were not fully taken into account. Dirac himself stressed[31] that, as a result, he missed a factor two in the relation (15.4). The systematic treatment of polarization begins with Jordan and Pauli's paper[19] of 1928.*

3. Scattering of photons; virtual states. The emission and absorption processes discussed above involve only one photon. In his first paper[25] Dirac remarked: 'Radiative processes of the more general type . . . , in which more than one light-quantum take part simultaneously are not allowed in the present theory'.[34] How young quantum mechanics still was. Early in 1927 Dirac did not yet know that these processes are perfectly well included in his theory. All one has to do is apply second- (and higher-) order perturbation theory to his H^1. So, in his second paper,[26] Dirac developed second-order perturbation theory by deriving the formula for the transition matrix element H_{fi} from an initial state i to a final state f:

$$H_{fi} = \sum_n \frac{H^1_{fj} H^1_{ji}}{E_i - E_j}, \tag{15.33}$$

* In his first paper Dirac raised the further question of what is the Schroedinger equation

$$H\Psi = i\hbar \frac{\partial \Psi}{\partial t} \tag{15.32}$$

corresponding to the Hamiltonian given by Eq. (15.15)? His answer is now found in standard textbooks.[32] Note that the term (photon) vacuum for that Ψ corresponding to all $n_j(\vec{k}) = 0$ is not yet present in Dirac's work. He refers to a 'zero state' of a photon in which its momentum and therefore its energy are zero, and looks upon emission/absorption as jumps out of/into the zero state respectively.[33] This picture is not useful; in particular it fails to apply to fields whose quanta have non-zero rest mass.

an equation which was to dominate quantum field theory in the nineteen-thirties. The transition $i \to f$ goes via 'intermediate states' j. Dirac applied this result to photon scattering by the bound electron which represented his atom. 'The scattered radiation', Dirac wrote,[26] 'thus appears as a result of the two processes $[i \to j$ and $j \to f]$ one of which must be an absorption, the other an emission, in neither of which the total proper energy is even approximately conserved'. As is well known, this does not imply an actual violation of the energy principle; rather, energy conservation cannot be given meaning here because of the transient existence of the intermediate state (remember that $\Delta E \Delta t > h$)—whence its alternative name: virtual state.

In Dirac's problem the summation over n involves two options (e_i means electron in the state i, etc.)

$$\begin{aligned} j=1:& \quad \gamma + e_i \to e_1 \to \gamma' + e_f \\ j=2:& \quad \gamma + e_i \to e_2 + \gamma + \gamma' \to \gamma' + e_f; \end{aligned} \tag{15.34}$$

see further Appendix, Note 4.

'Three physicists above all are prominent by their contributions to quantum electrodynamics in the first third of our century: Max Planck, Albert Einstein, and P. A. M. Dirac'.[27] We shall hear more about Dirac, but in this book we now say farewell, thank you, to Planck and Einstein.

(d) Quantum fields and quantum statistics

Born, Heisenberg, and Jordan had indicated, and Dirac had demonstrated, the close connection between quantum fields and quantum statistics. Second quantization guarantees that photons obey BE statistics. What about other particles which obey BE statistics?

The year 1927 was not over before Jordan and Klein addressed this question. 'Here we follow a paper by Dirac.'[36] This is what they did. Pick a field $\psi(\vec{x}, t)$, any field $\psi(\vec{x}, t)$, which satisfies some Schroedinger equation

$$H\psi(\vec{x}, t) = i\hbar \frac{\partial \psi(\vec{x}, t)}{\partial t}. \tag{15.35}$$

Expand ψ and its complex conjugate ψ^* in a complete set $u_k(\vec{x})$ of orthogonal normalized functions:

$$\psi(\vec{x}, t) = \sum_k a_k(t) u_k(\vec{x}) \tag{15.36}$$

$$\psi^*(\vec{x}, t) = \sum_k a_k^*(t) u_k^*(x). \tag{15.37}$$

Remember that $|a_k|^2$ is the probability of finding the system in the state k, and also that completeness means

$$\sum_k u_k^*(\vec{x}, t)u_k(\vec{x}', t) = \delta(\vec{x}-\vec{x}'). \tag{15.38}$$

The δ-function, another innovation due to Dirac (Appendix, Note 5), is a highly singular object, as is seen from its definition

$$\delta(\vec{x}) = 0 \quad \text{for } \vec{x} \neq 0, \qquad \int \delta(\vec{x})\, d\vec{x} = 1. \tag{15.39}$$

Second quantization: replace the numbers a and a^* by operators a and $a^\dagger$. Accordingly replace ψ^* by

$$\psi^\dagger(\vec{x}, t) = \sum_k a_k^\dagger(t) u_k^*(\vec{x}). \tag{15.40}$$

Imitate Dirac's equation (15.28) by postulating that

$$\begin{gathered}[a_k(t), a_l^\dagger(t)] = \delta_{kl} \\ [a_k(t), a_l(t)] = [a_k^\dagger(t), a_l^\dagger(t)] = 0.\end{gathered} \tag{15.41}$$

Follow Eq. (15.29) by setting

$$\begin{gathered}(a_k)_{n_k, n_k+1} = (a_k^\dagger)_{n_k+1, n_k} = \sqrt{n_k + 1} \\ n_k = 0, 1, 2, \ldots,\end{gathered} \tag{15.42}$$

so that the one particle probability $|a_k^2|$ (a_k is a number) is replaced by $(a_k^\dagger a_k)_{n_k, n_k} = n_k$. Interpret n_k as the number of particles in the state k. As before, these particles automatically satisfy BE statistics.

Again we have gone from a one-particle to a many-particle problem. Which problem? That depends on the choice of H in Eq. (15.35). In particular, Jordan and Klein stress, H *may include a potential* $V(\vec{x})$; second quantization is not confined to free particles only. As was to be clear later, Eqs. (15.41) rule the field theory of spinless mesons but, as Jordan and Klein already noted, one can quantize just as well the non-relativistic Schroedinger equation. In honor of these contributions the matrices in Eq. (15.42) have been named Jordan–Klein matrices. The authors also noted that Eqs. (15.36, 38, 40, 41) imply:

$$[\psi(\vec{x}, t), \psi^\dagger(\vec{x}', t)] = \delta(\vec{x}-\vec{x}') \tag{15.43}$$

$$[\psi(\vec{x}, t), \psi(\vec{x}', t)] = [\psi^\dagger(\vec{x}, t), \psi^\dagger(\vec{x}', t)] = 0. \tag{15.44}$$

These equations make it possible to formulate second quantization in configuration space, as was elegantly done by Vladimir Fock.[37]

Even before the completion of the Jordan–Klein paper, Jordan, inspired by Dirac's work, had raised[38] the question: is there a second quantization procedure which leads to FD statistics? Once again he was the first to ask the right question. He also found the correct answer: one needs a's and $a^\dagger$'s which are constructed from 2×2 matrices. Born referred[39] to this work at the Solvay conference in October 1927. Dirac was there and later recalled: 'At first I did not like this work of Jordan and Wigner . . . I did not appreciate . . . the importance of this other kind of second quantization'.[40] Here he referred to the improved treatment given shortly afterward by Jordan and Wigner.[41] In order to lend perspective to Dirac's remark, it should be recalled that in October 1927 he had not yet discovered his electron wave equation. The initial issue was how to express the exclusion principle in the language of second quantization.

Jordan and Wigner started again from Eqs. (15.36) and (15.40). Define the anti-commutator symbol

$$\{x, y\} = xy + yx. \tag{15.45}$$

Replace Eq. (15.41) by

$$\begin{gathered} \{a_k, a_l^\dagger\} = \delta_{kl} \\ \{a_k, a_l\} = \{a_k^\dagger, a_l^\dagger\} = 0, \end{gathered} \tag{15.46}$$

which are realized by the 'Jordan–Wigner matrices':

$$a_k = \eta_k \begin{pmatrix} 0 & 0 \\ 1 & 0 \end{pmatrix}, \qquad a_k^\dagger = \eta_k^* \begin{pmatrix} 0 & 1 \\ 0 & 0 \end{pmatrix}. \tag{15.47}$$

The phases η_k are introduced to guarantee that Eqs. (15.46) hold for $k \neq l$. For details on these phases I refer to textbooks.*

$a_k^\dagger$ and a_k are the creation and annihilation operators of the theory. The number operator is

$$n_k = a_k^\dagger a_k = \begin{pmatrix} 1 & 0 \\ 0 & 0 \end{pmatrix}, \quad \text{all } k. \tag{15.48}$$

In words: the number of particles in each state is either zero or one—the anticommutation relations provide a pretty way of implementing the exclusion principle. It follows from Eqs. (15.36, 38, 40, 46) that[41]

$$\{\psi(\vec{x}, t), \psi^\dagger(\vec{x}', t)\} = \delta(\vec{x} - \vec{x}') \tag{15.49}$$

$$\{\psi(\vec{x}, t), \psi(\vec{x}', t)\} = \{\psi^\dagger(\vec{x}, t), \psi^\dagger(\vec{x}', t)\} = 0. \tag{15.50}$$

As will be seen in the next chapter, the Jordan–Wigner formalism set the stage for the field theory of the electron–positron. Finally, new names: identical particles that satisfy BE (FD) statistics are called bosons and fermions respectively.

* See e.g. Ref. 24, pp. 507–8.

(e) Relativistic invariance and gauge invariance: beginnings

In the year 1928 the first steps were made along a road which, I am sure, has not yet been traveled to the end. The road sign reads: toward the incorporation of invariance principles in quantum field theory.

It began with relativistic invariance. To see what the question was, let us re-inspect the commutation relations encountered so far. There were those for mechanical systems, Eq. (15.14), and those for fields, Eqs. (15.43, 44). All are 'equal time commutators', the operators commuted refer to the same instant. From the point of view of relativity, Eqs. (15.43, 44) show an apparently awkward dissymmetry by referring to different points in space, but the same point in time. This is not to say that the relations violate the requirements of relativity, but that their compatibility with relativity needs to be proved.

That was first done by Jordan and Pauli[19] for the case of the free electromagnetic field in which the time dependence of operators is explicitly known so that one can compute explicitly the commutation relations between the various electric and magnetic field components at different space–time points. They verified that all is well with relativity and were the first to generalize the δ-function to unequal times (Appendix, Note 6).

This paper was Pauli's last while Privatdozent in Hamburg. Shortly thereafter, and for the rest of his life, he took up the post of professor in theoretical physics at the Eidgenössische Technische Hochschule in Zürich, made vacant by Debye's departure. Jordan succeeded Pauli in Hamburg.

In January 1928, within weeks of the completion of his work with Jordan, Pauli and Heisenberg decided[42] to attack the relativistic formulation of quantum electrodynamics in the presence of charges and currents.

In important respects their methods[43] are new. They begin by developing the analog for fields of the Lagrangian and Hamiltonian methods of mechanics.* They take certain fields $Q_\alpha(\vec{x}, t)$ themselves as an infinite set of dynamical variables, one for each point $\vec{x}$ and each label α. (α may refer to a vector component label for example.) In mechanics one starts from a typical Lagrangian $L = L(q, \dot{q})$ and defines the momentum conjugate to q by $p = \partial L/\partial \dot{q}$ (the dot denotes time differentiation). Likewise, they show, one can build a field theory on

$$L = \int \mathscr{L}\, \mathrm{d}\vec{x} \tag{15.51}$$

$$\mathscr{L} = \mathscr{L}\left(Q_\alpha, \dot{Q}_\alpha, \frac{\partial Q_\alpha}{\partial x_i}\right) \tag{15.52}$$

* For a simple presentation of their main points see for example Schiff's book.[44]

and define a 'field of momenta P_α conjugate to Q_α' by

$$P_\alpha(\vec{x}, t) = \frac{\partial \mathscr{L}}{\partial \dot{Q}_\alpha(\vec{x}, t)}. \tag{15.53}$$

Their commutation relations (which guarantee BE statistics) are the analogs of Eq. (15.14):

$$\begin{aligned} [P_\alpha(\vec{x}, t), Q_\beta(\vec{x}', t)] &= \delta_{\alpha\beta}\delta(\vec{x} - \vec{x}') \\ [Q_\alpha(\vec{x}, t), Q_\beta(\vec{x}', t)] &= [P_\alpha(\vec{x}, t), P_\beta(\vec{x}', t)] = 0. \end{aligned} \tag{15.54}$$

They proceeded to give a general proof that their formalism is compatible with relativity. (A slip in their argument was corrected one year later.*) This part of their work, which also includes a discussion of energy–momentum conservation laws, is not restricted to any specific dynamical system.

Next they turned to electrodynamics.

The application of Lagrangian methods to electrodynamics has a history in classical physics associated with illustrious names. Already in 1892, Helmholtz inquired whether 'we . . . can cast the empirically known laws of electrodynamics, as they are formulated in Maxwell's equations, in the form of a minimal principle' and made a first attempt[46] to do so. In 1903 Lorentz reviewed contributions to the same problem by Larmor, Poincaré, and himself.[47] In 1908 Born showed[48] that the Lagrangian for a pure radiation field:

$$\mathscr{L}_{\text{rad}} = \tfrac{1}{2}(\vec{E}^2 - \vec{H}^2) \tag{15.55}$$

is a relativistic invariant. In 1912, Mie pointed out that an invariant $\mathscr{L}$ leads to covariant field equations.[49]

Heisenberg and Pauli, desiring to exhibit the relativistic features of quantum electrodynamics, abandoned Eqs. (15.7–9), since div $\vec{A}^{\text{tr}} = 0$ does not look very covariant! Instead, they started from the Maxwell equations in their general form:

$$\vec{E} = -\frac{1}{c}\frac{\partial \vec{A}}{\partial t} - \nabla\phi \tag{15.56}$$

$$\vec{H} = \text{curl}\, \vec{A} \tag{15.57}$$

$$\text{div}\, \vec{E} = \rho \tag{15.58}$$

$$\text{curl}\, \vec{H} - \frac{1}{c}\frac{\partial \vec{E}}{\partial t} = \vec{j}. \tag{15.59}$$

* Their proof appeared to demand that the Hamiltonian satisfy a certain constraint, Ref. 43, Eq. (40). In 1930 it was shown by Rosenfeld[45] that this conclusion is due to a computational error; there is no such constraint. Much later Rosenfeld told me that he regarded his one-and-a-half-page article as his best contribution to physics. It is a pity that this paper was not included in the edition of his selected writings.

ϕ, ρ, and $\vec{j}$ are the Coulomb potential, charge, and current density respectively. The four fields $\vec{A}$, ϕ are taken as 'Q's'. The pure field part of their Lagrangian is given by Eq. (15.55), in which Eqs. (15.56, 57) are used as definitions.

Then, it seemed, the roof fell in over their heads. The P-field conjugate to ϕ is identically zero![50] This is evident from Eqs. (15.53, 55, 56). $\mathscr{L}_{\text{rad}}$ does not contain $\dot{\phi}$. (The other terms in the full $\mathscr{L}$ do not change this.)

To this day this is a highly nasty, but not an insurmountable complication, typical for vector fields associated with massless quanta. In 1929 it looked like such a catastrophe that it set Heisenberg and Pauli back a full year during which neither published any research. Then Heisenberg found a way out. What happened is indicated by Pauli in a letter[51] to Klein of February 1929: '[In the fall of 1928] I made, for my own amusement, a brief outline for a utopian novel . . . deluded by such dreams I suddenly received word in January [1929] from Heisenberg that he could eliminate the [above mentioned] difficulties by means of a formal trick . . . to my good fortune the utopian novel was buried deeply in my desk (where it still is) and the non-commuting space–time functions brought out from there'.

Heisenberg's trick (the first example of what today is called gauge fixing) consisted in adding to $\mathscr{L}$ certain terms* proportional to a small parameter ε and chosen such that ϕ does acquire a conjugate momentum field, then calculating whatever is desired, and letting $\varepsilon \to 0$ in the final answer. Their work could now proceed and was completed in March 1929.

Remark. This paper is also the first in which quantum electrodynamics with interactions is treated for the case that the electron obeys the meanwhile discovered Dirac equation. I return to that contribution in the next chapter.

In September 1929 Heisenberg and Pauli had completed a sequel[52] to their first paper. They had found a way to treat their problem without any ε-tricks. In order to see what they did next I need to introduce a new concept: gauge invariance.

The six fields $\vec{E}$ and $\vec{H}$ which encode electromagnetic phenomena registered in the laboratory can be expressed in terms of the four auxiliary fields $\vec{A}$ and ϕ by Eqs. (15.56, 57). The venerable age of ϕ is attested by its name: the Coulomb potential. The three fields $\vec{A}$ had long been known too. Precursors of Maxwell had used them.[53] So had Maxwell himself, calling them[54] electrotonic functions well before he wrote down his field equations, and calling them[55] electromagnetic momentum when, in 1864, he introduced the field concept.

I do not know who first used the name 'vector potential' for $\vec{A}$ (Helmholtz did so[46] in 1892) nor who first remarked that Eqs. (15.56–9) do not determine $\vec{A}$ and ϕ uniquely. Certain it is that Lorentz knew that last fact in

* Add the relativistic invariant $\varepsilon(\operatorname{div} \vec{A} + \dot{\phi}/c)^2$ to Eq. (15.55).

1903 (I use his text but modify his symbols):

'The potentials $\vec{A}$ and ϕ remain partially undetermined even though $\vec{E}$ and $\vec{H}$ are definite functions of $\vec{x}$ and t for every electromagnetic phenomenon. We eliminate this defect by the further condition

$$\operatorname{div} \vec{A} + \frac{1}{c}\frac{\partial \phi}{\partial t} = 0. \tag{15.60}$$

Indeed if $\vec{A}_0$ and ϕ_0 are some potentials compatible with [Eqs. (15.56, 57)] then every other admissible pair is of the form

$$\vec{A} = \vec{A}_0 - \nabla\chi, \qquad \phi = \phi_0 + \frac{1}{c}\frac{\partial \chi}{\partial t}. \tag{15.61}$$

Here the scalar function χ can be determined in such a way that [Eq. (15.60)] or

$$\Box\chi = \operatorname{div} \vec{A}_0 + \frac{1}{c}\frac{\partial \phi_0}{\partial t} \tag{15.62}$$

is satisfied.'[47]

A brief digression about physics today. The transformations (15.61) are gauge transformations. The invariance of Eqs. (15.56–9) under gauge transformations is called gauge invariance. Gauge invariance is a good thing. That was not appreciated until well into the twentieth century. Mie's electrodynamics[49] of 1912, for example, is not gauge invariant.[57] The gauge chosen by adopting Eqs. (15.60), (15.62) is called the Lorentz gauge; Eq. (15.60) is the Lorentz condition. The choice Eq. (15.9) is the Coulomb gauge. Either gauge leaves some freedom in the choice of potentials.

Back to history. The term 'gauge', which first entered physics in 1919, was introduced by Hermann Weyl in the context of a unified field theory of gravitation and electromagnetism which he had proposed the year before. The fundamental fields in his theory are the symmetric metric tensor $g_{\mu\nu}$ of gravitation and the electromagnetic four-vector potential A_μ,

$$A_\mu = (\vec{A}, i\phi).$$

The central requirement of the Weyl theory is invariance under the transformations ($x_\mu = \vec{x}, ict$)

$$g'_{\mu\nu} = e^{\chi} g_{\mu\nu} \tag{15.63}$$

$$A'_\mu = A_\mu - \frac{\partial \chi}{\partial x^\mu} \tag{15.64}$$

(the latter being the same as Eq. (15.61)) while the (contravariant) coordinates x^μ are unchanged. Accordingly the transformation (15.63) changes the (length element)^2 defined by $ds^2 = g_{\mu\nu}\, dx^\mu\, dx^\nu$ into $e^\chi\, ds^2$. Weyl had this change of *length* in mind when he chose the engineering expression 'change of gauge' for his transformation.* The main interest of Weyl's paper[58] is that for the first time conservation of electric charge is derived from an invariance principle:

$$\text{gauge invariance} \rightarrow \text{charge conservation.} \tag{15.65}$$

This, in Weyl's opinion was 'one of the strongest general arguments in favor of the theory presented here'. Later it became clear that a similar connection exists also for other unified field theories.** Further details need not concern us. Suffice it to say that Weyl himself soon abandoned his theory of 1918—but not the idea of gauge invariance to which he came back ten years later.

Meanwhile quantum mechanics had arrived. In his fourth communication† on wave mechanics, Schroedinger wrote down[62] the extension of Eq. (13.32) to the case that an electromagnetic field is present:

$$\left[\left(\frac{\partial}{\partial x_\mu} - \frac{ie}{\hbar c} A_\mu\right)^2 - \frac{m^2c^2}{\hbar^2}\right]\psi = 0. \tag{15.66}$$

In the first paper on the Dirac equation[63] we find Eq. (13.47). Schroedinger and Dirac may perhaps have noticed that their respective equations are invariant under the transformations (15.64) combined with

$$\psi' = e^{-(ie/\hbar c)\chi}\psi. \tag{15.67}$$

If they did, they did not say so. I believe that Walter Gordon, who independently discovered Eq. (15.66), was the first to mention this,[64] but only as a passing remark.†† From now on invariance under the combined transformations (15.64) and (15.67) will simply be called gauge transformations.

Weyl discussed the new style gauge transformations in his famous book of 1928: 'I now believe that . . . gauge invariance does not tie together electricity and gravitation but *electricity and matter*'.[66,67] Further reflection led to his paper[68] of March 1929 (further elaborated in May[69]) in which he made one of his lasting contributions to physics: 'The connection of [gauge] invariance with the conservation law of electricity remains exactly as before'; that is, the connection (15.65) holds again. To see this, it is sufficient, he noted, to consider the special case that χ does not depend‡ on $\vec{x}$ and t and is infinitesimal. The demand

$$\delta L = 0 \quad \text{for } \delta\psi = -\frac{ie}{\hbar c}\chi\psi \tag{15.68}$$

* In his first two papers[58,59] on the subject he called his new invariance *Maszstabinvarianz*; in a sequel[60] he introduced the term *Eichung*, gauging, for the first time.

** For the case of projective relativity theory see the general discussion in Ref. 61.

† In his earlier papers on non-relativistic wave mechanics the vector potential did not appear explicitly.

†† The issue is also raised in Ref. 65, where it is mixed with irrelevant geometrical arguments, however.

‡ This case is now called gauge invariance of the first kind.

yields current conservation.[70] He showed this for the case that ψ is a Dirac wave function. His conclusion is completely general, however.[71]

This introductory discourse on gauge invariance is not meant to convey all at once the extraordinary power of that concept. More will be said in what follows: first, in the next few lines; then in the discussion of renormalization (Chapter 18); and finally in the concluding pages (Chapter 21, Section (e)) where we shall see that generalizations of electromagnetic gauge invariance will lead us directly to the W- and Z-bosons.

The main message of Heisenberg and Pauli's second paper[52] can be stated succinctly: Get thee to the Coulomb gauge; that is, return to the Eqs. (15.7–9). Weyl's book[66] showed them the way. I quote their main points for the case of a pure radiation field (changing notation a bit): 'We can simply put ϕ equal to zero . . . the relativistic invariance of [this] scheme may seem dubious since ϕ is preferred to $\vec{A}$ . . . [however] it turns out that by means of a gauge transformation with a suitable function χ one can return to $\phi = 0$ . . . in [another] frame of reference'. The same procedure, they show, applies to the commutation relations (15.54). If ϕ can be put equal to zero then it does not matter that there is no momentum conjugate to ϕ. Hence they abandon their c-trick and prove covariance by again matching a Lorentz transformation with a gauge transformation. They write the interaction as

$$\text{interaction} = \text{static Coulomb terms} + \text{coupling to transverse waves (photons)}. \tag{15.69}$$

That is what Dirac had done in his non-relativistic theory. Heisenberg and Pauli added the important observation that this procedure may not look covariant but is covariant nevertheless—as is shown once again by suitable gauge transformations which from then on until the present are among the precious tools of field theory.

The procedure symbolized by Eq. (15.69) was used with very few exceptions (see, however, Appendix, Note 7) all through the 1930s and early 1940s. For example, the excellent first edition (1935) of Heitler's book[72] on quantum electrodynamics is written, one may say, in the Coulomb gauge. Thereafter (for reasons to be explained later) renormalization demanded: get thee to the Lorentz gauge.

(f) The positron

> I was reconciled to the fact that the negative energy states could not be excluded from the mathematical theory, and so I thought, let us try and find a physical explanation for them.
>
> P. A. M. Dirac in 1977[73]

Heisenberg once called the five years following the fifth Solvay conference, held in October 1927, 'das goldene Zeitalter der Atomphysik', the golden age of atomic physics.[74] Quantum mechanics was just over two years old, but much had already been done to provide a formal structure for the new theory. Its conceptual basis had been clarified by the uncertainty principle and by complementarity. Quantum mechanics and quantum statistics had consummated a happy union. There was also an emergent quantum electrodynamics, though not yet at a relativistic level. As the members of the Conference left Brussels there was a consensus, almost but not quite unanimous, that non-relativistic quantum mechanics was a well-established discipline. Heisenberg, the twenty-five-year-old bachelor, returned to Leipzig where earlier that month he had assumed his new position as the professor of theoretical physics. He and others now took up 'the innumerable problems, which, insoluble before, could be treated by the new methods'.[74]

The bliss of the next five years was not undivided, however.

To be sure, in the atomic periphery, and also in the solid state, all was going swimmingly. Things did not look as good elsewhere, however. As was seen in the previous chapter, during the late twenties the puzzlements about the atomic nucleus were such as to make grown men doubt whether quantum mechanics would suffice to lead them out of their miseries. The relativistic treatment of quantum electrodynamics raised complex technical problems. Yet, as is clear from all I have read and heard, another concern was more serious and more widespread. How could Dirac's relativistic wave equation for the electron be so successful yet so paradoxical? Nothing like it had been seen since the days when Bohr had presented to the world his splendid derivation of the Rydberg constant along with his brazen demand that certain closed electron orbits shall not radiate. When, in the 1960s, Heisenberg once spoke of the confusion in 1928 caused by the Dirac equation, he said: 'Up till that time I had the impression that in quantum theory we had come back into the harbor, into the port. Dirac's paper threw us out into the sea again'.[75]

I have already mentioned* the spectacular achievements contained in Dirac's two papers[63,76] of early 1928. Spin was a necessary consequence, the right magnetic moment was obtained, the Thomas factor appeared automatically, the Sommerfeld fine structure formula was derived with the correct Goudsmit–Uhlenbeck quantum numbers. I have also noted earlier that his equation gave twice as many states as, it seemed, were called for. There lay the rub.

In his first paper[63] Dirac, from the very outset fully aware of this apparent flaw, correctly diagnosed its cause. The energy–momentum–mass relation $E^2=c^2p^2+m^2c^4$ has two roots:

$$E=\pm c\sqrt{p^2+m^2c^2}. \qquad (15.70)$$

What to do with the negative E solutions? 'One gets over the difficulty on the classical theory by arbitrarily excluding those solutions that have a negative E. One cannot do this in the quantum theory, since in general a perturbation

* Chapter 13, Section (e).

will cause transitions from states with E positive to states with E negative'[63]—the ambiguity in the sign of E is the cause of the doubling of states. His paper also contains some speculations to the effect that negative energy solutions may be associated with particles whose charge is opposite to that of the electron. In that regard Dirac did not yet know as clearly what he was talking about as he would one and a half years later. This undeveloped idea led him to take the problem lightly, initially: 'Half of the solutions must be rejected as referring to the charge $+e$ of the electron'.[63] In a talk given in Leipzig, in June 1928, he no longer spoke of rejection, however. Transitions to negative energy states simply could not be ignored. 'Consequently the present theory is an approximation.'[77]

Even before Dirac's visit, Heisenberg must have been well aware of these difficulties. In May he had written to Pauli: 'In order not to be forever irritated with Dirac I have done something else for a change',[78] the something else being his quantum theory of ferromagnetism. In Leipzig, Dirac and Heisenberg discussed several aspects of the new theory.[79] Shortly thereafter Heisenberg wrote again to Pauli: 'The saddest chapter of modern physics is and remains the Dirac theory',[80] mentioned some of his own work which demonstrated the difficulties, and added that the magnetic electron had made Jordan *trübsinnig* (melancholic). At about the same time Dirac, not feeling so good either, wrote to Klein: 'I have not met with any success in my attempts to solve the $\pm e$ difficulty. Heisenberg (whom I met in Leipzig) thinks the problem will not be solved until one has a theory of the proton and electron together'.[81] Nor were Dirac and Heisenberg the only ones who kept brooding about the electron. Indeed, before the year 1928 was out, others made two important contributions to the subject. New characters now enter the story.

In October, Klein and Yoshio Nishina completed[82] their theory of Compton scattering based on the Dirac equation. But experimental information was not yet precise enough, nor had cosmic ray phenomena been sufficiently understood[83] to make clear at once that the Klein–Nishina formula ranks with Dirac's results earlier in the year as one of the great early successes of the new theory. However, by 1931 new experiments showed how well the formula works.[84]

Klein and Nishina replaced Dirac's non-relativistic expression Eq. (15.17) for H^1 by*

$$H^1 = -\int \vec{j}\vec{A}^{\text{tr}}\,\mathrm{d}\vec{x} \qquad \vec{j} = ie\bar{\psi}\vec{\gamma}\psi \tag{15.71}$$

and treated the radiation field semi-classically. The first quantum electrodynamic treatments, due independently to Ivar Waller from Uppsala[85] and Igor

* Notations are as in Chapter 13, Section (e).

Evgenievich Tamm from Moscow,[86] proceed by starting from the H_{fi} formula (15.33) applied to the case of a photon with initial (final) wave number $\vec{k}(\vec{k}')$ scattering off an electron at rest.[87] By momentum conservation the intermediate electron $e_1(e_2)$ (see Eq. (15.34)) has momentum $\vec{k}(-\vec{k}')$. Since energy is not conserved in the intermediate state the virtual electron can have positive *as well as negative* energy corresponding to its appropriate momentum. Waller and Tamm both made an important observation: *only* if one sums over *both* signs of energy does the Klein–Nishina formula reduce at low energy to the Thomson limit (Eq. 9.5) as of course it should. The situation became increasingly odd: negative energy states, those terrible nuisances, were necessary to obtain contact with the classical theory!

The last main event in 1928 occurred in December when Klein announced[88] his paradox, which caused such headaches to Bohr (Chapter 14, Section (d)) and others. Klein had noted that the Dirac theory leads to absurdities when one attempts to localize electrons with a precision $\lesssim \hbar/mc$. Consider electrons with energy E and density current j coming upon a potential barrier which varies by more than mc^2 over a distance $\sim \hbar/mc$. There is a reflection current j_1 at the barrier as well as a typically quantum mechanical transmission current j_2 into the barrier. One finds that $j=j_1+j_2$, as it should be, but that $j_1>j$, hence $j_2<0$, as it should not. More is reflected than what comes in. The later positron theory was to show that in actual fact, electron–positron pairs are produced at the barrier and all is well—but meanwhile one was stuck with more nasty problems. The following chain of events led to their unraveling.

May 1929. In one of his papers dealing with gauge invariance Weyl[69] suggests: 'It is plausible to anticipate that, of the two pairs of components of the Dirac quantity, one belongs to the electron, one to the proton'.

December 1929.[90] Dirac points out: 'One cannot simply assert that a negative energy electron *is* a proton', since this would violate charge conservation if an electron jumps from a positive to a negative energy state. Rather, 'let us assume . . . that all the states of negative energy are occupied except perhaps for a few of very small velocity', this occupation being one electron per state, as the exclusion principle demands. Imagine that one such negative energy electron is removed, leaving a hole in the initial distribution. The result is a rise in energy and in charge by one unit. This hole, Dirac notes, acts like a particle with positive energy and positive charge: 'We are . . . led to the assumption that the holes in the distribution of negative energy electrons are the protons'.

The identification of holes with particles is fine, but why protons? Dirac later remarked: 'At that time . . . everyone felt pretty sure that the electrons and the protons were the only elementary particles in Nature'.[91] Recall that in 1929 the proton–electron model of the nucleus was in trouble but had not yet been abandoned (Chapter 14, Section (b)). Just prior to submitting his paper,[90] Dirac wrote a letter[92] to Bohr which shows that he knew quite well that, at least in the absence of interactions, his holes should have the same mass as the electrons themselves. It was his hope (an idle one) that this equality would be violated by electromagnetic interactions: 'So long as one neglects interaction one has complete symmetry between electrons and protons; one could regard the protons as the real particles and the electrons as the holes in the distribution of protons of negative energy. However, when the interaction between the electrons is taken into account this symmetry is spoilt. I have not yet worked out mathematically the consequences of the interaction . . . One can hope, however, that a proper theory of this will enable one to calculate the ratio of the masses of proton and electron'. Actually the 'complete symmetry' of which Dirac wrote, now known as charge conjugation invariance, extends to the electromagnetic interactions as well. For want of a better procedure Dirac briefly considered the mass *m* in his equation to be the average of the proton and the electron mass.[93]

We just saw how important it is for the Klein–Nishina formula that intermediate state electrons can jump into negative energy states. That was the picture before hole theory came along. Would there not be a calamity since the exclusion principle forbids these jumps in the new version of the theory according to which negative energy states are almost completely occupied? Not so, Dirac (who knew of Waller's work) explained[94] to Bohr. 'On my new theory . . . there is . . . a new kind of double transition now taking place in which first one of the negative-energy electrons jumps to the proper final state with emission (or absorption) of a photon, and secondly the original positive-energy electron jumps down and fills up the hole, with absorption (or emission) of a photon. This new kind of process just makes up for those excluded and restores the validity of the scattering formulas derived on the assumption of the possibility of intermediate states of negative energy.' One sample of a 'new double transition': initial photon + initial positive energy electron + a negative energy electron → initial positive energy electron + final positive energy electron + hole, the later positron → final positive energy electron + final photon.*

Let us pause, take a deep breath and realize that this letter announces a monumental change in physical theory. *The simple problem of the scattering of a photon on an electron is no longer a two-body problem. It is an infinitely-many-body problem.* The mechanism which Dirac described to Bohr (before he had the positron) is now called virtual electron–positron pair formation. In one of the intermediate states there is not just one electron but two of them, plus a newly-formed hole (a positron), three charged particles in all. As the incoming momentum varies, so do the energies of the virtual particles. For a full description of the process as a function of energy one probes infinitely many filled negative energy states, one at a time. In order to calculate the Klein–Nishina formula one must follow the fate of infinitely many particles. (Accord-

* For more details on this change of picture see Ref. 95.

ingly the current $\vec{j}$ in Eq. (15.71) must be given a new meaning, as will be discussed in the next chapter.)

February 1930. Robert Oppenheimer points out[96] a difficulty with the new theory: it allows for the process proton + electron → two photons, so that a hydrogen atom could spontaneously annihilate into radiation: '[This] gives a mean life for ordinary matter of 10^{-10} seconds'. '[This] gives a way out', he suggests, 'if we return to the assumption of two elementary particles', that is, disconnect the proton states from the holes in the electron states, and if we further assume that 'no transitions to states of negative energy occur [because] *all* such states are occupied' (my italics). That is not quite to the point, since if all these states were filled at all times then (in modern terms) positrons and electrons could never appear.

April 1930. Independently, Tamm[86] reaches the same conclusion about the instability of matter. His only comment: 'This result constitutes a fundamental difficulty for the whole Dirac theory of protons'.

November 1930. Weyl takes a stand[97] in regard to the protons: 'However attractive this idea may seem at first it is certainly impossible to hold without introducing other profound modifications . . . indeed, according to [the hole theory] the mass of the proton should be the same as the mass of the electron; furthermore . . . this hypothesis leads to the essential equivalence of positive and negative electricity under all circumstances . . . The dissimilarity of the two kinds of electricity thus seems to hide a secret of Nature which lies yet deeper than the dissimilarity between past and future . . . I fear that the clouds hanging over this part of the subject will roll together to form a new crisis in quantum physics'.

May 1931. Dirac, having taken due note of all these objections, bites the bullet.[98] 'A hole, if there were one, would be a new kind of particle, unknown to experimental physics, having the same mass and opposite charge of the electron.'

Thus, a quarter of a century after the light-quantum did the second prediction of a new particle, the anti-electron, as Dirac initially called it, enter the scientific literature.*

September 1932. Carl Anderson submits a paper[99] to *Science* entitled: 'The apparent existence of easily deflectable positives'.

After receiving his Ph.D. at Caltech in June 1930 Anderson continued his work with cloud chambers, but shifted his attention to cosmic rays at the instigation of Millikan, his mentor.** To this end he built a cloud chamber with the highest magnetic field then in existence (up to 25 000 gauss). Before long he detected curious tracks in his chamber with a curvature that indicated

* Half a year earlier Pauli had suggested the neutrino, but in 1931 he was not yet prepared to publish (Chapter 14, Section (d)).

** See Ref. 100 for a brief sketch of Anderson's life and for his recollections.

positive charge if they were moving downward, negative charge if upward. Anderson has recalled[101] what happened next. 'Well, there seemed to be too many upward moving electrons, and I would tell Millikan "They can't be protons. They're minimum ionization, and a proton should have one and a half or two times the ionization or something like that, that would be clearly seen on the photograph". And he said, "Everybody knows that cosmic ray particles go down. They don't go up except in very rare circumstances". So then I got the idea of putting a lead plate across the center of the chamber to distinguish between upward and downward particles, whatever they might be . . . a quarter of an inch of lead would easily allow one to determine . . . the sign of the charge', the point being that if (for example) the particle travels downward and passes through the plate it will have lost energy so that its track will be curved more sharply on the downward side. His finding:[99] 'Some tracks . . . seem to be produced by positive particles, but if so the masses of these particles must be small compared to the mass of the proton'. Carefully sifting out alternatives, he reached the cautious conclusion: 'It seems necessary to call upon a positively charged particle having a mass comparable with that of an electron'.

This is as far as the paper goes. The first picture of one of his remarkable tracks was published[99] in *Science News Letter* of December 19, 1931. Anderson did not like[100] the suggestion by the Editor of the *Letter* to name the new particle 'positron', but adopted it nevertheless. It first appeared in print in his paper[102] of February 1933.

More later about subsequent work by Anderson and others which rapidly confirmed his initial conclusion. As far as this chapter is concerned, I shall conclude with Anderson's comments regarding the influence of Dirac's ideas on his own work:

'Yes, I knew about the Dirac theory . . . But I was not familiar in detail with Dirac's work. I was too busy operating this piece of equipment to have much time to read his papers[101] . . . [Their] highly esoteric character was apparently not in tune with most of the scientific thinking of the day . . . The discovery of the positron was wholly accidental.'[100]

(g) Appendix

1. Born and Jordan[9,10] derived the quantum version of the classical continuity equation for W:

$$\frac{\mathrm{d}W}{\mathrm{d}t}+\operatorname{div}\vec{S}=0 \tag{15.72}$$

where

$$\vec{S}=\frac{c}{2}\{[\vec{E},\vec{H}]-[\vec{H},\vec{E}]\} \tag{15.73}$$

is the quantum Poynting vector. Eq. (15.72) can be derived without using explicit representations for the fields. All one needs is to keep track of the ordering of operators in performing differentiations. They also gave an argument for the need to quantize the electromagnetic field in the presence of sources: a charged material linear oscillator with coordinate q emits an amount of energy $\int W \, \mathrm{d}\vec{x}$ which is proportional to $(\mathrm{d}^2q/\mathrm{d}t^2)^2$. This electromagnetic energy is necessarily a quantum mechanical operator, since the same is true for q.

2. Einstein's energy fluctuation formula[14] is

$$\langle E^2\rangle - \langle E\rangle^2 = \left(h\nu\rho + \frac{c^3}{8\pi\nu^2}\rho^2 \right) v \, \mathrm{d}\nu. \qquad (15.74)$$

v is the subvolume considered, ρ is Planck's spectral density function, $\langle E\rangle$ and $\langle E^2\rangle$ are the average energy and energy-squared respectively in the frequency interval $\mathrm{d}\nu$.

3. Debye's derivation of Planck's law.[17] The equilibrium energy density $\rho(\nu, T)\,\mathrm{d}\nu$ per unit volume at temperature T in the frequency interval $\mathrm{d}\nu$ is given by

$$\rho(\nu, T)\,\mathrm{d}\nu = \frac{8\pi\nu^2}{c^3}\langle E(\nu, T)\rangle \, \mathrm{d}\nu \qquad (15.75)$$

where $\langle E\rangle$ is the equilibrium energy of the oscillator with frequency ν. The cofactor of $\langle E\rangle$ is the oscillator density in the interval $\mathrm{d}\nu$. In equilibrium each E_n, see Eq. (15.10), is weighted with its Boltzmann factor y^n, where $y = \exp(-h\nu/kT)$. Thus

$$\begin{aligned} \langle E(\nu, T)\rangle &= \left(\sum_n E_n y^n \right) \Big/ \sum_n y^n \\ &= \frac{h\nu}{\mathrm{e}^{h\nu/kT} - 1}. \end{aligned} \qquad (15.76)$$

Eqs. (15.75) and (15.76) are Planck's law. One might define $\langle E^2\rangle$ in Einstein's Eq. (15.74) by using Eq. (15.76) with E_n replaced by E_n^2. This, however, does not reproduce Eq. (15.74).

4. As Dirac noted, H^1 Eq. (15.17), is not the only interaction to be considered. There is a further term

$$H^2 = \frac{e^2}{2mc^2}(\vec{A}^{\mathrm{tr}})^2; \qquad (15.77)$$

see Ref. 26, Eq. (10). The coherent sum of H^1 to second-order and H^2 to first-order 'gives just Kramers' and Heisenberg's dispersion formula when the incident frequency does not coincide with that of an absorption or emission line of the atom'. Resonance scattering occurs when such a coincidence does take place, a case also treated by Dirac.[26,103]

5. When Dirac introduced[104] his δ-function he remarked right away: 'Strictly, of course, $\delta(x)$ is not a proper function of x, but can be regarded only as the limit of a certain sequence of functions. All the same one can use $\delta(x)$ as though it were a proper function for practically all the purposes of quantum mechanics without getting incorrect results'. Rigorous treatments led to the theory of distributions.[105]

6. The manifestly covariant commutation relations of Jordan and Pauli are found, for example, in Schiff's book.[106] The physical interpretation of these relations and their associated uncertainty relations was discussed in detail by Bohr and Rosenfeld.[107] Their work was aimed, in part, at dispelling the suspicion[108] that radiative properties of an accelerated charged test body might affect the momentum–position uncertainty relation of that body.

Jordan and Pauli[19] also remarked that the infinite zero point energy of the radiation field (corresponding to all $n_j(\vec{k})=0$ in Eq. (15.30)) can be eliminated painlessly by a reordering of operators; see also Ref. 109. This is true in the vacuum but not in general. For example, the zero point energy leads to a zero point pressure between two conducting plates, the Casimir effect.[110]

7. The modern way of working in the Lorentz gauge rests on the following idea first mentioned by Fermi in 1929.[111]

Instead of using Eq. (15.55) start from

$$\begin{aligned}\mathscr{L} &= \tfrac{1}{2}(\vec{E}^2 - \vec{H}^2) - \tfrac{1}{2}\lambda^2 \\ \lambda &= \operatorname{din} \vec{A} + \frac{1}{c}\dot{\phi}.\end{aligned} \tag{15.78}$$

The field equations in covariant form are

$$\frac{\partial F_{\mu\nu}}{\partial x_\mu} + \frac{\partial \lambda}{\partial x_\nu} = 0. \tag{15.79}$$

P_ϕ, the momentum conjugate to ϕ is

$$P_\phi = -\lambda. \tag{15.80}$$

Dilemma: To get the Maxwell equations in their usual form one wishes the second term in Eq. (15.79) to vanish, which can be shown to amount to $\lambda=0$—the Lorentz gauge condition (15.60). But $\lambda=0$ means $P_\phi=0$ and one cannot quantize consistently, as seen earlier.

Fermi's resolution: only such states Ψ, the 'physical states', are realized in nature for which

$$\lambda\Psi = 0, \qquad \frac{d\lambda}{dt}\Psi = 0, \tag{15.81}$$

weaker demands than the operator identity $\lambda = 0$. (This procedure is evidently different from the way Heisenberg and Pauli handled their ε-term.) A nice exposé of this method is found in Wentzel's book.[112]

Other items:

It is possible to quantize the electromagnetic field without any constraint on A_μ. This, however, leads to a Hamiltonian which is not positive definite.[113]

The electromagnetic field without Lorentz condition is a mixture of spin-1 and spin-0 fields. The Lorentz condition serves to eliminate spin-zero states from the theory.[114]

No problems with ϕ and its conjugate momentum would arise if the photon had a mass $m = \hbar\kappa/c$. In that case a term $-\kappa^2 A_\mu^2/2$ must be added to $\mathscr{L}$ in Eq. (15.55). The equation div $\vec{E} = 0$ becomes div $\vec{E} + \kappa^2\phi = 0$ so that ϕ can be eliminated as a dynamical variable in favor of $\vec{E}$. One may try to treat quantum electrodynamics as the limit $\kappa \to 0$ of such a formalism. That limiting process introduces singularities, however, which are equivalent to those generated by a Lorentz transformation for velocity v followed by $v \to c$.[115]

Fermi applied Eq. (15.81) in the right way but his formalism was inconsistent. It turned out that his Ψ's have infinite norm.[116] This difficulty is overcome in the indefinite metric approach first developed for the free field case,[117] then extended to the presence of sources.[118] This formalism, based on Fermi's approach, contains a constraint on physical states given by

$$\lambda^+\Psi = 0 \tag{15.82}$$

where λ^+ is the positive frequency part of λ defined in Eq. (15.78). This equation implies

$$\langle\Psi|\lambda|\Psi\rangle = 0. \tag{15.83}$$

For details see e.g. Ref. 119.

Sources, quantum field theory prior to renormalization

Schwinger's source book[120] contains reprints of Dirac's first paper,[25] one of Fermi's papers,[111] and the Jordan–Wigner article.[41]

For reminiscences by participants in the developments see papers by Dirac,[40] Peierls,[121] Weisskopf,[122] and Wentzel.[123]

For historical essays see Jost,[27] Pais,[124] Weinberg,[125] and Wightman,[126] and especially 'Some chapters for a history of quantum field theory: 1938–1952', by Schweber, to be published. I had the privilege to see this article in reprint form.

The evolution of quantum field theory can be followed in the successive reviews by Fermi (1932),[111] Pauli (1933),[23] Pauli (1941),[71] and Pais (up to

December 1947),[127] and in the outstanding books by Heitler (1936)[72] and Wentzel (1943).[112]

Finally I mention Bromberg's interesting historical essay on particle creation.[128]

References

1. A. Einstein, *Phys. Zeitschr.* **18**, 121, 1917.
2. A. Pais, *Subtle is the Lord*, Chapter 5, Section (a), Oxford Univ. Press 1982.
3. J. C. Maxwell, *Theory of Heat*, Chapter 22, Longman, London 1872, repr. by Greenwood Press, Westport, Conn.
4. Ref. 2, Chapter 19, Section (f), and Chapter 22.
5. A. Einstein, *Verh. Deutsch. Phys. Ges.* **18**, 318, 1916.
6. A. Einstein, *Mitt. Phys. Ges. Zürich* **16**, 47, 1916.
7. Cf. Ref. 2, Chapter 21, Section (b).

7a. N. R. Hanson, *The concept of the positron*, Cambridge Univ. Press 1963.

8. W. Heisenberg, interview with T. Kuhn, February 28, 1963, transcript in Niels Bohr Library, Am. Inst. of Physics, New York.
9. M. Born and P. Jordan, *Zeitschr. f. Phys.* **34**, 858, 1925.
10. English transl. of Ref. 9 in B. L. van der Waerden, *Sources of quantum mechanics*, Dover, New York 1968.
11. W. Heisenberg, letter to W. Pauli, October 23, 1925, repr. in *Wolfgang Pauli, scientific correspondence*, referred to as *PC* below, Eds. A. Hermann, K. von Meyenn, and V. Weisskopf, Vol. 1, p. 251, Springer, New York 1979.
12. M. Born, W. Heisenberg, and P. Jordan, *Zeitschr. f. Phys.* **35**, 557, 1925; Engl. transl. in Ref. 10.
13. P. Jordan, *Zeitschr. f. Phys.* **30**, 297, 1924; cf. also A. Einstein, *Zeitschr. f. Phys.* **31**, 784, 1925.
14. A. Einstein, *Phys. Zeitschr.* **10**, 185, 817, 1909.
15. Cf. also Ref. 2, Chapter 21, Section (a).
16. A. Einstein, *Ann. der Phys.* **20**, 199, 1906; cf. also Ref. 2, Chapter 19, Section (d).
17. P. Debye, *Ann. der Phys.* **33**, 1427, 1910.
18. L. S. Ornstein and F. Zernike, *Proc. K. Ak. Amsterdam* **28**, 280, 1919; P. Ehrenfest, *Zeitschr. f. Phys.* **34**, 362, 1925.
19. P. Jordan and W. Pauli, *Zeitschr. f. Phys.* **47**, 151, 1928.
20. Cf. J. J. Gonzalez and H. Wergeland, *K. Nord. Vidensk. Skr.* No. 4, 1973.
21. P. A. M. Dirac, interview with T. Kuhn, May 7, 1963, transcript in Niels Bohr Library, American Institute of Physics, New York.
22. P. A. M. Dirac, *Proc. Roy. Soc. A* **112**, 661, 1926.
23. W. Pauli, in *Handbuch der Physik*, Vol. 24/1, Sections 15 and 16, Springer, Berlin 1933.
24. L. I. Schiff, *Quantum mechanics*, 2nd edn, Chapter 10, McGraw-Hill, New York 1955.
25. P. A. M. Dirac, *Proc. Roy. Soc. A* **114**, 243, 1927.
26. P. A. M. Dirac, *Proc. Roy. Soc. A* **114**, 710, 1927.
27. R. Jost, in *Aspects of quantum theory*, Eds. A. Salam and E. P. Wigner, p. 61, Cambridge Univ. Press 1972.
28. Ref. 26, p. 715.

29. Ref. 26, Eqs. (9) and (10).
30. Ref. 24, Sections 48, 46.
31. Ref. 25, p. 262, footnote.
32. Cf. Ref. 24, pp. 352–5.
33. Ref. 25, p. 260.
34. Ref. 25, p. 261.
35. Cf. Ref. 24, pp. 201–5.
36. P. Jordan and O. Klein, *Zeitschr. f. Phys.* **45**, 751, 1927.
37. V. Fock, *Zeitschr. f. Phys.* **75**, 622, 1932; **76**, 852, 1932.
38. P. Jordan, *Zeitschr. f. Phys.* **44**, 473, 1927.
39. M. Born, in *Electrons et photons*, Proc. 5th Solvay conference p. 176, Gauthier-Villars, Paris 1928.
40. P. A. M. Dirac, in *History of twentieth century physics*, p. 140, Academic Press, New York 1977.
41. P. Jordan and E. Wigner, *Zeitschr. f. Phys.* **47**, 631, 1928.
42. W. Pauli, letter to H. A. Kramers, February 7, 1928, *PC*, Vol. 1, p. 432.
43. W. Pauli and W. Heisenberg, *Zeitschr. f. Phys.* **56**, 1, 1929.
44. Ref. 24, Sections 45, 48.
45. L. Rosenfeld, *Zeitschr. f. Phys.* **63**, 574, 1930.
46. *Wissenschaftliche Abhandlungen von Hermann von Helmholtz*, Vol. 3, p. 476, Barth, Leipzig 1895.
47. H. A. Lorentz, in *Encyclopädie der mathematischen Wissenschaften*, Vol. 5, Chapter 14, Teubner, Leipzig 1904–22.
48. M. Born, *Ann. der Phys.* **28**, 571, 1909.
49. G. Mie, *Ann. der Phys.* **37**, 511, 1912; **39**, 1, 1912; **40**, 1, 1913.
50. Ref. 43, Eq. (46).
51. W. Pauli, letter to O. Klein, February 18, 1929, *PC*, Vol. 1, p. 488.
52. W. Pauli and W. Heisenberg, *Zeitschr. f. Phys.* **59**, 168, 1930.
53. Cf. E. T. Whittaker, *A History of the theories of aether and electricity*, 2nd edn, Vol. 1, pp. 189, 201, Nelson, London 1958.
54. J. C. Maxwell, 'On Faraday's lines of force', 1856, repr. in *The scientific papers of J. C. Maxwell*, Vol. 1, p. 477, Dover, New York.
55. J. C. Maxwell, papers, Ref. 54, Vol. 1, p. 555.
56. Ref. 47, Chapter 14, Section 3; also H. A. Lorentz, *The theory of electrons*, p. 239, Teubner, Leipzig 1909.
57. Cf. also W. Pauli, 'Relativitätstheorie', Section 64, in *Encyclopädie der mathematischen Wissenschaften*, Vol. 5, Teubner, Leipzig 1921.
58. H. Weyl, *Sitz. Ber. Preuss. Ak. Wiss. 1918*, p. 465.
59. H. Weyl, *Math. Zeitschr.* **2**, 384, 1918.
60. H. Weyl, *Ann. der Phys.* **59**, 101, 1919.
61. A. Pais, *Physica* **8**, 1137, 1941.
62. E. Schroedinger, *Ann. der Phys.* **81**, 109, 1926, Eq. (36).
63. P. A. M. Dirac, *Proc. Roy. Soc. A* **117**, 610, 1928, Eq. (14).
64. W. Gordon, *Zeitschr. f. Phys.* **40**, 117, 1927, esp. p. 119.
65. F. London, *Zeitschr. f. Phys.* **42**, 375, 1927.
66. H. Weyl, *Gruppentheorie und Quantenmechanik*, 1st edn, pp. 87, 88, Hirzel, Leipzig 1928.
67. English transl. of Ref. 66 (2nd edn): *The theory of groups and quantum mechanics*, transl. H. P. Robertson, pp. 100, 101, Dover, New York.

68. H. Weyl, *Proc. Nat. Ac. Sci.* **15**, 232, 1929.
69. H. Weyl, *Zeitschr. f. Phys.* **56**, 330, 1929.
70. Cf. also Ref. 67, p. 214.
71. W. Pauli, *Rev. Mod. Phys.* **13**, 203, 1941.
72. W. Heitler, *The quantum theory of radiation*, 1st edn, Clarendon Press, Oxford 1936.
73. Ref. 40, p. 144.
74. W. Heisenberg, *Der Teil und das Ganze*, Piper Verlag, Munich 1969; in English: *Physics and beyond*, transl. A. J. Pomerans, Chapter 8, Harper and Row, New York 1971.
75. W. Heisenberg, interview with T. Kuhn, July 12, 1963. Transcript in the Niels Bohr Library, American Institute of Physics, New York.
76. P. A. M. Dirac, *Proc. Roy. Soc. A* **118**, 351, 1928.
77. P. A. M. Dirac, *Phys. Zeitschr.* **29**, 561, 712, 1928.
78. W. Heisenberg, letter to W. Pauli, May 3, 1928, *PC*, Vol. 1, p. 443.
79. Ref. 77, p. 562, footnote 2.
80. W. Heisenberg, letter to W. Pauli, July 31, 1928, *PC* Vol. 1, p. 466.
81. P. A. M. Dirac, letter to O. Klein, July 24, 1928, copy in Niels Bohr Library, American Institute of Physics, New York.
82. O. Klein and Y. Nishina, *Zeitschr. f. Phys.* **52**, 853, 1929; Y. Nishina, ibid., 869.
83. W. Heisenberg, *Ann. der Phys.* **13**, 430, 1932.
84. L. Meitner and H. Hupfeld, *Zeitschr. f. Phys.* **67**, 106, 1931.
85. I. Waller, *Zeitschr. f. Phys.* **61**, 837, 1930.
86. I. Tamm, *Zeitschr. f. Phys.* **62**, 545, 1930.
87. See also Ref. 72, Section 16.
88. O. Klein, *Zeitschr. f. Phys.* **53**, 157, 1929.
89. Cf. e.g. C. Itzykson and J. B. Zuber, *Quantum field theory*, Chapter 2, McGraw-Hill, New York 1980.
90. P. A. M. Dirac, *Proc. Roy. Soc. A* **126**, 360, 1929; also *Nature* **126**, 605, 1930.
91. Ref. 40, p. 144.
92. P. A. M. Dirac, letter to N. Bohr, November 26, 1929, copy in Niels Bohr Library, American Institute of Physics, New York.
93. P. A. M. Dirac, *Proc. Cambr. Phil. Soc.* **26**, 361, 1930.
94. P. A. M. Dirac, letter to N. Bohr, December 9, 1929, copy in Niels Bohr Library, American Institute of Physics, New York.
95. Ref. 72, pp. 189, 190.
96. R. Oppenheimer, *Phys. Rev.* **35**, 562, 1930.
97. Ref. 67, pp. 263–4 and Preface.
98. P. A. M. Dirac, *Proc. Roy. Soc. A* **133**, 60, 1931.
99. C. D. Anderson, *Science* **76**, 238, 1932.
100. C. D. Anderson and H. L. Anderson, in *The birth of particle physics*, Eds. L. M. Brown and L. Hoddeson, p. 131, Cambridge Univ. Press 1983; also C. D. Anderson in *Nobel lectures in Physics 1922–41*, p. 365, Elsevier, New York 1965.
101. C. D. Anderson, interview with C. Weiner, June 30, 1966, transcript in Niels Bohr Library, American Institute of Physics, New York.
102. C. D. Anderson, *Phys. Rev.* **43**, 491, 1933.
103. Cf. also I. Waller, *Zeitschr. f. Phys.* **21**, 213, 1928.
104. P. A. M. Dirac, *Proc. Roy. Soc. A* **113**, 621, 1927, Section 2.

105. Cf. I. Halperin and L. Schwartz, *Introduction to the theory of distributions*, Toronto Univ. Press 1952.
106. Cf. Ref. 24, Eqs. (48.40–42).
107. N. Bohr and L. Rosenfeld, *Dansk. Vid-Selsk. Mat.-Fys. Medd.* Vol. 12, p. 8, 1933; *Phys. Rev.* **78**, 794, 1950.
108. L. Landau and R. Peierls, *Zeitschr. f. Phys.* **69**, 56, 1931.
109. L. Rosenfeld and J. Solomon, *J. de Phys.* **2**, 139, 1931.
110. H. B. G. Casimir, *Proc. K. Ak. Amsterdam* **51**, 793, 1948.
111. E. Fermi, *Rend. Acc. Lincei* **9**, 881, 1929; **12**, 431, 1930; *Rev. Mod. Phys.* **4**, 87, 1932.
112. G. Wentzel, *Einführung in die Quantentheorie der Wellenfelder*, Section 16, Deuticke, Vienna 1943; in English: *Quantum theory of fields*, transl. C. Houtermans and J. M. Jauch, Interscience, New York 1948.
113. L. Rosenfeld, *Zeitschr. f. Phys.* **76**, 729, 1932.
114. M. Fierz, *Helv. Phys. Acta* **12**, 3, 1939.
115. L. J. F. Broer and A. Pais, *Proc. K. Ak. Amsterdam* **48**, 3, 1945.
116. S. T. Ma, *Phys. Rev.* **75**, 535, 1949.
117. S. N. Gupta, *Proc. Phys. Soc. London A* **63**, 681, 1950.
118. K. Bleuler, *Helv. Phys. Acta* **23**, 567, 1950.
119. Ref. 89, Chapter 3.
120. J. Schwinger, *Selected papers on quantum electrodynamics*, Dover, New York 1958.
121. R. Peierls, in *The physicist's conception of nature*, Ed. J. Mehra, p. 370, Reidel, Boston 1973.
122. V. Weisskopf, Ref. 100, p. 56; *Physics Today*, November 1981, p. 69.
123. G. Wentzel, in *Physics in the twentieth century*, p. 48, Interscience, New York 1960, repr. in Ref. 120, p. 380.
124. A. Pais, Ref. 27, p. 79.
125. S. Weinberg, *Daedalus* **106**, No. 4, p. 17, 1976.
126. A. Wightman, Ref. 27, p. 95.
127. A. Pais, *Developments in the theory of the electron*, pamphlet, Princeton Univ. Press 1948.
128. J. Bromberg, *Hist. St. Phys. Sc.* **7**, 161, 1976.

16

Battling the infinite

I think that this discovery of antimatter was perhaps the biggest jump of all the big jumps in physics in our century.

Heisenberg in 1972[1]

(a) Introduction

Pauli's rhythmic body movements back and forth* were sometimes accompanied by a shaking of the head in the horizontal plane. The frequencies of both motions, increasing as they did with his level of agitation, must have been high when during his stay* at Ann Arbor in the summer of 1931 he attended a lecture on the Dirac equation by Robert Oppenheimer. In the middle of that talk (George Uhlenbeck told me) Pauli stood up, marched to the blackboard, and grabbed a piece of chalk. There he stood, facing the board, waving the chalk in his hand, then said: 'Ach nein, das ist ja alles falsch', . . . all that is wrong anyway. Kramers commanded his friend Pauli to hear the speaker out. Pauli walked back and sat down.

What exactly was on Pauli's mind at that time I do not know. The Ann Arbor summer school took place shortly after Dirac had proposed[2] his anti-electron (positron) but I am not sure whether Pauli knew that.** It would not have made much difference in any event. Well after Pauli had heard of this new idea but before Anderson's discovery he wrote: 'We do not believe that this way out [holes are positrons] should be seriously considered'.[3]

It goes without saying that the detection of the positron was considered by nearly everyone as a vindication of Dirac's theory. Yet its basic idea: a positron is a hole in an infinite sea of negative electrons remained unpalatable to some, and not without reason. Even the simplest state, the vacuum, was a complex consisting of infinitely many particles, the totally filled sea. Interactions between these particles left aside, the vacuum had a negative infinite 'zero point energy' and an infinite 'zero point charge'. Pauli did not like that. Even after the positron had been discovered he wrote to Dirac: 'I do not believe in your perception of "holes" even if the "anti-electron" is proved.'[4] That was not all, however. Pauli to Heisenberg one month later: 'I do not believe in the hole theory since I would like to have asymmetries between positive and

* Chapter 14, Section (d). ** Dirac's paper appeared in print in September.

negative electricity in the laws of nature (it does not satisfy me to shift the empirically established asymmetry to one of the initial state)'.[5] Regarding the zero point infinities, let it directly be said that these are actually innocuous and can be eliminated by a simple reformulation of the theory (as we shall see).

Other successes soon followed the discovery of the positron. At moderate energies the Klein–Nishina formula turned out to work well, as did the predictions for radiation loss (*Bremsstrahlung*) and the production and annihilation of electron–positron pairs. Nevertheless, leaders in the field continued to express grave misgivings. The cause for their concern (briefly alluded to in Chapter 15, Section (a)) was obvious. Even after the zero point energy and charge are eliminated, the positron theory is still riddled with infinities caused by interactions. All good results were obtained in leading approximation. All higher approximations were divergent. (The concept of 'approximations' is defined in Section (d).)

It is the main aim of this chapter to describe how these infinities came to be diagnosed and how theorists responded to what may fairly be called a new crisis: how to cope with a theory which works very well approximately but which makes no sense rigorously. As Pauli put it in 1936 during a seminar given in Princeton: 'Success seems to have been on the side of Dirac rather than of logic'.[6]

Since it is imperative to recall that some of the infinities to be encountered below are not uniquely characteristic for the positron theory but actually have a history dating back to classical times, I have devoted Section (c) of this chapter to the infinities before the positron theory.

Before defining the issues with greater precision I should like first to recapture some of the attitudes of those who were actively involved with these problems during the thirties. First there is Dirac's remark[7] (mentioned earlier*) made *a propos* certain doubts concerning the validity of the energy conservation law: 'The only important fact [of existing theory] that we have to give up is quantum electrodynamics ... we may give it up without regrets—in fact because of its extreme complexity, most physicists will be glad to see the end of it.'

These lines were written in 1936 by which time Dirac had made his last contributions of major importance: the theory of magnetic monopoles, the application of Hartree's self-consistent field method to hole theory, and the diagnosis of the polarization of the vacuum (Section (d)). Thereafter his strong discontent with quantum electrodynamics led him to the conviction that one should scrap the theory and seek a new starting point. Several times he made such new beginnings (see Section (e)) but without success.

Next there is Heisenberg[8] writing to Pauli in 1935: 'In regard to quantum electrodynamics we are still at the stage in which we were in 1922 with regard

* Chapter 14, Section (d), Part 5.

to quantum mechanics. We know that everything is wrong. But in order to find the direction in which we should depart from what prevails we must know the consequences of the prevailing formalism much better than we do.'*

Evidently Heisenberg shared Dirac's disbelief in the theory but, unlike him, took the position that progress might be possible by facing the issues squarely. He is one of that quite small band of theoretical physicists who had the courage to explore those aspects of quantum electrodynamics which remained in an uncertain state until the late 1940s, when renormalization would provide more systematic and more successful ways of handling the problems. Heisenberg's 1935 letter to Pauli is also of great interest because of its sharp contrast to his later opinion mentioned at the end of this section. Whether or not one agrees in detail with what he said in 1972, certain it is that the positron theory must be considered the greatest triumph for theoretical physics in the 1930s—but that became clear only in retrospect.

My final deponent is Wentzel whose recollections[9] of the thirties were written in 1960. 'In spite of all failures, the general confidence in quantum electrodynamics as a supreme, though as yet imperfect, tool of atomistic theory remained alive .... Nevertheless, the awareness of the basic difficulties weighed heavily on our minds.'

I believe this spectrum of opinions to be characteristic for that decade of uncertainty. Not that Pauli, Dirac, Heisenberg, and Wentzel would hold on to one immutable position during that period. Nor should it be forgotten that these men, though only in their thirties, were already seasoned campaigners, having participated in establishing the great successes of non-relativistic quantum mechanics. Those experiences must have colored their opinions in regard to the positron theory. In those times a variance in age of but a few years could amount to a generational difference. To mention but one example, Robert Oppenheimer, who made his mark in quantum field theory, was only four years younger than Pauli. Yet, when in later years I was on occasion together with both, I noted how Oppenheimer understandably treated Pauli as the veteran of battles waged in his childhood.

I conclude this sampling of views with a few remarks by experimentalists. As was mentioned at the end of the previous chapter, in September 1932 Anderson announced[10] his preliminary evidence for the positron. By February 1933 he had obtained more detailed results.[11] Meanwhile, earlier that month, independent confirmation had been obtained at the Cavendish, where, sometime in 1932, Occhialini, thoroughly familiar with Geiger counters, had joined Blackett, a cloud chamber expert. 'According to Blackett, at that time the Geiger counter was a very delicate instrument. As he put it: "In order to make

* 'Wir sind ... in Bezug auf der Quantenelectrodynamik noch in dem Stadium in dem wir bezüglich der Quantum Mechanik 1922 waren. Wir wissen das alles falsch ist. Aber um die Richtung zu finden in der wir das Bisherige verlassen sollen, müssen wir die Konsequenzen des bisherigen Formalismus viel besser wissen, als wir es tun.'

it work you had to spit on the wire on some Friday evening in Lent." They decided to pool their instrumental knowledge and investigate cosmic rays'.[12] The result was the invention of the counter-controlled cloud chamber, a device in which the expansion is triggered by the coincidence of pulses in one Geiger counter above the chamber with one below it.[13] 'As soon as they saw Anderson's article they immediately looked for [positron] events in their own plates and found them in great abundance',[12] including a number of cases in which an electron–positron pair appears.[14] They considered three possibilities for the origin of these pairs: they may have existed in the struck nucleus, or in the incident particle, or 'they may have been created during the process of collision ... it is reasonable to adopt the last hypothesis'. It is evident from their paper[15] that they had been in touch with Dirac. Indeed Blackett later recalled[12] that Dirac worked closely with them and was often in the laboratory.

Since Oppenheimer, expert on positron theory, spent a considerable time at Caltech, I wondered whether he likewise had suggested the pair formation mechanism to Anderson. When I recently asked Carl Anderson he replied: 'I am astounded to this day that [O.] did not come up with the idea of pair production* ... During the time he was in Pasadena he had daily access to any new results .... We were all familiar (even me) with Dirac's [hole theory] interpretation .... I first learned of the pair production idea from the Blackett–Occhialini paper ... and immediately recognized it as being correct'.[17]

I return once more to Blackett. Upon being asked in 1962 how long he and Occhialini had known about the Dirac theory at the time of their discoveries, he replied that he could not recall but that it did not matter anyway because nobody took Dirac's theory seriously.[12] I tend to think that at this late date we still see here the influence of Rutherford who in 1933 made one of his characteristic comments[18] on theoretical physics: 'It seems to me that in some way it is regrettable that we had a theory of the positive electron before the beginning of the experiments. Blackett did everything possible not to be influenced by the theory, but the way of anticipating results must inevitably be influenced to some extent by the theory. I would have liked it better if the theory had arrived after the experimental facts had been established'.

One last word about experimental physics in the early 1930s and its bearing on quantum electrodynamics. That period is marked not only by theoretical difficulties resulting from the infinities but also by confusion about what cosmic rays were revealing experimentally. For example, in April 1932 Millikan and Anderson concluded[19] in their report on cloud chamber observations of cosmic rays: 'Formulae like Klein–Nishina ... can have no validity in the cosmic ray field.'

It would take some years before it became clear that this apparent conflict between theory and experiment was caused by the presence of mesons in

* For Oppenheimer's recollections on this point see Ref. 16.

cosmic rays. Meanwhile one finds recurrent statements in the literature about a breakdown of quantum electrodynamics at energies of the general order of 100 MeV (see e.g. Section (d), Part 1, below).

1920–30 was the final decade in which physics at the frontiers was quintessentially European. Moving now into the 1930s we encounter fundamental contributions to quantum field theory originating in the United States and in Japan. Leaving the latter for Chapter 17, I turn in Section (b) to a brief sketch of the American physics scene with emphasis on Robert Oppenheimer's role in bringing quantum electrodynamics to the U.S. and making it flourish there. Section (c) deals, as said, with the early history of infinities. Section (d), devoted to quantum electrodynamics in the 1930s, begins with an enumeration of the good results obtained in second order. Thereafter the early battles with the infinities are treated. The final Section (e) deals with searches for alternatives to quantum electrodynamics resulting from dissatisfaction with that theory. These efforts sometimes side-stepped concurrent developments in the positron theory, thus lending a curious diversity to the theoretical efforts of that period. It is worthwhile to mention at least some of these fruitless attempts, if it were only to illustrate the varieties of opinion in regard both to how things stood and how to proceed next.

(b) Physics in America: the onset of maturity; the emergence of Oppenheimer

On 20 May, 1899, thirty-eight physicists,* gathered in Fayerweather Hall, Columbia University, to found the American Physical Society, elected from their midst two colleagues of international renown as senior officers: Henry Rowland as president, Albert Michelson (who some years later would become the first U.S. Nobel laureate in physics) as vice-president. *The Physical Review* founded by Edward Nichols had begun publication six years earlier. The title pages of its first fifteen volumes are marked 'Published for Cornell University'. In 1903 this was replaced by 'Conducted with the cooperation of the American Physical Society'.

Neither Rowland nor Michelson had formally earned the Ph.D. degree although the first American Ph.D. in physics was conferred (by Yale) in 1861.[21] Prior to that time American physics was principally represented by a small number of highly prominent figures each working in virtual isolation. In the closing decades of the nineteenth century a broader base began to emerge as increasing numbers of Americans went abroad for the completion of their physics studies. Already before 1900, leading European physicists had come for visits to the United States: Helmholtz, Rayleigh, Rutherford, W. Thomson

* Their names are listed in Ref. 20.

(Kelvin), J. J. Thomson, Tyndall. The list of distinguished visitors grew rapidly. Before 1910, J. J. Thomson, Rutherford, and Nernst were Silliman lecturers at Yale; Boltzmann and Rutherford taught at Berkeley; Lorentz and Planck at Columbia; Jeans and O. W. Richardson joined the faculty in Princeton.*

At home, educational opportunities and laboratory facilities improved, as did scientific productivity in experimental physics. American activity on the new frontiers in theoretical physics, the old quantum theory and special relativity, remained marginal, however. Still there were some notable contributions such as Kemble's** (Harvard) on band spectra,[24] van Vleck's[25] and Slater's[26] on the old quantum theory of radiation, and Lewis and Tolman's (MIT)[27] on the energy–momentum laws in relativity.†

During the 1920s a new generation was preparing at four main European schools for what was to come: Bohr's in Copenhagen, Born's in Goettingen, Rutherford's in Cambridge, and Sommerfeld's in Munich. 'In the winter of 1926', K. T. Compton has recalled, 'I found more than twenty Americans in Goettingen at this fount of quantum wisdom'.[29] In addition, by 1930 the new gospel had been preached in America by Bohr, Born, Dirac, Schroedinger, and Sommerfeld. From the 1920s also dates the establishment of the National Research Fellowships, the first national program for support of post-doctoral studies. The International Education Board and the John Simon Guggenheim Memorial Foundation, both newly founded, likewise made available research stipends.

American efforts in experimental physics continued at a high level. What is most characteristically new for that period is the steep rise in U.S. contributions to theoretical physics as the era of quantum mechanics began. Van Vleck later said: 'Fairly suddenly . . . America came of age in physics, for although we did not start . . . quantum mechanics, our young theorists joined it promptly.'[30]

A list of those who, by 1930, had begun to make their mark in quantum theory includes Breit, Condon, Dennison, Eckart, Houston, Loomis, Margenau, Ph. Morse, Mulliken, Oppenheimer, Pauling, Rabi, Slater, Stratton, van Vleck. 'By 1930 or so, the relative standing of *The Physical Review* and *Philosophical Magazine* were interchanged.'[30]

The rise of the natives†† was accompanied by an influx of European physicists, caused both by the dearth of academic positions in Europe and the hunger for young talent in the U.S. Several men encountered earlier in this book are

* For a list of European visitors to the U.S. from 1872 to 1935 see Ref. 22.
** In 1929 Kemble would publish the first review article[23] on quantum mechanics for *Reviews of Modern Physics*.
† There were also American contributions to the old quantum theory of magnetism.[28]
†† Among whom I count men like Breit and Rabi, born elsewhere but raised in the U.S. from an early age.

among those who found jobs in the U.S. before the Hitler period: Goudsmit, Herzfeld, Landé, von Neumann, L. H. Thomas, Uhlenbeck, Wigner.* This list swelled mightily after 1933 when the Nazis began their cruel attacks on the culture of Western Europe. 'These distinguished Europeans were responsible, not for giving us maturity, but rather for carrying us still further to pre-eminence.'[30] The most prominent of these later emigrés was, of course, Einstein, whose fame more than his later contributions added so much to the prestige in which the American scientific establishment came to be held. When he spoke at an AAAS meeting in 1934, admission was by ticket only and some newspaper reporters broke through the windows to get in.[30]

The impact of the summer symposia at Ann Arbor on the growth of theoretical physics in America deserves special mention. Begun in 1923, these gatherings acquired additional vigor some years later when four young men, Dennison, Goudsmit, Laporte, and Uhlenbeck, took principal charge. Among those who lectured at the summer school before the Second World War were Bethe (twice), Bohr, Dirac, Ehrenfest, Fermi (four times), Heisenberg, Kramers (twice), Pauli (twice), Sommerfeld, and Wigner. Of particular interest for the spreading of quantum field theory were Fermi's 1930 lectures on quantum electrodynamics which, edited by Uhlenbeck, were published soon thereafter.[32] Fermi's exposé made this new subject clear to a wider audience especially because it contained numerous applications. According to Uhlenbeck, Fermi had prepared himself for lecturing in English by taking a course at a Berlitz school in Rome, thus acquiring fluency which, however, was not quite accent-free. 'Finite' became 'feeneetay'. When a listener asked about something being infinite Fermi did not understand and Uhlenbeck had to explain. 'Ah', Fermi smiled, 'infeeneetay'.

Vigorous administrative efforts by Birge (Berkeley) and Millikan (Pasadena) did much to attract young talent in theoretical physics to California. Thus it came to pass that in 1929 J. Robert Oppenheimer betook himself to the Far West to teach how particles are made and how they disappear.

Already when Oppenheimer graduated *summa cum laude* in chemistry from Harvard College he had a great love for physics, in which he was widely read. In October 1925 he entered Christ's College, Cambridge, as a Fellow. Initially intending to do experimental physics, he hoped to be taken on by Rutherford; 'but Rutherford wouldn't have me',[33] so he worked briefly under J. J. Thomson. He was soon fascinated by the new theoretical developments. 'I didn't learn about quantum mechanics until I got to Europe . . . I formed the avid habit of reading everything as fast as I could lay my hands on it.'[33] Fowler became his first mentor in theoretical physics. He got to know his fellow student Dirac, as well as Bohr, Born, and Ehrenfest when they came for visits. 'When

* For a list of scientific emigrés to the U.S. up to 1936 see Ref. 31.

Rutherford introduced me to Bohr he asked me what I was working on. I told him and he said, "How is it going?" I said, "I'm in difficulties." He said, "Are the difficulties mathematical or physical?" I said, "I don't know." He said, "That's bad".'[33]

Stimulated by discussion with Fowler and Dirac, Oppenheimer published his first two papers, on quantum mechanics. His Cambridge period was therefore not without profit. Above all, however, it was for him a time of highly precarious emotional balance. Rutherford told Dirac that he once saw Oppenheimer fall in a faint to the floor in a laboratory. (Dirac witnessed a similar occurrence later in Goettingen.[34]) In January 1926 Oppenheimer attempted to strangle a close friend of his (no harm was done).[35,36] There were visits to psychiatrists, first one in Cambridge, then one in London.[37] The cause of these disturbances is not quite clear to me. I have the impression that Oppenheimer's struggles to make the change from experimental to theoretical physics played some role, but only a secondary one. In any event he pulled through. Yet, during the sixteen years I knew him in later times, I came to understand that vast insecurities lay forever barely hidden behind his charismatic exterior, whence an arrogance and occasional cruelty befitting neither his age nor his stature. Rabi's six-page[38] essay on Oppenheimer, which conveys more about the man than anything else written about him, says it best: 'In Oppenheimer the element of earthiness was feeble'.

In the summer of 1926 Oppenheimer worked several months with Ehrenfest. During that Leiden period he shared a pair of rooms with Uhlenbeck, who remembers his great need for intimate friendships at that time. From there he went to Goettingen where he stayed till the summer of 1927. He now began to flourish. His teacher, Born, remembered him later as a man already of great talent and conscious of his superiority. From that time also dates his habit, causing occasional annoyance, of explaining to someone what he was saying before the other had finished saying it.[39] In Goettingen he got his Ph.D. degree on the quantum mechanics of continuous spectra,[40] a difficult subject for its time. His contributions prior to those in quantum field theory are contained in seventeen papers, some of them very fine, which I shall not discuss in detail.* A few samples may suffice. With Born he developed approximation procedures in the theory of molecular spectra.[42] He was the first to note and evaluate the role of quantum mechanical tunneling in the field emission of electrons from metal surfaces,[43] and also the first to treat[44] exchange interference in electron collisions.**

After his return to the U.S. in the summer of 1927, Oppenheimer divided a year's NRC Fellowship among Harvard, Berkeley, and Caltech. His future plans began to take shape. 'I visited Berkeley and I thought I'd like to go to

* An analysis of this work is found in Ref. 41.

** The well-known scattering formula $\sigma(\vartheta) = (|f(\vartheta)+g(\vartheta)|^2 + 3|f(\vartheta)-g(\vartheta)|^2)/4$ is due to him.

Berkeley because it was a desert. There was no theoretical physics and I thought it would be nice to start something. I also thought it would be dangerous because I'd be too far out of touch so I kept the connection with Caltech . . . [which] was a place where I would be checked if I got too far off base.'[16] First, however, he intended to use another year's NRC Fellowship for a further stay in Europe. 'I had some T.B. . . . . It was never very serious but I did go to the mountains for the summer . . . . I was a little late in getting to Europe because of this illness.'[16]

In September 1928 Oppenheimer returned to Ehrenfest.* 'I went to Ehrenfest because he asked me to and I was a great admirer of his.'[16] Ehrenfest, who thought well of Oppenheimer but found him too quick and too learned,[45] directed his next move which was to be decisive for his scientific career: He dissuaded him from his intention of going to Copenhagen afterward. '[It was] Ehrenfest's certainty that Bohr with his largeness and vagueness was not the medicine I needed but that I needed someone who was a professional calculating physicist and that Pauli would be right for me.'[16] In November Ehrenfest wrote to Pauli: 'He [O.] always has very witty ideas . . . . I am really convinced that for the full development of his (great) scientific talent [he] should in time be a bit lovingly beaten into shape.'[46]**

Shortly after Oppenheimer's arrival in Zürich (January 1929) Pauli gave Ehrenfest his impressions[47] of him. His strength: many and good ideas, much inventiveness. His weakness: too quickly satisfied with imperfectly substantiated statements, lack of perseverance and thoroughness, leaves his problems in a half-digested state of belief or disbelief. 'I definitely hope, however, that in the course of time he can improve by energetic coaxing.' Pauli further noted that Oppenheimer treated him too much as an authority figure. 'I believe to know how this need for extraneous authorities comes about in him. They have to solve his problems and answer his questions, so that he does not have to do it himself. (This connection is of course not *consciously* clear to him, only latent in the unconscious.)' Pauli said to Rabi (who was also in Zürich at that time) that Oppenheimer seemed to treat physics as an avocation and psychoanalysis as a vocation.[38]

Pauli's hopes for improvement turned out to be altogether justified. Oppenheimer soon completed a paper on the radiation of electrons in a Coulomb field[48] about which Pauli wrote approvingly to Sommerfeld.[49] Then he went on to quantum electrodynamics. For that purpose the timing of his arrival in Zürich was perfect. In March, Heisenberg and Pauli part one[50] had been completed, after which its authors went to work right away on part two.[51] Recall† that this was the paper which laid the basis for the Coulomb gauge calculations of the 1930s. Thus Oppenheimer had the great good fortune to

* During this Dutch period he also spent a month with Kramers in Utrecht.
** '. . . noch rechtzeitig a bisserl liebevoll zurechtgeprügelt werden sollte.'
† See Chapter 15, Section (e).

enter practical quantum electrodynamics on the ground floor.* Of that he made good use and completed a piece of work on the self-energy problem to which I return in the next section. Heisenberg suggested to Pauli that this work be included in their part two with Oppenheimer as co-author.[52]

In actual fact Oppenheimer's contribution ('the third part of the Heisenberg–Pauli paper', he called it[16]) was published separately after he had gone back to the U.S. in June 1929. (He was not to return to Europe until after the Second World War.) His European sojourns may have been terribly difficult for him at times but, all told, their harvest had been excellent. Now he was to become America's leading teacher in quantum field theory.

In the fall of 1929 Oppenheimer commenced concurrent appointments as assistant professor in Berkeley and at Caltech.** In Berkeley, his principal headquarters, he settled down in Room 219, Le Conte Hall. Space for theoreticians was limited. Thus for example his later student Willis Lamb had a desk in Oppenheimer's own office.[53]

'From all I hear I was a very difficult lecturer', Oppenheimer recalled later.[16] Having heard him often in later years I would describe his lecturing as priestly. He did not teach in the common style; he initiated. Sometimes I have come away greatly moved by an Oppenheimer lecture, realizing only afterwards that I had not understood much. He spoke softly as is illustrated by an exchange between him and Ehrenfest during a Caltech seminar. E.: 'Louder, please, dear Oppenheimer.' O.: 'But this room is so big.' E.: 'You always adjust your voice so we can't hear. I couldn't hear you in a telephone booth.'[54] Yet substance conquered style as the following judgment[41] by Hans Bethe makes abundantly clear: 'In addition to [his] massive scientific work, Oppenheimer created the greatest school of theoretical physics that the United States has ever known. Before him, theoretical physics in America was a fairly modest enterprise, although there were a few outstanding representatives. Probably the most important ingredient he brought to his teaching was his exquisite taste. He always knew what were the important problems, as shown by his choice of subjects. He truly lived with these problems, struggling for a solution, and he communicated his concern to his group. In its heyday, there were about eight or ten graduate students in his group and about six postdoctoral fellows. He met his group once a day in his office, and discussed with one after another the status of the student's research problem . . . [His] lectures were never easy but they gave his students a feeling of the beauty of the subject and conveyed his excitement about its development.'

There is no greater tribute to Oppenheimer than the list of Ph.D.s he delivered, which includes Carlson, Christy, Dancoff, Kusaka, Lamb, Morrison, Snyder, and Volkoff, and the list of NRC Fellows who came to work with

* Oppenheimer's later reference to the Heisenberg–Pauli papers as 'a monstrous boo-boo'[16] is rather excessive, even with the hindsight of the advances due to renormalization.

** At both places he was promoted to associate professor in 1931 and to full professor in 1936.

him: Furry, Plessett, R. G. Sachs, Schiff, Schwinger, Serber, and Uehling among them. Together with Oppenheimer these young people worked on quantum electrodynamics, cosmic ray physics, mesons, nuclear physics, and astrophysics. During that period Oppenheimer continued to contribute importantly himself. (Some of his work will be discussed in the next section and the next chapter.) His calculations often contained errors, however. As Serber has recalled: 'His physics was good, but his arithmetic awful'.[41a]

As to the later years, there is general agreement that Oppenheimer was an outstanding director of the Los Alamos project. Much has been written about those war years as well as about his turbulent life thereafter. I do not think that physics research is the only road to salvation, but do believe that none of his contributions deserves higher praise and will longer endure than what he did for the growth of American physics in the decade before the Second World War.

(c) A prelude: infinities before the positron theory

Once upon a time, it was at the turn of the century, an electron was considered to be a little marble with charge e, mass m, and 'classical electron radius' a:

$$a = \frac{e^2}{mc^2}. \tag{16.1}$$

This value for a was derived from the dynamical assumption that the electrostatic energy of an electron at rest, the so-called self-energy, accounts fully for its mass: $e^2/a = mc^2$. Not that anyone knew what prevented this little sphere from exploding on account of the electrostatic repulsion between its parts. Nor that there existed any experimental evidence for this or any other value of the electron's size. The only place where a had entered in an observed quantity was the Thomson cross section $\sigma = 8\pi a^2/3$ for the scattering of soft X-rays on electrons (see Eq. (9.51)). Of course such long wavelength scattering says nothing about the size of the scatterer; one would need X-rays with energies $\sim 137 mc^2$ to explore lengths of order a—for which case the Thomson formula no longer holds, however.

Then came special relativity theory.

A new question now arose: how can a finite-sized electron be made to obey the general kinematic relation

$$E^2 = c^2p^2 + m^2c^4 \tag{16.2}$$

supposed to hold for a stable system which, experimentally, the electron was, but which, theoretically, it was not?*

* According to special relativity theory, the energy of any system is given by the space integral of the 44-component of its energy–momentum tensor. The transformation properties of this integral imply Eq. (16.2) if and only if the system is stable.[55]

Einstein, though vitally interested* in the experimental validity of Eq. (16.2), wisely never touched on the dynamical question of the electron's stability in any of his papers on special relativity. In fact, as far as I know, the term 'classical electron radius' never appeared in his writings. Not until 1919, after he had mastered gravitation, did he make a brief comment on the stability problem. In a paper entitled: 'Do gravitational fields play an essential role in the structure of material elementary particles?'[56] he suggested that gravitational attraction and electromagnetic repulsion might hold the electron together.** As happened so often in that phase of his career, the idea came and went without leaving any trace. Yet in all his strivings toward a unified field theory, the search for stable particle-like solutions of field equations remained forever one of his dominant concerns.[57]

Lorentz' response to the demands of special relativity was quite different. Ever since the 1890s a dynamical model for the electron had been foremost in his mind;[58] so it remained till his death in 1928. Lecturing at Columbia University in 1906, shortly after special relativity had arrived, he had this to say[58] on the stability question: 'In speculating about the structure of these minute particles we must not forget that there may be many possibilities not dreamt of at present; it may well be that other internal forces serve to ensure the stability of the system, and perhaps, after all, we are wholly on the wrong track when we apply to the parts of the electron our ordinary notion of force'.†

Apart from a serious but unsuccessful effort by Gustav Mie,†† the problem of the electron's structure receded into the background during the next two decades.‡ One can think of several reasons for this. First, the problem was intractable. Secondly, the discovery of the proton raised the question of how to find a model for not one but two fundamental particles. Thirdly, the old quantum theory occupied center stage more and more. When Bohr proposed his theory of the hydrogen atom, he introduced the coordinate $\vec{q}$ and momentum $\vec{p}$ of the electron without any reference to its finite size. One may consider this natural since the ground state orbit of hydrogen has a radius $(137)^2$ times larger than a, so that for practical purposes a would not play a role anyway. Nevertheless, *the old quantum theory eased the transition toward an electron without spatial extension*.

All the same the classical electron radius remained a concept hovering just barely off-stage, even though special relativity abhors finite-size rigid bodies.

* Cf. Chapter 4, Section (f) and Ref. 55.

** A non-zero cosmological constant was essential to his arguments.

† The issue arose in the course of Lorentz' discussion of Poincaré's idea that the electron is held in equilibrium by stresses of non-electromagnetic origin. Lorentz had found[59,60] that this equilibrium is not stable against deformations.[55]

†† Mie attempted to modify the Maxwell–Lorentz equations in the interior of the electron in such a way that this particle is held together by additional purely electromagnetic forces which counteract the Coulomb repulsion.[61]

‡ Pauli's review of developments up to 1921 includes a critical assessment of Einstein's and Mie's proposals.[62]

To face the alternative, a point electron, was abhorrent for other reasons: a zero radius for the electron means an infinite self-energy e^2/a, $a \rightarrow 0$. So, I would think, one rather suffered a particle of finite size than a particle with infinite mass.

As if individual electrons and protons weren't trouble enough, there was the added confusion caused by the picture of the atomic nucleus as consisting of protons and electrons held together by Coulomb forces.* In 1925, having tried in vain[63] to make sense out of this nuclear model (details are unimportant), Frenkel finally gave up on the little marbles: 'The inner equilibrium of an extended electron becomes . . . an insoluble puzzle from the point of view of electrodynamics. I hold this puzzle (and the questions related to it) to be a scholastic problem. It has come about by an uncritical application to the elementary parts of matter (electrons) of a principle of division, which when applied to composite systems (atoms etc.) just led to these very "smallest" particles. The electrons are not only indivisible physically, but also geometrically. They have no extension in space at all. Inner forces between the elements of an electron do not exist because such elements are not available. The electromagnetic interpretation of the mass is thus eliminated.'[64]

It may have been the first time that someone explicitly opted for the point model.

Then came non-relativistic quantum mechanics.

Just as in the old Bohr–Sommerfeld theory, the electron continued to be described by a $\vec{q}$ and a $\vec{p}$, now of course taken to be non-commuting operators. The Schroedinger equation for the hydrogen atom described the electron without any reference to its size; in so far as the electron behaved as a particle, it was a point particle. *Non-relativistic quantum mechanics continued to ease the transition toward a point model.*

Soon the question arose as to the electron's electrostatic self-energy in quantum mechanics, this quantity now being defined as the expectation value of the Coulomb energy operator for one electron in a zero momentum state. First Jordan and Klein[65] then Heisenberg and Pauli[66] noted that, just as in the classical theory, this energy behaves as the limit of e^2/a for $a \rightarrow 0$.

Then came quantum electrodynamics and the Dirac equation.

Let us first consider the time from early 1928 to late 1929. The Dirac equation had been discovered and there were the beginnings of quantum electrodynamics, but not yet a hole theory, let alone a positron theory. During that period Oppenheimer made the important discovery[67] of a new source of self-energy, a typical quantum effect without classical counterpart. (This is the work which had been considered for joint publication with Heisenberg and Pauli; see the previous section.)

* Chapter 11, Section (e).

Oppenheimer started from Dirac's second-order perturbation formula Eq. (15.33) for the matrix element H_{fi},

$$H_{fi} = \sum_j \frac{H^1_{fj} H^1_{ji}}{E_i - E_j}, \tag{16.3}$$

and adopted the expression (15.71) for the perturbation H^1:

$$H^1 = -\int \vec{j}\vec{A}^{\text{tr}}\,\mathrm{d}\vec{x}, \qquad j = ie\bar{\psi}\vec{\gamma}\psi. \tag{16.4}$$

Eq. (16.3) can be applied as well for $i=f$ in which case it represents a perturbation to the energy of the state i. Let this state be a single free* electron e with momentum $\vec{p}$, energy $E(p) = c(p^2 + m^2c^2)^{1/2}$ (and given spin). In this case the sequence of transitions $i \to j \to i$ is

$$e \to e' + \text{photon} \to e \tag{16.5}$$

where virtual states correspond to all momentum-conserving partitions of $\vec{p}$ between e' and the photon. *There are infinitely many such states*. Moreover, in the spirit of 1929, the negative energy states were still empty, so that for any momentum, e' can be in a negative as well as a positive energy state. The self-energy $W(\vec{p})$ is found to be

$$W(\vec{p}) \sim \frac{e^2\hbar^2c^2}{\hbar cE(p)} \int k\,\mathrm{d}k \tag{16.6}$$

plus smaller terms including, of course, the electrostatic self-energy. Quantum effects therefore made the situation seem even worse: whereas the classical contribution to W diverges linearly ($\sim 1/a$ as $a \to 0$), in quantum electrodynamics W diverges quadratically, since the integral in Eq. (16.6) extends to infinity! Note further that

$$W(\vec{p}_1) - W(\vec{p}_2) \text{ is also infinite} \tag{16.7}$$

so that, as Oppenheimer stressed, self-energy effects cause infinite displacements of spectral lines.

The battle with infinities continues to be waged to this day but prospects never again looked as bleak and confusing as in 1930. As will be seen in the next section, a few years later the positron theory brought major changes. Both the classical linear divergence and the quantum mechanical quadratic divergence turned out to be spurious. We shall see later that Oppenheimer's

* Oppenheimer actually considered the case of an electron in a bound atomic state. All his qualitative conclusions remain valid for a free electron, however, as was noted by Waller.[68] Shortly afterward Rosenfeld made a similar calculation for an electron bound in a harmonic oscillator potential.[69]

idea of displaced spectral lines is answered in finite terms with the help of the renormalization program.

(d) Quantum electrodynamics in the thirties

1. Second-order successes. From the first quantum electrodynamical calculation ever performed, Dirac's treatment in late 1926 of spontaneous emission,* until the present, all we know about quantum electrodynamics is based on perturbation theory. Even though this method has served us well we would like to be free of it. We don't know how.

The perturbation expansion is based (it would be too much to say 'is justified') on the treatment of the fundamental charge e as a small parameter or, expressed in more appropriate dimensionless terms, on

$$\alpha = \frac{e^2}{\hbar c} \ll 1, \tag{16.8}$$

perhaps the most important numerical relation, not only for quantum electrodynamics, but also for the constitution of ordinary matter.

As we have seen,** Dirac performed his calculation of Compton scattering with the help of Eq. (16.3). The scattering amplitude so obtained is approximate, however. There are further contributions $O(\alpha^n)$, $n = 2, 3, \ldots$ arising from higher-order terms† in the perturbation expansion. To order α there are only two virtual states; see Eq. (15.34). In all higher orders there are infinitely many virtual states for much the same reason as was explained in the derivation of Eq. (16.6). This leads to momentum integrations which invariably diverge. Here we have the dilemma: lowest-order effects are finite and work well. Higher-order contributions are small but infinite, small because of powers of α, infinite because of the integrations.

I turn next to a quick enumeration of the main leading-order results obtained by 1935, adding also a few comments on alternative computational schemes developed during those years. The calculations were made in the Coulomb gauge Eq. (16.4), the quantity $\vec{j}$ referring to a sum over all electrons present, including those in the filled negative energy states. (A more precise prescription will follow anon.)

March 1930. Dirac shows[71] that the Klein–Nishina formula is unaffected (to leading order) by the new hole theory postulates.†† He also gives the cross-section for the process

$$e^+ + e^- \rightarrow 2\gamma. \tag{16.9}$$

* Chapter 15, Section (c).
** Chapter 15, Section (c), and Appendix, Note 4.
† The first few of these are written out in Ref. 70.
†† See also Chapter 15, Section (f), especially Dirac's letter to Bohr of December 9, 1929, mentioned there.

Well, not quite. In 1930 Dirac still believed that the holes were protons.* His answer could immediately be adapted to Eq. (16.9), however, by changing the mass parameter he had used (average of proton and electron mass) to the electron mass.[72]

February 1931. Heisenberg proposes[73] a new calculational method: take the Maxwell equations considered as operator equations but with the usual (c-number) Dirac charge-current as a source, integrate them by means of retarded potentials, expand the Dirac ψ in a power series in e (he stopped at the first order), and in the final expression take matrix elements of the products of photon creation and/or annihilation operators.

(In 1933 Casimir used the method for computing line widths.[74] In 1934 this technique was extended to the quantized Dirac field by Heisenberg[75] and by Weisskopf.[76] This method, later cast in modern form by Yang and Feldman[77] and by Källén,[78] is fully equivalent to Hamiltonian procedures.)

May 1931. Dirac notes[79] the occurrence of the process

$$\gamma + \gamma \rightarrow e^{+} + e^{-} \tag{16.10}$$

but does not calculate its probability.

March 1932. In an attempt to improve on the Heisenberg–Pauli version of quantum electrodynamics Dirac proposes[80] a 'many-time formalism', ascribing individual time coordinates to each electron. Pauli writes Dirac: 'Your comments were—to put it mildly—certainly no masterpiece.'[81] Actually, this new version of the theory (equivalent to earlier ones[82]) is an important first step toward manifestly covariant procedures.

April 1932. Møller gives the invariant matrix element for electron–electron scattering ('Møller scattering') by patching together the Coulomb effect with the second-order contribution from Eq. (16.3).[83]

As a result of the experimental discoveries by Anderson, and by Blackett and Occhialini, the pace begins to quicken.

June 1933. Oppenheimer and Milton Spinoza Plesset point out: 'If we allow gamma rays of [sufficient] energy to fall upon a nucleus, we should expect pairs to appear'.[84] This effect, they note, can be seen as a photoelectric absorption of the γ by an electron in the filled negative energy states, the nucleus picking up some recoil momentum. This is an important observation since this pair formation mechanism is obviously far more efficient than the one of Eq. (16.10).**

June 1933. Casimir notes[87] that calculations with the Dirac equation can often be simplified by the introduction of projection operators and the taking of traces over products of Dirac matrices. This procedure, initially known as the

* So did Oppenheimer who, also in March 1930, calculated the same annihilation rate but whose answer was off[71a] by a factor $(2\pi)^4$.

** There are several errors in their results, corrected in Refs. 85, 86.

Casimir trick, has been used ever since as a means of avoiding the tedium of handling separately the four Dirac equations for each spin value.
August 1933. Fermi and Uhlenbeck point out[86] the possibility of one-quantum annihilation:

$$e^+ + e^- \rightarrow \gamma, \tag{16.11}$$

if the electron is bound to a nucleus which can take up the recoil momentum; they estimate the cross-section.
December 28, 1933. The first symposium on the positron held at an American Physical Society meeting. Uhlenbeck, Anderson, and Oppenheimer were the speakers. 'This proved to be a session of great interest and importance and the attendance was about five hundred.'[88]
February 1934. Bethe and Heitler compute the ***Bremsstrahlung*** (loss of energy by radiation) of an electron in a (screened) Coulomb field.[89]* They note that their formulae work well at low but not at high (cosmic ray) energies: 'The theoretical energy loss by radiation for high initial energy is far too large to be in any way reconcilable with the experiment of Anderson'.[92] Reviewing[93] the situation in 1936, Heitler remarked: 'It is highly probable that the theory breaks down at . . . high energies. The limit where this breakdown begins . . . lies between 150 and $300mc^2$'—just the order of energy where meson effects, not yet known at that time, obscure the comparison with experiment!
July 1934. The Pauli–Weisskopf theory to which I return below.
October 1934. Breit and Wheeler compute the cross section for the process (16.10).[94]
January 1935. Heitler points out that electromagnetic radiation in thermal equilibrium must be accompanied by a temperature-dependent number of electron–positron pairs and makes estimates of that number.[95]
October 1935. Bhabha evaluates[96] the cross-section for electron–positron scattering ('Bhabha scattering'), noting the importance of the one-photon intermediate state

$$e^+ + e^- \rightarrow \gamma \rightarrow e^+ + e^-. \tag{16.12}$$

This concludes the brief account of second-order results up till the mid-thirties. Some other important findings bearing on quantum electrodynamics, but not specifically on positron theory, also date from that period, among them: a procedure[97] in which the action of two moving charges on each other is approximated by the action of photon bunches, in a suitable coordinate system; and the Bloch–Nordsieck method[98,99] (1936) for describing multiple emission of very low energy photons.

* Shortly afterward Racah independently made the same calculation.[90] For other references to early papers on this effect see Heitler's book.[91]

2. The Dirac field quantized. In all mentioned second-order calculations the full complexity of the positron theory does not yet come to the fore. The next question is how the intrinsic many-body aspect of the theory came to be handled. That problem was first broached by Dirac in his address to the seventh Solvay conference, October 1933. That report may be said to mark the beginning of positron theory as a serious discipline.

Dirac adopted Hartree's self-consistent field method as an approximate way of dealing with the infinitely many bodies making up the (almost) fully occupied negative energy states and one or a few occupied positive energy states. 'We shall suppose that each electron has its own individual wave function in space-time (instead of there being an enormous number of variables to describe the whole distribution), and also we shall suppose that each electron moves in a definite electromagnetic field, which is the same for all the electrons. This field will consist of a part coming from external causes and a part coming from the electron distribution itself, the precise way in which the latter depends on the electron distribution being one of the problems we have to consider.'[100] Denote one of Dirac's 'individual wave functions' by $\psi_\alpha(k, x)$, where $k = 1, \ldots, 4$ enumerates the four solutions of the Dirac equation, $\alpha = 1, \ldots, 4$ refers to the components of each solution, and x represents the four space–time variables. Dirac next introduced an off-diagonal density matrix R defined by

$$(k', x'|R|k'', x'') = \sum_{\alpha=1,\ldots,4} \sum_{\text{occ}} \psi_\alpha(k', x')\psi_\alpha^*(k'', x''). \tag{16.13}$$

The 'occ' summation goes over all occupied states including the sea of negative energy states. With the help of his R-matrix he made his last major contribution to particle physics, a leading-order calculation of vacuum polarization. (I come back to that.) The method is somewhat messy and does not permit the development of a systematic approximation procedure. Dirac, of course, knew that and noted in a subsequent paper[101] that the method 'is incomplete in that the effect of the exclusion principle . . . has not been investigated'. This paper dates from February 1934. When Pauli had received it, he wrote a letter to Heisenberg signed: 'Your (drowned in Dirac's formulae) W. Pauli'.[102] Heisenberg replied: 'I regard the Dirac theory . . . as learned trash which no one can take seriously'.[103] He would change his mind about that.

Meanwhile it had begun to dawn on several people that the quantization of the Dirac field was a superior way of treating the positron theory systematically. Before turning to who did what, I first summarize the main points.

As was seen in Chapter 15, Section (d), in 1927 Jordan and Wigner had invented a second quantization method which incorporates the exclusion principle. That was even before Dirac's relativistic wave equation. By means of simple modifications of formulae given earlier, the Jordan–Wigner formalism can at once be adapted* to the quantization of the four-component Dirac ψ_j, $j=1,\ldots,4$. In the absence of interaction the energy $E(\vec{k},s)$ of an individual electron is given by

$$E(\vec{k},s)=(\hbar^2c^2k^2+m^2c^4)^{1/2}\cdot\begin{cases}+1 & s=1,2,\\ -1 & s=3,4.\end{cases} \tag{16.22}$$

It is easily verified[105] that the respective eigenvalues for the total energy H_0 and the total charge Q_0 are given by

$$H_0=\sum_{\vec{k},s} n(\vec{k},s)E(\vec{k},s) \tag{16.23}$$

$$Q_0=e\sum_{\vec{k},s} n(\vec{k},s). \tag{16.24}$$

So far the postulates of the hole theory have not been invoked. In particular it follows from Eqs. (16.22) and (16.23) that the eigenvalues of H_0 can be either positive or negative. Let us next turn to the hole theory.

Already in December 1929 Dirac had replied to a query by Bohr: 'I do not think the infinite distribution of negative energy electrons need cause any difficulty. One can assume that in Maxwell's equation

$$\operatorname{div}\vec{E}=\rho; \tag{16.25}$$

the ρ means the difference in the electric density from its value when the world is in its normal state, (i.e. when every state of negative energy and none of positive energy is occupied).'[106] Accordingly Dirac's first step in his Solvay report[100] was to define the physical energy and charge densities by subtracting the corresponding quantities referring to the completely filled negative energy

* Let $u_j(\vec{k},s;\vec{x})$ be a complete orthonormal set of solutions. $\vec{k}$ denotes momentum, $s=1,\ldots,4$ enumerates the two spin times two sign of energy states. Replace Eqs. (15.36, 40, 47–50) by the respective generalizations

$$\psi_j(\vec{x},t)=\sum_{\vec{k},s} a(\vec{k},s;t)u_j(\vec{k},s;\vec{x}) \tag{16.14}$$

$$\psi_j^\dagger(\vec{x},t)=\sum_{\vec{k},s} a^\dagger(\vec{k},s;t)u_j^*(\vec{k},s;\vec{x}) \tag{16.15}$$

$$\{a(\vec{k},s;t),a^\dagger(\vec{k}',s';t)\}=\delta_{\vec{k}\vec{k}'}\delta_{ss'} \tag{16.16}$$

$$\{a(\vec{k},s;t),a(\vec{k}',s';t)\}=\{a^\dagger(\vec{k},s;t),a^\dagger(\vec{k}',s';t)\}=0 \tag{16.17}$$

$$a(\vec{k},s;t)=\eta(\vec{k},s;t)\begin{pmatrix}0&1\\0&0\end{pmatrix};\qquad a^\dagger(\vec{k},s,t)=\eta^*(\vec{k},s,t)\begin{pmatrix}0&0\\1&0\end{pmatrix} \tag{16.18}$$

$$n(\vec{k},s)=a^\dagger(\vec{k},s;t)a(\vec{k},s;t)=\begin{pmatrix}1&0\\0&0\end{pmatrix}\quad \text{all } \vec{k},s \tag{16.19}$$

$$\{\psi_j(\vec{x},t),\psi_l^\dagger(\vec{x}',t)\}=\delta_{jl}\delta(\vec{x}-\vec{x}') \tag{16.20}$$

$$\{\psi_j(\vec{x},t),\psi_l(\vec{x}',t)\}=\{\psi_j^\dagger(\vec{x},t),\psi_l^\dagger(\vec{x}',t)\}=0. \tag{16.21}$$

These formulae hold whether or not the electron interacts with the electromagnetic field. See Ref. 104.

states, to begin with in the absence of interactions. We now call this step the zeroth-order subtraction. Departing from Dirac's own procedure let us recall next how that step is implemented by means of second quantization methods.

Replace H_0 and Q_0 by

$$H = H_0 - \sum_{\vec{k}} \sum_{s=3,4} E(\vec{k}, s) \tag{16.26}$$

$$Q = Q_0 - \sum_{\vec{k}} \sum_{s=3,4} e. \tag{16.27}$$

Write these expressions far more elegantly by making the substitutions

$$\begin{aligned} a(\vec{k}, 3; t) &= b^\dagger(\vec{k}, 1; t) \\ a(\vec{k}, 4; t) &= b^\dagger(\vec{k}, 2; t) \end{aligned} \tag{16.28}$$

and likewise for their conjugates. Then

$$H = \sum_{\vec{k}} \sum_{s=1,2} [n^{(-)}(\vec{k}, s) + n^{(+)}(\vec{k}, s)](\hbar^2 c^2 k^2 + m^2 c^4)^{1/2} \tag{16.29}$$

$$Q = e \sum_{\vec{k}} \sum_{s=1,2} [n^{(-)}(\vec{k}, s) - n^{(+)}(\vec{k}, s)] \tag{16.30}$$

with

$$n^{(-)}(\vec{k}, s) = a^\dagger(\vec{k}, s; t) a(\vec{k}, s; t) \tag{16.31}$$

$$n^{(+)}(\vec{k}, s) = b^\dagger(\vec{k}, s; t) b(\vec{k}, s; t). \tag{16.32}$$

The Eqs. (16.29, 30) show that $b^\dagger$ and b may be interpreted as creation and annihilation operators for positron states (with positive energy, of course). Note that according to Eqs. (16.14, 15) ψ is now a superposition of electron annihilation and positron creation operators; likewise for $\psi^\dagger$ with electron and positron interchanged. *In this new language all reference to an 'infinite sea of occupied negative energy states' is eliminated.*

Let us see how this second quantization method, standard procedure today, entered into the mainstream of physics.

The first to use the transcription Eq. (16.28) was Heisenberg—not for the positron theory, however, but in the course of his analysis[107] of 'a far-reaching analogy between the terms of an atomic system with n electrons and those of a system in which n electrons in a closed-shell are lacking'. In this paper (1931) Heisenberg uses Jordan–Wigner methods and replaces 'holes' ('Löcher'), as he calls them, by particles, using essentially Eq. (16.28). Fock was the first to note[108] that Heisenberg's procedure was tailor-made for the positron theory and applied it to the case of free electrons only. What happened next has been described by a contemporary[9] in these words: 'Furry and Oppenheimer[109] employed the Jordan–Wigner formalism ... but without much benefit.... Heisenberg[75] proceeded to generalize the subtraction rules ... making use of the Jordan–Wigner ... quantization which at this point starts to play its essential role'. (As was noted later by Wightman and Schweber,[109a] Furry and Oppenheimer's discussion of particle position, momentum, and spin in terms of Fock's second quantization methods is in fact incorrect.)

Right after the Solvay conference of 1933, Heisenberg immersed himself in the positron theory. There were distractions, however: a trip to Stockholm, thoughts on nuclear physics, and time needed to digest Fermi's theory of β-radioactivity. All these topics are touched on in the letters of that period exchanged between Heisenberg and Pauli. This correspondence shows, however, that from late 1933 until early 1936 nothing in physics preoccupied Heisenberg more than the positron theory. Pauli was a little less negative now: 'At this time my attitude toward the hole theory is like Bohr's and your own, neither completely disapproving nor negative',[110] but remained skeptical about the 'Limes–Akrobatik'.[111] During this period Pauli served as Heisenberg's sounding board and critic. At one point Heisenberg suggested a joint publication on positron theory. Pauli initially liked that idea and outlined a draft[112] but nothing ever came of this common enterprise (in part, I believe, because soon thereafter Pauli got involved with the Pauli–Weisskopf theory).

Heisenberg's first paper on subtraction techniques was completed in June 1934.[75,113] I leave the new physics it contains until a little later and turn to the new methodology found in its second part, entitled *Quantentheorie der Wellenfelder*. It is motivated by 'the necessity to formulate the fundamental equations of the theory in a way which goes beyond the Hartree–Fock approximation procedure'. Heisenberg continues to employ the R-matrix but now takes ψ to be quantized, replacing the summation in Eq. (16.13) by an expectation value. Then he leaves the R-matrix and introduces Hamiltonian methods: 'One can also construct a Hamilton function in the usual way and carry out perturbation theory'. In making this step Heisenberg became the first to cast the Dirac–Maxwell formalism in the form we use today. In essence he took Eq. (16.4) for the interaction with radiation, ψ now being a q-number field. Pauli realized at once how great an advance this was: 'In principle and in respect to physics your Ansatz has proved to be practicable, I believe, so that the route is given for liberating Dirac's Ansatz from the assumption of the applicability of the Hartree–Fock method and for calculating self-energies.'[114]

One last remark about general methods. A month after completing this major paper, Heisenberg presented a short note[115] on charge fluctuations in which he computed the vacuum expectation value Q^2 of

$$\int_v \int_v \rho(\vec{x})\rho(\vec{x}')\,\mathrm{d}\vec{x}\,\mathrm{d}\vec{x}' \tag{16.33}$$

$\rho = e\psi^\dagger\psi$, with ψ quantized; v is the spatial integration volume. He found Q^2 to have 'boundary divergences': Q^2 is infinite if v has a sharp boundary, but finite if the boundary is smoothed out over a thickness b. He discussed only the case $b \gg \hbar/mc$. Jost and Luttinger[116] treated the problem for $b \ll \hbar/mc$.

3. Charge conjugation; Furry's theorem. Already in his June 1934 paper[75] Heisenberg demanded 'that the symmetry of nature with respect to positive and negative charges is expressed from the outset in the fundamental equations of the theory', a requirement indeed satisfied by his formalism. It took a few years, however, before Heisenberg's demand was cast in the form of an invariance principle, invariance under charge conjugation (C-invariance) a term introduced by Kramers,[117] who was the first to implement this invariance in all generality.

Our collection of formulae indicates how to proceed. The theory should be invariant under exchange of electrons and positrons which, loosely speaking, amounts to an interchange of a's and b's, $a^\dagger$'s and $b^\dagger$'s, and thus of ψ and $\psi^\dagger$, see Eqs. (16.14, 15). This interchange should flip the sign of the current:

$$j_\mu \to -j_\mu. \tag{16.34}$$

But if the theory is to be invariant then the interaction $j_\mu A_\mu$ may not change, so that

$$A_\mu \to -A_\mu. \tag{16.35}$$

The precise transformations were found by Kramers:

$$\psi_\alpha \to C_{\alpha\beta}(\psi^\dagger \gamma_4)_\beta \tag{16.36}$$

where the 4×4 matrix $C_{\alpha\beta}$ must satisfy

$$(C^{-1}\gamma_\mu C)_{\alpha\beta} = -(\gamma_\mu)_{\beta\alpha}. \tag{16.37}$$

Eqs. (16.36, 37) yield Eq. (16.34) provided the anti-commutation relations (16.20) are used: the great novelty about C-invariance is that it can only be consistently formulated in a *quantized* field theory. It turns out that the matrix C is necessarily anti-symmetric (the symmetry properties of C depend in an interesting way on the number of space–time dimensions[118]). Kramers' argument holds for any realization of the γ_μ. Earlier Majorana had presented[119] the same reasoning for a special choice of γ's.

Already in 1936 Wendell Furry had noted curious cancellations in higher-order quantum electrodynamical calculations. These, he found,[120] can be understood in terms of a rule, now known as Furry's theorem, which shortly afterward was realized to be a special case of C-invariance: any matrix element with an odd number of external photons and no external $e^\pm$'s is zero if the theory is C-invariant. This follows at once from Eq. (16.35).

For an excellent pedagogical exposé of C-invariance I recommend a book by Sakurai.[121]

4. The polarization of the vacuum. In August 1933 Dirac wrote[122] to Bohr: 'Peierls and I have been looking into the question of the change in the distribution of negative energy electrons produced by a static electric field. We find that this changed distribution causes a partial neutralization of the charge producing the field .... If we neglect the disturbance that the field produces in negative energy electrons with energies less than $-137mc^2$, then the neutralization of charge produced by the other negative energy electrons is small and of the order 136/137 .... The effective charges are what one measures in *all* low-energy experiments, and the experimentally determined value of e must be the effective charge on an electron, the real value being slightly bigger ... One would expect some small alterations in the Rutherford scattering formula, the Klein–Nishina formula, the Sommerfeld fine structure formula, etc. when energies of the order mc^2 come into play'.

Transcribed in the modern vernacular, Dirac's effective charge is our physical charge; his real charge our bare charge; his neutralization of charge our charge renormalization. For a further understanding of Dirac's letter it is helpful to turn first to his Solvay report[100] which contains in more quantitative form the results he had mentioned to Bohr. I should point out right away, however, that the existence of vacuum polarization was also independently diagnosed by Furry and Oppenheimer.[123]

As mentioned before, Dirac began his report by describing zero-order subtractions: ψ and $\psi^\dagger$ on the rhs of Eq. (16.13) are taken to be free-particle wave functions. As long as $x_\mu = x'_\mu - x''_\mu \neq 0$ the summation in R gives finite results, but R develops singularities depending on x_μ^2 only* as $x_\mu \to 0$. Dirac's prescription for extracting finite results was: first subtract these singular terms, then let $x_\mu \to 0$.

Thereupon he raised the all-important question of how and what to subtract in realistic situations where electrons interact with each other and with external fields. He did not give a general answer but presented a calculation, the first of its kind, for the case that a static external source $\rho(\vec{x})$ is present which is supposed to vary slowly in space. Making the same subtractions as before he found (I paraphrase) that the presence of ρ induces an additional charge density $\delta\rho$, due to virtual creation and annihilation of pairs, which to order α is given** by

$$\delta\rho = \alpha\left\{C\rho - \frac{1}{15\pi}\left(\frac{\hbar}{mc}\right)^2 \Delta\rho\right\}, \tag{16.38}$$

where C is a logarithmically divergent integral. (Δ is the Laplace operator.) A new infinity had entered the theory, now known as the charge renormali-

* In modern parlance, we deal with a light cone expansion.

** A numerical error in the coefficient of the second term has been corrected.

zation term. The slow space variance of ρ permits the neglect of higher derivative terms of order $\Delta\Delta\rho$.

Concerning the C-term, Dirac opined that this infinity should not be taken seriously since 'we may not suppose the theory to be valid for energies larger than $137mc^2$'—whence the remark in his letter to Bohr about neglecting states below $-137mc^2$. Neither in that letter nor in his report did Dirac pay adequate attention to the $\Delta\rho$-term. In referring to 'alterations' of standard results he had in mind: 'In experiments involving energies of the order mc^2 it would be the real [i.e. the bare] charge, or some intermediate value of the charge which comes into play',[122] not a bad idea either.

Regarding other developments on vacuum polarization in the thirties:

(a) Heisenberg rederived (and corrected) Dirac's Eq. (16.38) but initially did not interpret the answer correctly: '[The second term in Eq. (16.38)] has no physical meaning ... the "polarization of the vacuum" first becomes a physical problem for external densities which vary with time ...'[124] Actually this term is measurable and (as we shall see later) agrees with experiment.

(b) Are Dirac's subtraction prescriptions unique? That question was independently raised by several. Already in 1934 it was realized that constraints imposed by invariance, so crucial for the further developments in the forties, are of the essence. Peierls emphasized[125] that subtractions may not destroy gauge invariance since otherwise one may risk 'an induced current in the vacuum even if there is no field but only a non-vanishing vector potential'. Similar comments were made by Furry and Oppenheimer.[109] Heisenberg showed that Dirac's subtractions are in fact compatible with the conservation laws for electric charge[126] and energy-momentum.[127]

(c) In 1935 two NRC Fellows from Berkeley extended Eq. (16.38). Robert Serber gave the generalization to time-dependent external sources.[128] Edwin Uehling obtained a formula which includes effects due to sources which vary strongly over space.[129] In 1936 Pauli and Rose simplified Serber's calculations,[130] while Weisskopf derived the Serber–Uehling results for the first time with a q-number Dirac field.[131]

Uehling also calculated the effect of the second term in Eq. (16.38) for an electron moving in a hydrogen-like atom with nuclear charge $-Ze$. He found that the nS levels are displaced by an amount $\Delta E(nS)$ proportional to $\alpha^5 Z^4$:

$$\Delta E(nS) = -\frac{8Z^2\alpha^3}{15\pi n^3} R \tag{16.39}$$

where R is the corresponding Rydberg constant (see Chapter 9). For hydrogen and $n=2$ the corresponding frequency displacement is

$$\Delta\nu = -27 \text{ megacycles per second.} \tag{16.40}$$

Right after the Second World War this result would set mighty forces in motion.

5. The self-energy of the electron, of the vacuum, and of the photon. The positron theory has taught us two new lessons regarding self-energies. First, all that was said on this subject earlier in the century is irrelevant. Secondly, new kinds of self-energies arise never dreamt of in those old times.

Consider the electrostatic self-energy to $O(\alpha)$ of the electron, first calculated by Weisskopf[76] in the positron theory. The corresponding operator is

$$\frac{1}{2}\frac{\int\{\rho(\vec{x})-\rho_0(\vec{x})\}\{\rho(\vec{x}')-\rho_0(\vec{x}')\}}{|\vec{x}-\vec{x}'|}\,\mathrm{d}\vec{x}\,\mathrm{d}\vec{x}'. \tag{16.41}$$

$\rho = e\psi^\dagger\psi$, ρ_0 is its vacuum expectation value to zeroth order, ψ is quantized and also zeroth order. The subtraction of ρ_0 is Dirac's zeroth-order subtraction. Compute the expectation value $W_2^{\text{stat}}(\text{vac}+1)$ for the state: one electron with zero momentum. The result is not yet what we want since, even after subtracting ρ_0, the operator (16.41) has a non-zero vacuum expectation value $W_2(\text{vac})$ because the vacuum expectation value of $\rho(\vec{x})\rho(\vec{x}')$ is not equal to the product of the vacuum expectation values of $\rho(\vec{x})$ and $\rho(\vec{x}')$. $W_2(\text{vac})$ is the electrostatic *second order self-energy of the vacuum*, an altogether new concept. The quantity we need is evidently

$$W_2^{\text{stat}} = W_2^{\text{stat}}(\text{vac}+1) - W_2^{\text{stat}}(\text{vac}). \tag{16.42}$$

The procedure described is a *second-order subtraction*. Now lo and behold: W_2^{stat} diverges only logarithmically!

We are not yet finished. In addition one must repeat Oppenheimer's calculation Eqs. (16.3–6) but now for the positron theory. By similar reasonings as for the static effect one finds that there is no quadratic divergence as in Eq. (16.6) but, instead, another logarithmic divergence!*

Whatever became of Lorentz' classical electron radius?

Calculate W_2, the total self-energy to order e^2 but cut off all virtual momenta at a value $P = \hbar/a$. Then

$$W_2 = \frac{3}{2\pi}\cdot\frac{e^2}{\hbar c}\cdot mc^2 \ln\left[\frac{\hbar}{mca}+\sqrt{1+\frac{\hbar^2}{m^2c^2a^2}}\right]$$

$$\cong \frac{3e^2}{2\pi a} \quad \text{if } a \gg \frac{\hbar}{mc}. \tag{16.43}$$

Thus the classical answer returns if one makes the absurd assumption of an

* This was first noted by Furry. As Weisskopf later wrote[132]: 'I made a mistake in the first calculation . . . Then I received a letter from Furry . . . who pointed out that the divergence is logarithmic . . . Instead of publishing the result himself he allowed me to publish a correction[133] quoting his intervention.'

electron radius large compared with the Compton wavelength. Whatever the future may hold, the tiny marble with energy e^2/a is gone forever.

Another novelty: the self-energy depends on m, the mass parameter—now called the bare mass—appearing in the Dirac equation. The physical mass equals $m + W_2/c^2$ + additional contributions to the self-energy of fourth-* and higher-order in e. The physical mass will be of the form

$$mf\left(\frac{\hbar}{mca}, \frac{e^2}{\hbar c}\right), \qquad a \to 0. \tag{16.44}$$

It was conjectured[135] that f may have a finite limit f_0 as $a \to 0$ and that, if so, the equation $m = mf_0$ would restore the possibility that all the electron's mass is electromagnetic; at the same time the fine structure constant would be fixed. This cannot be correct since other fields contribute to the electron's self-energy as well. The only important result about higher-order self-energies dating from that period is Weisskopf's proof that they all diverge at most logarithmically.[136]

Lastly, also new in the positron theory is the self-energy of the photon associated (to order e^2) with the virtual transitions $\gamma \to e^+ + e^- \to \gamma$, an effect first noted by Heisenberg.[75] His evaluation of the photon self-energy contains an instructive mistake. He found that this energy diverges (logarithmically) even before the off-diagonal distance x_μ is set equal to zero. Serber noted[137] that this impossible answer is due to an inappropriately performed contact transformation.

Gauge invariance demands that this energy must be strictly zero. Also in later years the photon self-energy would cause problems if manifest covariance and gauge invariance were not carefully maintained at all stages of the calculation.

In his paper[137] Serber introduced a new expression which is now part of the physics language: to *renormalize* the polarization of the vacuum. The physical idea of renormalization was definitely in the air. One also finds germs of it in the papers by Furry and Oppenheimer,[109] and by Weisskopf[131] who notes: 'A constant polarizability is in no way observable'.

6. In which the Maxwell equations become non-linear. According to Eq. (16.10) two γ's can give a pair. According to Eq. (16.9) a pair can give two γ's. Treating the pair as a virtual state one sees that two γ's can scatter into two γ's with an amplitude $O(\alpha^2)$. The existence of this phenomenon was first noted in October 1933 by Otto Halpern[138] from New York University and was for some time called Halpern scattering. The calculation of its amplitude by the methods of the thirties is extremely complicated (and not that simple even today) since one must consider all possible sequences of absorption and

* The earliest attempt to evaluate the fourth-order effect dates from the thirties.[134]

emission of the four photons. In 1934 Heisenberg set two of his students, Hans Euler and Bernard Kockel, to work on the problem. It is to their credit that they correctly obtained a finite answer even though one integration over virtual momenta occurs. Considering the case $\lambda \gg \hbar/mc$ (λ is the photon wavelength in the center of momentum system) they found[139] a cross-section $\sigma \sim \alpha^4(\hbar/mc)^8\lambda^{-6}$. For $\lambda \ll \hbar/mc$, $\sigma \sim \alpha^4\lambda^2$, a result first obtained by Aleksandr Il'ich Akhieser.[140]

Light-by-light scattering implies a breakdown of linear superposition for electromagnetic radiation. The effect is small, σ is at most $\sim\alpha^4(\hbar/mc)^2$, for $\lambda \sim \hbar/mc$. Nevertheless the implications are profound: the Maxwell equations are no longer linear. In fact Euler and Kockel showed* that their answers amount to the statement: for low frequencies and low field intensities light behaves *as if* the Lagrangian for the radiation field were

$$L = \tfrac{1}{2}(\vec{E}^2 - \vec{H}^2) + \frac{\alpha^2}{360\pi^2 mc^2}\left(\frac{\hbar}{mc}\right)^3 \{a(\vec{E}^2 - \vec{H}^2)^2 + b(\vec{E}\vec{H})^2\} \tag{16.45}$$

with

$$a = 1, \qquad b = 7. \tag{16.46}$$

Here we encounter the first use of what now is called an *effective Lagrangian*.

The α^2-terms describe the net effect of virtual pairs on pure radiation phenomena. L may also be used for the description of the coherent scattering of light by electrostatic fields ('Delbrück scattering').[142,143] The new terms in L are covariant and gauge invariant, as they should be. These invariance properties in fact make it possible to simplify very considerably the light-by-light scattering calculation for large λ. Invariance alone fixes the α^2-terms in L up to the values of a and b. To find these numbers one can use simpler field configurations, however. That is what Heisenberg and Euler did in 1936.[144] They inserted rigorous solutions of the Dirac equation in constant and parallel electric and magnetic fields into Eq. (16.13) and thus obtained not only a and b but the more general non-linearity in L valid for strong but slowly varying $\vec{E}$ and $\vec{H}$.** This calculation was subsequently simplified by Weisskopf.[131]

This paper with Euler, Heisenberg's last contribution to the positron theory, concludes with a recitation of difficulties: infinities occurring in the vacuum self-energy, in the fourth-order amplitudes for Compton scattering; in light-by-light scattering to sixth order. As they state in their conclusion: 'The theory of the positron and quantum electrodynamics must undoubtedly be considered provisional'.

* See also the detailed discussion in Euler's Ph.D thesis.[141]

** This calculation also makes clear[143] why light-by-light scattering is finite in spite of an infinity of virtual states.

So it was in the 1930s, everywhere. The tools were there to push the theory much further, but experimental incentive was lacking and there were so many other interesting things to do in theoretical physics: nuclear physics, β-radioactivity, cosmic rays, mesons....

7. The Pauli–Weisskopf theory. We have encountered Pauli as a creator of new physics, as a critic of other's ideas, but not yet as a teacher. In that last respect I cannot think of a better testimonial to his influential role than giving a list (chronologically ordered) of his assistants in the Zürich years: Kronig, Bloch, Peierls, Casimir, Weisskopf, Kemmer, Ludwig, Fierz, Jost, Schafroth, Thellung, Enz. All these men went on to make their own names in physics, several of them in quantum field theory. One among the latter is Victor (much better known as Viki) Weisskopf whose contributions to the positron have already been mentioned. I now turn to his joint work with Pauli on the quantum electrodynamics of spinless fields.

The first intimations of this theory are found, as usual, in a letter[145] from Pauli to Heisenberg: 'I have hit upon a sort of curiosum .... The application of our old formalism of field quantization to [the scalar] theory leads *without any further hypotheses* (without the "hole" idea, without *Limes Akrobatik*, without subtraction physics!) to the existence of positrons and to pair creation processes .... After field quantization the energy is automatically positive! Everything gauge invariant and relativistically invariant! ... It has pleased me that once again I could say something nasty about my old enemy the Dirac theory'.* In his Princeton seminar shortly thereafter Pauli referred to this work as the 'anti-Dirac theory'.[6] As he soon realized, however, the new theory was neither better nor worse than the positron theory in regard to the subtractions needed to cope with infinities.**

Pauli and Weisskopf had returned to the scalar wave equation (15.66) criticized by Dirac because its characteristic density given in Eq. (13.34) (for $e=0$) is not positive definite (see Chapter 13, Section (e)). They pointed out[147] that this causes no problems if one quantizes ψ and interprets $e\rho$ as the charge density operator rather than ρ as a probability density. Briefly, the formal steps are these:

The basic commutation relations of the theory are the Jordan–Klein relations (15.43), (15.44) for ψ and $\psi^\dagger$. In order to accommodate particles with both signs of charge replace Eqs. (15.36), (15.40) by

$$\psi=\frac{1}{\sqrt{V}}\sum_{\vec{k}}\sqrt{\frac{\hbar}{2\omega_{\vec{k}}}}\,(a_{\vec{k}}(t)+b^{\dagger}_{\vec{k}}(t))\,e^{i\vec{k}\vec{x}} \tag{16.47}$$

* 'Es hat mich gefreut, dass ich meiner alten Feindin ... wieder eins anhängen konnte.' Pauli's use of 'positron' must of course be understood to mean particle of positive charge.
** For Weisskopf's recollections about the origins of this theory see Ref. 146.

$$\psi^\dagger = \frac{1}{\sqrt{V}} \sum_{\vec{k}} \sqrt{\frac{\hbar}{2\omega_{\vec{k}}}} \, (a_{\vec{k}}^\dagger(t) + b_{\vec{k}}(t)) \, e^{-i\vec{k}\vec{x}}$$

$$\omega_{\vec{k}} = c\sqrt{k^2 + \kappa^2}, \qquad \kappa = \frac{mc}{\hbar} \tag{16.48}$$

where the a and $a^\dagger$ operators satisfy the commutation relations (15.41). So do the b and $b^\dagger$, while any cross-combination of a and b commute. Both the a- and b-matrices have representations as in Eq. (15.42). In the absence of interaction one finds for the Hamiltonian H and the charge $Q = e\int \rho \, \mathrm{d}\vec{x}$ (see any textbook on the subject)

$$H = \sum \hbar\omega_{\vec{k}}(n_{\vec{k}}^{(-)} + n_{\vec{k}}^{(+)} + 1) \tag{16.49}$$

$$Q = e\sum (n_{\vec{k}}^{(-)} - n_{\vec{k}}^{(+)}) \tag{16.50}$$

where both

$$n_{\vec{k}}^{(-)} = a_{\vec{k}}^\dagger a_{\vec{k}} \quad \text{and} \quad n_{\vec{k}}^{(+)} = b_{\vec{k}}^\dagger b_{\vec{k}} \tag{16.51}$$

have eigenvalues 0, 1, 2, ... Thus $n_{\vec{k}}^{(-)}$ and $n_{\vec{k}}^{(+)}$ are the number operators for particles with mass m and charge e and $-e$ respectively. Eq. (16.50) shows how Dirac's objection about a nondefinite density ρ has successfully been overcome in quantum field theory.

Pauli and Weisskopf went on to calculate the pair creation cross-section, and also the vacuum polarization, which is again logarithmically divergent. Some years later Weisskopf found[136] that the self-energy to order α is quadratically divergent.

In his letter[145] to Heisenberg, Pauli further wrote that unfortunately this new theory had little to do with reality. Soon after the subsequent discovery of spinless charged mesons it became evident, however, that the theory had a bright and lasting future.

(e) Scientific nostalgia: the search for alternatives

> I really spent my life trying to find better equations for quantum electrodynamics, and so far without success, but I continue to work on it.
>
> Dirac at age 75[148]

Today, half a century after the events just described, we are still battling the infinite but the nature of the attack has changed. The renormalization techniques of the 1940s have given us new weapons which have served to reduce all the many infinities in quantum electrodynamics to three fundamental ones, those of mass and charge and of non-measurable vacuum quantities. The result was a vastly increased predictive power of the theory which has stood experimental test extremely well. Furthermore, since the 1930s the battles have been waged on new fronts. To give but one example, it has definitely been established that the electron's self-energy is not purely electromagnetic in origin. Other contributions arise from its interaction with W- and Z-bosons and, less directly, with other fields as well.

Both these new techniques and these new dynamics received important stimuli from experimental information not yet available in the thirties. Already then, however, dissatisfaction with the status of quantum electrodynamics led to searches either for new computational methods or for modifications of the dynamics. All these attempts have left the darkness unobscured. There is at present no reason for believing that any of this work may contain germs for future progress.

These early exertions moved in a variety of directions, yet they have one common trait: a nostalgia for the past. Let us return, its proponents suggested, to the classical theory, modify it so as to be rid of infinities and thereupon revisit the quantum theory in the hope that also there all will now be well. It was, as I see it, a misjudgment comparable in style (though not in content) with Einstein's hopes for a revised classical theory which would yield quantum mechanics as an inevitable consequence.[149] In view of the discussion following Eq. (16.43) it rather seems that the quantum theory itself should be the point of departure for a much needed understanding of what lies beyond the infinities.

I conclude this chapter with a brief catalogue of some alternatives proposed in the thirties and early forties. These remarks can be skipped without loss of continuity.

(a) *λ-limiting process.* In 1933 Wentzel proposed[150] to consider the classical point electron as the limit in which a time-like vector rather than the usual space-like classical electron radius tends to zero. For example the electric field $\vec{E}(\vec{x}, t)$ at the position $\vec{x}$ of the electron is defined as

$$\vec{E}(\vec{x}, t) = \tfrac{1}{2} \lim_{\xi \to 0, \tau \to 0} (\vec{E}(\vec{x} + \vec{\xi}, t + \tau) + \vec{E}(\vec{x} - \vec{\xi}, t - \tau)) \qquad (16.52)$$

where the limit is taken such that $|\vec{\xi}| < c\tau$. He obtained in this way a zero self-energy of the electron. The only reaction of the self-field on the electron is Lorentz' damping term $-2e^2\dddot{x}3c^3$ in the equation of motion (in the electron's rest system).

Wentzel considered next the equivalent for his theory of the calculation which led to the quadratic divergence Eq. (16.6) (Dirac theory with negative energy states empty) and found that the self-energy remains infinite in this case.[151] In order to eliminate this infinity Dirac introduced[152,153] what amounts to photons of negative energy, as a result of which the divergence $\int_0^\infty k \, dk$ is replaced by $\int_{-\infty}^{\infty} k \, dk = 0$. He attempted to eliminate the physical paradoxes resulting from negative energy photons by introducing an indefinite metric.[154] This in turn leads to new difficulties critically analyzed by Pauli.[155] As far as

I know the λ-process never was applied to the positron theory. For a review of the method see Wentzel's book.[156]

(b) Marcel Riesz proposed a specific way of continuing the solutions of the classical inhomogeneous electromagnetic wave equations analytically in the complex plane, again for the purpose of eliminating the classical infinities. References to this work are found in a paper by Ma[157] who showed that this method is equivalent to the λ-limiting process.

(c) Non-linear modifications of the classical Maxwell equations (Born and coworkers). These were motivated by 'the conviction of the great superiority of the unitary idea ... [Here 'unitary' means: the mass of the electron is of electromagnetic origin]. The present theory (formulated by Dirac's wave equation) holds ... as long as the wavelengths ... are long compared with the "radius of the electron" but breaks down for a field containing shorter waves. The non-appearance of Planck's constant in the expression for the radius indicates that in the first place the electromagnetic laws are to be modified; the quantum laws may then be adapted to the new field equations'.[158] These ideas gave rise to a number of interesting classical studies. For a review and literature see a paper by Born.[159] Attempts to quantize the theory were never successful.[160]

(d) The next example, Heitler's quantum theory of radiation damping, does not follow the pattern of a revision of the classical theory. Here an algorithm is developed in which all those virtual transitions ('round-about transitions') leading to infinite self-energies and to infinite contributions to transition probabilities are discarded.[161] Bethe and Oppenheimer showed that this leads to serious difficulties in the low frequency region of electromagnetic processes.[162]

(e) Finally, Dirac's two attempts at a revised classical electron theory. The first one dates from 1938. 'A new physical idea is needed which should be intelligible both in the classical theory and in the quantum theory and our easiest path of approach is to keep within the confines of the classical theory.'[163] He did not attempt to modify the Maxwell–Lorentz equations but rather to seek a new interpretation for them. His method rests on the invariant separation of fields into 'proper' and 'external' parts, the latter being half the sum of the retarded and the advanced interaction. His final rigorous equation of motion for the electron is

$$m\dot{v}_\mu - \tfrac{2}{3}e^2\ddot{v}_\mu - \tfrac{2}{3}e^2\dot{v}_\lambda^2 v_\mu = ev_\mu F_{\mu\nu} \tag{16.53}$$

where v_μ is the electron four-velocity and $F_{\mu\nu}$ the external field. This equation looks the same as Lorentz' equation of motion, but actually it has different properties. In Lorentz' case terms proportional to positive powers of the electron radius a had to be neglected. There are no such terms here: a negative mechanical mass has been introduced so as to make the observed mass equal to m, after which the limit $a \to 0$ is taken. The occurrence of this sink of mechanical energy is closely connected with the appearance of 'runaway solutions' of Dirac's *exact* equation of motion, i.e. solutions corresponding to accelerations even in the absence of external fields. This can readily be seen from the general integration of the above equation with $F_{\mu\nu} = 0$. For that reason Dirac was obliged to impose a final condition on the acceleration (along with initial conditions on position and velocity) in order to single out the acceptable solutions $v_\mu = constant$. Dirac's subsequent concern with the quantization of his theory led to his involvement mentioned earlier with the λ-process and with the indefinite metric.

Unable to find a quantum version of his classical point electron he never mentioned this theory again in later years. He certainly had abandoned it by 1951 since at that time he started all over again a second time. Once again the point of departure was classical: 'The troubles . . . should be ascribed . . . to our working from the wrong classical theory'.[164] His new suggestion may be considered as the extreme opposite of what he had proposed in 1938. This time he began with a classical theory that does not contain discrete particles at all. 'The theory of electrons should be built up from a classical theory of the motion of a continuous stream of electricity rather than the motion of point charges. One then looks upon the discrete electrons as a quantum phenomenon.'[165] After 1954 this model, too, vanished from his writings without leaving a trace.

(f) I shall return later to Kramers' ideas, also dating from the thirties, which are the classical forerunners of renormalization.

Sources

On the positron theory: the sources mentioned at the end of Chapter 15. In addition: the notes[6] of a seminar conducted by Pauli in Princeton, in the academic year 1935–6; and an essay by the author on Heisenberg's contributions to the positron theory, to be published in Heisenberg's collected works.

In preparing the section on physics in America I have greatly benefited from books by Sopka,[20] by Kevles[166] and by Moyer.[167] For Oppenheimer's Berkeley period see especially an article by Serber.[41a]

References

1. W. Heisenberg in *The physicist's conception of nature*, Ed. J. Mehra, p. 271, Reidel, Boston 1973.
2. P. A. M. Dirac, *Proc. Roy. Soc. A* **133**, 60, 1931.
3. W. Pauli, *Handbuch der Physik*, Vol. 24/1, Part B, Section 5, Springer, Berlin 1933.
4. W. Pauli, letter to P. A. M. Dirac, May 1, 1933, repr. in *W. Pauli, scientific correspondence*, Eds. A. Hermann and K. von Meyenn, referred to below as *PC*, Vol. 2, p. 159, Springer, New York 1985.
5. W. Pauli, letter to W. Heisenberg, June 16, 1933, *PC*, Vol. 2, p. 169.

6. *The theory of the positron and related topics*, Report of a seminar conducted by W. Pauli, notes by B. Hoffmann, Inst. Adv. Study, Princeton 1935–6, mimeographed.
7. P. A. M. Dirac, *Nature* **137**, 298, 1936.
8. W. Heisenberg, letter to W. Pauli, April 25, 1935, *PC*, Vol. 2, p. 386.
9. G. Wentzel, in *Theoretical physics in the twentieth century*, p. 48, Interscience, New York 1960.
10. C. D. Anderson, *Science* **76**, 238, 1932.
11. C. D. Anderson, *Phys. Rev.* **43**, 491, 1933.
12. J. L. Heilbron, summary of an interview with P. M. S. Blackett, December 17, 1962, copy in the Niels Bohr Library, American Institute of Physics, New York.
13. P. M. S. Blackett and G. P. S. Occhialini, *Nature* **130**, 363, 1932.
14. P. M. S. Blackett and G. P. S. Occhialini, *Proc. Roy. Soc. A* **139**, 699, 1933.
15. Ref. 14, p. 715.
16. J. R. Oppenheimer, interviewed by T. Kuhn, Nov. 20, 1963, transcript in the Niels Bohr Library, American Institute of Physics, New York.
17. C. D. Anderson, letter to A. Pais, March 30, 1983.
18. E. Rutherford, *Proc. Solvay Conference*, p. 177, Gauthier-Villars, Paris 1934.
19. R. A. Millikan and C. D. Anderson, *Phys. Rev.* **40**, 325, 1932.
20. K. R. Sopka, *Quantum physics in America 1920–35,* Appendix 4, Arno Press, New York 1980.
21. Ref. 20, p. 1.76.
22. Ref. 20, Appendix 2.
23. E. C. Kemble, *Rev. Mod. Phys.* **1**, 157, 1929; E. C. Kemble and E. L. Hill, ibid. **2**, 1, 1930.
24. E. C. Kemble, *Phys. Rev.* **15**, 95, 1920.
25. J. H. van Vleck, *Phys. Rev.* **24**, 330, 347, 1924.
26. J. C. Slater, *Phys. Rev.* **25**, 395, 1925.
27. G. N. Lewis and R. C. Tolman, *Phil. Mag.* **18**, 510, 1909.
28. Cf. J. H. van Vleck, *The theory of electric and magnetic susceptibilities*, Chapter 5, Clarendon Press, Oxford 1932; also for further references.
29. K. T. Compton, *Nature* **139**, 238, 1937.
30. J. H. van Vleck, *Physics Today*, June 1964, p. 21.
31. Ref. 20, Appendix 3.
32. E. Fermi, *Rev. Mod. Phys.* **4**, 87, 1932.
33. Ref. 16, interview on November 18, 1963.
34. P. A. M. Dirac, private communication.
35. Francis Fergusson, private communication.
36. A. K. Smith and C. Weiner, *Robert Oppenheimer, letters and recollections*, p. 91, Harvard Univ. Press, Cambridge, Mass. 1980.
37. Ref. 36, pp. 92, 94.
38. I. I. Rabi, in *Oppenheimer*, p. 3, Scribner, New York 1969.
39. M. Born, *My life*, p. 229, Scribner, New York 1978.
40. R. Oppenheimer, *Zeitschr. f. Phys.* **41**, 268, 1927.
41. H. A. Bethe, *Biogr. Mem. Fell. Roy. Soc.* **14**, 391, 1968.
41a. R. Serber, in *The birth of particle physics*, Eds. L. M. Brown and L. Hoddeson, p. 206, Cambridge Univ. Press 1983.
42. M. Born and J. R. Oppenheimer, *Ann. der Phys.* **84**, 457, 1927.
43. J. R. Oppenheimer, *Phys. Rev.* **31**, 914, 1928; see also R. H. Fowler and L. Nordheim, *Proc. Roy. Soc. A* **119**, 173, 1928.

44. J. R. Oppenheimer, *Phys. Rev.* **32**, 361, 1928.
45. G. E. Uhlenbeck, private communication.
46. P. Ehrenfest, letter to W. Pauli, November 26, 1928, *PC*, Vol. 1, p. 477.
47. W. Pauli, letter to P. Ehrenfest, February 15, 1929, *PC*, Vol. 1, p. 486.
48. J. R. Oppenheimer, *Zeitschr. f. Phys.* **55**, 725, 1929.
49. W. Pauli, letter to A. Sommerfeld, May 16, 1929, *PC*, Vol. 1, p. 500.
50. W. Heisenberg and W. Pauli, *Zeitschr. f. Phys.* **56**, 1, 1929.
51. W. Heisenberg and W. Pauli, *Zeitschr. f. Phys.* **59**, 168, 1930.
52. W. Heisenberg, letter to W. Pauli, July 20, 1929, *PC*, Vol. 1, p. 514.
53. W. E. Lamb, private communication.
54. R. F. Bacher, *Proc. Am. Philos. Soc.* **116**, 279, 1972.
55. A. Pais, *Subtle is the Lord*, Chapter 7, Section (e), Oxford Univ. Press 1982.
56. A. Einstein, *Sitz. Ber. Preuss. Ak. Wiss. 1919*, pp. 349, 463; also *Math. Annalen* **97**, 99, 1927.
57. Ref. 55, Chapter 15, Section (f) and Chapter 17.
58. H. A. Lorentz, *The theory of electrons*, p. 215, Teubner, Leipzig 1909, repr. by Dover, New York 1952.
59. H. A. Lorentz, *Collected works*, Vol. 5, p. 314, Nyhoff, The Hague 1937.
60. Ref. 58, p. 335.
61. G. Mie, *Ann. der Phys.* **37**, 511, 1912; **39**, 1, 1912; **40**, 1, 1913.
62. W. Pauli, 'Relativitätstheorie', Chapter 5, *Encyclopädie der math. Wiss.*, Vol. 5, Part 2, Teubner, Leipzig 1921; in English: *Theory of Relativity*, transl. G. Field, Pergamon Press, Oxford 1958.
63. J. Frenkel, *Naturwiss.* **12**, 882, 1924.
64. J. Frenkel, *Zeitschr. f. Phys.* **32**, 518, 1925.
65. P. Jordan and O. Klein, *Zeitschr. f. Phys.* **45**, 751, 1927, esp. p. 762.
66. Ref. 50, esp. Eq. (115).
67. J. R. Oppenheimer, *Phys. Rev.* **35**, 461, 1930.
68. I. Waller, *Zeitschr. f. Phys.* **62**, 673, 1930.
69. L. Rosenfeld, *Zeitschr. f. Phys.* **70**, 454, 1931.
70. Ph. M. Morse and H. Feshbach, *Methods of theoretical physics*, Chapter 9, Section 1, McGraw-Hill, New York 1953.
71. P. A. M. Dirac, *Proc. Cambr. Phil. Soc.* **26**, 361, 1930.
71a. J. R. Oppenheimer, *Phys. Rev.* **35**, 939, 1930.
72. See Ref. 14, Section 6.
73. W. Heisenberg, *Ann. der Phys.* **9**, 338, 1931.
74. H. B. G. Casimir, *Zeitschr. f. Phys.* **81**, 496, 1933.
75. W. Heisenberg, *Zeitschr. f. Phys.* **90**, 209, 1934.
76. V. Weisskopf, *Zeitschr. f. Phys.* **89**, 27, 1934.
77. C. N. Yang and D. Feldman, *Phys. Rev.* **79**, 972, 1950.
78. G. Källén, *Ark. f. Fys.* **2**, 187, 371, 1950.
79. P. A. M. Dirac, *Proc. Roy. Soc. A* **133**, 60, 1931.
80. P. A. M. Dirac, *Proc. Roy. Soc. A* **136**, 453, 1932.
81. W. Pauli, letter to P. A. M. Dirac, September 11, 1932, *PC*, Vol. 2, p. 115.
82. P. A. M. Dirac, V. Fock, and B. Podolsky, *Phys. Zeitschr. Soviet Union* **2**, 468, 1932; L. Rosenfeld, *Zeitschr. f. Phys.* **76**, 729, 1932; F. Bloch, *Phys. Zeitschr. Soviet Union* **5**, 301, 1934.
83. C. Møller, *Ann. der Phys.* **14**, 531, 1932.
84. J. R. Oppenheimer and M. S. Plesset, *Phys. Rev.* **44**, 53, 1933.

85. W. Heitler and F. Sauter, *Nature* **132**, 892, 1933.
86. E. Fermi and G. E. Uhlenbeck, *Phys. Rev.* **44**, 510, 1933.
87. H. B. G. Casimir, *Helv. Phys. Acta* **6**, 287, 1933.
88. *Phys. Rev.* **45**, 284, 1934.
89. H. Bethe and W. Heitler, *Proc. Roy. Soc. A* **146**, 83, 1934.
90. G. Racah, *Nuov. Cim.* **11**, 461, 1934.
91. W. Heitler, *The quantum theory of radiation*, 1st edn, p. 161, Clarendon Press, Oxford 1936.
92. C. D. Anderson, *Phys. Rev.* **44**, 406, 1933.
93. Ref. 91, p. 228.
94. G. Breit and J. A. Wheeler, *Phys. Rev.* **46**, 1087, 1934.
95. W. Heitler, *Proc. Cambr. Phil. Soc.* **31**, 242, 1935.
96. H. J. Bhabha, *Proc. Roy. Soc. A* **154**, 195, 1936.
97. C. F. von Weizsäcker, *Zeitschr. f. Phys.* **88**, 612, 1934; E. J. Williams, *Kgl. Dansk Vid. Selsk. Mat.-Fys. Medd.* **13**, No. 4, 1935.
98. F. Bloch and A. Nordsieck, *Phys. Rev.* **52**, 54, 1937.
99. Ref. 98, repr. in J. Schwinger, *Selected papers on quantum electrodynamics*, p. 129, Dover, New York 1958.
100. P. A. M. Dirac, in *Rapports du Septième conseil de physique*, p. 203, Gauthier-Villars, Paris 1934.
101. P. A. M. Dirac, *Proc. Cambr. Phil. Soc.* **30**, 150, 1934; also *ibid.* **25**, 62, 1929; **26**, 376, 1929; **27**, 240, 1931.
102. W. Pauli, letter to W. Heisenberg, February 6, 1934, *PC*, Vol. 2, p. 274.
103. W. Heisenberg, letter to W. Pauli, February 8, 1934, *PC*, Vol. 2, p. 279.
104. L. I. Schiff, *Quantum mechanics*, 2nd edn, Section 47, McGraw-Hill, New York 1955.
105. Ref. 104, Eqs. (47.17), (47.18).
106. P. A. M. Dirac, letter to N. Bohr, December 9, 1929, copy in the Niels Bohr Library, American Institute of Physics, New York.
107. W. Heisenberg, *Ann. der Phys.* **10**, 888, 1931.
108. F. Fock, *Dokl. Ak. Nauk* **1**, 267, 1933.
109. W. H. Furry and J. R. Oppenheimer, *Phys. Rev.* **45**, 245, 343, 1934.
109a. A. S. Wightman and S. S. Schweber, *Phys. Rev.* **98**, 812, 1955.
110. W. Pauli, letter to W. Heisenberg, Sept. 29, 1933, *PC*, Vol. 2, p. 212.
111. W. Pauli, letter to W. Heisenberg, June 14, 1934, *PC*, Vol. 2, p. 327.
112. W. Pauli, letter to W. Heisenberg, January 21, 1934, *PC*, Vol. 2, p. 254.
113. W. Heisenberg, *Zeitschr. f. Phys.* **92**, 692, 1934.
114. W. Pauli, letter to W. Heisenberg, December 11, 1933, *PC*, Vol. 2, p. 235.
115. W. Heisenberg, *Ber. Sächs. Ak. Wiss.* **86**, 317, 1934, repr. in Ref. 99, p. 62.
116. R. Jost and J. Luttinger, unpublished; cf. also E. Corinaldesi, *Nuovo Cim.* **8**, 494, 1951; *Nuovo Cim. Suppl.* **10**, 83, 1953; W. Pauli, *Selected topics in field quantization*, pp. 41, 42, MIT Press, Cambridge, Mass. 1973.
117. H. A. Kramers, *Proc. Ak. Wet. Amsterdam* **40**, 814, 1937, repr. in *Collected scientific papers*, p. 697, North-Holland, Amsterdam 1956.
118. A. Pais, *J. Math. Phys.* **3**, 1135, 1962.
119. E. Majorana, *Nuovo Cim.* **14**, 171, 1937; cf. also G. Racah, *Nuov. Cim.* **14**, 322, 1937.
120. W. H. Furry, *Phys. Rev.* **51**, 125, 1937.
121. J. J. Sakurai, *Invariance principles and elementary particles*, Chapter 5, Princeton Univ. Press 1964.

122. P. A. M. Dirac, letter to Niels Bohr, August 10, 1933, copy in the Niels Bohr Library, American Institute of Physics, New York.
123. See Ref. 109, especially footnote 11, and also Ref. 88.
124. Ref. 75, p. 222.
125. R. Peierls, *Proc. Roy. Soc. A* **146**, 420, 1934, esp. Section 6.
126. Ref. 75, Eq. (20).
127. Ref. 75, Eq. (25).
128. R. Serber, *Phys. Rev.* **48**, 49, 1935.
129. E. Uehling, *Phys. Rev.* **48**, 55, 1935.
130. W. Pauli and M. Rose, *Phys. Rev.* **49**, 462, 1936.
131. V. Weisskopf, *Kg. Dansk. Vid. Selsk. Mat.–fys. Medd.* Vol. 14, No. 6, 1936; repr. in Ref. 99, p. 92.
132. V. Weisskopf, *Physics Today*, November 1981, p. 76.
133. V. Weisskopf, *Zeitschr. f. Phys.* **90**, 817, 1934.
134. A. Mercier, *Helv. Phys. Acta* **12**, 551, 1938.
135. G. Racah, *Phys. Rev.* **70**, 406, 1946.
136. V. Weisskopf, *Phys. Rev.* **56**, 72, 1939.
137. R. Serber, *Phys. Rev.* **49**, 545, 1936.
138. O. Halpern, *Phys. Rev.* **44**, 855, 1936.
139. H. Euler and B. Kockel, *Naturw.* **23**, 246, 1935.
140. A. I. Akhieser, *Phys. Zeitschr. Soviet Union* **11**, 263, 1937.
141. H. Euler, *Ann. der Phys.* **26**, 398, 1936.
142. M. Delbrück, *Zeitschr. f. Phys.* **84**, 144, 1933.
143. N. Kemmer and V. Weisskopf, *Nature* **137**, 659, 1936.
144. W. Heisenberg and H. Euler, *Zeitschr. f. Phys.* **98**, 714, 1936.
145. W. Pauli, letter to W. Heisenberg, June 14, 1934, *PC*, Vol. 2, p. 327.
146. V. Weisskopf, interview with T. Kuhn and J. Heilbron, July 10, 1963, transcript in the Niels Bohr Library, American Institute of Physics, New York.
147. W. Pauli and V. Weisskopf, *Helv. Phys. Act.* **7**, 709, 1934.
148. P. A. M. Dirac, *Report KFKI-1977-62*, Hungarian Ac. of Sci.
149. Ref. 55, Chapter 26.
150. G. Wentzel, *Zeitschr. f. Phys.* **86**, 479, 1933; **87**, 726, 1934.
151. G. Wentzel, *Zeitschr. f. Phys.* **86**, 635, 1933.
152. P. A. M. Dirac, *Ann. Inst. Poincaré* **9**, 13, 1939.
153. P. A. M. Dirac, *Comm. Dublin Inst. Adv. Studies* **A1**, 1943.
154. P. A. M. Dirac, *Proc. Roy. Soc. A* **180**, 1, 1942; cf. also J. Eliezer, *Proc. Roy. Soc. A* **187**, 197, 1946.
155. W. Pauli, *Rev. Mod. Phys.* **15**, 175, 1943.
156. G. Wentzel, *Quantum Theorie der Wellenfelder*, Section 19, Deuticke, Vienna 1943; in English: *Quantum theory of fields*, transl. C. Houtermans and J. M. Jauch, Interscience, New York 1948.
157. S. T. Ma, *Phys. Rev.* **71**, 787, 1947.
158. M. Born and L. Infeld, *Proc. Roy. Soc. A* **144**, 425, 1934.
159. M. Born, *Ann. Inst. Poincaré* **7**, 155, 1937.
160. M. H. L. Pryce, *Proc. Roy. Soc. A* **159**, 355, 1937; see also Ref. 6.
161. W. Heitler, *The quantum theory of radiation*, 2nd edn, pp. 240–52, Clarendon Press, Oxford 1944.
162. H. A. Bethe and J. R. Oppenheimer, *Phys. Rev.* **70**, 451, 1946.

163. P. A. M. Dirac, *Proc. Roy. Soc. A* **167**, 148, 1938.
164. P. A. M. Dirac, *Proc. Roy. Soc. A* **209**, 291, 1951; also ibid. *A* **212**, 330, 1952.
165. P. A. M. Dirac, *Proc. Roy. Soc. A* **223**, 438, 1954.
166. D. J. Kevles, *The Physicists*, Vintage, Random House, New York 1979.
167. A. E. Moyer, *American physics in transition*, Tomash, Los Angeles 1983.

17

In which the nucleus acquires a new constituent, loses an old one, reveals new forces with new symmetries, and is explored by new experimental methods

(a) Enter the neutron

I think we shall have to make a real search for the neutron.
Chadwick to Rutherford in 1924[1]

Every good joke has a short punch line. Some have a short set-up as well. (What is the origin of the Grand Canyon? A Scotsman lost a nickel.) The set-up of others is leisurely and labyrinthine, as with the tale of the atomic nucleus which begins* with ante-nuclear physics, continues with the discoveries of the nucleus,** of isotopes and the determination of nuclear masses, charges, and radii,† runs off the track with the proton–electron model,†† and ends with the punch line: in 1932 Chadwick discovered the neutron. As with all good stories, the importance of the punch line is emphasized by its brevity, particularly fitting in this instance since, off and on, Chadwick had kept looking for the neutron during a twelve-year period while his actual discovery took only 'a few days of strenuous work'.[2]

As is evident from the letter by Chadwick to Rutherford quoted above, the neutron concept had a long gestation. Already in his Bakerian lecture of 1920 Rutherford had suggested that 'it may be possible for an electron to combine much more closely with the H-nucleus [than is the case in the ordinary hydrogen atom] . . . It is the intention of the writer to test [this idea] . . . The existence of such atoms seems almost necessary to explain the building up of heavy elements'.[3] These remarks teach us three things.

* Chapters 2, 3, 6, 8. ** Chapter 9. † Chapter 11. †† Chapters 11 and 14.

First, Rutherford's 'atom' is not 'our' neutron, which has half-integer spin.

Secondly, Rutherford's proton–electron composite should have a mass approximately equal to the proton. It never seems to have bothered him or his coworkers that there was no room for such a state in Bohr's old quantum theory of the hydrogen atom.* That is not surprising, however. If, for reasons unknown, two protons and four electrons could tightly bind to form an α-particle (we are in 1920!) then why couldn't the same happen to one proton and one electron?

Thirdly, Rutherford's speculation sprang from a primitive yet profoundly sound idea about nucleosynthesis. Given that the mass of a nucleus is essentially that of a set of protons (again, we are in 1920) how can this set congeal without some agent that neutralizes electrostatic repulsion?

Rutherford, in typical fashion, right away set some of his boys to work on a search for the neutral object. One of them looked for its possible production in a discharge tube filled with protons and electrons, seeking to detect its formation by means of secondary collisions in a vessel filled with mercury vapor. In the report[5] of negative results we find mention of the neutral object 'to which the name *neutron* has been given by Professor Rutherford'. That, as far as I know, is the first occasion on which this name appears in a paper from the Cavendish. 'Neutron' reappears in a paper[6] by Rutherford and Chadwick in 1929, and also in the Rutherford–Chadwick–Ellis textbook.[7] It should be noted, however, that the name 'neutron' for some neutral composite particle had cropped up numerous times on earlier occasions.** (Remember also Pauli's initial name for the neutrino, Chapter 14.)

Another Rutherford student tried in vain to find the heat generated when a hydrogen atom collapses into a neutron.[11] Chadwick searched for γ-rays emitted in that same process.[2] As he later recalled, these were the more respectable attempts, 'others were so desperate, so far-fetched as to belong to the days of alchemy'.[2] He and Rutherford came close (but found nothing) when in 1929 they looked for neutrons produced in α-aluminum collisions.[6]

The final phase of the neutron search began in June 1930 when Walther Bothe and Herbert Becker reported[12] that the exposure of beryllium to α-particles from polonium generated 'new radiations . . . so hard that one can hardly doubt their nuclear origin'. Their assumption that this radiation consists of nuclear γ-rays later turned out to be part of the truth[13] but not the whole truth. There was something odd about this radiation process. Neither its energy balance[14] nor its angular distribution[15] readily fitted the assumption of γ-emission.

Then came the Joliot–Curies' communication[16] of January 28, 1932 which, in Chadwick's words,[2] had an electrifying effect. In this paper entitled 'The

* Chadwick did not raise this point until 1933.[4]

** Feather found[8] that 'neutron' was used as early as 1899. Stuewer[9] cites other examples. I further found the same term used in a book by Stark.[10]

emission of high energy photons from hydrogenous substances irradiated with very penetrating alpha rays' the authors reported that the alleged γ-rays from the α-beryllium reaction were capable of ejecting protons from paraffin. They further noted that these γ's should have an energy $\sim$50 MeV if the proton ejection were a Compton effect. They were aware that this was a peculiarly high energy but 'that is not a sufficient reason for rejecting the [Compton effect] hypothesis'. On further consideration they changed their mind, however—not about the γ-rays but about the Compton effect. In a follow-up communication (February 22)[17] they wrote that the mechanism for proton ejection 'corresponds to a new mode of interaction between radiation and matter'.

Meanwhile Chadwick had discovered the neutron.

When the January 28 paper reached the Cavendish, neither Chadwick nor Rutherford believed its conclusions.[2] Chadwick immediately went to work and on February 17 submitted a paper[18] entitled 'possible existence of a neutron' in which he proposed that the α–Be reaction is (n = neutron)

$$\alpha + \mathrm{Be}^9 = \mathrm{C}^{12} + n. \tag{17.1}$$

In support of this conclusion he presented two distinct arguments:

(1) From a comparison of recoil velocities produced by the radiation when it traverses a vessel filled either with hydrogen or with nitrogen he found that the radiation consists of particles with close to protonic mass.*

(2) Suppose nevertheless that the radiation consists of photons. Then the process would be $\alpha + \mathrm{Be}^9 \rightarrow \mathrm{C}^{13} + \gamma$. If so, then from known mass defects Chadwick estimated that the γ's could have an energy of at most 14 MeV—not enough. So, he concluded, 'all the evidence is in favor of the neutron [unless] the conservation of energy and momentum were relinquished at some point'. At any other time this final caveat might have seemed excessive. But this was February 1932, and the debate about the validity of the conservation laws in certain nuclear processes was still very much alive.[20]

Years later Joliot,** paying Rutherford a handsome compliment, said that he had not read the latter's Bakerian lecture and that, had he done so, it is possible or probable that he and his wife would have identified the neutron in place of Chadwick.[22] Fortunately Chadwick's discovery did not lead them to terminate their experiments with polonium α-particles. As a result he and his wife, too, made a major discovery. There is some poetry in that they could do so because mother Marie had provided them with the world's strongest source of an element she had discovered and named: polonium.

* Let the radiation consist of particles with mass M (in proton units) moving with a velocity at most equal to v. Let $v(H)$ and $v(N)$ be the respective maximum recoil velocities in hydrogen and in nitrogen. Then $v(H)/v(N) = (M+14)/(M+1)$. Comparing this formula with his measurements Chadwick found[19] $M \simeq 1.15$.

** For a biography of Joliot see Ref. 21.

In July 1933 the Joliot–Curies reported[23] that α-irradiation of aluminum as well as boron led to two kinds of final products:*

$$\alpha + \mathrm{Al}^{27} \rightarrow \mathrm{Si}^{30} + p \tag{17.2}$$

$$\rightarrow \mathrm{Si}^{30} + n + e^{+} \tag{17.3}$$

$$\alpha + \mathrm{B}^{10} \rightarrow \mathrm{C}^{13} + p \tag{17.4}$$

$$\rightarrow \mathrm{C}^{13} + n + e^{+}. \tag{17.5}$$

Initially they interpreted Eqs. (17.3) and (17.5) as a disintegration of a proton: 'The proton is complex and results from an association of a neutron with a positive electron'. That is, 'the transmutation with emission of a proton can sometimes lead to the latter's dissociation'.[24] That was the set-up. Fortunately again, they continued the experiment and on January 15, 1934 announced[25] the punch-line: all nuclear reactions known till then had been instantaneous, but when an aluminum foil is irradiated the emission of positrons does not cease immediately when the α-source is removed. The foil keeps emitting positrons; the process has a characteristic half-life ($\sim$3 minutes): 'These experiments demonstrate the existence of a new kind of radioactivity with positron emission'. Instead of Eqs. (17.3) and (17.5) they now proposed the chains

$$\alpha + \mathrm{Al}^{27} \rightarrow \mathrm{P}^{30} + n \quad \searrow \quad \mathrm{Si}^{30} + e^{+} \tag{17.6}$$

$$\alpha + \mathrm{B}^{10} \rightarrow \mathrm{N}^{13} + n \quad \searrow \quad \mathrm{C}^{13} + e^{+}. \tag{17.7}$$

Three weeks later they reported[26] the direct detection by radiochemical means of the respective phosphorus and nitrogen isotopes—which explains why the discovery of β^{+}-radioactivity was awarded a Nobel Prize in chemistry. (It may also have been a consideration to honor the Joliot–Curies and Chadwick in the same year.) That happened in 1935. Marie Curie just did not live long enough to witness the occasion.

Events followed each other very fast now. Another five weeks later Fermi submitted the first in a series of articles on radioactivity induced by neutron bombardment.[27] With this paper Fermi began his experimental studies in neutron physics which made him perhaps the world's leading expert on that subject during the thirties. He and his co-workers made experimental physics flourish in Rome during that period, as has been described in great detail by Gerald Holton[28] and by Emilio Segré.[29] Neutron physics also blossomed

* In the spirit of the original papers I do not mention neutrinos in Eqs. (17.2)–(17.7).

elsewhere. The scattering and the absorption of neutrons became a main source of information about nuclear structure. Neutron bombardment led to the discovery of nuclear fission. Nuclear fission changed the world in which we try to live. All these are topics of high interest but not part of the present story.

In April 1934 Rutherford congratulated Fermi 'on your successful escape from the sphere of theoretical physics'.[30] One must conclude that Rutherford had either failed to notice, or else had not paid much attention to, the fact that, only two months before Fermi began his experimental neutron work, he had used the neutron as a theorist when giving the first correct interpretation of β-radioactivity.

Two main themes dominate the physics of the thirties. As previously noted, it was a time of struggle with quantum electrodynamics. It was also a time which Bethe has called[31] the happy thirties, with good reason: the discovery of the neutron made possible for the first time a rational theory of nuclear phenomena. This major advance is the topic of the present chapter. In Section (c) it is described how the electron–proton model of the nucleus with its attendant horrors* made way for the neutron–proton model. That transition did not occur all at once. There were two years, from early 1932 to early 1934, during which it was evident that the neutron was an essential nuclear constituent but it was by no means obvious as yet how to get rid of the electron as a nuclear building block. In that period it was a common device to hide electrons inside neutrons; one might well call those years the time of the proton–electron model of the neutron. The most notable theoretical contributions during this transition are Heisenberg's articles on nuclear forces discussed in Section (d). Heisenberg's neutron was a bound proton–electron system. His work straddles two themes: find the consequences of this composite picture of the neutron; and find those consequences for the nucleus, conceived as a neutron–proton system, which are independent of this alleged neutron structure. Also in this last respect his theory leaves much to be desired. Nevertheless this work was instrumental in showing the way toward a phenomenological treatment of nuclear forces in the context of non-relativistic quantum mechanics.

The period of transition came to an end with Fermi's theory of β-decay (Section (e)) which made clear, finally, not only *that* but also *how* electrons should be excised as nuclear constituents. It is hard to overrate Fermi's contribution. He showed that β-radioactivity is due to a new force. He liberated the neutron from the vestiges of the proton–electron model. And he silenced Bohr's conjectures about non-conservation of energy.[32] It may well have been to Fermi's advantage that he did not move in Bohr's orbit. 'Fermi was fully

* Chapter 14, Section (b).

aware of his accomplishment and he said he thought he would be remembered for this paper, his best so far.'[33]

The attempts in dealing with nuclear forces, initiated by Heisenberg and rapidly developed by others, notably Wigner and Majorana, may be called the Coulomb phase. One tried to fit the as yet meager facts by means of some phenomenological static potential between the nuclear constituents. Next came the 'Maxwell phase'. Could the nuclear potential be the static manifestation of an interaction mediated by a field, analogous to the connection between the Coulomb potential and the Maxwell field? The early attacks on this question are described in Section (g) where we encounter the first mesons, introduced by Yukawa. The preceding section (f) covers charge independence and the introduction of isotopic spin. The chapter concludes with a tribute to Rutherford, who died in 1937.

It is fitting to follow this outline with a mention of some major topics that will *not* be discussed in this or later chapters. Being inward bound, it is not my intent to give anything like a history of nuclear physics beyond those brief glimpses which are indispensable for an understanding of an even more refined subsequent description of matter. Nor shall I treat in any detail the immense amount of important work, theoretical and experimenal, having to do with developments in β-decay following the enunciation of the Fermi theory. Even the story of single items such as the β-spectrum of radium E are worthy subjects for a separate essay, but such topics do not fit the design of the present enterprise.

It is imperative, however, to mention a number of experimental advances which contributed centrally to progress in the thirties and which drastically transformed the character of research in nuclear and in particle physics. These topics are sketched in the next section. Interested readers are urged to acquaint themselves with the reminiscences of participants in these very important developments; see the sources collected at the end of this chapter.

(b) Of the deuteron, of cosmic rays, and of accelerators

December 5, 1931. Urey, Brickwedde, and Murphy announce[34] the discovery of the deuteron (in the beginning also called deuton, diplon), the nucleus of the mass-2 isotope of hydrogen. They began (at the Bureau of Standards) by evaporating liquid hydrogen near the triple point and 'collecting the gas from the last fraction of the last cubic centimeter'. Then (at Columbia University) they examined the atomic spectrum of their sample. Next to each of the usual hydrogen lines H_β, H_γ, H_δ they found a weak satellite at the position expected* from the old Bohr theory applied to an isotope of mass 2.

* This is the same nuclear mass effect discussed earlier in connection with He^+; see Eqs. (9.19) and (9.20). The effect is obviously most pronounced for the case of deuterium.

The deuteron was discovered ten weeks before the neutron. I already mentioned* that in 1932 the deuteron binding energy was still estimated on either of two assumptions: the deuteron's constitution is $2p+e$; or it is $p+n$.

In nuclear physics the deuteron matches in importance but not in simplicity the hydrogen atom in atomic physics. As the unique bound two-body system of nuclear physics, the deuteron is a rich source of information about nuclear forces. Already in the thirties it began to be clear, however, that the nuclear force law is incomparably more complex than the Coulomb law which serves to give such an excellent first approximation to the hydrogen spectrum. So it is and so it will remain.

On the subject of isotopes, developments in the thirties were extremely rapid. By 1937 nearly a hundred new radioactive species had been discovered.[35]

December 19, 1931. The first publication** of a positron picture, by Anderson, marks but one example of major progress in cosmic ray research during the thirties. Until just shortly before Anderson's discovery, hardly anything was known about the origin and constitution of this radiation, though by then nearly twenty years had passed since its discovery.

Some years ago I visited Professor Berta Karlik at the *Institut für Radiumforschung* in Vienna. As we were mounting the stairs she stopped at a window, pointed to a parking lot, and said approximately: 'There the first steps were made toward the discovery of cosmic rays. It was a meadow then'. Much later I came upon an account of what happened in that meadow, written by Victor Franz Hess, and dating from 1911.

By that time numerous experiments had demonstrated that ionization chambers of electroscopes registered radiations even when strongly shielded. It had become increasingly difficult to regard these penetrating radiations as originating in the earth's interior or in its atmosphere. In Hess' 1911 paper[36] two series of experiments are reported. The first, done in the meadow, was designed to check whether theoretical estimates of absorption coefficients agreed with experiment. The horizontal distance between a strong radium source and an electrometer was made to vary up to 90 m. The absorption found agreed with the theory. In the second set of experiments, made with manned balloons rising up to 1000 m, the vertical intensity was measured. It was found that the penetrating radiation at this altitude was about as intense as at sea level—in conflict with the meadow experiment *if* the radiation emanated from the earth. Next, on 17 April 1912, Hess together with two companions began a series of manned balloon ascents, eventually reaching 5000 m altitude, at which height the radiation was found to be about nine times as intense as on earth.[37] New hypotheses were necessary: 'Either a substance hitherto unknown which is found predominantly at high altitude or . . . an extraterrestrial source of

* Chapter 11, Section (f). ** Chapter 15, Section (f).

penetrating radiation', mostly called ultraradiation until late 1925 when Millikan (always with a keen ear for the catchy phrase) proposed[38] the name 'cosmic rays'.

Thus began a new branch of particle physics. Unlike the case of radioactivity, no laborious preparation of sources was needed. Unlike accelerator physics, no particles needed speeding up by man. The key advances were tied almost uniquely to the evolution of detection apparatus, which explains why, during the first fifteen years following Hess' discovery, progress was rather slow. The history of what came next contains many a tale of ocean voyages, dives in deep waters, airplane rides, and adventures on top of the Eiffel Tower and in the high mountains. With some regret I pass these by and list next, ever so briefly, the highlights in the developments up to the end of 1933, continuing beyond that in Section (g).

December 1925. First report[39] by the Millikan group* of measurements with self-registering equipment 'carried up by two balloons eighteen inches across'. This new technique was to make possible observations at much higher altitudes than before.

April 1927. In a β-ray experiment Dimitry Skobeltzyn places a cloud chamber in a magnetic field and finds[41] that a few of his pictures show electron tracks so straight that they cannot be due to β-radioactivity—the first observation of cosmic ray *particles*.

February 1929. Skobeltzyn notices[42] that cosmic ray particles frequently occur in groups—first indication of showers.**

June 1929. Advances in counter techniques and coincidence circuitry enable Bothe and Kolhörster to show[44] that some of the cosmic rays incident at sea level must be *charged* particles. They found that the number of coincidences between two counters in a vertical plane diminishes only slightly when a 4 cm-thick lead absorber is placed between the counters. The result is impossible to understand if, as was then generally believed, cosmic rays at sea level were photons with energies $\lesssim$ a few hundred MeV. The effect had rather to be due to single fast charged particles traversing the lead with moderate energy loss. 'The impact of this experiment on the history of cosmic ray research is well known. Until then cosmic rays were believed to be high-energy γ-rays.'[45]

1927–33. Experimental and theoretical studies of the dependence of cosmic ray intensities on latitude and on direction for a given position on earth (east–west effect) reveal the important influence of the earth's magnetic field and make clear that cosmic rays consist *preponderantly* of charged particles.†

February 1933. Blackett and Occhialini's first report[47] on cosmic ray showers, studied with their counter-controlled cloud chamber.

* Already then Millikan had been interested in cosmic rays for several years.[40]

** For Skobeltzyn's reminiscences of those years see Ref. 43.

† See e.g. Ref. 46 for detailed references to this complex subject.

1932–3. First evidence for billion-volt energies of individual cosmic ray particles. The most compelling results are those of Bruno Rossi who repeated the Bothe–Kolhörster experiment with lead absorbers up to one meter thick.[48] As has become known since, cosmic rays can have energies even much, much higher.

After just about 1930 the number of those active in cosmic ray studies began a rapid rise. So did the number of papers on the subject, as is illustrated in the following table culled from a 1934 literature survey[49] claimed to be fairly complete. 'Total' and 'U.S.' refer respectively to the number of papers published world-wide and in the United States:

Year	'24	'25	'26	'27	'28	'29	'30	'31	'32	'33
Total	9	15	48	37	43	54	62	95	132	184
U.S.	1	2	8	3	7	7	9	16	44	69

February 17, 1932. Discovery of the neutron with the help of an α-particle source. Thereafter all basic particles are discovered either in cosmic rays, or with accelerators, or with reactors.

June 15, 1932. Cockcroft and Walton report[50] the first nuclear process initiated by particles speeded up in an accelerator:

$$p + Li^7 \rightarrow 2\alpha. \tag{17.8}$$

They had been able to accelerate protons up to 0.7 MeV although for the purpose of the lithium reaction 0.12 MeV was found to suffice.

Voltage differences of this order of magnitude had long been available. Rutherford recalled seeing an experiment at the St. Louis Exposition of 1904 in which a transformer generated about half a million volts.[51] Whatever industrial reasons made such pursuits worthwhile, it was quite something else to achieve such voltages on the terminals of a highly exhausted vacuum tube. Only with such an arrangement could one hope to accelerate particles without loss of energy due to scattering on the gas in the tube. Thus from the outset the achievements of effective high energy were linked to those of high vacuum. Well into the twenties the limit on useful acceleration was about 200 000 volts; higher potentials led to disturbing discharges.[52]

It appears that the drive toward high energy research started with the 'development of a program in nuclear physics begun in 1926 under the direction of Dr. Gregory Breit'[53] at the Department of Terrestrial Magnetism of the Carnegie Institution in Washington. According to Merle Tuve, the principal experimentalist in this effort, 'The broad aim was to understand how atomic nuclei, composed [1926] of protons and electrons could hold so stably together . . . We expected to find that some non-electrical short range forces were involved'.[54] Using high voltage X-ray tubes Breit and Tuve reached[55] peak potentials of ~5 MeV. In 1927 a team in Berlin embarked on a program (not without success) of using atmospheric electricity as a high voltage source.[56] In 1928, a high energy nuclear program (starting out with X-ray tubes) was initiated at Caltech.[57] In 1929 Van de Graaff, then an NRC Fellow at Princeton, began his work* on the belt-charged electrostatic generator since named after him. In this electrical equivalent of the bucket hoist, motor-driven silk belts transport charge so as to generate opposite potentials on a pair of electrodes. In 1931 Van de Graaff, reporting 1.5 MeV potentials, noted: 'The machine is simple, inexpensive, and portable. An ordinary lamp socket furnishes the only power needed'.[59]

Meanwhile, back at the Cavendish, Cockcroft and Walton had begun work** on their voltage multiplier device now known as 'the Cockcroft–Walton'. The impetus came from a discussion in which Cockcroft had learned from Gamow that protons of a few hundred keV could penetrate nuclear barriers ('inverse α-decay') at a sufficient rate to initiate nuclear reactions.[60] Cockcroft and Walton's efforts had the blessing of Rutherford, who shortly before had expressed himself publicly in favor of high energy physics. 'It has long been my ambition to have available for study a copious supply of atoms and electrons which have an energy far transcending that of the α- and β-particles from radioactive bodies.'[61] Cockcroft was able to persuade Rutherford to obtain a large grant for the work—one thousand pounds.[62] Some years of labor finally produced their gadget, a combination of condensers and rectifiers with which they could multiply the voltage of a 200 keV transformer to a direct current potential of 700 keV. Cockcroft has described[62] what happened next. 'At last we obtained a high energy proton beam and brought it out into the air through a window to check on its energy and range in air. After wasting a certain amount of time in this way, until prodded by Chadwick and Rutherford, we directed it on to a lithium target and at once observed, with a zinc sulphide screen, the bright scintillations which were obviously due to particle emission from lithium.' Thus was achieved the first experimental result in accelerator physics.

Not long after this event, Cockcroft and Walton each received[63] the following telegram: 'The Associated Press of America serving about 1400 newspapers in North and South America would be deeply grateful if you will grant us an explanatory interview on your recent experiments in splitting the atom'.

* McMillan has given an eyewitness account of these beginnings.[58]
** For an account of those days see the correspondence between McMillan and Walton.[58]

September 15, 1932. The first physics result obtained with an American accelerator: from Berkeley, Ernest Lawrence, Stanley Livingston, and Milton White report[64] on a study of the reaction (17.8) over a proton energy range of 100–700 keV.

I met Lawrence only once, shortly after the Second World War, in Oppenheimer's Princeton home. (The two were still on good terms at that time.) My vivid memory of that occasion is of a man radiating self-confidence and controlled power. We were talking about the 400 MeV cyclotrons then under construction. I asked him how much higher energy he thought could be reached with present technology. Well over a thousand MeV, he said, an awesome number for those times.

So far as the U.S. is concerned, Lawrence was the first of the new breed of machine-builders. Experimenting with fast particle beams was perhaps not his main concern, but his single-minded dedication to forging ahead to ever higher energies was outstanding, as was his leadership in rearing the next generation of machine-builders: Alvarez, Brobeck, K. Green, Livingood, Livingston, Lofgren, McMillan, Panofsky, Snell, Thornton, White, and R. R. Wilson among them.* In the post-war era their experience was to benefit the National Laboratories at Argonne, Batavia, Brookhaven, Oak Ridge, and Stanford.

In addition, Lawrence had a most uncommon talent for obtaining funds, not a trivial matter in the early thirties. 'Lawrence's optimism, even his boosterism, engaged the willing belief of ordinary people sick of Depression . . . There was little money despite Lawrence's success at grant gathering; while the machines consumed tens of thousands, and then hundreds of thousands of dollars, the staff made do with small salaries, if any, and none of the fringe benefits now common: medical insurance, secretaries, and paid travel to meetings.'[65]

Lawrence, a native of South Dakota (where he and his high school friend Tuve did ham radio together[58]) arrived in Berkeley in 1928. In 1930 he started work on resonance acceleration. This concept goes back to 1924 when Gustaf Adolf Ising from Stockholm's Tekniske Högskola proposed to accelerate particles by multiple traversal of a voltage difference.[66] His idea was to send charged particles through a tube containing a linear array of accelerating electrodes. When they come to the end of the tube the electric field reverses, and the particles turn around and double their energy as they return to their starting point, when the field again reverses, etc. Following Ising's suggestion, the Norwegian Rolf Wideröe, 'the first designer of accelerators',[67] received his degree at the Technische Hochschule in Aachen by achieving[68] energy doubling, in 1928. (Wideröe later served on the advisory committees for the building

* I am beholden to Ed Lofgren for his help in compiling this list.

of the first proton synchrotron at CERN and the electron synchrotron DESY at Hamburg.*)

Lawrence has recalled[69] how, in 1929, the reading of Wideröe's paper gave him the inspiration for his own design.** According to an eyewitness, 'The idea of bending the orbits in a magnetic field occurred to Ernest immediately on reading Wideröe'.[71] In outline† Lawrence's strategy went like this. Two flat semicircular hollow electrodes ('dees') are mounted in a vacuum chamber placed between the poles of an electromagnet which produces a uniform magnetic field B perpendicular to the dees. Charged particles (charge e, mass m) produced near the center of the chamber follow a semicircular path in each dee with a frequency ν of revolution given by

$$\nu = \frac{eB}{2\pi m}. \tag{17.9}$$

ν is independent (that is the joke) *of the particle velocity*. In a gap left between the dees the particles are acted on by B and also by an alternating electric field with frequency ν'. Resonance, established by adjusting ν' to equal ν, guarantees that during each full circuit the particle energy increases twice; and these increases are cumulative. The particles travel in widening semicircles until they reach the dee periphery.

In the spring of 1930 Lawrence's student Edlefsen built 'two crude models'[73] of what is now called a cyclotron. The vacuum chamber†† made at first of pieces of window pane, scraps of brass, and an overcoat of wax, fitted between the 4-inch-diameter poles of a spectroscopist's magnet. In September 1930, Lawrence gave a paper on this device at the National Academy of Science meeting in Berkeley.[74] Resonance was still rather diffuse. On April 30, 1931, Lawrence and Livingston reported[75] on the first cyclotron that worked, constructed by the latter as his Ph.D. assignment. 'Using a magnet with pole faces 10 cm in diameter and giving a field of 12 700 gauss, 80 000 volt hydrogen molecule ions have been produced using 2000 volt high frequency oscillators on the plates.' Three months later they reached the next milestone: with a nine-inch model they accelerated protons beyond 1 MeV.[76] In February 1932 they readied a paper[77] on the results with a newly installed 11-inch magnet: a current of 10^{-9} amperes of 1.22 MeV protons. In September they and White finished their first physics experiment, the lithium reaction mentioned earlier.[64]

It took a number of months to obtain this lithium result because, as Livingston has recalled,[78] they were unprepared at that time to observe disintegrations with adequate instruments. 'The planning of physics experi-

* A sketch of Wideröe's career is found in Ref. 67.

** In 1931 Wideröe's linear accelerator design was applied in Berkeley to accelerate mercury ions.[70]

† For details see the book by Livingston and Blewett.[72]

†† Now on display in the Lawrence Hall of Science in Berkeley, as is the next cyclotron built by Livingston.

ments had not paralleled the construction of instruments to perform them. This negative consequence of Lawrence's concentration on accelerator improvement was to recur throughout the thirties.'[65] The building went on. The 11-inch was followed by the 27-inch, then the 37-inch, then the 60-inch cyclotron.[79] By then proton and deuteron energies had reached 8 and 16 MeV respectively. Cyclotrons were constructed in other places as well. Physics kept pouring out. The number of publications grew speedily.[80] Lawrence started planning for a 100 MeV machine.

Then came the Second World War.

Anticipating what is to come later I note that Eq. (17.9), the cyclotron principle, has restricted validity. It neither takes account of the relativistic variation of mass with velocity, nor of energy loss by radiation as charged particles move in curved orbits. These restrictions were not important for the relatively low energies of the first generation of cyclotrons. Also, at higher energies, Lawrence's idea of curved orbits has remained fundamental. As McMillan, himself a leading authority on the subject, said[58] about half a century later: 'I consider this [cyclotron principle] to be the single most important invention in the history of accelerators; it brought forth a basic idea of great power, and one capable of later elaborations and variations, such as the use of phase stability and strong focussing. All the big proton synchrotrons are really just an extension of the cyclotron principle'.

(c) What is a neutron?

> The chief point of interest is how far neutrons can be considered as elementary particles (something like protons or electrons).
>
> D. Iwanenko, April 21, 1932[81]

> . . . A proton embedded in an electron.
>
> J. Chadwick, April 28, 1932[19]

Rutherford's vision of the neutron, so brilliantly vindicated by Chadwick, illustrates strikingly how the pursuit of a conjecture based on very good sense (Rutherford's idea about the synthesis of the elements) can open new vistas not anticipated by the best of physicists. In the present case the search for a tightly bound proton–electron system led to the neutron, which is actually neither more nor less elementary than the proton. This and the next section are devoted to the transition from the long-held preconceived notion of the neutron as a hydrogen-like structure to the modern view. As we shall see, more was at stake than a process of liberation from ideas about composite particles. In particular, Bohr's speculation about a possible breakdown of quantum mechanics at short distances did not help matters at all.

The course of events is presented in the form of a brief chronology. The year is 1932.

April 18. Francis Perrin[82] and Pierre Auger[83] suggest that light elements are exclusively. built of α-particles, protons, and neutrons, the latter being considered as bound pe-systems, but that, from the naturally radioactive isotope K^{41} on, additional electrons appear as separate nuclear constituents. This proposal was still discussed at the Solvay meeting in October 1933, where, somewhat unexpectedly, it found support from Dirac: 'If we consider protons and neutrons as elementary particles we would have three kinds of elementary particles [p, n, e] out of which the nucleus is made up. This number may seem large but, from that point of view, two is already a large number'.[84]
April 21. Dmitri Dmitriievich Iwanenko must be credited with the first suggestion that the neutron might be something like a proton (see quotation above). Not that he banished electrons from the nucleus. Rather, he suggested, these are '*all* packed in α-particles or neutrons . . . [This] sounds not so improbable if we remember that the nuclei [sic] electrons profoundly change their properties when entering into nuclei'. Thus he believed electrons to be tucked away as constituents of constituents of the nucleus—which does, of course, not solve the many problems of the nuclear pe-model discussed in Chapter 14, Section (b).
April 18. Another 'Discussion on the structure of atomic nuclei' at the Royal Society.[20] It is perfectly natural that on that day the men from the Cavendish would stick with the pe-model of the neutron. Rutherford: 'An electron cannot exist in the free state in a stable nucleus but must always be associated with a proton or other possible massive units. The indication of the existence of the neutron in certain nuclei is significant in this connection'. Chadwick: 'The neutron may be pictured as a small dipole, or perhaps better, as a proton embedded in an electron', his second alternative alluding to the smallness of the classical proton radius e^2/Mc^2 compared to its electron counterpart.
May 10. Chadwick: 'It is of course possible to suppose that the neutron may be an elementary particle. This view has little to recommend it at present, except the possibility of explaining the statistics of such nuclei as N^{14}'.[85]
June 7. The first of a series of three articles on nuclear forces by Heisenberg, to be discussed in the next section. Heisenberg sticks with the pe-model of the neutron.
June 18. Oppenheimer and his student Frank Carlson's paper[86] on 'The impact of fast electrons and magnetic neutrons'. Here magnetic neutrons are Pauli's neutrinos. Listening to Pauli's colloquium at the Ann Arbor summer school* gave the authors the idea[87] of studying the absorption of these hypothetical particles in cosmic ray events. For the present only the following is important. They first refer to the 'third element in the building of nuclei'—that is, neutrinos as proposed by Pauli* to solve both 'the anomalous spin and statistics

* Chapter 14, Section (d).

and the apparent failure of the conservation of energy in beta-particle disintegration'. Then they go on: 'One may, however, assume that the neutron has a mass very close to that of the proton, and that such neutrons are substituted for pairs of electrons and protons in certain nuclei, instead of being added to them; such neutrons would help explain the anomalous spin and statistics of nuclei, although they would throw no light on beta-ray disintegrations. The experimental evidence on the penetrating beryllium radiation suggests that neutrons of nearly protonic mass do exist . . .'. To my knowledge this is the first time that it is stated in the literature that the neutron saves spin and statistics *and* that energy conservation in β-decay is a separate issue.
August 2. Bacher and Condon attempt[88] to determine the neutron spin from the spins of Li^6, Li^7, and N^{14} under various assumptions of the (p, n, e)-content of these nuclei. They find it most probable that in the nucleus the spins are 1/2 for p and n, zero for e—not bad, since there are no electrons in the nucleus.

For a better appreciation of this work it should be recalled that not even the spin of the deuteron was known at that time. In May 1933, experimental analysis of band spectra* of the d–d molecule was to show[89] that 'the spin of the deuteron is neither zero, 1/2, nor 3/2. It may be stated without reasonable doubt that this spin is 2/2'. Not until March 1934 did similar measurements definitively show[90] the deuteron spin to be one. Once that was established, the neutron spin could be eiter 1/2 or 3/2. Everyone took the value 1/2 for granted of course. The first conclusive proof for neutron spin 1/2 was not given until 1937 when Schwinger showed[91] that 3/2 could be excluded from data on neutron scattering by para- and orthohydrogen.
August 17. Iwanenko[92] on the resolution of the spin-statistics paradoxes (Chapter 14, Section (b)): 'We do not consider the neutron as built up of an electron and a proton but as an *elementary particle*. Given this fact we are obliged to treat neutrons as possessing spin 1/2 and obeying Fermi–Dirac statistics . . . Nitrogen nuclei appear to obey Bose–Einstein statistics. This now becomes understandable since N^{14} contains just 14 elementary particles [7p+7n], that is, an even number, and not 21 [14p+7e]', the earliest explicit statement I have found on the resolution of the nitrogen problem.

The year 1932 ended with a stand-off between the old and the new picture of the neutron.** The year thereafter, proponents of the old view began to vacillate. One example: Chadwick in his Bakerian lecture[93] (May 1933): 'There can be no doubt that the mass of the neutron is distinctly less [!] than that of the hydrogen atom. This is consistent with the view that the neutron consists of a proton and an electron . . . If the neutron is a proton and an electron why does not the hydrogen atom transform into a neutron with release of

* At issue was the intensity alternation described in Chapter 14, Section (b), Part 3.
** More opinions are quoted by Stuewer.[9]

energy? . . . It seems necessary for the present to recognize these difficulties and, while retaining the hypothesis that the neutron is complex for some purposes, to retain it as an elementary unit in the structure of atomic nuclei'. Note the additional complication of an as yet inaccurate value of the neutron mass m_n. Actually the neutron is heavier than the proton (by $\simeq 1.3$ MeV). The best early value* for m_n was obtained in August 1934 by Chadwick and Goldhaber[95] from the application of energy conservation to the photo-effect of the deuteron, $\gamma + d \rightarrow n + p$.

As if the neutron did not cause trouble enough, early in 1933 the proton became an object of puzzlement as well. At the turn of the year Bloch wrote[96] to Bohr: 'Pauli told me that Stern has measured the proton [magnetic] moment and appears to have found about $3e\hbar/2Mc$. That would mean that the Dirac theory does not apply to the proton, to Pauli's great joy'. Shortly afterward Frisch and Stern announced[97] their result: a moment between 2 and 3 μ_p ($= e\hbar/2Mc$, present best value $\simeq 2.8\ \mu_p$) against the general expectations of one μ_p, by analogy with the electron. 'This result is extremely surprising', Stern wrote with good reason.[98] Later in the year Bohr, on the lookout for new physics at short distances, saw the neutron as the culprit: 'This interesting discovery . . . most undoubtedly finds its explanation in the fact that the diameter of the neutron, and consequently that of the proton, is considerably larger than $\hbar/Mc$'.[99] Pauli, on the lookout for trouble with the Dirac equation, wrote to Heisenberg: 'There are theoretical as well as empirical (Stern's experiment) reasons for the inapplicability of the relativistic Dirac equation to heavy particles'.[100] Neither Bohr nor Pauli were on the right track. As to the magnetic moment of the neutron, already in 1933 Bacher found indications[101] (from the magnetic moment of N^{14}) that the neutron's moment is about minus one μ_p. The first good direct measurement made[102] by Alvarez and Bloch gave $\simeq -1.9\ \mu_p$; but that was not until 1940. I shall return to magnetic moments in Section (g).

The two years following the discovery of the neutron were somewhat perplexing. Neither those who considered the neutron to be composite nor those who favored it to be elementary had compelling arguments to present. The neutron's mass and spin were still under advisement, the proton's magnetic moment was a new riddle. Above all else, however, there was the unresolved mystery: what is the dynamical origin of β-decay? Once this question was answered by Fermi in 1934 the proton–neutron model without hidden electrons became compelling. I turn to Fermi's theory in Section (e). First, however, we must attend to the birth of the theory of nuclear forces.

* For a survey of other m_n-determinations up to that time see Ref. 94.

(d) Nuclear forces: phenomenological beginnings

> The basic idea is: shove all fundamental difficulties on to the neutron and practice quantum mechanics inside the nucleus.
> Heisenberg to Bohr, June 1932[103]

Classical mechanics had played a crucial role in the discovery of the nucleus.* The old quantum theory had sufficed to show that hyperfine structure can serve as a diagnostic for nuclear magnetic moments.** Non-relativistic quantum mechanics had successfully explained α-decay† and had paved the way toward further progress by revealing the paradoxes of nuclear spins, magnetic moments, and statistics.†† None of these advances touched in much detail on the questions of nuclear constitution and nuclear dynamics—except for first indications that intra-nuclear forces are not purely electromagnetic‡ and for the negative conclusion that the proton–electron model of the nucleus was a source of aggravation.

Three papers by Heisenberg[104,105,106] completed in the latter half of 1932 mark the transition to the modern view on nuclear forces. These articles, important though they are, must not be considered as a clean break with the past, however. Heisenberg's nuclear theory is a hybrid of the old and the new. It has the virtue of being based on the proton–neutron model of the nucleus, but the drawback of a proton–electron model for the neutron.

In order to understand Heisenberg's position it is essential to remember that in 1932 the controversy‡‡ between Bohr and Pauli about the interpretation of the continuous β-spectrum was still unresolved. We have seen that this and other difficulties had led Bohr as far as speculating about a breakdown of quantum mechanics at short distances and doubting the validity of conservation laws in β-decay. Heisenberg had probably been the first to be informed of Pauli's counter proposal, the neutrino postulate. The key to understanding Heisenberg's 1932 papers is simply this: at that time he sided with Bohr.

Earlier in that year Heisenberg had, in fact, attempted to work out a theory (never published) along Bohr's lines of thought. It was a lattice model of space with a lattice constant equal to $\hbar/Mc$. That picture caused trouble with the energy–momentum conservation laws—but, perhaps, that was as it should be. 'A further interesting result would be that nuclei consist of protons and (slow) light quanta of mass M, not of electrons.'[103] What these quanta were exactly, Heisenberg did not say. Bohr's influence continues to be manifest in the published nuclear physics papers of 1932. 'The very existence of the neutron

* Chapter 9, Section (d). ** Chapter 13, Section (c). † Chapter 6, Section (f).
†† Chapter 14, Section (b). ‡ Chapter 11, Section (g). ‡‡ Chapter 14, Section (d).

contradicts the laws of quantum mechanics in their present form. The admittedly hypothetical validity of Fermi statistics for neutrons as well as the failure of the energy law in β-decay proves the inapplicability of present quantum mechanics to the structure of the neutron.'[107]

Heisenberg's neutron is a particle with spin 1/2 obeying Fermi–Dirac statistics. 'It will however be assumed that under suitable circumstances [the neutron] can break up into a proton and an electron in which case the conservation laws of energy and momentum probably do not apply.'[108] We now grasp Heisenberg's strategy: first, let us admit that we don't understand the neutron; secondly, let us see how far we can come with standard non-relativistic quantum mechanics in regard to a nucleus built of protons and neutrons only—the electrons somehow hiding inside the neutrons.

To begin with let us get out of the way Heisenberg's thoughts on the structure of the neutron. First, he proposed[109] to explain certain experimental deviations from the Klein–Nishina formula for γ-nucleus scattering in terms of γ-scattering off electrons bound to protons, a large effect because of the smallness of the electron's mass on the the nuclear scale.* The true explanation of this scattering anomaly came in 1933: the effect was caused by electron–positron annihilation.** Secondly, he suggested that β-decay may come about because 'a neutron, analogous to quantum mechanical systems, would occasionally decay spontaneously under the influence of a strong electric field'.[112] Thirdly, 'the circumstance that the neutron behaves in many respects as if it is composed of a proton and an electron [finds] . . . its expression in the position exchange (*Platzwechsel*)'.[113]

This last point, to be explained next, brings us to Heisenberg's theory of nuclear forces.

Along with the Coulomb forces between protons Heisenberg postulated new interactions of short range:

(a) Between p and n. This force is introduced 'in analogy with the H_2^+–*ion*'.[104] Just as an attractive potential in H_2^+ is generated by the exchange of an electron between two protons, so, Heisenberg argued, a similar exchange takes place in the deuteron, an electron now associating itself with one proton, forming a neutron, then with the other proton, doing the same. He cast this exchange mechanism in a form which to him must have been a mathematical convenience but which a few years later would become of momentous physical importance. This is what he did.†

Consider a particle described by a wave function $\psi(\vec{x}, s, w)$, where $\vec{x}=$ position, $s=\pm 1/2$ is the spin value in some given direction, and $w=\pm 1/2$ is

* In presenting estimates for this effect Heisenberg used the occasion to thank Bohr for clarifying discussions.[110]

** For a detailed account of this episode see an article by Brown and Moyer.[111]

† I change Heisenberg's notations to conform to later applications.

a new two-valued variable. We decree:

$$\text{If } w=+1/2\ (-1/2), \text{ the particle is a proton (neutron).} \tag{17.10}$$

Introduce three 2×2 matrices τ_i, which act on the w-variable

$$\tau_1=\begin{pmatrix}0 & 1\\ 1 & 0\end{pmatrix},\quad \tau_2=\begin{pmatrix}0 & -i\\ i & 0\end{pmatrix},\quad \tau_3=\begin{pmatrix}1 & 0\\ 0 & -1\end{pmatrix} \tag{17.11}$$

just as the Pauli spin matrices Eq. (13.36) act on s. Thus

$$\tau_3\begin{pmatrix}\psi(\vec{x}, s, 1/2)\\ \psi(\vec{x}, s, -1/2)\end{pmatrix}=\begin{pmatrix}\psi(\vec{x}, s, 1/2)\\ -\psi(\vec{x}, s, -1/2)\end{pmatrix}. \tag{17.12}$$

Put

$$\tau_\pm=\frac{\tau_1\pm i\tau_2}{2} \tag{17.13}$$

τ_+ turns a p into an n:

$$\tau_+\begin{pmatrix}\psi(\vec{x}, s, 1/2)\\ \psi(\vec{x}, s, -1/2)\end{pmatrix}=\begin{pmatrix}\psi(\vec{x}, s, -1/2)\\ 0\end{pmatrix} \tag{17.14}$$

τ_- turns an n into p. Consider next a two-particle system with wave function $\psi(\vec{x}^{(1)}, s^{(1)}, w^{(1)}; \vec{x}^{(2)}, s^{(2)}, w^{(2)})$. Introduce two triples $\tau_i^{(1)}, \tau_i^{(2)}$ of τ-matrices, the ones with superscript 1 (2) acting only on $w^{(1)}$ $(w^{(2)})$. Heisenberg writes his neutron–proton potential W_{12} as* $(r_{12}=|\vec{x}_1-\vec{x}_2|)$

$$W_{12}=P_{12}F(r_{12}) \tag{17.15}$$

$$P_{12}=\tau_+^{(1)}\tau_-^{(2)}+\tau_-^{(1)}\tau_+^{(2)} \tag{17.16}$$

W_{12} is non-zero only when acting on a two-particle state consisting of one p and one n. In particular

$$P_{12}\psi(\vec{x}^{(1)}, s^{(1)}, 1/2; \vec{x}^{(2)}, s^{(2)}, -1/2)=\psi(\vec{x}^{(1)}, s^{(1)}, -1/2; \vec{x}^{(2)}, s^{(2)}, 1/2). \tag{17.17}$$

Thus P_{12} changes a proton with spin $s^{(1)}$ at $\vec{x}^{(1)}$ into a proton with spin $s^{(2)}$ at $\vec{x}^{(2)}$; similarly for the neutron. In other words proton and neutron exchange charge and spin, whence the name charge–spin exchange for the interaction W_{12}.

(b) Between n and n: this interaction is written as

$$\frac{1-\tau_3^{(1)}}{2}\cdot\frac{1-\tau_3^{(2)}}{2}K(r_{12}). \tag{17.18}$$

Again appeal is made to a molecular analogy, this time to the neutral H_2

* In his third paper[106] Heisenberg added a non-exchange term to W_{12}.

molecule which, with its two protons and two electrons, bears similarity to a two-neutron system *à la* Heisenberg.
(c) Between p and p: Coulomb forces only, *no* new short-range interaction. There is no molecule to appeal to!

In summary, inspired by molecular analogies Heisenberg had introduced nuclear forces of the '*Platzwechsel*' kind. His potentials J and K were phenomenological; wisely he preferred to learn from experiment what they were rather than predict them.

In assessing the uses to which Heisenberg put his new forces, it is important to remember how very little was known in 1932 about nuclear two-body systems. Thus he wrote down the Schroedinger equation for the deuteron[114] but did not discuss its solutions. How could he? The deuteron spin was not known as yet (as mentioned earlier); nor do the papers contain any mention of a value for the deuteron binding energy. There is no discussion of neutron–proton scattering. True, the first few such scattering events had been reported[115] in March 1932 by the Joliot–Curies, cloud-chamber pictures of protons ejected from paraffin, but that was not enough for a comparison with theory. Nor was Heisenberg very successful with his calculations[106] on the saturation of nuclear binding energies; the property of these energies to grow approximately linearly with the total number of protons and neutrons. In fact he stated incorrectly that his forces would give the α-particle a saturated structure, as is desired. Shortly afterward Majorana pointed out[116] that the deuteron, not the α-particle, is fully saturated by Heisenberg's forces.

In spite of these and other shortcomings, Heisenberg's work represents nothing less than a breakthrough, for one single reason: his insistence that the nucleus, considered as a p–n system, is amenable to non-relativistic quantum mechanical treatment. The message rapidly sank in, as did the realization that Heisenberg's molecular analogies were instructive but not compelling. Before long proposals were made for n–p forces with other exchange properties, named after their proponents:

	Spin exchange	*Charge exchange*
Heisenberg forces (1932)[104]	yes	yes
Wigner forces (1932)[117]	no	no
Majorana forces (1933)[116]	no	yes
Bartlett forces (1936)[118]	yes	no

Fitting experimental information to combinations of these forces with various assumptions about the structure of the short-range potentials became a new industry. By early 1936 data were available on the energy of the ^{3}S-ground state of the deuteron and its excited ^{1}S-state as well as on n–p scattering up to ~5 MeV. In addition p–p scattering experiments up to ~900 keV had demonstrated[119] the existence of an attractive short-range p–p nuclear force.

All this work is reviewed in the celebrated article by Bethe and Bacher[120] from which people of my generation learned with such profit their first nuclear physics. That was the time of merrily calculating away with various shapes of short-range potentials, square wells, exponentials, Gaussians, and others. It could have been realized already then, however, that the available scattering data only fix the strength and the range of the nuclear potential but tell us nothing about its shape. Remarkably, this rather simple fact was not recognized for years.[121] As to later phenomenological developments: in 1936 the first suggestions appeared about a relation between np-, pp-, and nn-forces known as charge independence.* In early investigations it was generally assumed that nuclear two-body potentials depend only on the relative distance between the particles (central forces). The discovery[122] of the electric quadrupole moment of the deuteron in 1939 demonstrated[123] the existence of an additional noncentral force of the tensor type. Shortly after the War it was realized[124] that still another noncentral force is needed, the spin–orbit coupling. It also became clear that the radial dependence of the forces was not simply monotonic as had most often been assumed. These discoveries as well as the availability of scattering data at increasingly high energies have eventually led to a quite complicated representation for the effective two-body forces in the nucleus.[125]

(e) Fermi's tentativo

1. In which Fermi introduces quantized spin-1/2 fields in particle physics. Pauli found Heisenberg's np-model of the nucleus quite interesting but did not care at all for his ep-model of the neutron. In July 1933 he criticized[126] Heisenberg's mechanism for β-decay: 'If it is true that the neutron has spin 1/2 and Fermi statistics . . . then a neutron can never be decomposed [as H. had suggested] into an electron and proton in the presence of external fields (I will insist on this in Brussels)'. Pauli continued: 'A neutron . . . can decompose, however, in a more complicated way, for example into a proton, an electron, and a neutrino'.

When, at the Solvay meeting in Brussels in October of that year, Pauli commented on Heisenberg's report, 'General theoretical considerations on the structure of the nucleus', he did defend his neutrino hypothesis[127] but neither he nor anyone else present, Fermi included, stated the fundamental neutron decay just mentioned. At that time Bohr, subdued in his comments** on energy nonconservation, certainly did not yet fancy neutrinos. In March Gamow wrote[128] to Goudsmit, in his inimitable English: 'Bohr . . . (well, you know that he absolutely does not like this chargeless, massless little thing!) thinks that continuous β-spectra is compensated by the emmition of gravitational waves (!!!) which play the role of neutrino but are much more physical things.

* See further, Section (f). ** See Chapter 14, Section (d), Part 5.

It is . . . very difficult to put through'. I do not know the details of this idea. It is clear, though, that the theory of particles and fields belongs to the post-Bohr era.

In the same letter Gamow also mentioned a new β-decay theory of Fermi's which 'explaines excellently' a number of experimental facts.

Fermi must have gone to work right after the Solvay conference. In December he sent a note on the subject to *Nature*. It was rejected 'because it contained speculations too remote from reality to be of interest to the reader'.[129] An Italian version entitled 'Tentativo di una teoria della emissione di raggi β' fared better.[130] More detailed accounts[131,132] appeared early in 1934. Here, at long last, it is stated that the language appropriate to β-decay is the language of quantum field theory: 'Electrons (or neutrinos) can be created and can disappear . . . The Hamilton function of the system consisting of heavy and light particles must be chosen such that to every transition from neutron to proton there is associated a creation of an electron and a neutrino. To the inverse process, the change of a proton into a neutron, the disappearance of an electron and a neutrino should be associated'.

In making these assertions Fermi appealed to an analogy with the creation and annihilation of photons. Comparisons like these were in the air. Iwanenko, August 1932: 'The expulsion of an electron [in β-decay] is similar to the birth of a new particle'.[133] Perrin, December 1933: 'The neutrino . . . does not preexist in atomic nuclei [but] is created when emitted, like the photon'.[134] No one before Fermi had put these ideas in operational form, however. *In fact, Fermi was the first to use second quantized spin-1/2 fields in particle physics.* Remember* that Heisenberg's first use of such fields in the positron theory dates from June 1934!

Before describing how Fermi did all that, I must digress briefly.

2. De Broglie on antiparticles. Neutrino theory of light. The neutrino, a particle with spin 1/2, mass μ, and zero charge, is described by a four-component field $\psi_\nu(\vec{x}, t)$ (ν stands for neutrino, the four-valued Dirac index is suppressed) which satisfies the Dirac equation. Except for electromagnetic interactions, all that has been said about the electron holds for the neutrino as well. There are positive and negative energy states; and a hole in the occupied negative energy states behaves as a particle—a new particle. In a communication dated January 8, 1934 (almost exactly the time at which Fermi was busy)—Louis de Broglie stated[135] that this new particle is 'an *antiparticle* [my italics] which is related to [the neutrino] like the positive electron to the negative electron in the hole theory of Dirac'. In this note the terms 'antiparticle' and 'antineutrino' enter physics for the first time.

* Chapter 16, Section (d), Part 2.

De Broglie's interest in neutrinos had nothing to do with β-decay, however, but with a new theory of light. A photon, he suggested, might be a composite consisting of a $\nu-\bar{\nu}$ pair ($\bar{\nu}$ = antineutrino). In this he was stimulated by Perrin's remark at the Solvay conference* that the ν-mass might be zero. De Broglie used non-quantized ψ_ν's for his purpose. In later work by others these ψ_ν's were quantized and a more sophisticated theory was developed in which a photon is represented by a superposition of $\nu\bar{\nu}$-states.** These efforts came to an abrupt end when Maurice Pryce discovered[137] that all proposals for a neutrino theory of light suffered from a lack of invariance under spatial rotations.[138]

The following ditty (provenance unknown to me) about Jordan, who also worked on this theory, should be recorded for posterity, however. It is to be sung to the tune of 'Mac the Knife'.

Und Herr Jordan
Nimmt Neutrinos
Und daraus baut
Er das Licht.
Und sie fahren
Stets in Paaren
Ein Neutrino
Sieht man nicht.

Rendered in English:

Mister Jordan
Takes neutrinos
And from those he
Builds the light.
And in pairs they
Always travel
One neutrino's
Out of sight.

Thus fortified let us return to Fermi.

3. The tentativo. Fermi implemented the analogy with photons by mimicking the electromagnetic interaction between the electric matter current j_μ and the vector potential A_μ responsible for photon emission and absorption. The β-decay analog of A_μ should be, he argued, an expression bilinear in the quantized electron field $\psi_e^\dagger$ (e for electron) and the quantized $\psi_\nu^\dagger$. Recall the properties of ψ_e and its conjugate $\psi_e^\dagger$ collected in Eqs. (16.14–32), in particular that $\psi_e^\dagger$ contains electron creation operators $a^\dagger$ and positron annihilation

* See Chapter 14, Section (d).
** For reviews and guides to the literature see papers by Kronig[136] and by Pryce.[137]

operators b; see Eqs. (16.15, 28). The same formalism (with $e=0$) applies to neutrinos. Thus $\psi_\nu^\dagger$ contains ν-creation, $\bar{\nu}$-annihilation. Fermi therefore replaced A_μ by some bilinear form of the type $\psi_e^\dagger\psi_\nu^\dagger$.

Once again I must interrupt Fermi's argument.

In neutron β-decay is the electron accompanied by a ν or a $\bar{\nu}$? At this stage this is purely a matter of convention. Fermi chose ν; in fact there is no mention of the antineutrino in this paper. For future purposes and for a slight easing of the formalism* I shall at once go to the modern picture:

$$\mathrm{n}\rightarrow \mathrm{p}+\mathrm{e}^-+\bar{\nu} \tag{17.19}$$

in which case Fermi's wholly unobjectionable bilinear $\psi_e^\dagger\psi_\nu^\dagger$ is replaced by $\psi_e^\dagger\psi_\nu$, as was first done by Konopinski and Uhlenbeck.[139] Fermi, aware of course of the need for Lorentz invariance, replaced A_μ by[140] $\bar{\psi}_e\gamma_\mu\psi_\nu$.

What should be the analog of j_μ? 'Here we meet a difficulty which stems from the fact that the relativistic wave equation for the heavy particles is unknown.'[132] Remember, it was the time of bewilderment about the magnetic moment of the proton. Fermi, ever pragmatic, therefore chose to consider only the analog of the static, non-relativistic part of $j_\mu A_\mu$, that is, $\rho\phi$ (ρ = charge density, $\phi=-iA_4$ = static potential). Since in the electric case $\rho=e\delta(\vec{x}-\vec{x}_0)$, $\vec{x}_0$ = particle position, he introduced a corresponding density $g_V\delta(\vec{x}-\vec{x}_0)$ (he wrote g for g_V). But in β-decay the heavy particle density should induce transitions p $\leftrightarrow$ n. To that end he used Heisenberg's $\tau_\pm$-operators, Eq. (17.13), integrated over $\vec{x}$, and finally came forth with the following β-decay interaction (recall that $\bar{\psi}_e\gamma_4\psi_\nu=\psi_e^\dagger\psi_\nu$).

$$H_{\text{int}}=g_V[\tau_-\psi_e^\dagger(\vec{x}_0,t)\psi_\nu(\vec{x}_0,t)+\tau_+\psi_\nu^\dagger(\vec{x}_0,t)\psi_e(\vec{x}_0,t)], \tag{17.20}$$

the starting point for all further developments in β-decay theory.** Note that g_V has a dimension $\dim(g_V)$ distinct from $\dim(e)$, e = electric charge:

$$\dim(g_V)=L^5MT^{-2}=\dim(e^2)L^2 \tag{17.21}$$

(L = length, M = mass, T = time). Later this innocent-looking relation was to show the way to W- and Z-bosons.

Using Eq. (17.20) Fermi discussed the following problems:

a. Neutrino mass μ. In the process Eq. (17.19) the electron senses the nuclear Coulomb field, hence its wave functions (the u's in Eq. (16.14)) are Coulomb wave functions. Neglect the Coulomb effect for a moment. Then the matrix element of the operator Eq. (17.20) does not depend on the energy W of the electron and the probability $P(W)\,\mathrm{d}W$ for an energy between W and $W+\mathrm{d}W$

* With the conventions corresponding to Eq. (17.19) it is unnecessary to introduce Fermi's δ-matrix, Ref. 132, Eq. (14).

** H_{int} acts on two-component pn-wave functions as in Eq. (17.14).

is entirely *statistical*, that is, proportional to the corresponding phase-space volume. In that case[141]

$$P(W)\,\mathrm{d}W = \text{const. } W(W_0 - W)[W^2 - m^2c^4]^{1/2}[(W_0 - W)^2 - \mu^2c^4]^{1/2}\,\mathrm{d}W. \tag{17.22}$$

W_0 is the upper limit of the β-spectrum. Eq. (17.22), which is the quintessential expression of the long-mysterious continuous electron spectrum, shows that the dependence on the neutrino mass μ is most sensitive near $W = W_0$. This remains true if one includes Coulomb effects, as Fermi did. From Eq. (17.22) (which remains essentially the same near $W = W_0$ also in the presence of Coulomb effects) he found 'the greatest similarity with the empirical curves for $\mu = 0$'.

b. Allowed and forbidden transitions. Concepts introduced by Fermi. To obtain the matrix element of H_{int} for β-decay of a nucleus one must evaluate $\int \varphi_p^* H_{\text{inf}} \varphi_n \,\mathrm{d}\vec{x}_0$, where φ_p and φ_n are wave functions of heavy particles bound in nuclear matter. Assume (said Fermi) that the $\bar{\nu}$- and e-wave functions do not vary over the nuclear volume. Then the decay probability is proportional to

$$M_{\text{pn}} = \int \varphi_p^* \varphi_n \,\mathrm{d}\vec{x}_0. \tag{17.23}$$

He noted that the 'nuclear matrix element' M_{pn} is non-zero only if the initial and final nucleus have the same angular momentum.* He went on to distinguish two possibilities: $M_{\text{pn}} \sim 1$ or $= 0$, 'allowed' and 'forbidden' transitions respectively, and observed that the latter case means a slower rate of decay rather than no decay at all, since various assumptions made are not rigorous: ψ_e may vary over the nuclear volume; and the neglected terms $\sim v/c$ in the heavy particle velocity v may be important if $M_{\text{pn}} = 0$.

These rudimentary but far-seeing remarks have grown into a major study of favored, allowed, and forbidden transitions, with corresponding selection rules, beginning with the 1941 paper[142] of Konopinski and Uhlenbeck. Refined treatments of nuclear matrix elements have become an important source of information about nuclear structure.

c. The value of g_V. If M_{pn} were always ~ 1 then the decay rate would be a function of Z and W_0 only. Already in 1933 this was known[143] not to be the case. Fermi identified a subset of relatively short-lived cases as 'allowed'. From these he deduced the 'very crude estimate'

$$g_V = 4 \times 10^{-50} \text{ erg cm}^3. \tag{17.24}$$

The present best value is $\sim 1.4 \times 10^{-49}$.

* And the same parity, but that is not yet in his paper.

Thus began the modern era in β-decay theory. Fermi himself never wrote about the subject again (apart from a critical note[144] on a now forgotten alternative theory). Others, however, immediately began to elaborate these new ideas.

4. Further developments in the thirties.

(a) In November 1933 an experiment done with a radium source showed[145] that neutrinos have a mean free path >150 km in nitrogen gas at 75 atmospheres pressure.

(b) In March 1934 Gian-Carlo Wick applied[146] the Fermi theory to β^+-radioactivity, just discovered by the Joliot–Curies. He, and independently Bethe and Peierls,[147] noted the possibility of the capture process

$$e^- + p \rightarrow n + \nu. \tag{17.25}$$

First detailed capture calculations: Yukawa and Sakata (1935).[148] First observation of K-capture: Alvarez (1938).[149]

(c) In July 1934, Hugh Wolfe and Uhlenbeck completed a paper[150] on 'Spontaneous disintegration of proton or neutron according to the Fermi theory'. For the best value of the p–n mass difference (evidently still under debate!) they found the free neutron to be unstable with a life-time $\sim 10^3$ sec, 'improbably small but there is no evidence to the contrary'. First observation of n-decay: in 1948, in a reactor experiment.[151] Present best mean-life value: 925 ± 11 sec.

(d) Fermi had already noted[152] that compared with the data his $P(W)$ gave too few slow electrons. When Bethe and Peierls suggested[153] that this could be remedied by introducing derivatives of ψ_e, ψ_ν in the interaction, Fermi replied[154] that one should wait for improved experiments. In May 1935, Uhlenbeck and his Ph.D. student Konopinski, pursuing the same idea, introduced[139] the derivative of ψ_ν in the interaction.* As a result the factor $(W_0 - W)^2$ in $P(W)$ (see Eq. (17.22) with $\mu = 0$) becomes $(W_0 - W)^4$. This change fitted the current data very well. For the next five years the 'KU theory' was widely accepted. Then Jim Lawson came to see Uhlenbeck and told him (as Uhlenbeck told me) that all the β-spectra measurements were wrong. The spectra were distorted by secondary effects, in particular by absorption and scattering within the source itself and within its support material. Thinning the source (down to a few mg/cm^2) improved the agreement with the Fermi theory[155] and the KU theory vanished from the scene.

(e) Once one dared use Dirac wave functions for p, n, it was obvious how to write Fermi's interaction (17.20) in relativistic form, and equally obvious that all the von Neumann covariants Eqs. (13.48–53) could generate β-decay

* Coupling: $\psi_n \gamma_\mu \psi_p \bar{\psi}_e \partial \psi_\nu / \partial x_\mu$.

couplings. Thus the following generalized interaction emerged:

$$H_{\text{int}} = \sum_i g_i \bar{\psi}_p O_i \psi_n \bar{\psi}_e O_i \psi_\nu + \text{hermitian conjugate}, \qquad i = S, V, T, A, P,$$

$$O_S = 1, \qquad O_V = \gamma_\mu, \qquad O_T = \sigma_{\mu\nu}, \qquad O_A = i\gamma_\mu\gamma_5, \qquad O_P = i\gamma_5. \tag{17.26}$$

When I asked Uhlenbeck who first thought of this general form, he replied that it was in the air. The earliest published account I found is in Bethe and Bacher's review article.[156] The corresponding $P(W)$ contains terms $\sim g_i^2$ as well as interferences[157] between S and V, A and T. Of great importance are the Gamow–Teller couplings[158] between the $\vec{\sigma}$'s of the heavy and light particles, generated in the static limit by A and T. Eq. (17.26) also permits the incorporation of nuclear recoil upon $(e, \bar{\nu})$-emission. The first rather qualitative evidence for this effect dates[159] from 1936.*

(f) May 1935. First estimates[161] of the rate for double β-decay

$$(A, Z) \to (A, Z \pm 2) + 2e^{\mp} + 2\begin{pmatrix}\bar{\nu}\\ \nu\end{pmatrix}. \tag{17.27}$$

Typical lifetimes: $\sim 10^{20}$ yr.[162]

(g) *Majorana neutrinos*. The charge conjugation operation Eq. (16.36) which interchanges the role of e^- and e^+ can likewise be applied to ν and $\bar{\nu}$. In 1937 Majorana noted[163] an alternative to the Fermi theory: one can arrange for ν and $\bar{\nu}$ to become identical, to wit, by substituting (in Kramer's language)

$$\psi_\alpha = C_{\alpha\beta}(\psi^\dagger \gamma_4)_\beta$$

for Eq. (16.36). This implies a modification of the Fermi theory.[164] An important distinction[165] is the double β-decay process which on the Majorana picture becomes

$$(A, Z) \to (A, Z \pm 2) + 2e^{\mp},$$

with lifetimes $\sim 10^5$ faster[162] than for Eq. (17.27). Neither version of double β-decay has yet been observed.

(f) How charge independence led to isospin

In his first nuclear physics paper[104] Heisenberg had realized that the approximate equality between the number of protons and neutrons in nuclei implied that short-range forces between nn could not be very different from those between pp. Since he did not introduce such pp-forces at all, he had therefore to assume the nn-force to be weak compared to np. For the same reason, in early models[166] with both nn- and pp-interactions these two were taken equal. In 1935, it was shown[167] by variational methods that the difference between

* For a review of early recoil experiments see Ref. 160.

the binding energies of H^3 (pnn) and He^3 (ppn) can entirely be accounted for by Coulomb effects—which implies equality of short-range pp- and nn-forces. 'We know today how hard it is to get good values for [the binding energies of] these nuclei.'[168] However that may be, it was appreciated early that nuclear pp- and nn-forces, if present, should be equal, a property since named charge symmetry.*

In August 1936 it became clear that such forces actually exist, that they are strong, and that they are intimately related to pn-forces. Three papers received by *Physical Review* within two days in that month mark major strides forward in the understanding of nuclear interactions.

One, by Tuve and co-workers,[119] announced evidence from scattering experiments for strong pp-interactions. Another, by Breit, Condon, and Present,[169] contained an analysis of these data, a comparison with available pn-scattering information, and the following conclusion: 'The pp- and pn-interactions in ^{1}S-states are found to be equal within the experimental error. This suggests that interactions between heavy particles are equal also in other states.'

The third paper,[170] by Cassen and Condon, deals with a formalism that incorporates equality of pp-, pn-, *and* nn-forces in ^{1}S-states, a property called *charge independence*. Their tools are Heisenberg's matrices $\vec{\tau}$, Eq. (17.11), which they called the characteristic vector. In 1937 Wigner[171] gave $\vec{\tau}$ the name 'isotopic spin', not a fortunate choice.** Today the common term is 'isospin'.

Cassen and Condon emphasized analogies between isospin and Pauli spin, Eq. (13.36). Consider two spins 1/2, $\vec{\sigma}^{(1)}/2$ and $\vec{\sigma}^{(2)}/2$ and their resultant $\vec{S}=(\vec{\sigma}^{(1)}+\vec{\sigma}^{(2)})/2$ with magnitude 0 or 1. There are three states with $S=1$ ($S_3=1, 0, -1$); these are symmetric under an interchange of (1) and (2). Let the forces in these three states be equal. Then, quantum mechanics tells us, the forces are of the form $a+b\vec{\sigma}^{(1)}\vec{\sigma}^{(2)}$, where a and b may depend on other variables. These forces are unaltered, invariant, under rotations of the spins.

Now, Cassen and Condon say, let

$$\vec{T}=\tfrac{1}{2}(\vec{\tau}^{(1)}+\vec{\tau}^{(2)}) \tag{17.28}$$

be the analog of $\vec{S}$. The three symmetric states with $T=1$ are pp, $(\text{pn}+\text{np})/\sqrt{2}$, and nn. Next, 'We postulate that in an assembly of heavy particles the wave function has to be anti-symmetric in all particles with respect to the exchange of all five of their coordinates. We want to show that this gives a convenient formalism for working with nuclear problems.'

The five coordinates are $\vec{x}$, s, and w, met in Eqs. (17.12, 14). Applied to the ^{1}S states, antisymmetric in $(\vec{x}, s)$; this means that they are symmetric in w; they are an isotopic triplet, or isotriplet. Experiment tells us that the forces

* More precisely charge symmetry means that the Hamiltonian is symmetric in the n- and p-variables (excluding electromagnetic effects).

** The name isobaric spin, suggested with good reason by others, is occasionally found in the early literature.

W_{12} in these three states are equal. Hence

$$W_{12} = a + b\vec{\tau}^{(1)}\vec{\tau}^{(2)}. \tag{17.29}$$

Thus *charge independence implies that nuclear forces are invariant under isospin rotations*. These rotations form a symmetry group, the group SU(2). Thus did a new symmetry,* manifest only in the subatomic world, enter physics. Some additional comments:

(1) Once (17.29) is established, other two-particle states can be labeled as well: ^{3}P has $T=1$, ^{1}P has $T=0$, etc. For more particles: $\vec{T} = \sum \vec{\tau}^{(j)}$, where $\vec{\tau}^{(j)}$ is the isospin of the jth particle. In general

$$[T_i, T_j] = i\varepsilon_{ijk} T_k, \tag{17.30}$$

the analog of Eq. (12.14) for angular momentum; $\varepsilon_{ijk} = +1(-1)$ if i, j, k is an even (odd) permutation of 1, 2, and 3; $=0$ otherwise.

(2) Isospin has very important applications, not discussed here, in many branches of nuclear physics.[173]

(3) W_{12} *refers to strong interactions only*. Electromagnetic (and weak) interactions do not obey SU(2) symmetry; after all a proton is electrically charged, a neutron is not.**

(4) To the extent that the strong forces inside a nucleus depend neither on the orientation of spin nor of isospin, they obey a stronger symmetry, SU(4). This approximate symmetry, discovered in 1937 by Wigner[171] and independently by Hund, is of importance for a more detailed classification of nuclear spectra.[174,175] Later we shall encounter an analog of SU(4) for the spectra of particle physics; see Chapter 21, Section (b).†

(5) A final word about Cassen and Condon's starting point, the extension of the Pauli principle. As was seen for two particles, and as is true in general, once one accepts isospin invariance this extension becomes automatic. As to the second quantization of the p and n Dirac fields ψ_p, ψ_n, each satisfies Eqs. (16.14–21). One can choose ψ_p and ψ_n to commute. Alternatively, without changing the physical content, one can arrange[177] the phases η introduced in Eq. (15.47) in such a way that the eight-component quantity

$$\psi_\alpha = \begin{pmatrix} \psi_p \\ \psi_n \end{pmatrix}, \qquad \alpha = 1, \ldots, 8 \tag{17.31}$$

and its conjugate $\psi^\dagger_\alpha$ satisfy Eqs. (16.14–21).

* Isospin symmetry is a continuous symmetry. One cannot[172] associate the physics of isospin with a discrete subgroup of SU(2).

** Another non-isospin invariant effect, the n–p mass difference, is commonly believed to be a secondary effect of these non-invariant forces.

† See also Ref. 176.

(g) Quantum field theory encounters the nucleus. Mesons

1. In which the Fermi theory is taken too seriously and high energy neutrino physics makes a first brief appearance. Within days of reading Fermi's paper on β-decay Heisenberg realized that the interaction Eq. (17.20) taken to second order (see Eq. (15.33)) gives rise to a charge exchange force between n and p due to virtual exchange of electron–neutrino pairs via the chains $n+p \to p+e^- + \bar{\nu} + p \to p+n$ and $n+p \to n+e^+ + \nu + n \to p+n$. On 18 January 1934 he sent[178] his results to Pauli: for $r_{12} \neq 0$ the function $J(r_{12})$ in Eq. (17.15) becomes 'up to a factor of order one'

$$J(r_{12}) = \frac{g^2}{\hbar c} \cdot \frac{1}{r_{12}^5}, \tag{17.32}$$

as is easily verified by dimensional reasoning.* A potential that singular cannot be treated[179] by standard wave mechanical methods. Heisenberg realized that cutting off the potential at $r_{12} \sim 10^{-13}$ cm would give an np-force too small by a factor $\sim 10^{10}$ (use Eq. (17.24)). To get an interaction energy ~ 1 MeV one needs to cut off at $\sim 10^{-15}$ cm, an inadmissibly small value for a nuclear range. These same conclusions were published half a year later by Tamm,[180] Iwanenko,[181] and Nordsieck.[182]**

Thus did the quantum field theory of nuclear forces make its not very propitious beginnings. In their review of nuclear physics up to early 1936, Bethe and Bacher wrote: 'This highly unsatisfactory result [too weak nuclear forces] is of course due to the extremely small values of the constant [g_V] which governs the beta-emission. However, the general idea of a connection between beta-emission and nuclear forces is so attractive that one would be very reluctant to give it up.'[184] There was no alternative—at least none that they knew of.

A change for the worse occurred later in 1936 when charge independence was discovered. How could that property of nuclear forces be reconciled with the Fermi interaction which to $O(g_V^2)$ generates a force between np, but none between nn and pp? Several attempts were made to cope with this problem: studies[185,186,187] of like-particle forces due to the Fermi interaction taken to $O(g_V^4)$; of additional[188] Fermi-type (e^+e^-) and ($\nu\bar{\nu}$) forces; of the possible role[189] of hypothetical spinless heavy particles (n′, p′).† It all looked artificial or hopeless or both.

* Use Eq. (17.21) and neglect the dependence on heavy particle masses ($\to \infty$) and light particle masses ($\to 0$).

** Heisenberg did not publish on the subject till 1935 when he noted[183] that the Fermi interaction gives an infinite contribution to the self-energy of the heavy particles.

† The idea was to generate charge-independent forces by virtual transitions of the type $n \to n' + \nu$, $p' + e^-$; $p \to p' + \nu$, $n' + e^+$.

In 1935 the following related and quite interesting idea bearing on the magnetic moment μ_p of the proton was proposed[190] by Wick. For a fraction τ of the time a proton, he said, is dissociated into $n+e^++\nu$. Therefore its magnetic moment μ_p equals

$$\mu_p = \frac{\mu + \tau\mu'}{1+\tau} \tag{17.33}$$

where $\mu = e\hbar/2Mc$ and μ' is the magnetic moment of the dissociated state. This is the fundamentally correct explanation of the anomalous proton moment—though the relevant dissociated state is not the one Wick thought it to be. Moreover, as others[185,191] rapidly noted, had he actually calculated μ' on the basis of Fermi's Eq. (17.20) he would have found $\mu'=0$, since that interaction is spin-independent. The more general interaction (17.26) gives a non-zero result, however.[191,192]

All these considerations were quite soon to come to an end without leaving a trace, but another field theory question concerning the Fermi interaction, first raised by Heisenberg in 1936, would continue to preoccupy a later generation: what is its high energy behavior?

Heisenberg considered[193] specifically the cross sections σ for inelastic processes in which a high energy proton produces electrons, positrons, and neutrinos in the Coulomb field of a nucleus with charge Ze. By purely dimensional arguments (use Eq. (17.24)) he found:

$$p \to n + e^+ + \nu, \qquad \sigma \sim \left(\frac{Ze^2}{\hbar c}\right)^2 \left(\frac{g_V}{\hbar c}\right)^2 k^2 \tag{17.34}$$

$$p \to p + j(e^- + e^+ + \nu + \bar{\nu}), \qquad \sigma \sim \left(\frac{Ze^2}{\hbar c}\right)^2 \left(\frac{g_V}{\hbar c}\right)^{4j} k^{8j-2} \tag{17.35}$$

where the first factor in σ arises from the energy–momentum transfer in the Coulomb field and k is a typical wave number of the particles produced. Heisenberg observed that the higher-order processes became more important than the lower-order ones at a sufficiently high energy, which he estimated to be $\sim$1000 MeV; perturbation theory breaks down at such energies. 'In this way the Fermi theory gives a quantitative explanation for the creation of [cosmic ray] showers', which on this picture should be events in which many particles are created in a single explosive act.

A brief digression about showers.

By 1936 it had become experimentally clear that the more common type of showers consists of photons, electrons, and positrons. In the same December week of 1936, first Carlson and Oppenheimer,[194] then Bhabha and Heitler,[195] submitted papers in which they showed that this type of shower can be interpreted as a cascade phenomenon. For example, a photon with energy $\sim$1000 MeV traversing lead creates an e^+e^- pair within a mean distance $\lesssim$1 cm.

These particles suffer *Bremsstrahlung*, the newly created photons make pairs again, and so on. The theory of cascades, a wonderful problem in fundamental interactions combined with statistical considerations, clarified satisfactorily the nature of the so-called soft component of cosmic rays.

Carlson and Oppenheimer, familiar with Heisenberg's new ideas, commented: 'We believe that [the idea of explosions] rests on an abusive extension of the formalism of the electron neutrino field'.[194] When Heisenberg heard of the cascade theory he at once conceded[196] that its authors had a point, but noted that rarer showers might still leave room for his explosions, an idea which continued to intrigue him several more years.[197] I well recall discussions at early postwar conferences on the issue 'plural versus multiple production'.

It appears to me that Heisenberg's experiences with the Fermi theory made a deep and lasting impact on him, which changed the course of his thinking. The strong increase of cross sections with energy he had found led him to surmise that physics may have to be revised at short distances. Already in his 1936 paper[193] we find reference to 'the introduction of a universal length [which] perhaps must be connected with a new change of principle in the formalism, just as for example the introduction of the constant c led to a modification of prerelativistic physics'. Note also that 1936 was the last year Heisenberg worked on quantum electrodynamics. In 1938 he wrote a beautiful paper,[198] dedicated to Planck on his eightieth birthday, in which he compared the need for a universal length with the need for Planck's constant at an earlier time. From the Heisenberg–Pauli correspondence one sees how, in those years, he kept groping for ways to introduce his length by means of a lattice theory (as he had already tried in 1932; see Section (d)) which would eliminate the infinities of the continuum field theories. Observe, furthermore, that the last dedicated efforts in Heisenberg's life concerning a theory in which couplings of the Fermi type are central can be traced back to 1936. In a letter of that year to Pauli[199] he mentioned the equation $\Box\psi = \lambda\psi\psi^*\psi$ (a simplified version of a wave equation arising in the Fermi theory) and commented: 'The problem of the elementary particles is a mathematical one, it is simply the question of how to construct a nonlinear relativistically invariant quantized wave equation without any fundamental constants.' (He presumably meant that particle masses should be deducible from the one constant λ.)

Two final comments.

A paper by Fierz (February 1936)[200] contains the first calculation (as best I know) of neutrino scattering. At high energies the differential cross section $d\sigma$ for

$$\bar{\nu} + p \rightarrow n + e^+ \tag{17.36}$$

is found to be (p_ν is the center of mass neutrino momentum)

$$d\sigma = \frac{g_V^2}{\pi^2} p_\nu^2 \, d\Omega, \tag{17.37}$$

a result which Fierz noted without further comment. In 1937, Tomonaga and Tamaki, stimulated by Heisenberg's work, made similar calculations for $\nu + n \rightarrow p + e^-$ and $\rightarrow n + e^+ + e^- + \nu$, and noted[201] that 'there exists . . . a certain limit of energies beyond which . . . perturbation theory ceases to be valid'. I shall return later to this very important issue.

The Fermi interaction has remained useful to this day, but cannot be taken seriously at high energies and momentum transfers (of the order of the W-boson mass); the clues for a universal length perceived by Heisenberg in the high energy behavior of Fermi's theory are now known to be spurious; see Chapter 21, Section (e), Part 1.

2. A meson proposed. Yukawa. Yukawa and I often discussed quantum field theory during the academic year 1948–9 which he spent at the Institute for Advanced Study in Princeton[202] and also during the next four years when he would visit from New York where he was visiting professor at Columbia University. To me these conversations were stimulating and pleasant. His seminars were something else. In his autobiography[203] Yukawa quotes one of his students' description of his lecturing style: 'His voice was as gentle as a lullaby and he spoke with little emphasis—it was ideal as an invitation to sleep'; and another as noting that he would turn his back to his audience and address the blackboard. These portrayals agree in detail with my own experiences. I would ascribe this conduct far more to reserve than to shyness. Yukawa's self-perception: 'I am a solitary man, and also obstinate',[203] strikes me as accurate.

Among other topics of much interest in Yukawa's book are his comments on those Japanese physicists who made a particularly strong impression on him during his formative years: Nagaoka, 'the most important scientist of Japan [in the twenties]', Nishina, 'the kindly father figure', and Tomonaga, fellow student in high school and at the university (both in Kyoto), 'smarter than any other friend I had ever known'.

Nagaoka[204] had spent the years 1893–6 studying physics at German universities. Upon his return to Japan he was appointed professor at Tokyo University. It was there that he did his work (mentioned earlier*) on the Saturnian model of the atom. He must have been instrumental in sending several of his students to Europe, among them Ishiwara who became an expert on relativity theory, and Nishina who from April 1923 to October 1928 was at Niels Bohr's Institute where in all eight young Japanese physicists spent time[205] during the 1920s. Nagaoka must also have participated in arranging visits to Japan (all in the twenties) by Dirac, Einstein, Heisenberg, Laporte, and Sommerfeld. After the founding of Osaka Imperial University in 1931 he became its first president, and as such passed on Yukawa's appointment there, first as lecturer (1933), then as associate professor (1936).

* Chapter 9, Section (c), Part 5.

Nishina, founder of experimental nuclear and cosmic ray research in Japan was the principal figure of the next generation. He will be remembered both for his theoretical work with Oskar Klein and his supervising the construction of several accelerators at the Institute of Physical and Chemical Research in Tokyo. In an act mindless even by military standards, the most advanced of these, a 150 cm cyclotron, was destroyed after the War by the American occupation forces. Serber has told me how bitterly this enraged Lawrence.

For the present account a discussion in 1933 between the 43-year-old Nishina and the 26-year-old Yukawa is of particular interest. Yukawa had started his academic career in 1932. That was the year of the neutron and of Heisenberg's theory of nuclear forces, the importance of which Yukawa appreciated at once.* 'I decided to carry the theory one step further ... By confronting this difficult problem I committed myself to long days of suffering ... Is the new nuclear force a primary one? ... It seemed likely that [it] was a third fundamental force unrelated to gravitation and electromagnetism ... Perhaps the nuclear force could also find expression as a field ... If one visualizes the force field as a game of "catch" between protons and neutrons, the crux of the problem would be the nature of the "ball" or particle.' That had also been the line of thought of Heisenberg who, in developing his molecular analogies (Section (d)), had taken the 'ball' to be the electron. Yukawa at first pursued that same idea on which he reported at a physics meeting in April 1933, in a paper (never published) 'The electrons within nuclei'. It was after that talk that Nishina and Yukawa had a discussion in which 'Nishina suggested that I consider an electron with other statistics, namely Bose–Einstein statistics'.

Yukawa has recalled how at first this idea seemed too radical to him; how he then became acquainted with Fermi's paper and the subsequent electron–neutrino theory of nuclear forces with all its troubles. 'I was heartened by the negative result and it opened my eyes so that I thought: let me not look for the particle that belongs to the nuclear force among the *known* [my italics] particles ... If I focus on the characteristics of the nuclear force field, then the characteristics of the particle I seek will become apparent.' Then he continued with a daring and innovative thought which brought him just fame: 'The crucial point came to me in October [1934]. The nuclear force is effective at extremely small distances. My new insight was that this distance and the mass of the new particle are inversely related to each other.'

In present terms, Yukawa saw that the electrodynamic connection

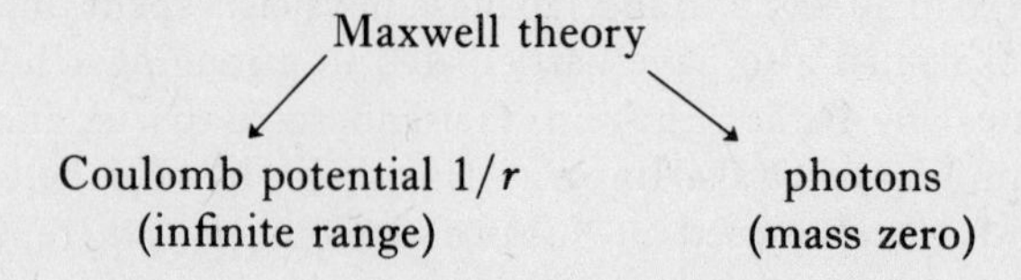

* All quotations in this and the next paragraph are from Yukawa's book.[203]

has the parallel

Yukawa theory ↘

Yukawa potential $\frac{e^{-\kappa r}}{r}$ (range $1/\kappa$) mesons (mass μ)

where

$$\frac{1}{\kappa}=\frac{\hbar}{\mu c}, \tag{17.38}$$

a relation which was to establish a link between nuclear and cosmic ray phenomena. The now current name 'meson' appears to have entered the literature[206] in 1939 (Yukawa used the term 'heavy particle'). Manuscripts presented that year to the cosmic ray symposium at Chicago contained no less than six names (including 'yukon'); the editors decided[207] upon 'mesotron'.

Of his first paper[208] on mesons, completed in November 1934, Yukawa wrote three years later: 'The mathematical formalism was preliminary and incomplete in many ways'.[209] His was indeed not yet a systematic treatment. He described mesons by a one-component complex field $U(\vec{x}, t)$ which, however, is not the scalar field discussed previously by Pauli and Weisskopf.* Rather, his U is supposed to be the fourth component of a vector field, the analog of the electrodynamic scalar potential. Thus his initial attempt is in terms of a vector meson theory without vector potential (in dropping the latter he must have been following Fermi's strategy described in the foregoing Section (e)). Yukawa, well aware that U should be quantized by the rules of Bose statistics, treated the nuclear force problem semi-classically. He started from the wave equation

$$\begin{gathered}(\Box-\kappa^2)U=g\rho,\\ \rho=\psi^*\tau_-\psi, \qquad \psi=\begin{pmatrix}p\\ n\end{pmatrix}.\end{gathered} \tag{17.39}$$

g is a constant with dimensions of electric charge. In the static limit (when ρ does not depend on t)

$$U(\vec{x})=-\frac{g}{4\pi}\int \rho(\vec{x}')\frac{e^{-\kappa|\vec{x}-\vec{x}'|}}{|\vec{x}-\vec{x}'|}\,\mathrm{d}\vec{x}' \tag{17.40}$$

By pursuing the electromagnetic analogy he ended up with a static pn-force of the Heisenberg type given by Eqs. (17.15, 16) with**

$$J(r_{12})=g^2\frac{e^{-\kappa r_{12}}}{r_{12}} \tag{17.41}$$

* Chapter 16, Section (d), Part 7.
** Yukawa incorrectly believed to have the freedom of replacing g^2 by $-g^2$.

By using the nuclear force range roughly known to be $1/\kappa = 2\times10^{-13}$ cm, he found a value $\sim 200\ m_e$ for the *U*-particle mass and $g \sim$ a few times e. Like-particle forces are not mentioned—but remember it was 1934, charge independence was not yet there.

Yukawa's paper is a primitive improvization, yet it contains three major lasting points: the dissociation of the strong nuclear force from the weak β-decay force; a sensible first estimate of the meson mass; and the remark that 'the massive quanta may also have some bearing on the shower produced by cosmic rays'.

Also in respect to β-decay the paper contains an ingenious new proposal: the *U*-field shall not only be coupled strongly to (p, n) but also weakly to (e, ν). Thus, he suggested, neutron β-decay would proceed via the chain of virtual processes:

$$\mathrm{n} \to \mathrm{p} + \mathrm{U}^-; \ \mathrm{U}^- \to \mathrm{e}^- + \bar{\nu} \tag{17.42}$$

We now know that β-decay does proceed via the virtual appearance of a Bose particle—which is not Yukawa's meson, however.

After November 1934, neither Yukawa nor anyone else wrote about mesons until June 1937 when an experimental discovery which, it seemed, had much to do with Yukawa's theory stirred considerable theoretical activity.

3. A meson discovered. By the mid-thirties it was known that *qua* absorptive properties cosmic rays consisted of a 'soft' component, well understood in terms of an electron–positron composition, and a much more penetrating 'hard' component, whose absorptive properties caused puzzlement. Neddermeyer and Anderson took up this problem, using the same magnet cloud chamber which had served in the discovery of the positron, and inserting a platinum bar in it to study absorption. Analyzing their data on the basis of received electrodynamic wisdom, they found that the hard component could not be protons since that would not fit the observed spectrum of electron secondaries. Nor could they be electrons, since that did not fit the observed *Bremsstrahlung*. On the other hand, again accepting standard quantum electrodynamics, they could understand their results reasonably well if the hard component were to consist of particles with unit charge and mass intermediate between p and e. A brief announcement to that effect appeared[210] in November 1936. A detailed paper was completed[211] in March 1937. Its conclusion: either the standard theory fails, a view which at that time had its distinguished proponents; or 'There exist particles of unit charge but with a mass (which may not have a unique value) larger than that of a normal free electron and much smaller than that of a proton'.

Shortly afterward the same conclusion was reached independently by Street and Stevenson[212] at Harvard and by the Nishina group[213] in Tokyo. None of these papers contain as yet an estimate of the mass of the new particle. A review[214] of data up to mid-1939 puts the mass between 100 and 400 m_e, about 200 m_e being most probable.

In his reminiscences[215] about this discovery, Anderson emphasized that it did not come abruptly, as with the positron. The absorption data had been odd for some time. The main block had been disbelief in quantum electrodynamics.

So now there was a theoretical meson and an experimental meson.

4. More on pre-war meson theory. The first prediction of a particle on symmetry grounds. On 1 June 1937, Oppenheimer and Serber sent a letter[216] to the Editor of *Physical Review* suggesting that the recently discovered 'heavy electron' (as Neddermeyer and Anderson had called their new particle) might in fact be Yukawa's meson.* This appears to be the first time the Yukawa theory is mentioned in a Western publication. Five days later a similar letter was sent from Geneva, by Stückelberg.[217] In July, Yukawa himself pointed out[209] the same connection as well.

As will be seen later, ten years were to go by before it was discovered that this cosmic ray particle actually is *not* Yukawa's meson. Because of this new pitfall of simplicity, quantitative theoretical conclusions obtained on mesons in the late thirties are no longer of much value. Nevertheless, it was a period of great progress. It became clear how charge independence could be incorporated in meson theory. New types of field equations were discovered, both for charged and neutral mesons. I content myself with recording the new equations, referring to Wentzel's book, still excellent reading,[218] for many omitted details.

(a) In 1936, the Rumanian-born French physicist Alexandre Proca proposed[219] a new type of field equations of great importance for the further development of meson theory. The Proca equations are

$$\begin{aligned} F_{\mu\nu} &= \partial_\mu A_\nu - \partial_\nu A_\mu \\ \partial_\mu F_{\mu\nu} - \kappa^2 A_\nu &= 0. \end{aligned} \tag{17.43}$$

These look rather like the Maxwell equations but differ in the following ways. The A_μ are complex. Polarization is no longer only transverse; there are additional longitudinal modes. Quantization leads to particles with charge $\pm e$ (quite like the Pauli–Weisskopf Eqs. (16.47, 48)), mass $\hbar\kappa/c$, and spin 1. The quantization is simpler for the non-zero mass case than for photons.**

(b) June 1937. In their letter[216] Oppenheimer and Serber criticize many details of Yukawa's paper, such as the absence of saturation (Heisenberg forces!) and the lack of charge independence.

* Serber told me that Oppenheimer knew of Yukawa's paper because the latter had sent him a reprint.

** See Chapter 15, Appendix, Note 7.

(c) September 1937. Yukawa and Sakata[220] abandon Yukawa's initial interpretation of the U-field and give a systematic treatment of charged scalar mesons quantized *à la* Pauli–Weisskopf. The field equations are*

$$\phi_\mu = \partial_\mu \phi + v_\mu$$
$$\partial_\mu \phi_\mu - \kappa^2 \phi = s \tag{17.44}$$

$$v_\mu = ig\bar{\psi}\tau_- \gamma_\mu \psi, \qquad s = f\bar{\psi}\tau_- \psi. \tag{17.45}$$

To order f^2 pn-forces, generated by virtual emission and absorption of mesons (use Eq. (15.33)), are given by Eqs. (17.15, 16, 41) with $g^2 \to -f^2$. It is noted that like-particle forces appear to $O(f^4)$. Attempts are made to get charge independence by playing with the values for f and κ. They are not sure this will work: 'It is not certain whether or not we have to introduce neutral heavy quanta also.'
(d) December 1937. Kemmer[221] and independently Serber[222] note that the scalar theory gives the wrong relative position of the deuteron ^{3}S and ^{1}S states. They, and also Bhabha,[223] introduce charged vector mesons. Bhabha further notes that Yukawa's mechanism for β-radioactivity implies that a free meson U^+ will decay spontaneously,

$$U^+ \to e^+ + \nu \tag{17.46}$$

(a remark which, oddly, is not found in Yukawa's first paper).

Vector meson field equations are Proca equations with sources:

$$F_{\mu\nu} = \partial_\mu A_\nu - \partial_\nu A_\mu = t_{\mu\nu} \tag{17.47}$$
$$\partial_\mu F_{\mu\nu} - \kappa^2 A_\nu = j_\nu$$

$$t_{\mu\nu} = ig\bar{\psi}\tau_- \sigma_{\mu\nu}\psi, \qquad j_\mu = if\bar{\psi}\tau_- \gamma_\mu \psi. \tag{17.48}$$

(e) February 1938. Kemmer introduces[224] two new types of meson: pseudoscalar mesons satisfying Eqs. (17.44) but with

$$v_\mu = if\bar{\psi}\tau_- \gamma_\mu \gamma_5 \psi, \qquad s = ig\bar{\psi}\tau_- \gamma_5 \psi \tag{17.49}$$

and pseudovector mesons satisfying Eqs. (17.47) but with**

$$t_{\mu\nu} = if\bar{\psi}\tau_- \gamma_5 \sigma_{\mu\nu}\psi, \qquad j_\mu = ig\bar{\psi}\tau_- \gamma_\mu \gamma_5 \psi. \tag{17.50}$$

(f) April 1938. Kemmer proposes[225] a neutral meson to save charge independence. Since his argument holds for any coupling of any of the mesons just encountered it suffices to illustrate the point by using Eqs. (17.44) with

* I systematically omit electromagnetic effects; Yukawa and Sakata only use the f-term; for $\bar{\psi}$ see Eq. (17.31). *Warning*! In Eq. (17.45) f and g cannot simultaneously be non-zero, since that would violate time-reversal invariance.
** *Warning*! In Eq. (17.50) f and g cannot be simultaneously non-zero, since that would violate time-reversal invariance.

s-coupling only. At this point I replace g by $g\sqrt{2}$ to conform to standard conventions. Thus the s-coupling corresponds to the interaction (use Eq. (17.13))

$$H_{\text{int}} = g\sqrt{2}\bar{\psi}(\tau_-\phi + \tau_+\phi^*)\psi = g\bar{\psi}(\tau_1\phi_1 + \tau_2\phi_2)\psi, \tag{17.51}$$

where

$$\phi_1 = \frac{\phi^* + \phi}{\sqrt{2}}, \qquad \phi_2 = \frac{i(\phi^* - \phi)}{\sqrt{2}}. \tag{17.52}$$

Kemmer enlarged this expression to

$$H_{\text{int}} = g\bar{\psi}(\tau_1\phi_1 + \tau_2\phi_2 + \tau_3\phi_3)\psi = g\bar{\psi}\vec{\tau}\psi\vec{\phi}. \tag{17.53}$$

The new field ϕ_3 shall have the same κ as in Eq. (17.44), hence its quanta have the same mass as the charged scalar mesons, but zero charge since $\bar{\psi}\tau_3\psi = \bar{\text{p}}\text{p} - \bar{\text{n}}\text{n}$. Proceed likewise for the other f- and g-interactions. Calculate to second order the coupling between two heavy particles. The result is of the form $\text{b}\vec{\tau}^{(1)}\vec{\tau}^{(2)}$, which is charge independent; see Eq. (17.28). Kemmer next makes a crucial observation: 'We can now further deduce that in any higher approximation of the perturbation theory the heavy particle interaction is almost as simple; it is in fact $[a + b\vec{\tau}^{(1)}\vec{\tau}^{(2)}$, Eq. (17.28)]'.

Hence *there is charge independence to any order in perturbation theory.*

Why? Since ψ is an isospinor ($T = 1/2$), $\bar{\psi}\ \vec{\tau}\psi$ is an isovector ($T = 1$) (remember once again the familiar analog with ordinary spin). Thus we can read Eq. (17.53) as the coupling of this isovector to another isovector: $\vec{\phi}$. That is, *we consider the mesons to be a mass degenerate* isotriplet* $T = 1$. This coupling is invariant under isospin rotations, and the same must therefore be true for all its consequences—whence the charge independence of nuclear force. The resulting theory came to be known as the 'symmetrical meson theory'.** There was, as yet, no evidence for a neutral meson but, as Sakata and Tanikawa noted (1940),[227] it could easily have escaped detection since it can decay into two photons with a lifetime $\sim 10^{-16}$ sec—which is exactly as it turned out to be.

These, then, were the formal advances in meson theory of the thirties. Further elaborations followed, such as work on the charged vector theory by Yukawa, Sakata, Taketani, and Kobayasi[228] and by Fröhlich, Heitler, and Kemmer.[229]† A few other theoretical and experimental results of that period will be found in the next chapter, where we shall see how the actual connection between

* Up to small electromagnetic corrections of course.

** An alternative, the 'neutral meson theory' which operates with a single (isosinglet) neutral meson was discussed shortly thereafter.[226]

† Both groups calculated the proton magntic moment without, however, considering the important role of virtual proton–antiproton and neutron–antineutron pairs.

the Yukawa meson and the first cosmic ray meson was established; how the choice was made among the four kinds of mesons encountered in the foregoing; and how the battle of the infinities flared up in meson field theories. Even when clarity on these issues had much improved, meson theory remained opaque. A remark by Fermi[230] made in his Silliman lectures of 1951 may serve as conclusion to this first phase of meson physics and as prelude to what is to come: 'It is fashionable to assume [that mesons are pseudoscalar] since one finds that it leads to nuclear forces less in disagreement with experiment. We shall follow this fashion.' Only in the 1970s did it become clear that the forces responsible for strong interactions are not those Yukawa had introduced; see Chapter 21.

It is Yukawa's great and lasting contribution that he has changed our thinking about forces. To this day his extension to strong interactions of the relation between forces and virtual particle exchange, his analogy with quantum electrodynamics, has remained our guide to thinking about all fundamental forces.

(h) Death of Rutherford

On 11 September 1933 Rutherford remarked in an address to the British Association's meeting at Leicester: 'The energy produced by the breaking down of the atom is a very poor kind of thing. Anyone who expects a source of power from the transformation of these atoms is talking moonshine'. When Lawrence was asked for his reaction to this statement he commented: 'Personally I have no opinion as to whether it can ever be done, but we're going to keep on trying to do it'.[231]

In October 1933, Rutherford and Chadwick met Lawrence at the Solvay conference in Brussels. Both liked him very much. Their discussions must have turned to the end of the hegemony of the α-particle, which had served Rutherford in discovering the atomic nucleus and artificial transmutation, and Chadwick in discovering the neutron. 'But this did not persuade Rutherford to begin the construction of a cyclotron at the Cavendish. Chadwick on the other hand was firmly convinced that a cyclotron was an essential tool for further research in nuclear physics.'[232] This divergence contributed to Chadwick's leaving the Cavendish in 1935 for Liverpool, where he did build a cyclotron. During the remaining almost forty years of his life he published only one more research paper. In 1936, Ellis left the Cavendish for King's College London. As for Rutherford himself, 'toward the end . . . he was increasingly occupied with outside responsibilities, but yet he kept absolute control of his laboratory. In that large rambling collection of buildings his personality was everywhere'.[233]

In his later years Rutherford suffered from a slight hernia at the umbilicus for which he wore a truss harness.[234] In mid-October 1937, he complained of indigestion. After examination he was hospitalized. 'On the way [to the hospital] he told his wife that his business and financial affairs were all in order. [She] told him not to worry, for his illness was not serious. To which he replied that life was uncertain.'[235] On the 15th he was operated on for a strangulated hernia. There were brief hopes for recovery. On 19 October in the evening, Rutherford died peacefully. He was sixty-six years old.

The news reached Bohr in Bologna where he was attending the Galvani Conference. On 20 October, Bohr gave a short address[236] about Rutherford which one eye-witness has called one of the most moving experiences in his life.

On 21 October, Rutherford's remains were cremated. On the 25th his ashes were interred in Westminster Abbey, in the presence of his widow, his son-in-law (his daughter had died before him), his grandchildren Peter and Elizabeth, a representative of the king, several members of the British cabinet, the High Commissioner of New Zealand, leaders in science (including Dirac) and industry; among others, a representative of the Professional Committee for German Jewish refugees.[237]

J. J. Thomson was not at the Abbey. He, the eighty-year-old Master of Trinity, was in Cambridge that day where, at the Trinity College Chapel memorial service* in remembrance of Ernest, Baron Rutherford of Nelson, he read the Lesson, verses of *Ecclesiastes* beginning with: 'I will make mention of the works of the Lord, and will declare the things that I have seen'.

It was a time of affectionate remembrance.

Several of those who had known him in New Zealand wrote of Rutherford the youth.[238] Of his extraordinary powers of concentration manifest already then. Of his high school days at Nelson College when he won prizes in English literature and a French scholarship; of his fondness for music; of keeping order with a wicket after becoming College librarian. How when teaching his younger sisters he would tie their pigtails together in order to hold their attention and keep them quiet. Of his stint as substitute teacher at the boys' high school in Christchurch. 'He was entirely hopeless as a schoolmaster . . . Disorder prevailed in his classes. The main characteristic of the mob mind of a schoolboy class is its ape-like cunning, and we certainly had him added up as a genial person whose interests were nothing to do with the keeping in order of small boys. He used to blurt suddenly into anger which was succeeded by a sort of desperate calm. This latter mood was hailed with real joy by all pupils.' And of his days at Canterbury College, where he was on the rugby team.

Andrade wrote[239] of his Manchester days: 'When things were going well and new discoveries were coming in at the rate of one a week, a tune recognizable by the elect as "Onward, Christian Soldiers" could be heard

* I am grateful to the Librarian of Trinity College for a copy of the Order of Service for that day.

accompanying the Professor's steps along the corridor: when things were going less well another tune, no less holy, held sway'.

Chadwick wrote[240] of the Cavendish days: 'Rutherford had no cleverness—just greatness. He treated his students, even the most junior, as brother workers in the same field'.

Eve, his principal biographer: 'He was always a charming blend of boy, man, and genius'.[241]

A correspondent to *The Times*[242] who preferred to remain anonymous: 'To speak of devotion to science would be to invite his sarcasm, for he was ever scornful of "these fellows who take themselves so seriously"'.

In 1938 nuclear fission was discovered.
In 1939 the Second World War broke out.
A new era was upon physics and mankind.

Sources

Personal reminiscences. Chadwick on the neutron.[2] Rossi[45] and Skobeltzyn[243] on cosmic rays. Livingston[67,78] and McMillan[58,79] on accelerators. On nuclear physics: a series of contributions to Stuewer's book.[29] Serber on Berkeley in the thirties.[244] Yukawa on mesons.[203] Anderson on the first cosmic ray meson.[245] Kemmer on isospin.[246]

Books. The Joliot–Curies[247] on the neutron in 1932. Miehlnickel[248] on the early phases of cosmic rays. Rossi[249] on cosmic rays. Livingston and Blewett[72] on accelerators. Brink[104] on nuclear physics. Segré's scientific biography of Fermi.[33] Wentzel on meson theory in the thirties.[218]

Articles. Feather[8] and Bromberg[250] on the neutron. Proceedings[207] of the 1939 Chicago symposium on cosmic rays. Heilbron *et al.* on the Lawrence years at Berkeley.[65] Konopinski[251] on the status of β-radioactivity in 1943. L. Brown[252] and Mukherji[253] on mesons. Rasche[254] on isospin.

References

1. J. Chadwick, letter to E. Rutherford, September 1924, quoted in A. S. Eve, *Rutherford*, p. 300, Cambridge Univ. Press 1939.
2. J. Chadwick, in *Proc. 10th Int. Congress on the History of Science, Ithaca, New York*, p. 135, Hermann, Paris 1964.
3. E. Rutherford, *Proc. Roy. Soc. A* **97**, 374, 1920.
4. J. Chadwick, *Proc. Roy. Soc. A* **142**, 1, 1933.
5. J. L. Glasson, *Phil. Mag.* **42**, 496, 1921.
6. E. Rutherford and J. Chadwick, *Proc. Cambr. Phil. Soc.* **25**, 186, 1929.

7. E. Rutherford, J. Chadwick, and C. D. Ellis, *Radiations from radioactive substances*, p. 523, Cambridge Univ. Press 1930.
8. N. Feather, *Contemp. Phys.* **1**, 191, 257, 1960.
9. R. H. Stuewer, in *Otto Hahn and the rise of nuclear physics*, p. 19, footnote 150, Reidel, Boston 1983.
10. J. Stark, *Atomstruktur und Atombindung*, p. 131, Seydel, Berlin 1928.
11. J. K. Roberts, *Proc. Roy. Soc. A* **102**, 72, 1922.
12. W. Bothe and H. Becker, *Naturw.* **18**, 705, 1930.
13. F. Rasetti, *Naturw.* **20**, 252, 1932; H. Becker and W. Bothe, ibid. **20**, 349, 1932.
14. W. Bothe, *Phys. Zeitschr.* **32**, 661, 1931.
15. H. C. Webster, *Proc. Roy. Soc. A* **136**, 428, 1932.
16. I. Curie and F. Joliot, *Comptes Rendus* **194**, 273, 1932.
17. I. Curie and F. Joliot, *Comptes Rendus* **194**, 708, 1932.
18. J. Chadwick, *Nature* **129**, 312, 1932.
19. J. Chadwick, *Proc. Roy. Soc. A* **136**, 735, 1932.
20. Chapter 14, Section (d).
21. M. Goldsmith, *Frédéric Joliot–Curie*, Lawrence and Wishart, London 1976.
22. Ref. 1, p. 360.
23. I. Curie and F. Joliot, *J. de Physique* **4**, 494, 1933.
24. I. Curie and F. Joliot, *Proc. Seventh Solvay Conference, October 1933*, p. 154, Gauthier-Villars, Paris 1934.
25. I. Curie and F. Joliot, *Comptes Rendus* **198**, 254, 1934; also *Nature* **133**, 201, 1934.
26. I. Curie and F. Joliot, *Comptes Rendus* **198**, 559, 1934.
27. E. Fermi, *Ric. Scient.* **5**, 283, 1934; repr. in *E. Fermi, collected papers*, Vol. 1, p. 645; Engl. transl., ibid., p. 674.
28. G. Holton, *Minerva* **12**, 159, 1974.
29. E. Segré, in *Nuclear physics in retrospect*, Ed. R. H. Stuewer, p. 35, Univ. Minnesota Press, Minneapolis 1979.
30. E. Rutherford, letter to E. Fermi, 23 April 1934, repr. in Ref. 27, Vol. 1, p. 641.
31. H. A. Bethe, in Ref. 29, p. 9.
32. See esp. N. Bohr, *Nature* **138**, 25, 1936.
33. E. Segré, *Enrico Fermi*, p. 72, Univ. of Chicago Press 1970.
34. H. Urey, F. G. Brickwedde, and G. M. Murphy, *Phys. Rev.* **39**, 164, 864, 1932.
35. T. E. Allibone, *Proc. Roy. Soc. A* **282**, 447, 1964.
36. V. F. Hess, *Phys. Zeitschr.* **12**, 998, 1911.
37. V. F. Hess, *Phys. Zeitschr.* **13**, 1084, 1912; **14**, 610, 1913.
38. R. A. Millikan, *Proc. Nat. Ac. Sci.* **12**, 48, 1926.
39. R. A. Millikan and I. S. Bowen, *Phys. Rev.* **27**, 353, 1926.
40. R. H. Kargon, *The rise of Robert Millikan*, Cornell Univ. Press, Ithaca 1982.
41. D. Skobeltzyn, *Zeitschr. f. Phys.* **43**, 354, 1927; esp. Fig. 12 and p. 372.
42. D. Skobeltzyn, *Zeitschr. f. Phys.* **54**, 686, 1929.
43. *The birth of particle physics*, Eds. L. M. Brown and L. Hoddeson, p. 111, Cambridge Univ. Press 1983.
44. W. Bothe and W. Kolhörster, *Zeitschr. f. Phys.* **56**, 751, 1929.
45. B. Rossi, *International colloquium on history of particle physics*, Paris, July 1982. *J. de Physique*, **43**, Suppl. to No. 12, 1982, p. 69.
46. D. J. X. Montgomery, *Cosmic ray physics*, Princeton Univ. Press 1949.
47. P. M. S. Blackett and G. P. S. Occhialini, *Proc. Roy. Soc. A* **139**, 699, 1933.

48. B. Rossi, *Zeitschr. f. Phys.* **82**, 151, 1933, esp. p. 158; also C. D. Anderson, *Phys. Rev.* **41**, 405, 1932; T. H. Johnson, *Phys. Rev.* **41**, 545, 1932.
49. E. G. Steinke, *Erg. Ex. Naturw.* **13**, 89, 1934.
50. J. D. Cockcroft and E. T. S. Walton, *Proc. Roy. Soc. A* **137**, 229, 1932.
51. Ref. 1, p. 337.
52. A. Brasch and F. Lange, *Naturw.* **18**, 16, 1930.
53. M. A. Tuve, *J. Franklin Institute* **216**, 1, 1933.
54. Ref. 29, p. 135.
55. G. Breit and M. A. Tuve, *Nature* **121**, 535, 1928.
56. A. Brasch and F. Lange, Ref. 52, and *Zeitschr. f. Phys.* **70**, 10, 1931.
57. C. C. Lauritsen and R. D. Bennett, *Phys. Rev.* **32**, 850, 1928.
58. E. M. McMillan, in Ref. 29, p. 113.
59. R. Van de Graaff, *Phys. Rev.* **38**, 1919, 1931.
60. J. D. Cockcroft, *Nobel lectures in physics 1942–62*, p. 167, Elsevier, New York 1967.
61. E. Rutherford, *Nature* **120**, 809, 1927; *Proc. Roy. Soc. A* **117**, 300, 1928.
62. M. Oliphant and Lord Penney, *Biogr. Mem. Fell. Roy. Soc.* **14**, 139, 1968.
63. Ref. 1, p. 360.
64. E. O. Lawrence, M. S. Livingston, and M. G. White, *Phys. Rev.* **42**, 15, 1932.
65. J. L. Heilbron, R. W. Seidel, and B. R. Wheaton, *Lawrence Berkeley Laboratory News Magazine*, **6**, No. 3, 1981.
66. G. A. Ising, *Ark. f. Mat. Astr. och Fys.* **18**, No. 30, 1924.
67. M. S. Livingston, *Adv. in Electronics and Electron Phys.* **50**, 1, 1980.
68. R. Wideröe, *Arch f. Elektrotechn.* **21**, 387, 1928.
69. E. O. Lawrence, Ref. 60, p. 430.
70. D. H. Sloan and E. O. Lawrence, *Phys. Rev.* **38**, 2021, 1931.
71. T. H. Johnson, letter to E. M. McMillan, 9 April 1977, repr. in Ref. 58.
72. M. S. Livingston and J. P. Blewett, *Particle accelerators*, McGraw-Hill, New York 1962.
73. Ref. 69, Fig. 2.
74. E. O. Lawrence and N. E. Edlefsen, *Science* **72**, 376, 1930.
75. E. O. Lawrence and M. S. Livingston, *Phys. Rev.* **37**, 1707, 1931.
76. E. O. Lawrence and M. S. Livingston, *Phys. Rev.* **38**, 834, 1931.
77. E. O. Lawrence and M. S. Livingston, *Phys. Rev.* **40**, 19, 1932.
78. M. S. Livingston, *Physics Today*, October 1959, p. 18.
79. E. M. McMillan, *Physics Today*, October 1959, p. 24.
80. C. Weiner, *Physics Today*, May 1972, p. 40.
81. D. Iwanenko, *Nature* **129**, 798, 1932.
82. F. Perrin, *Comptes Rendus*, **194**, 1343, 1932; also ibid. **194**, 2211; **195**, 236.
83. P. Auger, *Comptes Rendus*, **194**, 1346, 1932.
84. P. A. M. Dirac, Ref. 24, p. 328.
85. J. Chadwick, *Proc. Roy. Soc. A* **136**, 692, 1932.
86. J. F. Carlson and J. R. Oppenheimer, *Phys. Rev.* **41**, 763, 1932; also ibid. **39**, 864, 1932.
87. J. F. Carlson and J. R. Oppenheimer, *Phys. Rev.* **38**, 1787, 1931.
88. R. F. Bacher and E. U. Condon, *Phys. Rev.* **41**, 683, 1932.
89. G. N. Lewis and M. F. Ashley, *Phys. Rev.* **43**, 837, 1933.
90. G. M. Murphy and H. Johnston, *Phys. Rev.* **45**, 550, 761, 1934.
91. J. Schwinger, *Phys. Rev.* **52**, 1250, 1937.
92. D. Ivanenko, *Comptes Rendus*, **195**, 439, 1932.

93. J. Chadwick, *Proc. Roy. Soc. A* **142**, 1, 1933.
94. E. Ruth and A. E. Kempton, *Proc. Roy. Soc. A* **143**, 724, 1934.
95. J. Chadwick and M. Goldhaber, *Nature* **134**, 237, 1934; also M. Goldhaber, in Ref. 29, p. 83.
96. F. Bloch, letter to N. Bohr, 30 December 1932, copy in Niels Bohr Library, Am. Inst. Phys., New York.
97. R. Frisch and O. Stern, *Zeitschr. f. Phys.* **85**, 4, 1933; also R. Frisch, in Ref. 29, p. 65.
98. O. Stern, *Helv. Phys. Acta* **6**, 426, 1933.
99. N. Bohr, Ref. 24, p. 227.
100. W. Pauli, letter to W. Heisenberg, 17 April 1934, repr. in *W. Pauli, scientific correspondence*, Eds. A. Hermann and K. v. Meyenn, quoted as *PC* below, Vol. 2, p. 316, Springer, New York, 1985.
101. R. Bacher, *Phys. Rev.* **43**, 1001, 1931.
102. L. Alvarez and F. Bloch, *Phys. Rev.* **57**, 111, 1940.
103. W. Heisenberg, letter to N. Bohr, 20 June 1932, copy in Niels Bohr Library, American Institute of Physics, New York.
104. W. Heisenberg, *Zeitschr. f. Phys.* **77**, 1, 1932; Engl. transl. in D. M. Brink, *Nuclear forces*, p. 144, Pergamon, Oxford 1965.
105. W. Heisenberg, *Zeitschr. f. Phys.* **78**, 156, 1932.
106. W. Heisenberg, *Zeitschr. f. Phys.* **80**, 587, 1933; Engl. transl. in Ref. 104, p. 155.
107. Ref. 105, p. 162.
108. Ref. 104, pp. 1, 2.
109. Ref. 105, Section 2, and Ref. 106, Section 3.
110. Ref. 105, p. 160, footnote.
111. L. M. Brown and A. E. Moyer, *Am. J. of Phys.* **52**, 130, 1984.
112. Ref. 104, Section 3.
113. Ref. 105, Section 3.
114. Ref. 104, Eq. (4).
115. I. Curie and F. Joliot, *Comptes Rendus*, **194**, 876, 1932.
116. E. Majorana, *Zeitschr. f. Phys.* **82**, 137, 1933.
117. E. P. Wigner, *Phys. Rev.* **43**, 252, 1933; *Zeitschr. f. Phys.* **83**, 253, 1933.
118. J. H. Bartlett, *Phys. Rev.* **49**, 102, 1936.
119. W. H. Wells, *Phys. Rev.* **47**, 591, 1935; M. G. White, *Phys. Rev.* **49**, 309, 1936; M. A. Tuve, N. Heydenberg, and L. R. Hafstad, *Phys. Rev.* **50**, 806, 1936.
120. H. A. Bethe and R. F. Bacher, *Rev. Mod. Phys.* **8**, 82, 1936.
121. L. Landau and J. Smorodinsky, *J. Phys. USSR* **8**, 154, 1944; J. Schwinger, *Phys. Rev.* **52**, 1250, 1947; J. M. Blatt and J. D. Jackson, *Phys. Rev.* **76**, 18, 1949; *Rev. Mod. Phys.* **22**, 77, 1950; H. A. Bethe, *Phys. Rev.* **76**, 38, 1949.
122. J. M. B. Kellogg, I. I. Rabi, N. F. Ramsey, and J. R. Zacharias, *Phys. Rev.* **55**, 318, 1939; **57**, 677, 1940.
123. J. Schwinger, *Phys. Rev.* **55**, 235, 1939; cf. also H. A. Bethe, *Phys. Rev.* **55**, 1261, 1939.
124. J. H. D. Jensen, *Phys. Rev.* **75**, 1766, 1949; M. G. Mayer, *Phys. Rev.* **75**, 1969, 1949; cf. also K. M. Case and A. Pais, *Phys. Rev.* **79**, 185, 1950; **80**, 203, 1950.
125. Cf. e.g. A. Bohr and B. R. Mottelson, *Nuclear structure*, Vol. 1, Chapter 2, Section 5, Benjamin, New York 1969.
126. W. Pauli, letter to W. Heisenberg, 14 July 1933, *PC*, Vol. 2, p. 184.
127. W. Pauli, Ref. 24, p. 324.

128. G. Gamow, letter to S. A. Goudsmit, 8 March 1934.
129. F. Rasetti, in Ref. 27, Vol. 1, p. 450.
130. E. Fermi, *Ric. Scient.* **4**, 491, 1934; see Ref. 27, Vol. 1, p. 538.
131. E. Fermi, *Nuov. Cim.* **11**, 1, 1934; see Ref. 27, Vol. 1, p. 559.
132. E. Fermi, *Zeitschr. f. Phys.* **88**, 161, 1934; see Ref. 27, Vol. 1, p. 575.
133. D. Iwanenko, *Comptes Rendus* **195**, 439, 1932.
134. F. Perrin, *Comptes Rendus* **197**, 1625, 1933.
135. L. de Broglie, *Comptes Rendus* **198**, 135, 1934.
136. R. de L. Kronig, *Ann. Inst. H. Poincaré* 6, 215, 1936.
137. M. H. L. Pryce, *Proc. Roy. Soc. A* **165**, 247, 1938.
138. Cf. also K. M. Case, *Phys. Rev.* **106**, 1316, 1957.
139. E. J. Konopinski and G. E. Uhlenbeck, *Phys. Rev.* **48**, 7, 1935.
140. Ref. 132, Eq. (11).
141. E. J. Konopinski, *The theory of β-radioactivity*, Eq. (1.8), Clarendon Press, Oxford, 1966.
142. E. J. Konopinski and G. E. Uhlenbeck, *Phys. Rev.* **60**, 308, 1941.
143. B. W. Sargent, *Proc. Roy. Soc. A* **139**, 659, 1933.
144. E. Fermi, *Zeitschr. f. Phys.* **89**, 522, 1934; see Ref. 27, Vol. 1, p. 592.
145. J. Chadwick and D. E. Lea, *Proc. Cambr. Phil. Soc.* **30**, 59, 1934.
146. G.-C. Wick, *Rend. Acc. Lincei* **19**, 319, 1934.
147. H. A. Bethe and R. Peierls, *Nature* **133**, 532, 1934.
148. H. Yukawa and S. Sakata, *Proc. Phys. Math. Soc. Japan* **17**, 467, 1935.
149. L. Alvarez, *Phys. Rev.* **54**, 486, 1938.
150. H. Wolfe and G. E. Uhlenbeck, *Phys. Rev.* **46**, 237, 1934.
151. A. H. Snell and L. C. Miller, *Phys. Rev.* **74**, 1217, 1948.
152. Ref. 132, p. 176.
153. H. A. Bethe, *International Conference on Physics, London, 1934*, Vol. 1, p. 66, Cambridge Univ. Press 1935.
154. E. Fermi, Ref. 153, p. 67.
155. J. L. Lawson and J. M. Cork, *Phys. Rev.* **57**, 982, 1940; see also C. S. Wu, *Rev. Mod. Phys.* **22**, 386, 1950.
156. Ref. 120, pp. 190, 191.
157. M. Fierz, *Zeitschr. f. Phys.* **104**, 553, 1937.
158. G. Gamow and E. Teller, *Phys. Rev.* **49**, 895, 1936.
159. A. I. Leipunski, *Proc. Cambr. Phil. Soc.* **32**, 301, 1936.
160. R. H. Crane, *Rev. Mod. Phys.* **20**, 278, 1938.
161. M. Goeppert-Mayer, *Phys. Rev.* **48**, 512, 1935.
162. C. S. Wu and S. A. Moszkowski, *Beta-decay*, Section 5–2, Interscience, New York 1966.
163. E. Majorana, *Nuovo Cim.* **14**, 171, 1937; cf. also K. M. Case, *Phys. Rev.* **107**, 307, 1957.
164. G. Racah, *Nuovo Cim.* **14**, 322, 1937.
165. W. Furry, *Phys. Rev.* **56**, 1184, 1939.
166. K. Guggenheim, *J. de Phys. et le Rad.* **5**, 475, 1934; L. A. Young, *Phys. Rev.* **47**, 972, 1935.
167. E. Feinberg and J. G. Knipp, *Phys. Rev.* **48**, 906, 1935.
168. R. Peierls, in Ref. 29, p. 189.
169. G. Breit, E. U. Condon, and R. D. Present, *Phys. Rev.* **50**, 825, 1936.
170. B. Cassen and E. U. Condon, *Phys. Rev.* **50**, 846, 1936.

171. E. P. Wigner, *Phys. Rev.* **51**, 106, 1937.
172. K. M. Case, R. Karplus, and C. N. Yang, *Phys. Rev.* **101**, 874, 1956.
173. Cf. Ref. 125, Vol. 1, pp. 42 ff.
174. F. Hund, *Zeitschr. f. Phys.* **105**, 202, 1937; see also P. Franzini and L. Radicati, *Phys. Lett.* **6**, 322, 1963.
175. K. M. Case, R. Karplus, and C. N. Yang, *Phys. Rev.* **101**, 874, 1956.
176. A. Pais, *Rev. Mod. Phys.* **38**, 215, 1966.
177. O. Klein, *J. de Phys. et le Rad.* **9**, 1, 1938.
178. W. Heisenberg, letter to W. Pauli, 18 January 1934, *PC*, Vol. 2, p. 250.
179. K. M. Case, *Phys. Rev.* **80**, 797, 1950.
180. I. Tamm, *Nature* **133**, 981, 1934.
181. D. Iwanenko, *Nature* **133**, 981, 1934.
182. A. Nordsieck, *Phys. Rev.* **46**, 234, 1934.
183. W. Heisenberg, in *Pieter Zeeman*, p. 108, Nyhoff, The Hague 1935.
184. Ref. 120, p. 203.
185. B. Kahn, *Physica* **3**, 495, 1936.
186. D. Iwanenko and A. Sokolov, *Zeitschr. f. Phys.* **102**, 119, 1936; also *Nature* **138**, 246, 684, 1936.
187. B. Kahn, *Physica* **4**, 403, 1937.
188. G. Gamow and E. Teller, *Phys. Rev.* **51**, 289, 1937; N. Kemmer, *Phys. Rev.* **52**, 906, 1937.
189. G. Wentzel, *Zeitschr. f. Phys.* **104**, 34, 1936; **105**, 738, 1937.
190. G. Wick, *Rend. Acc. Lincei*, **21**, 170, 1935.
191. C. F. von Weizsäcker, *Zeitschr. f. Phys.* **102**, 572, 1936.
192. M. Fierz, *Zeitschr. f. Phys.* **104**, 553, 1936; contains a list of corrections to Ref. 191.
193. W. Heisenberg, *Zeitschr. f. Phys.* **101**, 533, 1936.
194. F. Carlson and J. R. Oppenheimer, *Phys. Rev.* **51**, 220, 1937; also J. R. Oppenheimer, *Phys. Rev.* **50**, 389, 1936.
195. H. J. Bhabha and W. Heitler, *Proc. Roy. Soc. A* **159**, 432, 1937.
196. W. Heisenberg, *Verh. Deutsch. Phys. Ges.* **18**, 50, 1937.
197. W. Heisenberg, *Rev. Mod. Phys.* **11**, 241, 1939; *Zeitschr. f. Phys.* **113**, 61, 1939.
198. W. Heisenberg, *Ann. der Phys.* **32**, 20, 1938.
199. W. Heisenberg, letter to W. Pauli, 23 May 1936, *PC*, Vol. 2, p. 442.
200. M. Fierz, *Helv. Phys. Acta* **9**, 245, 1936.
201. S. Tomonaga and H. Tamaki, *Inst. Phys. Chem. Res. Sci. Papers*, **33**, 288, 1937.
202. Cf. H. Yukawa, *Phys. Rev.* **76**, 300, 1949.
203. Cf. H. Yukawa, *Tabibito*, transl. L. Brown and R. Yoshida, World Scientific, Singapore 1979.
204. T. Kimura, *Jap. St. Hist. Sc.* **11**, 90, 1972.
205. P. Robinson, *The early years*, pp. 156–8, Akademisk Forlag, Copenhagen 1979.
206. H. J. Bhabha, *Nature* **143**, 276, 1939; cf. also C. G. Darwin, *Nature* **143**, 602, 1939.
207. *Rev. Mod. Phys.* **11**, 122, 1939.
208. H. Yukawa, *Proc. Phys. Math. Soc. Japan*, **17**, 48, 1935.
209. H. Yukawa, *Proc. Phys. Math. Soc. Japan*, **19**, 712, 1937.
210. *Science*, **84**, 1936, supplement following p. 464.
211. S. H. Neddermeyer and C. D. Anderson, *Phys. Rev.* **51**, 884, 1937.
212. J. C. Street and E. C. Stevenson, *Phys. Rev.* **51**, 1005, 1937.
213. Y. Nishina, M. Takeuchi, and T. Ichimaya, *Phys. Rev.* **52**, 1198, 1937.

214. S. H. Neddermeyer and E. C. Stevenson, *Rev. Mod. Phys.* **11**, 191, 1939.
215. C. D. Anderson, Ref. 43, p. 131.
216. J. R. Oppenheimer and R. Serber, *Phys. Rev.* **51**, 1113, 1937.
217. E. C. G. Stückelberg, *Phys. Rev.* **52**, 41, 1937.
218. G. Wentzel, *Einführung in die Quantentheorie der Wellenfelder*, Deuticke, Vienna, 1943; in English: *Quantum theory of fields*, transl. C. Houterman and J. M. Jaunch, Interscience, New York 1948.
219. A. Proca, *J. de Phys. et le Rad.* **7**, 347, 1936.
220. H. Yukawa and S. Sakata, *Proc. Math. Phys. Soc. Japan*, **19**, 1084, 1937.
221. N. Kemmer, *Nature* **141**, 116, 1938.
222. R. Serber, *Phys. Rev.* **53**, 211, 1938.
223. H. J. Bhabha, *Nature* **141**, 117, 1938.
224. N. Kemmer, *Proc. Roy. Soc. A* **166**, 127, 1938.
225. N. Kemmer, *Proc. Roy. Soc. A* **166**, 354, 1938.
226. H. J. Bhabha, *Proc. Roy. Soc. A* **172**, 384, 1939; H. A. Bethe, *Phys. Rev.* **55**, 1261, 1939.
227. S. Sakata and Y. Tanikawa, *Phys. Rev.* **57**, 548, 1940.
228. H. Yukawa, S. Sakata, and M. Taketani, *Proc. Phys. Math. Soc. Japan*, **20**, 319, 1938; the same and M. Kobayasi, ibid., 720.
229. H. Fröhlich, W. Heitler, and N. Kemmer, *Proc. Roy. Soc. A* **166**, 154, 1938.
230. E. Fermi, *Elementary particles*, p. 22, Yale Univ. Press, New Haven, Conn., 1951.
231. *New York Herald Tribune*, 12 September 1933.
232. H. Massey and N. Feather, *Biogr. mem. Fell. Roy. Soc.* **22**, 11, 1976.
233. C. D. Ellis, *Proc. Phys. Soc. London*, **50**, 463, 1938.
234. M. Oliphant, *Rutherford, recollections of the Cambridge days*, Elsevier, New York 1972.
235. Ref. 1, p. 425.
236. N. Bohr, *Nature* **140**, 752, 1937.
237. *The Times*, London, 26 October 1937, p. 17; also *Nature* **140**, 754, 1937.
238. *The Times*, London, 22 October 1937, p. 18.
239. E. N. da C. Andrade, *Nature* **140**, 753, 1937.
240. J. Chadwick, *Nature* **140**, 749, 1937.
241. A. S. Eve, *Nature* **140**, 746, 1937.
242. *The Times*, London, 25 October 1937, p. 20.
243. D. Skobeltzyn, Ref. 43, p. 111.
244. R. Serber, Ref. 43, p. 206.
245. C. D. Anderson, Ref. 43, p. 131.
246. N. Kemmer, Ref. 45, p. 359.
247. I. Curie and F. Joliot, *L'existence du neutron*, Hermann, Paris 1932.
248. E. Miehlnickel, *Höhenstrahlung*, Steinkopff, Dresden 1938.
249. B. Rossi, *Cosmic rays*, McGraw-Hill, New York 1964.
250. J. Bromberg, *Hist. St. Phys. Sc.* **3**, 307, 1971.
251. E. Konopinski, *Rev. Mod. Phys.* **15**, 209, 1943.
252. L. Brown, Introduction to Ref. 203 and *Centaurus* **25**, 71, 1981.
253. V. Mukherji, *Arch. Hist. Ex. Sc.* **13**, 27, 1974.
254. G. Rasche, *Arch. Hist. Ex. Sc.* **7**, 257, 1971.

Part Two

The Postwar Years: A Memoir

. . . it is a privilege to see so
much confusion.

Marianne Moore, *The steeple-jack*

18

Of quantum electrodynamics' triumphs and limitations and of a new particle's sobering impact

> The nearer we come to the present, of course, the more opinions diverge. We might, however, reply that this does not invalidate our right to form an opinion.
>
> Jakob Burckhardt[1]

(a) Shelter Island and other personal reminiscences

I first met I. I. Rabi in September 1946 at a meeting of the American Physical Society in mid-Manhattan. After a few pleasantries, he fired this question at me: 'Do you think the polarization of the vacuum can be measured?' It was my first week in the United States, and I recall my astonishment at being in a new land where experimentalists would know, let alone bother, about vacuum polarization. Nor did Rabi's question about Uehling's prewar formula, Eq. (16.39), stem from idle curiosity. A month later there appeared[2] in the *Quarterly Progress Report of the Columbia University Radiation Laboratory* a proposal entitled, 'Microwave physics. Experiment to determine the fine structure of the hydrogen atom (Lamb, Retherford)'.

For me that September week was rich in other experiences as well. I ate my first banana in six years. At the physics meeting I met Uhlenbeck, whom I had not seen since he left Holland in August 1939. After I had answered some of his questions with: yes professor, no professor, he said to me: 'Why don't you call me George'. Kramers, in New York for a meeting of the scientific and technological subcommittee of the United Nations' Atomic Energy Committee, attended the Physical Society meeting as well. We were sitting next to each other at one of the sessions when he handed me a slip of paper on which he had scribbled: 'Turn around and pay your respects to Robert Oppenheimer'. I did and there was the man, grinning pleasantly, whose face had up till then been only familiar to me from newspaper photographs. He wore a short-sleeved shirt, no tie, no jacket, unheard of for a professor in my native Holland, but then so was the heat. I have written elsewhere[3] about my meetings with Oppenheimer during the next few months which led to my long-term association with the Institute for Advanced Study in Princeton.

It is recorded[4] in the minutes of that Physical Society meeting that 'It was confined to papers on three topics: cosmic-ray phenomena, theories of elementary particles, and the design and operation of accelerators of nuclear particles and electrons. Disparate as these three subjects may appear to be, the trend of physics is rapidly uniting them'.

I gave an invited paper[4] dealing with work done during the war, while I lived in hiding. I had often discussed my ideas with Kramers, the only physicist who knew where I was and who would visit me from time to time, in my room in an attic in Amsterdam.

Soon after receiving my Ph.D. (1941),[5] I became interested in the question of how to eliminate the logarithmic infinities* of the electron's self-energy. I noted that this singular part of the self-energy would remain unaltered to all orders in $\alpha = e^2/\hbar c$ if the electromagnetic field were replaced by a massive neutral vector meson field coupled with strength e to the electron.** Thus it is possible to get a finite self-energy by coupling the electron both to photons and to neutral vector mesons and subtracting the latter's contribution from that of the photon (an idea with classical antecedents[6]). I remarked that this procedure was covariant but that it was unacceptable as a realistic physical theory: 'Subtractive fields are incompatible with a stable vacuum distribution as required by the hole theory. Thus they have to be discarded'.[7] Some years later this subtractive procedure resurfaced as a convenient computational device for locating infinite contributions in field theories (regularization).[8]

Next I asked: could a finite self-energy arise in a realistic theory by compensation between the electromagnetic infinities and those due to mesons of other than vector kinds? From a computation[9] of the second-order self-energy due to couplings with the four varieties of mesons mentioned earlier† it turned out[7] that to $O(\alpha)$ one can obtain a finite self-energy by assuming that the electron couples not only to photons but also to neutral scalar mesons.†† Moreover one can obtain[7] a sensible value for the proton–neutron mass difference by assuming that the proton, but not the neutron, couples to the same scalar field.[10] I was astonished to learn in 1947 that during the war the same scalar field had also been introduced by Sakata.[11] These attempts at a realistic theory did not lead far enough, however. The infinities in the vacuum polarization and in higher-order self-energies remained as obscure as before.[12]

Kramers and I had many discussions on these questions when he visited me in Amsterdam. He was interested in them but looked at the problem in a distinct way which he had first communicated[13] at the Galvani conference of 1937. His starting point was the classical non-relativistic theory of an electron in interaction with the electromagnetic field which is the sum of a 'proper

* Chapter 16, Section (d), Part 5.
** The field equations are Eq. (17.47) with A_μ real (self-adjoint), $f=e$, $g=0$.
† Eqs. (17.44, 45, 47–50).
†† Use Eq. (17.44) with ϕ real (self-adjoint), $g=0$, $e^2=f^2/2$.

field', the Coulomb field dragged along by the electron in uniform motion, and an 'external field'. Now, he reasoned, all the proper field does in this approximation is to contribute to the electron's mass. Therefore one may effect a trade-off. Replace the electron mass parameter in the equations which are the starting point of the calculations, the 'bare' mass, by the experimental mass, the sum of the bare and the electromagnetic mass, and drop the proper field. Kramers referred to this procedure as the 'structure independent' electron theory. In this strategy, the coupling between proper and external fields is neglected, but (he noted) that does not change low-frequency secular effects due to the external field. Finally, he quantized the *external* field in the standard way, hoping that his method would avoid such nasty effects as infinite displacements of spectral lines.[14]*

In our many conversations I would argue against Kramers' approach. A non-relativistic theory is not a good starting point, I would say. But we have no reliable relativistic theory, he would reply. I agreed but said the positron theory is still the best we have. Using such arguments as noted after Eq. (16.43), I would note that the self-energy problem is inherently a quantum problem and that his external field, once quantized, would generate new infinities all over again. He would not deny that, but would keep insisting that one should first remedy the physics of the low-frequency fields and then hope for the best. In his last paper on the subject[15] he put it like this: 'A relativistic treatment would ... hardly seem possible or promising ... one should not think too hard of the device: first quantizing the wrong Hamiltonian and trying to make amends later on'. That was in 1948. Kramers was by then already deeply weary. He, Holland's best, died a few years later at age 57. I owe him much and should have paid closer attention to his wartime ideas, which were nothing less than a mass renormalization program (albeit cast in an outdated framework), instead of searching for a realistic finite theory, an idea neither timely then nor even now.

Shortly after the war I was offered fellowships both at Bohr's Institute and at the Institute in Princeton (where I hoped to work with Pauli who, however, returned to Europe shortly afterward). Early in 1946, I went to Copenhagen where I got immersed in other questions. Making use of an earlier idea[16] of introducing real but continuous angular momentum as a variational parameter, Lamek Hulthén and I[17] calculated n–p scattering up to an extremely high energy: twenty-five million volts, for which the first few data, obtained with the Harvard cyclotron, had just become available.** Møller and I worked on a theory which produced elementary particle spectra. We considered a de Sitter world, a five-dimensional spherical shell with finite thickness and with periodic boundary conditions in the fifth direction. All known particles were

* See Chapter 16, Section (c).

** 25 MeV neutrons had been produced[18] in the reaction $Li^7 + d \rightarrow Be^8 + n$.

ground states of a spectral series (called towers these days). One such ground state was the proton-neutron for which Møller had invented (1941)[19] the collective name 'nucleon'; likewise for the meson and the electron-neutrino. The latter and their higher mass states needed also a collective name, we thought, for which we proposed[20] 'lepton' (from $\lambda\varepsilon\pi\tau\acute{o}\sigma^{-}$ = small). Last but not least, I spent part of my Copenhagen days in close contact with Bohr, as I have described elsewhere.[21]

In September 1946 I went to the United States, first attending the New York physics meeting mentioned earlier, then going on to the Institute in Princeton. Dirac was there that fall. Kramers came in the spring; so did Oppenheimer, on a short visit as Director-elect. As to Rabi's question about vacuum polarization, I did not hear any more about that subject until the morning of June 2, 1947.

The Shelter Island Conference grew out of an idea of Duncan McInnes of the Rockefeller Institute (now Rockefeller University) for organizing a set of small conferences aimed at assessing the current status and prospects for development in several fields. He and Karl Darrow, secretary of the American Physical Society, did the initial planning for the physics meeting, submitting to the National Academy of Sciences a cost proposal of $3100 for housing, travel, and sundries for 25 people. Funds were granted, Darrow was appointed as chairman, Kramers, Oppenheimer, and Weisskopf as discussion leaders.* The Ram's Head Inn on Shelter Island was to be the meeting place. An early list[22] of invitees included several who were unable to come, Einstein among them.

On June 1 most participants gathered at the American Institute of Physics (then on Manhattan's 55th Street), boarded a bus and set off for the long ride across Long Island. As we entered Nassau County, the bus was stopped by a police trooper on a motorcycle who asked: 'Are you the scientists?' Yes, we were the scientists. Follow me, he said to the driver, escorting us to the tune of sirens to Suffolk County where other troopers took over. We had no clue as to what this meant and speculated about security measures which in those days did tend to take bizarre forms. All was cleared up after a dinner offered us at Greenport when a kindly gentleman rose and said that the escort was appreciative of what 'the scientists' had meant in his own life. He had been a Marine in the Pacific and might well not have been there to thank us had it not been for the atomic bomb. After a ferry ride to Shelter Island we settled down.

Oppenheimer's comment on the meeting, written[23] on the day of its conclusion, summarizes fittingly, I think, the response of all participants: 'The three days were a joy to us and perhaps rather unexpectedly fruitful ... [we] came away a good deal more certain of the directions in which progress may

* Each of these prepared a brief working paper, reproduced in an article by Schweber.[22]

lie'. Beginning with Willis Lamb's report on the first morning it was clear to all that a new chapter in physics was upon us. Lamb himself has described[24] how familiarity with microwave technology, acquired during war work done at Columbia University, had enabled him to do the experiment on which he now reported. The result: the $2^2S_{1/2}$ level of hydrogen lies higher than the $2^2P_{1/2}$ level by about 1000 megacycles per second (MHz) or 0.033 cm^{-1}, in conflict with the Dirac theory for the electron–proton system with pure Coulomb interaction, according to which these two levels should be degenerate.* Here then was the answer to Rabi's question—except that the answer was of opposite sign and forty times larger than the vacuum polarization effect given by Eq. (16.40).

A remark in the paper[26,27] on the 'Lamb shift', completed two weeks later: 'The results [are] in essential agreement with Pasternack's hypothesis', serves as a reminder that the Lamb shift has a bit of prehistory. The June 1947 paper[26] was in fact Lamb's third on the subject of deviations from Coulomb's law. Indications of an anomaly in the hydrogen fine structure found in 1938 by spectroscopic methods[28] had been interpreted by Pasternack[29] in terms of an upward displacement of $2^2S_{1/2}$ by 0.03 cm^{-1}, possibly due to a repulsive deviation from the Coulomb potential.** Thereupon it was suggested[31] that this deviation might have the same explanation as Wick's† for the anomalous moment of the proton: partial dissociation $p \rightarrow n +$ positive meson. Lamb showed in two papers[32] that this suggestion is quantitatively incorrect.

Back to that first morning at Shelter Island, Rabi reported on his measurements with Nafe and Nelson[33] of the hyperfine structure of hydrogen and deuterium in the $1^2S_{1/2}$ ground state which gave results about 0.3% larger than those expected theoretically, and on experiments by Kusch and Foley[34] on the g-value of gallium in $^2P_{1/2}$ and $^2P_{3/2}$ states which showed similar deviations.

It was at once accepted by all that these new effects demanded interpretation in terms of radiative corrections to the leading-order predictions of quantum electrodynamics. It was therefore particularly timely that Kramers should present his program for coping with the infinities to the meeting. As Lamb has recalled:[24] 'No concrete method of calculation was described'. Bethe found this talk inspiring, however.[22] Kramers also participated in the discussions, but it was Oppenheimer who took charge. For the first time I saw him in action as the threefold master: stressing the important problems, directing the discussion, and summarizing the findings. He was of course particularly qualified to do so since not only had he been the first (1930)†† to note that radiative corrections will displace spectral lines, but in 1933 he had also noted[35]

* Similar results for the fine structure of He^+ were reported a few months later.[25]
** Conjectures about deviations from Coulomb's law on grounds of spectroscopic evidence go back to 1933.[30]
† See Eq. (17.43).
†† Chapter 16, Section (c).

that changes may subsequently occur in the fine structure for which, even then, there were slight indications. At the time of the meeting the quantum electrodynamic explanation was only an article of faith, however, without quantitative support. In particular, it was as yet by no means clear that the magnetic anomalies are due to a deviation of the electron's magnetic moment from the conventional value.[36] This lack of clarity is illustrated by a remark I found in my flimsy notes of the meeting: since the hyperfine splitting is proportional to the probability of finding the electron at the position of the nucleus and since the latter decreases because of a new repulsive term in the potential (Lamb shift!), so the splitting also decreases, whereas we want an increase . . .

Other recollections: late night talks on quantum electrodynamics with Schwinger, who was quiet throughout the sessions. Feynman trying to explain to me a new calculational method in field theory. Creation and annihilation operators were not in evidence. He would start by drawing some pictures. I asked him to derive in his way some results known to me. He did so, with lightning speed. Whatever he was doing had to be important—but I did not understand it.

Puzzling features of cosmic rays, the other main topic of discussion at Shelter Island, led Robert Marshak to suggest that two kinds of mesons were called for. In order to appreciate his point let us first go back a few years.

(b) Divine laughter: the muon

Between 1938 and 1943 experiments with cosmic ray mesons had yielded several important results. It had been found that these particles decay into electrons. (The first cloud chamber picture of such an event[37] dates from 1940). This, it seemed, was a confirmation of Yukawa's and Bhabha's ideas.* The meson's mean life had been pinned down** to 2.15 ± 0.07 microseconds (present best value: 2.198 ± 0.001), its mass lay near $200 m_e$ (still with fairly large errors†). There were also some data on meson scattering.

None of these results made meson theoreticians very comfortable.

Recall,†† this was the period of exploring which (or which combination) of the four kinds of mesons with coupling constants, generically called g in the next few paragraphs, would best fit the phenomena. Most calculations of that time were performed by means of the second-order perturbation formula (Eq. (15.33)), although there were already some doubts (about which more later) about whether this might work as well as in quantum electrodynamics, since

* See Eqs. (17.42) and (17.46).
** For a review of lifetime measurements of the period, see Rossi.[38]
† In 1946 the mass value 202 ± 5 was obtained.[39]
†† Chapter 17, Section (g), Part 4.

$g^2/\hbar c$ was much larger than $\alpha = e^2/\hbar c$. From the many results obtained, I single out those* which bear directly on what follows next.
(a) Nucleon–nucleon interaction. It was noted that a mass bigger than $200m_e$ would better fit scattering data.[41]
(b) Meson lifetime. According to Eq. (17.42) the effective Fermi coupling is proportional to gg', where g' is the meson–lepton coupling constant. Thus, the value of g' can be estimated from data on β-decay and nucleon–nucleon forces. Hence the meson lifetime, depending only on g', Eq. (17.46), can be deduced. The result: theory demands a lifetime ~100 times shorter than observed.[42]
(c) Meson–nucleon scattering. Just as e enters both in the Coulomb potential and in Compton scattering, so g plays the dual role of determining the strength of nuclear potentials and of meson–nucleon scattering. Considering the former as input, one has a piece of theoretical information about the latter. Result: experimentally, mesons scatter less by about two orders of magnitude than is theoretically indicated.[43]

The response to these difficulties ranged from unease to worry. In particular, some hoped that a theoretical adjustment of parameters (several g's? several ranges?) might save the day.
(d) Meson absorption, the cause of a real crisis. Theoretically,[44] slow positive Yukawa mesons traversing matter should strongly prefer to decay rather than be absorbed by a nucleus, Coulomb repulsion mitigating against the meson reaching the nucleus. Negative Yukawa mesons, on the other hand, should strongly prefer absorption to decay. These predictions were blown to bits by the Conversi–Pancini–Piccioni experiment (December 1946)[45]: positive cosmic ray mesons behaved the way the theory said, and negative cosmic ray mesons were absorbed in lead, also as expected, *but not in carbon*! A simple analysis of this result[46] showed that the negative cosmic ray meson's interaction with nuclei was ten to twelve orders of magnitude weaker than that of the Yukawa's meson.

It was during the Shelter Island discussion of this obviously grave paradox that Marshak proposed a way out: there is the Yukawa meson, strongly absorbable as estimated, which decays into another, weakly-absorbable meson, the cosmic ray meson seen at low altitudes.[47] At the time he made this suggestion (well received), neither he nor anyone else present knew that the same idea had been put forward earlier, nor that the May 24, 1947 issue of *Nature* contained preliminary evidence supporting the two-meson hypothesis.

As to earlier predictions: Tanikawa (in 1942)[48] and Sakata and Inoue (in 1943)[49] had made the same proposal,** publication of which had been much delayed because of the war. The first experimental 'evidence which suggests

* More details are found in an article by Mukherji.[40]
** The three independent suggestions differed in spin assignments, obviously open at the time. Sakata and Inoue made the correct guess.

that two types of mesons exist of different mass' came from Powell and his group in Bristol[50] who had developed to high perfection a method which goes back to Rutherford's early days[51] and which Powell had already exploited[52] before the war: when a heavy ionizing particle passes through a photographic emulsion it leaves a succession of developed grains, a track. In their May 1947 paper[50] the Bristol group reported two cases of a particle track coming to the end of its range, then showing a big kink, indicating that a charged decay product is taking off. The masses before and after the kink were of mesonic magnitude, though not measurable with much precision. More such events reported in their next paper (October)[53] made their conclusion firmer: 'There is . . . good evidence for the production of secondary mesons constant in mass and kinetic energy . . . . It is convenient to refer to this process . . . as the μ-decay. We represent the primary meson by the symbol π, and the secondary by μ.'

Thus came about the vindication of the two-meson idea, and the resolution of earlier difficulties. The π-meson (now called pion), the Yukawa meson, is heavier than the μ-meson (muon); that is good for nuclear forces. The pion, strongly coupled to nucleons, is produced copiously. Its daughter, the muon's, interaction is mainly electromagnetic, whence its small scattering and its negligible absorption. The decay electrons seen at sea level originate from muons, not pions. Thus the observed two microseconds lifetime pertains to the muon.

These discoveries and their early elaboration came at just about the time when a new generation of accelerators began complementing cosmic rays. Powell's 1950 summary[54] of progress on both fronts contains extensive references to the literature of that period as well as the following data (m is expressed in electron masses, τ in seconds; modern values in parentheses)

$$\mu^+: \quad m = 212\,(207), \qquad \tau = 2.1\,(2.2) \times 10^{-6}, \qquad \mu^+ \to e^+ + 2\nu \tag{18.1}$$

$$\pi^+: \quad m = 276\,(277), \qquad \tau = 1.6\,(2.6) \times 10^{-8}, \qquad \pi^+ \to \mu^+ + \nu \tag{18.2}$$

where 'ν represents any particle of small rest mass not a photon'.

I first heard the terms 'π-mesons', 'μ-mesons', in September 1947 when attending a lecture by Powell in Copenhagen, which he began with an altogether fitting quotation from Maxwell: 'Experimental science is continually revealing to us new features of natural processes and we are thus compelled to search for new forms of thought appropriate to these features'. The new punchline, 'There is a muon', may have caused laughter in the heavens, but man was, and still is, ignorant of the joke. What else was the muon good for other than being the pion's favorite decay product? To be sure, the discovery of the electron had also been totally unexpected, but its universal use as an ingredient of the atomic periphery was recognized rapidly. The neutron, less of a surprise, made it possible almost at once to develop theories of nuclear structure and β-decay. But the muon? Now, forty years later, the divine laughter continues unabated.

(c) Quantum electrodynamics: The great leap forward

> I hold that the chief merit of a theory is, that it shall guide experiment, without impeding the progress of the true theory when it appears.
>
> Maxwell in *On Faraday's lines of force*[55]

1. From the week after Shelter Island to the month after Pocono. Quantum electrodynamics was twenty years old, quite old by current standards, when, beginning with Shelter Island, it entered a new phase. In later years, veterans of the 1930s battle with the infinite would occasionally and wistfully remark that this change should already have occurred in prewar days. They would remind themselves that, after all, the theory so masterfully developed further in the late forties was based on the same dynamical equations, subject to the same rules of quantum theory and relativity, with which they had struggled before. More than that, several of the main themes marking this new phase had originated in those earlier days:* displacement of spectral lines due to radiative corrections; insistence on relativistic and gauge invariance in performing subtractions; renormalization of mass and charge; the identification of an observable vacuum polarization effect; a finite displacement $\Delta E(nS)$ of nS levels in a hydrogen atom given by (see Eq. (16.39))

$$\Delta E(nS) = \frac{-4Z^4\alpha^5}{15n^3}\, mc^2. \tag{18.3}$$

All these earlier ideas and results now returned, though in vastly improved form. Some veterans might even remember a calculation (1939)[56] of radiative corrections $O(\alpha)$ to the scattering of a Dirac electron in a fixed external Coulomb field. If the necessary mass and charge renormalizations are performed properly, the result is finite. A slip in the mass renormalization gave an infinite answer, however. As Tomonaga said[57] much later, had this calculation been performed correctly, the history of renormalization theory would have been completely different.

The development of physics in the twentieth century, and earlier as well, is replete with examples of 'it could have been done earlier', each with its own circumstances. As a matter of fact, in the present case the delay (if one even wishes to call it that) was not all that long, especially if one remembers the diversions due to the advent of nuclear physics in the thirties, and to the not

* Cf. Chapter 16, Section (d).

unrelated necessity of physicists' participation in war work. As far as quantum electrodynamics is concerned, the experiments first reported at Shelter Island were, of course, the great stimulus to creating not a new theory but a far more mature version of an earlier one. In the broadest terms, the developments which started in the late forties, with a largely new cast of characters, and which are being elaborated to this day, have shown that quantum electrodynamics, a source of despair in earlier times, has extraordinary vitality.

It began in a quite old-fashioned way.

Five days after Shelter Island, Bethe circulated the participants of that meeting with his theoretical results[58] for the Lamb shift. His calculation for the nS-level shift, in which the electron was treated as unrelativistically as it had been by Dirac in his first attempts,* went like this. The leading term in the electron's self-energy diverges linearly, Lorentz' old e^2/a. This term is the same for a free electron as for a bound electron with the same average kinetic energy. Subtract that term: mass renormalization, and identify the remainder as the level displacement. Answer:

$$\Delta E(nS) = \frac{4Z^4\alpha^5}{3\pi n^3} \cdot mc^2 \ln \frac{K}{\bar{E}} \tag{18.4}$$

in the limit $K \to \infty$: after subtraction the result is still divergent. $\bar{E}$ is the average excitation energy for the nS state. Next Bethe said with great aplomb: a relativistic treatment ought to justify cutting off at

$$K = mc^2. \tag{18.5}$$

Doing so and calculating $\bar{E}$ he obtained 1040 MHz, 'in excellent agreement with the observed value'.

While hardly a calculation from first principles and manifestly incomplete,** it was an exceedingly encouraging result showing that the direction was right and also, once again, that Bethe was, and is, a physicist's physicist.

The first 'clean' result of the new era was Julian Schwinger's evaluation of the electron's anomalous magnetic moment (December 1947).[59] In physical terms, the idea is quite similar to Wick's for the anomalous moment of the proton; see Eq. (17.33): an external magnetic field sees the electron in a state of dissociation into electron plus photon for part of the time, during which the electron has a different momentum and hence a different magnetic interaction than in its undissociated state. Schwinger's calculation by the rules of positron theory, with renormalizations properly performed, made clear beyond doubt that the effects reported by Rabi at Shelter Island are due to radiative

* See Eqs. (15.15)–(15.17) and (15.78).

** Not only because of the cutoff but also because of the omission of such effects as vacuum polarization and P-level displacements.

corrections, as is seen from a comparison* of his answer with data then available:

$$a=\tfrac{1}{2}(g-2)=\frac{\alpha}{2\pi}=1\,162\times 10^{-6}\,(\text{theory}) \tag{18.6}$$

$$=1\,18(3)\times 10^{-6}\,(\text{expt.}). \tag{18.7}$$

Schwinger realized that his result (18.6) also enabled him to guess a new contribution to the Lamb shift, the full evaluation of which was not completed until late 1948. For the present purposes it suffices to recount how ultimately it came to be agreed that to leading order the nS levels of the hydrogen atom are displaced upward by an amount

$$\Delta E(nS)=\frac{4Z^4\alpha^5}{3\pi n^3}\left(\ln\frac{mc^2}{\bar{E}}-\ln 2+\frac{11}{24}+\frac{3}{8}-\frac{1}{5}\right)mc^2. \tag{18.8}$$

(There is also a small downward displacement of the $P_{1/2}$-level.[60]) Two of the terms in parentheses have been seen before: the first and biggest is Bethe's ingenious estimate; the last, the Uehling term Eq. (18.3).

Schwinger's guess concerned the 3/8-term. He argued that his $\alpha/2\pi$ result implied that one should add the expression

$$\frac{\alpha}{2\pi}\frac{e\hbar}{2mc}\bar{\psi}(\vec{\sigma}\cdot\vec{H}+i\vec{\alpha}\vec{E})\psi \tag{18.9}$$

to the effective Hamiltonian for the electromagnetic interaction of electrons, the magnetic term being the one he had just calculated, the electric term then being necessary for ensuring relativistic invariance ($\vec{\alpha}$ is Dirac's velocity matrix). The latter modifies the Coulomb interaction and gives the 3/8-contribution. Schwinger's direct calculation of the electric term produced[61] a most unpleasant surprise, however: it was too small by a factor 1/3! A year later he would write[62] in a footnote (it is remarkable how much of the drama of that time is recorded in footnotes): 'This difficulty is attributable to the incorrect transformation properties of the electron self-energy in the conventional Hamiltonian treatment and is completely removed in the covariant formalism now employed'.

How can a fully covariant theory yield non-covariant results? Because during the calculation one has to subtract infinity from infinity, which in general is not a well-defined step. How can one hope to avoid non-covariant answers? By computing in such a way that covariance is manifest at every stage; and likewise for gauge invariance. Take, for example, Heisenberg and Pauli's treatment of quantum electrodynamics in the Coulomb gauge which, as noted earlier,** may not look covariant but is covariant nevertheless. It is not, however,

* g was defined in Eq. (13.8). Numbers in parentheses indicate uncertainties in the last decimal(s).
** Chapter 15, Section (e).

manifestly covariant at every stage. Thus the Coulomb gauge does not lend itself (readily) to the evaluation of radiative corrections. Likewise the second-order perturbation formula (Eq. (15.33)) and its higher-order partners, though actually covariant, are not manifestly so. One can, however, cast them in an equivalent manifestly covariant form, as in fact Stückelberg had already shown[63] in 1934.

At a meeting, the sequel to Shelter Island, held from 30 March to 2 April 1948, at Pocono Manor, a large inn in Pennsylvania's Pocono Mountains, the main event was Schwinger's marathon performance on the first day, when he presented to us his 'covariant formalism now employed'. So packed was the agenda that during a session that same evening Rabi said: 'As so and so pointed out yesterday . . . '.

I have now presented the evolution of quantum field theory from Dirac's first paper up to the Pocono's. While my presentation of the subject is not quite finished, I am coming close to the stage at which 'what happened next' is part of the current physics scene. I therefore believe it to be not only helpful but also proper to begin steering the reader to recent publications, references to which are collected in the sources found at the end of this section. I am convinced that the general reader who has come this far would not be helped, that the experts keen to learn about the early beginnings of their subject would not be further enlightened, and that justice would not be done to the original authors, if I were to compress into a few pages the highly technical aspects, so readily accessible elsewhere, of the radiation theories of Tomonaga, Schwinger, and Feynman. Thus in what follows, the reader will not even find displayed those most ubiquitous and useful modern tools in field theory, the Feynman diagrams, the Feynman rules, and the Feynman integrals.

Back to the Poconos. Schwinger's report (the starting point of a series of papers[64,65]), replete with technical novelties, was a manifestly covariant and gauge invariant reformulation of familiar quantum electrodynamics. Among the results were a proof that to $O(\alpha)$ the photon self-energy vanishes, as gauge invariance demands (recall that this problem had plagued Heisenberg years earlier*); and a preliminary answer for the Lamb shift in which everything was in order except for the 11/24-term. If Schwinger's arguments were hard to keep track of because of their wealth of technicalities, Feynman's shorter presentation which came next was difficult to understand (I don't think anyone did) because of its unfamiliar way of thinking. There were path integrals, positrons moving backward in time,** and closed loops (vacuum polarization) which, at that time, he was not quite sure how to handle. I had heard some of this before during the private conversations at Shelter Island. Again I was

* Chapter 16, Section (d), Part 5, cf. also Ref. 66. ** An idea foreseen by Stückelberg.[67]

struck, as were we all, by the speed and apparent efficiency of his methods; he practically calculated the Lamb shift before our eyes. Everything was in order except for the 11/24-term.

A sidelight: at one point Feynman said, let's forget about the Pauli principle, or words to that effect, whereupon Niels Bohr—perhaps not quite aware that while Feynman might like to clown he was dead serious about physics—strode to the blackboard and delivered himself of an encomium on the exclusion principle.

From Feynman's summary of the meeting, written[68] shortly afterward: 'A major portion of the conference was spent in hearing and discussing [the] results of Schwinger. They represent a real advance in our understanding of physics. There was also presented by Feynman a theory in which the equations of electrodynamics are artificially altered [by a cut-off] so that all quantities, including the inertia of the electron, turn out finite. The results of this theory are in essential agreement with those of Schwinger, but they are not as complete'.

A few weeks had passed when we received from Oppenheimer a copy[69] (dated April 5) of a letter Tomonaga had written to him.

2. About Tomonaga and his group. I saw a good deal of Tomonaga when—on leave from the Tokyo University of Education—he spent academic '49–'50 at the Institute in Princeton. The year 1948, during which he had been frenetically active, was behind him and he had turned his attention to sound waves in a Fermi gas.[70] I remember him as soft-spoken, serene, ascetic in appearance and as the most profound of the Japanese physicists I have known.

Early in his research career, Tomonaga had been assistant to Nishina and had also spent time with Heisenberg in Leipzig, working on the liquid drop model of the nucleus.[71] Conversations with the master on the status of quantum field theory stimulated the program he set himself upon his return to Japan. His first major paper[72] entitled 'On a relativistically invariant formulation of the quantum theory of wavefields', published in Japanese in 1943, in English in 1946, is devoted to the extension of the many-time formalism for n particles* to a manifestly covariant 'super-many-time' formalism for fields, including the proof of its equivalence with the Heisenberg–Pauli theory.** Schwinger acknowledged[73] in his first article of 1948 that Tomonaga's formulation was the same as his 'interaction representation'. Neither knew that the starting point of these considerations was already to be found in a 1938 paper[74] by Stückelberg, whose originality commands high respect.

Old-fangled methods, rather than Tomonaga's new one, were still employed in 1947 when he and his group embarked on a series of investigations of the scattering of a Dirac electron in a Coulomb field, a problem which, as

* Cf. Chapter 16, Section (d), Part 1.

mentioned before, had been unsuccessfully tackled in 1939. They began with the combined photon–scalar meson theory (Section (a))[7,11] which to $O(\alpha)$ gives a finite self-energy of the free electron and asked: Will this theory also give finite mass effects in this scattering problem? At first they made the same mistake as in the 1939 paper and found the answer to be no.[75] Then they discovered the error in that earlier work and found the answer to be yes.[76] At just about that time the papers by Lamb and Retherford and Bethe reached Japan, leading them to inquire whether the same scattering process would also have a finite answer without the scalar field but with the exclusive help of renormalizations or, as they called it, self-consistent subtractions. Again the answer was yes,[77] as had likewise been noted a few weeks earlier in Princeton.[78] In May 1948 they obtained[79] the same result once again, now for the first time using Tomonaga's manifestly covariant formalism.[72] Tomonaga's letter to Oppenheimer summarized all these conclusions.*

In September 1948 the Tomonaga group[81] reported the results of its Lamb shift calculation performed with the same covariant technique. Everything was in order, including the 11/24-term and the $2P_{1/2}$-displacement. And so they became the first to record the correct formula for the $2S_{1/2}$–$2P_{1/2}$ splitting due to radiative corrections.

Weeks later, Kroll and Lamb[81a] and, right thereafter, French and Weisskopf[82] submitted their papers on the Lamb shift. Deftly these authors had managed to obtain correct answers by non-covariant methods, imposing as additional constraints that the electric term in Eq. (18.9) shall have the coefficient dictated by covariance. Early in 1949 Schwinger[83] and Feynman[84,85] published their Lamb shift calculations, also with the correct answer. If it had taken them longer, that was largely because they were after bigger fish: their respective manifestly covariant formulations of quantum electrodynamics.

What about the considerable confusion that had been caused by the 11/24-term? That had nothing to do with subtleties of renormalization. The point was that the calculation had been divided into two parts. The electron in the hydrogen atom moves in a Coulomb potential which can be handled as a perturbation for high virtual electron energies but which must be treated rigorously for low frequencies. (One has since learned how to circumvent this two-part procedure.[86]) All the trouble arose from the proper joining of these two parts, a question about which all parties concerned kept in constant touch.[87] Schwinger later said: 'Both Feynman and I were careless about the low energy'.[61] As to agreement with experiment: in 1949 the theory gave 1051 MHz, the experimental value was slightly higher (1062 ± 5),[88] but the difference was nothing much to worry about—yet.

* At that time the Tokyo group still had difficulties with the photon self-energy problem.[80]

3. 1949–84 in a nutshell. With the successful completion of the magnetic moment and Lamb shift calculations, a solid beachhead had been established in uncharted terrain, the physics of radiative corrections. Meanwhile an outstanding problem had been resolved by Dyson's demonstration[89] of the equivalence of Feynman's approach with the formulations (almost identical) of Tomonaga and Schwinger. The simpler and far easier to apply Feynman version now began its rapid and never waning rise in popularity. It was the main topic of the next small conference, held from 11 to 14 April 1949, at Oldstone-on-the-Hudson, 60 km north of New York City. By then it was clear that to $O(\alpha)$ the renormalization program suffices to make *all* predictions of quantum electrodynamics finite (except of course for the electron's mass and charge). My only other recollection of that gathering is sitting in on an illuminating evening's poker game with Oppenheimer, von Neumann, and Teller.

After that third meeting, Oppenheimer decided that the original purpose, to assess the current status and the prospects for further developments, had been accomplished. Thereupon Marshak took a new initiative, to the lasting benefit of the high energy physics community. It was his belief that the profitable meetings of the kind just ended should be continued, but be more international in scope and have a better mix of experimentalists and theorists, especially in view of rapid developments in physics done with accelerators. The result was a series of seven annual conferences, beginning in 1950, with ever-growing attendance, all held in Rochester, New York, with financial support from local industry. In connection with the subject here under discussion, be it recorded that during Rochester I, the 1950 meeting, the subject of quantum electrodynamics did not even come up (although Feynman was there).*

In 1957 the responsibility for these meetings was transferred to a commission of the International Union for Pure and Applied Physics. The next conferences were held in Geneva (1958), Kiev (1959), Rochester (1960), and on and on, all of them being referred to as Rochester Conferences. Eventually these meetings were convened bi-annually and are now the high energy physicists' principal assemblies.

Let us return to quantum electrodynamics. After the beachhead had been established, the next task was to break out of it in conquest of still more territory. Some of these attacks have been spectacularly successful; others, no

* From Rochester II on, the proceedings were published, at first in ditto form, in later years as books. Since there exists no such record for Rochester I it may be of some use to state that it was a one-day meeting held on 16 December 1950, with an attendance of about 50; that it consisted of a morning, afternoon, and evening session chaired respectively by Pais, Oppenheimer, and Bethe, and that the topics discussed were: accelerator results on the interaction of pions and nucleons with matter, muon physics, and cosmic ray physics. I am grateful to Robert Marshak for making available to me an unedited transcript of that meeting.

less pertinent, were stymied. About these subjects tons has been written, some of it mentioned hereafter in the sources. The reader will need to consult those and other post-war physics texts should she or he wish to have more information on modern developments than is contained in the topics, ordered in a series of six brief entries, which I have selected for further and final discussion.

(1) To begin with, consider a world containing electrons and photons only. After a book-keeping had been found which to order α gives finite predictions for all processes, the obvious next question was: does the renormalization program also guarantee finite answers for the radiative corrections associated with the higher-order terms in the perturbative expansion in powers of α? Yes, it does. According to a fundamental theorem stated by Dyson in 1949[90] three and only three renormalizations suffice to each order: mass, charge, and wave function renormalization,* the latter not manifesting itself directly in terms of a physical parameter. The theorem's complicated proof was initiated by Dyson,[90] emended by Salam,[92] refined by Weinberg,[93] and elaborated by others.[94] The vast literature on this subject which has developed since bears witness to the emergence of a new highly technical branch of mathematical physics, advanced renormalization theory.

(2) The need for renormalizing mass and charge implies that the theory cannot predict their values but borrows those from experiment. Mass and charge are what one calls *phenomenological parameters*. Quantum electrodynamics is nevertheless a theory with enormous predictive power. A *finite* number of parameters treated phenomenologically make possible an infinite number of predictions. Such a theory is called renormalizable. Theories which would demand renormalizations of new physical parameters (of scattering, or particle production amplitudes, for example) to each order of a perturbative expansion in some coupling constant have, in all, an infinity of renormalizations and are called unrenormalizable. Such theories are useless, since infinitely many data must now be borrowed from experiment. Are there such theories? Yes, most of the meson field theories met before** are of this kind. More about that in the next chapter.

The nature of the infinities encountered in quantum field theory is closely related to the fact that integral (half-integral) spin fields satisfy partial differential equations of the second (first) order.† Several attempts were made to

* Wave function renormalization, a not too well-chosen name, refers to a rescaling of field operators. In zeroth order ($e = 0$), the free Dirac field operators $\psi_0(\vec{x}, t)$, $\psi_0^\dagger(\vec{x}', t)$ satisfy the anti-commutation relations (16.21). For $e \neq 0$ but for point pairs far from the regions of interaction one must require that the corresponding field operators $\psi(\vec{x}, t)$, $\psi^\dagger(\vec{x}', t)$ satisfy the same relations (16.21). One finds, however, that in the latter case the δ-function on the rhs is multiplied by an infinite constant. Wave function renormalization is a rescaling of the ψ and $\psi^\dagger$ such that this constant becomes 1. Likewise for the photon field operators.[91]

** Eqs. (17.44)–(17.50).

† And to the fact that our world has $3+1$ dimensions.

lessen, or even eliminate, the infinities of standard theories by supposing these field equations to be of higher than the conventional order. Uhlenbeck and Pais showed (1950), however, that this leads to other serious difficulties[95] which indicate that the common field theories, while not perfect, have fewer afflictions than others. It was found that field theories considered up to that time either contain infinities, or that it is impossible to define a positive definite energy for free (uncoupled) fields, or that the state vectors of physical systems show non-causal behavior. Quantum electrodynamics does not suffer from these last two diseases which are fatal for interacting systems.

(3) Back once again to the theory of photons and electrons. It cannot predict the electron's mass m *regardless* of the renormalization issue simply because m is its only mass scale. m is therefore as uncalculable in this theory as is Planck's $\hbar$ constant in quantum theory, or the light velocity c in relativity. It is not unreasonable to hope, however, that theories with larger particle content may ultimately explain mass ratios such as that of the muon to the electron, or mass differences such as that of the proton and neutron. All these mass questions are presently unresolved, nor is it clear whether they will be answered one at a time or all at once.

The world of photons and electrons does contain the scale $\sqrt{\hbar c}$ for the electron charge e. Should one therefore aspire to predict the dimensionless number $\alpha = e^2/\hbar c$ the current best value[96] of which is

$$\alpha^{-1} = 137.035\,963(15)? \tag{18.10}$$

There is no harm in aspiring but no one has a clue.

(4) All results in quantum electrodynamics obtained to date are based on the presumed validity of a perturbative power series expansion in α. Is this expansion legitimate? Three main problems arise.

Problem 1. Consider the renormalized amplitude A for some process, expanded as

$$A = \sum_1^\infty A_n \alpha^n \tag{18.11}$$

with finite A_n. Does this series converge? We do not know but, especially as a result of lessons learned[97] from simpler but non-trivial model field theories, we believe that the answer probably is no.

Problem 2. According to the α-expansion, the mass and charge renormalizations are of the form

$$\delta m = \sum_1^\infty \delta m_n \alpha^n, \qquad \delta e = \sum_1^\infty \delta e_n \alpha^n \tag{18.12}$$

where δm_n and δe_n are infinite for each n. Does this mean that δm and δe are infinite? That is, could rigorous rather than perturbative methods yield finite answers for δm and δe? No definitive answer has been found in spite of strenuous efforts.

Problem 3. Are unrenormalizable theories bad? Yes, on the face of it. Note, however, that unrenormalizability was defined in the context of a perturbative expansion in some coupling constant. Could a non-perturbative treatment of such a theory nevertheless make sense? Perhaps in some cases,[97a] but no general answer is known.

These three problems which illustrate the strong need for other than perturbative methods are among those which spurred the activities of the axiomatic field theorists (Chapter 15, Section (a)) whose counsel is sorely needed.

(5) For the last time in this book, back to quantum electrodynamics, not now to the world of photons and electrons, however, but to the real world.

Let us follow the fate of a virtual photon as it contributes to the radiative correction of the electron's magnetic moment μ_{el} because an external magnetic field sees the electron e part of the time in the dissociated state $e+\gamma$. In turn, the γ can dissociate part of its time into a pair: $\gamma \rightarrow e^+ + e^- \rightarrow \gamma$. Counting powers of e one sees that this contributes to μ_{el} in $O(\alpha^2)$. The γ can, however, also dissociate into muons: $\gamma \rightarrow \mu^+ + \mu^- \rightarrow \gamma$ and this too contributes to μ_{el} in $O(\alpha^2)$. More than that, γ can dissociate as $\gamma \rightarrow x + \bar{x} \rightarrow \gamma$ where x is any charged particle and $\bar{x}$ its antiparticle. x may be coupled to other fields, meson fields for example, and dissociated accordingly. All this contributes to μ_{el}. What has been illustrated with the help of μ_{el} holds of course for any electron–photon process. Hence:

Beginning with $O(\alpha^2)$ one finds in the guts of the radiative corrections contributions from all species of charged particles in the physical world.

This shows first that the answer to the question of whether the α-expansion in the photon–electron world converges or not will not say much about the real world, secondly that radiative corrections of order α^2 and higher contain low-energy information about the full particle spectrum in the universe.

(6) Introducing his final attempt[98] to revert to a new classical basis for a theory of the electron, Dirac wrote (1951): 'Recent work by Lamb, Schwinger, and Feynman and others has been very successful . . . but the resulting theory is an ugly and incomplete one'. Quantum electrodynamics was indeed incomplete at that time. As the preceding comments aimed to demonstrate, it still is. So, of course, is classical mechanics, a part of physical theory which stands as a monument of perfection in its domain of applicability, that is where quantum and relativity effects can be ignored. By comparison quantum electrodynamics fares very well in so far as its successes are considered. What is lacking, on the other hand, is a clear definition of how far it goes. A final judgement on the merits of that theory can only come if we can define its limits. That, at this time, is a matter for conjecture. Physics may well look different when we know the answer. I shall leave aside these deep questions

and conclude my account of quantum electrodynamics with brief remarks on its spectacular successes in the post-war era.

We have already seen the beginnings of that story, how faith in the theory was greatly bolstered by agreements between experiment and the theory of corrections to $O(\alpha)$. That, however, was by no means the end. Since the early fifties, experimental precision has reached new heights while theoretical analysis of radiative corrections has been pushed to higher and higher orders, having reached $O(\alpha^4)$ at the time of writing. Even at that level there is no evidence for any failure of quantum electrodynamics.

We owe these advances to the devoted labor of relatively few experimentalists and theorists. Many of those involved are named and their results quoted in recent reviews by Drell[99] and by Kinoshita.[100] The development in this highly specialized field is a tale of mutual stimulus between theory and experiment. For example, the theoretical studies of $O(\alpha^2)$ effects began in 1949 with a pioneering evaluation[101] of $O(\alpha^2)$ corrections to the electron's magnetic moment. It is a mark of the complexity of these calculations that the answer remained unchallenged until 1957 when an apparent discrepancy[102] between improved experiment and theory was eliminated after it was found that the theoretical result was not quite right.[103] The two reviews just mentioned are rich in details concerning the evolution of this subject, covering such topics as fine structure, hyperfine structure, positronium (bound e^+e^- systems) and muonium (bound μ^+e^-). I conclude my own account with two examples.

The magnetic moments of the electron and positron. The present best experimental values are[104] (a was defined in Eq. (18.6))

$$\left.\begin{aligned} a(\mathrm{e}^-) &= 1\,159\,652\,200(40)\times 10^{-12} \\ a(\mathrm{e}^+) &= 1\,159\,652\,222(50)\times 10^{-12}. \end{aligned}\right\} \quad \text{(expt.)} \tag{18.13}$$

The theoretical answer consists of two parts,

$$a(\mathrm{e}^\pm) = a^{(1)}(\mathrm{e}^\pm) + a^{(2)}(\mathrm{e}^\pm). \tag{18.14}$$

The first term contains the contributions from the electron–photon world only and is given by[105]

$$a^{(1)}(\mathrm{e}^\pm) = \frac{\alpha}{2\pi} - 0.328\,478\,966\left(\frac{\alpha}{\pi}\right)^2 + 1.176\,5(13)\left(\frac{\alpha}{\pi}\right)^3 - 0.8(1.4)\left(\frac{\alpha}{\pi}\right)^4 \tag{18.15}$$

where the theoretical error expresses the best estimate of inaccuracies in numerical evaluations. The contribution from other particles which (as was just mentioned) feed into $a(\mathrm{e}^\pm)$ from $O(\alpha^2)$ on is given by[106]

$$a^{(2)}(\mathrm{e}^\pm) \simeq 4\times 10^{-12}. \tag{18.16}$$

Using Eq. (18.10) one finds

$$a(\mathrm{e}^{\pm}) = 1\,159\,652\,460(44)(127)\times 10^{-12}\,\text{(theory)}. \tag{18.17}$$

The first error is theoretical, the second is due to the error in α, Eq. (18.10).

The magnetic moment of the muon. Best experimental[107] values:

$$\left.\begin{aligned} a(\mu^-) &= 11\,659\,370(120)\times 10^{-10} \\ a(\mu^+) &= 11\,659\,110(110)\times 10^{-10}. \end{aligned}\right\}\text{(expt.)} \tag{18.18}$$

An improvement of the experimental accuracy by a factor 30 is actively being considered.[108] The best theoretical value is contained in a publication[109] by Kinoshita, expert among experts, and co-workers which reached my desk in March 1984:

$$a(\mu^{\pm}) = a^{(1)}(\mu^{\pm}) + a^{(2)}(\mu^{\pm}). \tag{18.19}$$

Here

$$a^{(1)}(\mu^{\pm}) = \frac{\alpha}{2\pi} + 0.765\,858\,10(10)\left(\frac{\alpha}{\pi}\right)^2 + 24.073(11)\left(\frac{\alpha}{\pi}\right)^3 + 140(6)\left(\frac{\alpha}{\pi}\right)^4 \tag{18.20}$$

contains the contribution from the world of photons, electrons, muons, and τ-particles (about which more later in Chapter 21), while

$$a^{(2)}(\mu^{\pm}) \simeq 722\times 10^{-10} \tag{18.21}$$

comes from all other known sources. Again using Eq. (18.10) one finds

$$a(\mu^{\pm}) = 11\,659\,204(20)\times 10^{-10}. \tag{18.22}$$

All these magnetic moment calculations took years of hard work including hundreds of hours of numerical integrations on a CDC-7600 computer.[100]

The agreement between experiment and theory shown by these examples, the highest point in precision reached anywhere in the domain of particles and fields, ranks among the highest achievements of twentieth-century physics.

Meanwhile the battle with the infinite continues.

Sources

Reminiscences. On quantum electrodynamics: by Lamb,[2,24] Schwinger,[61] Tomonaga,[57] Weisskopf;[110] also contributions to Ref. 22. On the muon: by Marshak,[111] Peyrou,[112] Piccioni,[113] and Rossi.[38] Also Schwinger on Tomonaga,[114] Dyson on Feynman,[115] Marshak on the Rochester conferences.[116]

Books. An anthology of papers on quantum electrodynamics;[27] selected papers of Schwinger[117] and of Feynman;[85] the scientific papers of Tomonaga.[118]

Articles. Galison on the muon.[119] Drell (1979)[99] and Kinoshita (1983)[100] on the status of quantum electrodynamics. Wheeler's notes on the Pocono conference (unpublished) distributed to the participants.

Textbooks on quantum electrodynamics. Jauch and Rohrlich (1955),[60] Bjorken and Drell (1965),[120] and Itzykson and Zuber (1980)[121] give clear pictures of the state of the art at various times.

References

1. J. Burckhardt, *Reflections on history*, Liberty Classics, Indianapolis 1979. (*Weltgeschichtliche Betrachtungen*, transl. M. D. Hottinger.)
2. W. E. Lamb, repr. in *A Festschrift for I. I. Rabi*, Ed. L. Motz, p. 82, *Trans. N.Y. Ac. Sc. Series II*, Vol. 38, 1977.
3. A. Pais, *Physics Today*, Oct. 1967, p. 42, repr. in *Oppenheimer*, p. 31, Scribner, New York 1969.
4. *Phys. Rev.* **70**, 784, 1946.
5. A. Pais, *Projective theory of meson fields and electromagnetic properties of atomic nuclei*, Noord Holl. Uitg., Maatschappy 1941; also *Physica* **8**, 1137, 1941; **9**, 267, 407, 1942.
6. F. Bopp, *Ann. der Phys.* **38**, 345, 1940; A. Landé and L. H. Thomas, *Phys. Rev.* **60**, 121, 574, 1940.
7. A. Pais, *Phys. Rev.* **68**, 227, 1945.
8. F. Villars, in *Theoretical physics in the twentieth century*, Eds. M. Fierz and V. Weisskopf, p. 78, Interscience, New York 1960.
9. A. Pais, *Trans. Kon. Ak. Wet. Amsterdam* **19**, 1, 1947; earlier similar calculations by N. Kemmer, *Proc. Roy. Soc. A* **166**, 127, 1938, contain several inaccuracies.
10. See however A. S. Wightman, *Phys. Rev.* **71**, 447, 1947.
11. S. Sakata, *Progr. Th. Phys.* **2**, 145, 1947; S. Sakata and O. Hara, ibid. **2**, 30, 1947.
12. Cf. D. Ito, Z. Koba, and S. Tomonaga, *Progr. Th. Phys.* **2**, 217, 1947.
13. H. A. Kramers, *Nuov. Cim.* **15**, 108, 1938; repr. in *Collected scientific papers*, p. 831, North Holland, Amsterdam 1956.
14. See further J. Serpe, *Physica* **7**, 133, 1940; W. Opechowski, *Physica* **8**, 161, 1941.
15. H. A. Kramers, *Proc. Solvay Conference 1948*, p. 241; repr. in *Collected works*, p. 845, North Holland, Amsterdam 1956.
16. A. Pais, *Proc. Cambr. Phil. Soc.* **42**, 45, 1946.
17. L. Hulthén and A. Pais, *Proc. international conf. on fundamental particles and low temperatures*, Vol. 1, p. 177, Taylor and Francis, London 1947.
18. R. Sherr, *Phys. Rev.* **68**, 240, 1945.
19. C. Møller, *Kgl. Dansk Vid. Selsk. Math.-Fys. Medd* **18**, No. 6, 1941, p. 3, footnote.
20. C. Møller and A. Pais, reported by C. Møller, Ref. 17, Vol. 1, p. 184.
21. A. Pais, in *Niels Bohr*, Ed. S. Rozental, p. 215, North Holland, Amsterdam 1967.
22. S. Schweber, in *Shelter Island II*, Eds. R. Jackiw, N. N. Khuri, S. Weinberg, and E. Witten, p. 301, MIT Press, Cambridge, Mass. 1985.
23. J. R. Oppenheimer, letter to F. B. Jewett, June 4, 1947, copy in Rockefeller University Archives.

24. W. E. Lamb, in *The birth of particle physics*, Eds. L. M. Brown and L. Hoddeson, p. 311, Cambridge Univ. Press 1983.
25. J. E. Mack and N. Austern, *Phys. Rev.* **72**, 972, 1947.
26. W. E. Lamb and R. C. Retherford, *Phys. Rev.* **72**, 241, 1947.
27. Ref. 26, repr. in *Quantum electrodynamics*, Ed. J. Schwinger, p. 136, Dover, New York 1958.
28. R. C. Williams, *Phys. Rev.* **54**, 558, 1938.
29. S. Pasternack, *Phys. Rev.* **54**, 1113, 1938.
30. E. C. Kemble and R. D. Present, *Phys. Rev.* **44**, 1031, 1933.
31. W. Fröhlich, W. Heitler, and B. Kahn, *Proc. Roy. Soc. A* **171**, 269, 1939; *Phys. Rev.* **56**, 961, 1939; B. Kahn, *Physica* **8**, 58, 1941.
32. W. E. Lamb, *Phys. Rev.* **56**, 384, 1939; **57**, 458, 1940.
33. J. E. Nafe, E. B. Nelson, and I. I. Rabi, *Phys. Rev.* **71**, 914, 1947, and later papers by Nafe and Nelson; also D. E. Nagel, R. S. Julian, and J. R. Zacharias, *Phys. Rev.* **72**, 971, 1947.
34. P. Kusch and H. M. Foley, *Phys. Rev.* **72**, 1256, 1947, and later papers.
35. See W. V. Houston and Y. M. Hsieh, *Phys. Rev.* **45**, 263, 1934, esp. p. 272.
36. Cf. also G. Breit, *Phys. Rev.* **72**, 984, 1947.
37. E. J. Williams and G. E. Roberts, *Nature* **145**, 102, 1940.
38. B. Rossi, Ref. 24, p. 183.
39. W. B. Fretter, *International colloquium on the history of particle physics*, p. 191, *J. de Phys.* **43**, suppl. to No. 12, 1982.
40. V. Mukherji, *Arch. Hist. Ex. Sci.* **1**, 27, 1974.
41. L. E. Hoisington, S. S. Share, and G. Breit, *Phys. Rev.* **56**, 884, 1939.
42. Cf. e.g. H. Yukawa and S. Sakata, *Nature* **143**, 761, 1939; L. W. Nordheim, *Phys. Rev.* **55**, 506, 1939; H. A. Bethe and L. W. Nordheim, *Phys. Rev.* **57**, 998, 1940.
43. Cf. J. G. Wilson, *Proc. Roy. Soc. A* **174**, 73, 1940.
44. S. Tomonaga and G. Araki, *Phys. Rev.* **58**, 90, 1940.
45. M. Conversi, E. Pancini, and O. Piccioni, *Phys. Rev.* **71**, 209, 1947.
46. E. Fermi, E. Teller, and V. Weisskopf, *Phys. Rev.* **71**, 314, 1947.
47. R. E. Marshak and H. A. Bethe, *Phys. Rev.* **72**, 506, 1947.
48. Y. Tanikawa, *Progr. Theor. Phys.* **2**, 220, 1947.
49. S. Sakata and T. Inoue, *Progr. Theor. Phys.* **1**, 143, 1946.
50. C. M. G. Lattes, H. Muirhead, G. P. S. Occhialini, and C. F. Powell, *Nature* **159**, 694, 1947.
51. E. Rutherford, *Phil. Mag.* **10**, 163, 1905.
52. C. F. Powell and G. E. F. Hertel, *Nature* **144**, 115, 1939; the same and W. Heitler, ibid. p. 283.
53. C. M. G. Lattes, G. P. S. Occhialini, and C. F. Powell, *Nature* **160**, 453, 486, 1947.
54. C. F. Powell, *Rep. Progr. Phys.* **13**, 350, 1950.
55. J. C. Maxwell, 1856, *Collected works*, Vol. 1, p. 208, Dover, New York.
56. S. M. Dancoff, *Phys. Rev.* **55**, 959, 1939.
57. S. Tomonaga, in *The physicist's conception of nature*, Ed. J. Mehra, p. 404, Reidel, Boston 1973.
58. H. A. Bethe, *Phys. Rev.* **72**, 339, 1947, repr. in Ref. 27, p. 139.
59. J. Schwinger, *Phys. Rev.* **73**, 416, 1948, repr. in Ref. 27, p. 142.
60. J. M. Jauch and F. Rohrlich, *The theory of photons and electrons*, p. 359, Addison-Wesley, Reading, Mass. 1955.
61. J. Schwinger, Ref. 24, p. 329.

62. J. Schwinger, *Phys. Rev.* **75**, 898, 1949, repr. in Ref. 27, p. 143, footnote 8.
63. E. C. G. Stückelberg, *Ann. der Phys.* **21**, 367, 1934.
64. J. Schwinger, *Phys. Rev.* **74**, 1439, 1948.
65. J. Schwinger, *Phys. Rev.* **75**, 651, 1949; **75**, 898, 1949 (repr. in Ref. 27, p. 143); **76**, 790, 1949 (repr. in Ref. 27, p. 169).
66. G. Wentzel, *Phys. Rev.* **74**, 1070, 1948.
67. E. C. G. Stückelberg, *Helv. Phys. Act.* **15**, 23, 1942.
68. R. P. Feynman, *Physics Today*, **1**, June 1948, p. 8.
69. Reproduced in Ref. 22.
70. S. Tomonago, *Progr. Theor. Phys.* **5**, 544, 1950.
71. S. Tomonago, *Zeitschr. f. Phys.* **110**, 573, 1938.
72. S. Tomonaga, *Progr. Theor. Phys.* **1**, 27, 1946, repr. in Ref. 27, p. 156; sequels: Z. Koba, S. Tati, and S. Tomanaga, ibid. **2**, 101, 198, 1947.
73. Ref. 64, footnote 11.
74. E. C. G. Stückelberg, *Helv. Phys. Acta* **11**, 225, 1938, Section 5.
75. D. Ito, Z. Koba, and S. Tomonaga, *Progr. Theor. Phys.* **2**, 216, 1947.
76. The same, ibid. **2**, 217, 1947.
77. Z. Koba and S. Tomonaga, *Progr. Theor. Phys.* **2**, 218, 1947; **3**, 391, 1948; the same and D. Ito, ibid. **3**, 276, 325, 1948.
78. H. W. Lewis, *Phys. Rev.* **73**, 173, 1948; S. T. Epstein, ibid., p. 179.
79. T. Tati and S. Tomonaga, *Progr. Theor. Phys.* **3**, 391, 1948.
80. S. Tomonaga, *Phys. Rev.* **74**, 224, 1948.
81. H. Fukuda, Y. Miyamoto, and S. Tomonaga, *Prog. Theor. Phys.* **4**, 47, 121, 1949.
81a. N. M. Kroll and W. E. Lamb, *Phys. Rev.* **75**, 388, 1949.
82. J. B. French and V. Weisskopf, *Phys. Rev.* **75**, 1240, 1949.
83. J. Schwinger, *Phys. Rev.* **75**, 898, 1949, repr. in Ref. 27, p. 143.
84. R. P. Feynman, *Phys. Rev.* **76**, 769, 1949, repr. in Ref. 27, p. 236.
85. Also repr. in R. P. Feynman, *Quantum electrodynamics*, p. 167, Benjamin, New York 1961.
86. G. W. Erickson and D. R. Yennie, *Ann. of Phys.* **35**, 271, 1965.
87. Cf. Ref. 82, Section 6; Ref. 84, footnote 13.
88. W. E. Lamb, *Rep. Progr. Phys.* **14**, 19, 1951.
89. F. J. Dyson, *Phys. Rev.* **75**, 486, 1949, repr. in Ref. 27, p. 275.
90. F. J. Dyson, *Phys. Rev.* **75**, 1736, 1949, repr. in Ref. 27, p. 292.
91. Cf. Ref. 60, p. 221.
92. A. Salam, *Phys. Rev.* **82**, 217, 1951; **84**, 426, 1951.
93. S. Weinberg, *Phys. Rev.* **118**, 838, 1960.
94. Cf. N. N. Bogoliubov and D. V. Shirkov, *Introduction to the quantum theory of fields*, Interscience, New York 1959; K. Hepp, *Théorie de la rénormalisation*, Springer, Berlin 1969.
95. A. Pais and G. E. Uhlenbeck, *Phys. Rev.* **79**, 145, 1950.
96. E. R. Williams and P. T. Olsen, *Phys. Rev. Lett.* **42**, 1575, 1979.
97. Cf. e.g. J. Glimm and A. Jaffe, *Proc. Int. School of Physics 'Enrico Fermi'*, Ed. R. Jost, Academic Press, New York 1969; C. de Calan and V. Rivasseau, *Commun. Math. Phys.* **83**, 77, 1982.
97a. Cf. D. J. Gross and A. Neveu, *Phys. Rev.* **D10**, 3235, 1974.
98. P. A. M. Dirac, *Proc. Roy. Soc. A* **209**, 291, 1951.
99. S. D. Drell, *Physica* **96A**, 3, 1979.
100. T. Kinoshita, Ref. 22, p. 278.

101. R. Karplus and N. M. Kroll, *Phys. Rev.* **77**, 536, 1949.
102. P. Franken and S. Liebes, *Phys. Rev.* **104**, 1197, 1956.
103. C. M. Sommerfield, *Phys. Rev.* **107**, 328, 1957; *Ann. of Phys.* **5**, 26, 1957; A. Petermann, *Helv. Phys. Act.* **30**, 407, 1957.
104. P. B. Schwinberg, R. S. van Dyck, and H. G. Dehmelt, *Phys. Rev. Lett.* **47**, 1679, 1981.
105. T. Kinoshita and W. B. Lindquist, *Phys. Rev. Lett.* **47**, 1573, 1981; *Phys. Rev.* **D27**, 867, 877, 1983; M. Caffo, S. Turrini, and E. Remiddi, *Phys. Rev.* **D30**, 483, 1984.
106. Ref. 105 and T. Kinoshita, in *New frontiers in high energy physics*, p. 127, Plenum, New York 1978.
107. J. Bailey *et al.*, *Phys. Lett.* **68B**, 191, 1977; also F. J. M. Farley and E. Picasso, *Ann. Rev. Nucl. Sci.* **29**, 243, 1978.
108. T. Kinoshita, quoting V. W. Hughes, Ref. 100.
109. T. Kinoshita, B. Nižić, and Y. Okamoto, *Phys. Rev. Lett.* **52**, 717, 1984.
110. V. Weisskopf, Ref. 24, p. 56.
111. R. E. Marshak, Ref. 24, p. 376.
112. C. Peyrou, Ref. 39, p. 7.
113. O. Piccioni, Ref. 24, p. 222; Ref. 39, p. 207.
114. J. Schwinger, Ref. 24, p. 354.
115. F. J. Dyson, *Disturbing the universe*, Harper and Row, New York 1979.
116. R. E. Marshak, *Bull. At. Sci.* **26**, June 1970, p. 92.
117. *Selected Papers of Julian Schwinger*, Reidel, Boston 1979.
118. *Scientific papers of S. Tomonaga*, Ed. T. Miyazima, Misuzo Shobo, Tokyo 1971.
119. P. Galison, *Centaurus* **26**, 282, 1983.
120. J. D. Bjorken and S. D. Drell, *Relativistic quantum mechanics* and *Relativistic quantum fields*, McGraw-Hill, New York 1964 and 1965 respectively.
121. C. Itzykson and J. B. Zuber, *Quantum field theory*, McGraw-Hill, New York 1980.

19

In which particle physics enters the era of big machines and big detectors and pion physics goes through ups and downs

(a) Of new accelerators and physics done by consortium

> Gone was the reluctance to do big things, gone the sometimes valuable, sometimes hampering isolation of the research worker.
>
> Ph. Morrison, *Physics in 1946*[1]

Back from the wars they came, the physicists, those who had contributed to victory, those who had been unable to stave off defeat, and those who, whether by design or circumstance, had been otherwise engaged. Some returned with many qualms about their participation in war work, others with none. International communication, severely disrupted for several years, was re-established rather quickly. Past events were by no means easily forgotten, yet in scientific circles there was little of that lingering vindictiveness which had marked the years following the First World War.

For better or for worse, the chain of events spanning only thirteen years which directly linked Chadwick's table-top discovery of the neutron at the Cavendish to the first test explosion of the atomic weapon at Trinity's tower in New Mexico had created new options for post-war civilization. From politics to poetry it was a changed world. Science itself is now perceived differently. Just as in earlier times, scientific figures continue to appear who, because of individual accomplishment, are held in highest esteem beyond their circle of peers, in their community, in their nation, internationally. The fairly widely held additional new perception is that of the scientist not as an individual but as a member of a caste, the shamans of the atomic age, believed capable of dispensing good and evil to our planet.

Immediate post-war research benefited very considerably from the military's new respect for science. Laboratories at Los Alamos, Oak Ridge, and Chicago (Argonne), initially established for war-related work, were allowed to continue and gradually developed pure research programs. Lawrence made good use of the high esteem in which he was held by General Groves, the head of the Manhattan Engineer District (MED), for furthering the cause of the Radiation Laboratory at Berkeley (now the Lawrence Berkeley Laboratory) which had been exclusively devoted to the war effort in the years just passed. In 1946, the Berkeley electron synchrotron project was authorized at a cost of $500 000, while the Army was still in charge, a directive to this effect having been issued by the MED office at Oak Ridge.[2] Groves also authorized $170 000 for the completion of the 184-inch cyclotron* and saw to the transfer of several hundred thousand dollars' worth of surplus radar sets and capacitors from other Army projects.[3] The availability of such large funds was a heady new experience for those who remembered so well the tight budgets of the Depression days.

Another novelty dating back to 1946 is the formation of the first consortium of universities for the purpose of jointly managing costly research enterprises. Early that year representatives from nine universities in the eastern United States gathered to discuss how to participate in reactor and accelerator physics, radiochemistry, and nuclear medicine, given the impossibility of adequately doing so within the financial constraints of their respective institutions. As a result of that meeting a letter dated 19 January 1946 was transmitted to Groves, proposing a regional laboratory in the nuclear sciences operated with Federal funds but managed by the joint universities. On 18 July 1946, the group started its corporate life as Associated Universities Incorporated. The laboratory site eventually decided on was Camp Upton, an Army installation, now known as Brookhaven National Laboratory (BNL). When a civilian authority, the U.S. Atomic Energy Commission (AEC), took over MED's responsibilities, one of its first acts was to grant formal permission[4] for establishing BNL (January 1947). This institution grew into a multipurpose laboratory with programs in biology, chemistry, engineering, medicine, and physics.

The next joint venture after Brookhaven was international in character.

In June 1950 the American delegation to the UNESCO conference in Florence sponsored a resolution which begins[5] as follows: 'The Director-General is authorized to assist and encourage the formation and organization of regional research centers and laboratories in order to increase and make more fruitful the international collaboration of scientists in the search for new knowledge in fields where the effort of any one country in the region is insufficient for the task'.

In speaking to the resolution, Rabi, who had been instrumental in its formulation, and who had also played a key role in the establishment of BNL,

* Chapter 17, Section (b).

stressed that Europe would be a particularly suitable region and physics an appropriate field of endeavor.[6]

Already before 1950 that same idea had been on the minds of those leading European physicists who had become convinced that European science and technology could keep in step with developments elsewhere and that the brain drain could be slowed down only by joining forces. In addition, the idea had immense appeal as an expression of European unity. Complex and sometimes tense negotiations led to an agreement among eleven countries, signed in Geneva in February 1952, for setting up a provisional CERN, a European Center for nuclear research.

I first visited CERN in the summer of 1954 when it was still temporarily headquartered in the Villa Cointrin at the Geneva airport. The previous May ground had been broken at nearby Meyrin, the permanent site. The formal date of birth of CERN is 29 September 1954, when sufficient ratifications of the convention had been obtained from member states.

The first CERN accelerator (600 MeV) went into operation in 1957. Now, thirty years later, CERN ranks among the very few leading centers in the field, has produced the highest effective energy reached thus far in a laboratory, and stands as the most successful example of what Europe can achieve by pooling its resources.

In the same month that CERN began to exist officially, fourteen U.S. institutions joined in forming the Midwestern Universities Research Association. Relations among consortiums have on the whole been amicably and respectfully competitive.

I return to the immediate postwar years for brief comments on advances in accelerator design. That was the time in which the U.S. was far ahead in the development of 'the great new tools which are the legacies of war'.[1] In mid-1948 the number of high energy (from several keV up) installations was about 60 in the U.S., about 50 in the rest of the world. The Soviet Union, hurt worse in its own land than any of the other Allies, had four van de Graaffs and one cyclotron producing 1.8 MeV deuterons. Berkeley was in the lead with 380 MeV α-particles.[7]

New ideas for accelerators date back to the years of war. A paper,[8] completed in December 1945, describes 'six devices [which] show promise of attaining energies well over 200 MeV'. All of these* are described in textbooks.[9] In this chapter I confine myself largely to the principal advances in circular proton accelerators, the instruments which dominated experimental research until well into the sixties, leaving linear accelerators and storage rings for Chapter 21.

* They are: betatron, synchrotron, microtron, linear resonance accelerator, linear wave guide accelerator, and relativistic ion cyclotron.

1. The synchrocyclotron (SC). Consider again* Lawrence's cyclotron principle:

$$\nu = \frac{eB}{2\pi m} \tag{19.1}$$

which governs the motion of a particle with charge e and effective mass m spiralling outward from the center of a cyclotron. Recall: B is the constant magnetic field perpendicular to its orbit, ν is the frequency of the rf power source which gives the particles an accelerating kick as they pass the gap between the dees. Particles further along in the spiral beam have higher energy than those nearer the source, yet all particles obey Eq. (19.1): they do not all have the same energy yet remain in phase with the accelerating frequency.

As mentioned before, all this holds as long as m may be considered to be energy-independent, as is very nearly the case for protons $\lesssim 25$ MeV. At higher energies the relativistic dependence of m on the particle velocity v (m_0 is the rest mass)

$$m = m_0/\sqrt{1 - v^2/c^2} \tag{19.2}$$

becomes disturbing. Higher energy particles arrive at the gaps too late to get their extra kick. The remedy is clear: take particles with some fixed m at a certain point in their orbit, and fiddle with ν and B so as to keep those particles in phase. For example fix B, lower ν appropriately. That is the principle of the synchrocyclotron or frequency-modulated cyclotron: decreasing ν, constant B, spiralling orbits.

One would not get very far if that were all, however. The precise tuning of ν to m would mean that extremely few particles would reach high energy. The situation is saved by the principle of phase stability stated by Veksler[10] in 1944 and independently by McMillan[11] in 1945.** The idea is about like this. The rf power source produces a potential $V = V_0 \sin \phi(t)$ at the gap. Consider three particles, 1, 2, 3, all with select mass m, which reach the gap at respective phases $\pi/2 < \phi_1 < \phi_2 < \phi_3 < \pi$. Let 2 be synchronous, that is, it will return with the same phase ϕ_2. 1 came too early, gets a larger kick than 2, therefore moves to a larger orbit, hence arrives a little later next time, with a phase closer to ϕ_2. 3 came too late, gets a smaller kick than 2, moves to a smaller orbit, hence arrives a little earlier next time, with a phase *also* closer to ϕ_2. Thus the phases tend to bunch, the particles tend to synchronize.

After a bunch of particles has been delivered by progressively lowering ν, one has to raise this frequency again and go through the same procedure. Thus, unlike the cyclotron, the SC delivers particles in pulses. Evidently the

* See Chapter 17, Section (b), esp. Eq. (17.9). ** Others claim early awareness of this idea.[12]

usefulness of this accelerator depends not only on the energy it can reach but also on its time-average yield (current) of particles (down by a typical factor ~1000 compared to a cyclotron) and its duty cycle, the fraction of the time that a usable beam is available for research.

That is still not all. For a variety of reasons the orbits wiggle around their ideal path in their plane, and wobble in the direction vertical to that plane. As long as the frequencies of these various oscillations are lower than the revolution frequency of the particles then (it can be shown) the orbits can be held in check by a suitable weak decrease of B in the radial direction, a trick now called weak focussing. These frequency constraints are one of the two major factors that limit the energy which can be reached by an SC. The other is the weight of the magnet, which grows with the third power of the energy, and which, in the GeV region, would soon demand enough iron for building a battleship. The effective SC limit is about 700 MeV.

Berkeley's 184-inch machine, planned as a (single dee) cyclotron—which would not have worked!—was converted to an SC which operated at almost its first trial, and already in November 1946 produced 190 MeV deuterons and 380 MeV α-particles. It was later upgraded to 720 MeV. In that period several other SC's in the U.S. as well as one at CERN (600 MeV) and one at the Joint Institute for Nuclear Research in Dubna (680 MeV) also went into operation.

2. The weak focussing synchrotron. Principle: ν increases, B increases, approximately circular orbits. One again uses Eq. (19.1), now matches both ν and B to some m, and profits from phase stability. The rise of B is arranged so as to keep particles in orbits of almost constant radius instead of in the earlier spiral pattern. This has the major economic advantage of eliminating the huge magnet at the center of the earlier cyclotrons. Instead one surrounds a doughnut-shaped vacuum chamber by much smaller magnets. Unlike in earlier devices the particles must now be injected from the outside.

Since it is very difficult to control B precisely near zero field, due to small residual fields in iron and other magnet imperfections, one must inject particles pre-accelerated in a simpler apparatus. Thus injection as well as extraction of the beam cause new technical problems.

In 1946 the first application to electrons was made, in England.[13] In 1949 McMillan's electron synchrotron, conceived[11] in 1945, came in at 320 MeV. The first experiment[14] was the production of mesons by high-energy X-rays which are part of the so-called synchrotron radiation emitted by the electrons in their curved path. The attendant energy loss can be held low by judicious choice of machine parameters. The effect is proportional to $(E/m)^4/R$, where R is the radius of curvature of the orbit. By choosing R large, B low (~3k Gauss) and rf power high one has been able to reach energies ~20 GeV.

The first proton synchrotron was the Cosmotron (3 GeV) followed by similar accelerators in Birmingham (1953, 1 GeV), Berkeley (1954, 6 GeV), Dubna (1957, 10 GeV), Saclay (1958, 2 GeV), and later by others. Meanwhile a crucial new insight had opened the way toward reaching even much higher energies.

3. Strong focussing; the alternating gradient synchrotron (AGS). The abovementioned frequency constraints on the oscillations around the ideal orbits become harder and harder to maintain—and correspondingly beam intensity suffers—the higher the energy, that is, the higher the revolution frequency of the beam. Weak focussing does not work when the conditions are violated. In that case focussing the wiggles defocusses the wobbles and vice versa. In 1952 Courant, Livingston, and Snyder hit upon[15] a marvelous way, now called strong focussing,* of overcoming this problem. They found that wiggles and wobbles could both be damped by using a very high gradient of B which alternates in clever ways between horizontal and vertical directions. In other respects the AGS is similar to the weak focussing version (B and ν increase with time; orbits have a nearly constant radius). Strong focussing also has great economic advantages: the strong damping of the oscillations makes possible an even smaller doughnut cross section, hence smaller magnets and lower cost per GeV.**

Once again the new technique can be applied to electrons as well as protons. The first electron machine of this kind was completed at Cornell (1954, 1 GeV), followed by the CEA (MIT–Harvard, 1962, 6 GeV). Other such installations are in Hamburg (Desy, 1965, 7 GeV), Daresbury near Liverpool (Nina, 1967, 4 GeV), Erevan (Arus, 1967, 6 GeV), and Cornell (1967, 10 GeV).

The highest energy accelerators to date are the AG proton synchrotrons. The first two of these, at CERN and BNL, have energies about 30 GeV and were completed around 1960. By then, thought had already been given[17] to building accelerators going well beyond 100 GeV. These came into being in the 1970s, but not without political maneuvering which, in the United States, involved the U.S. Congress, Governors of States, and the White House. That story should be told by participants. I do wish, however, to note here some facts and dates in order to illustrate how physics has changed since the time when Becquerel, all by himself, discovered radioactivity with a bit of a uranium salt and a photographic plate as his only tools.

1963. The 'Ramsey panel', appointed jointly by the General Advisory Committee of the AEC and the President's Scientific Advisory Committee, releases its recommendation for the construction of a 200 GeV accelerator.

* Subsequently it turned out that this idea had been foreseen by Christofilos in an unpublished report.[16]

** For example the Cosmotron doughnut had a 6×26-inch cross-section and was surrounded by 8×8-ft magnets with 2000 tons total weight. For the Brookhaven AGS (33 GeV) these parameters are 3×7 inches, 3×3 feet, 4000 tons, respectively. Thus only twice the amount of iron was needed to reach a ten times higher energy.

1965. February: the Joint Committee on Atomic Energy of the U.S. Congress releases a report[18] on national policy in high energy physics.* March: hearings on the high energy physics program before a congressional committee. June: incorporation of a new consortium, the Universities Research Associates (URA) which presently comprises more than fifty institutions from the U.S. and Canada.

1966. From among more than 200 proposals, the Site Evaluation Committee of the National Academy of Sciences chooses the 6800-acre site which (since 1974) has been known as the Fermi National Accelerator Laboratory (FNAL).

1968. January: after authorization by the U.S. Congress for the AEC to proceed, a contract is signed between the AEC and the URA. June: Robert Wilson assumes his duties as director. December: ground is broken for the first building.

1972. The accelerator, completed within the 250-million-dollar construction budget, reaches 200 GeV.

The following incomplete list** of AG proton machines shows the highest energies (in GeV) reached at various times (the length in parentheses denotes the machine's approximate circumference).

CERN, 1959 (PS):	28	(0.8 km)
BNL, 1960 (AGS):	33	(0.8 km)
Serpukhov, 1967†:	76	(1.6 km)
FNAL††, 1972:	200	(6.4 km)
CERN, June 17, 1976 (SPS):	400	(6.4 km)
FNAL, May 14, 1976:	500	
1984:	800	

The last entry, now known as the Tevatron (earlier as the energy doubler/saver) is a project initiated by Wilson[19] and brought to its present stage under Leon Lederman, his successor. It is a new ring, installed directly below the 'old' ring, consisting of a doughnut surrounded by 990 superconducting magnets (the last one was emplaced on March 18, 1983) which are cooled by a refrigerator system capable of pumping 'a river of sub-cooled liquid helium at 4.5 K'[19] through the ring. The machine is planned to reach 1 TeV, or tera-electron volt, which is 1000 GeV or 1.6 erg.

As was mentioned earlier, to reach high energies it is necessary to inject pre-accelerated particles. That has become a complex sequence of events. One

* This report includes the Ramsey Panel report as well as reports of earlier similar Panels.

** Others: PPA at Princeton (3 GeV, 1962); Nimrod at Harwell (7 GeV, 1962); ZGS at Argonne (12.5 GeV, 1962), a variant (using 'edge focussing'), KEK in Japan (12 GeV, 1976).

† This accelerator was completed just in time to help celebrate the 50th anniversary of the October Revolution.

†† The FNAL and SPS machines use dipole magnets for bending the beam, quadrupole magnets for focussing it.

generation's highest energy accelerator has become the next generation's injector. This is what happens at the Tevatron. A Cockcroft–Walton accelerates protons from an ion source to 0.75 MeV—just the energy attained forty years earlier by Cockcroft and Walton in their first experiment on nuclear transmutation. These protons are injected into a linear accelerator which brings them to 200 MeV. Next follows injection into an 8 GeV synchrotron, whereafter the protons enter the 'old' ring, tuned down to 150 GeV. Finally the beam enters the new ring. Already now there are plans to use the Tevatron as an injector for future projects.

This rapid glance at the growth of accelerators does not remotely suffice to convey the complexities of the developments which have led to the modern high energy laboratories. To do so would demand all by itself a volume comparable in size to the present one. I must content myself here with a few scattered additional remarks.

It cannot be emphasized enough that bigness is inevitable but precision is the key. In 1960, one of the AG machine-builders wrote: 'Particular attention must be given to the placement of the [magnet] units . . . Root mean square errors in position may not exceed about 0.5 mm in machines such as for 25 GeV . . . Foundations are required to be stable to a degree previously unheard of in the architectural profession'.[20] Even with all due regard for precision, the coaxing of a high energy beam through a doughnut remains part craft, part art. There does exist a vast body of mathematical analysis of stability against wiggles and wobbles. The problem is so delicate and complex, however, that most often unforeseen perturbations arise during the first test runs of a new machine. To subdue those—there lies the art.

Accelerators produce a mixed bag of particles. To separate the various species becomes increasingly difficult with increasing energies. Separation devices can reach lengths of 1 km or more.

As the energy increases, the task of shielding against radiations dangerous to people and disturbing to experiments becomes ever more demanding. That problem, negligible when Lawrence began, has long since required that accelerator rings be buried in underground tunnels.

As the machines evolved, so did the detectors needed for coping with their outpour. The first new type of post-war detector would have pleased Rutherford both because of its modest size and its use of scintillations. In his days, scintillations were observed in solids of a kind which rapidly absorb the light flashes. When it was found that certain organic crystals (like naphthalene, anthracene) were transparent to their own fluorescent radiation, one was able to use a large amount of such substances and send the radiation into a photomultiplier. Thus was the scintillation counter created.[20a] Cerenkov counters and diffusion cloud chambers came next. Much larger and far more complex new detectors were to follow. More on that in Section (e) below.

(b) Home-made pions

When in May 1947 the Powell group announced their discovery of pions in cosmic rays* they did not mention that these new particles had already since November 1946 been produced in the laboratory, to wit, by the α-particles accelerated to 380 MeV in Berkeley. Nor was that fact discussed at the Shelter Island conference in June 1947. The reason for this silence is simple: pions were being produced but no one knew that yet. So it remained until March 1948. There are at least four reasons why it took nearly a year and a half before pions were seen.

First, as Serber told me, in the true Lawrence style,** the Berkeley group had concentrated mostly on readying the SC and had paid little attention to the building of detection apparatus.

Secondly, even if they had had detectors, they would have needed time to become familiar with phenomena in a newly accessible energy regime. This is a perennial problem in the progress of experimental particle physics. A new machine is an indolent monster without the equipment for analyzing what it produces. Nearly every advance in energy poses new detector problems.

Thirdly, while one can do certain experiments inside the machine it is obviously preferable, for reasons of background and flexibility, to get the beam out, another problem which demands special devices for nearly every new accelerator. For the Berkeley SC all this was not in hand[21] until well into 1948.

Finally there was a qualitative theoretical point that remained to be clarified. A 380 MeV α-particle actually does not have enough energy (95 MeV per nucleon) for producing a pion upon hitting a nucleon at rest. When it collides with a sufficiently heavy nucleus, however, the internal kinetic energy of a nucleon inside it suffices for pion production. This effect of what is now called Fermi motion, was not diagnosed until mid-1947, by W. G. McMillan and Teller.[22] Serber told me how, at about that time, Teller came to ask him whether the Berkeley machine should not produce pions. Serber replied that he was sure it did but that no one yet knew how to find them.

Thereupon Eugene Gardner betook himself to stick a carbon target as well as a stack of photographic emulsions inside the SC, hoping to find pion tracks similar to those seen by the Powell group. Still no pions. Then, early in 1948 Lattes, a member of the Bristol group and thus familiar with what to look for, came to Berkeley. In no time pions were found[23] in copious numbers. The first account I heard of this discovery was Serber's talk on March 30, 1948, the opening item on the agenda of the Pocono Conference.

These early experiments were not yet quantitatively accurate. For example the pion mass was quoted to be $313 \pm 16m$ (m = electron mass). Comparison

* Chapter 18, Section (b). ** Chapter 17, Section (b).

with the best value[24] of that period from cosmic rays: $283 \pm 7m$ and with the present best value: about $273m$ provides but one illustration that the accelerator physicists were not quite ready yet to take the lead, even though they had vastly more pions to play with (by a typical factor $\sim 10^8$). That changeover occurred in 1950 when, for the first time, a new particle was discovered in an accelerator experiment.

That was the neutral pion, π^0. Recall* that already in 1940 it had been proposed that this particle should decay as

$$\pi^0 \rightarrow \gamma + \gamma \tag{19.3}$$

with a lifetime $\sim 10^{-16}$ sec. Shortly after the war it was noted[25] that short-lived neutral mesons might possibly help explain large cosmic air showers. Next, in 1949, a Berkeley SC experiment[26] on high energy photon production by proton bombardment showed that the photons' spectral and angular distribution could not be due to *Bremsstrahlung*. The production of a neutral meson with mass $\sim 300m$, halflife $< 10^{-11}$ sec, decaying by Eq. (19.3) would fit all the data, however. The problem was clinched[27] in April 1950 when Steinberger, Panofsky, and Steller, using X-rays from the Berkeley electron synchrotron as a source, observed $\gamma\gamma$-coincidences which could be ascribed to π^0-production followed by 2γ-decay with a lifetime $<10^{-13}$ sec. Later that year[28] the discovery of the reaction

$$\pi^- + \text{p} \rightarrow \pi^0 + \text{n} \tag{19.4}$$

made it possible to deduce a charged pion mass of $275.2 \pm 2.5m$ and a π^0-mass smaller by $10.6 \pm 2m$ (present value $9m$). The isotriplet predicted in 1938** had been found. Pion physics was now in the hands of the accelerator physicists for good. In later times it has become a field of specialization for 'meson factories', accelerators which produce pions with intensities typically $\sim 10^3$–10^4 times larger than all-purpose machines.

By 1950, times were ripe for determining the spins and parities of pions. Note that Eq. (19.3) implies[29] even spin for the π^0. It was pointed out[27] in 1950 that this fact, combined with data on X-ray-produced charged and neutral pions *and* with the assumption that $\pi^\pm$ and π^0 have the same spin-parity, permitted the conclusion that pions are pseudoscalar: zero spin, odd parity. That last assumption is of course implied by isospin. At that time isospin invariance was not yet[30,31] as universally an article of faith, however, as it would be shortly thereafter. Accordingly—which was just as well—one looked for independent ways of determining the spin and parity of charged and of neutral pions.

* Chapter 17, Section (g), Part 4. ** Chapter 17, Section (g), Part 4.

The case of the charged pion spin provides a good example of the style of reasoning.* Consider the reactions

$$p+p \rightarrow \pi^{+}+d \tag{19.5}$$

$$\pi^{+}+d \rightarrow p+p \tag{19.6}$$

with respective cross sections σ_1, σ_2. Since the probabilities of the two processes are the same under identical kinematical conditions (detailed balancing) one has[30,31]

$$\frac{\sigma_1}{\sigma_2}=\frac{3(2S+1)}{2}\frac{k^2}{K^2} \tag{19.7}$$

where K (k) is the momentum of the protons (pion) in either reaction and S the pion spin. Eq. (19.7) is *independent of any details of pion dynamics*; no knowledge whatever of learned quantum field theory equations is called for. The moral, most important for what follows later in this chapter, is that one can learn a lot about pions from simple general principles, in this instance detailed balancing and other simple rules of ordinary quantum mechanics.

Experiments[33] completed in June 1951 showed that the $\pi^{\pm}$ has zero spin. For other pretty arguments and experiments which settled the pseudoscalar nature of all pions, I refer the reader to textbooks.[34]

(c) In which meson field theories fall upon hard times

1. Nuclear forces. In 1948, the physics of particles and fields looked most promising. Schwinger had just announced his theory of the anomalous electron moment as the year began. Artificially produced pions had been detected for the first time in March. Agreement on the theoretical value of the Lamb shift had been reached in the autumn. Experimentalists were obviously eager to do pion physics with their new machines. Their efforts were to be richly rewarded. Equally obviously, theorists were ready to attack pion physics with the new tools that had scored such impressive successes in quantum electrodynamics. They, however, were soon in deep trouble.

As we have seen,** meson theory originated from attempts to understand nuclear forces in terms of fields, the main early data to be interpreted being deuteron properties and low energy nucleon–nucleon scattering. Some qualitative insights had emerged. The short-range nature, strengths, and spin dependence of the forces could be incorporated. The deuteron's cigar shape, manifested by its quadrupole moment, demanded a non-central force

* For early attempts to fix meson spins from cosmic ray data see Ref. 32.
** Chapter 17, Section (g).

component. That too could be explained, though less easily,* by a particular choice of theory: a pseudoscalar isotriplet $\vec{\pi}$ of meson fields with pseudoscalar coupling** H_{int} to nucleons:

$$H_{\text{int}} = ig \int \bar{\psi}\vec{\tau}\gamma_5\psi\vec{\pi}\, \mathrm{d}\vec{x}. \tag{19.8}$$

Here the operators (π^+, π^-, π^0) are given by

$$\pi^+ = \frac{\pi_1 + i\pi_2}{\sqrt{2}}, \qquad \pi^- = \frac{\pi_1 - i\pi_2}{\sqrt{2}}, \qquad \pi^0 = \pi_3. \tag{19.9}$$

(π^+, π^-) create (π^+, π^-)—and annihilate (π^-, π^+)—mesons; π_3 contains both creation and annihilation terms for neutral pions. Thus the interaction (19.8) was looked upon with some favor already before the pion spin-parity was known, in fact even before the pion had been discovered.

In spite of these promising qualitative features, the meson theory of nuclear forces was not at all a comfortable subject in those early years, even after the confusion stemming from the ignorance about two kinds of mesons, π and μ, had been resolved. The main cause of worry concerned the magnitude of g. It is readily seen that Eq. (19.8) implies an interaction between nucleons essentially proportional to their relative velocity/c, a ratio which is small ($<1/10$) in the deuteron. Therefore g must be large for the forces to be sufficiently strong—very large. Estimates in the 1940s gave

$$\frac{g^2}{\hbar c} \text{ roughly equals } 15. \tag{19.10}$$

(In 1949 Bethe quoted[36] a value of 40.) That is very bad. Consider the logic. Eq. (19.10) was derived from Dirac's good old perturbation formula (15.33), that is, on the assumption that nuclear forces are due to virtual exchange of single mesons. That assumption was based on the analogy with the Coulomb force, which is due to virtual single photon exchange proportional to (in dimensionless units)

$$\alpha = \frac{e^2}{\hbar c} \simeq \frac{1}{137}. \tag{19.11}$$

Now (as we have seen) even before the days of renormalization it was intuitively clear that the numerous successes of quantum electrodynamics to order α had to imply that multiple virtual photon effects ought to be small, and that this

* Non-central forces derived from pseudo-scalar theory have the right sign but suffer from a $1/r^3$-singularity which is inadmissible in quantum mechanics. That complication was coped with either by arbitrarily cutting off the r^{-3} force at some finite distance or by mixing pseudo-scalar with vector mesons.[35]

** See Eqs. (17.52), (17.53), where a similar structure was derived for the scalar case.

smallness was somehow related to the smallness of α. By sharpest contrast, if one believed Eq. (19.10), obtained in second order in g, then higher-order effects had to be so huge as to render meaningless the very concept of 'higher order': *The Maxwell–Yukawa analogy breaks down*. The least one could say is that perturbation theory in powers of $g^2/\hbar c$ made no sense. The best one could hope for was to save the theory by using non-perturbative methods which, however, have not been found to this day.

Thus began a quarter-century of uncertainty about the dynamics of strong interactions. That question is still not fully resolved today, yet we now have excellent reasons for believing that the Yukawa meson fields are not as fundamental as the electromagnetic field, a subject to which I return in Chapter 21. As to the 1940s, the mood is perhaps best summarized by the quotation from Macbeth chosen by Rosenfeld[37] as an epigraph for his report 'Problems in nuclear forces' to the Solvay conference in the fall of 1948: 'Such welcome and unwelcome things at once/'Tis hard to reconcile'.
Also in later years nuclear forces never became an inspiring source of information about fundamental interactions, in spite of heroic efforts. The nuclear two-body problem is just too complicated.

2. The proton and neutron magnetic moments. As early as 1938 it had been noted[38] that the virtual sequence

$$\text{nucleon} \rightarrow \text{nucleon} + \text{pion} \rightarrow \text{nucleon} \tag{19.12}$$

provides a Wick mechanism* for the anomalous magnetic moment of nucleons, an idea which recurred[39] during the next few years. It became an obvious topic for further pursuit after the successful evaluation of the electron's anomalous moment. By then an effect had been discovered which likewise ought to have an explanation in terms of Eq. (19.12): the electron–neutron interaction, another novelty at Shelter Island where Rabi reported on the first experimental evidence.[40]

Between late 1948 and the middle of 1949 at least six papers appeared[41] reporting on second-order calculations of nucleon moments (some also including a treatment of electron–neutron scattering) by the new methods. The essentially identical results were terrible. I confine myself to the answers for the interaction (19.8). Once again, the neutron and proton moments demanded huge values for $g^2/\hbar c$. Much worse, it was impossible to fit the two moments with the same value of g. The neutron demanded that $g^2/\hbar c$ equal 7, the proton 52.[42] While through the years the radiative corrections to the electron moment have proved successful at least to order α^4, the corresponding meson calculations were already in 1949 a flop to order g^2.

* Chapter 17, Section (g), Part 1.

3. Meson dynamics and renormalizability. At about the same time that these second-order calculations were in progress, the question was raised as to whether the general renormalization theory of quantum electrodynamics also applied to the variety of possible meson field theories. Answer: yes for a few, no for most.

The first partial result, due to Case,[43] was that a neutral vector meson theory with tensor interaction* has additional infinities beyond mass and charge. This conclusion, he noted, remains true for zero meson mass. Thus quantum electrodynamics with an interaction $f\bar{\psi}\sigma_{\mu\nu}\psi F_{\mu\nu}$ in addition to $j_\mu A_\mu$ is more viciously divergent for $f \neq 0$ than for $f = 0$. In fact, as was soon to be clear, for $f \neq 0$ the theory is unrenormalizable.**

General renormalization properties of meson theories, first stated by Matthews,[44] can be summarized as follows:

(a) All vector and pseudovector theories involving charged mesons are unrenormalizable in the presence of either or both g- and f-coupling; see Eqs. (17.47–50). Neutral vector meson theory with g-coupling only is renormalizable.

(b) Scalar and pseudoscalar theories involving charged mesons are unrenormalizable in the presence of f-coupling; see Eqs. (17.44, 45, 49).

(c) Scalar and pseudoscalar theories with g-couplings only are renormalizable. One and only one further renormalization is necessary, however, in addition to those for mass, coupling constant,† and wave function. The amplitude for meson–meson scattering exhibits a new infinity (not present in the corresponding light by light scattering because of gauge invariance[45]) and must accordingly be treated as an additional phenomenological parameter (at some fixed energy).††

As to the symmetrical pseudoscalar theory with the coupling Eq. (19.8) the fact that it turned out to be one of the few renormalizable theories was of no help at all. The mathematical property of renormalizability is (to repeat one last time) of no physical use when the 'expansion' parameter $g^2/\hbar c$ is large. In Fermi's words:[47] '[Pseudoscalar theory with pseudovector coupling] is not renormalizable, as [pseudoscalar theory with pseudoscalar coupling] is, and it is therefore frowned upon by the high priests of field theory. There is a hope somewhere between ~5 and 95% that [the latter theory] is the truth'.

These phrases are found in Fermi's last (posthumous) publication, notes of his lectures delivered in the summer of 1954 at a school since named after

* Eq. (17.47) with real (self-adjoint) A_μ and $f \neq 0$.

** That term was defined in Chapter 18, Section (c).

† Coupling constant renormalization is the more fitting name used for what is called charge renormalization in quantum electrodynamics.

†† In a neutral scalar meson theory the three-meson vertex demands still a further renormalization.[46]

him. When, a few months later, he died of stomach cancer, just 53 years old, physics lost one of its great twentieth-century figures, a man at home in the instrument shop, in the laboratory, and in theoretical physics. Pion–nucleon scattering was the last of the very many subjects spanning the spectrum of physics to which he made seminal contributions. His teaching inspired two generations on two continents. 'On a heroic scale was his acceptance of death.'[48]

4. *Free meson processes.* The final part of this litany. During the late 30s and all through the 40s, considerable effort had gone into second-order perturbation calculations of the production, scattering, and absorption of pions. Marshak's excellent and extensive review[49] of this work was completed when the stream of experimental results had barely begun to come in. A book[34] based on lectures by Bethe is the best source of comparison between these calculations and experiments up to the end of 1954. I cite some representative conclusions concerning data for pion laboratory energies $\lesssim 100$ MeV.

The energy dependence for $\pi^+ p \to \pi^+ p$ scattering is 'hopelessly wrong'. The cross-section ratio $(\pi^+ p \to \pi^+ p/(\pi^- p \to \pi^- p)$, expected to be $\simeq 1$ at 50–100 MeV, is found to be $\simeq 7$ at 100 MeV. For $(\pi^- p \to \pi^0 n)/(\pi^- p \to \pi^- p)$: expected ~2–3 per cent, observed ~2 at 100 MeV.[50] Angular distributions for scattering do not fit.[51] Photomeson production is likewise full of contradictions.[52]

Perturbative meson field theory was wrong wherever one looked. New theoretical strategies had to be devised. The remainder of this chapter includes responses in the fifties and early sixties to this awkward situation.

(d) Symmetry saves, up to a point

1. *Isospin as a free-floating invariance. A new spectroscopy.* When I met Fermi on the evening preceding the second Rochester conference (January 11–12, 1952), he told me with enthusiasm of the work he and his collaborators were then engaged in at the 450 MeV Chicago synchrocyclotron: the pioneering pion–nucleon scattering experiments.[53] He was particularly intrigued by indications that at their highest pion energies (~140 MeV) the scattering appeared to proceed predominantly in the isospin $T = 3/2$ state. When I asked him how he knew that, he replied that the cross-sections for the processes $\pi^+ p \to \pi^+ p(\sigma_1)$, $\pi^- p \to \pi^0 n(\sigma_2)$, and $\pi^- p \to \pi^- p(\sigma_3)$ stood in ratios close to

$$\sigma_1 : \sigma_2 : \sigma_3 = 9:2:1 \tag{19.13}$$

and that these numbers follow from isospin considerations alone as long as $T = 3/2$ dominates. I excused myself shortly afterward and went to my room to check this statement.

Since a nucleon (N) has $T = 1/2$ and a pion $T = 1$, a πN system has $T = 1/2$ or 3/2 by angular momentum composition. Since isospin is conserved (electromagnetic effects are dropped) any πN-scattering amplitude A is a linear combination of transitions $1/2 \to 1/2$ (amplitude $A(1/2)$) and $3/2 \to 3/2$ ($A(3/2)$). The amplitudes for the three processes just mentioned turn out to be, in that order

$$\begin{aligned} A_1 &= A(3/2) \\ A_2 &= \sqrt{2}/3(A(3/2) - A(1/2)) \\ A_3 &= 1/3(A(3/2) + 2A(1/2)). \end{aligned} \tag{19.14}$$

I doubt whether I knew then that these relations had first been given by Heitler[54] in 1946, when it was too early to do much with them. Now the Chicago experiments had brought them to life: drop $A(1/2)$, remember that each σ is proportional to its $|A|^2$, and find Eq. (19.13).

When, in the late thirties, Kemmer incorporated isospin into meson theory he had insisted that the consequences of isospin invariance hold regardless of dynamical approximations.* In those days calculations were so dominated by second-order perturbation theory, however, that the real generality of isospin arguments receded into the background of, I dare say, nearly everyone's mind. All that changed in 1951, when Berkeley data on π-production in N–N and N-nucleus collisions provided an incentive for analyzing these reactions by isospin methods, a program initiated by Watson and Brueckner.[55] Discussing this work with Watson I suggested that isospin might constrain the average numbers $\bar{n}^+$, $\bar{n}^-$, $\bar{n}^0$ of created pions with charges as indicated. He soon produced[56] the 'Watson theorem'

$$\bar{n}^+ + \bar{n}^- = 2\bar{n}^0 \tag{19.15}$$

for any nucleon–nucleon collision process. This relation was one of the first exact predictions of isospin, independent of the dominance of any particular state, as is the case for Eq. (19.13).

In 1952, papers appeared in rapid succession on isospin in single pion production,[57] in multiple pion production,[58] and in nuclear physics.[59]** I believe that this widespread interest stemmed to a large extent from the discussion (to which I return) of Eq. (19.13) at the second Rochester conference.

At least as important as the experimental stimulus for this focus on isospin was the disarray in regard to meson theory. Relations like Eqs. (19.13), (19.15) came as a blessing; theorists had at least something to offer their experimental colleagues. Moreover, since isospin has nothing to do with perturbation theory,

* Chapter 17, Section (g), Part 4.

** A review of isospin in nuclear physics is found in Ref. 60.

it could serve as a reliable guide to *what* needed explanation by alternative theoretical methods. Isospin does not, of course, suffice to inform *how* it should be explained that, for example, a certain state dominates at a certain energy, as in the Chicago experiments. Symmetry saves, but only up to a point.

The new trend in particle theory, which began to emerge in the early 50s as the result of the experiences with isospin, is perhaps best explained by comparing Lorentz invariance as abstracted from Maxwell–Lorentz theory with isospin invariance as abstracted from meson theory. In the former case the equations which had revealed the invariance were treated with new respect. They became easier to interpret, simpler to handle. In the meson case, on the other hand, isospin survived even when the equations which had served as a means of abstraction became highly dubious. Isospin invariance, first introduced in phenomenological nuclear potentials, then built into Yukawa equations, became a free-floating invariance in search of a badly needed appropriate dynamics. That type of development was to recur in subsequent years: start with dynamical equations, abstract some of their general features, forget the starting point.

As to πN-scattering, later experiments showed[61] that the cross-sections peak strongly at ~ 180 MeV, where the scattering is nearly pure $T = 3/2$, after which they drop, another sure sign of the failure of perturbation theory, which cannot account for such peaks. In addition, information was coming in about the angular dependence at the peak. As the Chicago group had noted[53] this distribution should be isotropic or behave as $1 + 3\cos^2\theta$ if an angular momentum $J = 1/2$ or $3/2$ state dominates. The latter turned out to be close to the truth.[62] Thus at the peak, scattering is predominantly in a state with $T = J = 3/2$, later called the 33-resonance and denoted by Δ, a multiplet with charges (check the πN combinations)

$$\Delta^{++}, \Delta^{+}, \Delta^{0}, \Delta^{-}. \tag{19.16}$$

This brief account of the discovery of Δ and its properties is grossly oversimplified. Detailed analysis demands knowledge of phase shifts for scattering in various (T, J)-states, a problem which initially caused a fair amount of argument concerning possible ambiguities.[63] Yet it was soon agreed that the (3/2, 3/2)-phase dominates at the peak where it reaches its *resonance** value of 90°.

The role of Δ in πN-scattering can be represented as a (real, not virtual) formation: $\pi + \mathrm{N} \rightarrow \Delta$, followed by a decay: $\Delta \rightarrow \mathrm{N} + \pi$. The state Δ can be assigned not only a T, a J, and a parity (even), but also a mass ~ 1230 MeV/c^2 and a lifetime $\sim 10^{-23}$ sec, as follows respectively from the position and width (~ 115 MeV) of the peak. Thus Δ has all the attributes of an unstable particle such as the neutron except for the quantitative difference of being exceedingly

* See standard texts for resonance in scattering processes.[64]

short-lived. It took quite a few years before physicists became comfortable with the idea that there is no real difference between a resonance and an unstable particle.*

2. Antinucleons; nucleon conservation; charge and mass formulae; violation rules. The Berkeley bevatron was designed for accelerating protons up to 6 GeV laboratory kinetic energy, more than the 6 Mc^2 (M = nucleon mass) necessary for producing antiprotons ($\bar{p}$) by reactions such as

$$p+p \rightarrow p+p+p+\bar{p}. \tag{19.17}$$

All went according to plan. The $\bar{p}$ was found[66] in 1955 and in 1956 served as a source for discovering[67] the antineutron ($\bar{n}$) via the reaction

$$p+\bar{p} \rightarrow n+\bar{n}. \tag{19.18}$$

Assign a 'nucleon number', call it B, equal to +1 (−1), to nucleons (antinucleons). Eqs. (19.17), (19.18) satisfy the rule that the sum of B's on the left equals that on the right. That rule is the conservation of nucleons.

A faint first glimpse of this conservation principle appeared in 1929 when Weyl, conjecturing** that the negative energy states of the Dirac equation were somehow associated with protons, stated: 'There should be two conservation laws of electricity which state (upon quantization) that the numbers of electrons as well as protons remain constant'.[68] Weyl's world only contained protons, negative electrons, and photons as fundamental particles, so that the physical content of nucleon conservation: *protons are stable*, was at that time true in any case. It was Stückelberg[69] who, in 1938, formulated the principle in a more realistic setting: 'Apart from . . . the conservation law of electricity there exists evidently a further conservation law: No transmutations of heavy particles (neutron and proton) into light particles (electron and neutrino) have yet been observed in any transformation of matter. We shall therefore demand a conservation law of heavy charge'.†

The apparent similarities between the additive conservation laws of electric charge and nucleon number noted by Weyl and Stückelberg led to speculations[71] that there might exist a neutral vector field, similar to the electromagnetic potentials, which associates nucleon conservation with a local gauge principle. Those ideas never got very far. In fact it is now believed that nucleon conservation may not strictly hold (see Chapter 22). If such violations exist, they are bound to be so weak, however, that they will not measurably affect conclusions concerning strong and electromagnetic couplings, where B may be treated as a quantum number as good as the charge quantum number Q.

* See especially the discussion following a paper presented by Anderson[65] in 1980.
** Chapter 15, Section (f).
† A similar formulation was given later by Wigner.[70]

Note that antinucleons have $T=1/2$, while Eqs. (19.17), (19.18) imply $T_3=+1/2$ $(-1/2)$ for $\bar{n}$ ($\bar{p}$). Hence

$$Q = T_3 + \frac{B}{2}, \qquad (T_3 = -T, -T+1, \ldots, +T) \tag{19.19}$$

applies to nucleons ($B=1$, $T=1/2$), antinucleons ($B=-1$, $T=1/2$), Δ ($B=1$, $T=3/2$), and pions ($B=0$, $T=1$).

Since B is an isoscalar, Eq. (19.19) implies a specific 'violation rule' for the way electromagnetic interactions disobey rigorous isospin invariance: they are a coherent mixture of isovector and isoscalar couplings. This knowledge suffices to draw important consequences from isospin even in the presence of isospin violation. The key is that, to a good approximation, one is allowed to treat electromagnetic effects to lowest non-vanishing order in e. *Exercise 1*: With only that last restriction, single photon transitions between two nuclei are possible only[72] if their isospin difference $=0, \pm 1$. *Exercise 2*: with only that last restriction, the ratio of cross sections for $\gamma p \to p\pi^0$ and $\gamma p \to n\pi^+$ equals 2 if $T=3/2$ dominates, a result which was important[73] for diagnosing the Δ. *Exercise 3*: To order e^2 the masses $m(T_3)$ in a given isomultiplet are split according to

$$m(T_3) = \alpha T_3^2 + \beta T_3 + \gamma, \tag{19.20}$$

a mass formula first used in particle physics by Weinberg and Treiman.[74]

Isospin symmetry saves, up to a point. In particular its predictive power decreases with increasing numbers of particles participating in a reaction. $\bar{p}p$ annihilation into n pions is a case in point. Consider for example $n=6$, not at all an extravagant number. Even if one of the two possible isospin transitions ($0\to 0$ or $1 \to 1$) dominates, we still know little about the 6π-system since it can be in any of 15 (36) linearly independent states with $T=0$ (1). The situation is slightly improved because, for given n, large blocks of these states have the same branching ratios into the various possible charge partitions. This makes it possible[75] to set bounds on the number of charged prongs for given n. Likewise[76] for e^+e^--annihilation via a single photon, $e^+e^- \to \gamma \to n\pi$. Multi-pion systems are also of some use for observing Bose statistics at work: pions of the same charge like to stay closer together than those of unlike charge.[77]

3. G-parity. In 1952 Jost and I noted[78] a new selection rule based on the validity of isospin and charge conjugation but otherwise free of assumptions. In its simplest, though not its most general, form it says that an even number of neutral pions cannot go into an odd number and vice versa. Our attention was drawn to this problem by several earlier papers[79] in which related results had been found in lowest non-vanishing order of perturbation theory.

The point is elementary. Under charge conjugation* C: $\pi_3 \rightarrow \pi_3$, $\pi^+ \rightarrow \pi^-$ or $\pi_2 \leftrightarrow -\pi_2$ by Eq. (19.9). Thus C is a reflection in the isotopic 13-plane. Rotate by 180° around the isotopic 1-axis (operation R_1). Then, again by (19.9), $\pi^+ \leftrightarrow \pi^-$ and $\pi_3 \rightarrow -\pi_3$. So the product CR_1 sends π_3 into $-\pi_3$ and the mentioned result follows. The neutral pion rule combined with charge conservation implies that, for whatever charge, an even number cannot go into an odd number. Michel next noted[80] that this can also be formulated as invariance under C times R_2, a 180° rotation around the isotopic 2-axis, which yields $\pi \rightarrow -\pi$ for all charges. He applied CR_2 to nucleon–antinucleon annihilation from states with definite spin-parity. So did Lee and Yang[81] who gave the operation CR_2 its present name: G-parity.

(e) Of a new spectroscopy and new detectors

I have dwelt at length on the 33-resonance because, in 1952, it played such a crucial role in changing the thinking on pion physics. It is now known that this resonance is but one of a horde. (Note: Strange particles are left for the next chapter.)

π-nucleon scattering experiments rapidly moved to higher energies. Already in late 1953, Cool, Madansky, and Piccioni reported[82] on data up to 1.5 GeV obtained at the Cosmotron.** These showed evidence for a new scattering peak, at ~1 GeV, $T=\frac{1}{2}$. Thus began the hunt for higher nucleon resonances, an ongoing process.† At the last count this new spectroscopy has yielded some 25 $T=\frac{1}{2}$ resonances and a similar number with $T=\frac{3}{2}$. J-values range upward to $\frac{11}{2}$, masses go beyond 3 GeV. The analysis of high mass resonances is complicated by the low relative weight at high energies of states with any given J and parity, and by inelastic processes. One example: at ~1900 MeV there is a $T=\frac{3}{2}$, $J=\frac{5}{2}$ resonance which decays with 80 per cent probability into $2\pi+\mathrm{N}$. To this day there does not exist a theory remotely precise enough to account quantitatively for this wealth of spectral lines. Models must suffice.[85]

Yet another chapter of the new spectroscopy began in 1961 with the discoveries in rapid succession of three meson resonances. First[86] came the ρ-meson ($T=J=1$) seen as a peak with mass $\mathrm{M}\simeq 770$ MeV and width $\Gamma \simeq 150$ MeV in the $(\pi^+\pi^-)$ distribution produced in the reaction $\pi+\mathrm{N}\rightarrow 2\pi+\mathrm{N}$. (The invariant definition of the rest mass M is:

$$M^2c^4=(E_1+E_2)^2-c^2(\vec{p}_1+\vec{p}_2)^2 \tag{19.21}$$

where the E's and $\vec{p}$'s are the respective energy–momentum of the pions observed in the laboratory.) Then followed[87] the ω-meson ($T=0$, $J=1$, $M\simeq 783$, $\Gamma\simeq 10$) first seen as a 3π-peak in $\bar{\mathrm{p}}\mathrm{p}\rightarrow 5\pi$; finally there was[88] the

* Chapter 16, Section (d), Part 3. ** For a tabulation of data up to late 1954 see Ref. 83.
† For a recent review see Ref. 84.

η ($T = J = 0$; $M \simeq 549$, $\Gamma \simeq 1$ keV), a 3π-peak in $\pi^+ d \to p + p + \pi^+ + \pi^- + \pi^0$. All these discoveries were made possible by a major advance in detection technique: Glaser's invention of the bubble chamber.

Roughly his idea was this. A liquid enclosed in a vessel is put under pressure and heated above its normal boiling point. Upon expansion the liquid becomes thermodynamically unstable against the formation of vapor bubbles which can be nucleated by the passage of an electrically charged particle. Such particles leave photographable tracks. Glaser proposed[89] this idea in 1952; a year later he published[90] his first pictures of tracks made in a two-cubic-centimeter chamber.* The bubble chamber is an imaging detector, like the cloud chamber. It has enormous advantages over the latter, however, in respect of sensitivity, rapid cycling, spatial resolution, and high stopping power. For accelerator physics it has the particular advantage that the times of beam arrival and of chamber expansion can be matched.

The first high energy bubble chamber experiment[92] was published in 1957, at the Cosmotron. Liquid propane served as fluid. It was obvious to all concerned, however, that one should aim for liquid hydrogen and deuterium, the ideal targets for studying fundamental interactions. The first small prototype of this kind which produced tracks, completed[93] in 1954, was built in Berkeley where Alvarez, used to thinking big as a result of his wartime experiences with radar and the atomic bomb, had decided upon a large-scale chamber. In 1955 a proposal was drafted for a 500 litre hydrogen chamber, later known as 'the 72-inch'. The planning for a detector that large was based on the sound physical argument that it would make possible the observation, in sufficient numbers, of the production and subsequent decay of the newly discovered long-lived particles (see the next chapter). Four years, and two and a half million dollars, later, the chamber was operational, complete with elaborate cryogenics for keeping hydrogen at −250°C, a 15 000 gauss magnetic field for bending tracks, its own separate building, and its elaborate safety precautions. (Elsewhere, some time later, an exploding hydrogen chamber caused a death.)

By that time the Berkeley group was also prepared for applying new techniques to the scanning of their photographs and the reduction of their data. As Alvarez had stressed[94] already in 1955: '[Data analysis] is the bottleneck in cloud chamber and emulsion work all over the world . . . The rate at which interesting events will occur in the large bubble chamber is . . . very much greater. Certainly, one day of bubble chamber operation could keep a group of "cloud chamber analysts" busy for a year'. Thus, he concluded, track scanning and analysis needed to become semi-automatic; computers should be used for analysis and storage of data. And that is exactly what happened. Such were the techniques which led to the discovery of the ρ-meson in BNL

* For a review of the early years see Ref. 91.

hydrogen bubble chamber exposures, and of the ω and η in pictures from the 72-inch.

I well recall the initial reservations of quite a few members of the profession in regard to Alvarez' managerial approach which, they felt, would remove physicists yet another step from experimentation done in the old ways. They were right and so was Alvarez. It came to be recognized that either one accepts the new style as indispensable, or particle physics will languish. How else could the Berkeley group alone have measured 1.5 million events in 1968?

Large accelerators had changed the style of doing experiments. Large, costly, and essentially immovable detectors changed the style once again. In many instances one could no longer bring along home-built detection equipment for experimentation at an accelerator but had to use permanent on-site detectors instead.

With increasing energies, an ever greater number of possible final states opened up, all recorded somewhere on the exposed film. Since one can therefore ask many questions of the same roll of film, either at once or later, these records had to be kept as a library: 'An event catalog becomes an important part of the data analysis scheme . . . whenever necessary, the master list is updated with new scanning information, measurement requests, modifications or results'.[95] One could borrow film from the library and take it home for analysis, using scanning devices and computers available at one's own or nearby institutions. That is how, for example, the η was discovered at Johns Hopkins in 100 000 pictures borrowed from Berkeley.

High energy experiments, never easy, never cheap, were becoming ever more complex and expensive—which brings me to the final entry in the list of changes in style during the fifties. Earlier it had been necessary to form consortia in order to build accelerators. Now one had to form consortiums in order to finance, perform, and analyze experiments. An early example is the case of the seven European institutions which, in 1958, jointly undertook the analysis of tracks in a stack of emulsions after a group from Berkeley had seen to its exposure at the bevatron.[96]

As to the meson resonances: by now over forty such isomultiplets are known. As to bubble chambers: even bigger ones followed. Gargamelle, heavy liquid, 12 000 litres, was completed in 1970; BEBC, the Big European Bubble Chamber, hydrogen, 35 000 litres, in 1975; both at CERN. As to detectors, many new types are now in use: spark chambers; multi-wire proportional chambers; drift chambers; streamer chambers; others. At this point I desist and direct the reader to the literature.[97]

(f) What use quantum field theory? The years of uncertainty

1. The age of diversity. I return to quantum field theory in the early fifties when, as we have seen, electrodynamics looked better and meson theory less tractable than they had ever done before. The remainder of this chapter is devoted to an account of the theorists' efforts to get on with the meson problems.

It was an unusual situation. As I try to illustrate the status of pion dynamics at that time with a parallel to earlier events in physics, I find myself unable to produce one. Perhaps there is none. It was not like classical mechanics where the equations are known but outstanding important problems remain, for lack of acumen in manipulating them. In the pion case, by contrast, one could well question whether the equations had more than the most qualitative meaning. It was not like the days of the old quantum theory when the fundaments of physics clearly needed rebuilding. In the pion case, by contrast, there was no reason to conclude from failure that general principles, or even the particular Yukawa equations, were in need of overhaul. It was not like the dispiriting days of quantum electrodynamics in the thirties when there was much to be concerned about, yet when the predictions made in perturbative approximations scored one success after another. In the pion case, by contrast, there was not a single prediction on a par with those good results. Typical for that period are statements such as the following (author anonymous) made[98] during the 4th Rochester conference (January 25-7, 1954): 'You cannot make any general predictions about what the situation is with regard to what the theory says'.

Those were not just years of uncertainty, however. Exciting experimental discoveries, of new particles, of parity non-conservation (to be discussed later), and so on, drew many theorists into much needed new branches of phenomenology. In regard to pion–nucleon phenomena it became increasingly evident that these were but part of a larger design which should include more particles, more fields. I am now prepared to give a general characterization of the period. Phenomena were exhibiting a marked increase in complexity; their description in terms of new phenomenological rules became an important and fruitful new activity, but their dynamical explanation was trailing behind and causing frustration. So it remained until well into the sixties when a new description of old as well as new strongly interacting particles in terms of quarks and gluons caused a drastic change.

Meanwhile a striking early response to complexity and frustration was the emergence in the 1950s of new levels of specialization. Up till that time, those engaged in particle/field theory could and would cover the whole range of questions it presented. In the early fifties one witnesses the creation of subspecialties each with its own practitioners and terminology. This was not (or only quite rarely) a matter of factionalism. Rather, specialists in one or another area would keep up with each other's progress without immersion in each other's problems, while a smaller contingent would continue to move actively from one area to another. Within this new diversity one recognizes efforts at finding new inspiration from older problems. In non-relativistic quantum mechanics the question was considered whether the perturbation series (the Born expansion) for the scattering of a particle by a center of force converges.[99] In relativistic quantum mechanics the axioms of field theory were re-examined and refined.* The relation between dispersion and absorption of light, a subject with 19th century origins, became the spur to dispersion relations. More diversity was added because some would insist on mathematical rigor, as in axiomatic field theory, while others would be willing to entertain educated guesses, as in analytic S-matrix theory. In the latter instance, there were those who considered these guesses to be tentative, others who accepted them as articles of a new faith.

Nor was the period at hand one of consensus concerning the future of relativistic quantum field theory. In regard to strong interactions, some thought the theory was plainly wrong, others apologetically borrowed some of its concepts. Meanwhile a search had begun for implications of the theory that do not depend on perturbative expansions** which, everybody of course agreed, were useless for strong interactions. In general terms, the problem posed was to identify those aspects of field theory (including interactions) which do not depend on the magnitudes of coupling constants and to ask whether these suffice for making at least some limited predictions concerning strong interaction phenomena. This program produced answers of lasting value, among which I shall single out shortly low energy theorems and dispersion relations. The rigorous proofs of the latter are in large measure due to axiomatic field theorists.

The discovery of a new class of field theories, the non-Abelian gauge theories, also falls in this period. I shall return later to this subject, the far-reaching consequences of which were not immediately realized.

Some went this way, others went that way in the theory of particles and fields. So it always was, but never more so in my time than in the 50s. All told the harvest of that period was a deeper insight not so much into the structure of matter as in the structure of theories. The rest of this section contains brief sketches† of some main issues without pretence to completeness. Among the subjects I pass by is axiomatic field theory, the early history of which, I hope, will some day be systematically recorded.

2. Fermi's assessment in 1951. In October 1951 the American Institute of Physics organized a symposium on contemporary physics as part of its twentieth-anniversary celebrations. Fermi spoke on 'The nucleus' and concluded his talk[101] with his views on meson physics: 'Of course, it may be that someone will come up soon with a solution to the problem of the meson, and that experimental results will confirm so many detailed features of the theory that

* Chapter 15, Section (a).
** This search included studies of simple model theories which allow for a comparison between exact solutions and approximate methods.[100]
† These can be skipped without loss of continuity with later chapters.

it will be clear to everybody that it is the correct one. Such things have happened in the past. They may happen again. However, I do not believe that we can count on it, and I believe that we must be prepared for a long hard pull . . . '. His reservations stemmed only in part from the difficulties described earlier in this chapter: 'When the Yukawa theory first was proposed there was a legitimate hope that the particles involved, protons, neutrons and pi-mesons, could be legitimately considered as elementary particles. This hope loses more and more its foundation as new elementary particles are rapidly being discovered'.

Fermi was right on all counts. The pull was long and hard. Protons, neutrons, and pions are currently no longer considered as elementary but as composites. The new particles and the compositeness question will be left to later chapters except for one remark. In 1949 Fermi was the first[102] to suggest that pions might be composites of nucleons and antinucleons. This idea, quite novel for its time,[103] was worked out by him and Yang.[104] The notion that nucleons and antinucleons, but not pions, are fundamental has since been modified: none of them are fundamental.

3. The semi-classical nucleon. This is a topic which brings us back for the last time to prewar meson theory. During the lively discussion following the presentation of π-nucleon scattering data at the January 1952 Rochester meeting, Oppenheimer remarked: 'It is the first time that there is any smell at all of phenomena which have been on the books for ten years, namely the existence of some loosely bound systems of isotopic spin 3/2 and spin 3/2'.[105] This was in reference to the so-called strong coupling theory, initiated[106] by Wentzel in 1940, in which a nucleon is treated as a classical extended source of classical (unquantized) meson fields. Recall* that in those pre-muon days the question had arisen as to how to reconcile strong nuclear forces with weak meson (that is, muon!) scattering. Strong coupling was designed largely as an attempt to cope with this fictitious problem. For present purposes its sole interest is the fact that it led to guesses about the 33-resonance.**

The idea is to consider the nucleon as a classical extended static source with density $\rho(\vec{x})$ and radius† a coupled to an isotriplet ϕ_λ of pseudoscalar mesons by

$$H_{\rm int} = -\frac{f}{\mu}\sum_{\lambda=1}^{3}\tau_\lambda\int \mathrm{d}\vec{x}\,\rho(\vec{x})\vec{\sigma}\vec{\nabla}\phi_\lambda \tag{19.22}$$

This expression is the non-relativistic semi-classical limit of the γ_5-coupling

* Chapter 18, Section (b).

** For details on strong coupling see Ref. 107. For the specific case of the pseudoscalar symmetric theory see Ref. 108.

† a is defined by $\iint \rho(\vec{x})\rho(\vec{x}')/|\vec{x}-\vec{x}'|\cdot \mathrm{d}\vec{x}\,\mathrm{d}\vec{x}' = 1/8\pi a$.

Eq. (19.8) provided that (μ and M are the pion and nucleon mass respectively)

$$\frac{f^2}{\hbar c} = \left(\frac{\mu}{2M}\right)^2 \cdot \frac{g^2}{\hbar c} \tag{19.23}$$

and that, essentially, $\bar{\psi}\psi$ is replaced by ρ. Nucleon recoil and nucleon–antinucleon pair formation are evidently neglected. The self-field carried by the blob $\rho(\vec{x})$ generates a self-energy $\sim f^2/a$, the Lorentz term, and in addition a spectrum of excited nucleon states which are due to the gyrations of the quantized spin and isospin. If

$$\frac{f^2}{\hbar c} \gg 1 \gg \frac{\mu c a}{\hbar}, \tag{19.24}$$

that is, if the coupling is strong and the source is small then, it turns out, the total spin and the total isospin have equal values $j = \frac{1}{2}, \frac{3}{2}, \ldots$ for low lying excitations. $j = \frac{1}{2}$ is the physical nucleon, the other j's have a higher energy than $j = \frac{1}{2}$, given by

$$E_j = \frac{12\pi a(\mu c^2)^2}{f^2}\left[j(j+1) - \tfrac{3}{4}\right], \qquad j \geq \tfrac{3}{2}. \tag{19.25}$$

$E_j \to 0$ as $f \to \infty$ whence the name nucleon isobars for these states.

In the case of π-nucleon scattering only the isobar $j = \frac{3}{2}$ can be excited—which brings us back to Oppenheimer's remark: that was just the state under discussion at Rochester. Actually, already two years earlier, data on photomeson production had led to speculations[109] and calculations[110] on the possible role of isobars. Also, the strong coupling picture had led Brueckner to guess at Rochester[111] that the apparent resonance had quantum numbers $T = J = \frac{3}{2}$, even before J had been determined.

Thus strong coupling had provided a certain stimulus—yet an isobar is not necessarily a resonance. Eq. (19.25) shows that for large f (and fixed a) the isobars are actually bound states, stable against decay into π + nucleon. It is clear that strong coupling overshoots the mark: from Eqs. (19.10) and (19.23) one finds that

$$\frac{f^2}{\hbar c} \simeq 0.08, \tag{19.26}$$

in conflict with Eq. (19.24). Nor do the quantum numbers of the next nucleon resonance fit the picture.

And so strong coupling faded out* but not the study of the interaction (19.22) which continued during 1952–5, now, however, with mesons treated

* Nor is 'intermediate coupling' much in evidence any longer. This variational method of the Hartree type was initiated in 1941 by Tomonaga whose papers on that topic have been collected in Ref. 112.

as quantized fields. The foremost result of these efforts, initiated by Chew,[113] is the Chew–Low equation,[114] which is best appreciated in the context of dispersion relations (see below).*

4. Low energy theorems. The simplest theorem of this kind concerns the Compton scattering cross section σ in the limit of zero photon frequency ω, given classically by Thomson's good old formula (9.5)

$$\sigma \to \frac{8\pi}{3}\left(\frac{e^2}{mc^2}\right)^2 \quad \text{as } \omega \to 0. \tag{19.27}$$

The theorem says that this equation remains valid in quantum electrodynamics to any order in α. All that radiative corrections do is renormalize e and m. This result (it follows from gauge invariance only, in particular from an important identity[116] due to Ward) implies that σ $(\omega = 0)$ provides a means of defining and measuring the physical charge. It was initially proved[117] for an electron interacting with the electromagnetic field but is also true for the proton with its additional meson interactions, as long as these are renormalizable.[118] A further theorem holds for the scattering amplitude to $0(\omega)$. Calculate it in lowest (first) order in α, replace e, m, and the magnetic moment, which appear in the answer, by their physical values, and the result is again exact.[119] Similar theorems hold for higher intrinsic multipole moments carried by particles with higher spin.[120]

It would be all-important if these non-perturbative theorems had analogs for meson couplings. The situation was indeed so confused at that time that there did not even exist an unambiguous prescription for defining the magnitude of g. Another low energy theorem which helped much had meanwhile been found:[121] in photo-meson production the rigorous amplitude A at threshold is related to the perturbative amplitude ($\sim eg$) A_{pert} by

$$A = A_{\text{pert}}\left[1 + C \cdot \frac{\mu}{M} + O\left(\left(\frac{\mu}{M}\right)^2\right)\right] \tag{19.28}$$

where C is a model-dependent constant. One could therefore only draw rigorous conclusions in the unphysical limit $\mu \to 0$. Extrapolating data in this simple, but nevertheless ad hoc way, one found[122] that $f^2/\hbar c$ has about the value given by Eq. (19.26): the early guesses for f were not so bad after all. Dispersion relations would confirm this.

5. The S-matrix. Dispersion relations. Well before the troubles with meson theories had been fully appreciated, Heisenberg already had expressed serious reservations about the unlimited validity of quantum field theory. His doubts, dating from the thirties, were caused less by the difficulties of the positron theory**

* For a detailed discussion of this equation see Ref. 115. ** Chapter 16, Section (d).

than by his inability to construct a sensible field theory out of Fermi's β-decay model. This had led him to wonder about the need for 'the introduction of a universal length which perhaps must be connected with a new change of principle in the formalism'.* Such speculations made him search for a new 'correspondence principle': what large distance aspects of the theory would survive a change of its short distance properties? In his S-matrix theory[123] of 1943 he assumed these to be the quantum mechanical superposition principle for non-interacting particles, unitarity, the standard conservation laws, relativistic invariance and symmetries such as isospin.**

Let us recall the definition of an S-matrix and what is meant by unitarity. Suppose two or more particles interact via forces of sufficiently short range. Then in the remote past they were in a free particle state, call it n. Let a collision event produce two or more particles. Long after the event these are likewise free, in a state m. The S-matrix is the collection S_{mn} of transition amplitudes $n \rightarrow m$, normalized such that $|S_{mn}|^2$ is the probability for this transition. The total probability that a given n ends up in whatever m must be unity:

$$\sum_m |S_{mn}|^2 = 1, \qquad (19.29)$$

or

$$SS^\dagger = S^\dagger S = 1 \qquad (19.30)$$

that is, S is unitary. In quantum mechanics S can in principle be calculated from the Hamiltonian H which describes in full the space–time evolution of states, including their behavior at short relative particle distances. Heisenberg suggested that the large distance information coded by S remains meaningful even for a theory in which H can no longer be defined.

Unitarity is the first important constraint on S; relativistic invariance and symmetries provide others.† All these are precious and useful conditions, yet at this stage the S-matrix is little more than a vessel containing well-arranged experimental data. The real fun with S began in the fifties when theorists began to constrain S further by borrowing general conditions from quantum field theory (thereby departing from Heisenberg's program). This strategy revealed important analyticity properties of S-matrix elements. In order to explain by what route these were discovered I must briefly digress.

Consider a bundle of light passing perpendicularly through a slab of dielectric with thickness l. The electric fields $E_{\text{in}}(\omega)$ and $E_{\text{out}}(\omega)$ associated with light of frequency ω as it enters and leaves the slab respectively are related by

$$E_{\text{in}}(\omega) = G(\omega)E_{\text{out}}(\omega) \qquad (19.31)$$

* Chapter 17, Section (e), Part 4.

** Heisenberg's applications of the theory to the definition of bound states by analytic continuation (not as simple as he thought) and to specific S-matrix models are of no interest for what follows.

† Constraints on S were first considered by Wheeler, for nuclear physics purposes.[123a]

$$G(\omega)=\exp i\omega\left[n(\omega)+\frac{i\alpha(\omega)c}{2\omega}\right]l. \tag{19.32}$$

The phenomenological parameters $n(\omega)$ and $\alpha(\omega)$ are respectively the index of refraction and the absorption coefficient. The idea that n and α are related goes back to the 1870s when the formula

$$n(\omega)-1=\sum_i \frac{\alpha(\omega_i)}{\omega_i^2-\omega^2} \tag{19.33}$$

was derived[124], for visible light, in terms of a model in which dispersion and absorption are due to the action of light on particles which are harmonically bound with proper frequencies ω_i. In 1926 Kramers,[125] and independently Kronig,[126] proposed to extend this formula to the continuous X-ray region, as follows:

$$n(\omega)-1=P\int_0^\infty \frac{\alpha(\omega')\,\mathrm{d}\omega'}{\omega'^2-\omega^2} \tag{19.34}$$

where P denotes the Cauchy principal value.

I do not know who first stated that this Kramers–Kronig relation does not depend on any dynamical detail of the interaction between light and matter and that the necessary and sufficient condition for its validity is causality, an extremely general requirement.* This reasoning goes as follows.[127] Suppose a 'cause' C at time $t-\tau$ contributes to an 'effect' E at time t, and that C and E are linearly related:

$$E(t)=\int_{-\infty}^{\infty} F(\tau)C(t-\tau)\,\mathrm{d}\tau. \tag{19.35}$$

Causality, 'E cannot precede C', is expressed by

$$F(\tau)=0, \qquad \tau<0. \tag{19.36}$$

A general mathematical theorem says that this condition is equivalent to the following two statements: the Fourier component $G(\omega)$ defined by

$$G(\omega)=\int_{-\infty}^{\infty} F(t)\,e^{i\omega t}\,\mathrm{d}t \tag{19.37}$$

can be continued analytically to complex values of ω with $Im\omega>0$, and has no singularities in this region; and $G(\omega)$ satisfies

$$Re\,G(\omega)=\frac{1}{\pi}P\int_{-\infty}^{\infty}\frac{ImG(\omega')\,\mathrm{d}\omega'}{\omega-\omega'}. \tag{19.38}$$

* This statement is not found in the original papers, Refs. 125, 126. A 1929 paper by Kramers[127] gives some indication that he may have known this.

This relation, known by physicists as a dispersion relation, can be shown[128] to lead to Eq. (19.34) if $G(\omega)$ satisfies Eq. (19.32), provided one adds the definitions

$$n(-\omega)=n(\omega), \qquad \alpha(-\omega)=-\alpha(\omega). \tag{19.39}$$

This first example of 'crossing symmetry' was introduced in 1929 by Kramers.[127]

The digression ends here and the S-matrix story now continues in 1946 when Kronig,[129] aware by then of the general argument just given, suggested adding causality to the S-matrix postulates. Shortly afterward, this proposal was applied in non-relativistic quantum mechanics where, however, the definition of causality turns out to be delicate.[130] Relativity, on the other hand, leads to a simple formulation of this requirement: a cause at the space–time point $(\vec{x}, t)$ cannot lead to an effect at $(\vec{x}', t')$ if a light signal from $\vec{x}$ cannot reach $\vec{x}'$ in a time span less than $|t-t'|$. Translated into quantum theory, any two field observables* A and B must obey

$$[A(\vec{x}, t), B(\vec{x}', t')]=0 \quad \text{if } (\vec{x}-\vec{x}')^2-c^2(t-t')^2>0. \tag{19.40}$$

This condition (also known as the locality condition), valid in the presence of interactions and independent of their strengths, was now borrowed from field theory, and its consequences for S-matrix elements was examined. With similar generality crossing symmetry, the analog of Eq. (19.39), was also abstracted from general field theory.[132]

Armed with these ingredients Goldberger (1955)[133] obtained dispersion relations for π-nucleon scattering in the forward direction. His conclusions, arrived at in part by heuristic arguments, were substantiated soon afterward by rigorous arguments[134] based on the 'LSZ formalism', so named after its authors, Lehmann, Symanzik, and Zimmerman.[135] One sample of the answers is found via the following main steps.
(a) Introduce a matrix T related to S by

$$S_{\mathrm{ab}}=\delta_{\mathrm{ab}}+2\pi i\delta(p_\mu^{\mathrm{a}}-p_\mu^{\mathrm{b}})T_{\mathrm{ab}}, \tag{19.41}$$

where the δ-function implements overall energy–momentum conservation in the transition $\mathrm{a}\rightarrow\mathrm{b}$. It is the T-matrix whose analyticity is at issue.
(b) For forward scattering of pions with frequency ω (in the laboratory system) on nucleons, T reduces to a 3×3 matrix $T_{\alpha\beta}(\omega)$ in the nucleon isospin space,

$$\begin{gathered} T_{\alpha\beta}(\omega)=\delta_{\alpha\beta}T^{(1)}(\omega)+\tfrac{1}{2}[\tau_\alpha, \tau_\beta]T^{(2)}(\omega) \\ T^{(1)}=\tfrac{1}{2}(T^{(-)}+T^{(+)}),\ T^{(2)}=\tfrac{1}{2}(T^{(-)}-T^{(+)}), \end{gathered} \tag{19.42}$$

where $T^{(\pm)}$ refer to the processes $\pi^\pm+\mathrm{p}\rightarrow\pi^\pm+\mathrm{p}$.

* In the Heisenberg representation. The relevance of Eq. (19.40) for dispersion relations was first stated by Gell–Mann, Goldberger, and Thirring.[131]

(c) Sample result:*

$$Re\, T^{(2)}(\omega)=\frac{2\omega f^2}{\omega^2-(\mu^2/2M)^2}+\frac{\omega}{\pi}P\int_{\mu}^{\infty}\frac{k'\,d\omega'[\sigma_-(\omega')-\sigma_+(\omega')]}{\omega'^2-\omega^2}. \tag{19.43}$$

(d) $\sigma_{\pm}(\omega)$ are the forward cross sections for $\pi^{\pm}+p\rightarrow$ all possible final states. Their appearance results from the proportionality of the imaginary part of T and $\sigma_{\pm}$ (up to explicit kinematic factors). In turn, this proportionality (the optical theorem) is a consequence of unitarity.
(e) The integral in Eq. (19.43) extends over the physically allowed region $\omega=\mu-\infty$. According to Eq. (19.37) (and crossing) the unphysical region $\omega=0-\mu$ must be included as well, however. This leads to the 'pole term', proportional to f^2, which corresponds to the unphysical process $\pi^-+p\rightarrow n$ for which $\omega=\mu^2/2M$ and k is imaginary. By appealing once more to field theory (but not perturbation theory) f can be identified with the *renormalized* coupling constant introduced earlier.**

Following the announcement of the forward π-nucleon dispersion relations, which caused considerable stir and was a highlight of the sixth Rochester conference (April 1956),[137] there began a period of intense and widespread analysis of their applications and extensions, more generally of the analytic properties of the S-matrix combined with causality, unitarity, and crossing. I list a few main points.
(a) π-nucleon forward dispersion relations. Nothing in their derivation guarantees that dispersion integrals like the one in Eq. (19.43) converge; in that example it actually does. In known cases where it does not, meaningful though weaker statements can still be made. Theorems[138] on the high energy behavior of cross sections give some control over the situation. The most famous of these is the Pomeranchuk theorem[138] according to which collisions between particle and target and antiparticle and the same target become asymptotically equal at high energy. The validity of the theorem depends on assumptions that do not all follow from general quantum field theory, however.

In the example, $T^{(2)}$ can be measured (in the interference between nuclear and Coulomb effects), as can $\sigma_{\pm}(\omega)$, at least over some ω-region. Making sensible extrapolations to large ω, a region with relatively low weight, one has a one-parameter (f) fit to the data which continues to be good.[139] Alternatively one can use Eq. (19.43) for determining f. Already in 1956[140] this yielded

$$\frac{f^2}{\hbar c}=0.082\pm0.015. \tag{19.44}$$

This derivation of a value for f from a rigorous equation (with μ-dependence explicit, unlike the low energy theorems) is perhaps the greatest achievement

* Eq. (19.42) is expressed in units $\hbar=c=1$; $k^2=\omega^2-\mu^2$. ** For more details see Ref. 136.

of the whole subject.* It vindicates earlier crude estimates for f and marks the happy ending of at least one chapter in pion physics.

Other chapters remained untouched, as is already clear from the example Eq. (19.43): resonances are contained in $\sigma_{\pm}(\omega)$ but dispersion relations do not claim to explain their presence. Interesting results (sum rules) can be obtained by approximating $\sigma_{\pm}(\omega)$ in terms of resonance contributions only. The Chew–Low equation (mentioned earlier) which must be considered as a precursor to dispersion relations is the oldest example. This equation can be retrieved[142] by retaining only p-wave contributions to the dispersion integrals, cutting off those integrals, and dropping terms $0(\mu/M)$.

(b) Other applications. The usefulness of analyticity is much more obscure for other processes such as nucleon–nucleon scattering, where complications stemming from the unphysical region remain unresolved. A measure of success was achieved for electron–nucleon scattering due (to $O(\alpha)$) to the electron shaking off a virtual photon which sees the nucleon as a complicated blob because of its surrounding virtual meson cloud. This blob or vertex satisfies a dispersion relation[143] containing a dispersion integral, somewhat as in Eq. (19.43). Estimates for this integral did not fit the massive amount of important data gathered in the fifties by Hofstadter and co-workers.[144] The situation was saved[145] by assuming an enhancement of the integral due to an as yet unknown resonance, which in fact was found soon thereafter: the ρ-meson. These same data also served to predict[146] the ω-meson by a more phenomenological argument.**

(c) Non-forward dispersion relations. Considerable effort has gone into finding dispersion relations for other than forward scattering. These are bound to be harder to interpret because there is no optical theorem to provide aid and succor. Crucial to such studies is the choice of convenient variables. For the scattering

$$1+2\rightarrow 3+4 \tag{19.45}$$

these turn out to be $s=(p_1+p_2)^2$ and $t=(p_1-p_3)^2$, the invariant energy and momentum transfer respectively (p^2 stands for $E^2-\vec{p}^2c^2$). For a number of specific reactions proofs of dispersion relations for the scattering amplitude $A(s, t)$ have been given[134] for t held fixed and confined to limited regions of its allowed range. It has also been shown[134] that $A(s, t)$ is analytic in certain regions of the t-plane, for fixed s. It is not known whether these various (s, t)-regions can be extended. Applications include the establishment of bounds on pion–pion scattering amplitudes.[148]

* Dispersion relations were also of use in discriminating between various options for π-nucleon phase shifts.[141]

** Another application[147] of dispersion relations to electromagnetic processes deals with the decay $\pi^0\rightarrow 2\gamma$.

A number of conjectures concern the simultaneous consideration of (19.45) and

$$1+\bar{3}\rightarrow\bar{2}+4 \tag{19.46}$$

$$1+\bar{4}\rightarrow\bar{2}+3 \tag{19.47}$$

where the bar denotes the antiparticle. For reaction (19.46) t is now the energy variable, s the momentum transfer but of course in other ranges of values than for Eq. (19.45). Similarly for (19.46): $u=(p_1+p_4)^2$ represents energy, s and t momentum transfer ($u+s+t$ equals a constant). The physical parameter regions for (s, t, u) are disjoint for the three reactions. It has been conjectured that there exists a function $A\,(s, t, u)$ which is sufficiently analytic to permit continuation from one physical region to another and which represents the physical scattering amplitude in each of these regions. This is the consummate form of crossing symmetry. Expressions for A have been suggested[149] but no general proofs exist. The ultimate conjecture in this line of thought, much debated in the fifties, is the bootstrap idea[150] according to which the constraints of analyticity, unitarity, and crossing, later supplemented with Regge pole ideas (see below) will suffice to give a self-consistent theory of strong interactions.

Interest in these conjectures has since receded. The active study of the analytic S-matrix by rigorous methods has continued up to the present, however.*

6. Regge poles. Eq. (19.43) exemplifies the occurrence of poles in scattering amplitudes at energies corresponding to bound states or to resonances (the latter contained in $\sigma_\pm$), each pole occurring in a specific partial wave. Such poles are long familiar from potential scattering in non-relativistic quantum mechanics. For that case, new and fundamental results were obtained[153] in 1959, when Regge showed that these poles are interconnected. He proved for a class of potentials that the partial wave amplitudes $f_l(s)$ in the expansion

$$f(s, \theta)=\sum_l (2l+1)f_l(s)P_l(\theta) \tag{19.48}$$

(the P_l are Legendre polynomials) are linked by an 'interpolating function' $f(l, s)$, defined for complex l, which has these main properties:

$$f(l, s)=f_l(s) \quad \text{for } l=0, 1, 2, \ldots; \tag{19.49}$$

$f(l, s)$ is analytic for all $Re\, l>-\frac{1}{2}$ except for a finite number of poles—the Regge poles; and $f(l, s)$ is unique provided it satisfies certain further analyticity properties and is sufficiently bounded as $|l|\rightarrow\infty$. The positions of

* For books carrying the subject into the sixties see Refs. 148, 151. For much more recent contributions see Ref. 152.

the poles are solutions of an equation

$$l = \alpha(s) \tag{19.50}$$

with a calculable function α. As s varies the poles trace a 'Regge trajectory'.

To those engaged in the study of the analytic S-matrix, this totally unexpected development offered prospects for adding new criteria to unitarity, causality, and crossing. In 1961 the appealing hypothesis was put forth[154] that all poles of the S-matrix for strongly interacting particles are Regge poles. That is to say all such particles, whether stable or not (resonances), can be located as follows. Divide the particles of whatever spin into groups distinguished by specific values of quantum numbers such as isospin, nucleon number, etc. Subdivide the meson groups further into even and odd spin. Enter each group into a plot[155] (a projection of Regge trajectories) of $Re\,\alpha(s)$ versus $(\text{mass})^2$ (which is the real part of s). Note that half-integer and integer values of $Re\,\alpha(s)$ just correspond to spin values of real physical states. Now behold: the members of each group lie on lines which are quite straight for mesons, and approximately straight for half-integer spins.[84,156] This is obviously an important message.* It is not known, however, why these lines are straight, nothing that simple is indicated from potential scattering (for Yukawa potentials the corresponding plot is quite curvaceous).[158] More generally, Regge poles in S-matrix context can in fact not be handled with a rigor comparable to Regge's original treatment for the simple reason that whereas Regge could study specific potentials, their relativistic analogs, interactions at short distances, are not at all understood with comparable precision.

In order to make headway, additional assumptions and approximations had therefore to be made. About these it was written in 1967: 'The uncertainties about some of the assumptions and approximations in Regge theory necessitate detailed experimental study of the predictions that result from various approximations. It seems unlikely that relativistic theory will by itself provide a rigorous justification of the wide range of approximation schemes within Regge theory that are required for experimental comparisons. The corollary to this theoretical impasse is that "predictions" of Regge theory are likely to be of a heuristic nature for some time to come', a situation which has not changed since.[159]

Regge poles have, however, led to an interesting phenomenology for 'soft scattering', processes in which the produced particles carry small momenta in directions perpendicular to the beam. These include reactions of the types $A+B \to C+D$, $A+B \to C+$whatever (inclusive reactions), and diffraction scattering. By combining the trajectory concept with crossing, one has been able to discuss the data in terms of trajectory parameters rather than individual particle parameters (masses, coupling constants). These analyses are most

* For a popular account of Regge trajectories and related plots see Ref. 157.

often exceedingly complex—but so are the phenomena. The degree of success of this phenomenology has shown marked ups and downs in the course of time and continues to do so, as recent summaries[160] indicate. An assessment of the status of Regge phenomenology in just a few lines does not appear possible at this time.

As we shall see later, the new physics of the sixties and seventies originated from hard scattering processes about which Regge analysis has little to say. It is devoutly to be hoped that current efforts[161] at finding a synthesis between the new field theories and Regge poles will lead to a better understanding of this interesting subject.

Sources

Reminiscences. McMillan on the synchrocyclotron;[11] Livingston on strong focussing;[12] Ramsey on the birth of Brookhaven National Laboratory;[4] Kowarski on the birth of CERN;[5] R. Wilson on Fermilab;[19] Goldwasser on the growth of big science;[162] Lattes on pion physics in Berkeley;[163] Anderson on pion physics in Chicago;[65] Glaser[164] and Alvarez[165] on bubble chambers; Nishijima, Michel, and Yang on G-parity;[166] Wightman on axiomatic field theory;[167] Gell–Mann on the S-matrix;[168] Chew on the bootstrap.[169]

References

1. Ph. Morrison, *J. Appl. Phys.* **18**, 133, 1947.
2. E. M. McMillan, *Physics Today*, February 1948, p. 31.
3. R. W. Seidel, *Hist. St. Phys. Sc.* **13**, 375, 1982.
4. N. F. Ramsey, *Report BNL 992, T-421*, 1966.
5. See L. Kowarski, *An account of the origins and beginnings of CERN*, *CERN Report 61-10*, 1961.
6. J. Krige, *CERN Report CHS-1*, 1983.
7. BNL *Report BNL-L-101*, 1948.
8. L. I. Schiff, *Rev. Sci. Instr.* **17**, 6, 1946.
9. See e.g. M. S. Livingston and J. P. Blewett, *Particle accelerators*, McGraw-Hill, New York 1962.
10. V. I. Veksler, *Dokl. Ak. Nauk* **43**, 329, 1944; **44**, 365, 1944; also *J. of Phys. USSR* **9**, 153, 1945.
11. E. M. McMillan, *Phys. Rev.* **68**, 143, 1945.
12. Cf. M. Livingston, *Adv. in electronics and electron physics* **50**, 1, 1980, esp. p. 41.
13. F. K. Goward and O. E. Barnes, *Nature* **158**, 413, 1946.
14. E. M. McMillan and J. M. Petersen, *Science* **109**, 438, 1949.
15. E. D. Courant, M. S. Livingston, and H. S. Snyder, *Phys. Rev.* **88**, 1190, 1952; also J. P. Blewett, *Phys. Rev.* **88**, 197, 1952; J. B. Adams, M. G. N. Hine, and J. D. Lawson, *Nature* **171**, 926, 1953.
16. E. D. Courant, M. S. Livingston, H. S. Snyder, and J. P. Blewett, *Phys. Rev.* **91**, 202, 1952.

17. D. L. Judd, in *Proc. int. conf. on high energy accelerators, CERN 1959*, p. 6, CERN Sci. info. Svce, 1959.
18. Joint Committee on Atomic Energy, print 42-613, U.S. Govt. Printing Office, Washington D.C., February 1965.
19. R. R. Wilson, *Physics Today*, October 1977, p. 23.
20. J. J. Livingood, *Principles of cyclic particle accelerators*, p. 200, Van Nostrand, New York, 1961.

20a. Cf. G. B. Collins, *Sci. Am.* **189**, November 1953, p. 36.

21. W. M. Powell *et al.*, *Rev. Sci. Instr.* **19**, 506, 1948; **20**, 887, 1949; Ref. 9, p. 388.
22. W. G. McMillan and E. Teller, *Phys. Rev.* **72**, 1, 1947.
23. E. Gardner and C. M. G. Lattes, *Science* **107**, 270, 1948; also J. Burfening, E. Gardner, and C. M. G. Lattes, *Phys. Rev.* **75**, 382, 1949.
24. F. W. Brode, *Rev. Mod. Phys.* **21**, 37, 1949.
25. Cf. e.g. H. Bridge, B. Rossi, and R. Williams, *Phys. Rev.* **72**, 257, 1947; H. W. Lewis, J. R. Oppenheimer, and S. A. Wouthuysen, *Phys. Rev.* **73**, 127, 1948.
26. R. Bjorklund, W. E. Crandall, B. J. Moyer, and H. F. York, *Phys. Rev.* **77**, 213, 1950.
27. J. Steinberger, W. K. H. Panofsky, and J. Steller, *Phys. Rev.* **78**, 802, 1950.
28. W. K. H. Panofsky, R. L. Aamodt, and J. Hadley, *Phys. Rev.* **81**, 565, 1951.
29. L. D. Landau, *Dokl. Ak. Nauk* **60**, 207, 1948; C. N. Yang, *Phys. Rev.* **77**, 242, 1950.
30. R. E. Marshak, *Phys. Rev.* **82**, 313, 1951; *Rev. Mod. Phys.* **23**, 137, 1951.
31. W. B. Cheston, *Phys. Rev.* **83**, 1181, 1951.
32. R. F. Christy and S. Kusaka, *Phys. Rev.* **59**, 414, 1941.
33. R. Durbin, H. Loar, and J. Steinberger, *Phys. Rev.* **83**, 646, 1951; D. L. Clark, A. Roberts, and R. Wilson, *Phys. Rev.* **83**, 649, 1951.
34. H. A. Bethe and F. de Hoffmann, *Mesons and fields*, Vol. 2, Section 28, Row and Petersen, Elmsford, N.Y. 1955.
35. Cf. L. Rosenfeld, *Nuclear forces*, Chapter 16, North Holland, Amsterdam 1948; G. Wentzel, *Rev. Mod. Phys.* **19**, 1, 1947.
36. H. A. Bethe, *Phys. Rev.* **76**, 190, 1949.
37. L. Rosenfeld, *Proc. 8th Solvay Conf.*, p. 179, Coudenberg, Brussels 1950.
38. H. Yukawa, S. Sakata, and M. Taketani, *Proc. Phys. Math. Soc. Japan* **20**, 319, 1938; H. Fröhlich, W. Heitler, and N. Kemmer, *Proc. Roy. Soc. A* **166**, 154, 1938.
39. J. M. Jauch, *Phys. Rev.* **36**, 334, 1943; G. Araki, *Progr. Th. Phys.* **1**, 1, 1946.
40. W. W. Havens, I. I. Rabi, and J. Rainwater, *Phys. Rev.* **72**, 634, 1947; E. Fermi and L. Marshall, *Phys. Rev.* **72**, 1139, 1947.
41. J. M. Luttinger, *Helv. Phys. Acta* **21**, 483, 1948; K. M. Case, *Phys. Rev.* **74**, 1884, 1948; **76**, 1, 1949; M. Slotnick and W. Heitler, *Phys. Rev.* **75**, 1645, 1949; S. D. Drell, *Phys. Rev.* **76**, 427, 1949; S. Borowitz and W. Kohn, *Phys. Rev.* **76**, 818, 1949; K. Sawada, *Progr. Th. Phys.* **4**, 383, 1949.
42. S. Borowitz and W. Kohn, Ref. 40.
43. K. M. Case, *Phys. Rev.* **75**, 1440, 1949.
44. P. T. Matthews, *Phys. Rev.* **76**, 1254, 1949; more details in *Phil. Mag.* **41**, 185, 1950; corrections in *Phys. Rev.* **80**, 293, 1950. See also A. Salam, *Phys. Rev.* **82**, 217, 1951, and the review by P. T. Matthews and A. Salam, *Rev. Mod. Phys.* **23**, 311, 1951.
45. J. C. Ward, *Phys. Rev.* **77**, 293, 1950.

46. P. T. Matthews, *Phys. Rev.* **81**, 936, 1951; cf. also A. Salam, *Phys. Rev.* **82**, 217, 1951.
47. E. Fermi, *Nuovo Cim.* Suppl. to Vol. 2, 1955, p. 33.
48. E. P. Wigner, *Symmetries and reflections*, p. 252, Indiana Univ. Press, Bloomington 1967.
49. R. E. Marshak, *Meson physics*, McGraw-Hill, New York 1952.
50. Ref. 34, Section 29c.
51. Ref. 34, Sections 29b, d.
52. Ref. 34, Section 35c.
53. H. L. Anderson *et al.*, *Phys. Rev.* **85**, 934, 936; **86**, 793, 1952.
54. W. Heitler, *Proc. Ir. Ac. Sc.* **51**, 33, 1946.
55. K. M. Watson and K. A. Brueckner, *Phys. Rev.* **83**, 1, 1951.
56. K. M. Watson, *Phys. Rev.* **85**, 852, 1952, cf. footnote 10.
57. R. L. Garwin, *Phys. Rev.* **85**, 1045, 1952; A. M. L. Messiah, *Phys. Rev.* **86**, 430, 1952; J. M. Luttinger, *Phys. Rev.* **86**, 571, 1952; M. Ruderman, *Phys. Rev.* **87**, 383, 1952.
58. L. van Hove, R. E. Marshak, and A. Pais, *Phys. Rev.* **88**, 1211, 1952.
59. R. K. Adair, *Phys. Rev.* **87**, 1041, 1952; N. M. Kroll and L. L. Foldy, *Phys. Rev.* **88**, 1177, 1952.
60. *Isotopic spin in nuclear physics*, Ed. D. H. Wilkinson, North Holland, Amsterdam 1969.
61. Cf. L. Yuan and S. J. Lindenbaum, *Phys. Rev.* **92**, 1578, 1953; J. Ashkin *et al.*, *Phys. Rev.* **93**, 1129, 1954.
62. Cf. Ref. 34, Section 34c.
63. H. L. Anderson, W. C. Davidon, and U. E. Kruse, *Phys. Rev.* **100**, 339, 1955; cf. also Ref. 34, Sections 34, 35.
64. E.g. M. L. Goldberger and K. M. Watson, *Collision theory*, Chapter 8, Wiley, New York, 1964.
65. H. L. Anderson, 'International colloquium on the history of particle physics', p. 160, *J. de Phys.* **43**, suppl. to No. 12, 1982.
66. O. Chamberlain, E. Segré, C. E. Wiegand, and T. Ypsilantis, *Phys. Rev.* **100**, 947, 1955.
67. B. Cook, G. R. Lambertson, O. Piccioni, and W. A. Wentzel, *Phys. Rev.* **104**, 1193, 1956.
68. H. Weyl, *Zeitschr. f. Phys.* **56**, 330, 1929.
69. E. C. G. Stückelberg, *Helv. Phys. Act.* **11**, 299, 1938, esp. p. 317.
70. E. P. Wigner, *Proc. Am. Phil. Soc.* **93**, 521, 1949; *Proc. Nat. Ac. Sci.* **38**, 449, 1952.
71. T. D. Lee and C. N. Yang, *Phys. Rev.* **98**, 1501, 1955; A. Pais, *Phys. Rev.* **D8**, 1844, 1973.
72. L. Radicati, *Phys. Rev.* **87**, 521, 1952.
73. Ref. 34, Section 37.
74. S. Weinberg and S. B. Treiman, *Phys. Rev.* **116**, 465, 1959.
75. A. Pais, *Ann. of Phys.* **9**, 548, 1960; **22**, 274, 1963.
76. A. Pais, *Phys. Rev. Lett.* **32**, 1081, 1974; *Phys. Rev.* **D10**, 2147, 1974.
77. G. Goldhaber, S. Goldhaber, W. Y. Lee, and A. Pais, *Phys. Rev.* **120**, 300, 1960; for recent experiments see T. Akesson *et al.*, *Phys. Lett.* **129B**, 269, 1983; W. A. Zacj *et al.*, *Phys. Rev.* **C29**, 2173, 1984.
78. A. Pais and R. Jost, *Phys. Rev.* **87**, 871, 1952.

79. H. Fukuda and Y. Miyamoto, *Progr. Th. Phys.* **4**, 389, 1949; C. B. van Wyk, *Phys. Rev.* **80**, 987, 1950; K. Nishijima, *Progr. Th. Phys.* **6**, 614, 1951; L. Michel, *Progr. Cosm. Ray Phys.* **3**, p. 142, Interscience, New York 1952. (For a loophole in the last paper see Ref. 78, footnote 21.)
80. L. Michel, *Nuovo Cim.* **10**, 319, 1953.
81. T. D. Lee and C. N. Yang, *Nuovo Cim.* **3**, 749, 1956.
82. R. L. Cool, L. Madansky, and O. Piccioni, *Phys. Rev.* **93**, 249, 637, 1954.
83. Ref. 34, pp. 113, 115.
84. L. Montanet, *AIP Conference Proc.* **68**, 1213, 1981.
85. Cf. N. Isgur and G. Karl, *Phys. Rev.* **D18**, 4187, 1978; R. Konink and N. Isgur, *Phys. Rev.* **D21**, 1868, 1980.
86. D. Stonehill *et al.*, *Phys. Rev. Lett.* **6**, 624, 1961; A. R. Erwin *et al.*, ibid. **6**, 628, 1961.
87. B. Maglić *et al.*, *Phys. Rev. Lett.* **7**, 178, 1961.
88. A. Pevsner *et al.*, *Phys. Rev. Lett.* **7**, 421, 1961.
89. D. A. Glaser, *Phys. Rev.* **87**, 665, 1952.
90. D. A. Glaser, *Phys. Rev.* **91**, 496, 762, 1953.
91. H. Bradner, *Ann. Rev. Nucl. Sc.* **10**, 109, 1958.
92. J. Brown, D. Glaser, M. Perl, and J. van der Velde, *Phys. Rev.* **107**, 906, 1957.
93. J. G. Wood, *Phys. Rev.* **94**, 731, 1954.
94. L. Alvarez, 'The bubble chamber program at UCRL', unpublished report dated April 18, 1955.
95. A. H. Rosenfeld and W. E. Humphrey, *Ann. Rev. Nucl. Sc.* **13**, 103, 1963.
96. B. Bhowmik *et al.*, *Nuovo Cim.* **13**, 690, 1959.
97. E.g. P. Rice-Evans, *Spark, streamer, proportional and draft chambers*, Richelieu Press, London 1974; *Physics Today* **31**, 1978, October issue.
98. *Proc. 4th Rochester Conference* (Jan. 25–7, 1954), p. 44, unpublished copies distributed to participants.
99. R. Jost and A. Pais, *Phys. Rev.* **82**, 840, 1951.
100. Cf. e.g. T. D. Lee, *Phys. Rev.* **95**, 1329, 1984.
101. E. Fermi, *Physics Today* **5**, March 1952, p. 6.
102. See C. N. Yang, *Selected papers 1945–1980*, p. 7, Freeman, San Francisco 1983.
103. See however N. Rosen, *Phys. Rev.* **74**, 128, 1948; H. M. Moseley, *Phys. Rev.* **76**, 197, 1949.
104. E. Fermi and C. N. Yang, *Phys. Rev.* **76**, 1739, 1949.
105. R. Oppenheimer, *Proc. 2nd Rochester Conference* (January 1952), p. 38, Univ. of Rochester Report NYO-346, 1952.
106. G. Wentzel, *Helv. Phys. Acta* **13**, 269, 1940; **14**, 633, 1941; cf. also W. Heisenberg, *Zeitschr. f. Phys.* **113**, 61, 1939.
107. W. Pauli, *Meson theory of nuclear forces*, Interscience, New York 1946; G. Wentzel, *Rev. Mod. Phys.* **19**, 1, 1947; also in *Theoretical physics of the 20th century*, Eds. M. Fierz and V. Weisskopf, p. 71, Interscience, New York 1960; C. J. Goebel, *Quanta*, p. 20, Univ. Chicago Press 1970.
108. W. Pauli and S. M. Dancoff, *Phys. Rev.* **62**, 85, 1942; A. Pais and R. Serber, *Phys. Rev.* **113**, 955, 1959.
109. Y. Fujimoto and H. Miyazawa, *Progr. Theor. Phys.* **5**, 1052, 1950; R. E. Marshak, *Phys. Rev.* **78**, 346, 1950.
110. K. A. Brueckner and K. M. Case, *Phys. Rev.* **83**, 1141, 1951.
111. K. A. Brueckner, Ref. 105, p. 27; *Phys. Rev.* **86**, 106, 1952.

112. S. Tomonaga, *Progr. Theor. Phys. Suppl. No. 2*, 1955.
113. G. F. Chew, *Phys. Rev.* **89**, 591, 1953; **94**, 1748, 1755, 1954; **95**, 1, 1669, 1954.
114. G. F. Chew and F. E. Low, *Phys. Rev.* **101**, 1571, 1579, 1956.
115. G. C. Wick, *Rev. Mod. Phys.* **27**, 339, 1955; S. S. Schweber, *An introduction to relativistic quantum field theory*, Section 12d, Row, Peterson, Elmsford, N.Y. 1961.
116. J. Ward, *Phys. Rev.* **77**, 293; **78**, 182, 1950.
117. W. Thirring, *Phil. Mag.* **41**, 1193, 1950.
118. M. Gell-Mann and M. L. Goldberger, *Phys. Rev.* **96**, 1433, 1954.
119. F. E. Low, *Phys. Rev.* **96**, 1428, 1954, and Ref. 118.
120. A. Pais, *Phys. Rev. Lett.* **19**, 544, 1967; *Nuovo Cim.* **53**, 433, 1968.
121. N. M. Kroll and M. A. Ruderman, *Phys. Rev.* **93**, 233, 1954; cf. also A. Klein, *Phys. Rev.* **99**, 998, 1955.
122. G. Bernardini and E. L. Goldwasser, *Phys. Rev.* **95**, 857, 1954.
123. W. Heisenberg, *Zeitschr. f. Phys.* **120**, 513, 673, 1943.
123a. J. A. Wheeler, *Phys. Rev.* **52**, 1107, 1937.
124. W. Sellmeyer, *Ann. der Phys.* **143**, 272, 1871; **145**, 339, 520; **147**, 380, 525, 1872.
125. H. A. Kramers, *Atti del Congr. di Como* **2**, 545, 1927; *Collected works*, p. 333, North Holland, Amsterdam 1956.
126. R. de L. Kronig, *J. Am. Opt. Soc.* **12**, 547, 1926.
127. H. A. Kramers, *Phys. Zeitschr.* **30**, 522, 1929; Ref. 125, p. 347.
128. J. Toll, Ph.D. Thesis, Princeton 1952, unpublished; N. G. van Kampen, *Ned. Tydschr. Natuurk*, **24**, 1, 29, 1958.
129. R. de L. Kronig, *Physica* **12**, 543, 1946.
130. See esp. N. G. van Kampen, *Phys. Rev.* **91**, 1267, 1953.
131. M. Gell-Mann, M. L. Goldberger, and W. Thirring, *Phys. Rev.* **95**, 1612, 1954; M. L. Goldberger, *Phys. Rev.* **97**, 508, 1954.
132. M. Gell-Mann and M. L. Goldberger, Ref. 98, p. 36; G. C. Wick, Ref. 115.
133. M. L. Goldberger, *Phys. Rev.* **99**, 975, 1955; M. L. Goldberger, H. Miyazawa, and R. Oehme, *Phys. Rev.* **99**, 986, 1955.
134. Cf. H. Lehmann, *Nuovo Cim.* **14**, Suppl. 153, 1959; N. N. Bogoliubov and D. V. Shirkov, *Introduction to the theory of quantized fields*, Chapter 9, Interscience, New York 1959.
135. H. Lehmann, K. Symanzik, and W. Zimmerman, *Nuovo Cim.* **11**, 342, 1954; **1**, 205, 1955.
136. Ref. 64, Chapter 10.
137. M. L. Goldberger, *Proc. 6th Rochester Conference* (April 3–7, 1956), p. 1, Interscience, New York 1956.
138. I. Ya. Pomeranchuk, *Soviet Phys. JETP* **7**, 499, 1958; cf. also R. J. Eden, *High energy collisions of elementary particles*, Chapter 8, Cambridge Univ. Press, 1967.
139. R. E. Hendrick and B. Lautrup, *Phys. Rev.* **D11**, 529, 1975.
140. U. Haber-Schaim, *Phys. Rev.* **104**, 1113, 1956.
141. H. L. Anderson, W. C. Davidon, and U. E. Kruse, *Phys. Rev.* **100**, 339, 1955.
142. G. F. Chew, M. L. Goldberger, F. E. Low, and Y. Nambu, *Phys. Rev.* **106**, 1337, 1957.
143. P. Federbush, M. L. Goldberger, and S. B. Treiman, *Phys. Rev.* **112**, 642, 1958.
144. See R. Hofstadter, *Nuclear and nucleon structure*, Benjamin, New York 1963.
145. W. R. Frazer and J. R. Fulco, *Phys. Rev. Lett.* **2**, 365, 1959.
146. Y. Nambu, *Phys. Rev.* **106**, 1366, 1957.
147. M. L. Goldberger and S. B. Treiman, *Nuovo Cim.* **9**, 451, 1958.

148. A. Martin, *Scattering theory, analyticity and crossing*, Springer, New York 1969.
149. S. Mandelstam, *Phys. Rev.* **112**, 1344, 1958; **115**, 1741, 1752, 1959.
150. G. F. Chew, *S-Matrix theory of strong interactions*, Benjamin, New York 1961.
151. R. J. Eden, P. V. Landshoff, D. I. Olive, and J. C. Polkinghorne, *The analytic S-matrix*, Cambridge Univ. Press 1966.
152. D. Iagolnitzer, *The S-matrix*, Elsevier, New York 1978; J. Bros, *Mathematical Physics VII*, p. 145, North Holland, Amsterdam 1984.
153. T. Regge, *Nuovo Cim.* **14**, 951, 1959; **18**, 947, 1960.
154. R. Blankenbecler and M. L. Goldberger, *Phys. Rev.* **126**, 766, 1962, esp. footnote p. 766; G. F. Chew and S. C. Frautschi, *Phys. Rev. Lett.* **7**, 394, 1961.
155. G. F. Chew and S. C. Frautschi, *Phys. Rev. Lett.* **8**, 41, 1962.
156. P. D. B. Collins, *An introduction to Regge theory and high energy physics*, Cambridge Univ. Press 1977.
157. G. F. Chew, M. Gell-Mann, and A. H. Rosenfeld, *Sci. Am.* **210**, February 1964, p. 74.
158. Cf. Ref. 138, p. 126.
159. Ref. 138, p. 230.
160. A. C. Irving and R. P. Worden, *Phys. Rep.* **34**, 117, 1977; S. N. Gauguli and D. R. Roy, *Phys. Rep.* **67**, 201, 1980; J. K. Storrow, *Phys. Rep.* **103**, 317, 1984.
161. P. D. B. Collins and A. Martin, *Rep. Progr. Phys.* **45**, 335, 1982.
162. E. L. Goldwasser, Ref. 65, p. 345.
163. C. M. G. Lattes, in *The birth of particle physics*, Eds. L. M. Brown and L. Hoddeson, p. 307, Cambridge Univ. Press 1963.
164. D. A. Glaser, in *Nobel lectures in physics 1942–62*, p. 529, Elsevier, New York 1964.
165. L. Alvarez, in *Nobel lectures in physics 1963–70*, p. 241, Elsevier, New York 1972. See also P. Galison in *Experiment and observation in modern science*, MIT Press, Cambridge, Mass. 1985.
166. Ref. 65, pp. 483 ff.
167. A. Wightman, *Physics Today* **22**, September 1969, p. 53.
168. M. Gell-Mann, *Proc. conf. 'Symmetries in physics'*, Sant' Felice, Spain, September 1983.
169. G. F. Chew, *Physics Today* **17**, April 1964, p. 30; **23**, October 1970, p. 23; *Science* **161**, 762, 1968.

20

Onset of an era: new forms of matter appear, old symmetries crumble

(a) 'Four τ-mesons observed on Kilimanjaro'

In late September 1947, on the way home by slow freighter, there was ample time to reflect on Powell's report about π-mesons and μ-mesons which I had just heard* in Copenhagen. Was the muon telling us that there was a lepton spectrum, that the electron is some sort of ground state, that mass is quantized, as Møller and I had speculated** the year before? Perhaps. The pursuit of these preconceived but not necessarily relevant notions raised the question, however, why there was no sign of mass states built on the pion or on the nucleon as respective ground states, as our little model had suggested. During the next few years I did not forget the curious muon (no particle physicist ever forgets the muon, curious to this day) but dropped all thoughts about particle models and turned to other interests.

In December 1947, an article[2] by Rochester and Butler appeared entitled 'Evidence for the existence of new unstable particles'. Working in Blackett's laboratory in Manchester, they had discovered two unusual events in their cloud chamber cosmic ray pictures. One, found on 15 October 1946, showed a forked track which they interpreted as the spontaneous decay of a neutral particle into a pair of charged particles. The other, found on 23 May 1947,† showed a track with a marked kink, most probably the decay of a charged particle into another charged particle plus one or more neutrals. In both cases the mass of the parent particle lay somewhere between 770 and 1600m (m = electron mass). This range was very wide, not only because the identity of the decay products was unknown but also because the masses of the pion and muon, likely candidates, were not at all well established yet. Remember, we are only a few months after the discovery of π–μ decay.

In later years a discovery like Rochester and Butler's, however preliminary the data, would at once have caused excited corridor talk and would have set

* Chapter 18, Section (b). ** Chapter 18, Section (a); also Ref. 1.
† These dates are given by Rochester.[3]

telephones ringing across continents. In 1948, I must have heard of the Manchester pictures but do not recall their causing a stir or immediate awareness of a new era being upon us in respect to the structure of matter. It was certainly unlike the arrival of the muon, a few months earlier, which had been greeted with surprise but also with relief because it resolved paradoxes of meson scattering and absorption. The Manchester 'V-particles' (the early name for these new objects), on the other hand, served no purpose at all. Another contrast: after the first few muons had been found more kept coming, slowly but steadily. Regarding the V-particles, Rochester later wrote: 'The two years following 1947 were tantalising and embarrassing to the Manchester group because no more V-particles were found'.[4] Nor, apparently, did others at once re-examine earlier cloud chamber pictures for V-particle evidence.[5] Years later I raised this issue with Wilson Powell, a cosmic ray expert, asking him in particular why V-particles had not been discovered earlier as well they might have.* He promised an answer by the following day. When next we met he had with him a handful of cloud chamber pictures he had taken in the thirties, each one showing either a fork or a kink. He told me that some time in the early fifties he had convinced himself that these were V-particles, adding with a smile that, because of the great interest in showers in the thirties, experimentalists would load their cloud chambers with many metal plates in order to obtain high particle multiplicities and would pay little attention to individual tracks. Rochester and Butler had noticed their two events in a chamber containing only a single plate . . .

As said, in 1948 there was hardly any reaction to the news. The discussion on cosmic rays at the Pocono Conference (30 March–1 April 1948) does not contain[9] any mention of V-particles. Rochester's own report[10] on cosmic ray work to the symposium in honor of Millikan's eightieth birthday does not include V-particles in the list of main results. When I recently queried George Rochester about this he replied: 'It is true that there is meagre reference in the printed version . . . I did, however, say more about the V-particles in my talk and I spoke privately with a number of the participants, notably, Anderson, Brode, and Rossi'.[11]**

At that same symposium Brode from Berkeley mentioned[12] almost as an aside the observation in cosmic rays of eight particles with masses between 500 and $800m$. Leprince-Ringuet was more emphatic in announcing[13] an example of 'very heavy mesons which we call τ-mesons'. This event, found

* In 1944 a single event[6] had given a hint of a particle with mass $\sim 1000m$. However, 'till today [that] event remains a bit of a puzzle'.[7] See also Ref. 8.

** I am further indebted to Professor Rochester for sending me copies of letters (all from 1947) to him by Fermi (3 December), Heitler (23 November), Powell (24 November), Rossi (28 November), and Wheeler (9 December), all expressing great interest in the preprint they had received of the Rochester–Butler paper.[2] A discussion on these particles did take place during the eighth Solvay conference (Brussels, October 1948). Its proceedings were not published until 1950.

in emulsions, was a nuclear disintegration caused by a particle with mass at least $700m$. As the year drew to a close, the Bristol group found[14] the first example of a charged particle with mass between $870m$ and $985m$ decaying into three particles believed to be probably all pions, although $\pi+2\mu$ or $2\pi+\mu$ 'cannot be excluded'. This is the first example of 'τ-decay':

$$\tau^+ \to \pi^+ + \pi^+ + \pi^-. \tag{20.1}$$

It took more than a year before two more events of this kind were found.[15]

Nothing worthy of note happened in the first few months of 1950. In April, Fermi, lecturing at Yale on 'Elementary Particles',[16] made no mention of V's or τ's. Then, later that spring, I read a Caltech report on results obtained with 'the famous cloud chamber of Anderson that discovers all the new particles'.[17] Thirty forks and four kinks had been observed, most of them on Mount Wilson (3200 m). 'One must come to the same remarkable conclusion as that drawn by Rochester and Butler . . . these events . . . represent . . . the spontaneous decay of neutral and charged unstable particles of a new kind.'[18] No mass values were quoted. The decay products of the neutrals could be two π's, a π and a μ, or a proton and a meson (π or μ). I was startled. Were there meson and nucleon families after all?

Now, slowly, the new particles began to gain attention. During the Harwell conference (September 1950) Blackett reported[19] that the Manchester group, having moved their cloud chamber to the observatory at the Pic-du-Midi (2900 m), had found ten more events. In October the first case was found[20] of a V-particle, now called[22] Λ, with a proton as one identified decay product, the other being a π or a μ. During the first Rochester conference (December 16, 1950) Oppenheimer suggested[21] a discussion on τ-mesons. Nothing came of that, the day was spent on π-nucleon and muon physics. In October 1951, Fermi was taking the new particles quite seriously, as we have seen.* During the two-day long second Rochester conference (11–12 January 1952) one full day was devoted to the new particles; they have remained an important agenda item ever since. Also in early 1952, the first theoretical ideas on the new particles emerged, as will be seen in the next section. Marshak's textbook[23] of that year, *Meson Physics*, was the first to contain a chapter on the new particles.

Let us recapitulate the early years. The era of the new particles began in the 1940s with cosmic ray discoveries by groups at Manchester, the Ecole Polytechnique, and Bristol. The detectors used were either cloud chambers or emulsions. The latter technique received a great boost with the introduction[24] (1953) of stripped emulsions, sheets removed from plates and packed into solid blocs which eventually grew to sizes over 10 litres and weights over 60 kg. All early experiments were of high quality but suffered severely from

* Chapter 19, Section (f), Part 2.

low statistics, as I have endeavored to illustrate by naming this section after the title of a paper[25] published in 1954. It should be stressed that these early discoveries were part of a process of *gradual discovery*. Initially particles in a new mass range were observed. Their masses, decay products, and lifetimes, only barely defined, had yet to be given precision by experimental elaboration, an undertaking to which many contributed.

The early fifties witnessed a rapid expansion of experimental activities. The proceedings[26] of the Bagnères-de-Bigorre conference, held in July 1953, contains contributions from about twenty groups* active in the field. How could so many participate so rapidly in the new venture? Because they could use worldwide available beams, cosmic rays, and because they did not need special new instrumentation. All they needed to do was pay careful attention to rare events. I turn next to a brief summary of their findings, referring for many more details to Peyrou's excellent recent historical survey[7] of that period. In regard to literature I confine myself to the first mention of new phenomena. More detailed important references are found in Peyrou's paper.

At this point it is convenient to begin using some terminology introduced during the next few years. At Bagnères the following names[27] were adopted: 'L-mesons' for π and μ collectively; 'K-mesons' for particles with mass intermediate between the pion and the proton; 'hyperons'[28] for new particles 'with mass intermediate between those of the neutron and the deuteron (this definition to be revised if "fundamental" particles heavier than the deuteron are found)'. In 1954, the name 'baryon' was proposed[29] to denote nucleons and hyperons collectively. Symbols for specific baryons which came into use at that time:

Ξ^-	Ξ^0		(1320)
Σ^-	Σ^0	Σ^+	(1190)
	Λ		(1115)
	n	p	(940)

The numbers in parentheses are rounded-off average mass values (in MeV) of the particles to their left; more detailed specifications of these states will be discussed in Section (b). The symbols Σ and Ξ were proposed[30] in 1954.**

The best record on the status of the new particles in the early fifties is undoubtedly the proceedings[26] of the Bagnères conference.† In closing that

* These were located at Bern, Bombay, Bristol, Brussels, Chicago, Cork, Ecole Polytechnique, Genoa–Milan, Goettingen, Indiana University, Manchester, M.I.T., Naval Research Laboratory (Washington, D.C.), Paduva, Pasadena, Princeton, Rehovoth, Rochester, Rome, and St. Louis.

** After Gell-Mann and I had thought of the symbol Σ, we needed a name for the next higher state. At that point the name of the society $\Sigma\Xi$ came to my mind. That is how the cascade particle got its name.

† For more easily accessible reviews dating from that time see Ref. 31.

meeting, Blackett called it 'in many respects the best I ever attended'. Rossi had the difficult task of presenting a summary* which could not quite reflect a consensus. (A few reported results turned out later to be incorrect.) The following items and quotations are lifted from his report.
(a) Λ-decays:

$$\Lambda \to \mathrm{p} + \pi^- + (\text{m.p.})37\ \text{MeV} \tag{20.2}$$

(m.p. = most probably), an excellent value; mean life: 3 (present best value 2.6) $\times 10^{-10}$ sec.
(b) Some evidence[33] for

$$\begin{aligned}\Sigma^+ &\to \mathrm{p} + (\text{m.p.})\pi^0,\\ &\to \mathrm{n} + \pi^+.\end{aligned} \tag{20.3}$$

(c) 'One of the big surprises': evidence[34] for 'cascade particles' decaying into $\Lambda + \mathrm{L}$, later understood to be

$$\Xi^- \to \Lambda + \pi^-. \tag{20.4}$$

(d) First report[35] of Λ-particles bound to nuclear matter, 'hyperfragments'.
(e) Neutral K-mesons. The decay

$$\theta^0 \to \pi^+ + \pi^- \tag{20.5}$$

(θ^0 was later written as $K^0_{\pi 2}$) is well established; mass 495 ± 2 MeV (on the dot), lifetime $\simeq 10^{-10}$ sec (now 0.9 times that); some evidence for $\tau^0(K^0_{\pi 3}) \to \pi^+ + \pi^- + \pi^0$.
(f) Charged K-mesons, 'the most difficult part of my task'. τ-decay Eq. (20.1) well established. Mass: 495 MeV (very close); mean life 'conservatively' between 10^{-8} and 10^{-10} sec (now 1.2×10^{-8}). Furthermore there was evidence[36] for

$$\kappa^\pm \to \mu^\pm + 2\ \text{neutrals}, \tag{20.6}$$

then thought to be $\kappa^\pm \to \mu^\pm + \nu + \gamma$, now known to be

$$K^\pm_{\mu 3} \to \mu^\pm + \nu + \pi^0. \tag{20.7}$$

There were also weak indications[37] for

$$\chi^\pm \to \pi^\pm + 1\ \text{neutral}, \tag{20.8}$$

now known to be

$$\theta^\pm(K^\pm_{\pi 2}) \to \pi^\pm + \pi^0. \tag{20.9}$$

Commenting on this variety of processes, Rossi in his wisdom 'would like to take the point of view that two particles are equal until they are proved different', and proceeded to make the following most important comments:

* For a much abbreviated version see also Ref. 32.

'You find the same proportion of χ- and κ-decays and of τ-decays in photographic emulsions and in the cloud chambers ... [This] means that the κ-, χ-particles group and the τ-particles have very approximately the same lifetime .... *We can then reduce the number of particles to one* ... [There is a] very close similarity between the masses of two of the best established particles, I mean the charged τ-particle with a mass of 970 [m] and the θ^0-particle with a mass of 971. *This looks hardly like an accident* ...' [my italics].

His points were well taken. To begin with one should treat κ, χ, and τ as potentially non-identical objects. The lifetime argument showed, however, that these were probably alternative decay modes of a single particle, the K^+ (or the K^-). The mass argument showed that charged and neutral K's probably formed a multiplet. As time went by these probabilities turned into certainties.

Thus, with limited statistics and without prodding by theorists, did the cosmic ray physicists usher in the era of the new particles, and, by their conjecture that there exists only one charged K-particle, pave the way toward the discovery of non-conservation of parity.

In his closing remarks at Bagnères, C. F. Powell said: 'Gentlemen, we have been invaded ... the accelerators are here'.[26]

In April 1953 a report had been submitted[38] of the first new particles produced by the big machines: two V-events had been seen at the Cosmotron. These were discussed[39] at Rochester 4 (25–7 January 1954). At Rochester 5 (31 January–2 February 1955) the good news[40] was that a dozen V's had been found at the bevatron. Thereafter the accelerators really took off.

The first of the man-made V-particles were found with bubble chambers and emulsions. Soon, however, new detectors such as Cerenkov counters made their appearance. Transistors, just discovered, were simplifying electronics equipment. Then came the bubble chambers. Next a new breed, beam builders, managed to create electromagnetically separated beams with high enrichment factors for some desired species of particles.* Cosmic ray studies did not come to a halt but by the mid-fifties 'everybody knew that the cosmic rays had finished their task in elementary particle physics'.[7] The rays from the skies simply could not compete with the beams from the accelerators as the following crude numbers illustrate. The flux of cosmic ray protons with energies in the few GeV range reaching the top of a medium high mountain is something like one per cm^2 per minute per steradian. About $1\frac{1}{2}$ V-events (decays) per day are generated in a medium-sized bubble chamber installed at this altitude. The Cosmotron (3 GeV) deposits $\sim 10^{11}$ protons per pulse (every 3 secs) on the target. These produce $\sim 10^9\, K^+$. After beam separation one is left with $\sim 10^3\, K^+$ (always per pulse). For the AGS (30 GeV) these numbers are: $\sim 4 \times 10^{12}$ protons per pulse (every 3 sec) $\rightarrow \sim 10^{11}\, K^+ \rightarrow \sim 10^6\, K^+$ after beam separation (K^- is less by a factor 3–10). In fact, in later years the main problem

* These various innovations are surveyed in Ref. 41.

with K^- (and other) beams shifted to keeping event rates low enough so as not to swamp detectors.

As to contributions from cosmic ray studies to elementary particle processes, they will probably be right back at the frontiers, sooner or later.

(b) Early theoretical ideas

1. Associated production. In recent years, the April issue of *Reviews of Modern Physics* has contained an updated 'Review of particle properties'. Entries recently listed under the heading 'Stable particles' include γ (the photon), W, Z, leptons, K-particles, and all the baryons previously listed. At first acquaintance it must strike the uninitiated as bizarre to find particles with mean lives $\sim 10^{-10}$ sec listed as stable. Even the particle physicist will concede that this stretches things a little. He (short for he or she) has good reasons, however, for calling a Λ or a K very long-lived. His scale of lifetimes is set by the number 10^{-23} sec, the time it takes for a resonance* like a Δ (also listed as a particle in the 'Review') to disintegrate into π + nucleon. That time has never been directly measured; it is too short. It can be estimated, however, from the observed width $\Delta E \sim 100$ MeV of the Δ-resonance and from the uncertainty relation

$$\Delta E \Delta t \gtrsim h \simeq 4 \times 10^{-21} \text{ MeV sec.} \tag{20.10}$$

A Λ, too, is strictly speaking a resonance, it also decays into π + nucleon, but no one has ever directly observed its width, $\sim 10^{-5}$ eV, corresponding to its readily measurable lifetime, $\sim 3 \times 10^{-10}$ sec. Following a slightly arbitrary yet understandable usage, the particle physicist likes to call the Δ a resonance, the Λ a particle.

Why do Λ's and K's, and also Σ's and Ξ's, live so long? What makes $\Lambda \to p + \pi^-$ go so very slow compared to $\Delta^0 \to p + \pi^-$? That problem, by far the most intriguing novelty of the new particles, needs to be stated more precisely. Before doing so I give away the punch line. In the 'Review' just quoted, stable particles are defined as 'All particles stable under the strong interactions'.

It was known[42] (we are in about 1951) that the production rate of V-particles is $\gtrsim$ one per cent of that of pions. This copious production had therefore to be due to a strong interaction, roughly comparable in strength to the π-nucleon coupling. Suppose there were a coupling of a Λ to nucleon and π of comparable strength. That would explain copious Λ-production but would also lead to $\Lambda \to p + \pi^-$ as a fast process, and the Λ would be a 'resonance' which it is not. The problem therefore was how to reconcile copious production with slow decay.

* Chapter 19, Section (d).

Some time early in 1951 it was suggested* that a steep potential barrier between nucleon and pion might suppress Λ-decay, a low energy process, without inhibiting Λ-production, a high(er) energy process, as would be the case if the Λ had a spin** of, say, 13/2. I did not much care for that idea. As indicated earlier I had the vague notion that the Λ should be related to the nucleon as the muon is to the electron. If so, one should seek a solution in which 'the heavy fermion [Λ] is as elementary as the nucleon'.[42] Thus I began to look for a model in which 'the emphasis will be ... on the role of selection rules', rules which had to be of a new kind.

Being familiar with the difficulties of pion physics, I knew that perturbation theory would be of no use. In addition there was the further problem of electromagnetic decay. Further prohibition was required to slow down processes like $K^0 \rightarrow 2\gamma$ by at least six orders of magnitude. Thus I looked for *selection rules which would hold for strong and electromagnetic but not for weak processes*, meaning by weak processes reactions 'similar to ... neutrino processes where a coupling constant occurred which is very small'. The selection rule I proposed, the 'even–odd rule', can in simplest form be stated as follows. Assign a number 0 to all 'old' particles (π, N, γ, leptons) and a number 1 to the new particles Λ, K (the others were not there yet). In any process, first add these numbers for initial-state particles, then for final-state particles, the respective sums being n_i and n_f. Then in all strong and electromagnetic processes n_i and n_f must be both even or both odd; in weak decays of the new particles one shall be odd, the other even. Thus $\pi^- + p \rightarrow \Lambda + \pi^0$ is strongly forbidden, $\pi^- + p \rightarrow \Lambda + K^0$ strongly allowed. In general new particles come in pairs, a mechanism later named 'associated production' (I do not know who coined this term). Electromagnetic decays like $K^0 \rightarrow 2\gamma$ are forbidden. $\Lambda \rightarrow p + \pi^-$, $K^0 \rightarrow 2\pi$, etc. proceed by weak interactions. Later I learned† that several Japanese colleagues[46] had been considering a series of options which included the idea of a strong interaction selection rule.

I noted at that time[42] that 'there is perhaps an indication of the existence of families†† of elementary particles like a nucleon and an electron family in which, not unlike the levels in a given kind of atom, the members of a given family are distinguished from each other through a quantization process, but one of a new kind', adding that 'the search for ordering principles at this moment may ultimately have to be likened to a chemist's

* According to Feynman (note added on 7 June 1951, to his lectures[43] on high energy physics) this idea had originated with W. Fowler and Fermi. Cf. also Ref. 44.

** A critique of this idea is found in Ref. 30.

† I lectured on my ideas in the autumn of 1951 at the Institute for Advanced Study, and again at the second Rochester conference. (I am not responsible for the term 'megalomorphs' in the title of that talk. That was an invention of Oppenheimer, made because 'Fermi had become bored with the name "elementary particles".'[45]) The Japanese papers are referred to in the paper[42] prepared shortly thereafter.

†† In a table I drew on the board during the Rochester conference I made parallels between three families, then called nucleons, bosons, and leptons. (This table was recently reproduced by Marshak.[47])

attempt to build up the periodic system if he were only given a dozen odd elements'.

Early in 1953, I made the first try at enlarging the isospin group so as to find room for an additional quantum number which should obey an additional selection rule. In other words, the plan was to give the needed additional selection rule a group theoretical foundation. The new particles should be grouped into isospin multiplets, like the old ones. Isospin should be conserved in *all* strong interactions[48] but not in electromagnetic interactions, while the new rule should hold for both these forces but not for weak interactions: 'The present picture seems to involve a hierarchy of interactions corresponding to the symmetry classes of the [intrinsic] variables'.[49] That picture has survived. The first example of an enlarged group[49] was richly premature, however. For one thing, there still was no cascade particle! Further possibilities, including several variants of the Fermi–Yang model (Chapter 19, Section (f)),[50] were explored during the next few years* but the correct answer was not found until the early sixties (Chapter 21).

As to associated production, cosmic ray evidence[53] seemed at first against it. Cyclotron experiments at energies too low for production of the new particles in pairs gave no evidence[54] for their single production. The issue was settled when accelerators in the GeV range became available. A Cosmotron experiment (November 1953)[55] yielded the first convincing results and also provided the first example of one of the new particles being discovered at an accelerator: the Σ^-, decaying into $n+\pi^-$.

2. The strangeness scheme. The next step in the reconciliation of strong production and weak decay was made in August 1953 when Murray Gell-Mann made the ingenious proposal[56] of assigning integer isospin to hyperons (there was *still* no cascade particle) and half-integer values to K-particles. Later that year the same scheme was put forward[57] independently by Tadao Nakano and Kazuhiko Nishijima. (Recall once again: isospin, T, works like angular momentum, see Eq. (17.30). A particle is labeled by T and its third component T_3.)

Let us see how it works.** Σ's shall have $T=1$, $T_3=1, 0, -1$ for plus, zero, minus charge; $T=0$ for Λ. Further (K^+, K^0) and $(\bar{K}^0, K^-)$ each have $T=\frac{1}{2}$ and, in the order written, $T_3=\frac{1}{2}, -\frac{1}{2}$. The introduction of *two* neutral K's is another novelty about which more shortly. In the presence of only strong interactions, all Σ's are mass degenerate, as are all K's. Just a reminder, the nucleons are as usual, (p, n), $T=\frac{1}{2}$, $T_3=\frac{1}{2}, -\frac{1}{2}$ respectively. All multiplets will be slightly split by electromagnetic effects.

What does this do? Consider $\Lambda \to p+\pi^-$ with $T_3=0$ on the left, $T_3=\frac{1}{2}-1=-\frac{1}{2}$ on the right. The Λ survives strong and electromagnetic† interactions

* For a review of these see Refs. 51 and 52.

** For ease I use modern particle symbols and add Λ to the particles introduced in Refs. 56, 57.

† One can construct fundamental interactions that do not conserve T_3. Those, however, are absent in any sensible version of the theory.[58,59]

which conserve T_3 but decays because of weak interactions which don't. Production processes can or cannot proceed via strong interactions depending on whether or not T, T_3 are conserved. Examples of the first kind:

$$\pi^- + p \to \Lambda + K^0, \quad \text{or} \quad \Sigma^- + K^+ \tag{20.11}$$

and of the second kind:

$$n + n \to \Lambda + \Lambda. \quad \text{or} \quad \pi^- + p \to \Sigma^+ + K^-. \tag{20.12}$$

Every allowed process complies with associated production but the converse is not true. As time went by, all these consequences turned out to be good things experimentally, as did the assignments, made shortly afterward,[60] for the Ξ-doublet: $T = \frac{1}{2}$, $T_3 = \frac{1}{2}$ $(-\frac{1}{2})$ for Ξ^0 (Ξ^-).

In 1954 Gell-Mann and I prepared a joint report 'Theoretical views on the new particles' in which all ideas on the subject known to us were summarized.[30] This paper was presented in July at a conference in Glasgow. The Σ^0 and Ξ of the Gell-Mann scheme were as yet hypothetical. These particles were discovered some years later in bubble chambers exposed to accelerators: Σ^0 in 1956,[61]* Ξ^0 in 1959.[62] In regard to decays Λ, $\Sigma \to \pi + \text{nucleon}$, $\Xi \to \Lambda + \pi$ which go with a change $|\Delta T_3| = \frac{1}{2}$, it was suggested[30] that weak interactions obey the stronger 'violation law'**

$$|\Delta T| = \tfrac{1}{2} \tag{20.13}$$

which, it turned out later, has many good consequences.†

Detailed and streamlined accounts of this scheme, including more of its consequences, were given in 1955 by Nishijima[64] and by Gell-Mann.[59] Both papers contain the extension of Eq. (19.19) needed to include the new particles:

$$Q = T_3 + \frac{S}{2} + \frac{B}{2}. \tag{20.14}$$

S is the 'strangeness' quantum number (Nishijima called it η, the 'η-charge'), conserved in strong and electromagnetic interactions. Assignments for S: 0 for (p, n); -1 for Λ, Σ, $\bar{K}^0$, K^-; -2 for Ξ; $+1$ for K^+, K^0. Weak interactions obey to leading order the violation rule

$$|\Delta S| = 1. \tag{20.15}$$

The strangeness scheme was a faithful guide through the fifties. Its implications have been described in accounts, all written by experts, ranging from a

* Σ^0 decays into $\Lambda + \gamma$, allowed by T_3-conservation and T-violation by one unit.
** Chapter 19, Section (d).
† Example: according to Eq. (20.13) the ratio of rates $\Lambda \to p\pi^-$, $\Lambda \to n\pi^0$ equals 2. Eq. (20.13) is not rigorous. For example weak $|\Delta T| = 3/2$ transitions are possible when this rule is combined with the electromagnetic rule $|\Delta T| = 0, 1$ (Chapter 19, Section (d), Part 2). Such further suppressed transitions can be seen to be important for the process $K^+ \to \pi^+\pi^0$. A detailed analysis of these and other weak decays is found in Ref. 63.

popular article[65] to a simple book[66] to a more advanced book[52] to a technical treatment.[63] Its usefulness has remained undiminished in later years when it was found necessary to introduce still more quantum numbers.

3. The neutral K-particle complex. In the course of preparations for the Glasgow report, considerable time was devoted to discussions of a question raised by Fermi: If K^0 ($S=1$) and $\bar{K}^0$ ($S=-1$) are non-identical particles, how does one see that in the laboratory? The resulting new concept[67] of 'particle mixtures' was so unfamiliar that it was thought best not to report it at the Glasgow conference. It was written up only several months later. These ideas emerged before the invariance under charge conjugation (C) (and parity) had been challenged, and will be dealt with next in that initial context (using updated particle symbols). The subsequent modifications, small but most important, will be discussed shortly.

Consider the decay (20.5):

$$K^0 \rightarrow \pi^+ + \pi^-, \quad \text{with amplitude } A, \tag{20.16}$$

in which the final state goes into itself* under the operation of charge conjugation C (defined in Chapter 16, Section (d), Part 3). The charge conjugate of Eq. (20.16) is

$$\bar{K}^0 \rightarrow \pi^+ + \pi^-, \quad \text{also** with amplitude } A, \tag{20.17}$$

where the final state is the same as before but not the initial state which has $S=+1$ in the first, -1 in the second process. How can the operation C transform the final but not the initial state into itself? It cannot do so as long as S is a good quantum number (strong interactions) but it can, and does, when S is not a good quantum number (weak interactions). In fact weak interactions, acting twice, mix K^0 and $\bar{K}^0$ via

$$K^0 \rightleftarrows \pi^+ + \pi^- \rightleftarrows \bar{K}^0. \tag{20.18}$$

Leave this small mixing aside for a moment and introduce the combinations of one-particle states

$$K_1 = \frac{K^0 + \bar{K}^0}{\sqrt{2}}; \qquad K_2 = \frac{K^0 - \bar{K}^0}{\sqrt{2}}. \tag{20.19}$$

According to (20.17), (20.19) the state K_1^0 can decay into $\pi^+ + \pi^-$ (with amplitude $A\sqrt{2}$) but K_2^0 cannot decay at all that way (amplitude zero). More generally K_1 (K_2) can only decay into states even (odd) under C. These two states will therefore have different lifetimes, τ_1 and τ_2 respectively. Since we

* Since K^0 has zero spin the two pions have zero orbital angular momentum so that the state is unchanged when $\pi^+ \leftrightarrow \pi^-$.

** Up to a phase which we are free to choose to be +1.

should properly reserve the name 'particle' for an object with a unique lifetime, the K_1 and K_2 are true particles. K^0 and $\bar{K}^0$ must be considered as 'particle mixtures' given by

$$K^0 = \frac{K_1 + K_2}{\sqrt{2}}; \qquad \bar{K}^0 = \frac{K_1 - K_2}{\sqrt{2}}. \tag{20.20}$$

It was at once suspected[67] that $\tau_2 \gg \tau_1$, K_2 should live much longer than K_1, simply because for a variety of reasons the rates of decay into $\pi^+\pi^-$ and $2\pi^0$ (both inaccessible to K_2) are so much larger than the rates for alternative decays. The first evidence for such a long-lived neutral K-particle was obtained[68] two years later in one of the last cloud chamber experiments at accelerators; the present best value for τ_2 is 5.2×10^{-8} sec, longer than τ_1 by a factor ~ 500. In 1957 it was found[69] that not more than half of the particles in a beam consisting only of K^0's can decay into two pions, another prediction[67] of particle mixing.

After I had given a seminar on this subject at Columbia University sometime early in 1955, Piccioni came up to talk to me. Out of our long discussion about the fate of a freshly created K^0 beam* in the course of time grew the idea of 'regeneration' which we illustrated[70] by the following example.

High energy particles bombarding a thin plate in a cloud chamber produce at time zero a pure K^0 beam (for ease supposed to be monoenergetic) with density ρ. t seconds later the beam has changed into a mixture of K^0 and $\bar{K}^0$ since, meanwhile, more K_1 than K_2 have decayed; t is chosen such that $\tau_2 \gg t \gg \tau_1$. Straightforward quantum mechanics says that the probabilities $P_\pm(t)$ for finding a K^0 or a $\bar{K}^0$ at time t are

$$P_\pm(t) = \frac{\rho}{4}\left[e^{-t/\tau_1} + e^{-t/\tau_2} \pm 2\, e^{-t/2\tau_1 - t/2\tau_2} \cos \frac{\Delta c^2 t}{\hbar}\right] \tag{20.21}$$

where + (−) refers to K^0 ($\bar{K}^0$), and

$$\Delta = m(K_2) - m(K_1) \tag{20.22}$$

is the tiny mass difference generated[67] by weak interaction effects such as Eq. (20.18).

Suppose next that at time t the beam, now almost purely K_2, hits a second plate thick enough to absorb the $\bar{K}^0$ component of K_2 by the allowed strong processes $\bar{K}^0 + \text{nucleon} \to \Lambda$ or $\Sigma + \pi$. The rules of strangeness forbid analogous processes for K^0. Thus absorption enriches the K^0-component of the beam so that a regeneration of K_1 occurs, manifested by reappearance of $K_1 \to 2\pi$. Scattering of the beam in the second plate alters the relative phase of its K^0- and $\bar{K}^0$-components; this effect too will regenerate K_1. In all, 2π-decays must

* Neutral K-particle production, a strong interaction process, generates necessarily K^0 or $\bar{K}^0$ states.

reappear at a rate at most one fourth of the rate after the first plate.* Regenerations by this method, as well as by K_2-diffraction off complex nuclei and off nucleons were all observed[73] in the early sixties.

Thus far t was taken large compared to τ_1. As Serber noted,[70] the interference term involving Δ in Eq. (20.21) may give observable effects for t comparable to τ_1. Theoretical proposals soon appeared for actually measuring Δ by following aging K^0 beams, using as criteria either[74] the decays into lepton + π^0 + ν, or[75] the distinct strong interactions of the K^0 and $\bar{K}^0$ components.

Δ is extremely small. From Eq. (20.10) one estimates it (putting $t = \tau_1$) to be $\simeq 10^{-5}$ eV, the value actually observed. It is due to the possibility of Δ-effects building up in time that this quantity, the smallest mass difference between particles ever detected, is observable. Also the sign of Δ is known: positive.**

There are many more 'bizarre manifestations of the mixing of K^0 and $\bar{K}^0$'.[77] My favorite one concerns the annihilation $\bar{p} + p \rightarrow \bar{K}^0 + K^0$ which at low energies proceeds predominantly from the ^{1}S-state. It can be shown[78] that of the three conceivable combined decays, K_1K_1, K_1K_2, and K_2K_2, only the first one is allowed. Experiments have verified this.[79] This is a further striking example of how the neutral K-system serves as a testing ground, not only for particle properties but also for such quite general concepts as the quantum mechanical superposition principle. As Feynman put it: 'It is not based on an elegant hocus-pocus .... We have taken the superposition principle to its ultimate logical conclusion .... It works'.[80] For a detailed survey of these phenomena I direct the reader to a review by Lee and Wu.[81]

(c) In which the invariance under reflections in space and conjugation of charge turns out to be violated

1. The Dalitz plot thickens. The first strange particle observations at accelerators date from 1953 (as noted) but it was not until 1955 that machine physics began to dominate the subject. Meanwhile cosmic rays continued to yield important information, including confirmations[82] of the $\theta^{\pm}$ decay mode Eq. (20.9), of importance for what follows next. As accelerators began to take over, high priority was given to checking whether the cosmic ray physicists had been correct in surmising (see Section (a)) that all charged K-decay modes exhibit the same lifetime. In 1953, these earlier results were corroborated in a series of experiments[83] with much improved statistics. This evidence for the existence of only one K^+ (and one K^-) particle with many decay options received further credence when it was found[84] that the corresponding branching ratios remain unaltered under changing conditions of K^+-production; and from the observed[85] equality of $\theta^{\pm}$ and $\tau^{\pm}$ scattering. Mass measurements

* These mechanisms were analyzed in more detail by Case[71] and by Good.[72]
** References to measurements of Δ are collected in Ref. 76.

reaffirmed another conclusion drawn from cosmic ray data: the near equality[86] of charged and neutral K-masses. Thus it appeared that the situation was as simple as could be: one charged K-particle of each sign forming an approximate multiplet with neutral K-particles.

There was one hitch, however.

The individual pions produced in $\tau^{\pm}$ decay (Eq. (20.1)) do not, of course, emerge with a unique energy. Their possible energy ranges can be coded in a finite two-dimensional region such that each point therein marks uniquely the energy configuration of the three pions. At the Bagnères conference of 1953, Dalitz presented[87] a very convenient way of mapping this region. This 'Dalitz plot' is described in every good text on particle physics.[88] Suffice it to say here that equal areas in the plot correspond to equal volumes of (covariant) phase space, a very useful property. The number of τ-events which, in the course of time, Dalitz registered in his plot tells the story of the period. He had access[87] to 13 points (all from cosmic rays) in 1953; 53 (42 from cosmic rays) at the fifth Rochester conference (31 January–2 February 1955)[89]; over 600, most of them from accelerators, at the sixth Rochester conference (3–7 April 1956).[90] The more points appeared, the stronger the evidence became that their distribution is uniform. As can readily be seen[88] this strongly indicates* that the $\tau^{\pm}$ have zero spin. That, however, raised a serious issue, known in its day as the tau–theta puzzle.

Straightforward reasoning[88] shows that, given the spin (zero) and parity (odd) of pions, a τ with zero spin must have odd parity; it is a pseudoscalar. If, as looked plausible, τ^{+} and θ^{+} (Eq. (20.9)) are decay modes of the same particle, then θ^{+} should of course also have zero spin, odd parity. But that cannot be! It is equally straightforward[88] to show, using again the pion properties, that Eq. (20.9) implies: zero spin goes with even parity. A particle cannot have a parity which is now even, then odd.** What to do?

There appeared to be three options. First, perhaps the τ and θ are two distinct particles[91] rather than alternative decay modes of one particle. If so, their near mass degeneracy was a mystery, their lifetime equality distinctly unpalatable. Secondly, perhaps the spin was not zero. For example, spin 2, even parity, would permit both 2π- and 3π-decay. That option became less and less attractive the more the Dalitz plot thickened. These two ways out (rapidly abandoned shortly afterward) represented attempts to save the situation by conventional means. Not so option three: there is only one charged

* The distribution depends not only on the spin but also on the interactions among the decay pions. The distribution *is* uniform (for large numbers of points) if the spin is zero and if these interactions are feeble. No evidence to the contrary was ever found.

** Before θ^{+} was well established one used to compare τ^{+} with θ^{0}, Eq. (20.5). This led to the same conclusion if one assumed that these two particles belong to the same doublet. That reasoning is sound but less direct than the τ^{+}–θ^{+} comparison, which came into use in 1956.

K-particle, its spin is zero but parity is not conserved in its decay, so that nothing prevents 2π- and 3π-decays from being alternative modes.

I do not know when the subject of parity-nonconservation in K-decay was first mentioned; it may well be that this thought had occurred to several independently. The issue was raised, more as a logical possibility than as an idea to be embraced, at least as early as 1954, at the Kyoto conference.* It was not until the sixth Rochester conference, however, that serious discussion of this option arose for the first time, during sessions as well as over coffee. Oppenheimer noted[93] after Dalitz' report: 'The τ-meson will have either domestic or foreign complications. It will not be simple on both fronts'. Shortly afterward I wrote: 'Be it recorded here that on the train back from Rochester to New York, Professor Yang and the writer each bet Professor Wheeler one dollar that the theta- and the tau-meson were distinct particles; and that Professor Wheeler has since collected two dollars'.[94]

I shall relate shortly what happened next but need to raise first several related issues bearing on earlier times.

2. P, T, and C prior to 1956. Early in 1924, Otto Laporte, a student of Sommerfeld, discovered[95] that the energy levels of iron atoms consist of two subsets that do not intercombine. A few months later the same observation[96] was made independently by Henry Norris Russell while studying titanium. Later their results turned out to be special cases of a general quantum mechanical rule according to which all spectral lines produced by electric dipole radiation are generated by transitions between atomic (or molecular) states of opposite parity. In 1924, quantum mechanics was not yet there, however, hence parity could not even be defined. Heisenberg, in his last paper (April 25)[97] prior to his discovery of quantum mechanics, attempted to formulate a rationale for the 'Laporte rule'. That was still too early for a correct interpretation. In May 1927, two years after quantum mechanics, Wigner found the right answer.[98] He divided atomic states into 'normal terms' (we would say even-parity states) and 'reflected (*gespiegelte*) terms' (odd-parity) and noted that for electric dipole radiation only transitions normal $\leftrightarrow$ reflected are allowed. In February 1928, he returned to this problem in his seminal paper 'On the conservation laws of quantum mechanics'.[99] There he noted that these laws are associated with the existence of unitary operators P that commute with the Hamiltonian H. As a result states can be chosen such that P and H are simultaneously diagonal: P is conserved. Taking P to be the reflection operator (*Spiegelung*), its eigenvalues, the parity quantum number, or parity** for short, can be either $+1$ or -1. This led Wigner[99] to remark:

* Brueckner: 'If one assigns spin higher than zero to these particles . . . then the same particle can decay into 2 or 3 pions without any violation of parity conservation.'[92]

** Weyl[100] in 1931 and Pauli[101] in 1933 still use the early name 'signature'. The name 'parity' is found[102] in 1935. I do not know whose invention that was.

'Only rarely will one be able to use [parity] since it has only two eigenvalues (±1) and has therefore too little predictive power [!].'*

More importantly, Wigner stressed that '[Parity] has no analog in classical mechanics'. By way of comparison, invariance under spatial rotations yields a conservation law; angular momentum is an integral of the motion, whether in classical or quantum mechanics. Invariance under spatial reflections, on the other hand, while well defined (and used, as in the classification of crystal types) classically, yields the associated concepts of parity and its conservation only in quantum mechanics. Thus did Wigner for the second time introduce a discrete symmetry with ramifications that have no classical counterpart. (The first time had been his treatment of permutation symmetry for identical particles.**)

The parity selection rule is the first instance of a multiplicative rule. Example: the parity of a two-particle system (AB) in a state of definite relative orbital angular momentum is the product of the intrinsic parities of A and B respectively and the orbital parity. Nevertheless we say that the parity of a state of the H-atom is identical with the orbital parity. The intrinsic parity of the proton and the electron do not come into question, because for all of physics one may introduce the *convention* that these two parities are plus. Had we chosen any other convention, we might have had to change our language somewhat without changing any observable conclusions. As was particularly stressed by Wick, Wightman, and Wigner[103] the question of whether the relative parities of two states are measurable or just conventional is related to the possibility of observing, in any experiment whatsoever, the relative phase of these states. Note also that in quantum field theory there is no distinction between free spin zero (or one) fields of even and odd parity. *The physical content of parity is bound to interactions*, a fact which is unnecessarily obscured by the term 'intrinsic' parity. Parity is violated when H cannot be made invariant under reflections by whatever choice of phases.

Invariance under time reversal (T), the next discrete symmetry to enter quantum physics, had played a prominent role in the classical physics of earlier times, especially in the reconciliation of apparently irreversible thermodynamic properties (the second law) with the classical mechanical laws on which thermodynamics is based and which themselves are T-invariant. The extension to classical electrodynamics is due to Boltzmann who noted[104] in 1897 that the Maxwell equations remain unchanged 'if one at a given moment leaves all electric forces and polarizations unchanged but reverses the time direction and the magnetic forces and polarizations', this in refutation of an argument to the contrary by Planck.[105]

* 'Man wird ihn aber nur selten gebrauchen können da er nur zwei Eigenwerte (±) hat und so zu wenig auszusagen vermag.'
** Chapter 13, Section (a).

The first treatment of T-invariance in quantum mechanics is once again due to Wigner.[106] The stimulus came from Kramers' degeneracy theorem[107] which says (in its most general form) that the energy eigenstates of an odd number of spin-$\frac{1}{2}$ particles are at least doubly degenerate in the absence of an external magnetic field. Kramers used specific dynamical considerations. Wigner reduced the theorem to its essence, a consequence of T-invariance. The theorem exemplifies a general feature of T-invariance implications: these always involve relations between different states, never an intrinsic property of one state, quite unlike the situation for P. This distinction stems from the fact that the operation of time reversal (in Wigner's words[106]) 'is not quite trivial because of its non-linear character'. Specifically, in the simple Schroedinger theory T sends a wave function into its complex conjugate (T is 'anti-unitary'). Accordingly there is never a quantum number related to T-invariance. Note further that in applying T we must not only reverse momenta but spins as well (unlike the situation in detailed balance arguments[108]).

The next discrete operation to appear, charge conjugation (C-) invariance, has been discussed earlier.* It was noted that C can properly be treated only in the context of quantum field theory; and that the electromagnetic current and vector potential change sign under C. The C-operation can likewise be applied to all five covariants which appear in the general β-decay interaction Eq. (17.26) which I copy:

$$H_1 = \sum_i g_i \bar{\psi}_{\mathrm{p}} \mathrm{O}_i \psi_{\mathrm{n}} \cdot \bar{\psi}_{\mathrm{e}} \mathrm{O}_i \psi_\nu + \text{hermitian conjugate}$$

$$i = S, V, T, A, P, \tag{20.23}$$

$$\mathrm{O}_S = 1, \qquad \mathrm{O}_V = \gamma_\mu, \qquad \mathrm{O}_T = \sigma_{\mu\nu}, \qquad \mathrm{O}_A = i\gamma_\mu\gamma_5, \qquad \mathrm{O}_P = i\gamma_5,$$

with the following result (a, b = p, n or e, ν)

$$\bar{\psi}_a \mathrm{O}_i \psi_b \rightarrow \varepsilon \bar{\psi}_b \mathrm{O}_i \psi_a, \tag{20.24}$$

$$\varepsilon = \begin{cases} 1 & \text{for } \mathrm{O}_i = S, A, P, \\ -1 & \text{for } \mathrm{O}_i = V, T. \end{cases} \tag{20.25}$$

P and C invariance provide excellent illustrations for the distinction between invariance of physical laws and non-invariance of initial conditions. P is not violated because our heart is on the left or because polarized light is twisted to the left when passing through an aqueous sugar solution. Nor is C violated because, in our environment, protons and electrons dominate over their anti-particles. All these situations are created by locally asymmetric initial conditions.

The reader will enjoy numerous applications of P, C, T in the books by Sakurai[52] and T. D. Lee.[109]

* Chapter 16, Section (d), Part 3.

3. Spin and statistics; the CPT theorem. As will be recalled shortly, experiment has revealed that nature does not strictly respect the invariances under C, P, and T. To this day, however, no violation has been found of the invariance under the application (in any order) of these three symmetries combined. Theoretically, this 'CPT invariance' occupies a unique position. When the invariances under C, P, and T separately (and under the product of any two of them) were found wanting, no other postulate of relativistic quantum field theory was thereby affected. Violation of CPT invariance, on the other hand, would be much more serious in that it would demand revision of other general principles. That is the content of the 'CPT theorem' which can remain valid even though C, P, and T are separately violated.

In order to appreciate the generality of that theorem, it is useful to begin with some remarks on a closely related subject, the connection between spin and statistics. That takes us back to the 1930s.

In June 1934, Pauli sent Heisenberg a copy of the Pauli–Weisskopf paper on the quantization* of the spin-zero field. Recall (Eqs. (16.47), (16.48), (15.41)) that this field had been quantized by the rules of Bose–Einstein statistics. An accompanying letter to Heisenberg[110] contains the first comment on the connection between spin and statistics: 'One cannot quantize the scalar wave equation by the rules of the exclusion principle [that is according to Fermi–Dirac statistics**] in such a way that simultaneously: (1) Relativistic and gauge invariance hold; (2) The eigenvalues of the energy are positive, while both requirements are satisfied if one quantizes according to Bose statistics.' Out of this remark grew, by and by, the spin-statistics theorem which says, loosely stated, that half-integer (integer) spins can only be consistently quantized according to Fermi–Dirac (Bose–Einstein) statistics. Fierz' paper[111] on fields with spin >1, followed by Pauli's paper[112] of 1940, were important steps on the way.† More and more the theorem was freed from unnecessary assumptions, until in 1958 Burgoyne phrased it in the most general way[115] in which no constraints on the form of field equations or interactions are needed. (Gauge invariance is not a requisite.) If a field theory satisfies the conditions: (1) Invariance under proper (no space reflections) orthochronous (no time reversal) inhomogeneous (space–time translations included) Lorentz transformations; (2) No states of negative energy††; (3) The metric in Hilbert space is positive definite; (4) Distinct fields either commute or anti-commute for space-like separations (cf. Eq. (19.40)), then no field can have the 'wrong' connection between spin and statistics; and this is true for any spin.

* Chapter 16, Section (d), Part 7.

** See Chapter 16, Section (d), Part 2, and also Eq. (15.46).

† Many more details on the stages of evolution of the spin statistics as well as of the CPT theorem are found in Refs. 113, 114, where one will also find references to all the pertinent papers.

†† Or better (to avoid any ideas that 'hole theory' is excluded) a state of lowest energy can be defined, and is given zero energy by convention.

The *CPT* theorem can now be simply stated: the conditions just mentioned are sufficient* for showing that the invariance under the combined *CPT* symmetry: right ↔ left, particle ↔ anti-particle, past ↔ future, *must* hold. Pauli again played a key role[116] in the evolution of *CPT*-invariance. The most general proof, due to Jost[117] is based on** Wightman's axiomatic formulation of quantum field theory. This, and the previous theorem, are among the finest examples of the meetings between rigor and practical application.

The following example[118] shows how the theorem 'works'. Consider the interaction

$$H = H_1 + H_1' \tag{20.26}$$

where H_1 is given by Eq. (20.23) and

$$\begin{gathered} H_1' = \sum_i g_i' \bar{\psi}_p O_i \psi_n \bar{\psi}_e O_i' \psi_\nu + \text{hermitian conjugate} \\ O_S' = i\gamma_5, \quad O_V' = i\gamma_\mu \gamma_5, \quad O_T' = \gamma_5 \sigma_{\mu\nu}, \quad O_A' = \gamma_\mu, \quad O_P' = 1 \end{gathered} \tag{20.27}$$

If P is conserved then either all g_i or all g_i' vanish. If in addition C is conserved then all g_i, or all g_i', are real, as is also independently required by T-invariance: 'P yes and C yes imply T yes'. If P is not conserved, then C demands g_i to be imaginary relative to g_i', while T requires them to be relatively real: 'P no and C yes imply T no'; etc. Anyone who has played with these invariances knows that it is an orgy of relative phases. *CPT*-invariance is the minimal sufficient ground for the existence of anti-particles to particles† and for the equalities of masses and lifetimes of these pairs of objects.[119] How delicate the argument is can be exemplified by the fact that if a particle can branch into two or more decay modes, then *CPT* does *not* imply the same branching ratios for these reactions and the corresponding anti-reactions. Most delicate of all, as usual, is the K^0–$\bar{K}^0$ system. These are particles-antiparticles only if weak interactions are ignored. Their inclusion leads to the strongest test to date for *CPT*: a bound[120] on the mass difference of K^0 and $\bar{K}^0$.

4. Toward the universality of weak interactions. Conservation of leptons. It is convenient to preface the next stage of developments regarding the discrete symmetries with a brief partial update on weak interactions.

As has been noted previously,†† by about 1940 it had become difficult to reconcile the lifetime and scattering of the cosmic ray meson (the later muon) with those of the Yukawa meson (the later pion). As a result, the possibility had been considered[121] as early as 1941 that the muon might be a spin-$\frac{1}{2}$

* Actually the commutation conditions can be weakened slightly (to 'weak local commutativity').[113,114]
** His proof is based on the fact that in the complex extension of the Lorentz group *PT* can be generated continuously from the identity element.
† Excepting of course self-conjugate particles like π^0 and γ.
†† Chapter 18, Section (b).

particle with possible decay modes

$$\mu \rightarrow e + 2\nu \tag{20.28}$$

$$\mu \rightarrow e + \gamma. \tag{20.29}$$

(Recall, this was well before the distinction between π and μ had been clarified.) The first attempts, dating from 1947–8, at finding the monochromatic γ-ray in the process, Eq. (20.29), yielded negative results.[122] As will be seen in the next chapter, this is, we think, as it should be.* Let us continue here with the discussion of the reaction (20.28).

During 1948–9 this decay as well as μ-capture: $\mu^+ + n \rightarrow p + \nu$ were the subject of many theoretical studies, all of them starting from interactions analogous to Eq. (20.23). Write the latter symbolically as $(\bar{p}n)(\bar{e}\nu)$ and consider the triple of couplings

$$(\bar{p}n)(\bar{e}\nu), (\bar{p}n)(\bar{\mu}\nu), (\bar{\nu}\mu)(\bar{e}\nu), \tag{20.30}$$

which generate β-decay, μ-capture, and μ-decay respectively (as well as various inverse and conjugate processes). Analysis of experiments revealed that the coupling constants for μ-decay and μ-capture were of the same order of magnitude as those for β-decay. This led to the hypothesis of a universal Fermi-interaction: all three couplings in (20.30) should be of the same type (S, V, T, etc.) and essentially of the same strength.[124]

The electron spectrum in μ-decay depends in general on five g_i's; cf. Eq. (20.23). The next advance was Louis Michel's proof[125] that this dependence can be lumped** into a single dimensionless combination ρ of these constants, since known as the Michel parameter. Michel noted[125] that the possible range of ρ depends on whether the reaction (20.28) produces two neutrinos, or two antineutrinos, in which case $0 \leq \rho \leq 3/4$, or one of each: $0 \leq \rho \leq 1$. It was to be clear a few years later that the latter holds true:

$$\mu^{\pm} \rightarrow e^{\pm} + \nu + \bar{\nu}. \tag{20.31}$$

The option Eq. (20.31) is dictated by the principle of conservation of leptons, a concept that originated with the 1953 paper[127] by Konopinski and Mahmoud: ascribe by convention a 'lepton number' +1 to e^-, and add the assignments: +1 for μ^-, ν; −1 for e^+, μ^+, $\bar{\nu}$; 0 for all other particles. Then in all reactions lepton number is conserved (in the above example, for $\mu^-: 1 = 1 + 1 - 1$). Lepton conservation implies the absence of neutrinoless β-decay Eq. (17.28), and vice versa.[128]

* In 1977 there was an intense flurry of excitement when evidence for this decay was rumored to have been found.[123] It passed.

** This is true only if one sums over all polarizations and neglects small terms proportional to the electron mass. These conclusions remained unaltered[126] after the discovery of parity violation when, in general, ten coupling constants could be introduced in μ-decay.

Other weak decays were discussed in the framework of Eq. (20.30) as well, such as $\pi^+ \to \mu^+ + \nu$, conceived as the two-stage process $\pi^+ \to p + \bar{n}$ (strong) followed by $p + \bar{n} \to \mu^+ + \nu$ (Fermi interaction). According to (20.30) $\pi^+ \to e^+ + \nu$ should proceed by a similar chain. Calculations showed that (as experiment demands) this decay is suppressed compared to the μ-mode only if the pion is pseudoscalar and if the universal β-decay interaction in the chain is dominantly of type A. In that case the rate for $\pi^+ \to l^+ + \nu$, $l = e$ or μ is given by

$$R(\pi^+ \to l^+ + \nu) = \frac{f_{\pi l}^2 m_l^2}{4\pi m_\pi^3}(m_\pi^2 - m_l^2)^2 \tag{20.32}$$

where all strong interaction features are lumped in the constant $f_{\pi l}$. Universality demands that (up to electromagnetic radiative corrections)

$$f_{\pi e} = f_{\pi\mu} \equiv f_\pi \tag{20.33}$$

Accordingly the ratio of e/μ rates is $\sim 10^{-4}$. For a number of years the experimental search for the electron mode at this rate remained fruitless, causing great concern as late as the 8th Rochester conference (July 1958).[130] To everyone's great relief $\pi \to e\nu$ was found[131] two months later at about the expected rate.

An argument similar to that for pion decay was applied shortly afterwards[132] to K-decay.

5. Late 1956: P and C are violated. On 16 January 1957, the *New York Times* carried on its front page an article headlined: 'Basic concept in physics is reported upset in tests/Conservation of parity in nuclear theory challenged by scientists at Columbia and Princeton Institute'.

Explicit experimental proof of parity violation demands measurement of the distribution of a 'screw', a quantity that is odd under spatial reflections. Example: choose a definite coordinate system in which to define the quantity H:

$$H = \vec{J} \cdot \hat{p} \tag{20.34}$$

where $\vec{J}$ is an angular momentum in units $\hbar$ (even under P) and $\hat{p}$ a unit momentum vector (odd). Measure the probability distribution of the angle θ between $\vec{J}$ and $\hat{p}$. If this distribution is asymmetric with respect to the plane perpendicular to $\vec{J}$ (that is, with respect to $\theta \to \pi - \theta$) then parity is violated. Observe that the processes $K \to 2\pi$ and 3π—which had caused the parity issue to be raised in the first place—do not provide enough independent vectors for defining a screw of the type (20.34) or of any other kind.

One further remark on Eq. (20.34). H is called helicity if $\vec{J}$ and $\hat{p}$ refer to spin and momentum of one and the same particle. The expected value of helicity depends on the choice of coordinate system except for the case of massless particles. For neutrinos, for example, spin and momentum can be seen to be necessarily parallel or anti-parallel to the momentum, hence $H = \pm\frac{1}{2}$ in any frame of reference. By convention we speak of a right- (left)-handed neutrino if $H = +\frac{1}{2}$ $(-\frac{1}{2})$.

Let us now return to our story. What happened next was a stunning theoretical discovery: weak processes other than $K \rightarrow 2\pi$, 3π do provide possibilities for measuring screw distributions—but no such experiment had ever been performed. Parity had never been tested in any weak interaction!!

Late in 1957, when memory was still fresh, I wrote[94] of the events following the sixth Rochester conference: 'It seems fair to say that two years ago (to fix a reasonable point in time) nearly all theoretical physicists believed that the general validity of space reflexion invariance had firmly been established by observation.* It seems fair to say that at that time few experimentalists thought of devising experiments which might challenge the universal validity of parity conservation.** It seems fair to say that the main contribution to [the parity] question which Lee and Yang have made . . . has been to point out that parity conservation had never been checked in a domain of physics of which much, but not enough, was known two years ago (β-decay, π-decay, μ-decay); and to discuss a series of experimental conditions under which such checks could be made.[134]

'The incentive for these investigations came from the puzzling properties of the K-mesons . . . Lee and Yang faced the challenge. Immediately after the [Rochester] conference they started a systematic investigation of the then status of experimental knowledge concerning the verification of space reflexion invariance and charge conjugation invariance. Their conclusion was that for one group of interactions neither invariance had so far been established. These reactions are all characterized by their weakness and to these belong the three types of decay processes mentioned above, as well as K-particle and hyperon disintegration. Thus the attention became focussed on a whole class of phenomena instead of on an exciting but rather isolated puzzle. Soon a theoretical investigation followed[119] together with Oehme, on the question of time reversal invariance and on the interrelations between possible violations of 'C-, P- and T-invariance' with the help of the CPT theorem. And then came the great news: neither P- nor C-invariance holds true in β-decay, nor in π-, nor in μ-decay. It will be remembered how the Co^{60}-experiment of Chien-Shiung Wu (Mrs Yuan) and coworkers[135] provided the first evidence. More recently it has been established that parity is not conserved in Λ-decay either, and so, via only a slight theoretical detour, the puzzle may be considered to be solved.

* For a notable case of misgivings see Ref. 103, footnote 9.
** For a notable exception see Ref. 133.

'Thus Lee and Yang's suggestions have led to a great liberation in our thinking on the very structure of physical theory. Once again principle has turned out to be prejudice . . .

'Their more recent work deals with the two-component formulation of neutrino theory,[136] a topic to which independent contributions had been made by Salam[137] and by Landau;[138] and with the concept of lepton conservation.[136,139] At the time of writing the experimental situation in β-decay, after having gone through a phase of utter confusion last summer, is still too unclear to enable one to judge the tenability of these appealing ideas. Most recently [Lee and Yang and coworkers] have been analyzing the information obtainable from hyperon decay.[140]

'The work of T. D. and of Frank, as they are affectionately called, is characterized by taste and ingenuity, by physical insight and formal power. Their counsel is sought by theorist and experimentalist alike. In this they have more than a touch of the late Fermi . . .'.

Several items just mentioned demand elaboration. This is done in the next sections which continue the account of weak interactions through 1964, when the next startling news arrived: the violation of *CP*- and of *T*-invariance.

(d) Being selecta from the exploding weak interaction literature

We have now reached the years in which the particle physics literature begins its tremendous growth. On the subject of weak interactions alone, 1000 experimental and 3500 theoretical articles as well as 100 reviews (all in round numbers) appeared between 1950 and 1972, written by authors from 50 countries.[141] The distribution in time shows a large peak immediately following the discovery of parity violation. As announced in Chapter 1, I must from here on confine myself to brief indications whereafter I direct the reader to the literature, which includes recollections of those intense years by a number of participants, both from the experimental and the theoretical side (see the sources below), as well as a number of very good textbooks.* What follows next is a synopsis of the main events.

1. The pioneering experiments. The first of these,[135] a study of β-decay of polarized Co^{60} nuclei, yielded an H distribution (with $\vec{J}$ the cobalt spin, $\hat{p}$ in the direction of the electron momentum) of the form $1+a\cos\theta$ $(a<0)$, a clear signal of parity violation. It did not take anti-Co^{60} to show that *C*-invariance is violated as well. From a more detailed reasoning, having to do with the relative phases of coupling constants (see the discussion following Eq. (20.27)), it could be shown that the observed magnitude of a implied *C*-violation. Immediately thereafter two other experiments demonstrated[146]

* See Refs. 52, 77, 109, 142, 143, 144. A number of important papers were reprinted in Ref. 145.

P- and C-violation, both in $\pi^+ \to \mu^+ + \nu$ and in $\mu^+ \to e^+ + \nu + \bar{\nu}$. This can be seen[134] to follow from the observed form $1 + b\hat{p}_\mu \cdot \hat{p}_e (b \neq 0)$ of the electron distribution observed in the π–μ–e sequence.* Next it was noted[147] that nonzero helicity of an electron in β-decay from *un*polarized nuclei also implies P-violation. The experimental detection of that effect followed quickly.[148] As the spring of 1957 arrived it had been established that P and C are violated in weak processes wherever one looked. Moreover the effects were *large*.

2. CP-invariance; K^0–$\bar{K}^0$ revisited. Even before the experimental observation of P-violation, the question had been posed: suppose P is violated; is there a weaker but nevertheless elegant symmetry that could take its place? It had occurred to several theorists that combined CP-invariance might serve that purpose: look in a mirror *and* flip the charge, and all will be symmetric.[103,149]

It was at once noted[138] that CP-invariance sufficed to maintain the main features of the K^0–$\bar{K}^0$ system as initially discussed on the assumption of C-invariance (see the preceding Section (b), Part 3). Redefine $K_1(K_2)$ as states with $CP = +1\ (-1)$ instead of $C = +1\ (-1)$, as done earlier. Since zero angular momentum states $\pi^+\pi^-$ and $2\pi^0$ are even under P and C, they have $CP = +1$. Thus it remains true that K_1 can, K_2 cannot, decay into 2π, so that the earlier argument for a long-lived K_2 continues to hold. The switch from C to CP importantly affects the 3π-decays of neutral K's however.[150]

The neutral K complex now acquired new significance as a testing ground for CP-invariance. Was it indeed true that K_2 never decayed into 2π? By 1961, an upper limit of 0.3 per cent for the branching ratio of this mode was reported.[151] Another important test concerned the ratio (R = rate)

$$\delta_l = \frac{R(K_2^0 \to \pi^- l^+ \nu) - R(K_2^0 \to \pi^+ l^- \bar{\nu})}{R(K_2^0 \to \pi^- l^+ \nu) + R(K_2^0 \to \pi^+ l^- \bar{\nu})}, \qquad l = \text{e or } \mu. \qquad (20.35)$$

Any deviation of δ_l from zero would violate C and T, hence CP by the CPT theorem.[152] By 1964 experiment showed[153] no deviation from $\delta_l = 0$. Nor had earlier tests[154] for T-violation using polarized neutrons shown any effect. As a result, 'even though parity had been overthrown a few years before, one was quite confident about CP-symmetry'.[155]

3. The two-component theory of the neutrino. Like CP-invariance, this theory was first proposed[137,156] before P-violation had been seen, and was further elaborated right thereafter.[147] The idea goes back to Weyl's attempt** to assign two of the four components of the Dirac wave function to electrons, the other two to protons. Since this picture violates parity it was thought (1933)[157] to be 'inapplicable to physical reality'. Interest in two-component theories revived

* $\hat{p}_\mu$ is the muon direction in the π-rest system; $\hat{p}_e$ is the electron direction in the μ-rest system.
** Chapter 15, Section (f).

once P-conservation was put in doubt, particularly in regard to applications to massless neutrinos (ν) and antineutrinos ($\bar{\nu}$). It should be remembered, however, that such theories can be consistently formulated whether or not the particles have zero mass.[158]

Let us stay with zero ν-mass and start with the four-component version described by

$$\gamma_\mu \frac{\partial \psi}{\partial x_\mu} = 0 \tag{20.36}$$

(see Eq. (13.41); ψ is quantized). Introduce the 'chirality' operator* γ_5 which anticommutes with γ_μ; hence $\gamma_5\psi$ also satisfies Eq. (20.36) and so do

$$\psi_\pm = \frac{1 \pm \gamma_5}{2}\psi. \tag{20.37}$$

γ_5 can be chosen diagonal with eigenvalues ± 1, each twice. $\psi_\pm$ are then two-component wave functions. The two-component theory in its simplest form asserts that *either* ψ_+ *or* ψ_- enters in *all* neutrino interactions. The two alternatives are distinguished by the possible values $\pm\frac{1}{2}$ of helicity. One can show that

$$\psi_+ \leftrightarrow H_\nu = -H_{\bar{\nu}} = -\tfrac{1}{2} \quad \text{only}, \tag{20.38}$$

$$\psi_- \leftrightarrow H_\nu = -H_{\bar{\nu}} = +\tfrac{1}{2} \quad \text{only}. \tag{20.39}$$

These two options have in common: a) a 'γ_5-invariance' $\psi_\pm = \pm\gamma_5\psi_\pm$ which implies zero mass[137] and induced magnetic moment; b) γ_5-invariance is a special case of CP-invariance. P is violated since it flips helicity while (anti)particle stays (anti)particle; C is violated since it interchanges particle and antiparticle without change of helicity; and CP is satisfied; c) combined with lepton conservation the Michel parameter is predicted** to equal 3/4, in agreement with experiment.

4. The universal V–A theory. The two-component neutrino looked good. Lepton conservation looked good. Yet (as already alluded to at the end of Section (c)), by the summer of 1957 weak interaction theory was in a state of utter confusion. At issue was a third and equally attractive concept: the universality of Fermi interactions. By the end of that year that problem, too, had been resolved.

The principal difficulties during this well-documented period[161] were the following. Measurement of electron–neutrino correlations (having no bearing on parity) in He^6 β-decay appeared to have given strong evidence for interactions predominantly of types S and T (see Eqs. (20.23), (20.26)), with

* See Eq. (13.53): $\gamma_5 = \gamma_1\gamma_2\gamma_3\gamma_4$. The name chirality was adopted[159] in this context in 1957.
** Up to radiative corrections. It took some time[160] before this parameter settled down to its present best value 0.752 ± 0.003.

comparable strength. It could be deduced from the $\pi\mu e$ chain experiments[146] that the μ-decay interaction is a combination of V and A. As mentioned, the decay ratio $\pi \to \mu\nu/\pi \to e\nu$ demanded[129] a dominant role for A. Universality, on the other hand, required the same coupling type for all these interactions. Confusion: something had to be wrong either with universality or with some experiments.

Several participants[162,163] have recalled the complex sequence of events that followed. As early indications suggested (later fully confirmed) that the He^6 experiment was incorrect, they began opting[164,165,166] for what has become known as the V–A theory.* All Fermi interactions are of type V and A; lepton conservation is incorporated; neutrinos are two-component with $\nu(\bar{\nu})$ left-(right)-handed corresponding to the choice ψ_+, Eq. (20.38). Improved β-decay data all fell into line with this picture. In addition, a brilliant experiment[167] demonstrated directly that ν is left-handed.

In summary, by the late fifties the symbolic expressions Eq. (20.30) had take on the concrete form

$$H = \frac{G}{\sqrt{2}}[J_\lambda^\dagger(j_\lambda(\mathrm{e}) + j_\lambda(\mu)) + j_\lambda^\dagger(\mu)j_\lambda(\mathrm{e})] + \text{hermitian conjugate} \tag{20.40}$$

$$J_\lambda = \bar{n}(1 + a\gamma_5)\mathrm{p} \tag{20.41}$$

$$j_\lambda(l) = \bar{l}\gamma_\lambda(1 + \gamma_5)\nu, \qquad l = \mathrm{e}, \mu \tag{20.42}$$

where particle symbols refer to the corresponding quantized wave fields. G is the Fermi constant, hitherto called $g_V\sqrt{2}$:

$$G \simeq 10^{-49}\ \text{erg cm}^3 = 6.25 \times 10^{-44}\ \text{MeV cm}^3, \tag{20.43}$$

not much different from Fermi's own initial estimate, Eq. (17.24). a is the axial to vector ratio, g_A/g_V; modern value:

$$a = 1.2539 \pm 0.0063. \tag{20.44}$$

I leave until the next chapter the extension of H needed to incorporate strange particle decays, except for noting that tests for P-violation in those processes had been contemplated from the start.[134] Indeed, the first reference[168] to P-violation in the experimental literature concerns the decay $\Lambda \to \mathrm{p} + \pi^-$.

5. *The interplay of weak and strong forces: local action of lepton pairs, conserved vector current, partially conserved axial current.* As the foundation of weak interactions became much more firm, the time was at hand for raising the hitherto neglected issue of how strong forces affect weak processes. It is clear that they must. The β-decay interaction Eq. (20.26) is local; that is, all four

* The appellation 'V minus A' harks back to earlier conventions for the signs of the constants g_V and g_A. For the definitions of g_A and g_V see Eq. (17.26); also Eqs. (20.43), (20.44).

particles interact at the same space–time point. The strong processes of virtual pion exchange introduce non-local modifications insofar as nucleons are concerned. The leptons, on the other hand, untouched by these strong influences, keep interacting as a local pair* with the smeared-out nucleons. Because of this circumstance, weakly interacting lepton pairs can serve as precious probes of strongly interacting systems.

The earliest example of such probing is $\pi \to l\nu$ decay. The explicit lepton mass dependence in Eq. (20.32) follows exclusively from the fact that the action of lepton pairs is local, in the form of an axial current, and is independent of whether the pion 'sees' this pair via virtual nucleon pair formation (as was the case in the initial model).[129] The very same argument concerning the e/μ decay ratio also demonstrated that $K \to l\nu$ proceeds via axial vector coupling. Further examples: the local action of lepton pairs applied[169] to $K^+ \to \pi^0 + e^+ + \nu$ has made it possible to demonstrate experimentally[170] that the strangeness changing weak interactions contain not only an A but also a V component; applied[171] to $K^+ \to \pi^+ + \pi^- + e^+ + \nu$, it has yielded information[172] on pion–pion scattering, a strong interaction process; applied to high energy neutrino–nucleon collisions, it simplifies the analysis of final states independently of their complexity.[173]

The property of weak interactions that acquires particular importance when confronted with strong interactions is their universality. Why should the Fermi constant for β-decay be so closely equal to the one for μ-decay, given that the former will in general be renormalized by strong interactions, the latter not? The answer was found by borrowing from quantum electrodynamics where the proton's electric charge is not renormalized by strong interactions, because the electric current j_λ^{elm} is conserved. It was therefore proposed that V_λ, the vector part of J_λ, is likewise conserved. When this suggestion was first made (1955),[174] it was considered 'of no practical significance but only of theoretical interest'. That changed when it was proposed[164] more specifically that V_λ, $V_\lambda^\dagger$ and the isovector part of j_λ^{elm} (in units e) form an isotriplet.** Consequently, the net weak vector current should contain terms additional to the $\bar{n}p$-terms of Eq. (20.41). These include terms bilinear in pion fields[164] from which the rate for $\pi^+ \to \pi^0 + e^+ + \nu$ can at once be predicted, and 'weak magnetism' terms[177] with prescribed strength, the weak counterparts of the induced magnetic (isovector) terms in quantum electrodynamics that give rise to the anomalous magnetic moments of nucleons. All these implications have been well verified.†

* Here much smaller electromagnetic effects are neglected.
** Conservation of V_λ alone allows this current to have[175] a part even (first-class current) and a part odd (second-class current) under the G-parity operation (Chapter 19, Section (d), Part 3). The isotriplet picture (where V_λ-conservation holds up to electromagnetic corrections) permits only first-class currents as, after much travail, appears indeed to be the case.[176]
† For comparisons with experiment see Ref. 178.

The conservation law for V_λ has no counterpart for A_λ, the axial part of J_λ with strength $G_A = Ga$. In fact A_λ cannot be conserved, since if it were, one can show[179] that the decays $\pi \to$ e or $\mu + \nu$ would be forbidden. Moreover, even in a fictitious world in which A_λ were conserved, we still would not be able to infer anything about G_A since a non-zero static vector charge (like e) happens not to have an axial analog. Nevertheless it was found possible to establish a remarkable link between strong and axial weak interactions, to wit the Goldberger–Treiman relation[180]

$$mG_A = f_\pi f \tag{20.45}$$

where m is the nucleon mass, f_π the pion-decay coupling constant defined by Eqs. (20.32), (20.33), and f^2 (in units $4\pi\hbar c$) the strong interaction π-nucleon coupling constant defined in Eq. (19.23). The initial derivation of this relation was conceived as an exercise in applying dispersion techniques to decay amplitudes, and was executed by considering the chain then in vogue: $\pi^+ \to \mathrm{p} + \bar{\mathrm{n}} \to \mathrm{e}^+ + \nu$. In spite of these and further approximations, the numerical consistency of the relation was, and has remained, uncanny—currently to within six percent.[181] Eq. (20.45) has been a stimulus for further theoretical developments,* including the study of dynamical models which yield the relation in a more rigorous fashion; extension of the reasoning to K-particles; attempts to account for the missing six per cent[181,183]; and the observation[184] that Eq. (20.45) can be seen as a consequence of the operator relation ($\pi(x)$ is the pion field operator)

$$\frac{\partial A_\mu}{\partial x_\mu} = f_\pi m_\pi^2 \pi(x). \tag{20.46}$$

This simultaneous constraint on strong and weak interactions implies (modulo delicacies) that A_μ is conserved in the limit $m_\pi \to 0$. Accordingly Eq. (20.46) is known as the partially conserved axial current (PCAC) condition.

None of the above explains as yet why a in Eq. (20.44) has the value it does. Brief comments on that point will be found in Chapter 21.

(e) Unfinished story of a near miss: the violations of *CP*- and of time-reversal-invariance

One morning, in the spring of 1964, I was having breakfast in the Brookhaven National Laboratory cafeteria when Jim Cronin and Val Fitch came to the table to talk to me about their recent experiment. That was the first time I had heard of their discovery, made together with Christenson and Turlay,[185] that long-lived neutral K-particles decay into $\pi^+ + \pi^-$ at a rate of about 2×10^{-3} of all charged modes (a number close to the modern branching ratio: $0.203\pm$

* See Ref. 182 for numerous details.

0.005 per cent). How can that be? I asked; it violates *CP*-invariance. They knew that, they said, but there it was. Why was the effect not due to regeneration of short-lived K's, I wanted to know. Because, they said, that effect was far too small in the helium bag where the 2π events had been found. I asked many more questions, why were they not seeing $2\pi\gamma$ decays with a soft photon, or $\pi\mu\nu$ decays with a soft neutrino and perhaps some confusion about mass. They had thought long and hard about these and other alternatives (the actual experiment had been concluded the previous July) and had ruled them out one by one. After they left I had another coffee. I was shaken by the news. I knew quite well that a small amount of *CP*-violation would not drastically alter[119] the earlier discussions, based on *CP*-invariance, of the neutral K-complex. Also, the experience of seeing a symmetry fall by the wayside was not new to anyone who had lived through the 1956–7 period. At that time, however, there had at once been the consolation that *P*- and *C*-violation could be embraced by a new and pretty concept, *CP*-invariance. What shook all concerned now was that with *CP* gone there was nothing elegant to replace it with. The very smallness of the 2π rate, *CP*-invariance as a near miss, made the news even harder to digest.

During 1964–5 the widespread unease initially caused by the new effect (which was rapidly confirmed[186]) led to a variety of efforts at explaining it in a *CP*-conserving way. A list of such inevitably exotic alternatives includes the introduction of new very weakly-coupled long-range fields; of *ad hoc* new particles; of a shadow universe coupled to ours by weak interactions; of deviations from the exponential decrease of decay rates with time; of deviations from the quantum mechanical superposition principle; and others.* Most of these proposals could be experimentally disproved; others faded away. *CP*-violation has turned out to be unavoidable. Accordingly, the role of K_1 ($CP=+1$) and K_2 ($CP=-1$), defined in Eq. (20.20) as the observed physical states, is now taken over by K_S and K_L respectively, new particle states to be defined precisely in a moment.

The novel situation created in 1964 raised many further problems. Are T and CPT, or even both, violated? What can one learn from K_L-decays other than $\pi^+\pi^-$? And from processes other than those associated with neutral K's? Experiments to probe these questions are continuing to this day; others are in preparation. Without exception they are subtle, lengthy, and difficult. As Cronin noted[188] in 1980, six years of his professional life had already been spent in the pursuit of *CP*-violation in $K_L \rightarrow 2\pi^0$.

Let us first summarize what K_L has revealed to date (progress on these matters can be followed via a series of status reports).** The phenomenological analysis needed for this purpose is based on careful use of simple tools: the

* These suggestions are discussed further in Ref. 187.
** See for example Refs. 81 (1966), 189 (1968), 190 (1976), 188 (1981), 191 (1984), 192 (1985).

quantum mechanical concepts of superposition and unitarity, and a few properties of isospin. Watch out for phases! The entire detailed discussion is riddled with some phases defined by convention and others that have physical significance.

The main points are these. As in the old K_1, K_2-days, K_S and K_L are defined as those linear combinations of K^0 and $\bar{K}^0$ that each have a unique lifetime. They satisfy a pair of coupled Schroedinger equations with solutions

$$\begin{aligned} K_S &= [(1+\varepsilon+\Delta)K^0+(1-\varepsilon-\Delta)\bar{K}^0]/N_+ \\ K_L &= [(1+\varepsilon-\Delta)K^0-(1-\varepsilon+\Delta)\bar{K}^0]/N_- \\ N_\pm &= \{2(1+|\varepsilon\pm\Delta|^2)\}^{1/2} \end{aligned} \tag{20.47}$$

where (one proves) ε and Δ are complex numbers signifying *CP*-violation and, furthermore: $\Delta=0$, $\varepsilon\neq 0$: *T*-violation, *CPT* good; $\Delta\neq 0$, $\varepsilon=0$: *CPT*-violation, *T* good. Note that the states K_S, K_L are not orthogonal. The projection $\langle K_S|K_L\rangle$ of K_S on K_L satisfies:

$$\begin{aligned} N_+N_-\langle K_S|K_L\rangle &= 2(\varepsilon^*+\varepsilon)\text{: real, if } \Delta=0, \\ &= 2(\Delta^*-\Delta)\text{: imaginary, if } \varepsilon=0. \end{aligned} \tag{20.48}$$

These reality conditions are observable and are at the root[188,193] of the proof, based for the rest on experiment, that *CPT* remains good, within the errors, while there is a 10 standard deviation from *T*-invariance. No evidence for *CPT* violation has turned up anywhere else either.[196]

Assume next that Δ is strictly zero and focus on

$$\varepsilon = |\varepsilon| \exp i\phi. \tag{20.49}$$

One shows (M and Γ refer to mass and total width)

$$\begin{aligned} \phi &= -\tan^{-1}\left(-\frac{2(M_S-M_L)}{\Gamma_S-\Gamma_L}\right) \\ &= (43.7\pm 0.2)^0. \end{aligned} \tag{20.50}$$

CP-violation allows for nonzero values of the quantities δ_l defined in Eq. (20.35). These charge asymmetries have indeed been observed: nature prefers l^+ over l^- (at least in our corner of the universe). One can show that δ_l are related to ε by*

$$\delta_e = \delta_\mu = 2|\varepsilon| \cos\phi. \tag{20.51}$$

* Here the additional assumption is made that the so-called $\Delta S/\Delta Q$ rule holds good. See Chapter 21 for this rule and its experimental validity.

Using Eq. (20.50) and the modern data on $\pi l \nu$ decays one finds

$$\varepsilon = (2.27 \pm 0.08) \times 10^{-3} \exp i(43.7 \pm 0.2)^0. \tag{20.52}$$

Turn to *CP*-violation in 2π-decays which is conventionally described in terms of the two complex η-parameters (A = amplitude)

$$\begin{aligned} \eta_{+-} &= |\eta_{+-}| \exp i\phi_{+-} = \frac{A(\mathrm{K}_L \to \pi^+\pi^-)}{A(\mathrm{K}_S \to \pi^+\pi^-)}, \\ \eta_{00} &= |\eta_{00}| \exp i\phi_{00} = \frac{A(\mathrm{K}_L \to 2\pi^0)}{A(\mathrm{K}_S \to 2\pi^0)}. \end{aligned} \tag{20.53}$$

If these processes obey the $|\Delta I| = \frac{1}{2}$ rule *strictly* (no $|\Delta I| = 3/2$ contribution) then, one shows,

$$\eta_{+-} = \eta_{00} = \varepsilon. \tag{20.54}$$

These relations appear to be satisfied within present experimental accuracies[191,194]

$$\eta_{+-} = (2.279 \pm 0.026) \times 10^{-3} \exp i(44.6 \pm 1.2)^0 \tag{20.55}$$

$$\eta_{00} = (2.33 \pm 0.08) \times 10^{-3} \exp i(54.5 \pm 5.3)^0 \tag{20.56}$$

Thus twenty years of hard experimental labor following the discovery of *CP*-violation have reduced the phenomena to a single complex number, ε, the 'impurity parameter'. At the phenomenological level the appearance of ε corresponds to a so-called 'superweak'[197] $|\Delta S| = 2$ mixing of the states K^0, $\bar{\mathrm{K}}^0$, which violates *CP*.

(f) Final comments on discrete symmetries

At the time of writing, the neutral K-complex remains the unique source of information on *CP*- and *T*-violation. For this reason, if for no other, further experiments aiming at an increased precision of the parameters just discussed are under way. One is particularly interested in possible small $|\Delta I| = 3/2$ contributions to $\mathrm{K}_L \to 2\pi$; and in possible *CP*-effects in $\mathrm{K}_L \to 3\pi$. Searches for *CP*-violations elsewhere have so far yielded either negative or inconclusive results. Particularly substantial efforts have been devoted to detecting electric dipole moments of fundamental particles. These, as Landau was the first to note,[149] signal violations of P and T. Even the best experiments of this kind—and they are very good—have not yet revealed an effect. Current status:[198] this moment is $< 3 \times 10^{-24}$ for the electron, $(2.3 \pm 2.3) \times 10^{-25}$ for the neutron (both in units $e \times$ cm). Other natural sources for *CP*-violation are neutral complexes associated with quantum numbers newer than strangeness, such as the D^0–$\bar{\mathrm{D}}^0$ mesons connected with charm (Chapter 21). For a number of technical reasons CP-violation for these complexes need not be a repeat of what happened for K-particles.[199,200] No results are available as yet.

The weak interactions have left us with questions that, so far, remain unanswered.

We do not understand why P, C, and T are violated if, and only if, weak interactions intervene. None of these effects cause any strain on other current postulates—as long as CPT-invariance is maintained. We do not understand the hierarchical nature of interaction symmetries (already remarked on in Section (b)), the curious links between strength of interaction and presence or absence of symmetry. Nor have the great advances of unified gauge theories (Chapter 21) shed any light on these questions. Those theories incorporate the violations but as yet do not explain them.

It is thought by many that the headiest challenge is the weakness of CP-violating effects even on the scale of weak interactions. As Cronin has put it: 'We must continue to seek the origin of CP-violation by all means at our disposal. We know that improvements in detector technology and quality of accelerators will permit even more sensitive experiments in the coming decades. We are hopeful, then, that at some epoch, perhaps distant, this cryptic message from nature will be deciphered'.[188]

Sources

Reminiscences. On new particles in cosmic rays: Rochester[3,4,201]; Butler;[202] Leprince-Ringuet;[28] C. D. Anderson;[203] Fretter;[204] Peyrou.[7] Marshak[47] on the years 1947–52. Gell–Mann[205] and Nishijima[206] on the strangeness scheme. Dalitz on the τ-meson.[207] On the days of parity violation: Yang;[208] Lee;[209] and Pais.[94] On the pioneering parity experiments: Wu;[210] Garwin;[211] and Telegdi.[212] On the V–A theory: Feynman[213] Gell–Mann;[162] and Sudarshan and Marshak.[214] On CP-violation: Fitch;[215] and Cronin.[155,188]

Status reports on weak interactions covering the period discussed in this chapter: proceedings of Rochester conferences; also Lee and Yang;[216] Gell–Mann and Rosenfield;[63] Wu;[217] Pais;[218] Lee and Wu.[219]

References

1. C. M. G. Lattes, G. P. S. Occhialini, and C. F. Powell, *Nature* **160**, 486, 1947, footnote 3.
2. G. D. Rochester and C. C. Butler, *Nature* **160**, 855, 1947.
3. G. D. Rochester, *Yearbook of the Phys. Soc. London* 1957, p. 61.
4. G. D. Rochester, *Early history of cosmic ray studies*, Eds. Y. Sekido and H. Elliot, p. 299, Reidel, Boston 1985.
5. See however Ref. 2, footnotes 3 and 4.
6. L. Leprince-Ringuet and M. Lhéritier, *Comptes Rendus* **219**, 618, 1944.

7. Ch. Peyrou, 'Colloq. hist. particle phys.', *J. de Phys.* **43**, suppl. to No. 12, p. 7, 1982.
8. H. A. Bethe, *Phys. Rev.* **70**, 821, 1946.
9. Dittoed notes of the Pocono Conference, unpublished.
10. G. D. Rochester, *Rev. Mod. Phys.* **21**, 20, 1949.
11. G. D. Rochester, letter to A. Pais, November 25, 1984.
12. R. B. Brode, *Rev. Mod. Phys.* **21**, 37, 1949.
13. L. Leprince-Ringuet, *Rev. Mod. Phys.* **21**, 42, 1949; also L. Leprince-Ringuet *et al.*, *Comptes Rendus* **226**, 1897, 1948.
14. R. Brown *et al.*, *Nature* **163**, 82, 1949.
15. J. B. Harding, *Phil. Mag.* **41**, 405, 1950.
16. E. Fermi, *Elementary particles*, Yale Univ. Press, New Haven, Conn. 1951.
17. W. B. Fretter, *Proc. 2nd Rochester conf.* (*January* 11–12, 1952), p. 56, Univ. of Rochester Report NY0–3046.
18. A. J. Seriff, R. B. Leighton, C. Hsia, E. W. Cowan, and C. D. Anderson, *Phys. Rev.* **78**, 290, 1950.
19. P. M. S. Blackett, *Proc. Harwell nucl. phys. conf.* 1950, p. 20, Ministry of Supply Report A.E.R.E., G/M 68; see further R. Armenteros *et al.*, *Nature* **167**, 501, 1951.
20. V. D. Hopper and S. Biswas, *Phys. Rev.* **80**, 1099, 1950.
21. First Rochester conference, December 16, 1950. Unedited and unpublished transcript.
22. E. Amaldi *et al.*, *Nature* **173**, 123, 1954.
23. R. E. Marshak, *Meson physics*, Chapter 9, McGraw-Hill, New York 1952.
24. C. F. Powell, *Phil. Mag.* **44**, 219, 1953.
25. R. Dixit, *Zeitschr. f. Naturf.* **9a**, 355, 1954.
26. Congrès sur le rayonnement cosmique, Bagnères-de-Bigorre, July 1953, proceedings unpublished.
27. Ref. 26, p. 269; also Ref. 22.
28. L. Leprince-Ringuet, Ref. 7, p. 165; also Ref. 21.
29. A. Pais, in *Proc. int. conf. on theor. phys. Kyoto, 1954*, p. 157, Science Council of Japan, Tokyo 1955.
30. M. Gell-Mann and A. Pais, *Proc. Glasgow conf. on nuclear and meson phys.*, July 1954 Eds. E. H. Bellamy and R. G. Moorhouse, p. 342, Pergamon, Oxford 1955.
31. L. Leprince-Ringuet, *Ann. Rev. Nucl. Sc.* **3**, 39, 1953; G. D. Rochester and C. C. Butler, *Rep. Progr. Phys.* **16**, 365, 1953; Royal Society discussion, *Proc. Roy. Soc. A* **221**, 277, 1954; C. Dilworth, G. P. S. Occhialini, and L. Scarsi, *Ann. Rev. Nucl. Sc.* **4**, 271, 1954.
32. L. Leprince-Ringuet and B. Rossi, *Phys. Rev.* **92**, 722, 1953.
33. A. M. York *et al.*, *Phys. Rev.* **90**, 167, 1953; A. Bonetti *et al.*, *Nuovo Cim.* **10**, 1736, 1953.
34. R. Armenteros *et al.*, *Phil. Mag.* **43**, 597, 1952; C. D. Anderson *et al.*, *Phys. Rev.* **92**, 1089, 1953.
35. M. Danysz and J. Pniewski, *Phil. Mag.* **44**, 348, 1953.
36. C. O'Ceallaigh, *Phil. Mag.* **42**, 1032, 1951.
37. M. G. K. Menon and C. O'Ceallaigh, *Proc. Roy. Soc. A* **221**, 292, 1954.
38. W. B. Fowler *et al.*, *Phys. Rev.* **90**, 1126, 1953; also *Phys. Rev.* **91**, 1287, 1953.
39. A. Thorndike, *Proc. 4th Rochester conf.*, p. 82, unpublished.
40. G. Goldhaber, *Proc. 5th Rochester conf.*, p. 104, Interscience, New York 1955.
41. D. M. Ritson, *Machine techniques of high energy physics*, Interscience, New York 1961.

42. A. Pais, *Phys. Rev.* **86**, 663, 1952.
43. R. P. Feynman, 'High energy phenomena and meson theories', p. 76, notes of lectures given at Caltech, January–March 1951, unpublished.
44. R. G. Sachs, *Phys. Rev.* **84**, 305, 1951.
45. J. R. Oppenheimer, Ref. 17, p. 52.
46. Y. Nambu, K. Nishijima, and Y. Yamaguchi, *Progr. Th. Phys.* **6**, 615, 619, 1951; K. Aizu and T. Kinoshita, ibid., 630; H. Miyazawa, ibid., 631; S. Oneda, ibid., 633.
47. R. E. Marshak, in *The birth of particle physics*, Eds. L. M. Brown and L. Hoddeson, p. 376, Cambridge Univ. Press 1983.
48. Cf. also D. C. Peaslee, *Phys. Rev.* **86**, 127, 1952.
49. A. Pais, *Physica* **19**, 869, 1953, esp. p. 885; also Ref. 29.
50. S. Sakata, *Progr. Theor. Phys.* **16**, 686, 1956; M. Goldhaber, *Phys. Rev.* **101**, 433, 1956.
51. B. D'Espagnat and J. Prentki, *Progr. elementary particle and cosmic ray phys.*, Vol. 4, Part 3, 1958.
52. J. Sakurai, *Invariance principles and elementary particles*, Chapter 11, Princeton Univ. Press 1964.
53. R. B. Leighton *et al.*, *Phys. Rev.* **89**, 148, 1953; W. B. Fretter *et al.*, ibid., p. 168.
54. R. L. Garwin, *Phys. Rev.* **90**, 274, 1953; A. H. Rosenfeld and S. B. Treiman, *Phys. Rev.* **92**, 727, 1953.
55. W. B. Fowler *et al.*, *Phys. Rev.* **93**, 861, 1953.
56. M. Gell-Mann, *Phys. Rev.* **92**, 833, 1953.
57. T. Nakano and K. Nishijima, *Progr. Theor. Phys.* **10**, 581, 1953.
58. Ref. 42, Eq. (6).
59. M. Gell-Mann, *Nuovo Cim.* **4**, Suppl., p. 848, 1956.
60. M. Gell-Mann, 'On the classification of particles', Univ. of Chicago preprint, unpublished.
61. R. Plano *et al.*, *Nuovo Cim.* **5**, 216, 1957; L. W. Alvarez *et al.*, ibid., 1056.
62. L. W. Alvarez *et al.*, *Phys. Rev. Lett.* **2**, 215, 1959.
63. M. Gell-Mann and A. H. Rosenfeld, *Ann. Rev. Nucl. Sc.* **7**, 407, 1957.
64. K. Nishijima, *Progr. Theor. Phys.* **13**, 285, 1955.
65. M. Gell-Mann and E. P. Rosenbaum, *Sci. Am.* **197**, July 1957, p. 72.
66. R. K. Adair and E. C. Fowler, *Strange particles*, Interscience, New York 1963.
67. M. Gell-Mann and A. Pais, *Phys. Rev.* **97**, 1387, 1955.
68. K. Lande, E. T. Booth, J. Impeduglia, and L. M. Lederman, *Phys. Rev.* **103**, 1901, 1956; also M. Bardon, K. Lande, L. M. Lederman, and W. Chinowsky, *Ann. of Phys.* **5**, 156, 1958; R. Ammar *et al.*, *Nuovo Cim.* **5**, 1801, 1957; M. Baldo-Ceolin *et al.*, *Nuovo Cim.* **6**, 130, 1957; W. B. Fowler *et al.*, *Phys. Rev.* **113**, 928, 1959.
69. F. Eisler, R. Plano, N. Samios, M. Schwartz, and J. Steinberger, *Nuovo Cim.* **5**, 1700, 1957.
70. A. Pais and O. Piccioni, *Phys. Rev.* **100**, 1487, 1955.
71. K. M. Case, *Phys. Rev.* **103**, 1449, 1956.
72. M. L. Good, *Phys. Rev.* **106**, 591, 1957.
73. R. H. Good *et al.*, *Phys. Rev.* **124**, 1221, 1961.
74. S. B. Treiman and R. G. Sachs, *Phys. Rev.* **103**, 1545, 1956.
75. W. F. Fry and R. G. Sachs, *Phys. Rev.* **109**, 2212, 1958.
76. *Rev. Mod. Phys.* **56**, No. 2, Part II, p. S107, 1984.

77. J. D. Jackson, *The physics of elementary particles*, p. 75, Princeton Univ. Press 1958.
78. M. Schwartz, *Phys. Rev. Lett.* **6**, 556, 1961; B. d'Espagnat, *Nuovo Cim.* **20**, 1217, 1961.
79. R. Armenteros *et al.*, *Proc. Cern conf.*, p. 417, Cern, Geneva 1962.
80. R. P. Feynman, *Theory of fundamental processes*, p. 50, Benjamin, New York 1961.
81. T. D. Lee and C. S. Wu, *Ann. Rev. Nucl. Sc.* **16**, 511, 1966.
82. A. L. Hodson *et al.*, *Phys. Rev.* **96**, 1089, 1954; H. S. Bridge *et al.*, *Nuovo Cim.* **1**, 874, 1955; D. Keefe, *Nuovo Cim.* **10**, suppl., p. 412, 1956.
83. G. Harris *et al.*, *Phys. Rev.* **100**, 932, 1955; L. W. Alvarez *et al.*, *Phys. Rev.* **101**, 503, 1956; D. M. Ritson *et al.*, *Phys. Rev.* **101**, 1085, 1956; J. Crussard *et al.*, *Nuov. Cim.* **3**, 731, 1956; R. Motley and V. Fitch, *Phys. Rev.* **105**, 265, 1957.
84. T. F. Hoang *et al.*, *Phys. Rev.* **102**, 1185, 1956.
85. M. Widgoff *et al.*, *Phys. Rev.* **104**, 811, 1956.
86. R. W. Birge *et al.*, *Phys. Rev.* **100**, 430, 1955.
87. R. H. Dalitz, Ref. 26, p. 236; also *Phil. Mag.* **44**, 1068, 1954; *Phys. Rev.* **94**, 1046, 1954; E. Fabri, *Nuovo Cim.* **11**, 479, 1954.
88. See e.g. Refs. 52, 63, 66.
89. R. H. Dalitz, *Proc. 5th Rochester Conference*, 1955, p. 140, Interscience, New York 1955.
90. R. H. Dalitz, *Proc. 6th Rochester Conference*, 1956, p. V111–19; Interscience, New York 1956.
91. T. D. Lee and J. Orear, *Phys. Rev.* **100**, 932, 1955; T. D. Lee and C. N. Yang, *Phys. Rev.* **102**, 290; **104**, 822, 1956.
92. K. A. Brueckner, Ref. 29, p. 272.
93. J. R. Oppenheimer, Ref. 90, p. V111–20; also discussions, ibid., pp. 27ff.
94. A. Pais, *Nucl. Phys.* **5**, 296, 1958.
95. O. Laporte, *Zeitschr. f. Phys.* **23**, 135, 1924.
96. H. N. Russell, *Science* **49**, 512, 1924.
97. W. Heisenberg, *Zeitschr. f. Phys.* **32**, 841, 1925, Section 5.
98. E. P. Wigner, *Zeitschr. f. Phys.* **43**, 624, 1927, esp. Sections 15, 25.
99. E. P. Wigner, *Goett. Nachr.* 1927, p. 375.
100. H. Weyl, *The theory of groups and quantum mechanics*, transl. H. P. Robertson, Methuen, London, 1931.
101. W. Pauli, in *Handbuch d. Phys.* **24/1**, p. 185, Springer, Berlin 1933.
102. E. U. Condon and G. H. Shortley, *The theory of atomic spectra*, Macmillan, New York, 1935.
103. G. C. Wick, A. S. Wightman, and E. P. Wigner, *Phys. Rev.* **88**, 101, 1952; cf. also G. Feinberg and S. Weinberg, *Nuovo Cim.* **14**, 571, 1959.
104. L. Boltzmann, *Berl. Ber.*, 1897, p. 660; also ibid., 1897, p. 1016; 1897, p. 182. Repr. in *L. Boltzmann*, *Wissenschaftliche Abhandlungen*, Vol. 3, pp. 615, 618, 622, repr. by Chelsea, New York 1968.
105. M. Planck, *Berl. Ber.*, 1897, pp. 55, 715; repr. in *Max-Planck*, *Physikalische Abhandlungen und Vorträge*, Vol. 1, pp. 493, 505, Vieweg, Braunschweig 1958.
106. E. P. Wigner, *Goett. Nachr.* 1932, p. 546.
107. H. A. Kramers, *Proc. Ak. v. Wet. Amsterdam*, **33**, 959, 1930; repr. in *H. A. Kramers*, *collected scientific papers*, p. 522, North Holland, Amsterdam 1956.
108. Cf. J. Blatt and V. Weisskopf, *Theoretical nuclear physics*, p. 531, Wiley, New York 1952.

109. T. D. Lee, *Particle physics and an introduction to field theory*, Harwood, New York 1982.
110. W. Pauli, letter to W. Heisenberg, June 28, 1934, repr. in *Wolfgang Pauli, scientific correspondence*, eds. A. Hermann and K. von Meyenn, Vol. 2, p. 334, Springer, New York 1985.
111. M. Fierz, *Helv. Phys. Act.* **12**, 3, 1939.
112. W. Pauli, *Phys. Rev.* **58**, 716, 1940.
113. R. F. Streater and A. S. Wightman, *PCT, spin and statistics, and all that*, Benjamin, New York 1964.
114. R. Jost, in *Theoretical physics in the twentieth century*, Eds. M. Fierz and V. Weisskopf, p. 107, Interscience, New York 1960.
115. N. Burgoyne, *Nuovo Cim.* **8**, 607, 1958.
116. W. Pauli, in *Niels Bohr and the development of physics*, Ed. W. Pauli, p. 30, McGraw-Hill, New York 1955.
117. R. Jost, *Helv. Phys. Act.* **30**, 409, 1957, and Ref. 114.
118. See Ref. 116; also C. N. Yang and J. Tiomno, *Phys. Rev.* **79**, 495, 1950.
119. T. D. Lee, R. Oehme, and C. N. Yang, *Phys. Rev.* **106**, 340, 1957.
120. Ref. 76, p. S35; Ref. 109, p. 329.
121. L. W. Nordheim, *Phys. Rev.* **59**, 555, 1941.
122. E. P. Hincks and B. Pontecorvo, *Phys. Rev.* **73**, 257, 1948; R. D. Sard and E. J. Althaus, *Phys. Rev.* **73**, 1251, 1948; O. Piccioni, *Phys. Rev.* **74**, 1754, 1948.
123. Cf. A. Pais, in '*Five decades of weak interactions*', Ed. N. P. Chang, p. 58, *Ann. N. Y. Ac. Sci.* **294**, 1977.
124. B. Pontecorvo, *Phys. Rev.* **72**, 246, 1947; O. Klein, *Nature* **161**, 897, 1948; G. Puppi, *Nuovo Cim.* **5**, 587, 1948; **6**, 194, 1949; J. Tiomno and J. A. Wheeler, *Rev. Mod. Phys.* **21**, 144, 1949; T. D. Lee, M. Rosenbluth, and C. N. Yang, *Phys. Rev.* **75**, 905, 1949.
125. L. Michel, *Proc. Phys. Soc. London* **63A**, 514, 1950.
126. C. Bouchiat and L. Michel, *Phys. Rev.* **106**, 170, 1957.
127. E. J. Konopinski and H. M. Mahmoud, *Phys. Rev.* **92**, 1045, 1953; see also W. Pauli, *Nuovo Cim.* **6**, 204, 1957.
128. T. D. Lee and C. S. Wu, *Ann. Rev. Nucl. Sc.* **15**, 381, 1965.
129. M. A. Ruderman and R. J. Finkelstein, *Phys. Rev.* **76**, 1458, 1949.
130. R. P. Feynman, *Proc. CERN conf.*, p. 216, CERN, Geneva 1958.
131. T. Fazzini *et al.*, *Phys. Rev. Lett.* **1**, 247, 1958; G. Impeduglia *et al.*, *Phys. Rev. Lett.* **1**, 249, 1958.
132. N. Dallaporta, *Nuov. Cim.* **1**, 962, 1953; M. Gell-Mann, Ref. 90, p. V111–23; also Ref. 63.
133. E. Purcell and N. F. Ramsey, *Phys. Rev.* **78**, 807, 1950.
134. T. D. Lee and C. N. Yang, *Phys. Rev.* **104**, 254, 1956.
135. C. S. Wu, E. Ambler, R. W. Hayward, D. D. Hoppes, and R. P. Hudson, *Phys. Rev.* **105**, 1413, 1957.
136. T. D. Lee and C. N. Yang, *Phys. Rev.* **105**, 1671, 1957.
137. A. Salam, *Nuovo Cim.* **5**, 299, 1957.
138. L. D. Landau, *Nucl. Phys.* **3**, 127, 1957; R. Gatto, *Phys. Rev.* **106**, 168, 1957.
139. T. D. Lee, *Proc. Seventh Rochester conf.* (*April* 15–19, 1957), P. V11–1, Interscience, New York 1957.
140. T. D. Lee, J. Steinberger, G. Feinberg, P. K. Kabir, and C. N. Yang, *Phys. Rev.* **106**, 1367, 1957; T. D. Lee and C. N. Yang, *Phys. Rev.* **108**, 1645, 1957.

141. D. H. White and D. Sullivan, *Physics Today* **32**, April 1979, p. 40.
142. E. J. Konopinski, *The theory of beta radioactivity*, Oxford Univ. Press 1966.
143. C. S. Wu and S. A. Moszkowski, *Beta decay*, Interscience, New York 1966.
144. R. E. Marshak, Riazuddin, and C. P. Ryan, *Theory of weak interactions in particle physics*, Wiley, New York 1969.
145. P. K. Kabir, *The development of weak interaction theory*, Gordon & Breach, New York 1963.
146. R. L. Garwin, L. M. Lederman, and M. Weinrich, *Phys. Rev.* **105**, 1415, 1957; J. I. Friedman and V. L. Telegdi, *Phys. Rev.* **105**, 1681, 1957.
147. L. D. Landau, Ref. 138; T. D. Lee and C. N. Yang, *Phys. Rev.* **105**, 1671, 1957; J. D. Jackson, S. B. Treiman, and H. W. Wyld, *Phys. Rev.* **106**, 517, 1957; R. B. Curtis and R. R. Lewis, *Phys. Rev.* **107**, 543, 1957.
148. H. Frauenfelder *et al.*, *Phys. Rev.* **106**, 386, 1957.
149. C. N. Yang, *Rev. Mod. Phys.* **29**, 231, 1957; L. D. Landau, *Soviet Phys. JETP* **32**, 405, 1957.
150. A. Pais and S. B. Treiman, *Phys. Rev.* **106**, 1106, 1957.
151. D. Neagu *et al.*, *Phys. Rev. Lett.* **6**, 552, 1961.
152. Ref. 119; H. W. Wyld and S. B. Treiman, *Phys. Rev.* **106**, 169, 1957; also R. G. Sachs and S. B. Treiman, *Phys. Rev. Lett.* **8**, 137, 1962; R. G. Sachs, *Phys. Rev.* **129**, 2280, 1963.
153. D. Luers *et al.*, *Phys. Rev.* **133**, 1276, 1964.
154. M. A. Clark *et al.*, *Phys. Rev. Lett.* **1**, 100, 1958; M. T. Burgy *et al.*, *Phys. Rev. Lett.* **1**, 324, 1958.
155. J. Cronin, *Physics Today* **35**, July 1982, p. 38.
156. L. D. Landau, *Soviet Phys. JETP* **32**, 407, 1957, and Ref. 138.
157. W. Pauli, *Handbuch der Physik*, Vol. 24/1, p. 226, Springer, Berlin 1933.
158. K. M. Case, *Phys. Rev.* **107**, 307, 1957.
159. S. Watanabe, *Phys. Rev.* **106**, 1306, 1957.
160. Ref. 128, Fig. 6.
161. See Refs. 52, Section 7.4; 63, Sections 4.2–7; 139; 142, Chapter 1; 143, Chapters 1, 3; 144, Section 1.4A.
162. M. Gell-Mann, *Proc. Int. Conf. Symmetries in Physics*, Sant Felice de Guixols, Spain 1983.
163. R. E. Marshak and E. C. G. Sudarshan, *Int. Conf.* '50 *years of weak interactions*', Racine, Wisc. 1984.
164. M. Gell-Mann and R. P. Feynman, *Phys. Rev.* **109**, 193, 1958.
165. E. C. G. Sudarshan and R. E. Marshak, *Phys. Rev.* **109**, 1860, 1958.
166. J. Sakurai, *Nuovo Cim.* **7**, 649, 1958.
167. M. Goldhaber, L. Grodzins, and A. W. Sunyar, *Phys. Rev.* **109**, 1015, 1958.
168. R. Budde *et al.*, *Phys. Rev.* **103**, 1827, 1956, Section 6.
169. A. Pais and S. B. Treiman, *Phys. Rev.* **105**, 1616, 1957.
170. G. E. Kalmus and A. Kernan, *Phys. Rev.* **159**, 1187, 1967; also contains references to other experimental papers on this subject.
171. A. Pais and S. B. Treiman, *Phys. Rev.* **168**, 1858, 1968; cf. also *Phys. Rev.* **178**, 2365, 1669.
172. L. Rosselet *et al.*, *Phys. Rev.* **D15**, 574, 1977.
173. T. D. Lee and C. N. Yang, *Phys. Rev.* **119**, 1410, 1960, Section 11; **126**, 2239, 1962; A. Pais, *Phys. Rev. Lett.* **9**, 117, 1962; A. Pais and S. B. Treiman, in *Problems of theoretical physics*, p. 257, Nauka, Moscow 1969.

174. S. S. Gershtein and Ia. B. Zel'dovich, *Soviet Phys. JETP* **2**, 576, 1976.
175. S. Weinberg, *Phys. Rev.* **112**, 375, 1978.
176. C. S. Wu, in *Unification of elementary forces*, Eds. D. B. Cline and F. E. Mills, p. 549, Harwood, London 1978.
177. M. Gell-Mann, *Phys. Rev.* **111**, 362, 1958.
178. Refs. 143, Sec. 7.3; 144, Sec. 4.3; 145, Sec 12.5; C. S. Wu, *Rev. Mod. Phys.* **36**, 618, 1964.
179. J. C. Taylor, *Phys. Rev.* **110**, 1216, 1958.
180. M. L. Goldberger and S. B. Treiman, *Phys. Rev.* **110**, 1178, 1958; **111**, 354, 1958.
181. M. D. Scadron, *Rep. Progr. in Phys.* **44**, 213, 1981.
182. J. Bernstein, *Elementary particles and their currents*, Chapters 11, 12, Freeman, San Francisco, 1968.
183. H. R. Pagels, *Phys. Rep.* **16**, 219, 1975.
184. M. Gell-Mann and M. Levy, *Nuovo Cim.* **16**, 705, 1960.
185. J. H. Christenson, J. W. Cronin, V. L. Fitch, and R. Turlay, *Phys. Rev. Lett.* **13**, 138, 1964.
186. A. Abashian *et al.*, *Phys. Rev. Lett.* **13**, 243, 1964; see also A. Abashian, *Physics Today* **36**, February 1983, p. 101.
187. J. Prentki, *Proc. Intern. Conf. on elementary particles, Oxford* 1965, p. 47, Rutherford Laboratory, 1966; A. Franklin, *Hist. St. Phys. Sci.* **13**, 207, 1983.
188. J. W. Cronin, *Rev. Mod. Phys.* **53**, 373, 1981.
189. P. K. Kabir, *The CP puzzle*, Academic Press, New York 1968.
190. K. Kleinknecht, *Ann. Rev. Nucl. Sc.* **26**, 1, 1976.
191. B. Winstein, *Proc.* 11*th int. conf. on neutrino physics and astrophysics*, Dortmund 1984.
192. I. I. Bigi and A. I. Sanda, *Comm. Nucl. Ptcle Phys.* **14**, 149, 1985.
193. K. R. Schubert *et al.*, *Phys. Lett.* **B31**, 662, 1970.
194. Ref. 76, p. S113.
195. L. Wolfenstein, *Phys. Rev. Lett.* **13**, 562, 1964.
196. K. Winter, *Proc. Amsterdam conference on elementary particles*, p. 333, North Holland, Amsterdam 1972.
197. L. Wolfenstein, *Phys. Rev. Lett.* **13**, 569, 1964.
198. Ref. 76, p. S37.
199. A. Pais and S. B. Treiman, *Phys. Rev.* **D12**, 2744, 1975.
200. L. B. Okun, V. I. Zakharov, and B. M. Pontecorvo, *Lett. Nuov. Cim.* **13**, 218, 1975; also Ref. 192.
201. G. D. Rochester, Ref. 7, p. 169.
202. C. C. Butler, Ref. 7, p. 177.
203. C. D. Anderson, Ref. 47, p. 131.
204. W. B. Fretter, Ref. 7, p. 191.
205. M. Gell-Mann, Ref. 7, p. 395.
206. K. Nishijima, Ref. 7, p. 403.
207. R. H. Dalitz, Ref. 7, p. 195.
208. C. N. Yang, *Selected papers* 1945–80, Freeman, San Francisco 1983.
209. T. D. Lee, *Selected papers*, Ed. G. Feinberg, Birkhaüser, Boston 1981.
210. C. S. Wu, *Adventures in expt. phys.*, Ed. B. Maglich, Vol. γ, p. 101, World Sci. Education, Princeton 1973.
211. R. Garwin, Ref. 210, p. 124.
212. V. L. Telegdi, Ref. 210, p. 131; Ref. 162.

213. R. F. Feynman, *Surely you're joking Mr Feynman!*, pp. 247–53, Norton, New York 1985.
214. E. C. G. Sudarshan and R. E. Marshak, *Proc. int. conf. on* 50 *years of weak interactions*, Racine, Wisc. 1984.
215. V. Fitch, *Rev. Mod. Phys.* **53**, 367, 1980; ref. 162.
216. T. D. Lee and C. N. Yang, *Brookhaven Nat'l Lab. Report BNL* 443 (*T*–91), 1957.
217. C. S. Wu, Ref. 114, p. 249.
218. A. Pais, *Proc.* 12*th Solvay Conf.*, 1961, p. 101, Interscience, New York 1962.
219. T. D. Lee and C. S. Wu, Refs. 81, 128, and *Ann. Rev. Nucl. Sc.* **16**, 471, 1966.

21

Essay on modern times: 1960–83

Essay . . . . originally implying want of finish.
Oxford English Dictionary.

(a) In which it is explained why a very rich period is treated with such brevity

In this final essay (it is to be followed by an epilog), the bridge between the X of Roentgen and the W and Z of UA1 and UA2 is closed. What was said in Chapter 1 applies most particularly to the present chapter: that the number of pages devoted to a given topic is not necessarily proportional to its importance—footnotes could easily be extended to essays, pages to monographs; and that more participants should be heard before making a historical assessment.

The period to which we now turn, from the early 1960s to the early 1980s, offers a rich variety of advances, in experimental techniques and discoveries, and in theoretical insights. In the course of the previous three chapters, principally devoted to the years from 1945 through the 1950s, some developments bearing on later years have already been sketched: the evolution of fixed-target accelerators;* quantum electrodynamics;** S-matrix theory;† Regge poles;†† and the violations of CP and T.‡ There is much more, however. Linear accelerators, colliders, and new detectors enter the picture. Experimental discoveries of new particles abound: new types of hadrons, the collective name for strongly interacting particles, baryons and mesons, adopted in 1962[1]; new leptons; and of course the W and the Z. Secondary beams of neutrinos became an important new source of experimental information. Also charged and neutral hyperons, moderately rare commodities in the 1950s, are now available as highly useful secondary beams.[2] High energy lepton–hadron scattering has revealed unexpected new properties of matter. A new class of weak reactions was discovered, the 'neutral current processes'. In theoretical physics new symmetries made their appearance. A new picture emerged in

* Chapter 19, Section (a). ** Chapter 18, Section (c), Part 3. † Chapter 19, Section (f), Part 5.
†† Chapter 19, Section (f), Part 6. ‡ Chapter 20, Section (e).

which hadrons, all those described earlier as well as new ones, are conceived as composites of finer subunits, quarks. Above all, there were breakthroughs in the understanding of weak and strong forces in terms of fundamental vector fields, advances largely made possible by the discoveries of new classes of renormalizable field theories and of 'asymptotic freedom'.

This brings me to the main message of the present chapter: relativistic quantum field theory is much healthier and much richer in new options than had been thought during the fifties and much of the sixties when, to be sure, quantum electrodynamics looked increasingly successful but the status of meson field theories remained highly problematical. As we now see it, the Yukawa-type interactions, unalterably important for low energy phenomena such as nuclear forces, are actually secondary manifestations of an underlying field theory, called quantum chromodynamics, not unlike the way Van der Waals forces are secondary consequences of electrodynamics. Furthermore, the Fermi interaction for weak processes, unalterably important for low energy phenomena such as β-decay, is also a secondary manifestation of an underlying field theory—whence the W and the Z.

There are numerous reasons why I have chosen to cram this plethora of novelties into a single chapter. It adds emphasis to the speed and intensity of these developments. It will serve neither the general reader nor the expert to go into too much technical detail, particularly plentiful in these areas. The topics to be discussed are extensively covered in recent books, anthologies, and reports; the reader interested in further particulars will find references to these at every step of the way. Also, there is perhaps some virtue in brevity when discussing events so recent that possible pitfalls of simplicity may yet lurk beyond our vision.

This chapter is laid out as follows. Section (b), devoted to higher symmetries, begins with comments on strange particle spectroscopy. Thereupon it is described by what steps one arrived at SU(3), the symmetry group that successfully unifies strangeness with isospin, and how this step led to the hypothesis that all baryons and mesons known till then were built of quarks, a hypothetical triplet of particles with fractional electric charges, 2/3, −1/3, −1/3 times e, respectively. Next follows a discussion of static SU(6), a symmetry group related to SU(3) just as Wigner's SU(4) for nuclei is related* to the isospin group SU(2). Then the reasons are recalled why static SU(6) led to the introduction of a new three-valued quantum number, 'color', for quarks; and how this in turn led to the further assumptions that color is a manifestation of a new SU(3) group, 'SU(3) color', distinct from the SU(3) mentioned just previously; that customary baryons and mesons and the photon are 'colorless' (singlets with respect to SU(3) color); and that SU(3) color (unlike the previous SU(3)) manifests itself dynamically through interactions

* Chapter 17, Section (f).

between quarks mediated by vector mesons—all these assumptions being later integrated into quantum chromodynamics. There follows a brief discussion of current algebra, a formalism that again leads to quarks, but via an alternate route. Section (b) concludes with a snapshot of particle physics as it looked in the mid-sixties.

Section (c) is devoted to machines other than the (earlier discussed) circular proton accelerators, and contains first glimpses of the new physics they helped produce: nuclear reactors made possible the direct observation of neutrinos produced in β-decay; neutrino beams revealed the existence of a second kind of neutrino, intimately related to the muon; SLAC, the world's major linear electron accelerator, produced a great surprise with far-reaching consequences; large-angle inelastic electron scattering was far more intense than had been anticipated; colliders appeared on the scene, beginning in the 1960s with electrons hitting electrons or positrons, followed in the 1970s by the first proton–proton collider, and in the eighties by the first proton–antiproton collider. In Section (d) we follow the route from the just mentioned SLAC results to scaling laws, and from there to the parton model.

Section (e), a conspectus of progress during the roaring seventies and early eighties, deals with the discoveries of quantum chromodynamics, the new road to the strong interactions; and of a unified field theory of weak and electromagnetic interactions. It is recalled how this unification stimulated the experimental discovery of a new type of current, the weak neutral current; and of the W- and Z-bosons. The theoretical proposal of a new hadronic quantum number, charm, and of a corresponding new fourth type of quark and its links to the discovery of a new class of massive hadrons with remarkably long lifetimes are discussed. Other fundamental experimental enrichments not bargained for by theorists include the discoveries of a new lepton, τ, and of still further new types of hadrons linked to the presence of a fifth and very probably a sixth type of quark.

(b) Higher symmetries take off

1. Further extensions of hadron spectroscopy. What is called high energy in one decade becomes medium energy in the next and low energy thereafter. The fifties were dominated* by weak focussing proton synchrocyclotrons in the 3–10 GeV region; the sixties by their strong successors, 30–70 GeV, and by the emergence of linear accelerators and electron colliders (see below); the seventies brought us circular fixed-target machines up to 500 GeV and the first hadron collider (again see below).

Strong and weak strange particle phenomena were at the frontiers during the fifties. Beginning in the sixties, these processes became the subject of ever

* Chapter 19, Section (a).

more refined medium energy studies. The increasing intensity of available K-beams has made it possible to analyze old and new decay modes with improved precision. Examples: the branching ratio for the dominant decay of $K^{\pm}$ into $\mu^{\pm}+\nu$ is now known within about 0.1 per cent accuracy. Seventeen other kinds of decays[3] of these particles have been detected. It has been possible (but not easy) to find rare modes with branching ratios of one part in 10^7. Likewise, thirteen decay modes of K_L are now under control.[3]

In regard to strong interactions, the sixties mark the beginning of strange particle spectroscopy. As noted before* the discoveries in 1961 of the first non-strange meson resonances, ρ, ω, η, were made possible by the advent of hydrogen bubble chambers. Strange particle resonances were first seen at about that same time and with the help of that same new type of detector. First[4] came the Σ (1385) resonance,** in late 1960. The year thereafter the first K-resonance,[5] K^- (892), and the first Λ-resonance,[6] Λ (1405), were reported. Ξ (1530), the first cascade resonance, was found[7] in 1962. All these resonances decay strongly into the metastable ground state, conserving strangeness (e.g. K^- (892) $\rightarrow \bar{K}^0\pi^-$, Σ (1385) $\rightarrow \Lambda\pi$, Λ (1405) $\rightarrow \Sigma\pi$, Ξ (1530) $\rightarrow \Xi\pi$). Every one of these and other resonances has its own individual history, how a multiplet was completed, how spins, parities, isospins were determined. Praised be the 'Review of particle properties'[3] for not only giving such data but also the pertinent literature.

Since the early sixties, the number of strange resonances has proliferated. By now 13 of type Λ, 9 of type Σ, 3 of type Ξ, and 8 of type K are known, with masses ranging up to about 3 GeV, the particle symbols indicating that they have the same quantum numbers as the first particles so named—except for spin which ranges up to 9/2, 7/2, 3/2, and 4 respectively.[3] It stands to reason that one would apply Regge pole ideas† to these new breeds. Once again plots of $(\text{mass})^2$ versus the real part of complex angular momentum showed a nearly linear behavior. Experts differ in opinion, however, as to the precision with which that linearity has been established.[8]

Meanwhile links had been established between strange and non-strange hadrons.

2. SU(3).†† The idea that strange particles call for an extension of isospin to a more embracing higher symmetry dates from the early fifties, as we have seen.‡ Recall‡‡ that, in those years, even the simple isospin rules were a

* Chapter 19, Section (e).
** Originally called Y_1^*. From here on I follow the usage of labelling a resonance by its modern symbol and mass value (averaged over a multiplet) in MeV.
† Chapter 19, Section (f), Part 6.
†† For an anthology of basic papers on SU(3) see Ref. 9.
‡ Chapter 20, Section (b), Part 1. Early attempts at finding such a symmetry are reviewed in Ref. 10.
‡‡ Chapter 19, Section (d), Part 1.

blessing since their validity was not tied to detailed assumptions about meson dynamics—a subject over which darkness reigned supremely. That situation had not appreciably changed when, in the early sixties, higher symmetries began taking off. In the first instance the strategy remained the same as for isospin: to find symmetries that yield new predictions and new constraints on a strong interaction dynamics yet to come. In order to specify in a little more detail what was at issue it may help to recall a few points about isospin.*

Isospin is not an exact symmetry, it is broken by electromagnetic (and weak) interactions. Nevertheless, isospin remains useful even in the presence of electromagnetism, because the violations appear as small corrections to the strong interactions. It is this smallness which makes it possible to formulate useful 'violation laws' such as the mass formula (19.20). This comes about because the breaking picks the specific preferred direction '3' in isospin space, as follows from the structure of the charge operator Q, in units e (Eq. (20.14))

$$Q = T_3 + \frac{Y}{2} \tag{21.1}$$

where

$$Y = B + S \tag{21.2}$$

is the so-called hypercharge, a quantity convenient for what follows.

The components T_i of the isospin operator satisfy the commutation relations (17.30) in straight analogy with angular momentum, Eq. (12.14). For a given isomultiplet, $T^2 = T_1^2 + T_2^2 + T_3^2$ equals $t(t+1)$ where $t = 1/2$ (nucleon) or 1 (pion) or $3/2(\Delta)$ ... To $t = 1/2$, 1, 3/2, ... correspond a multiplet of states with 2, 3, 4, ... members. The same, in mathematical language: T_i are the three generators of the group SU(2); a multiplet is an irreducible representation of that group, and is fully characterized by the number t. The number $2t+1$ is the dimension of the representation. Two isospins 1/2 can be composed to either isospin 0 or 1. Mathematically, in terms of dimensions: the product of two 2's can be decomposed into a 1 and a 3 according to the 'reduction formula' for dimensions

$$2 \times 2 = 3 + 1. \tag{21.3}$$

In the search for a higher strong interaction symmetry the following questions arose:

(a) Which group? Its generators should at least comprise T_i and Y. One 16should be able to label hadrons by T^2, T_3, and Y; hence Y should commute with T^2 and T_3.

(b) Which representations? As was true for isospin these representations must contain particles with the same spin and parity.

* Chapter 17, Section (f).

(c) The aim is to lump together more baryons, or mesons, than had already been grouped by isospin alone. If the symmetry were exactly valid then all particles in a given representation would have the same mass. This shows that the higher symmetry is necessarily a broken symmetry. That was not new, as witness isospin broken by electromagnetism. In the case of the higher symmetry, the breaking must be due to another, stronger, cause, however. To take but a mild case, if one wishes to lump together Λ with the nucleon (N) then one must account for the fact that Λ weighs 10 per cent more than N. How then is the higher symmetry broken down to SU(2)? Are there violation laws?

All such questions are familiar to today's graduate student taking a course in theoretical particle physics. He or she is exposed to Lie groups, knows that the rank of such a group equals the number of additively conserved quantum numbers, and has been briefed on other assorted links between group theory and particle physics. It was not so in, say, 1960. Whatever the average well-educated theorist then knew about groups concerned rotations, Lorentz transformations, and various discrete symmetries. There were exceptions, of course, most notably Racah and his school, who already, in the 1940s, had successfully applied much more advanced group theoretical methods[11] to atomic and nuclear spectroscopy. I well recall Racah's clear and interesting lectures in the spring of 1951 at the Institute for Advanced Study. 'From the start it was evident that here we physicists were being taught the real craft. I found Eugen Merzbacher and David Park willing to take notes out of which grew the widely known Racah lectures[12] on group theory and spectroscopy which have been of such great use ever since.'[13] Yet at that time I, among others, failed to distill those essentials from these lectures which, later on, would be of use in particle physics.

Let us return to the higher symmetry question. Note that, because of large mass splittings, it was not at once obvious which—if in fact any!—particles to lump together. For example, in 1960, a respectable attempt was made[14] to put (N, Σ, Ξ) in one representation, Λ by itself in another, with respective dimensions 7 and 1. Even earlier the idea of SU(3) as the looked-for symmetry had made its appearance. Taking their cue from a generalization of the Fermi–Yang model,* due to Sakata,[15] in which hadrons are conceived as composites of p, n, and Λ and their antiparticles, Ikeda, Ohnuki, and Ogawa had suggested[16] (1959) lumping (p, n, Λ) in one of the three-dimensional representations, called 3, of the group SU(3). ($\bar{p}$, $\bar{n}$, $\bar{\Lambda}$) were assigned to 3*, the other, distinct,** three-dimensional representation of SU(3). They had worked out (I transcribe their equations in modern notation) the commutation

* Chapter 19, Section (f), Part 2.

** This distinction has no analog for SU(2). Cf. also Ref. 17 where isospin is enlarged by a discrete invariance for the interchange n ↔ Λ.

relations for the eight operators F_i, later called unitary spin:

$$[F_i, F_j] = if_{ijk}F_k; \qquad i, j, k = 1, \ldots, 8 \tag{21.4}$$

which are the analogs of Eq. (17.30) for isospin, where the f_{ijk} are totally antisymmetric in i, j, k, with prescribed numerical values. They had further recognized that (F_1, F_2, F_3) can be identified with isospin because that triplet satisfies Eq. (17.30); and that Y is proportional to F_8:

$$Q = F_3 + \frac{F_8}{\sqrt{3}}; \qquad [F_8, F_{1,2,3}] = 0. \tag{21.5}$$

Considering pseudoscalar mesons as baryon–antibaryon composites à la Fermi–Yang, they had emplaced the three π's and four K's in an 8-dimensional representation, an 8, according to the reduction formula analogous to Eq. (21.3):

$$3 \times 3^* = 8 + 1. \tag{21.6}$$

One member of the 8, with $T = Y = 0$ (the later η) was still missing. These same results had also been found independently by Wess.[18]

As so often happens in physics, an unwarranted assumption leads to partial progress. Here: starting from (p, n, Λ) as a 3, attention is drawn to SU(3) which is good. Pseudoscalar mesons are an 8, also good. Yet (p, n, Λ) do not form a 3. (In the mentioned papers Σ and Ξ are not correctly assigned either.)

Early in 1961 the correct answer: N, Λ, Σ, Ξ form an 8 of SU(3), their antiparticles form another 8, was found independently by Gell-Mann, Ne'eman, and Speiser and Tarski. These authors had substantially, but not identically, the same motivations and initially pursued SU(3) to different levels of detail. For more on these matters the reader should turn to the respective reminiscences[19,20,21] and original papers.[22,23,24] Here I only note a few main points.

(1) ρ, ω, and K (892) form[22,23] another 8.

(2) About interactions. Remember the isospin structure of π-nucleon coupling which is essentially $\bar{N}\vec{\tau}N\vec{\pi}$ (Eq. (19.8)). It is an invariant, a scalar, with respect to SU(2). What is the SU(3) analog? An invariant coupling between the antibaryon 8, the baryon 8, and the meson 8? It turns out[22] that in general there are not one but two scalars that can be made of three 8's, named D- and F-coupling. The D/F ratio of strengths of these two interactions is not dictated by SU(3).

(3) Violation laws. The breakdown SU(3) → SU(2) which must preserve isospin and strangeness is effected most simply by the further assumption that the breakdown mechanism transforms like that component of an octet which lies in the hypercharge direction '8'. Assume further that the 'perturbation' causing

this breakdown is taken to lowest order. This leads* to the Gell-Mann–Okubo[25] mass formula for SU(3) multiplets of baryons:

$$M = M_0 + aY + b\left(T(T+1) - \frac{Y^2}{4}\right) \tag{21.7}$$

from which follows[22]

$$M_N + M_\Xi = \tfrac{1}{2}(M_\Lambda + 3M_\Sigma) \tag{21.8}$$

which is very well satisfied.[3]

It needs to be stressed that Eq. (21.7) follows exclusively from the transformation properties (octet) of the symmetry breaking under SU(3) and is *independent from whatever may be the dynamical origins of that breaking*. This is in apparent contrast to isospin breaking, where long-known electromagnetic effects are the prime candidate. In the later discussion of electroweak theory it will become clear that actually the symmetry-breaking mechanisms for isospin and for SU(3) are not all that fundamentally different.

The early SU(3) papers were received with interest but not at once with general acclaim. An authoritative review[26] on higher symmetries in early 1962 lists SU(3) as one among several options.** The next phase began at the eleventh Rochester conference (July 1962) where the resonance Ξ (1530) was first reported.[28] Gell-Mann observed[29] that if Δ (1232), Σ (1385), and this new resonance were all to have spin 3/2 (certain for Δ, not yet for the others) then one would have $4+3+2=9$ states that could be grouped in a 10 of SU(3). This proposal was supported by the fact that, one shows, for a 10 the special relation $T = 1 + Y/2$ holds, hence $M = \alpha + \beta Y$ by Eq. (21.7). The resulting equal mass spacing for the Δ, Σ, Ξ-resonances was well satisfied. A tenth state, named[29] Ω^-, with $T=0$, $Y=-2$, M about 1680 MeV, was not there, however. When early in 1964 one such particle† with the right mass was found[31] there was no longer any doubt, SU(3) was in.†† A number of other SU(3) multiplets have since been identified.[33]

I return a while later to the implications of SU(3) for weak interactions.

3. Quarks.‡ Nature appears to keep things simple but had bypassed the fundamental 3's in favor of 8's.

Or had it?

* The fact that Eq. (21.7) contains two symmetry-breaking parameters (a, b) is another consequence of the circumstance that three 8's make two scalars. For mesons one uses the same formula for (mass)^2.

** One implication of the octet picture: even relative ΣΛ-parity was still a subject of some debate at that time.[27]

† Some events observed earlier were probably of that same type.[30]

†† A semi-popular account of the status of particle physics in 1964 is found in Ref. 32.

‡ Anthology: Ref. 34. Reviews: Ref. 35.

In the early days of 1964 it was remarked by Gell-Mann[36] and by Zweig[37] that one can imagine baryons and mesons to be made up of 3's (quarks, q) and 3*'s (antiquarks, $\bar{q}$) if one allows these particles to have non-integral electric charges. These particles are, by name and attributes (Q in units e):

q	B	T	$T_3 = F_3$	Y	S	Q
u	$\frac{1}{3}$	$\frac{1}{2}$	$\frac{1}{2}$	$\frac{1}{3}$	0	$\frac{2}{3}$
d	$\frac{1}{3}$	$\frac{1}{2}$	$-\frac{1}{2}$	$\frac{1}{3}$	0	$-\frac{1}{3}$
s	$\frac{1}{3}$	0	0	$-\frac{2}{3}$	-1	$-\frac{1}{3}$

To go from q to $\bar{q}$, change all signs (except for T). All q, $\bar{q}$ must of course have the same spin, taken to be 1/2.

What does one buy with this? Mesons are taken to be bound $q\bar{q}$ states, $^1S \leftrightarrow$ pseudoscalar, $^3S \leftrightarrow$ vector. One readily recognizes familiar quantum numbers, $u\bar{d} \leftrightarrow \pi^+$ or ρ^+; $u\bar{s} \leftrightarrow K^+$ or K^+ (892); etc. Baryons are qqq states, evidently with the correct baryon number, $B = 1$. According to the corresponding reduction formula

$$3 \times 3 \times 3 = 1 + 8 + 8 + 10 \tag{21.9}$$

there is room for both the 8 and the 10 with the right spins, 1/2 and 3/2 respectively. Samples: uud $\leftrightarrow$ p or Δ^+; uuu $\leftrightarrow \Delta^{++}$; sss $\leftrightarrow \Omega^-$.

Important: SU(3) breaking can be accommodated by a mass difference between s and (u, d). This propagates to the mass formula (21.7) for other multiplets.

'The reaction of the theoretical physics community to the [quark] model was generally not benign . . . The idea that hadrons were made of elementary particles with fractional quantum numbers did seem a bit rich.'[38] A question not asked since the days when the reality of atoms was at issue now returned: is this a mnemonic device or is this physics? Particles with fractional charge, not easy to swallow for theorists, were a boon to experimentalists, however. A search for such objects has now been going on for twenty years. The simplest summary of this hunt is: no firm evidence for free quarks.* As we shall see, that negative result accords with later theoretical views.

Other obvious questions were raised at once. How big are the quark masses? What forces bind $q\bar{q}$ and qqq systems? Why should these forces not also bind other (q, $\bar{q}$) complexes? Why does the deuteron not collapse into a tight 6q state?** It is astonishing that it would take only one decade until quantum chromodynamics provided sensible views on these questions.

* Detailed references covering the period 1964–84 are found in Ref. 39.
** For opinions in the late sixties on these questions see Refs. 34, 40, 41.

Data analysis on the basis of q and $\bar{q}$ bound systems is continuing. Further elaborations* include the discussion of higher resonances, bound either in S- or in higher orbital angular momentum states; and, of 'exotic' systems such as $qq\bar{q}\bar{q}$ states (baryonium).[43] 'In a naive qualitative comparison the [quark] model holds up quite well.'[44]

*4. SU(6).*** In the early fall of 1964, I jotted down some dates and facts bearing on the hectic summer just past which I had spent at Brookhaven National Laboratory. Most of June had been devoted to preparing a rapporteur's talk on weak interactions for the forthcoming Dubna conference (August), an assignment which demanded particular care because of the great novelty: CP-violation, found at Brookhaven, first publication submitted July 10. Another subject to be discussed was a recent paper[47] (May) by Gell-Mann on current algebra (see Part 6 of this section). In that article there is mention of a group SU(6) which if 'of any use as an approximate symmetry ... would arrange particles of different spins and parities in super-super-supermultiplets'. I mentioned this remark to Gürsey (also at Brookhaven), suggesting he might look into this further. On 1 July Radicati arrived. He had only recently been working[48] on Wigner's supermultiplet theory of the thirties, based on SU(4). From this curious combination of circumstances was born the generalization of Wigner's SU(4) for nuclei to SU(6), later given the better name 'static SU(6)' (which has a physical content distinct from the group Gell-Mann had speculated about). On 8 July Gürsey and Radicati told me about[49] the 56 and the 35. I was taken by their remark, went to work on SU(6)—and never got to Dubna.† A few weeks later we found out that Sakita[50] had also come upon SU(6). Only later did I learn that Zweig[51] had had the same idea as well, also in that August.

Static SU(6) is an example of a type of symmetry that had entered physics twice before: Russell-Saunders coupling in atoms, where spin is conserved in the absence of spin–orbit coupling; and Wigner's SU(4) valid†† in the approximation that the Hamiltonian is spin- and isospin-independent. In static SU(6) one has independence of spin (generators S_i, $i=1,2,3$), unitary spin (F_i, $i=1,\ldots,8$), and also of S_iF_j, $3+8+3\times 8=35$ operators in all, collectively denoted by Λ_A, the generators of an SU(6). The Λ_A satisfy the commutation relations similar to Eq. (21.4): $[\Lambda_A, \Lambda_B]=f_{ABC}\Lambda_C$, with totally antisymmetric f_{ABC}.

In order to manipulate with SU(6), a bit of abstract group theory cannot do any harm.‡ I can attest, however, that the early practitioners of SU(6) were

* For a semi-popular account (1983), including references, see Ref. 42.
** Anthology: Ref. 45. Review: Ref. 46.
† Treiman was good enough to give the rapporteur's talk.
†† Chapter 17, Section (f).
‡ Cf. e.g. Ref. 52 where rules for deriving Eqs. (21.9–14) are collected.

'thinking quark' a good deal of the time, not because they were sold on the reality of these odd objects, but because they gave such handy and concrete pictures. In that spirit one starts out with the 6, the simplest representation:

$$6\text{:}\quad (u_\uparrow, u_\downarrow, d_\uparrow, d_\downarrow, s_\uparrow, s_\downarrow)$$

where the arrows denote spin up or down. One recognizes at once that SU(6) contains: (a) SU(3), transforms quark species for fixed spin; (b) SU(2), transforms spin for fixed species. Symbolically

$$6 = (3, 2), \tag{21.10}$$

the 6 contains an SU(3) triplet of spin doublets (all states with the same parity). Antiquarks form a 6*.

Let us transpose a few SU(3) relations to SU(6). First for mesons, à la Eq. (21.4)

$$6 \times 6^* = 35 + 1. \tag{21.11}$$

This is nice:

$$35 = (8, 1) + (8, 3) + (1, 3), \tag{21.12}$$

35 contains an octet of spin-zero mesons, another 8 of spin 1, just the two octets encountered before. In addition there is a (1, 3), an SU(3) singlet vector meson for which a candidate, the ϕ (1205), discovered[53] in 1962, was at hand. Next à la Eq. (21.9) for baryons:

$$6 \times 6 \times 6 = 20 + 56 + 70 + 70, \tag{21.13}$$

where 56 has the content

$$56 = (8, 2) + (10, 4), \tag{21.14}$$

an 8, spin 1/2, and a 10, spin 3/2. Thus 35, 56 contain all the particles now called the classical hadron states.

Static SU(6) has found applications in strong, electromagnetic, and weak interactions. I mention a few highlights:

(a) Symmetry breaking. Example: 56. Split 8 from 10 by adding a term depending on $J(J+1) = 3/4$ (for 8), 15/4 (for 10) to the mass formula Eq. (21.7) where now the coefficients a and b should be *the same* for 8 and 10. Thus the mass splittings within the 8 and the 10 become interrelated. The resulting new mass relations[54] work well (also for the 35), within 10 per cent accuracy.

(b) As mentioned earlier, SU(3) contains a dimensionless number, the D/F ratio, that is undetermined. A general argument[55] shows that this number is fixed as a result of embedding SU(3) into SU(6). This leads to new and stronger predictions, the most notable[56,57] being the ratio of magnetic moments μ for neutron and proton:

$$\frac{\mu(\mathrm{n})}{\mu(\mathrm{p})} = -2/3 \quad \text{(experimentally } -0.685). \tag{21.15}$$

Thus we learn something valuable about the good old proton and neutron via the roundabout way $SU(2) \to SU(3) \to SU(6)$. Other new and good predictions concern decays like[56] $\Delta^+ \to p + \gamma$ and[58] $\omega \to \pi^0 + \gamma$.

(c) Variants of static SU(6) such as 'collinear SU(6)'[59] have led to useful new predictions[60] about meson–baryon scattering.

(d) Brief but considerable confusion was created by conjectures[61] that static SU(6) pointed to an extension of the Lorentz group. The salutary effect was the derivation of several potent theorems demonstrating that non-trivial extensions of this kind are impossible.*

(e) In the long run the most important consequence of static SU(6) is that it pointed the way to a new dynamical degree of freedom now known as color.

*5. Color**: double SU(3) symmetry.* The next phase in the evolution of quark physics grew out of two types of discomfort. The first, which may be called aesthetic, concerned the fractionality of the quark charges. Could one retain the idea of quarks but have them be integrally charged? That is indeed possible provided one introduces more than one quark triplet. Two in fact will do.[64] The second reason for doubts about the quark picture was more directly physical and concerned an implication of static SU(6). This symmetry applies at best only when orbital angular momenta are effectively neglected (no spin–orbit couplings). Thus the quarks making up a baryon are in *S*-states, hence their three-quark wave function is symmetric in the orbital variables. As it happens, it is a specific property of the 56-representation, chosen for baryons, that it is totally symmetric in the spin and unitary spin variables. Thus these 3q-composites are symmetric in *all* variables—in apparent conflict with Fermi statistics (the exclusion principle). This problem caused a good deal of discussion; various remedies were proposed.

(a) SU(6) is good but the baryon octet is[50] in the totally antisymmetric representation $20 = (8, 2) + (1, 4)$. That was not to be the way.

(b) The 56 'ground state' is space-antisymmetric but intraquark forces nevertheless lead[65] to a lowest state with zero orbital angular momentum—not the way either.

(c) Quarks do not satisfy Fermi statistics but a so-called parastatistics in which a state can be occupied by a finite number larger than one (here: three) of particles (Greenberg[66]). Such statistics date back[67] to 1940 and had interested[68] Sommerfeld in his day.

* For summaries and references see Ref. 62. ** For a review of color models see Ref. 63.

(d) Each of the u, d, s quarks carries a *new* additional three-valued degree of freedom with respect to which the 3q 56-wave function is totally antisymmetric, thus saving the exclusion principle *and* having *S*-state wave functions (Han and Nambu[69]). Almost at once it became part of the jargon to speak of red, white, and blue quarks of type u or d or s, nine quarks in all. 'If the baryon is made up of a red, a white, and a blue quark, they are all different fermions and you thus get rid of the forced antisymmetry of spatial wave functions [1965].'[70]

Han and Nambu pushed this picture much further. They associated a *new* symmetry group, also an SU(3), with the new degree of freedom, thus arriving at a 'double SU(3) symmetry'. I shall at once introduce* the distinguishing symbol $SU(3)_c$ for this new group, c for 'color'. Each individual member of the old SU(3) triplet q is replaced by an $SU(3)_c$ triplet. It is further assumed that $SU(3)_c$ generates an additional contribution to the electric charge in such a way that now three integrally-charged triplets emerge with respective (u, d, s) charges $(0, -1, -1)$; $(1, 0, 0)$; $(1, 0, 0)$. Otherwise all consequences of SU(3) and SU(6) mentioned before continue to hold.

The most memorable remark in the Han–Nambu paper[69] concerns the origin of the strong interactions: 'We introduce . . . eight gauge vector fields which behave . . . as an octet in $SU(3)_c$, but as singlets in SU(3) . . . . The *superstrong* interactions for forming baryons and mesons have the symmetry $SU(3)_c$ . . . . The lowest [baryon and meson] mass levels will be $SU(3)_c$ singlets.'

In other words the ordinary hadrons do not see color, they are all color-neutral.

(e) It is noted (Greenberg and Zwanziger,[71] 1966) that the parastatistics model and the tricolored triplet model are equivalent, ***provided the three triplets carry the same charges***, that is, if there are three (u, d, s) sets each with the old charges $(2/3, -1/3, -1/3)$. In other words: color does not contribute to electric charge, the operator Q of Eq. (21.5) remains as it is, ***electromagnetism is color-neutral***.

The group $SU(3)_c$, its associated eight gauge vector fields, color-neutral hadrons, color-neutral electromagnetism—these are basic (though not complete) ingredients of quantum chromodynamics. Having traced the origins of these concepts I polled many experts on whether these facts had registered with them twenty years ago when these papers appeared. It was with them as it was with me; the answer was no. Three triplet models were duly noted, to be sure, but only as a means for hadron building. So do incomplete but profound ideas seep barely noticed into the body of physics.

*6. Quarks, the second route: current algebra.*** Quarks as discussed thus far were conceived as things subject to forces, of a kind not yet understood then,

* In Ref. 69 the old SU(3) is called SU(3)′, the new one SU(3)″.

** Annotated anthology: Ref. 72. Review: Ref. 73. Books: Refs. 74, 75.

that strongly bind them into and only into integrally charged hadronic states. There is a second logically independent reasoning that likewise led to quarks as fundamental entities. Here the point of departure is the weak interactions, a subject that I need to update in respect of strange particles.

So far we have encountered* non-leptonic decays ($\Lambda \to N\pi$, $K \to 2\pi, \dots$) and leptonic decays, the latter only for K particles, however ($K \to \mu\nu$, $\pi e \nu, \dots$). Baryon decays like Λ_β:

$$\Lambda \to p + e^- + \bar{\nu} \tag{21.16}$$

had not been seen as late as mid-1958 by which time their absence began to cause some concern.[77] A first guess[78] for the rate of (21.16), based on the assumption that the current J_λ in Eq. (20.41) should contain a term $\bar{\Lambda}p$ with about the same strength G as $\bar{n}p$, yields a branching ratio ~2 per cent, and it began to look as if the actual rate was much smaller than that. The first Λ_β decays were seen[78] in the fall of 1958; more followed before long. It soon became clear that the Λ_β rate *was* smaller than expected from naive universality.** Meanwhile another deviation from universality had been diagnosed[79]: the vector constant for β-decay was about 3 per cent smaller than for μ-decay. Something was not quite right with universality.

These matters were cleared up in 1963 when Cabibbo proposed[80] an extension to SU(3) of the earlier idea† that the weak vector current and the isovector electromagnetic current form an isotriplet. To see how this works recall the content of the meson octet: π, K, η. Pions: $T=1$, $S=0$, have the internal quantum numbers of the isotriplet just mentioned. (K^+, K^0): $T=1/2$, $S=1$ have the quantum numbers of a current that would induce processes like Λ_β. η: $T=0$, $S=0$ is like the isoscalar electromagnetic current. Whence Cabibbo's hypotheses:

(a) The electromagnetic and the weak vector currents are members of one and the same octet, call it $j_{a\mu}$, $a=1,\dots,8$. In particular (in units e)

$$j_\mu^{elm} = j_{3\mu} + \frac{1}{\sqrt{3}} j_{8\mu}. \tag{21.17}$$

(b) Axial weak currents are members of another octet, call it $j_{a\mu}^{(5)}$, '5' in honour of γ_5.

(c) Deviations from naive universality are parametrized by a single new number, the 'Cabibbo angle' θ. The entire weak interactions are††

$$H_{wk} = \frac{G}{\sqrt{2}} J_\lambda^\dagger J_\lambda \tag{2.18}$$

* Chapter 20, Section (b).
** Chapter 20, Section (d), Part 5.
† Chapter 20, Section (d), Part 5.
†† θ is introduced in Ref. 80, where Eq. (21.18) does not yet appear because of the focus on leptonic decays.

$$J_\lambda = \cos\theta(j_{1\lambda} + ij_{2\lambda} - j^{(5)}_{1\lambda} - ij^{(5)}_{2\lambda})$$
$$+ \sin\theta(j_{4\lambda} + ij_{5\lambda} - j^{(5)}_{4\lambda} - ij^{(5)}_{5\lambda})$$
$$+ \bar{e}\gamma_\lambda(1+\gamma_5)\nu + \bar{\mu}\gamma_\lambda(1+\gamma_5)\nu. \qquad (21.19)$$

The indices 1, 2, 4, 5 are according to standard usage. '$1+i2$' corresponds to a current ('π^+') with $Q=1$, $S=0$, '$4+i5$' ('K^+') has $Q=S=1$. Main implications:

(a) Suppression of $\Delta S=1$ versus $\Delta S=0$ β-decay rates by $\tan^2\theta$. A large class of processes (K vs. π-decays, Λ, Σ^- vs. neutron decays) yields consistently[81]

$$\theta \simeq 0.26 \text{ radians}. \qquad (21.20)$$

(b) Suppression of $n \to pe^-\bar{\nu}$ versus $\mu^- \to e^-\nu\bar{\nu}$ by $\cos^2\theta$. Eq. (21.20) accounts for the data.[82]

(c) All $|\Delta S|=1$ β-processes satisfy $|\Delta T|=1/2$. Thus, since the *hadronic* charge Q satisfies $Q=T_3+(S+B)/2$, we have the rule*

$$\frac{\Delta S}{\Delta Q} = +1 \text{ for } \beta\text{-processes} \qquad (21.21)$$

where ΔQ is the change of total *hadronic* charge. Therefore processes with $\Delta S/\Delta Q=-1$, like $\Sigma^+ \to n+e^+ +\nu$, are forbidden. During 1962–3 there was a brief but intense stir caused by evidence for violation of (21.21). Subsequent experiments showed** that there was no cause for concern, however.

(d) Eq. (21.18) also contains a mechanism for non-leptonic decays, to wit via the coupling of (1, 2)- to (4, 5)-components. The former has $|\Delta\vec{T}|=1$, the latter $|\Delta\vec{T}|=1/2$, hence their coupling has $|\Delta\vec{T}|=1/2$ and 3/2. In view of the success of the non-leptonic $|\Delta\vec{T}|=1/2$ rule, the 3/2 part ought to be strongly suppressed (experimentally the 1/2 to 3/2 amplitude ratio is of typical order 20). No convincing mechanism† for this suppression exists to this day. That is not good.

Let us see next how further reflection†† on the octets containing weak/electromagnetic currents opened an alternate road to quarks. The reasoning goes like this.

Denote by ρ_a, $\rho_a^{(5)}$ the charge density (fourth) components of $j_{a\mu}$, $j^{(5)}_{a\mu}$ and define

$$F_a = \int \rho_a \, d\vec{x}, \qquad F_a^{(5)} = \int \rho_a^{(5)} \, d\vec{x}. \qquad (21.22)$$

* First stated in Ref. 77, where it was also noted that violations of the rule would lead to disastrous $|\Delta S|=2$ processes.

** For a history of this episode see Ref. 83.

† For a sample of recent efforts see e.g. Ref. 84.

†† This reasoning is due to Gell-Mann who from the start[36] focussed more on quarks as current constituents, while Zweig[37] concentrated on quarks as hadron constituents. The idea of a current algebra actually predates the quark picture. Its initial formulation[85] borrowed heavily from the Sakata model. See further Ref. 47.

F_a, $F_a^{(5)}$ are called vector and axial 'charges' in analogy to the connection between Q, Eq. (21.5) and the electromagnetic density (21.17). Consider first the case of strict SU(3) symmetry when all F_i are time-independent (conserved). They satisfy of course Eq. (21.4) which I repeat:

$$[F_i, F_j] = if_{ijk}F_k. \tag{21.23}$$

The $F_a^{(5)}$ are not conserved, as witness for example the PCAC condition Eq. (20.46); thus they depend on t. They satisfy

$$[F_a, F_b^{(5)}(t)] = if_{abc}F_c^{(5)}(t), \tag{21.24}$$

same t on both sides. This equation is nothing but the mathematical expression of the fact that $F_b^{(5)}$ is an octet representation of SU(3). Now comes a striking new assumption

$$[F_a^{(5)}(t), F_b^{(5)}(t)] = if_{abc}F_c. \tag{21.25}$$

Form

$$F_a^{\pm} = \tfrac{1}{2}(F_a \pm F_a^{(5)}) \tag{21.26}$$

and find that all F_a^+ commute with all F_a^-, while F_a^+ satisfy SU(3) relations like (21.23), and so do F_a^-. Thus one is dealing with two independent SU(3)'s, in mathematical terms with SU(3)×SU(3), a structure* called chiral SU(3)× SU(3) in this instance.

Eq. (21.25) is a constraint on the dynamics, it does not follow from symmetry arguments. There are innumerable instances where it does not hold, such as for currents made from baryon and pseudoscalar meson octets that do satisfy Eqs. (21.23), (21.24), but not (21.25). Can it work at all? Yes, if currents are made[47] from quarks:

$$j_{a\mu} = i\bar{\psi}^\alpha \gamma_\mu (F_a)_\alpha^\beta \psi_\beta, \qquad j_{a\mu}^{(5)} = i\bar{\psi}^\alpha \gamma_\mu \gamma_5 (F_a)_\alpha^\beta \psi_\beta, \tag{21.27}$$

where one sums over the SU(3) indices α, $\beta = 1, 2, 3$.

ψ_β are three Dirac spinors, the quarks, $\bar{\psi}^\alpha$ are the conjugates. For each a $(F_a)_\alpha^\beta$ is a 3×3 matrix which represents F_a for the special case that it acts on 3's (just as the 2×2 matrices in Eq. (17.11) were a special representation of the isospin T_i, Eq. (17.30)). One verifies that these currents satisfy the relations (21.23) by using Eq. (21.22) and the *equal time* anticommutation relations Eqs. (16.2), embellished with SU(3) indices:

$$\begin{aligned} \{\psi_{\alpha j}(\vec{x}, t), \psi_l^{\dagger\beta}(\vec{x}', t)\} &= \delta_\alpha^\beta \delta_{jl} \delta(\vec{x} - \vec{x}'), \\ \{\psi_{\alpha j}(\vec{x}, t), \psi_{\beta l}(\vec{x}', t)\} &= \{\psi_j^{\alpha\dagger}(\vec{x}, t), \psi_l^{\beta\dagger}(\vec{x}', t)\} = 0. \end{aligned} \tag{21.28}$$

Thus the chiral group leads one to consider currents as quark currents. This consideration was initially presented[47] before color had appeared. It was quickly

* The possibility that this group might form Part of an SU(6) led to the remark[47] on supermultiplets quoted in Part 4 of this section.

seen, however, that color can be added with impunity.[86] At that stage the current J_λ in Eq. (21.19) becomes:

$$J_\lambda = \sum_{i=1}^{3} \{(\bar{d}^i \cos\theta + \bar{s}^i \sin\theta)\gamma_\lambda(1+\gamma_5)u_i\}$$

$$+\bar{e}\gamma_\lambda(1+\gamma_5)\nu + \bar{\mu}\gamma_\lambda(1+\gamma_5)^\nu \qquad (21.29)$$

where i is the color index.

There is more. As was noted after Eq. (16.21), 'the anticommutation relations hold whether or not the electrons interact with the electromagnetic field',—*always for equal times*. The same is true here.* Thus chiral SU(3)×SU(3) holds much more generally. In particular one can take the s-quark to have a different mass from u, d, and Eqs. (21.13–25) (with some F_i now t-dependent) still persist. These three relations therefore give dynamical meaning to SU(3) even when that symmetry is not obeyed.

All this is pretty but does it do any good? It does.

7. *Particle physics in the mid-sixties.* What was particle physics like in the mid-sixties?

Continued interest in dispersion relations and visions of the bootstrap** led to interesting theorems but little physical novelty.

Regge analysis,† partially successful, partially dubious, was going through its ups and downs.

Then there were SU(3), SU(6), and quark models, far less sophisticated than dispersion and Regge theory, not always a bed of roses, yet vastly more successful. The heady impact of SU(3) and SU(6) mark the beginning of an era, still ongoing, of a very strong emphasis on guidance by symmetry reasoning. Additional and amazingly elementary theoretical concepts made contact with the real world. Thus the assumption that the forward scattering amplitude for hadron–hadron scattering is the sum of all 2-body qq or q$\bar{q}$ scattering amplitudes of the respective constituents gave quite useful relationships between total cross sections.[87] The equally simplistic notion that quark systems making up hadrons may, up to a point, be treated as *non-relativistic* quantum mechanical systems with forces described by static potentials[88] has turned out to be very fruitful. In a 1983 review[89] of the achievements of that model it was written: 'The [constituent quark] model still lacks a convincing theoretical basis in quantum field theory .... Nonetheless, on purely phenomenological grounds, the model is now spectacularly successful in accounting for almost all (all?) major features of the baryon spectrum .... The ideas of spin-1/2 quarks ... must surely rate as the most significant achievement of baryon

* Provided only that interactions between quarks do not depend on derivatives of ψ, $\bar{\psi}$.
** Chapter 19, Section (f), Part 5.
† Chapter 19, Section (f), Part 6.

spectroscopy, perhaps comparable to the role of atomic spectroscopy in leading to the Bohr model and quantum mechanics.'

While the picture of 'constituent quarks', defined as quarks insofar as they appear bound by non-relativistic potentials, is not (yet) derivable from quantum field theory, 'current quarks', quarks as they appear in currents get much closer to field theory, as witness the role of Eq. (21.28). Alternative approaches to hadron structure, relativistic shell models known as 'bag models'* are inspired by (but not (yet) derived from) the strong interaction field theories of the 1970s. Hadrons are conceived as bubbles, bags, characterized by a small number of phenomenological parameters (bag pressure, bag surface tension), inside which 'bag quarks' move relativistically.

This is a good place for commenting on quark masses. Constituent quarks are taken to have masses of about 300 (u, d) and 500 (s) MeV. On the scale of the proton mass current quark masses are very small (<10 MeV) for u, d, and about 100 MeV for s. Bag quarks are also light. There is nothing necessarily paradoxical about all that. The *effective* mass of a particle may depend on its environment, as is familiar from the free-electron theory of metals.[92] Ideally one would like to control the problem: if the mass is this much in environment a, then it is that much in environment b. That has not yet proved possible for quarks—nor for electrons in metals. For electrons one can of course operationally define 'the' mass (0.5 MeV) as the value of that parameter when the electron moves freely. If our present ideas hold up, that situation has precisely no analog for quarks, however—there *are* no free quarks. Thus one is restricted to environmental definitions of quark mass, a subject under continued study.[93]

I return to particle physics in the mid-sixties, in particular to current algebra. 'The application of [current algebra] languished for a few years until [1965 when] . . . Adler[94] and Weisberger[95] . . . derived a useful sum rule relating the Gamow–Teller coupling in β-decay to sums over $\pi^{\pm}$–p scattering cross sections. This successful result produced an explosion of activity.'[73] By mid-1967 over five hundred papers on current algebra had appeared.[74]

The Adler–Weisberger relation reads

$$1-\frac{1}{G_A^2}=\frac{4m^2}{\pi f^2}\int_{m+\mu}^{\infty}\frac{E\,\mathrm{d}E}{E^2-m^2}[\sigma^+(E)-\sigma^-(E)], \tag{21.30}$$

where G_A is the Gamow–Teller constant (see Eqs. (20.40)–(20.45) for its definition), m (μ) is the nucleon (pion) mass, f the strong interaction constant defined in Eq. (20.45). $\sigma^{\pm}(E)$ are the total cross sections for the scattering of a zero-mass $\pi^{\pm}$ on a proton at center of mass energy E. The main ingredients that go into the derivation of the relation are: Eq. (21.25), specifically the

* These come in variants. For a review (1983) and detailed references see Ref. 90. Yet another quark variant is the 'Regge pole quark.'[91]

isospin equation $[F_1^{(5)}+iF_2^{(5)}, F_1^{(5)}-iF_2^{(5)}]=2F_3$, a non-linear relation that evidently sets the scale for the axial relative to the vector (Fermi) constant; and the PCAC relations (20.45, 46) that (as further examination shows) link the amplitude for the radiation of a 'soft' (low energy) massless pion in some process to the corresponding amplitude where this pion is absent. This soft pion theorem has physical content only if transition amplitudes can be extrapolated smoothly from physical to zero pion mass. Assuming this to be possible one can put in numbers for $\sigma^{\pm}$ and find a very good value for G_A (within a few per cent of Eq. (20.44)).

A few closing comments on current algebra.

(a) Eq. (21.30) is based in part, to repeat, on Eq. (21.25), *not* on any particular realization of that commutator, in terms of quarks or otherwise. That was considered a virtue during the 1960s when the emphasis on the quark structure of currents was less than pronounced. There are several reasons for this. First, we are still in the period (though the end is near now) that had begun in the 50s, during which it was high fashion to treat symmetries, including chiral SU(3)×SU(3), as free-floating invariances.* Secondly, considerable effort was going into the study of another type of model, less radical (it seemed) than the one with quarks, that sheds interesting light both on chiral SU(3)×SU(3) and on PCAC, and in which symmetry breaking can be studied in detail.**

(b) Applications of current algebra and PCAC to strange particles leads to important relations[98] between $K^+\to\pi^0 l^+\nu$ and $K^+\to l^+\nu$; and between $K^+\to\pi^+\pi^- l^+\nu$ and $K^+\to\pi^0 l^+\nu$ ($l=$ e or μ).

(c) Eqs. (21.23–5) have been extended to commutators of integrals over other current density components besides (as in Eq. (21.22)) the fourth one; and to the densities themselves. Most of these extensions are not straightforward, the reason being that naive application of the relations (21.28) to commutators of space-and-time components of current densities leads to wrong answers, as Schwinger has taught us.[99] There have been further advances in this direction (to which I shall return briefly) which, however, involve additional assumptions; see further Ref. 100.

(d) Finally, and slightly at a tangent, a remark on how careful application of field theory to products of current densities led to a wholly unexpected first piece of evidence for color. It was discovered (1969)† that the PCAC relation for the neutral axial vector current A_μ^0, $T=1$, $T_3=0$ is not what one would guess from the Eq. (20.46) for charged currents ($T=1$, $T_3=\pm1$). Instead ($\vec{E}$, $\vec{H}$ are the electromagnetic field operators, $\alpha=1/137$)

$$\frac{\partial A_\mu^0}{\partial x_\mu}=fm_\pi^2\pi^0+\frac{\alpha}{\pi}\vec{E}\cdot\vec{H}. \tag{21.31}$$

* In the sense indicated before, Chapter 19, Section (d), Part 1.

** This is the σ-model. Book: Ref. 96. Review: Ref. 97.

† Detailed discussions and references in Ref. 101.

It was further found that the 'anomaly' $\sim\alpha$ fully determines the rate for $\pi^0 \rightarrow 2\gamma$ up to a factor S having to do with the fermion content in the fundamental Hamiltonian of the theory. If those are quarks then, it turns out, $S = [\Sigma(2Q-1)]^2$, where Q is the charge of a u-quark and the summation is over all u-quarks. For a single u-quark ($Q = 2/3$), $S = 1/9$. For three u-quarks as desired by color, S = 1 whether the charges are either à la Han–Nambu or the same (u, d, s) three times over. Well, sir, $S = 1$ fitted the data on the dot.

That was quite amazing in its time.

(c) New tools—new physics

The sketch* given earlier of the evolution of high energy machines up to the present dealt with circular proton accelerators only. It is high time now for introducing other types of tools.

*1. Reactors—the first neutrino.*** Let us go back to the early 1950s to begin with, and recall that the neutrino postulate had been widely accepted by then. Energy, momentum, and spin balance all had fallen in line. No one, however, had yet seen a neutrino, 'seeing' meaning that, after its alleged creation in β-decay, a neutrino is observed doing something elsewhere, for example causing the reaction:

$$\bar{\nu} + \mathrm{p} \rightarrow \mathrm{n} + \mathrm{e}^+, \tag{21.32}$$

an inverse of the β-process. That, of course, is a tall order. A 1 MeV neutrino passing through a light year's thickness of lead interacts on the average just once. Thus it demands fierce fluxes of neutrinos to see some action.

As Reines, then on the staff at Los Alamos, was reflecting on whether a nuclear bomb detonation could produce any fallout of interest to pure science, the thought occurred to him (1951)[103] that here might be the flux needed for seeing neutrinos. Some time after he, together with Cowan, had begun to think of suitable detectors, they realized that steady fission reactors served their purpose much better than bombs, and proceeded accordingly. By 1953 they had the first hints of a signal[104] but it was not until 1956 that their experiment at the Savannah River reactor (flux: $10^{13}\ \bar{\nu}/\mathrm{cm}^2\,\mathrm{sec}$) gave them sufficient conviction[105] for sending a telegram† to Pauli announcing the experimental discovery of the neutrino.

Schematically their strategy for observing reaction (21.32) went like this. Large liquid scintillators, high in hydrogen content and loaded with a cadmium compound, are used as detectors, the trick being that e^+-annihilation leads to prompt scintillations while γ-rays from absorption of neutrons in cadmium give delayed pulses with predictable energy and time delay spectra. One needs

* Chapter 19, Section (a). ** Review: Ref. 102. † Reproduced in Ref. 103.

information on the flux and spectrum of the initial $\bar{\nu}$'s, knowledge of cross sections, checks on the event rate as a function of proton and cadmium densities, shielding from other radiations (fast neutrons), etc. After some vicissitudes it all worked out well in the end. The neutrino had been seen. The first type, that is.

2. Neutrino beams—the second neutrino. Serious interest in weak interactions at high energies dates from the late fifties. The initial experimental question was to find the optimal strategy for detecting such weak effects. Direct hadron beams are impractical since the huge noise, created by the preference of a proton (for example) for strong interaction processes, overwhelms the tiny signal of weak interactions such as $p + e \rightarrow n + \nu$. (Note: non-leptonic high energy reactions like $\pi^- + p \rightarrow \Lambda + \pi^0$ remain to be seen.) Neutrinos, particles that do not interact strongly or electromagnetically, and that can be made as beams by protons → pions → neutrinos via the dominant mode $\pi \rightarrow \mu\nu$, were the only chance. Several estimates looked promising, especially those by Schwartz.[106] Here it is a great help that neutrinos become more interactive the higher their energy. Where an MeV neutrino needs a light year of lead to do something, a GeV neutrino only needs a few million miles. Correspondingly one can make do with more modest fluxes at higher energies.

Theorists were much interested in these considerations, because of speculations on the possible existence of W-bosons (about which more later), and also because of the persistent absence of the decays $\mu^\pm \rightarrow e^\pm + \gamma$, processes that had been looked for in vain* since 1947–8. That problem became more serious when it was estimated[107] that $\mu \rightarrow e\gamma$ *had* to occur at a higher rate than the experimental upper limit then known. The argument goes roughly like this. Consider $\mu \rightarrow e\nu\bar{\nu}$ as a virtual step in a process during which a photon is also emitted: $\mu \rightarrow e\nu\bar{\nu}\gamma$. Since ν and $\bar{\nu}$ can annihilate into nothing, compatible with all conservation laws, it follows that $\mu \rightarrow e\gamma$ must occur. There was only one way out. The $\nu\bar{\nu}$ pair produced in $\mu \rightarrow e\nu\bar{\nu}$ must be forbidden to annihilate. Since a ν and a $\bar{\nu}$ can only do so if one is the antiparticle of the other, one must evade that connection. That can be arranged by assuming that μ decays mainly via

$$\mu^- \rightarrow e^- + \bar{\nu}_e + \nu_\mu \qquad (21.33)$$

where ν_e and ν_μ are *distinct* neutrinos (a possibility that had been conjectured earlier[108] for less direct reasons). More generally, μ is always accompanied by ν_μ, e by ν_e. Thus

$$\pi^+ \rightarrow \mu^+ + \nu_\mu \qquad (21.34)$$

$$\pi^+ \rightarrow e^+ + \nu_e \qquad (21.35)$$

* Chapter 20, Section (c), Part 4.

and likewise for other decay processes. This hypothesis can be verified by using reaction (21.34) as a neutrino source and verifying whether or not

$$\nu_\mu + p \to n + \mu^+ \text{ is allowed,} \tag{21.36}$$

$$\to n + e^+ \text{ is forbidden.} \tag{21.37}$$

A huge experiment mounted at Brookhaven by a Columbia group[109] showed that this was indeed the case. A major technical challenge in isolating these rare processes was that no particles other than neutrinos should enter the detection area. The required amount of shielding, unheard of till then, was provided by the obsolete U.S. battleship *Missouri* which was cut into iron plates that were shipped to Outer Mongolia, the name given to the area of Brookhaven where the neutrino experiment was set up. The result of eight months of hard experimentation: muons, twenty-nine; electrons, none. There *are* two neutrinos.

The Columbia experiment will be remembered not only for this beautiful result, but also for being the first to demonstrate the feasibility of physics done with neutrino beams, a subject to which Brookhaven, CERN, and Fermilab were to make fundamental contributions in later years.

Stay tuned for more leptons.

3. SLAC—hard scattering.* While at Los Alamos during the war, Alvarez had been asked to design a device for measuring the yield of a nuclear bomb explosion. Thinking this over, it occurred to him that he might profit from some results he had seen earlier in a report on blast effects of bullets. So he tracked down the report's author, a man named Panofsky, to an army camp somewhere back in the Sierras. Alvarez once said to me that he ranks the discovery of Panofsky among his finest achievements in science.

After war's end Alvarez immediately began collecting a large research group to join him in his project for building a proton linear accelerator (linac). Pief (as Panofsky is called by everyone), who was on top of his most wanted list, joined the Berkeley linac project in November 1945.

Linacs are the oldest devices for accelerating charged particles. In a cathode ray tube, the earliest instrument in that category, electrons gain energy by falling down, just once, a steep potential hill. That method can of course not be extended to high energies, since it would cause more trouble to build the hill than to have the electron, or ion, fall down the slope. The modern linac principle dates from the 1920s when Ising suggested, and Wideröe implemented,** resonance acceleration: let an ion successively fall down many small

* Brief history of SLAC up to 1983: Ref. 110; detailed history up to 1968: Ref. 111; the bible on linear accelerators up to 1970: Ref. 112.
** Chapter 17, Section (b).

hills in synchronism with the varying electric field of a radio frequency (rf) source. As we have seen, this idea had been Lawrence's inspiration for inventing the cyclotron. Lawrence also participated in linac development; in the early 1930s the rf method was developed a bit further in Berkeley, yielding[113] mercury ion beams up to 3 MeV. The main emphasis shifted entirely to cyclotrons, however.

When, in 1945, Alvarez returned to proton linacs, his arguments for doing so were mainly based[114] on cost versus energy. The price of a cyclotron or synchrocyclotron varies with the third power of the energy, principally because of the needed magnet,* whereas the cost of a linac rises linearly with energy. Therefore, it seemed, there had to be a crossover point beyond which linacs are cheaper. In any event, Alvarez and his group went ahead and for a large part of 1947 held the energy record for directly produced protons: 32 MeV. Two subsequent developments changed the cost picture in favor of circular machines: the advents of the synchrotron and of strong focussing.* As a result the principal emphasis on circular proton accelerators has persisted, but proton linacs do play their important role at lower energies. They serve as injectors for every single proton synchrotron. LAMPF,** at 800 MeV the highest energy proton linac, accelerates more protons per second above the pion production threshold than the sum of the proton intensities of all other accelerators combined.[115] The possibility that proton linacs may be relatively more cost-effective at ultra-high energies cannot be excluded either.

There came a turbulent time during which the University of California system required that its staff members take a loyalty oath. Panofsky would have none of that and accepted (1951) a professorship at Stanford, where, meanwhile, important strides had been made in developing electron linacs.† It had been realized early that energy loss by radiation strongly limits circular acceleration of electrons (as compared to protons), the more so as the desired energy increases;†† that problem is negligible, on the other hand, in linear acceleration. Another advantage of an electron over a proton linac is that synchronization of particle velocity with rf frequency is so much easier because after only a foot or two of accelerating path electrons move with nearly constant (light) velocity. Beam focussing is also relatively simple because to electrons of (say) 20 GeV a 3 km pipe appears about 1 m long. High rf power requirements could be met as the result of further development of the klystron (a pre-war Stanford invention‡), a device that delivers microwave rf power in intense short bursts.

In 1948, after pilot models had shown promising results, a proposal for a 1 GeV electron linac was submitted to, and soon accepted by, the Office of Naval Research. It was to be a 50 m-long machine, powered by 16 klystrons

* Chapter 19, Section (a). ** Los Alamos Meson Physics Facility.
† History of that early period: Ref. 116. †† Chapter 19, Section (a). ‡ History: Ref. 117.

in series, each to deliver 30 megawatts of pulsed power; cost: one million dollars. This became the 'Mark III', used in the mid-fifties by Hofstadter and his group for the study of nucleon structure by means of electron scattering.*

In 1956, encouraged by the successes of Mark III, a group at Stanford started planning SLAC (Stanford Linear Accelerator Center), a 3-km-long 20 GeV electron linac. It was to be a high intensity, low duty cycle traveling wave machine; that is, electrons surfboard down the tube on the crest of an electromagnetic wave. In 1957, a proposal was submitted to Federal agencies. In 1961—the year Panofsky became director of the project—Congress authorized $114 million for design and construction, making SLAC the most expensive accelerator yet. It became one of the world's finest, delivering to the community electrons of (at indicated year): 17 (1966), 20 (1967), 33 (1975), and, if all goes well, 50 (1986) GeV. SLAC is a national facility sustaining experimentalists in-house and elsewhere, for example by generating five million bubble chamber pictures per annum in the mid-1970s.

The early experiments** scheduled at SLAC included a project (executed in late 1967) by a SLAC–MIT collaboration for measuring inclusive, also called 'deep', inelastic scattering:

$$e + (p \text{ or } n) \rightarrow e + X \tag{21.38}$$

where X stands for 'anything'. It was anticipated that these reactions would give information on nucleon resonances; and so they did, a number of peaks were observed for an effective mass of X in the region up to about 2 GeV. It was further thought that the cross sections would fall off rapidly at higher energies and at large scattering angles, an expectation largely based on the earlier Hofstadter results which had shown that a nucleon, including its surrounding meson cloud, behaves qualitatively like smooth and soft jelly.

It came out quite otherwise.[119]

Electrons were found to scatter strongly over large angles by factors typically about 30 larger than anyone had predicted.† The jelly is only superficial. These new deeper probes inside the nucleon showed a behavior more like a box filled with hard nuggets. This result (about which more shortly), the experimental jackpot of the 1960s, provided the crucial missing piece of information which, a few years later, would lead to the theory of strong interactions.

4. Colliders. All accelerators encountered thus far are fixed-target machines: a beam of particles with energy E_0 hits a target at rest. Part of E_0 goes to waste. The center of momentum energy of (projectile + target particle) is of zero interest, all the physics depends on the remaining energy E of relative

* Chapter 19, Section (f), part 5.
** A report on the initial SLAC program is found in Ref. 118. † Popular account: Ref. 120.

motion. The percentage of waste increases with E_0. For a relativistic particle hitting a proton one has very nearly (E, E_0 in GeV)

$$E = \sqrt{2E_0}. \qquad (21.39)$$

Thus for $E_0 = 100$, only 14 per cent; for $E_0 = 1000$, only 4.5 per cent is useful energy.

It had of course long been obvious that colliders, accelerators in which two beams of particles with (for example) equal and opposite momenta hit head on, are free of waste. Closer scrutiny of such devices did not start until the 1950s, however. The beginnings of the subject can be traced to a study by Kerst and collaborators (1956)[121] on the efficiency of a system consisting of two tangent or interlaced beams; and to the observation of Touschek (1959)[122] that a single ring surrounded by magnets and rf cavities suffices as a collider, for counter-rotating particles of opposite charge. The first two-ring collider, constructed by a Princeton–Stanford group, was filled with 300 + 300 MeV electrons from the Mark III; results were reported[123] in 1966. The first single-ring device, called AdA,* was built at Frascati. The first report on e^+e^--interactions at 200 + 200 MeV appeared in 1964.

The economy of colliders with respect to useful energy is bought at a price: low event rate. The target for one collider beam is the other beam which contains a density of material comparable to a pretty good vacuum—ever so much less than for fixed targets. A main part of the art is therefore to do as well as possible on 'luminosity'** L. L is enhanced by injecting many pulses of particles into the ring(s) before making the two beams interact—whence the other name for colliders: storage rings. That introduced a crucial new parameter: the beam lifetime; one should be able to keep beams circulating over long periods of time without appreciable energy loss. That, in turn, imposed unprecedented demands on the quality of the vacuum, and on the stability of the accelerating rf voltage. Furthermore, the special geometry of colliders called for novel detector design.[124]

For these and further reasons, collider physics has brought new technology to the high energy physics laboratories. It has also profoundly affected their sociology. The primary protons or electrons accelerated in fixed-target machines can generate a variety of secondary beams that can be fed into separate experimental areas, allowing for a spread both in space and in choice of experiment. Collider experiments, on the other hand, are exclusively performed with primary beams, and must be executed right at the handful of beam intersection regions. The number of experiments that can be accommodated at any given time and, correspondingly, the number of options for

* Anelli di Accumulazione.

** L is defined by $\mathrm{d}n/\mathrm{d}t = L\sigma$, where σ is the cross section for the reaction at hand, n the number of events at time t.

participation by the experimental physics community, is therefore quite restricted. Further limitations arise because low event rates necessitate long running times for each experiment. The resulting problems of finding a healthy balance between fixed-target and collider programs are a very serious concern to all engaged in planning for the future of an experimental community with diverse means and interests.

I return to physics. More and improved e^+e^--colliders appeared on the scene from the later 1960s on. It was not until the seventies and early eighties, however, that storage rings began to occupy center stage. This was due to spectacular results (to which I return) obtained in e^+e^- physics* as well as to the arrival of the first two hadron colliders: the ISR (intersecting storage rings), mainly a pp-collider, and the $\bar{p}p$-collider, both at CERN.

The ISR, two interlaced storage rings with six intersection regions available for physics, was authorized in 1965, came on the air in 1971, and was shut down in 1983 to free resources for still more advanced projects. It was a marvel of precision engineering. The machine was fed by two counter-rotating 28 GeV proton beams from the PS.** Further acceleration in the ISR itself resulted in 31 + 31 GeV protons, the world's highest useful energy of the 1970s. The pressure inside the rings could be held at $\sim 3 \times 10^{-12}$ torr, by far the largest ultra-high vacuum system built to date. Typical filling times were an hour's worth of PS pulses, but beams could circulate for three days without appreciable deterioration (as became important when, later, the ISR briefly served as $\bar{p}p$ collider). In the course of time, luminosities improved by more than three orders of magnitude.

The history of physics at the ISR, brilliantly and concisely recorded meanwhile,[126] shows that already in its early years this machine began producing new physics, most notably the discoveries of a rise with energy of the total pp cross section[127] and of hard inelastic proton–proton scattering. Soon after large-angle inelastic electron–hadron effects had been discovered at SLAC, it was anticipated (1970)[128] that the forthcoming ISR might reveal analogous behavior for secondary hadrons produced with high transverse momentum p_T relative to the beams. 'It was nevertheless a big surprise to see such an effect taking place.'[126] Prior to the ISR, the p_T distribution of pions produced in $p + p \rightarrow \pi + X$ was known to drop exponentially for $p_T \lesssim 1$ GeV/c. ISR experiments first reported[129] at the Fermilab conference, September 1972, for π^0 production[130] up to $p_T \sim 7$ GeV/c, and $\pi^{\pm}$ production[131] up to ~ 5 GeV/c showed pion yields larger by several orders of magnitude compared to the extrapolation of the exponential drop seen earlier. 'Evidence for a marked departure from the exponential fall-off was a big event of 1972.'[126] Other ISR results as well as the meaning of hard scattering will be discussed in Section (e).

* A survey complete up to 1983 of e^+e^--colliders and their characteristics is found in Ref. 125.
** Chapter 19, Section (a).

The ISR set new records for useful energy during the week in August 1980 when it served[132] as collider for α on p, and α on α, at total energy 88 and 125 GeV respectively.

Then the $\bar{p}p$ collider took over the lead.

This device (already mentioned in Chapter 1) came into being because of the following set of circumstances. As I have yet to explain, in the mid-1970s strong theoretical arguments had been adduced for the existence of a charged boson, $W^{\pm}$, and a neutral companion, Z, with respective masses about 80 and 95 GeV. It would evidently be a big catch to find these hypothetical particles. No existing accelerator generated sufficient energy for doing so, however. In 1976 a group of physicists, including Rubbia, submitted a proposal[133] to CERN and to Fermilab* with the message: 'chances are excellent for finding the W and Z if you convert your high energy proton synchrotron into a $\bar{p}p$-collider'. For a complicated set of reasons Fermilab did not go along, but CERN accepted the proposal.[135]

I repeat in a bit more detail what was the strategy (already briefly outlined in Chapter 1). 3.5 GeV protons generated in the SPS** are collected in the antiproton accumulator (AA), a new ring built expressly for that purpose. As the antiprotons arrive in the AA, their motion is quite random relative to the desired storage orbit. A sensor measures the average deviation of a small slice of actual orbit from the storage orbit and, via a chord of the AA (that short cut is crucial) alerts a 'kicker' downstream with a message somewhat to the following effect. 'Sensor to kicker. A disorderly bunch of antiprotons is on its way to you. Apply corrective measure with the following electric field . . . .' So, bit by bit, are the antiprotons moved into storage orbit. This 'stochastic cooling' technique, invented by van der Meer and worth a Nobel Prize, is far more subtle and complex than need be discussed here.† After 24 hours of accumulation in the AA, the antiprotons are returned to the SPS, accelerated to 270 GeV, and made to collide with a counter-rotating proton beam of that same energy.

The construction of the AA was begun in 1979 and completed in 1980. Antiproton acceleration in the SPS was achieved in February 1981; simultaneous $\bar{p}$ and p injection in the SPS occurred the following April. The first $\bar{p}p$ collisions were detected[137] in July.

We now leave experiment for a while to catch up with theoretical developments.

(d) Scaling. Partons

Already in their first communications[119] the SLAC–MIT group had noted that their surprising data on hard scattering could be fitted reasonably well

* A popular account of the proposed scheme is found in Ref. 134.
** Chapter 19, Section (a). † For a detailed account see Ref. 136.

by a scaling law. In order to explain what that means, we need a master formula that applies not only to inelastic electron- and muon- but also to neutrino-scattering—soon found to obey similar regularities.

To a good approximation, the reaction (21.38) comes about because the electron emits a virtual photon that in turn shakes up the nucleon into the final complex X. The corresponding neutrino reactions

$$(\nu_\mu \text{ or } \bar{\nu}_\mu)+(\text{p or n})\rightarrow(\mu^- \text{ or } \mu^+)+\text{X} \tag{21.40}$$

proceed via the interaction (21.18). The partial cross sections in the nucleon rest frame can be represented by*

$$\frac{\mathrm{d}\sigma}{\mathrm{d}\Omega\,\mathrm{d}E'}=AE'^2\left[W_2(\nu,q^2)\cos^2\frac{\theta}{2}+2W_1(\nu,q^2)\sin^2\frac{\theta}{2}\right.$$
$$\left.+\varepsilon\frac{E+E'}{m}W_3(\nu,q^2)\sin^2\frac{\theta}{2}\right] \tag{21.41}$$
$$\nu=E-E',\qquad q^2=4EE'\sin^2\frac{\theta}{2}.$$

$E(E')$ is the initial (final) lepton energy, θ the lepton scattering angle, $\mathrm{d}\Omega$ the element of solid angle, m the nucleon mass, q^2 is minus the (lepton momentum change)^2. Finally

$$A=\frac{4\alpha^2}{(q^2)^2}\quad\text{for } e \text{ or } \mu,\qquad \alpha=\tfrac{1}{137}, \tag{21.42}$$

$$=\frac{G^2}{2\pi}\quad\text{for } \nu \text{ or } \bar{\nu}.\quad (G \text{ is the Fermi constant})^{**} \tag{21.43}$$

$$\varepsilon=\begin{cases}1, & \bar{\nu},\\ 0, & \text{e or } \mu,\\ -1, & \nu.\end{cases} \tag{21.44}$$

All information on the dynamical behavior of the nucleon is contained in the 'structure functions' W. In general these functions depend on the two independent† variables ν and q^2; and on what lepton and what target is being considered. Studies of properties of the W's (using dispersion theory and Regge arguments) began[139,140] in 1965.

Now to scaling. The experimental observation[119] of W_2 (the structure function most easily measurable) showed that, within the errors, νW_2 could be

* Units: $\hbar=c=1$. Eq. (21.41) applies upon averaging over initial spins and summing over the final lepton spin. The lepton mass is neglected. For a discussion of polarization, mass effects, and the role of T-invariance, see e.g. Ref. 138.

** This relation is modified when q^2 becomes of the order (W-boson mass)^2.

† Reactions (21.38), (21.40) depend on three independent variables, for example, ν, q^2, and the mass of the complex X.

represented as a function of a single dimensionless combination of ν and q^2:

$$\nu W_2^{ep} \cong F_2^{ep}(x); \qquad \nu W_2^{en} \cong F_2^{en}(x) \tag{21.45}$$

$$x = \frac{q^2}{2m\nu}, \qquad 0 \leq x \leq 1. \tag{21.46}$$

There was a good reason why the experimentalists at once plotted their data in terms of the not so obvious variable x. A year before the completion of their work, Bjorken, also from SLAC, had derived[141] the scaling relations

$$\begin{aligned} mW_1(\nu, q^2) &\to F_1(x), \\ \nu W_2(\nu, q^2) &\to F_2(x), \\ \nu W_3(\nu, q^2) &\to F_3(x), \end{aligned} \tag{21.47}$$

by combining not entirely unquestionable current algebra methods with brilliant intuition. The arrow means: in the limit $q^2 \to \infty$, x held fixed. A few weeks later, Callan and Gross found[141a] an additional asymptotic relation which specifically tests the current algebra assumption that currents consist of spin-1/2 particles, current quarks:

$$F_2(x) = 2xF_1(x). \tag{21.48}$$

These early experimental and theoretical results set in motion a variety of intense theoretical activities. First, when deviations from the scaling relations were found[142] in certain field theories that take account of quark interactions but treat those as perturbations, a heated debate arose as to whether the fault lay with scaling or with the use of perturbative methods. Secondly, important new mathematical tools were developed* for handling scaling phenomena. Thirdly, additional relationships between the F-functions were derived.[148] On the experimental side there developed an extensive and demanding program of studying electron, muon, and neutrino scatterings in the scaling region.** These experiments showed, in the most succinct terms, that the various scaling predictions are a good approximation to the actual state of affairs but that deviations do exist. This being an essay, I shall not enlarge on any of these issues, all carefully reviewed by others, but shall turn instead to the intermediate step that led from scaling to the new field theory: the parton picture.

Shortly after the discovery of scaling at SLAC, along comes Feynman with an incredibly simple-minded intuitive model[151,152] that nevertheless, as it

* These include the following interconnected topics: the treatment of scaling as an approximate invariance (reviews: Refs. 143, 144); light-cone algebra, a variant of current algebra, and the related problem of expanding products of operators whose respective arguments approach lightlike separations (reviews: Ref. 145); and further elaborations of the long-known renormalization group (review: Ref. 146), culminating in the Callan–Symanzik equation.[147]
** Reviews: Ref. 149 for electrons and muons; Ref. 150 for neutrinos.

turned out, contained key ingredients for a deepened understanding of the strong interactions. Let us assume, he said, that a high energy lepton beam sees the nucleon as a box filled with long-lived *structureless* (that is, pointlike) nuggets, partons, and that the *inelastic* lepton–nucleon scattering cross section in the scaling region equals the *incoherent* sum of *elastic* lepton–parton cross sections. This leads to scaling by the following argument. Go to a coordinate system where the nucleon moves with momentum $P \to \infty$. Let there be a probability P_N that the nucleon consists of N partons with individual masses that are negligible compared to q^2. The individual momentum of the ith parton ($i = 1, \ldots, N$) is $x_i P$, $0 \le x_i \le 1$. The allocation of parton momenta is governed by a distribution function $f_N(x_i)$. Individual parton momenta perpendicular to P are neglected, which is intuitively all right as long as x_i is not too small.* Let the electron scatter off a specific parton with momentum xP. q^2 and ν are *not* independent in that process since it is an elastic one, by assumption; one finds** from energy–momentum conservation that $q^2 = 2xm\nu$—x is just the scaling variable of Eq. (21.46)! Now, per prescription, add up the lepton–parton contributions. This yields

$$\nu W_2^e(\nu, q^2) = \sum_N P_N \left(\sum_{i=1}^{N} Q_i^2 \right) x f_N(x) \tag{21.49}$$

where Q_i is the electric charge of the ith parton. Similar reasonings apply to other structure functions, including those for neutrinos.[153] Scaling has been achieved. The final hadron state X (which does not contain free quarks) is reached upon redistribution of the shaken up box filled with partons—how *that* comes about the model does not and need not tell.

What are partons?

The simplest idea: partons are just the nucleon's three constituent quarks, does not agree with the observed behavior of the F's as functions as x. It was therefore suggested[154] that partons are the three 'valence quarks' just mentioned plus a 'sea' or core of $q\bar{q}$ pairs in an overall SU(3) singlet state: that still does not quite work. Decent fits could be obtained, however, by further adding neutral 'gluons', quanta of the force fields that hold the quarks together[155] and that do not directly interact with leptons.

With the help of the parton model it is possible to rederive relations in the scaling region that had been found earlier by more abstract methods. Two examples†: the Bjorken rule,[140]

$$A_1 \equiv \int_0^1 \mathrm{d}x \, [F_1^{\bar{\nu}p} - F_1^{\nu p}] = 2 \tag{21.50}$$

* Delicate points related to small x_i ('wee partons') are here ignored.

** Use the invariant definitions $q^2 = -(q_1 - q_2)^2$, $\nu = P(q_1 - q_2)/m$, where q_1, q_2, P are the four momenta of initial, final lepton and nucleon respectively.

† Here only the case is considered that the complex X does not contain strange particles.

and the Gross–Llewellyn Smith rule[156]

$$A_2 \equiv \int_0^1 \mathrm{d}x\, [F_3^{\bar{\nu}\mathrm{p}} + F_3^{\nu\mathrm{p}}] = 6. \tag{21.51}$$

The success of the parton model raised a difficult question. The incoherence of lepton–parton scattering means that the lepton sees the nucleon as an assembly of *freely moving* constituents. Then why does the nucleon not fall apart? I shall turn shortly to the answer: asymptotic freedom.*

(e) Conclusion: quantum field theory redux

Redux. Indicating the return of an organ to a healthy state.
Oxford English Dictionary.

1. Preamble. One third of the twentieth century had passed when Fermi took the initial steps toward a theory of weak interactions and Yukawa did likewise for strong interactions. In retrospect the next third appears as a time of probing the strengths and weaknesses of their concepts. Let us have a last look at that middle period, by way of introducing the final third (half-way gone at this writing) during which the theories of weak and strong interactions underwent a metamorphosis.

From the very beginning one encounters the idea that weak interactions are linked to field theories.** Fermi had constructed his theory by drawing an analogy with the electromagnetic process of virtual emission and absorption of a single photon. In his first paper on mesons, Yukawa had suggested that β-decay is mediated by his charged meson, see Eq. (17.52); a proposal he later withdrew. In the later thirties Heisenberg had begun to explore 'Fermi field theories': local couplings between four fermions treated as a perturbation in a fundamental Hamiltonian. To first order in G, the Fermi constant, this gives what Fermi taught us. Higher-order terms in G behave violently, Heisenberg had found. As became clear later,[158] this is because Fermi field theories are badly unrenormalizable; expansions in powers of G make no sense. Attempts at evading these unpleasant features of Fermi field theories never bore fruit.

The discovery in the late forties of the universality of weak interactions† at once provided incentives for drawing new parallels with field theory: 'One can perhaps explain the equality of [β- and μ-decay] interactions in a manner analogous to that used for Coulomb interactions, i.e. by assuming these interactions to be transmitted through an intermediate field with respect to

* For reviews of early attempts to incorporate partons in a field theory see Ref. 157.
** In this paragraph I summarize some points made in Chapter 17, Section (e), Part 2, and Section (g), Parts 1 and 2.
† Chapter 20, Section (c), Part 4.

which all particles have the same "charge".'[159] Speculations like these went into a temporary decline during the early 1950s when the β-decay interaction was believed* to be a mixture of S and T. Forces of that type would demand two different kinds of bosons with distinct spin. Then, late 1957, came the universal V–A theory.* From that time date the first suggestions[160] that weak interactions are mediated by charged *vector bosons*** $W^{\pm}$, that is, the *effective* interaction

$$H_{\mathrm{wk}} = \frac{G}{\sqrt{2}} J^{\dagger}_{\lambda} J_{\lambda} \qquad (21.18)$$

is supposed to be derivable from a *fundamental* interaction†

$$H_{\mathrm{W}} = g(J_{\lambda} W^{*}_{\lambda} + J^{\dagger}_{\lambda} W_{\lambda}). \qquad (21.52)$$

An example will clarify the connection between Eqs. (21.18) and (21.52). Consider the scattering

$$\mathrm{e}^{-}(p_1) + \nu_{\mu}(p_2) \to \mu^{-}(p_3) + \nu_{\mathrm{e}}(p_4) \qquad (21.53)$$

(the p's are the respective four-momenta). To second order in g this reaction proceeds via the virtual chains

$$\begin{aligned} &\mathrm{e}^{-} + \nu_{\mu} \to \nu_{\mathrm{e}} + \nu_{\mu} + \mathrm{W}^{-}(q) \to \mu^{-} + \nu_{\mathrm{e}}, \\ &\mathrm{e}^{-} + \nu_{\mu} \to \mathrm{e}^{-} + \mu^{-} + \mathrm{W}^{+}(-q) \to \mu^{-} + \nu_{\mathrm{e}}, \end{aligned} \qquad (21.54)$$

where $q = p_1 - p_4$. Apart from (important!) spin factors the matrix element M_{if} equals

$$M_{\mathrm{if}} = \frac{g^2}{q^2 + M^2_{\mathrm{W}}} \qquad (21.55)$$

where M_{W} is the W-mass. Consider the case $q^2 \ll M^2_{\mathrm{W}}$ so that $M_{\mathrm{if}} \simeq g^2/M^2_{\mathrm{W}}$. Now calculate the same transition based on H_{wk}. The results match if

$$\frac{G}{\sqrt{2}} = \frac{g^2}{M^2_{\mathrm{W}}}, \qquad (21.56)$$

or, using Eq. (20.43),

$$g^2 \simeq \frac{1}{\sqrt{2}} \left(\frac{M_{\mathrm{W}}}{m}\right)^2 \times 10^{-5}. \qquad (21.57)$$

Suppose that one *illegally* had dropped q^2 in Eq. (21.55), regardless of its magnitude. Then for the reaction at hand, as for all similar reactions, H_{W} to

* Chapter 20, Section (d), Part 4.

** Already in 1957 Schwinger had conjectured[161] that this is the case, for reasons to which I shall return.

† W_{λ} is a non-hermitian four-vector field, $W^{*}_{\lambda} = W^{\dagger}_{\lambda}$, $\lambda = 1, 2, 3$; $= -W^{\dagger}_{\lambda}$, $\lambda = 4$.

order g^2 gives the same results as H_{wk}. Equivalently, to second order in g the W-interactions yield all the Fermi results in the limit $M_W \to \infty$, g/M_W fixed.

Eq. (21.56) constrains g and M_W but we don't know enough to predict the value of M_W. The following argument* gives a rough idea what one might expect. Start from Eq. (21.18), calculate the cross section σ for (21.53) in the center of momentum frame, and find $\sigma = 4G^2\vec{p}^2/\pi$. Scattering via the Fermi interaction proceeds in a pure s-wave, hence σ may not exceed $\pi/2\vec{p}^2$ (the (sine)^2 of the phase shift is ≤ 1). It follows that *the Fermi interaction breaks down* for

$$|\vec{p}| \geq \left(\frac{\pi^2}{8G^2}\right)^{1/4} \simeq 300\,\text{GeV}. \qquad (21.58)$$

This, of course, proves nothing about the validity of the W-theory, but gives a hint that $M_W = 300$ GeV or less if the W-idea is right.

From the early sixties experimentalists went after the W. That was one of the main reasons for pushing neutrino beams, since these were considered relatively likely tools for producing W's in a detectable way. As Eq. (21.55) illustrates, one might also hope to catch glimpses of the new weak interaction physics even without actually producing W's, to wit, from high-precision measurements at relatively low energy that could reveal deviations $0\,(q^2/M_W^2)$ from the standard Fermi theory. The experimentalists' pursuits were matched by $0\,(g^2)$ calculations of various effects. By 1965 a lower limit $\simeq 2$ GeV for M_W had been established.**

There was one very serious hitch, however. The interaction (21.52) is once again unrenormalizable; theoretical estimates $0\,(g^2)$ could therefore not be trusted. (As it happened they turned out to be essentially all right.) Furthermore the interaction of W's with the electromagnetic field is unrenormalizable as well.[164]

Thus by the mid-sixties the theory of weak interactions had reached an impasse. The Fermi approach was as reliable as ever at low energies but had to break down at high energies. TheW-idea was tempting—but it was not a workable theory.

As to the status of the strong interactions at that same time, progress had been considerable since Yukawa's meson theory. Symmetry arguments had brought a degree of order in a wealth of new phenomena. Basic traits of quantum field theory had never been abandoned, and had in fact contributed to the evolution of dispersion theory, Regge pole models, and current algebra. Yet, at the level of fundamental dynamics, the situation was essentially the same it had been throughout all of the middle third of the century: obscure.

Then, very nearly simultaneously, a new weak interaction dynamics and a new strong interaction dynamics burst forth. An era began that, these days, is often called the new physics.

* Which I learned from T. D. Lee[162] in the summer of 1961.
** A review of theory and experiment up till June 1965 is found in Ref. 163.

The origins of these developments go back to the early fifties. Earlier* I remarked about that time: 'The discovery of a new class of field theories, the non-Abelian gauge theories, also falls in that period. I shall return later to that subject, the far reaching consequences of which were not immediately realized'. The point of return is here. As I try to sketch next how field theory regained center stage I imagine being interrupted once in a while by a curious questioner.

2. Non-Abelian gauge theories. My first encounter with non-Abelian gauges** (a term I had not yet heard of then) dates from the Lorentz–Kamerlingh Onnes centenary conference in Leiden, June 1953. At that meeting I had presented a paper[166] on the extension of isospin to a larger group, assuming that 'the element of space–time is not a point but a manifold . . . ,'† unaware of course that I was talking about a fibre bundle of the most trivial kind. Pauli, in the audience, was intrigued and in the subsequent discussion raised[167] the following question: 'I have a particular question regarding the interaction between mesons and nucleons . . . I would like to ask . . . whether the transformation group with constant phases can be amplified in a way analogous to the gauge group for the electromagnetic potentials in such a way that the meson–nucleon interaction is connected with the amplified group.'

In order to appreciate Pauli's question it will help to digress on the electromagnetic gauge group, a subject that was touched on earlier, but only in passing.†† For that purpose I again write down the Dirac equation (13.47) for an electron (changing to units $\hbar = c = 1$, used throughout hereafter, and putting $\partial_\mu = \partial/\partial x_\mu$),

$$[\gamma_\mu(\partial_\mu - ieA_\mu) + m]\psi = 0, \tag{21.59}$$

and the electromagnetic field equations,

$$F_{\mu\nu} = \partial_\nu A_\mu - \partial_\mu A_\nu \tag{21.60}$$

$$\partial_\nu F_{\mu\nu} + j_\mu = 0, \qquad j_\mu = ie\bar{\psi}\gamma_\mu\psi. \tag{21.61}$$

Consider the gauge transformation

$$\psi = S\psi' \qquad S = e^{-i\alpha} \simeq 1 - i\alpha \quad (\text{for small‡ } \alpha), \tag{21.62}$$

and distinguish two cases.

(a) Global gauge invariance: α is a constant, A_μ is unchanged. This implies current conservation: $\partial_\mu j_\mu = 0$; see Eq. (15.68).

* Chapter 19, Section (f), Part 1.

** I did not know at that time of a 1938 paper[165] by Oskar Klein that does not contain non-abelian gauge transformations, but nevertheless does bear on the subject at hand.

† A two-dimensional sphere.

†† See Eqs. (15.56)–(15.68).

‡ Infinitesimal gauge transformations will suffice here and below for the limited objectives of this chapter.

(b) Local gauge invariance: α depends on $\vec{x}$ and t; this can only be implemented if three conditions are met: (1) $e \neq 0$, since, for $e = 0$, ψ' cannot satisfy the same equation as ψ; (2) Eq. (21.62) must be completed by

$$A'_\mu = A_\mu + \frac{1}{e}\partial_\mu \alpha. \tag{21.63}$$

(3) Eqs. (21.60), (21.61) are also invariant under (21.62), (21.63). Note, however, that an important condition has sneaked in through the back door: imagine that the photon had mass μ. That would have led to an extra term $\mu^2 A_\mu$ on the lhs of Eq. (21.61), spoiling the invariance under Eq. (21.63): local gauge invariance demands the presence of a compensating *massless* vector field. The transformations (21.62), (21.63) form the 'local gauge group' named $U(1)$. The group is Abelian, which means that the result of two transformations does not depend on the order in which they are performed.

Consider next a nucleon and assume for a moment that it is described by an eight-component Dirac equation, four for the proton (ψ_p) and neutron (ψ_n) each:

$$(\gamma_\mu \partial_\mu + m)\psi = 0, \qquad \psi = \begin{pmatrix} \psi_p \\ \psi_n \end{pmatrix}. \tag{21.64}$$

Q. Why 'assume for a moment'?
A. Today we look upon the nucleon as a triquark composite and no longer use Eq. (21.64); just as we do not use it for hydrogen 3 or helium 3. But remember, we are in 1953.

Eq. (21.64) is invariant under the isospin rotation (τ_a are the isospin matrices; see Eq. (17.11))

$$\psi = S\psi' = (1 - i\tau_a \alpha_a)\psi', \tag{21.65}$$

$a = 1, 2, 3$ (summation over a understood). The way we have done business so far, α_a are constants and Eq. (21.65) may be called global isospin gauge transformations. Such transformations are non-Abelian; that means, the product $S_1 S_2$ of two such transformations is not generally equal to $S_2 S_1$. Pauli's question can now be rephrased: can one extend Eq. (21.65) to local isospin gauge transformations, with $\vec{x}$, t-dependent α_a?

Pauli kept pondering that question.* Right after the Leiden conference he wrote me[169] about a possible answer in terms of a generalized Kaluza–Klein theory.** A few weeks later, he sent me a manuscript entitled 'Meson nucleon interaction and differential geometry' that begins: 'Written down July 21 till

* I believe his interest may have stemmed from his earlier work on a unified theory of gravitation and electromagnetism in which a five-dimensional tangent bundle is hung at every space–time point.[168]
** Another version of a unified field theory.[170]

25 [1953] in order to see how it is looking'.* In these pages Pauli noted that local isospin gauge invariance demands the introduction of a compensating isotriplet of gauge field potentials, called $B_{a\mu}$ below, and finds, as his 'main result' the correct expression for the corresponding field strengths, Eq. (21.66) below. He did not, however, write down the associated dynamical field equations, Eq. (21.69) below.

Later in 1953 Pauli's enthusiasm began to wane. 'If one tries to formulate field equations . . . one will always obtain *vector mesons with rest mass zero* [his italics]. One could try to get other meson fields—pseudoscalars with positive rest mass— . . . But I feel that is too artificial.'[171] Meanwhile, Yang and Mills had independently tackled the same problem. In February 1954 Yang reported on their results in a seminar at the Institute for Advanced Study in Princeton. Pauli was in the audience and I well recall his critical and negative reaction.** Pauli never published;† Yang and Mills did. In two short brilliant papers[174] they founded modern gauge theory.

As a final remark on the origins of non-Abelian gauge theories, I mention the unpublished Cambridge dissertation of Ronald Shaw, submitted in August 1955. A footnote in this dissertation reads: 'The work described in this chapter was completed except for its extension [to a four-dimensional case] in Section 3, in January 1954, but was not published. In October 1954, Yang and Mills adopted independently the same postulate and derived similar consequences.'

Q. What in summary was the influence of Pauli and other major prewar figures on post-war particle physics?

A. Pauli made fundamental contributions to quantum field theory, in particular to the renormalization program and the *CPT* theorem. Fermi died soon after completing major work on pion physics (d. 1954). Schroedinger's later physics interests centered on unified field theories in the style of Einstein (d. 1961). During the last fifteen years of his life, Bohr's role in particle physics was that of spectator rather than actor (d. 1962). Dirac continued his solitary pursuits of an alternative to standard quantum electrodynamics (d. 1984). From 1954 to the end of his life, Heisenberg (d. 1976) was immersed in attempts at deriving all of particle physics from a fundamental non-linear wave equation. This program must be seen as a new variant of his old 'Fermi field theory'. Shortly after his work on gauge fields Pauli joined Heisenberg in this effort. Once again his initial interest rapidly waned, and shortly before his death (in 1958) he gave a just critique of this program.[176] Heisenberg was not swayed.†† His articles on this subject‡ contain several quite interesting

* A copy of this manuscript as well as of an important *Mathematischer Anhang*, written in December 1953, are in the Pauli Letters Collection, CERN Archives.

** See also Yang's own recollection of that occasion.[172]

† Notes of his lectures on this subject, given in the autumn of 1953, later appeared in print.[173]

†† For Heisenberg's reactions to Pauli's criticism see Ref. 177.

‡ These are found reprinted in his collected works.[178]

remarks, notably on symmetry breaking, but were not influential in the long run. His last published paper[179] which appeared shortly after his death concludes with this message: 'The particle spectrum can be understood only if the underlying dynamics of matter is known; dynamics is the central problem.'

Returning to the Yang–Mills paper[174] I give their main equations. The fields $G_{a\mu\nu}$ are defined in terms of the potentials $B_{a\mu}$ by ($a = 1, 2, 3$)

$$G_{a\mu\nu} = \partial_\nu B_{a\mu} - \partial_\mu B_{a\nu} - 2g f_{abc} B_{b\mu} B_{c\nu} \tag{21.66}$$

where g is the coupling constant of the theory and $f_{abc} = +1$ (-1) if a, b, c, is an even (odd) permutation of 1, 2, 3. Along with Eq. (21.65) $B_{a\mu}$ transforms as*

$$B'_{a\mu} = B_{a\mu} + 2 f_{abc} B_{b\mu} \alpha_c + \frac{1}{g} \partial_\mu \alpha_a. \tag{21.67}$$

The nucleon equation becomes

$$[\gamma_\mu(\partial_\mu - ig\tau_a B_{a\mu}) + m]\psi = 0. \tag{21.68}$$

The B-field equations are

$$\partial_\nu G_{a\mu\nu} + J_{a\mu} = 0, \tag{21.69}$$

$$J_{a\mu} = j_{a\mu} + j^{(B)}_{a\mu}, \tag{21.70}$$

$$j_{a\mu} = ig\bar{\psi}\tau_a\gamma_\mu\psi, \tag{21.71}$$

$$j^{(B)}_{a\mu} = 2g f_{abc} B_{b\nu} G_{c\mu\nu}. \tag{21.72}$$

The isotopic current is conserved,

$$\partial_\mu J_{a\mu} = 0. \tag{21.73}$$

Q. What about sources other than spin-1/2 fields?
A. The $B_{a\mu}$ play the same universal role here as do the A_μ in electromagnetism.[180,181]

The parallels with electromagnetism are striking—but so are the differences. The fields are non-linear in the potentials; and they are part of their own source. These non-linearities make the problem of enumerating even the classical solutions of these equations (even in the absence of ψ fields) extremely hard. This question has in fact not been completely answered to date.

These non-linearities also profoundly complicate the quantization of the B-field. While it was at once evident[174] that B quanta have unit spin and isospin, and charges 0, $\pm e$, their mass was initially a puzzlement. As Yang recalled[172] later: 'We found we were unable to conclude what the mass of the

* Unlike the electromagnetic case, $G_{a\mu\nu}$ is not invariant under Eq. (21.67) but the field Lagrangian $(G_{a\mu\nu})^2/4$ is, and that is all that matters.

gauge particle should be'. Here he was in good company: the non-linearities of the gravitational field had at one time caused protracted debates[182] as to whether gravitational waves, corresponding to zero-mass 'gravitons', do or do not exist. It was soon, and correctly, suspected, however, that the B-quanta have zero mass.

Almost twenty years had to go by before the B-quantum field theory was sufficiently understood: in 1971 it was shown by 't Hooft[183] that the theory is renormalizable, a result that was the culmination of a series of gradual advances. This proof has since found its way into textbooks.[184] In his important historical essay[185] of this evolution, Veltman, 't Hooft's teacher and himself one of the pioneers in the development of the quantum theory of gauge fields, has called the developments during the intervening years 'slow and painful'. It makes no sense to discuss these matters in a book of this kind. I do wish to stress, however, that the handling of the intrinsically non-linear non-Abelian gauge theories demands skills of a high level that had not yet been called for in, say, quantum electrodynamics.

Back to the fifties, the Yang–Mills theory was received with considerable interest. What to do with these recondite ideas was another matter, however. There were no strongly interacting vector mesons at that time, much less vector mesons of that kind with zero mass.

Q. Is that zero-mass property not a disaster?

A. It is not, as you will see shortly.

When, in the early sixties, massive strongly-interacting mesons were discovered,* it was at once suspected that these were somehow manifestations of non-Abelian gauge fields. First, the ρ-meson was tentatively identified with B-quanta.[186] When the vector octet (ρ, ω, K^*) appeared on the scene, it was conjectured by several[187] that these were associated with a generalization of the above gauge theory to SU(3). The straightforward extension[181] of the preceding formalism to that case goes as follows. $B_{a\mu}$ and $G_{a\mu\nu}$ are SU(3) octets, $a=1,\ldots,8$; f_{abc} are given by Eq. (21.4); otherwise Eqs. (21.66), (21.67), (21.72) are taken over unmodified. It was further widely believed that the octet of observed vector mesons had to be massive because of some breaking of the gauge symmetry.

Q. Was that not a sensible point of view?

A. Yes, it was sensible but not correct in so far as the strong interactions are concerned. It became clear in the seventies that the vector mesons of quantum chromodynamics, the theory of the strong interactions, are in fact strictly massless, a most startling new idea. It may help to consider the fundamental equations of that theory now, and only thereafter give reasons for the increasing conviction that these contain essentially the correct description of the strong interactions.

* Chapter 19, Section (e); this chapter, Section (b), Part 1.

*3. Quantum chromodynamics** (QCD) is a gauge field theory, the gauge group is SU(3) color, or $SU(3)_c$ for short. As we saw earlier in this chapter (Section (b), Part 5), the first conjecture that this group governs the strong interactions dates from 1965; in the early seventies that same idea re-emerged in various quarters.[192] The force fields are an $SU(3)_c$ octet $B_{a\mu}$ of massless vector fields (indices $a, b, \ldots$ all run from 1 to 8); their quanta are called gluons. These forces act on quarks, $SU(3)_c$ triplets, represented by Dirac fields $\psi_{\alpha\kappa}$. $\alpha = 1$, 2, 3, is the color index, κ denotes the species of quark (u, d, s, ...) or, in current terminology, κ is the *flavor* label. For each κ

$$[(\gamma_\mu \partial_\mu + m_\kappa)\delta_{\alpha\beta} - ig\gamma_\mu(\lambda_a)_{\alpha\beta}B_{a\mu}]\psi_{\beta\kappa} = 0. \qquad (21.74)$$

$(\lambda_a)_{\alpha\beta}$ are eight 3×3 matrices,[22] the $SU(3)_c$ analogs of the τ-matrices for isospin. Eqs. (21.66, 67, 69, 70, 72, 73) hold true with

$$j_{a\mu} = ig \sum_\kappa \bar{\psi}_{\alpha\kappa}(\lambda_a)_{\alpha\beta}\gamma_\mu\psi_{\beta\kappa}. \qquad (21.75)$$

This total set of equations constitute** QCD.

Q. Thank you for showing that the QCD equations can be written on the back of an envelope. Why is $SU(3)_c$ used and not the earlier SU(3) that governs hadron mass spectra and the like?

A. That other SU(3), now called SU(3) flavor or $SU(3)_f$, cannot be used because its associated quark current octet has already been spoken for in the electromagnetic and weak interactions (Cabibbo currents, Eq. (21.19)). One cannot hope to get strong interactions by coupling this same current octet to a *B*-octet since that would mess up the distinctions between the symmetries of the strong and those of the weak/electromagnetic interactions.

Q. Then how do the strong interaction symmetries of isospin and $SU(3)_f$ arise in QCD?

A. Isospin: because both the u- and the d-quark mass turn out to be quite small compared to hadronic masses.[192a] The old electromagnetic symmetry breaking of isospin is unalterably enforced. Approximate $SU(3)_f$: because of the mass difference between the s and the (u, d) quarks. Equal u, d, s mass would yield exact $SU(3)_f$ invariance.

Q. What does QCD predict for the quark masses m_κ in Eq. (21.74)?

A. Nothing, these are inputs.

Q. So isospin and $SU(3)_f$ are put in by hand?

* Theoretical status: Refs. 188 (1974); 189 (1978); 190 (1982); experimental evidence: Ref. 191 (1981).

** That statement is not quite precise. Additional terms ('θ-terms') arise because of an anomaly of the kind encountered in Eq. (21.31). This anomaly appears to be unwanted. Several ways of eliminating it are currently under advisement.[190]

A. That is the best we can do as long as we cannot predict anything about u, d, s masses. More about that later.
Q. What is known about solutions to the QCD equations?
A. A good deal but not remotely enough.
Q. What about deriving nucleon and meson properties in terms of bound qqq and $q\bar{q}$ systems?
A. It has not, not yet, been possible to calculate this binding from first principles.
Q. So there is nothing yet that can compare in numerical accuracy with the many decimal places computed in quantum electrodynamics?
A. Correct.
Q. Have people not tried to calculate binding energies of quark systems?
A. They are trying at this very moment, and have promising, preliminary, partial results. At this point we have reached the outer edge of history, however. I shall therefore not discuss these efforts that involve mathematical techniques with which I am not sufficiently familiar anyway.
Q. You are evidently very taken with QCD. Why?
A. First, because at this time there are no other viable approaches to the strong interactions. That in itself is not a very convincing argument, however. Far more compelling, QCD successfully explains a whole group of most important recent experimental results which, in broad terms, all carry a common message: *All the difficulties with the strong interactions encountered in earlier chapters reside in the low energy* ($\lesssim$ a few GeV) *régime*. The very difficult part of QCD that aims at coping with these problems is known as non-perturbative QCD. Here belong the questions you just asked about binding energies of quark systems. *Strong interactions diminish in strength as the energy gets higher*. The difficult part of QCD that deals with the high energy aspects is perturbative QCD. Here belongs the problem posed by the parton model: why does a nucleon behave in the scaling region as an assembly of freely moving constituents?

The response to that question marks the beginning of QCD.

Can the parton model be incorporated in a field theory? The lessons from quantum electrodynamics did not look encouraging. Consider the modifications due to quantum effects of the classical Coulomb interaction between two electrons: α/r in coordinate space, or α/q^2 in momentum space ($\alpha = e^2$ in units $\hbar c$). To leading order the interaction comes about via exchange of a virtual photon (four-momentum q). Higher-order effects arise (in part) because that photon is dissolved part of the time in electron–positron pairs. Write the corrected interaction as $\alpha(r)/r$ or $\alpha(q^2)/q^2$. It had long been known that $\alpha(r)$ grows with decreasing r or equivalently that $\alpha(q^2)$ increases as q^2 increases. In order to understand the parton picture one needs just the opposite: a coupling, call it $\alpha_c(q^2)$, such that $\alpha_c(q^2) \to 0$ as $q^2 \to \infty$, a behavior called asymptotic freedom.

In 1973 Gross and Wilczek,[193] and independently Politzer,[194] published their marvelous discovery: $SU(3)_c$ is asymptotically free!* Furthermore, Coleman and Gross showed[195] that *only* non-Abelian gauge theories have a chance to be asymptotically free.

The point is this: gluons, mediating quark interactions, are part of the time dissolved in $q\bar{q}$ pairs. That by itself would lead to an $\alpha_c(q^2)=g^2(q^2)$ that once again rises with q^2. However: because of the non-linearities in the B-field, gluons are also part of the time dissolved in gluons—and that does the trick. It overcompensates the $q\bar{q}$ effect, and the net result is**

$$\alpha_c(q^2)=\frac{12\pi}{(33-2n_f)\ln q^2/\Lambda^2}, \quad \text{for } \frac{q^2}{\Lambda^2}\gg 1, \tag{21.76}$$

which decreases with q^2 as long as n_f, the number of flavors, is less than sixteen.

The parameter Λ, 'the hidden scale of QCD', enters the theory via the quantum mechanical machinery of g-renormalization. The pure B-field (forget about quarks for the moment) contains non-linearities that induce B–B scattering—but not a scale for measuring that effect, the only constant being the dimensionless g^2 (in units $\hbar c$). The B–B scattering amplitude needs renormalization, which is performed for some fixed $(\text{mass})^2=-\Lambda^2$ of *virtual* gluons. One cannot take $\Lambda^2=0$ (real gluons) because there the amplitude is infra-red divergent—there are no low energy theorems in QCD! The renormalized g^2 will depend on Λ^2—whence the appearance of Λ in Eq. (21.76).

In the large q^2 region QCD makes predictions that include as well as sharpen those of the more naive parton model. Samples: Eqs. (21.50) and (21.51) become, respectively,

$$A_1=2\left(1-\frac{2\alpha_c(q^2)}{3\pi}\right) \tag{21.77}$$

$$A_2=-6\left(1-\frac{\alpha_c(q^2)}{\pi}\right). \tag{21.78}$$

Q. How big is Λ?
A. The examples just given provide but two of numerous independent ways for experimentally determining Λ, all of which should of course give the same answer. All these methods yield consistently $\Lambda\simeq 200$ MeV, with moderate precision.[191]
Q. Does QCD predict what the structure functions are in the scaling region?
A. It does not (yet) since these functions depend on the intractable behavior of low-energy bound three-quark systems. One can, however, predict many properties† of these functions, with satisfactory results.[191]

* This result was known to 't Hooft in 1972.[185]
** This formula is not rigorous. Corrections do not spoil asymptotic freedom. For a review of higher-order effects see Ref. 196.
† To wit, 'moments' of the type $\int dx\, F(x)x^n$.

Q. Does the masslessness of the gluons not cause trouble? Why aren't they seen?
A. I have been waiting for you to ask that question. First of all, you should know that the discoverers of asymptotic freedom were not at once convinced that gluons have zero mass, especially because it had meanwhile been discovered (I come back to that) that massive non-Abelian gauge fields can be renormalized provided that mass is introduced in a clever way. Only after asymptotic freedom did various physicists begin* to speculate that gluons are strictly massless, hence that $SU(3)_c$ is an ***unbroken*** symmetry.

One speculation led to another: going from high q^2 to not so high q^2, $\alpha_c(q^2)$ increases, and so therefore does $\alpha_c(r)$. Suppose this continues to hold all the way down to low q^2, large r—in other words that the attractive quark potential keeps growing with r. Then quarks cannot get away from each other, they are *confined*.[197,198,200]** That, in turn, led to speculations that all states with non-zero color, including single gluons, are confined.[197,198,200] That interdiction does not apply to colorless ($SU(3)_c$ singlet) states made up of two or more gluons. Such 'glueballs' are being sought.[202]

Moving from small to large r one goes from perturbative (weak coupling) to non-perturbative (strong coupling) QCD where the still outstanding problem† is a rigorous proof of confinement. There are many indications, however, that confinement does occur. The first of these date from early 1974: 't Hooft's example[204] of a one (space) plus one (time) dimensional gauge model that contains color and flavor and that exhibits both asymptotic freedom and confinement; and K. Wilson's demonstration[205] of confinement on a lattice. In such theories the space–time continuum is replaced by a discrete Euclidean lattice; quanta hop from one site to another. (The idea of treating strong coupling by lattice methods goes back to Wentzel's work[206] of 1940.) In more recent times this approach, sometimes referred to as the Euclidean revolution, has yielded increasingly powerful results.[207]

The beginnings of QCD have now been followed up to the summer of 1974. At the London conference on high energy physics (July 1974) the theory was awarded honorable mention—no less, no more. Then, in the fall of 1974, came great upheavals. I shall come back to what these did to QCD but first follow the weak interactions up to that same time.

4. The electroweak unification.†† Two seemingly independent pursuits transformed the description of the weak interactions from phenomenology into field theory. One, already touched on in the preamble to this section, was the search for a field theory that is renormalizable. The other, to be discussed

* As best I know the order of appearance is as in Refs. 197, 198, 199.
** The idea of confinement appeared in 1973–4, but not always in the context of QCD.[201].
† For a status report (1981) see Ref. 203.
†† Anthology: Ref. 208. Reviews: Refs. 209 (1972), 210 (1973), 211 (1974), 212 (1982). Books: Refs. 213, 214. Popular account: Ref. 215.

next, was the desire for a theory that unifies the weak with the electromagnetic interactions. The aesthetically appealing quest for unification and the technically compelling demand for renormalizability have, as it turned out, a common resolution.

The beginnings of the modern unification program demonstrate once again the great variety in patterns of discovery. During a ten-year period, elements of the correct answer would appear now here now there, but in those times the pieces would never quite fit, as is seen from the following sequence of main events: (1) it is suggested that the photon (γ) and the $W^{\pm}$ form a triplet with respect to a *new* group SU(2) ('new' in the sense of physical content, of course), and that the huge γ–W splitting in this triplet demands the introduction of an auxiliary scalar field (Schwinger[216]); (2) it is proposed[217] that $W^{\pm}$ form a triplet not with γ but with a new neutral heavy vector boson, called Z from now on. Z interacts weakly with a current, called 'the neutral current' from now on. Like the electric current, the neutral current does not induce changes of charge. It is further suggested that $W^{\pm}$ and Z are described by massive Yang–Mills fields; in which case lack of renormalizability causes serious problems (Bludman 1958); (3) the Yang–Mills theory is applied[218] to a γ–W-triplet; the origin of W-mass is obscure (Salam and Ward 1959); (4) it is noted[219] that SU(2) does not permit the unification* of parity-violating properties of weak currents with parity-conserving properties of the electromagnetic current. The remedy proposed is the introduction of an Abelian group U(1) in addition to SU(2), the resulting group being denoted by the conventional symbol SU(2)×U(1). There are now four vector bosons: $W^{\pm}$, coupled as always to a charge changing current, Z and γ, each coupled to neutral currents; the origins of the W- and Z-masses remaining obscure.

Looking back on those struggles one recognizes two ideas that were to survive: the use of SU(2)×U(1) which, if not the last word in unification, is indisputably a major stride in the right direction; and the conception of this group as a local gauge group. Nothing sensible had yet been said about the origins, let alone about the magnitudes, of the W- and Z-masses, however. And—we are now early 1967—renormalizability remained *the* stumbling block. The next and crucial step came when the unification program was wedded with another important notion that had sufficiently matured meanwhile: spontaneous symmetry breaking, an idea that had originated in solid state physics and had been injected into particle physics by Nambu and coworkers.[221]

This idea will be illustrated by a standard example treated in textbooks,[222] where one will find proofs of unsubstantiated assertions made in the next two paragraphs.

* It became clear later[220] that actually SU(2) is a logical possibility provided hypothetical fermions are introduced.

Step 1. Consider a complex spinless neutral field ψ that satisfies

$$\partial_\mu^2\psi - \mu^2\psi - \lambda\psi^*\psi\cdot\psi = 0. \qquad (21.79)$$

For $\lambda \neq 0$ the field is self-interacting. To insure stability (a finite lowest energy) λ must be ≥ 0. The energy density H is

$$H = \frac{\partial\psi^*}{\partial t}\frac{\partial\psi}{\partial t} + \vec{\nabla}\psi^*\cdot\vec{\nabla}\psi + V; \qquad V = \mu^2\psi^*\psi + \frac{\lambda}{2}(\psi^*\psi)^2. \qquad (21.80)$$

Eqs. (21.79), (21.80) are invariant under the global gauge transformation $\psi \to e^{i\alpha}\psi$ with constant α. For $\mu^2 > 0$ the theory is quantized in the standard way,* yielding particles with mass μ interacting via the λ-term. We shall take $\mu^2 < 0$, however, in which case μ still has the dimension but no longer the physical meaning of mass. Eq. (21.79) now has an infinity of space–time independent classical solutions

$$\psi = a\,e^{i\beta}, \qquad a = \left(\frac{-\mu^2}{\lambda}\right)^{1/2}, \qquad 0 \leq \beta < 2\pi, \qquad (21.81)$$

all with the same energy, the lowest energy attainable: the ground state, or vacuum, is degenerate, clearly a consequence of the global invariance.

Now declare: 'our' vacuum is going to be the single state corresponding to a fixed β, say $\beta = 0$. H is still globally gauge invariant but our choice of vacuum is not. That is called spontaneous symmetry breaking. This has a remarkable consequence: write the *general* solutions of Eq. (21.79) as

$$\psi = a + \varphi, \qquad \varphi = u + iv, \qquad (21.82)$$

(u, v) real (hermitian), and do a little algebra to find the separate wave equations for u and v. Result: $\partial_\mu^2 u + 2\mu^2 u + \cdots = 0$, $\partial_\mu^2 v + \cdots = 0$, where $\cdots$ stands for terms quadratic and cubic in (u, v). What does that mean? The term $2\mu^2 u$ implies that u-particles have mass $(-2\mu^2)^{1/2}$, the absence of a similar term in v means that *v-particles have mass zero*! The procedure has led to massless particles. This is a special case of the Goldstone theorem[223] according to which the spontaneous symmetry breaking of any continuous symmetry G is accompanied by the appearance of one or more species of massless particles, Goldstone bosons.

Notes. (1) The theorem goes further.** Treat φ (not ψ) as a quantized field and show that the theorem remains valid if quantum corrections may be treated perturbatively. It is less clear that the same is true independent of perturbation theory. (2) If (as can happen) the potential V by itself has a symmetry larger than the symmetry of the theory as a whole (accidental symmetry[225]), or if, more generally, the sector of V on which the potential is minimal has a larger such symmetry (vacuum symmetry[226]) then 'pseudo

* Chapter 16, Section (d), Part 7. ** For details see Ref. 224.

Goldstone bosons' appear, particles with zero mass at the classical but not at the quantum level. (3) Pseudo Goldstone bosons (not Goldstone bosons) can appear if a *discrete* symmetry is spontaneously broken.[227]

Step 2. Go back to Eq. (21.79), but demand *local* gauge invariance, $\psi \to e^{i\alpha(x)}\psi$. We need a compensating massless vector field:*

$$(\partial_\mu - igB_\mu)^2\psi - \mu^2\psi - \lambda\psi^*\psi \cdot \psi = 0 \tag{21.83}$$

with $B_\mu \to B_\mu + \partial_\mu\alpha/g$. Write out[222] the energy density of the B, ψ-system and find a coupling term $g^2B_\mu^2\psi^*\psi/2$ associated with the $B_\mu^2\psi$ term in Eq. (21.83). Break the symmetry by (21.82). Part of this coupling turns into $g^2a^2B_\mu^2/2$—which is the same as saying that the B-field has acquired a mass ga! However: a coupling survives which is of the form $gB_\mu\,\partial_\mu v/2$ and which contains that miserable massless scalar field.

Step 3. Hallelujah! A gauge transformation decouples[222]** v from all other fields. Therefore neither the creation nor annihilation of v-particles can occur, so that v can be ignored. What has happened is a trade-off in the number d of internal degrees of freedom. Start out with a complex scalar field ψ, $d=2$, one for u and one for v, and a massless vector field, $d=2$, two transverse polarization states. End up with a real scalar field (u), $d=1$, and a massive vector field, $d=3$, an additional longitudinal polarization appears. Several physicists contributed to the evolution of these ideas, see again Veltman.[185] The most complete treatment is due to Higgs.[228]

The three-step procedure, known as the Higgs phenomenon, can be applied to non-Abelian gauge theories as well. The surviving scalar particle(s) (u in our example) are commonly called Higgses. The Higgs potential V cannot depend on powers higher than the fourth in the scalar fields (as in Eq. (21.80)), since otherwise the theory is unrenormalizable. The modification of V due to quantum effects can give rise to subtleties, as has become especially clear from the analysis by Coleman and E. Weinberg.[229]

There came a time when it occurred to Weinberg[230] and to Salam[231] to unify weak and electromagnetic interactions in terms of a rigorously SU(2)×U(1) invariant gauge theory with massless gauge bosons, combined with a Higgs mechanism for generating W- and Z-masses (for the time sequence of events see their papers and recollections[232,233]). Let us see how this is done.

Associate a 'weak isospin' T with SU(2) and a 'weak hypercharge' Y with U(1). The electric charge in units e is

$$Q = T_3 + \frac{Y}{2} \tag{21.84}$$

* Since Eq. (21.82) will be used once more, ψ is *not* carrying charge (the vacuum must be electrically neutral), so do *not* think of B_μ as the electromagnetic field.

** The choice of gauge that causes v to decouple is called the unitary gauge.

just as in Eq. (21.1). There are four gauge bosons: $\vec{W}_\mu$ ($T=1$, $Y=0$) and B_μ ($T=Y=0$). Introduce a complex Higgs doublet ψ ($T=1/2$, $Y=1$) and its conjugate $\psi^\dagger$ ($T=1/2$, $Y=-1$):

$$\psi=\begin{pmatrix}\psi^+\\ \psi^0\end{pmatrix}, \qquad \psi^\dagger=(\bar{\psi}^+, \bar{\psi}^0) \tag{21.85}$$

(analogs of K, $\bar{K}$). ψ shall satisfy Eq. (21.83) with

$$-igB_\mu \to -ig\frac{\vec{\tau}}{2}\vec{W}_\mu-\frac{ig'}{2}B_\mu. \tag{21.86}$$

The theory contains two independent coupling constants g, g'. Look for vacuum solutions analogous to Eq. (21.82),

$$\psi=a\chi+\varphi, \qquad \psi^\dagger=a\chi^\dagger+\varphi^\dagger, \qquad \chi=\begin{pmatrix}0\\ 1\end{pmatrix}. \tag{21.87}$$

As before locate the candidate mass terms, here $\psi^\dagger(g\vec{\tau}\vec{W}_\mu+g'B_\mu)^2\psi/4$. The masses due to the χ-part of ψ are all contained in

$$\begin{aligned}a^2\chi^\dagger(g\vec{\tau}\vec{W}_\mu+g'B_\mu)^2\chi/4&=\frac{a^2}{4}[g^2\{(W_\mu^1)^2+(W_\mu^2)^2\}+(gW_\mu^3-g'B_\mu)^2]\\ &=M_W^2W_\mu^*W_\mu+\tfrac{1}{2}M_Z^2Z_\mu^2\end{aligned} \tag{21.88}$$

where the properly normalized fields are given by

$$W_\mu=\frac{W_\mu^1-iW_\mu^2}{\sqrt{2}}, \qquad Z_\mu=\cos\theta_W W_\mu^3-\sin\theta_W B_\mu, \tag{21.89}$$

$$\tan\theta_W\equiv\frac{g'}{g}. \tag{21.90}$$

The masses are

$$M_W^2=\frac{g^2a^2}{2}, \qquad M_Z^2=\frac{(g^2+g'^2)a^2}{2}, \tag{21.91}$$

hence

$$\frac{M_W^2}{M_Z^2}=\cos^2\theta_W. \tag{21.92}$$

What has happened? Three of the four scalar fields have been swallowed by W_μ, W_μ^*, and Z_μ, giving them masses.* One neutral vector field remains

* The fields Eq. (21.85) are the *minimal* but not the unique set of fields that can serve to generate the requisite masses. All the above formulae remain valid for more general choices, except for Eq. (21.92) which becomes $M_W^2=\rho M_Z^2\cos^2\theta_W$, where ρ is an adjustable dimensionless parameter. Experimentally,[234] ρ lies very close to one.

massless, the one orthogonal to Z_μ:

$$A_\mu = \sin\theta_W W^3_\mu + \cos\theta_W B_\mu, \tag{21.93}$$

to be identified with the electromagnetic vector potential. The spontaneous symmetry breaking introduces an angle θ_W, the Weinberg angle, in the (B_μ, W^3_μ)-plane, which fixes the mass eigenvalue directions in that plane.

Before symmetry breaking the interaction of the four-vector fields with matter has the invariant form

$$H_{int} = g\vec{T}_\mu(x)\vec{W}_\mu(x) + g'Y(x)B_\mu(x) \tag{21.94}$$

where $\vec{T}_\mu(X)$, $Y(X)$ are respectively the SU(2) and U(1) source densities. In terms of the physical fields, Eqs. (21.89), (21.93):

$$H_{int} = \frac{g}{\sqrt{2}}(J_\lambda W^*_\lambda + J^\dagger_\lambda W_\lambda) + \frac{gZ_\mu}{\cos\theta_W} J^{(0)}_\mu(x) + eA_\mu J^{elm}_\mu \tag{21.95}$$

with

$$J_\mu(x) = T^1_\mu(x) - iT^2_\mu(x), \tag{21.96}$$

$$J^{elm}_\mu(x) = T^3_\mu(x) + \frac{Y_\mu(x)}{2}, \tag{21.97}$$

$$J^{(0)}_\mu(x) = T^3_\mu(x) - J^{elm}_\mu(x)\sin^2\theta_W. \tag{21.98}$$

The electric charge e is *defined* by

$$e = g\sin\theta_W. \tag{21.99}$$

J_λ is the charged weak current first introduced in Eq. (21.19), coupled to W-bosons in a form foreshadowed in Eq. (21.52). $J^{(0)}_\lambda(x)$ is the new neutral current.

The earliest discussions[230,231] of the current contents were confined to leptons, which have to be introduced as SU(2)×U(1) multiplets, such that the parity properties of currents come out right. That is achieved by the assignments*

$$L_e = \begin{pmatrix} \nu^L_e \\ e^L \end{pmatrix}, \qquad L_\mu = \begin{pmatrix} \nu^L_\mu \\ \mu^L \end{pmatrix}, e^R, \mu^R$$
$$e^L = \tfrac{1}{2}(1+\gamma_5)e, \qquad e^R = \tfrac{1}{2}(1-\gamma_5)e, \tag{21.100}$$

and likewise for the other particles. Thus the L (left-handed) states form doublets, $T = 1/2$, $Y = -1$, the R (right-handed) states are singlets, $T = 0$,

* A particle symbol denotes the corresponding wave field. In Eq. (21.101) 'μ' means: corresponding muon terms.

$Y=-2$. Write out the currents and find

$$J_\mu^{elm}=\bar{e}\gamma_\mu e+\text{`}\mu\text{'},$$
$$J_\mu=\tfrac{1}{2}\bar{e}\gamma_\mu(1+\gamma_5)\nu_e+\text{`}\mu\text{'}, \qquad (21.101)$$
$$J_\mu^3=\tfrac{1}{4}[\bar{\nu}_e\gamma_\mu(1+\gamma_5)\nu_e-\bar{e}\gamma_\mu(1+\gamma_5)e]+\text{`}\mu\text{'},$$

from which $J_\mu^{(0)}$ can be read off using Eq. (21.98). We now have enough information to relate g to the Fermi constant G: compute μ-decay and follow the reasoning* that led to Eq. (21.56). Result:

$$\frac{G}{\sqrt{2}}=\frac{g^2}{8M_W^2}, \qquad (21.102)$$

hence, using also Eq. (21.99):

$$M_W=\frac{2^{-5/4}eG^{-1/2}}{\sin\theta_W}=\frac{37.3\text{ GeV}}{\sin\theta_W}; \qquad (21.103)$$

so from Eq. (21.92):

$$M_Z=\frac{74.6\text{ GeV}}{\sin\theta_W}. \qquad (21.104)$$

To complete this corner of the theory we need to ask: how do e and μ acquire mass? The e-mass term is of the form $m_e\bar{e}e=m(\bar{e}^Le^R+\bar{e}^Re^L)$—not allowed since it is not SU(2)×U(1) invariant. The Higgses help:[230,231] $f_e(\bar{L}_e\psi)e^R$ + conjugate *is* invariant and, upon symmetry breaking, contributes $f_ea(\bar{e}^Le^R+\bar{e}^Re^L)$, hence $m_e=f_ea$; likewise $m_\mu=f_\mu a$. Again a trade-off has been effected, here a coupling constant (f) for a mass (m).

Q. I have two questions. Why did one not introduce ν_e^R and ν_μ^R? And what is the fuss about the electron mass? What is so bad if SU(2)×U(1) symmetry is a little bit broken?

A. Both your questions relate to renormalizability. In 1967 it was conjectured[230,231] that the above theory is renormalizable if, before spontaneous symmetry breaking, it is rigorously SU(2)×U(1) invariant. The lack of proof thereof was the principal reason why that work initially drew little if any attention. It was not at all quoted in the literature[235] during 1967–9. Rapporteur talks at the biannual Rochester conferences in Vienna (1968) and Kiev (1970), as always devoted to what is currently fashionable, did not even mention SU(2)×U(1).

The great change came when 'at the Amsterdam conference [3 June–6 June 1971] . . . a young Dutch physicist, G. 't Hooft, not yet out of graduate school,

* The reader will note additional numerical factors on the rhs of Eqs. (21.101) and (21.102) as compared with Eqs. (21.19) and (21.56). These arise because of changes in the normalization procedures.

presented a paper[236] which would change our way of thinking on gauge field theory in a most profound way.'[237]* In addition to rediscovering the Higgs mechanism by himself, he presented a formulation of spontaneously broken gauge theories which is manifestly renormalizable. That development, which causes a great stir, made unification into a central research theme.

Now to your two questions. The proof of renormalizability rests on the condition that the theory is rigorously invariant prior to symmetry breaking; accordingly the introduction of e- or μ-mass terms by hand is prohibited. A further condition for the proof is that one must introduce in the Hamiltonian *every* possible term compatible with invariance and renormalizability (condition of 'strict renormalizability') in order to have the means of executing renormalizations when needed. Thus if one introduces ν_e^R, ν_μ^R one cannot avoid[239] non-zero neutrino masses—just as for e and μ. The non-occurrence of ν_e^R, ν_μ^R guarantees zero neutrino mass.

Q. Do we now have an understanding of the origin of particle masses?

A. Not really. We have realized that masses may arise from spontaneous symmetry breaking but, as said, this has so far only resulted in trading an adjustable mass value for an adjustable coupling constant. Quark masses m_κ occurring in the QCD Eq. (21.74) also originate from certain Higgs couplings in the electroweak sector; see what follows. None of these masses is predicted. Cleverly rigged combinations of Higgs fields shed, perhaps, some light on particle mass ratios.[240]

Q. About this one surviving Higgs, what is its mass?

A. We do not know, though there are sensible bounds on its value.[241]

Q. Has that Higgs been seen?

A. No. People have of course considered ways to detect it, if it exists, what are its decays, etc. All that belongs to the future. Do consult a good review on that subject.[242]

Q. Why 'if it exists'?

A. We are happy that the Higgs mechanism allows us to work with renormalizable massive vector boson theories. One is not sure at this time, however, whether that mechanism is the last word in symmetry breaking. After all we do wish to predict particle masses. For this and other reasons, one is very interested in alternatives to the Higgs description, such as 'technicolor' where no fundamental Higgs field is introduced, yet the dynamics causes Higgs-like spinless particle states to occur. This is still an open subject; again I refer you to the literature.[243]

Q. You mentioned that the parity properties of the various currents come out right. Does that mean that we now begin to understand why parity is violated in weak but not in other interactions?

* This paper as well as several others that refined the initial proof are all reprinted in Ref. 238.

A. It has been shown that parity can only be introduced in gauge theories in a manner consistent with observation. That, of course, is important but does not deepen our understanding of parity violation.
Q. Does that indicate a shortcoming of the theory?
A. I would rather defer answering that question until we know more.
Q. What is known about higher-order effects in this theory?
A. Two examples may illustrate important insights gained. μ–e universality has been implemented to second order in the gauge coupling constant g—that is, up to that order the couplings g_e, g_μ of $\bar{e}\nu_e$, $\bar{\mu}\nu_\mu$ to the W satisfy $g_e = g_\mu = g$ or

$$\frac{g_e}{g_\mu} = 1; \tag{21.105}$$

see Eqs. (21.95), (21.101). Inclusion of higher-order effects leads to a renormalization of the *one* constant g (along with g'). But we are dealing with *two* measurable constants g_e, g_μ. Hence even after renormalization there must exist a relation between the two which, to order g^4, turns out to be[244]

$$\frac{g_e}{g_\mu} = 1 + 0(\alpha), \qquad \alpha = \tfrac{1}{137}, \tag{21.106}$$

where the $0(\alpha)$ term is finite (and depends further on the ratio m_e/m_μ). *The expansion parameter for higher-order effects is the small fine structure constant!* The finiteness of the correction is a special example of a general theorem: write the theory in *strictly* renormalizable form. *Then any relation* (called a natural relation) *that holds to lowest order is also true in higher orders up to finite and small calculable corrections*. One further example[244]: corrections* to Eq. (21.92) are:

$$\frac{M_W^2}{M_Z^2} = \cos^2\theta_W + 0(\alpha) \tag{21.107}$$

where $0(\alpha)$ is again finite. This result depends on the specific assumption that the Higgs system is a doublet and thus illustrates that calculability can depend on the particle content of the theory.

Naturalness and calculability rank among the most important topics in gauge theories.[246] Here one also encounters the interplay of strong and weak interactions, specifically the question of whether radiative corrections could possibly mess up the respective symmetries of these two forces. All is well[197] if one assigns W, Z, and Higgses to colorless QCD states, gluons to states with zero weak isospin and hypercharge.

Finite higher-order weak corrections computed thus far are compatible with experiment, and in some cases are nearing the level of experimental

* Eq. (21.107) of course demands a precise *operational* definition[245] of what is meant by θ_W.

verification.[247] In any event, here lies the future of gauge theories both experimentally and theoretically.

Q. How big is θ_W?

A. SU(2)×U(1) does not predict its value. Neutral currents are a main key to the experimental answer, as Eqs. (21.95), (21.98) show (results are quoted in Eqs. (21.114) and (21.115) below).

5. The neutral current; charm.*** The search for neutral currents had been on the agenda from the very inception of neutrino beam experiments.[254] The first report of an upper limit for the rate of the neutral current process $\nu_\mu + p \rightarrow \nu_\mu + p$ dates[255] from 1963. At that time, neutral currents were not a high-priority item; theory had not as yet provided strong incentives. Experimentalists therefore contented themselves with fairly crude background estimates, the problem being how to distinguish $\nu_\mu + X \rightarrow \nu_\mu + Y$ from $\nu_\mu + X \rightarrow \mu + Y$, the muon escaping detection, and from neutron $+X \rightarrow Y$. It seems likely[249] that neutral currents might already have been diagnosed in the mid-sixties had the difficult background analysis been performed with greater precision.

In those years important low energy information about neutral currents did already exist, however. It was known then that the strength of the strangeness-changing ($|\Delta S| = 1$) part of the neutral current had to be very much weaker than the charged weak current. For example, two-thirds of all K^+'s decay into $\mu^+\nu_\mu$; but only a fraction $\sim 10^{-8}$ of K_L^0's decay† into $\mu^+\mu^-$. The same conclusion follows from the magnitude of K^0–$\bar{K}^0$ mixing, an effect at most of *second* order in the Fermi constant G; see Eq. (20.18). These two arguments concerning the $|\Delta S| = 1$ part of the neutral current imply nothing, of course, about its $\Delta S = 0$ component. Nevertheless 'by the end of the 1960s it was taken for granted that neutral weak currents were either absent in nature or else extremely rare'.[251] That was also the time when scaling was discovered at SLAC and when neutrino physicists turned their attention to that phenomenon in neutrino reactions.

Then came the theorists' great excitement of 1971 about gauge theories. Now neutral currents at once received high priority, especially because existing techniques sufficed to get it going at once. In particular the detection of $\nu_\mu + e \rightarrow \nu_\mu + e$, to order g^2 *purely* a neutral-current effect, was a ready challenge.

From the theorists' side the time was now more than ripe for adding quark contributions to the various currents. That appeared to be straightforward enough: as for leptons, introduce an L-doublet and R-singlets (I omit the

* Reviews: Ref. 248; historical essays: Refs. 249, 250; popular account: Ref. 251.
** Review: Ref. 252; popular account: Ref. 253.
† This fraction represents only an upper bound to the weak neutral current contribution since this decay may well be due primarily to electromagnetic effects.

color index for the moment; θ is the Cabibbo angle)

$$Q_1 = \begin{pmatrix} \mathrm{u}^L \\ \mathrm{d}^L \cos\theta + \mathrm{s}^L \sin\theta \end{pmatrix}; \qquad \text{singlets: } \mathrm{u}^R, \mathrm{d}^R, \mathrm{s}^R. \tag{21.108}$$

That indeed gives the long-known charged current (21.29). As before, quark masses are introduced by the Higgs device. What are the corresponding neutral current terms? A glance at Eqs. (21.95), (21.98) shows that $J_\lambda^{(0)}$ contains a term of type $g\ (\bar{\mathrm{d}}\mathrm{s}+\bar{\mathrm{s}}\mathrm{d}) \sin\theta \cos\theta/\cos\theta_{\mathrm{W}}$. To second-order in g, hence to first-order in G, this leads, via virtual exchange of a Z, to the process $\bar{\mathrm{s}}\mathrm{d} \leftrightarrow \bar{\mathrm{d}}\mathrm{s}$ or equivalently to $\mathrm{K}^0 \leftrightarrow \bar{\mathrm{K}}^0$. That spells disaster since, as just said, this mixing is at most of second order in G.

Enter charm.

In 1963–4 a flurry of papers had appeared[256] in which it was proposed to extend the flavor symmetry group $SU(3)_f$ to $SU(4)_f$. The motivations were various: to avoid quarks with fractional charge; to find new constraints in hadron spectroscopy; and to implement a baryon–lepton symmetry: four leptons (e, μ, ν_e, ν_μ) matched by four quarks u, d, s, and a fourth species carrying no (strong) isospin and strangeness, but a new quantum number $C = 1$, called[257] charm (u, d, s have $C = 0$), supposed to be conserved in strong and electromagnetic, but not in weak, interactions.

By and large these suggestions were politely ignored until early 1970 when Glashow, Iliopoulos, and Maiani ('GIM') pointed out[258] that the introduction of a fourth quark, call it c, provides a natural means for eliminating the highly unwanted terms in the neutral current.* Here c appears for the first time with fractional charge:

$$\mathrm{c}: \quad Q = \tfrac{2}{3}, \qquad B = \tfrac{1}{3}, \qquad S = 0, \qquad C = 1 \tag{21.109}$$

corresponding to the generalization of Eqs. (21.1), (21.2) to

$$Q = T_3 + \tfrac{1}{2}(S + B + C). \tag{21.110}$$

GIM discussed this problem in terms of an unrenormalizable version of $SU(2)\times U(1)$ (no Higgs mechanism). As Glashow recalled[260] later: 'Neither I, nor my coworkers, nor Weinberg sensed the connection between the two endeavors.' It took another year before Weinberg[261,262] brought the two ideas together: introduce a second quark doublet

$$Q_2 = \begin{pmatrix} \mathrm{c}^L \\ \mathrm{d}^L \sin\theta - \mathrm{s}^L \cos\theta \end{pmatrix} \tag{21.111}$$

and a further singlet, c^R, and verify that the contribution to $J_\lambda^{(0)}$ from Q_2

* Alternative proposals[259] to achieve this were short-lived.

exactly cancels the previously found $|\Delta S|=1$ terms, as was essentially known to GIM.

That is good, but not sufficient. The branching ratio 10^{-8} for $K_L \to \mu^+\mu^-$ is so small that one must worry about the magnitude of $|\Delta S|=1$ effects due to radiative corrections. Such contributions from Q_1 alone yield a decay amplitude $0(G\alpha)$, hence a branching ratio $\sim 10^{-4}$—inadmissible. Again charm saves the day. Inclusion of Q_2 leads to an amplitude (B. W. Lee, Primack, and Treiman, 1972[263])

$$\sim \sin\theta \cos\theta \, G\alpha \frac{m_u^2 - m_c^2}{M_W^2}, \qquad (21.112)$$

which gives the right order of magnitude as long as charmed quarks are not too heavy: $m_c/M_W \lesssim 0.1$.

Indications of the need for more than three quarks came from an altogether different direction when, in 1972, a loophole was discovered[264] in the proof of renormalizability of the $SU(2)\times U(1)$ theory when fermions are present. The point is quite technical* but its resolution can be stated very simply. A sufficient condition for the restoration of renormalizability is

$$\sum_i Q_i = 0 \qquad (21.113)$$

where Q_i is the electric charge of the ith fermion species, lepton or quark. (e, μ) contribute -2; (u, d, s), whether colored or not, give nothing $(2/3-1/3-1/3=0)$; a c-quark gives $+2/3$, not enough. Give it color: $3\times 2/3=2$ and Eq. (21.113) is satisfied. That resolution is of course not unique. Yet *it is the first instance of a confluence between color, charm, and unification*. That point was not yet emphasized in 1972, for obvious reasons: asymptotic freedom had not yet arrived, hence color had not yet taken its central position; nor had charm yet been discovered. With the discovery of the condition (21.113) the basics of $SU(2)\times U(1)$ were complete.

Meanwhile calculations had begun of individual hadronic neutral-current processes such as elastic ν_μ-proton scattering (Weinberg, 1971[265]). The experimental tasks were eased when it was realized that the ratio of inclusive $\Delta S=0$ cross sections $\sigma(\nu_\mu + T \to \nu_\mu + X)/\sigma(\nu_\mu + T \to \mu + X)$ had to be at least of order 20 per cent (Pais and Treiman, 1972[266]).**

The experimental search for neutral $\Delta S=0$ currents got under way in early 1972. Preliminary results reported[267] at the Batavia conference (September 1972) were discouraging. 'The upper bounds [for $\Delta S=0$ neutral currents] have recently diminished sufficiently to make serious trouble for certain models which feature neutral currents.'[268] Theorists responded by proposing alterna-

* The difficulty is that the presence of anomalies of the kind encountered in Eq. (21.31) spoil renormalizability. Eq. (21.113) forces anomaly cancelation in $SU(2)\times U(1)$.

** T = isoscalar target. The bound $|\theta_W|<35°$, then current, was assumed.

tive unification schemes based either on other gauge groups, or on other particle contents, or on both, in which neutral currents were either suppressed or absent.[269]

In early July 1973 the first sighting of a neutral-current event was reported[270] by a bubble chamber group working at CERN: one case of elastic ν_μe-scattering. Three weeks later the same group communicated[271] their observations of over a hundred inclusive hadronic events. Meanwhile, a neutral-current experiment in progress at Fermilab caused much uncertainty among its participants about background and cuts in the data, and hence about the interpretation of their results. Those were the days of the 'alternating neutral current'. Lengthy internal debates followed, accompanied by informal exchanges with CERN. Eventually it was decided to redesign the experiment. Galison has given a documented account[250] of these events, leading up to the publication[272] of the positive Fermilab results in 1974.

At the Bonn conference (August 1973) the situation was summarized[273] like this: 'There is now, for the first time, positive evidence for neutral currents in neutrino reactions.' At the London conference (August 1974) more data were reported, and a value

$$\sin^2 \theta_W = 0.39 \pm 0.05 \quad (1974) \tag{21.114}$$

was quoted. The rapporteur concluded[274] as follows: 'Overall conclusions: the field of neutrino physics is completely open and essentially unexplored!'

Thus did the neutral current, one of the fundamental scientific contributions of the twentieth century, enter physics. The subsequent maturation of the subject can be followed in the quoted reviews.[248] By 1981, θ_W had settled down to[248]

$$\sin^2 \theta_W \approx 0.22. \tag{21.115}$$

What is even more striking than the precision reached by then is the consistency between distinct types of experiments. These include the beautiful results[275] obtained at SLAC on asymmetries in longitudinally-polarized electron scattering off deuterium.*

We have now reached the summer of 1974, in respect of both strong and weak interactions. QCD could lay claim to asymptotic freedom as the explanation of scaling; and the confinement conjecture. Electroweak theory had the neutral current to its credit. Charm was still a guess.

Further progress lay just ahead.

6. Years of synthesis; the detection of charm, bottom, and, perhaps, top, of a new lepton, and of jets. In November 1974, a period in particle physics began of a kind rarely witnessed earlier in the post-war era. It was not just the rapid

* For reviews, including parity violation effects in atomic physics, see Ref. 276.

sequence of spectacular experimental discoveries that, by itself, made the difference. Above all, the novelty lay in the intensity and immediacy of the interplay between experimental progress and theoretical predictions based on fundamental dynamical principles rather than, as in earlier years, on phenomenological regularities ingeniously guessed at.

It is par for the course that this period would be preceded by months of great confusion, caused this time by the behavior of the cross section ratio

$$R(E)=\frac{\sigma(\mathrm{e}^+\mathrm{e}^-\rightarrow\gamma\rightarrow\text{hadrons})}{\sigma(\mathrm{e}^+\mathrm{e}^-\rightarrow\gamma\rightarrow\mu^+\mu^-)} \tag{21.116}$$

as observed in $\mathrm{e}^+\mathrm{e}^-$-colliders with energy E per beam. The data, reviewed by Richter[277] at the 1974 London conference, were to such a degree at variance with theoretical expectations that theorists spoke[278] of 'the gravity of this theoretical debacle'. Twenty-three predictions for R were presented at the meeting.[278] None of them fitted the facts. Let us see what experiment had revealed, and why theorists were genuinely worried.

During the preceding few years the favored theoretical anticipations about R had been linked to some variant or other of the quark–parton picture: $\mathrm{e}^+\mathrm{e}^-$ annihilates into a single virtual photon with mass $2E$ which turns into a quark–antiquark pair $\bar{q}_i q_i$ with quark charge Q_i and quark spin 1/2. Thereupon $\bar{q}_i q_i$ converts into conventional hadrons by complicated secondary dynamical processes. If E is large compared with quark masses, then

$$\sigma(\mathrm{e}^+\mathrm{e}^-\rightarrow\gamma\rightarrow\bar{\mathrm{q}}_i\mathrm{q}_i)=\frac{\pi\alpha^2}{3E^2}Q_i^2, \tag{21.117}$$

just the expression for $\mu^+\mu^-$, where $Q_i=1$. Thus it was thought that, for large E, R would tend to a constant[279]

$$R(E)=\sum_i Q_i^2 \tag{21.118}$$

where the summation extends over quarks with mass $m_i<E$. The value of R further depends on the choice of quark model. Three quark species without color give $R=(2/3)^2+(1/3)^2+(1/3)^2=2/3$; including color[280]: $R=3\times2/3=2$; including the color-carrying charmed quark:[280] $R=2+3\times4/9=10/3$. Eq. (21.118) was later derived[281] from QCD, including a correction factor $(1+\alpha_c/\pi)$; cf. Eq. (21.76).

The first data on R in the GeV region, taken at Frascati, Orsay, and Novosibirsk,* showed that R equals about $2\frac{1}{2}$ for $E\lesssim1\frac{1}{2}$ GeV, reasonably close to expectations for three colored quarks. The first indications that something odd was happening came from CEA, the collider at Cambridge (Mass.): $R=4.7\pm1.1$ and 6.0 ± 1.5 at $E=2$, $2\frac{1}{2}$ GeV respectively.[283] The general reac-

* For complete references see Ref. 282.

tion was something like: let us wait for results from Spear at SLAC which has a 200-times higher reaction rate and a far more advanced detector.

The driving force behind the successful completion of Spear (= Stanford positron electron asymmetric ring) was in the first instance Burt Richter (with strong support from Panofsky) who, since the late fifties,[123] had been pushing first e^-e^-, then e^+e^- colliders at Stanford.* The first formal proposal to the AEC for building Spear dates from 1964. Funds were not made available until 1970. (Richter speaks with some feeling about the intervening years.) Spear was ready to go in 1972, at $2\frac{1}{2}$ (later upgraded to 4) GeV per beam.

In 1973, a SLAC–Lawrence Berkeley Laboratory team initiated a program for measuring R by increasing E in 100 MeV steps from 1.2 to 2.4 GeV. First results, available by late 1973, were refined and included in Richter's London review.[277] The CEA points were confirmed; above about 2 GeV total energy R keeps rising, linearly on the face of it. That behavior was the London 'debacle'.

The understanding of what was really happening began in the autumn of 1974 when re-analysis of the Spear data showed that an anomalously high value for R at $2E = 3.1$ GeV did not want to go away. Further measurements near that point led to a startling conclusion.[285] There was a giant and very narrow resonance centered at 3.1 GeV!

Amazingly, the same result had been found at practically the same time in a quite different experiment. Sam Ting, proceeding on the strong hunch that there should exist other, more massive, vector meson resonances of the type** ρ, ω, ϕ, had assembled a group at MIT for the purpose of pursuing this idea. The strategy was to look for peaks in the distribution of the total energy of e^+e^- pairs produced in hadronic collisions, the peak being a signature for a resonance decaying into e^+e^-. The process $p + Be \rightarrow e^+ + e^- + X$, studied at the AGS in Brookhaven, revealed a huge peak[286]—at 3.1 GeV!

Formal announcement of the two independent discoveries was made on 11 November 1974. For details of the events preceding that date see the recollections of Ting,[287] Gerson Goldhaber,[288] and Richter.[289] Two names for the resonance were proposed: J (MIT) and ψ (SLAC–LBL). Eventually the name J/ψ was adopted, bringing into physics a symbol that had first appeared much earlier on the label of Valdespino, the distinguished Spanish sherry. The J/ψ was rapidly confirmed at Frascati[290] and at DESY (Hamburg).[291] Ten days after the unveiling of the J/ψ a second major and very narrow resonance with mass 3.695 GeV, named ψ', was found[292] at SLAC. Evidently the linear rise of R had been faked by the two resonances.

The general pandemonium following these discoveries compares only, in my experience, to what happened during the parity days of late 1956. By 1974,

* A very useful review of e^+e^--colliders up to mid-1976, including a detailed listing of earlier reviews, is found in Ref. 284.
** Chapter 19, Section (e).

resonances were of course known aplenty, but here was something new. A run of the mill 3 GeV hadronic resonance could be expected to have a width of several hundred MeV—vastly larger than the widths of J/ψ and ψ', since found to be $\simeq 0.06$ and 0.2 MeV respectively. What was going on? Theorists delved in their grab bag of exotica. I recall early discussions on three options: it was a weakly interacting Z-like boson; or a resonance carrying free color; or charmonium.

The answer was charmonium.

Shortly before the discoveries just recounted, an influential review[293] had been prepared on the properties of as yet hypothetical hadrons that contain charmed quarks c, or their antiparticles $\bar{c}$, or both, on the continued assumption that all quark structures are $\bar{q}q$ for mesons, qqq for baryons. New spectroscopy and decay schemes were foretold, essentially correctly as it would turn out. As to mesons, one anticipated either 'hidden charm': $c\bar{c}$, or 'naked charm': $\bar{d}c$, later called D^+ (spin 0) or D^{*+} (spin 1); $\bar{u}c$ (D^0, D^{*0}), $\bar{s}c$ (F^+, F^{*+}), and their antiparticles. It was argued qualitatively that the width of a $c\bar{c}$-resonance with spin parity 1^- must be narrow if its mass is less than twice that of naked charm mesons.

Also conceived prior to the J/ψ was the idea of treating the bound $c\bar{c}$ system, for which the name charmonium was proposed, by using QCD (Appelquist and Politzer[294]). The c mass, m_c, was assumed to be large (correctly; it is ~ 1500 MeV, about half the J/ψ mass), hence their Compton wavelength is small. Since binding holds particles at distances small compared to that length, it was argued that one is in the short-distance QCD régime, $\Lambda/m_c \ll 1$ (see Eq. (21.76)), so that *perturbative* QCD should apply. Furthermore the bound particles move with low mean velocity, since they are heavy, so that *non-relativistic* quantum mechanics should apply. In that situation QCD yields a potential that is Coulomb-like at short distances, $V(r) = -4\alpha_c/3r$, with a relatively small running constant α_c. Thus we have a hydrogen spectrum! Indeed, it has turned out that, in standard atomic notation, J/ψ and ψ' are the 1^3S_1 and 2^3S_1 states respectively of $c\bar{c}$. Additional QCD arguments led to narrow widths for these states.

The year 1974 was not over before it was argued that $V(r)$ could not possibly be Coulombic, since that potential does not confine. By more or less plausible arguments it was proposed[295] to use

$$V(r) = -\frac{4\alpha_c}{3r} + \sigma r, \tag{21.119}$$

α and σ to be fixed by experiment.

That model has successfully passed severe tests. Nine charmonium states are known at this time, the additional seven having been discovered either at SLAC or DESY (Hamburg).[296]* Some refinement of Eq. (21.119) has been

* Charmonium reviews: Refs. 296, 297.

found necessary (spin-dependent terms), but on the whole the picture works well. Charmonium comes in fact close to being the hydrogen atom of QCD. That analogy should not be exaggerated, however: the structure σr of the confining potential, a fair guess, is unsupported by proof; and no precision comparable to hydrogen levels has been reached, nor is this expected to happen soon. What is overridingly important in this development is its synthetic nature. Charmonium strongly reinforced belief in QCD *and* in electroweak theory. Note in particular that the magnitude $m_c \simeq 1500$ MeV is of just the right order to justify GIM suppression, Eq. (21.112). Accordingly, in 1975, the inner consistency of SU(2)×U(1) with four quarks earned it the name 'standard model'.[298]

Charm, naked charm that is, had still not been seen, however. The story of its discovery is somewhat complex. It is possible that a cosmic ray event registered[299] in 1971 contains a charmed particle with mass ~2–3 GeV. The most plausible explanation of a bubble chamber event found[300] at Brookhaven is the decay of a charmed baryon (1975). Further indications came from a series of neutrino experiments[301] that revealed an anomaly in the reactions ν_μ (or $\bar{\nu}_\mu$) + target $T \to \mu^+ + \mu^- + X$: the dimuon rate could not be explained by conventional mechanisms. Two possibilities remained. Either $\nu_\mu + T \to L^0 + X$, where L^0 is a neutral heavy lepton decaying into $\mu^+ + \mu^-$; or $\nu_\mu + T \to \mu^- + C + X$, where C is a new kind of hadron decaying into $\mu^+ + \cdots$. The L^0 option could be eliminated by a clever theoretical argument.[302] The C option survived. It was of course tempting to identify C with a charmed hadron. As a result, the *New York Times* of 1 January 1976, carried a front-page article announcing that charm had been discovered both at Fermilab and at CERN. That conclusion was tentative, however (though later it proved to be correct[303]). As yet it was neither possible to assign a definite mass value to C nor to show that the principal criterion: naked charm decays preferentially* into channels carrying non-zero strangeness was satisfied.[304]

All eyes were now on Spear. In 1975 a report came:[305] no evidence. Gerson Goldhaber has told me what happened next. In 1976 he went to a conference in Wisconsin where theoreticians told him that charm was inevitable. He was so impressed that, upon returning to SLAC, he and his co-workers dropped everything else. They found it: neutral D's in June,[306] charged D's in July.[307] Charmed hadron spectroscopy was on its way.** So was the development of needed new technology for measuring lifetimes $\sim 10^{-13}$ sec, typical for weak decays of charmed states.[310]

I conclude with a telegram-style account of most recent developments which it would be inexcusable to omit, and premature to expatiate upon in a book of this kind.

* That follows from Eq. (21.111) according to which the transition rates $c \to s$ and $c \to d$ are in the ratio $\cot^2 \theta \simeq 14$ see Eq. (21.20).
** Reviews: Refs. 308, 309.

(α) Having passed the wobbly threshold region for naked charm production, R settled down to a fairly constant behavior, as hoped, at a value of about 5, which, however, was noticeably higher than the 10/3 expected for four colored quarks. The resolution came when Perl and coworkers at SLAC made the entirely unanticipated discovery[311] of a new charged lepton, baptized τ. That particle adds one unit to R. The firm identification of τ as a lepton took some doing, especially because it arrived before the sought-for $D^{\pm}$. Because of its large mass (1784 MeV) the τ exhibits not only leptonic decays ($\tau^+ \to e^+ \nu_e \bar{\nu}_\tau$, $\mu^+ \nu_\mu \bar{\nu}_\tau$) but also decays $\tau^+ \to \bar{\nu}_\tau +$ hadrons ($\pi, \rho, K, \cdots$). The mass of the associated neutrino is not yet well known (at the time of writing $\lesssim 143$ MeV).[312] In the context of SU(2)×U(1), (τ, ν_τ) are treated as (e, ν_e), (μ, ν_μ); see Eq. (21.100).*

(β) The arrival of τ upset the equation of balance (21.113) needed for renormalizable SU(2)×U(1). Relief was forthcoming. In mid-1977, Lederman and his group at Fermilab announced[314] the discovery of a new type of resonance, the Υ, found in the reaction $p + (\mathrm{Cu}, \mathrm{Pt}) \to \mu^+ + \mu^- + X$: The (μ^+, μ^-) mass distribution peaked sharply at about 9.5 GeV. Two months later the same group[315] had resolved their peak into three resonances: Υ (9.4), Υ' (10.0), and Υ'' (10.4). The family has now ten members[316] and keeps growing even as these lines are written.[317] These further levels were found at the e^+e^- colliders at DESY, and CESR at Cornell.**

The Υ's are 'hidden bottom' states $b\bar{b}$ where b, for bottom, is a new quark species with $Q = -1/3$. The corresponding spectroscopy, richer and simpler than for ψ's because the b-mass ($\simeq 5$ GeV) is much larger than m_c, is again well described by the potential model Eq. (21.119).[297] The first evidence for 'naked bottom' B-mesons was obtained[319] in 1980 at CESR.

In the spirit of SU(2)×U(1) it is to be expected that the b has a brother, t for top, with $Q = 2/3$. States containing t have perhaps been seen at the CERN $p\bar{p}$ collider.[320] The presence of (b, t) quarks carrying color would restore Eq. (21.113).

(γ) As early as 1972, Kobayashi and Maskawa observed[321] that SU(2)×U(1) with three quark pairs and with Higgs fields as in Eq. (21.85) offers an interesting way of introducing CP-violation in the theory. The reason is that with more quarks one will have to replace Eqs. (21.108), (21.111) for Q_1, Q_2 by

$$Q_1 = \begin{pmatrix} u^L \\ \alpha_{11} d^L + \alpha_{12} s^L + \alpha_{13} b^L \end{pmatrix}, \qquad Q_2 = \begin{pmatrix} c^L \\ \alpha_{21} d^L + \alpha_{22} s^L + \alpha_{23} b^L \end{pmatrix},$$
$$Q_3 = \begin{pmatrix} t^L \\ \alpha_{31} d^L + \alpha_{32} s^L + \alpha_{33} b^L \end{pmatrix} \tag{21.120}$$

* Reviews of τ-physics: Ref. 313. ** Reviews of Υ-states: Ref. 318.

where α_{ij} is a unitary matrix satisfying $\alpha^\dagger\alpha = 1$. The α_{ij} are complex numbers. After their phases are absorbed as much as possible in redefinitions of d, s, b, one phase remains along with three mixing angles—instead of one mixing angle (Cabibbo) and no phase in the earlier 4-quark case. The presence of the phase signals *CP*-violation! This is an elegant, perhaps correct, but not unique way of introducing the mysterious *CP*-violation into electroweak theory. It does not give any clue, however, to the magnitude of the *CP*-effect. More details in Ref. 322.

(δ) In 1975, particle physics entered the jet age when it was found at Spear[323] that hadron systems produced in e^+e^- annihilation at $2E = 3.1$ and 3.7 GeV emerge more or less as two back-to-back jets—just what one would expect for the mechanism $e^+e^- \to q\bar{q}$ followed by the fragmentation of q and $\bar{q}$ into hadrons! The data analysis depends sensitively on the definability of jet axes, a delicate undertaking that is not quite unambiguous at the relatively modest quoted energies. The fragmentation step continues to be treated by empirical rules.

It was particularly exciting that the distribution in the angle θ between a jet axis and the beam direction was found to be approximately $1+\cos^2\theta$—just what one would expect for ***spin-$\frac{1}{2}$ quarks***! Jet studies have been extended in beautiful experiments at newer e^+e^- colliders: Petra at DESY* which began operations in late 1978, and which has reached energies $2E$ over 20 GeV; and Pep at SLAC (1980, $14\frac{1}{2}$ GeV).** At these higher energies jets are better defined. Among further results was the proof (from correlation effects between jets) that the primary q, $\bar{q}$ are electrically charged;[325] and that R behaves reasonably. Three jet events, first seen[324] at Petra in 1979, have been found by the thousands at Petra and Pep.

Jets are a boon to QCD according to which three-jet events are caused by gluon *Bremsstrahlung* of either q or $\bar{q}$ followed by gluon $\to$ hadrons. Jet phenomena provide numerous QCD tests; for example the ratio of three- jet to two-jet events gives a measure for α_c. Within the limits of accuracy, the results are consistent with predictions.†

(ε) After 'a long, patient effort which extended over a full decade',[126] conclusive proof was established at CERN for jets in hadronic processes, produced at the ISR. These are of two kinds: π-jet events in which one high p_T pion (see the preceding section (c)) and a jet go in opposite directions; and jet–jet events. Compelling evidence for a two-jet structure dates from 1976. 'It was, however, realized that triggering on a single high p_T particle was inefficient . . . It was only by 1982 that calorimetric studies [which collect the full jet] could claim evidence for jet dominance in processes with large transverse energy and

* For a review of Petra physics see Ref. 324.
** Petra = positron electron tandem ring accelerator. Pep = positron electron project.
† Reviews: Ref. 326.

measure a jet cross section, in agreement with expectations based on QCD.'[126] Similar agreement was found for production of prompt photons at large p_T (gluon Compton effect: gluon + q → photon + q); and muon pair production ($q+\bar{q}\rightarrow\gamma\rightarrow\mu^{+}+\mu^{-}$).[326] Finally, spectacularly clean jets have been found by UA1 and UA2 at 540 GeV center of mass $p\bar{p}$ energy. Jet physics, still young, shows great promise.

I turn to the final item.

7. Years of synthesis. The detection of W and Z. In December 1984, not quite two years after I first heard (Chapter 1) that the Z had been found, van der Meer, 'a soft spoken and gifted inventor with a high order of analytical ability',[329] and Rubbia, 'an exuberant extrovert, famous in his circle for unlimited energy and enthusiasm combined with a broad ranging and deep understanding of physics',[329] lectured in Stockholm. Both men emphasized the collective nature of the efforts that had gone into overcoming many difficulties encountered on the way to readying the CERN $\bar{p}p$ collider, the huge yet delicate detectors, and the data handling systems.

The search for the W and Z had not been a surprise party. It is in fact dubious whether this enormous enterprise would have been funded and executed had there not been excellent theoretical reasons for knowing where to look: near 80 GeV for M_W, near 90 GeV for M_Z. This follows from the relations (21.103), (21.104), along with the information on θ_W from neutral currents, e.g. Eq. (21.115). Thus the existence of the W and Z with well-predicted masses was the supreme test for the consistency of the electroweak theory.

As was noted in the opening pages of this book, the search was successful. The following results were reported at Stockholm.*

$$M_W = 80.9 \pm 1.5 \pm 2.4 \qquad \text{UA1}$$

$$83.1 \pm 1.9 \pm 1.3 \qquad \text{UA2}$$

$$M_Z = 95.6 \pm 1.4 \pm 2.9 \qquad \text{UA1}$$

$$92.7 \pm 1.7 \pm 1.4 \qquad \text{UA2}$$

Also

$$\left(\frac{M_W}{M_Z \cos\theta_W}\right)^2 = 0.968 \pm 0.045,\ \text{UA1};\ 1.02 \pm 0.06,\ \text{UA2},$$

in agreement with Eq. (21.107); and

$$\sin^2\theta_W = 0.226 \pm 0.015,\ \text{UA1};\ 0.216 \pm 0.010 \pm 0.007,\ \text{UA2},$$

in agreement with neutral-current results.

* Masses in GeV. The first error is statistical, the second systematic.

Also:

The spin of W equals one.

Observed W-decay distributions are consistent with the V–A theory. (V–A cannot be distinguished from $V+A$, in the absence of polarization experiments.)

Decay modes detected (hadronic modes expressed in quark language): $W^+ \to e^+\nu_e, \mu^+\nu_\mu, \tau^+\nu_\tau, \bar{d}u, \bar{s}c$, and, perhaps, $\bar{t}b$; $Z \to e^+e^-, \mu^+\mu^-$.

Having traveled from Roentgen to Rubbia, we have now arrived at the innermost point, about 10^{-16} cm (the Compton wavelengths of W and Z) to which, at this time, theory and experiment have jointly penetrated. Since it has been my intent to move inward as far as possible, but not further than where theory and experiment confirm each other, this book is therefore finished—but not, of course, the inward bound journey called particle physics.

Sources

Reminiscences. Experimental: Reines on the e-neutrino;[103] M. Schwartz on neutrino beams;[330] Panofsky on the Berkeley linac[331] and on SLAC;[110] Ting,[287] G. Goldhaber,[288] and Richter[289] on the J/ψ; Lederman[332] on the Υ. Theoretical: Gell-Mann,[19] Néeman,[20] and Speiser[21] on SU(3); Gell-Mann,[19] and Zweig[38] on quarks; Yang[172] on the Yang–Mills theory; Veltman[185] on renormalizability of Yang–Mills theories; Weinberg,[232] Salam,[233] Glashow,[260] and B. W. Lee[333] on electroweak theory.

References*

1. L. B. Okun, *Proc. int. conf. high energy physics* 1962, p. 845, CERN, Geneva 1962.
2. J. Lach and L. Pondrom, *ARNP* **29**, 203, 1969; M. Bourquin and J. P. Repellin, *Phys. Rep.* **114**, 99, 1984.
3. *Rev. Mod. Phys.* **56**, No. 2, Part II, April 1984.
4. M. Alston *et al.*, *Phys. Rev. Lett.* **5**, 520, 1960.
5. M. Alston *et al.*, *Phys. Rev. Lett.* **6**, 300, 1961.
6. M. Alston *et al.*, *Phys. Rev. Lett.* **6**, 698, 1961.
7. G. M. Pjerrou *et al.*, *Phys. Rev. Lett.* **9**, 114, 1962; L. Bertanza *et al.*, *Phys. Rev. Lett.* **9**, 180, 1962.
8. Cf. A. C. Irving and R. P. Worden, *Phys. Rep.* **34**, 118, 1977, Fig. 2.1; J. G. Hey and R. L. Kelly, *Phys. Rep.* **96**, 71, 1983, Figs. 3.2 and 3.3.
9. M. Gell-Mann and Y. Ne'eman, *The eightfold way*, Benjamin, New York 1964.
10. B. d'Espagnat and J. Prentki, *Progr. elem. particle and cosmic ray phys.*, Vol 4, Part 3, 1958; G. Morpurgo, *ARN* **11**, 41, 1961, Section 4; A. Pais, *Rev. Mod. Phys.* **33**, 493, 1961.

* In these references *ARN*(*P*) stands for *Annual Reviews of Nuclear* (*and Particle*) *Sciences*.

11. Cf. G. Racah, *Phys. Rev.* **61**, 186, 1941; **62**, 438, 1942; **63**, 367, 1943; **76**, 1352, 1949; A de Shalit and I. Talmi, *Nuclear shell theory*, Academic Press, New York 1963.
12. G. Racah, *Group theory and spectroscopy*, Mimeographed notes, Institute for Advanced Study, 1951; repr. in *Springer Tracts in modern physics*, Vol. 37, p. 28, Springer, New York 1965.
13. A. Pais, in *Spectroscopic and group theoretical methods in physics*, Ed. F. Bloch, p. 317, North-Holland, Amsterdam 1968.
14. R. E. Behrends and A. Sirlin, *Phys. Rev.* **121**, 324, 1961.
15. S. Sakata, *Progr. Th. Phys.* **16**, 686, 1959.
16. M. Ikeda, S. Ogawa, and Y. Ohnuki, *Progr. Theor. Phys.* **22**, 715, 1959.
17. Y. Yamaguchi, *Progr. Th. Phys.* 1959, Suppl. No. 11.
18. J. Wess, *Nuovo Cim.* **15**, 52, 1960.
19. M. Gell-Mann, in *Symmetries in physics*, Proc. Conf. held at San Felia de Guixols, Spain, 1983, forthcoming.
20. Y. Ne'eman, in Ref. 19.
21. D. Speiser, in Ref. 19; cf. also T. D. Lee and C. N. Yang, *Phys. Rev.* **122**, 1960, 1961.
22. M. Gell-Mann, *The eightfold way*, Caltech Report CTSL-20 (1961), repr. in Ref. 9, p. 11.
23. Y. Ne'eman, *Nucl. Phys.* **26**, 222, 1961, repr. in Ref. 9, p. 58.
24. D. Speiser and J. Tarski, *J. Math. Phys.* **4**, 588, 1963.
25. S. Okubo, *Progr. Theor. Phys.* **27**, 949, 1962; **28**, 24, 1962.
26. R. E. Behrends, J. Dreitlein, C. Fronsdal, and W. Lee, *Rev. Mod. Phys.* **34**, 1, 1962.
27. Cf. G. A. Snow, Ref. 1, p. 795.
28. G. M. Pjerrou *et al.*, Ref. 1, p. 289.
29. M. Gell-Mann, Ref. 1, p. 805, repr. in Ref. 9, p. 87.
30. Ref. 3, p. S134.
31. V. E. Barnes *et al.*, *Phys. Rev. Lett.* **12**, 204, 1964, repr. in Ref. 9, p. 88.
32. G. F. Chew, M. Gell-Mann, and A. H. Rosenfeld, *Sci. Am.* **210**, February 1964, p. 74.
33. S. D. Protopopescu and N. P. Samios, *ARNP* **29**, 339, 1979.
34. J. J. J. Kokkedee, *The quark model*, Benjamin, New York 1969.
35. G. Morpurgo, *ARN* **20**, 105, 1970; H. J. Lipkin, *Phys. Rep.* **8**, 173, 1973.
36. M. Gell-Mann, *Phys. Lett.* **8**, 214, 1964, repr. in Ref. 9, p. 168.
37. G. Zweig, CERN preprint 8182/Th 401, January 1964, unpublished; CERN preprint 8419/Th412, February 1964, unpublished, but repr. in D. B. Lichtenberg and S. P. Rosen, *Developments in the quark theories of hadrons*, Vol. 1, p. 22, Hadronic Press, Nonamtum, Mass. 1980.
38. G. Zweig, *Proc. fourth int. conf. on baryon resonances*, Ed. N. Isgur, p. 439, Univ. Toronto Press 1980.
39. Ref. 3, pp. S136–138.
40. R. H. Dalitz, *Proc. int. conf. high energy physics*, 1966, p. 215, Univ. California Press 1967.
41. G. Morpurgo, *ARN* **20**, 105, 1970.
42. N. Isgur and G. Karl, *Physics Today* **36**, November 1983, p. 36.
43. L. Montanet, G. C. Rossi, and G. Veneziano, *Phys. Rep.* **63**, 149, 1980.
44. Ref. 33, p. 340.

45. F. J. Dyson, *Symmetry groups in nuclear and particle physics*, Benjamin, New York 1966.
46. A. Pais, *Rev. Mod. Phys.* **38**, 215, 1966.
47. M. Gell-Mann, *Physics* **1**, 63, 1964, repr. in Ref. 9, p. 172.
48. P. Franzini and L. Radicati, *Phys. Lett.* **6**, 322, 1963, repr. in Ref. 45, p. 154.
49. F. Gürsey and L. Radicati, *Phys. Rev. Lett.* **13**, 173, 1964, repr. in Ref. 45, p. 159.
50. B. Sakita, *Phys. Rev.* **136B**, 1756, 1964, repr. in Ref. 45, p. 168.
51. G. Zweig, in *Symmetries in elementary particle physics*, Ed. A. Zichichi, p. 219, Academic Press, New York 1965.
52. Ref. 46, Section 3.
53. L. Bertanza *et al.*, Ref. 7.
54. A. Pais, *Phys. Rev. Lett.* **13**, 175, 1964, repr. in Ref. 45, p. 162; cf. further M. A. B. Bég and V. Singh, *Phys. Rev. Lett.* **13**, 418, 1964.
55. A. Pais, Ref. 54; see further F. Gürsey, A. Pais, and L. Radicati, *Phys. Rev. Lett.* **13**, 299, 1964, repr. in Ref. 45, p. 165.
56. M. A. B. Bég, B. W. Lee, and A. Pais, *Phys. Rev. Lett.* **13**, 514, 1964.
57. B. Sakita, *Phys. Rev. Lett.* **13**, 643, 1964.
58. V. V. Anisovich *et al.*, *Phys. Lett.* **16**, 194, 1965.
59. H. J. Lipkin and S. Meshkov, *Phys. Rev. Lett.* **14**, 670, 1965.
60. K. Johnson and S. B. Treiman, *Phys. Rev. Lett.* **14**, 189, 1965.
61. Ref. 49, esp. footnote 5.
62. Ref. 45, paper 3, 18–25; Ref. 46, Section 7.
63. O. W. Greenberg and C. A. Nelson, *Phys. Rep.* **32C**, 70, 1977.
64. H. Bacry, J. Nuyts, and L. van Hove, *Phys. Lett.* **9**, 279, 1964, repr. in Ref. 34, p. 120.
65. For references see Ref. 46, p. 237.
66. O. W. Greenberg, *Phys. Rev. Lett.* **13**, 598, 1964, repr. in Ref. 34, p. 122.
67. G. Gentile, *Nuov. Cim.* **17**, 493, 1940.
68. A. Sommerfeld, *Ber. Deutsch. Chem. Ges.* **75**, 1988, 1943.
69. M. Y. Han and Y. Nambu, *Phys. Rev.* **139B**, 1006, 1965, repr. in Ref. 34, p. 127.
70. A. Pais, in *Recent developments in particle symmetries*, Ed. A. Zichichi, p. 406, Academic Press, New York 1966.
71. O. W. Greenberg and D. Zwanziger, *Phys. Rev.* **150**, 1177, 1966.
72. S. L. Adler and R. F. Dashen, *Current algebras*, Benjamin, New York 1968.
73. J. D. Bjorken and M. Nauenberg, *ARN* **18**, 229, 1968.
74. B. Renner, *Current algebras*, Pergamon, Oxford 1968.
75. S. B. Treiman, R. Jackiw, and D. J. Gross, *Lectures on current algebra*, Princeton Univ. Press 1972.
76. Cf. M. Gell-Mann, *Proc. int. conf. high energy physics*, 1958, p. 260, CERN, Geneva 1958.
77. R. P. Feynman and M. Gell-Mann, *Phys. Rev.* **109**, 193, 1958.
78. F. Crawford *et al.*, *Phys. Rev. Lett.* **1**, 377, 1958; P. Norden *et al.*, *Phys. Rev. Lett.* **1**, 380, 1958.
79. Cf. R. P. Feynman, *Proc. int. conf. high energy physics*, 1960, p. 501, Interscience, New York 1960.
80. N. Cabibbo, *Phys. Rev. Lett.* **10**, 531, 1963, repr. in Ref. 9, p. 207.
81. Cf. N. Cabibbo, *Proc. int. conf. high energy physics*, 1966, p. 29, Univ. California Press 1967; L. M. Chounet, J. M. and M. K. Gaillard, *Phys. Rep.* **4**, 199, 1972.

82. C. S. Wu and S. A. Moszkowski, *Beta decay*, Chapter 7, Interscience, New York 1966.
83. D. H. Whyte and D. Sullivan, *Physics Today* **32**, April 1979, p. 40.
84. A. I. Vainshtain, V. I. Zakharov, and M. A. Shifman, *Soviet Phys. JETP* **45**, 670, 1977.
85. M. Gell-Mann, *Phys. Rev.* **125**, 1067, 1962, repr. in Ref. 9, p. 216.
86. S. Okubo, *Progr. Theor. Phys. Suppl.* **37**, 114, 1966.
87. H. J. Lipkin and F. Scheck, *Phys. Rev. Lett.* **16**, 71, 1966.
88. Cf. G. Morpurgo, *Physics* **2**, 95, 1965, repr. in Ref. 34, p. 132.
89. J. G. Hey and R. L. Kelly, Ref. 8, pp. 73, 193.
90. C. E. De Tar and J. F. Donoghue, *ARNP* **33**, 235, 1983.
91. Cf. V. Zakharov, in *High energy physics 1980*, AIP conf. proc. No. 68, p. 1235, AIP, New York 1981.
92. Cf. F. Seitz, *The modern theory of solids*, McGraw-Hill, New York 1940.
93. Cf. J. Gasser and H. Leutwyler, *Phys. Rep.* **87**, 77, 1983; also H. J. Lipkin, Ref. 38, p. 461.
94. S. L. Adler, *Phys. Rev. Lett.* **14**, 1051, 1965; *Phys. Rev.* **140B**, 736, 1965, repr. in Ref. 72, p. 87.
95. W. I. Weisberger, *Phys. Rev. Lett.* **14**, 1047, 1965; *Phys. Rev.* **143**, 1302, 1966, repr. in Ref. 72, p. 100.
96. B. W. Lee, *Chiral dynamics*, Gordon and Breach, New York 1972.
97. H. R. Pagels, *Phys. Rep.* **16**, 219, 1975.
98. C. G. Callan and S. B. Treiman, *Phys. Rev. Lett.* **17**, 336, 1966, repr. in Ref. 72, p. 197.
99. J. Schwinger, *Phys. Rev. Lett.* **3**, 296, 1959, repr. in Ref. 72, p. 235.
100. R. Jackiw, D. J. Gross, contributions to Ref. 75.
101. S. L. Adler, in *Lectures on elementary particles in quantum field theory*, Brandeis 1970, p. 1, M.I.T. Press, Cambridge, Mass. 1970; R. Jackiw, in Ref. 75.
102. F. Reines, *ARN* **10**, 1, 1960.
103. F. Reines, in 'Proc. int. colloq. history of particle phys.', *J. de Phys.* **43**, suppl. to No. 12, p. C8–237, 1982.
104. F. Reines and C. L. Cowan, *Phys. Rev.* **92**, 830, 1953.
105. C. L. Cowan *et al.*, *Science* **124**, 103, 1956.
106. M. Schwartz, *Phys. Rev. Lett.* **4**, 306, 1960; cf. also B. Pontecorvo, *Soviet Phys. JETP* **10**, 1236, 1960.
107. G. Feinberg, *Phys. Rev.* **110**, 1482, 1958.
108. E. J. Mahmoud and H. M. Konopinski, *Phys. Rev.* **92**, 1045, 1953; J. Schwinger, *Ann. of Phys.* **2**, 407, 1957.
109. G. Danby *et al.*, *Phys. Rev. Lett.* **9**, 36, 1962.
110. W. K. H. Panofsky, *Physics Today* **36**, October 1983, p. 34.
111. R. Neal (Ed.), *The Stanford two mile accelerator*, Benjamin, New York 1968.
112. *Linear accelerators*, Eds. P. M. Lapostolle and A. L. Septier, North-Holland, Amsterdam 1970.
113. D. H. Sloan and E. O. Lawrence, *Phys. Rev.* **38**, 2021, 1931; D. H. Sloan and W. M. Coates, *Phys. Rev.* **46**, 539, 1934.
114. L. W. Alvarez *et al.*, *Rev. Sci. Instr.* **26**, 111, 1955.
115. G. Lubkin, *Physics Today* **30**, February 1977, p. 17.
116. E. L. Ginzton, W. W. Hansen, and W. R. Kennedy, *Rev. Sci. Instr.* **19**, 89, 1948; M. Chodorow *et al.*, *Rev. Sci. Instr.* **26**, 134, 1955; E. L. Ginzton and W. Kirk, *Sci. Am.* **205**, November 1961, p. 49.

117. E. L. Ginzton, *Sci. Am.*, **190**, March 1954, p. 84.
118. W. K. H. Panofsky, *Proc. int. high energy conf. Vienna, 1968*, p. 23, CERN, Geneva 1968.
119. E. D. Bloom *et al.*, *Phys. Rev. Lett.* **23**, 930, 1969; M. Breidenbach *et al.*, *Phys. Rev. Lett.* **23**, 935, 1969.
120. H. W. Kendall and W. K. H. Panofsky, *Sci. Am.* **224**, June 1971, p. 61.
121. D. W. Kerst *et al.*, *Phys. Rev.* **102**, 590, 1956.
122. Cf. C. Bernardini, *Scientia* **113**, 27, 1978.
123. W. C. Barber *et al.*, *Phys. Rev. Lett.* **16**, 1127, 1966.
124. Cf. W. J. Willis, *Physics Today* **31**, October 1978, p. 32.
125. R. D. Kohaupt and G. A. Voss, *ARNP* **33**, 67, 1983.
126. M. Jacob and K. Johnson, CERN Rep. 84–13, LEP Division, Geneva 1984.
127. U. Amaldi *et al.*, *Phys. Lett.* **44B**, 112, 1973.
128. Cf. e.g. S. M. Berman and M. Jacob, *Phys. Rev. Lett.* **25**, 1683, 1970.
129. M. Jacob, *Proc. int. high energy physics conf., 1972*, Vol. 3, pp. 378, 427, National Accelerator Lab., Batavia, Ill. 1972.
130. F. W. Büsser *et al.*, *Phys. Lett.* **46B**, 471, 1973.
131. B. Alfer *et al.*, *Phys. Lett.* **44B**, 521, 1973; M. Banner *et al.*, *Phys. Lett.* **44B**, 537, 1973.
132. M. A. Faessler, *Phys. Rep.* **115**, 1, 1984.
133. D. B. Cline, P. McIntyre, F. Mills, and C. Rubbia, *Fermilab Report TM 687*, Chicago 1976.
134. D. B. Cline, C. Rubbia, and S. van der Meer, *Sci. Am.* **246**, March 1982, p. 48.
135. See F. Bonnaudi *et al.*, CERN report DG-2, 1977; A. Astbury *et al.*, CERN reports SPSC/78-06: SPSC/P92, 1978.
136. D. Möhl *et al.*, *Phys. Rep.* **58**, 73, 1980; popular account: D. Cline and C. Rubbia, *Physics Today* **33**, August 1980, p. 44.
137. The staff of the CERN proton–antiproton project, *Phys. Lett.* **107B**, 306, 1981.
138. A. Pais, *Ann. of Phys.* **63**, 361, 1971.
139. S. L. Adler, *Phys. Rev.* **143**, 1144, 1966.
140. J. D. Bjorken, *Phys. Rev.* **163**, 1767, 1967.
141. J. D. Bjorken, *Phys. Rev.* **179**, 1547, 1969.
141a. C. G. Callan and D. Gross, *Phys. Rev. Lett.* **22**, 156, 1969.
142. R. Jackiw and G. Preparata, *Phys. Rev. Lett.* **22**, 975, 1969; S. L. Adler and W. K. Tung, *Phys. Rev. Lett.* **22**, 978, 1969.
143. S. Coleman, Dilatations, in *Properties of fundamental interactions*, Ed. A. Zichichi, p. 359, Ed. Compositori, Bologna 1973.
144. *Scale and conformal invariance in hadron physics*, Ed. R. Gatto, Wiley, New York 1973.
145. H. Fritzsch and M. Gell-Mann, *Proc. Coral Gables conf. 1971*, Vol. 2, p. 1, Gordon and Breach, New York 1971; Y. Frishman, Ref. 129, Vol. 4, p. 119.
146. K. G. Wilson and J. Kogut, *Phys. Rep.* **12**, 75, 1974.
147. C. G. Callan, *Phys. Rev.* **D2**, 1541, 1970; K. Symanzik, *Comm. Math. Phys.* **18**, 227, 1970.
148. Review: C. H. Llewellyn Smith, *Phys. Rep.* **3**, 261, 1972.
149. F. J. Gilman, *Phys. Rep.* **4**, 95, 1972; J. I. Friedman and H. W. Kendall, *ARN* **22**, 203, 1972.
150. P. Musset and J. Vialle, *Phys. Rep.* **39**, 279, 1978; B. C. Barish, *Phys. Rep.* **39**, 279, 1978.

151. R. P. Feynman, articles: *Phys. Rev. Lett.* **23**, 1415, 1969, and in *High energy collisions*, Ed. C. N. Yang, p. 237, Gordon and Breach, New York 1969; book: *Photon–hadron interactions*, Benjamin, New York 1972; popular account: *Science* **183**, 601, 1974.
152. Review: J. Kogut and L. Susskind, *Phys. Rep.* **8**, 75, 1973.
153. J. D. Bjorken and E. A. Paschos, *Phys. Rev.* **D1**, 3151, 1970.
154. J. D. Bjorken and E. A. Paschos, *Phys. Rev.* **185**, 1975, 1969.
155. Cf. e.g. J. Kuti and V. Weisskopf, *Phys. Rev.* **D4**, 3418, 1971.
156. D. J. Gross and C. H. Llewellyn Smith, *Nucl. Phys.* **B14**, 337, 1969.
157. P. V. Landshoff and J. C. Polkinghorne, *Phys. Rep.* **5**, 1, 1972; T. M. Yan, *ARN* **26**, 199, 1976.
158. S. Kamefuchi, *Prog. Theor. Phys.* **6**, 175, 1951.
159. T. D. Lee, M. Rosenbluth, and C. N. Yang, *Phys. Rev.* **75**, 905, 1949.
160. R. P. Feynman and M. Gell-Mann, *Phys. Rev.* **109**, 193, 1958; E. C. G. Sudarshan and R. E. Marshak, *Phys. Rev.* **109**, 1860, 1958.
161. J. Schwinger, *Ann. of Phys.* **2**, 407, 1957.
162. See *Particle physics*, CERN report 61–30, Geneva 1961.
163. T. D. Lee and C. S. Wu, *ARN* **15**, 381, 1965, Chapter 3.
164. Cf. T. D. Lee and C. N. Yang, *Phys. Rev.* **128**, 885, 1962.
165. O. Klein, in *New theories in physics* (Warsaw Conference, 30 May–3 June 1938), p. 77, Nyhoff, The Hague 1939.
166. A. Pais, *Physica* **19**, 869, 1953.
167. W. Pauli, *Physica* **19**, 887, 1953.
168. W. Pauli and J. Solomon, *J. de Phys.* **3**, 452, 582, 1932; *Collected papers*, Vol. 2, p. 461, Interscience, New York 1964.
169. W. Pauli, letter to A. Pais, 3 July 1953.
170. See A. Pais, *Subtle is the Lord*, Chapter 17, Oxford Univ. Press 1982.
171. W. Pauli, letter to A. Pais, 6 December 1953.
172. C. N. Yang, *Selected papers 1945–1980*, p. 19, Freeman, San Francisco 1983.
173. P. Gulmanelli, *Su una teoria dello spin isotopico*, Publicazioni sezione di Milano dell'Istituto Nazionale di Fysica Nucleare, Casa Editrice Pleion, Milano, undated (probably 1954).
174. C. N. Yang and R. L. Mills, *Phys. Rev.* **95**, 631; **96**, 191, 1954.
175. R. Shaw, 'The problem of particle types and other contributions to the theory of elementary particles', Cambridge dissertation, 1955, unpublished.
176. W. Pauli, *Proc. int. conf. high energy physics 1958*, p. 122, CERN, Geneva 1958.
177. W. Heisenberg, *Physics and beyond*, Chapter 19, Harper and Row, New York 1971.
178. W. Heisenberg, *Collected works*, Eds. W. Blum, H. P. Dürr, and H. Rechenberg, Vol. C, pp. 524 ff., Springer, New York 1984.
179. W. Heisenberg, *Physics Today* **29**, March 1976, p. 32.
180. R. Utiyama, *Phys. Rev.* **101**, 1597, 1956.
181. S. L. Glashow and M. Gell-Mann, *Ann. of Phys.* **15**, 437, 1961.
182. Ref. 170, p. 281.
183. G. 't Hooft, *Nucl. Phys.* **B33**, 173, 1971.
184. C. Itzykson and J. B. Zuber, *Quantum field theory*, Chapter 12, McGraw-Hill, New York 1980.
185. M. Veltman, *Proc. 6th int. symp. electron and photon interactions at high energies, Bonn*, p. 439, North-Holland, Amsterdam 1974.
186. J. Sakurai, *Ann. of Phys.* **11**, 1, 1960.
187. A. Salam and J. Ward, *Nuovo Cim.* **19**, 167, 1961; M. Gell-Mann, Ref. 22; Y. Ne'eman, Ref. 23.

188. H. D. Politzer, *Phys. Rep.* **14**, 129, 1974.
189. W. Marciano and H. Pagels, *Phys. Rep.* **36**, 138, 1978.
190. F. Wilczek, *ARNP* **32**, 177, 1982.
191. P. Söding and G. Wolf, *ARNP* **31**, 231, 1981.
192a.D. J. Gross, S. B. Treiman, and F. Wilczek, *Phys. Rev.* **D19**, 2188, 1979.
192a. D. J. Gross, S. B. Treiman, and F. Wilczek, *Phys. Rev.* **D19**, 2188, 1979.
193. D. J. Gross and F. Wilczek, *Phys. Rev. Lett.* **30**, 1343, 1973.
194. H. D. Politzer, *Phys. Rev. Lett.* **30**, 1346, 1973.
195. S. Coleman and D. J. Gross, *Phys. Rev. Lett.* **31**, 851, 1973; also A. Zee, *Phys. Rev.* **D7**, 3630, 1973.
196. A. J. Buras, *Rev. Mod. Phys.* **52**, 199, 1980; also B. Adeva *et al.*, *Phys. Rev. Lett.* **54**, 1750, 1985.
197. S. Weinberg, *Phys. Rev. Lett.* **31**, 494, 1973.
198. D. J. Gross and F. Wilczek, *Phys. Rev.* **D8**, 3633, 1973.
199. H. Fritzsch, M. Gell-Mann, and H. Leutwyler, *Phys. Lett.* **47B**, 365, 1973.
200. H. Georgi and S. L. Glashow, *Phys. Rev. Lett.* **32**, 438, 1974.
201. Cf. D. Amati and M. Testa, *Phys. Lett.* **48**, 227, 1974; P. Olesen, *Phys. Lett.* **50B**, 255, 1974; Ref. 188, p. 154.
202. S. J. Lindenbaum, *Comm. Nucl. Ptcle Phys.* **13**, 285, 1984; P. M. Fishbane and S. Meshkov, ibid. **13**, 325, 1984.
203. M. Bander, *Phys. Rep.* **75**, 205, 1981.
204. G. 't Hooft, *Nucl. Phys.* **B75**, 461, 1974.
205. K. G. Wilson, *Phys. Rev.* **D10**, 2445, 1974.
206. G. Wentzel, *Helv. Phys. Acta.* **13**, 269, 1940; **14**, 633, 1941.
207. Reviews: C. Rebbi, *Phys. Rep.* **67**, 55, 1980; J. B. Kogut, *Phys. Rep.* **67**, 67, 1980; M. Creutz, *Phys. Rep.* **95**, 201, 1983.
208. *Gauge theory of weak and electromagnetic interactions*, Ed. C. H. Lai, World Scientific, Singapore 1981.
209. B. W. Lee, Ref. 129, Vol. 4, p. 249.
210. E. S. Abers and B. W. Lee, *Phys. Rep.* **9**, 1, 1973.
211. M. A. B. Bég and A. Sirlin, *ARN* **24**, 379, 1974.
212. M. A. B. Bég and A. Sirlin, *Phys. Rep.* **88**, 1, 1982.
213. J. C. Taylor, *Gauge theories of weak interactions*, Cambridge Univ. Press 1976.
214. T. P. Cheng and L. F. Li, *Gauge theory of elementary particle physics*, Clarendon Press, Oxford 1984.
215. S. Weinberg, *Sci. Am.* **231**, July 1974, p. 50.
216. J. Schwinger, Ref. 161, repr. in Ref. 208, p. 34; cf. also S. L. Glashow, *Nucl. Phys.* **10**, 107, 1959.
217. S. A. Bludman, *Nuovo Cim.* **9**, 433, 1958; cf. also S. L. Glashow, Ref. 216.
218. A. Salam and J. C. Ward, *Nuovo Cim.* **11**, 568, 1959.
219. S. L. Glashow, *Nucl. Phys.* **22**, 579, 1961, repr. in Ref. 208, p. 171; cf. also A. Salam and J. C. Ward, *Phys. Lett.* **13**, 168, 1964, repr. in Ref. 208, p. 181.
220. H. Georgi and S. L. Glashow, *Phys. Rev. Lett.* **28**, 1494, 1972.
221. Y. Nambu, *Phys. Rev. Lett.* **4**, 380, 1960; Y. Nambu and G. Jona Lasinio, *Phys. Rev.* **122**, 345, 1961; **124**, 246, 1961; cf. also W. Heisenberg, *Zeitschr. f. Naturf.* **14**, 441, 1959, Section 2.
222. E.g. Ref. 184, pp. 612 ff.; Ref. 214, Section 5.3.
223. J. Goldstone, *Nuovo Cim.* **19**, 154, 1961; J. Goldstone, A. Salam, and S. Weinberg, *Phys. Rev.* **127**, 965, 1962, repr. in Ref. 208, pp. 93, 118.
224. G. S. Guralnik, C. R. Hagen, and T. W. Kibble, *Adv. in Particle Phys.* **2**, 567, 1968.

225. S. Weinberg, *Phys. Rev. Lett.* **29**, 1698, 1972.
226. H. Georgi and A. Pais, *Phys. Rev.* **D12**, 508, 1975.
227. H. Georgi and A. Pais, *Phys. Rev.* **D10**, 1246, 1974.
228. P. W. Higgs, *Phys. Rev. Lett.* **12**, 132, 1964; *Phys. Rev.* **145**, 1156, 1966, repr. in Ref. 208, p. 133.
229. S. Coleman and E. Weinberg, *Phys. Rev.* **D7**, 1888, 1973.
230. S. Weinberg, *Phys. Rev. Lett.* **19**, 1264, 1967, repr. in Ref. 208, p. 185.
231. A. Salam, in *Elementary particle theory*, Ed. N. Svartholm, p. 367, Almqvist and Wiksell, Stockholm 1968, repr. in Ref. 208, p. 188.
232. S. Weinberg, *Rev. Mod. Phys.* **52**, 515, 1979, repr. in Ref. 208, p. 1.
233. A. Salam, *Rev. Mod. Phys.* **52**, 525, 1979, repr. in Ref. 208, p. 9.
234. Ref. 3, p. S293.
235. S. Coleman, *Science* **206**, 1290, 1979; D. Sullivan, D. Koester, D. H. White, and K. Kern, *Scientometrics* **2**, 309, 1980; D. Koester, D. Sullivan, and D. H. White, *Soc. St. of Sc.* **12**, 73, 1982.
236. G. 't Hooft, *Nucl. Phys.* **35**, 167, 1971.
237. B. W. Lee, Ref. 129, Vol. 4, p. 251.
238. Ref. 208, pp. 273 ff.
239. See e.g. Ref. 214, p. 417.
240. Cf. e.g. H. Fritzsch and P. Minkowski, *Phys. Rep.* **73**, 1967, 1981.
241. A. D. Linde, *JETP Lett.* **23**, 64, 1976; S. Weinberg, *Phys. Rev. Lett.* **36**, 294, 1976.
242. J. Ellis, M. K. Gaillard, G. Girardi, and P. Serba, *ARNP* **32**, 443, 1982.
243. E. Farhi and L. Susskind, *Phys. Rep.* **74**, 277, 1981; also Ref. 212, Section 7.2.3; Ref. 214, Chapter 13.
244. T. W. Appelquist, J. R. Primack, and H. R. Quinn, *Phys. Rev.* **D7**, 2998, 1973; C. G. Bollini, J. J. Giambiaggi, and A. Sirlin, *Nuovo Cim.* **16A**, 423, 1973; W. J. Marciano and A. Sirlin, *Phys. Rev.* **D8**, 3612, 1973.
245. Cf. W. J. Marciano and A. Sirlin, *Phys. Rev.* **D29**, 945, 1984.
246. S. Weinberg, *Phys. Rev. Lett.* **29**, 388, 1972; H. Georgi and S. L. Glashow, *Phys. Rev.* **D6**, 2977, 1972; **D7**, 2457, 1973; H. Georgi and A. Pais, *Phys. Rev.* **D10**, 539, 1974.
247. Ref. 212, Sec. 4.2.2.
248. P. Q. Hung and J. Sakurai, *ARNP* **31**, 375, 1981; J. E. Kim, P. Langacker, M. Levine, and H. H. Williams, *Rev. Mod. Phys.* **53**, 211, 1981; S. M. Bilenky and J. Hošek, *Phys. Rep.* **90**, 73, 1982.
249. F. Sciulli, *Progr. Ptcle Nucl. Phys.* **2**, 41, 1979.
250. P. Galison, *Rev. Mod. Phys.* **55**, 477, 1983.
251. D. B. Cline, A. K. Mann, and C. Rubbia, *Sci. Am.* **231**, December 1974, p. 108.
252. T. W. Appelquist, R. M. Barnett, and K. Lane, *ARNP* **28**, 387, 1978.
253. R. F. Schwitters, *Sci. Am.* **237**, October 1977, p. 56.
254. Cf. T. D. Lee and C. N. Yang, *Phys. Rev. Lett.* **4**, 307, 1960.
255. H. H. Bingham *et al.*, *Proc. Siena conf., 1963*, Vol. 1, p. 555, Soc. Ital. di Fysica, Bologna 1963.
256. P. Tarjanne and V. L. Teplitz, *Phys. Rev. Lett.* **11**, 447, 1963; W. Krolikowski, *Nucl. Phys.* **52**, 342, 1964; Y. Hara, *Phys. Rev.* **134B**, 701, 1963; Z. Maki and Y. Ohnuki, *Progr. Theor. Phys.* **32**, 144, 1964; J. D. Bjorken and S. L. Glashow, *Phys. Lett.* **11**, 255, 1964; D. Amati, H. Bacry, J. Nuyts, and J. Prentki, *Nuovo Cim.* **34**, 1732, 1964; L. B. Okun, *Phys. Lett.* **12**, 250, 1964.
257. J. D. Bjorken and S. L. Glashow, Ref. 256.

258. S. L. Glashow, J. Iliopoulos, and L. Maiani, *Phys. Rev.* **D2**, 1285, 1970.
259. Cf. S. Weinberg, *Rev. Mod. Phys.* **46**, 255, 1974, Section 2.
260. S. L. Glashow, *Rev. Mod. Phys.* **52**, 539, 1980.
261. S. Weinberg, *Phys. Rev. Lett.* **27**, 678, 1971.
262. S. Weinberg, *Phys. Rev.* **D5**, 1412, 1972.
263. B. W. Lee, J. R. Primack, and S. B. Treiman, *Phys. Rev.* **D7**, 510, 1972.
264. C. Bouchiat, J. Iliopoulos, and Ph. Meyer, *Phys. Lett.* **38B**, 519, 1972; D. J. Gross and R. Jackiw, *Phys. Rev.* **D6**, 477, 1972.
265. S. Weinberg, *Phys. Rev.* **D5**, 1412, 1972.
266. A. Pais and S. B. Treiman, *Phys. Rev.* **D6**, 2700, 1972; refined by E. A. Paschos and L. Wolfenstein, *Phys. Rev.* **D7**, 91, 1972.
267. D. H. Perkins, Ref. 129, Vol. 4, p. 189.
268. B. W. Lee, Ref. 129, Vol. 4, p. 255.
269. See the reviews Refs. 209, 210, 211, 259.
270. F. J. Hasert *et al.*, *Phys. Lett.* **46B**, 121, 1973.
271. F. J. Hasert *et al.*, *Phys. Lett.* **46B**, 138, 1973.
272. A. Benvenuti *et al.*, *Phys. Rev. Lett.* **32**, 800, 1974; B. Aubert *et al.*, *Phys. Rev. Lett.* **32**, 1454, 1974.
273. G. Myatt, Ref. 185, p. 389.
274. D. C. Cundy, *Proc. int. conf. high energy physics, London (1974)*, p. IV–131, Science Res. Council, London 1974.
275. C. Y. Prescott *et al.*, *Phys. Lett.* **77B**, 347, 1978.
276. E. D. Commins and P. H. Bucksbaum, *ARNP* **30**, 1, 1980; E. N. Fortson and L. L. Lewis, *Phys. Rep.* **113**, 289, 1984.
277. B. Richter, Ref. 274, p. IV–37.
278. J. D. Ellis, Ref. 274, p. IV–20.
279. N. Cabibbo, G. Parisi, and M. Testa, *Nuovo Cim. Lett.* **4**, 35, 1970.
280. W. A. Bardeen, H. Fritzsch, and M. Gell-Mann, Ref. 144, p. 139.
281. T. Appelquist and H. Georgi, *Phys. Rev.* **D8**, 4000, 1973; A. Zee, ibid., p. 4038.
282. R. F. Schwitters, *Proc. int. symp. lepton and photon physics (1975)*, p. 5, SLAC, 1975.
283. A. Litke *et al.*, *Phys. Rev. Lett.* **30**, 1189, 1973; G. Tarnopolsky *et al.*, *Phys. Rev. Lett.* **32**, 432, 1974.
284. J. P. Perez-Yorba and F. M. Renard, *Phys. Rep.* **31**, 1, 1977.
285. J. E. Augustin *et al.*, *Phys. Rev. Lett.* **33**, 1406, 1974.
286. J. J. Aubert *et al.*, *Phys. Rev. Lett.* **33**, 1404, 1974.
287. S. C. C. Ting, in *Adventures in experimental physics*, Ed. B. Maglich, **5**, 115, 1976, World Sci. Education, Princeton, N.J. 1976.
288. G. Goldhaber, Ref. 287, **5**, 131, 1976.
289. B. Richter, Ref. 287, **5**, 143, 1976.
290. C. Bacci *et al.*, *Phys. Rev. Lett.* **33**, 1408, 1649, 1974.
291. W. Braunschweig *et al.*, *Phys. Lett.* **53B**, 393, 1974.
292. G. S. Abrams, *Phys. Rev. Lett.* **33**, 1453, 1974.
293. M. K. Gaillard, B. W. Lee, and J. L. Rosner, *Rev. Mod. Phys.* **47**, 277, 1975.
294. T. Appelquist and H. D. Politzer, *Phys. Rev. Lett.* **34**, 43, 1975.
295. B. J. Harrington, S. Y. Park, and A. Yildiz, *Phys. Rev. Lett.* **34**, 168, 1975; E. Eichten, K. Gottfried, T. Kinoshita, J. Kogut, K. Lane, and T. M. Yan, ibid., p. 369.
296. T. Appelquist, R. M. Barnett, and K. Lane, *ARNP* **28**, 387, 1978.

297. C. Quigg and J. L. Rosner, *Phys. Rep.* **56**, 167, 1979; H. Grosse and A. Martin, *Phys. Rep.* **60**, 341, 1980.
298. A. Pais and S. B. Treiman, *Phys. Rev. Lett.* **35**, 1556, 1975.
299. K. Niu, E. Mikumo, and Y. Maeda, *Progr. Theor. Phys.* **46**, 1644, 1971.
300. E. G. Cazzoli *et al.*, *Phys. Rev. Lett.* **34**, 1125, 1975.
301. A. Benvenuti *et al.*, *Phys. Rev. Lett.* **35**, 1199, 1203, 1249, 1975; also B. C. Barish *et al.*, *Phys. Rev. Lett.* **36**, 939, 1976; M. Holder *et al.*, *Phys. Lett.* **69B**, 377, 1977.
302. A. Pais and S. B. Treiman, *Phys. Rev. Lett.* **35**, 1206, 1975.
303. B. C. Barish, *Phys. Rep.* **39**, 279, 1978; C. Baltay, *Proc. int. conf. high energy physics, Tokyo (1978)*, p. 894, Phys. Soc. Japan, Tokyo 1979.
304. See however J. von Krogh *et al.*, *Phys. Rev. Lett.* **36**, 710, 1976; J. Bleitschau *et al.*, *Phys. Lett.* **60B**, 207, 1976.
305. A. Boyarski *et al.*, *Phys. Rev. Lett.* **35**, 196, 1975.
306. G. Goldhaber *et al.*, *Phys. Rev. Lett.* **37**, 255, 1976.
307. I. Peruzzi *et al.*, *Phys. Rev. Lett.* **37**, 569, 1976.
308. G. J. Feldman and M. L. Perl, *Phys. Rep.* **33**, 285, 1977.
309. G. Goldhaber and J. E. Wiss, *ARNP* **30**, 337, 1980; A. Kernan and G. van Dalen, *Phys. Rep.* **106**, 297, 1984.
310. G. Bellini, *Phys. Rep.* **83**, 1, 1982.
311. M. L. Perl *et al.*, *Phys. Rev. Lett.* **35**, 1489, 1975; *Phys. Lett.* **63B**, 466, 1976.
312. C. Matteuzzi *et al.*, *Phys. Rev. Lett.* **52**, 1869, 1984.
313. Ref. 308; M. L. Perl, *ARNP* **30**, 299, 1980.
314. S. W. Herb *et al.*, *Phys. Rev. Lett.* **39**, 252, 1977.
315. W. R. Innes *et al.*, *Phys. Rev. Lett.* **39**, 1240, 1977.
316. Ref. 2, pp. S190–193.
317. Cf. R. Nernst *et al.*, *Phys. Rev. Lett.* **54**, 2195, 1985.
318. P. Franzini and J. Lee-Franzini, *ARNP* **33**, 1, 1983; K. Berkelman, *Phys. Rep.* **98**, 145, 1983.
319. D. Andrews, *Phys. Rev. Lett.* **45**, 219, 1980; G. Finocchiaro *et al.*, ibid., p. 222; full references to B-mesons up to 1984: Ref. 2, p. S120.
320. G. Arnison *et al.*, *Phys. Lett.* **147B**, 493, 1984.
321. M. Kobayashi and T. Maskawa, *Progr. Theor. Phys.* **49**, 652, 1973.
322. L. Wolfenstein, *Comm. Nucl. Ptcle Phys.* **14**, 135, 1985; I. I. Bigi and A. I. Sanda, ibid., p. 149.
323. G. Hanson, *Phys. Rev. Lett.* **35**, 1609, 1975.
324. S. L. Wu, *Phys. Rep.* **107**, 59, 1984.
325. R. Brandelik, *Phys. Lett.* **100B**, 357, 1981.
326. E. Reya, *Phys. Rep.* **69**, 195, 1981; G. Altarelli, *Phys. Rep.* **81**, 1, 1982.
327. M. Banner *et al.*, *Phys. Lett.* **118**, 203, 1982; G. Arnison *et al.*, *Phys. Lett.* **B123**, 115, 1983; **B132**, 214, 1983.
328. A. Pais, Proc. of the conf. 'Gauge theories and modern field theory', Northeastern Univ., p. 211, MIT Press, Cambridge, Mass. 1975.
329. L. M. Lederman and R. F. Schwitters, *Science* **227**, 131, 1985.
330. M. Schwartz, Ref. 287, **1**, 82, 1972.
331. W. K. H. Panofsky, in *Selected works of Luis W. Alvarez*, Ed. W. P. Trower, Chapter 10, Freeman, San Francisco 1986.
332. L. M. Lederman, *Sci. Am.* **239**, October 1978, p. 72.
333. B. W. Lee, *Phys. Rep. Repr. Book Series*, Vol. 2, p. 148, North Holland, Amsterdam 1978.

22

Being a conclusion that begins as epilog and ends as prolog

In 1968, the monthly *Physics Today* celebrated its twentieth anniversary by publishing an issue devoted to developments in physics during the preceding two decades. I contributed an article 'Particles'[1] in which I wrote: 'If there are W mesons they are most probably heavier than 2 GeV/c^2 . . . There is no a priori clue to how heavy the W's should be . . . Experimental studies of weak interactions at high energies and especially the search for W quanta constitute some of the most important future problems'. I went on to summarize the general situation at that time:

'The state of particle physics . . . is . . . not unlike the one in a symphony hall before the start of a concert. On the podium one will see some but not all of the musicians. They are tuning up. Short brilliant passages are heard on some of the instruments; improvizations elsewhere; some wrong notes too. There is a sense of anticipation for the moment when the concert starts.'

Seventeen years later, as again I reflected on how things stood, I asked myself: would I still subscribe to what I wrote then about the period 1948–68? And would I maintain the same about the subsequent years? I answered yes to the first question, yes and no to the second.

No—because two concepts meanwhile interjected, gauge theories and unification, while in need of elaboration and refinement, are not just brilliant improvizations but, demonstrably, pieces of reality of Maxwellian stature. Perhaps least expected was the transition of strong interaction dynamics from a state of chaos to one of discipline during that period. In this evolution the criterion of renormalizability (the necessity of which remains unclear) has been an extremely useful guide.

Yes—because improvizations, theoretical ideas that have not yet been validated by experiment, are in evidence even within these developments. Masses of leptons, quarks, and gauge bosons, supposed to be generated by spontaneous symmetry breaking, are arbitrary parameters. So is the mass of Higgs quanta. So are the angles that mix d, s, and b quarks. In all, color SU(3) and SU(2)×U(1) contain at least eighteen independent adjustable parameters. That does not violate any known principle. The presumption is very strong, however, that the present theory contains too much arbitrariness.

Other queries raised by the theory as it stands: it consistently incorporates non-invariance of parity and charge conjugation in weak interactions but leaves open the question of why these violations occur in weak, but not in other, interactions. Similarly for *CP*-violation, that puzzling small effect whose magnitude remains obscure. The theory currently operates with six flavors, u, d, s, c, b, t, but does not explain why Nature chooses this or any larger specific number of flavors. The relevance of all these questions is widely appreciated. Many and varied are the improvizations aimed at coping with them. A first example: some see, in the recent proliferation of quarks and leptons, indications for the existence of a small number of even more fundamental subunits out of which all these particles are built up.*

The theory in its present stage poses a number of outstanding problems of a more technical character. Foremost among these is the proof of confinement in QCD, virtually certainly true. Studies of non-perturbative QCD, and of other gauge theories, have led to potent topological results. These have not been discussed in the foregoing because, so far, their experimental implications remain to be verified.

Meanwhile particle physicists, stimulated by the spectacular successes in particle dynamics described in the preceding chapter, are in pursuit of new and more ambitious goals. At the center of current attention are efforts at achieving higher levels of unification. Those are not really topics for this book, but let us have a quick look anyway.

Grand unification, the program aiming at the union of color SU(3) with electroweak SU(2)×U(1), had its beginnings in the early seventies.[3] As my friend Howard Georgi, one of the pioneers, told me, he began thinking about that problem the week after he had understood confinement. He remembered that strong forces are only strong at low energies and, as asymptotic freedom tells us, get progressively weaker at higher energies, or, which amounts to the same thing, at smaller distances. In that second régime there exists therefore the possibility, needed for grand unification, of describing the forces of color, electromagnetism, and weak interactions by a common coupling strength. The distance scale at which these strengths become equal could be estimated to be about 10^{-29} cm—thirteen orders of magnitude smaller than the scale set by electroweak theory ($\hbar/M_W c \simeq 10^{-16}$ cm).

The search was on for the GUT (grand unified theory) gauge group. Its set of gauge bosons has to contain the eight massless gluons of $SU(3)_c$, the $W^\pm$, Z, and γ of SU(2)×U(1), and others, called X's (at least twelve of them) with masses M_X, thirteen orders of magnitude higher than the W and Z masses. Spontaneous symmetry-breaking can be rigged so as to produce those mass values but does not give a natural explanation for the occurrence of the

* Ref. 2. The few references in this chapter are exclusively to reviews, as much as possible of a popular kind.

dimensionless number 10^{13}. These traits are general, independent of the choice of gauge group. Other properties of almost equal generality: GUT provides a means for understanding why electric charge is quantized; and it fixes the value of the Weinberg angle θ_W.

GUT has profound implications for cosmology. It directs attention to the supreme high energy laboratory, the universe in its very earliest stages of development, since only at those extremely early times can particles with mass $M_X \sim 10^{15}$ GeV have been created. What happens to an X? It decays into a quark + lepton, a *baryon number-violating* process that (via an X in a virtual state) causes the proton to decay (into $e^+ + \pi^0$, for example). Thus GUT makes the most striking prediction that all atoms are unstable. This provides a dynamical basis for a speculation dating back to the sixties: that the matter–antimatter asymmetry seen here and now may have developed from symmetric beginnings of the universe, provided that baryon number is violated (and that the known violations of C and CP are taken into account). Thus the greatest single seriously thought-out leap ever to occur in particle physics, by thirteen powers of ten, has transformed the theorists' outlook by linking very small-scale with very large-scale phenomena.[4]

Which is the GUT group? We do not know, though there are favored candidates. On its choice depends the predicted value of θ_W and of the proton lifetime, assured to be long since it is proportional to M_X^4/M^5, M = proton mass. Information on proton decay is clearly crucial for the future of these speculations based on solid theorizing. There are a few candidate decay events at this time, but no firm claim as yet to proton instability. 'Nucleon decay, if it is ever discovered, will have to be based on unimpeachable evidence from several independent techniques. We are a long, long way from such a goal.'[5] Present limits on the proton lifetime have already served to eliminate the simplest possible GUT group (a version of SU(5)).

If there is truth to the strongly appealing GUT idea, then the entire subject matter of this book, from electron to Z, will some day be referred to as low energy particle physics.

Which brings us to the inwardmost scale of length yet contemplated in particle theory, the Planck length $(\hbar G/c^3)^{1/2} \simeq 10^{-33}$ cm (G is Newton's gravitational constant), corresponding to a mass $\sim 10^{19}$ GeV, at which quantum effects due to gravitation become important.

General relativity has not been mentioned up till now because gravitational effects may be ignored on the scales discussed thus far. When studies in gravitational quantum field theory, already for quite some time an important subject in its own right, revealed that the standard theory is unrenormalizable, it appeared that there was not much one could do.

New options made their appearance in the seventies. It began with the discovery of global supersymmetry, an elegant enlargement of the group of Lorentz transformations and translations, in which fermions and bosons appear jointly in specific mass-degenerate multiplets.[6] Known particles acquire 'superpartners'. This results to a remarkable degree in suppression of the field theoretical infinities. The symmetry must necessarily be broken to be of physical use. Ways of detecting superpartners have been suggested;[7] they have not as yet been seen. The transition from the global to the local version of the theory brings in a generalized version of the gravitational field, in the sense that the gravitational quantum, the graviton, is accompanied by superpartners of its own.[6] It was hoped that this formalism would lead to a consistent renormalizable quantum field theory of gravity. That, however, does not appear to be the case.

The most recent phase in the development of supersymmetry is superstring theory, a development that once again has its roots in the seventies.[8] Here the fundamental operator fields are no longer functions of a point that tracks a world line in space-time, but of a one-dimensional structure, a string, that sweeps a world sheet. Theories of this kind can be formulated consistently only on manifolds with more than three spatial dimensions (there is always only one time dimension). By means of a spontaneous symmetry-breaking mechanism ('compactification') the extra dimensions are supposed to shrivel to extremely tiny domains.[9] Theories of this kind are generalizations of the Kaluza–Klein theory, also conceived in its day as a unified field theory.[10] It appears that this approach may lead to a consistent renormalizable theory of gravitation, that, in fact, it is a candidate for TUT, the totally unified theory of all forces, the marriage between GUT and gravity. These theories contain super-high mass excitations, of the order of the Planck mass, necessary for consistency but negligible on the mass levels of the physics of known particles. In the limit where these supermasses are dropped, string theory reduces to point theory. In this picture even the GUT scale appears as a medium energy scale.

Intense explorations of these avenues are proceeding at this time. The principal reason for briefly mentioning them here is to give a sense of the current era of new diversity. Many implications of these new formalisms are beyond experimental reach. At the same time there is a keenly sensed need for new experimental stimulus. The general mood is positive, but there is no broad consensus as to where we stand and where we may be going.* In the words of Pogo, we are faced with insurmountable opportunities.

Experimentalists are busy as well.

While experimentation with existing accelerators progresses, vigorously, the entire high energy physics community is looking forward to the completion of new machines, among them SLC (SLAC), LEP (CERN), and HERA

* For comments on prospects see Ref. 11.

(Hamburg), all under construction. A huge American project, the SSC, is in the design stage and has not as yet been funded. It is a 20 + 20 trillion electron volt proton–proton collider,[12] a ring somewhere between 100 and 160 km in circumference, estimated to cost about $6 billion. There are infrequent but regular discussions[13] about the possibility of a VBA, an international super high energy project.*

There is strong awareness and concern in the community about the growing costs of these projects, not only of the machines but also of the needed new generation of detectors, larger and more complex than ever before. Indeed there is an urgent need for new ideas on accelerator design that will cut down their expense, and also their size.

It is obvious, yet bears saying anyway, that higher energy is but one of the two frontiers of new experimental knowledge. The other is improved measurement precision. Current example: searches for possibly non-zero neutrino masses.[14]

As I write this page in the summer of 1985, the discoveries of W and Z seem ages ago. That is not in the least unusual. There is a saying among experimental high energy physicists (my respects to whoever first thought of that): yesterday's sensation is today's calibration and tomorrow's background. Physicists, impressed by, but never content with, one glorious moment, move on to the next, inward bound.

Many of the most recent theoretical activities are highly speculative. New theories of the creation of the universe abound. These and other new mathematical structures are erected with few though important data to back them up. I cannot avoid thinking of Einstein in his late period, relying too much on formal elegance and too little on fact.

In these first post-W, Z years theorists, in their search for greater unity, are for the first time outrunning experiment, though it would of course be preposterous to say that they know it all in so far as lower energies are concerned. One discerns a developing gap between fact and fancy, accentuated by the current waiting for new machines, the fate of some of which is uncertain.[15] There appears to be somewhat of a pause in the dialog between experiment and theory. May it not last long.

To end, a story.

As a visitor to Washington D.C. was being driven along Pennsylvania Avenue, he noticed at the back of the National Archives a statue of a seated woman holding an open book on her lap. He was puzzled by the inscription

* SLC = Stanford Linear Collider; LEP = Large Electron Project; HERA = Hadron Elektron Ring Anlage; SSC = Superconducting Super Collider; VBA = Very Big Accelerator.

on the statue's base:

WHAT IS PAST IS PROLOGUE

What does that mean, he asked his cab driver.

To which the cabbie replied: 'It means you ain't heard nothing yet'.

References

1. A. Pais, *Physics Today* **21**, May 1968, p. 24.
2. H. Harari, *Sci. Am.* **248**, April 1983, p. 56.
3. H. Georgi, *Sci. Am.* **244**, April 1981, p. 48.
4. F. Wilczek, *Sci. Am.* **243**, December 1980, p. 82; H. R. Pagels, *Perfect symmetry*, Simon and Schuster, New York 1985; technical: E. W. Kolb and M. S. Turner, *ARNP* **33**, 645, 1983.
5. D. H. Perkins, *ARNP* **34**, 1, 1984.
6. D. Z. Freedman and P. van Nieuwenhuizen, *Sci. Am.* **238**, February 1978, p. 126; technical: H. P. Nilles, *Phys. Rep.* **110**, 1, 1984.
7. G. L. Kane, *Comm. Nucl. Ptcle. Phys.* **13**, 313, 1984.
8. M. B. Green, *Nature* **314**, 409, 1985.
9. A. Chodos, *Comm. Nucl. Ptcle. Phys.* **13**, 171, 1984.
10. A. Pais, *Subtle is the Lord*, Chapter 19, Oxford Univ. Press 1982.
11. L. van Hove, *Phys. Rep.* **104**, 87, 1984.
12. Cf. M. M. Waldrop, *Science* **225**, 490, 1984.
13. R. R. Wilson, *Physics Today* **37**, September 1984, p. 9.
14. W. Marciano, *Comm. Nucl. Ptcle. Phys.* **9**, 169, 1981; A. K. Mann, ibid. **10**, 155, 1981; F. Boehm and P. Vogel, *ARNP* **34**, 125, 1984.
15. Cf. also *Nature* **315**, 689, 1985; D. Dickson, *Science* **228**, 1509, 1985.

Appendix

A synopsis of this book in the form of a chronology

(A number and letter in parentheses refer to the chapter and section in which the subject is discussed. A date refers to time of receipt by a journal.)

1815–16	Prout claims that specific gravities of atomic species are integral multiples of the value for hydrogen (9b).
1819	Dulong and Petit present a list of twelve atomic weights (11a).
1830–50	Multiple discovery of macroscopic energy conservation (6b).
1833	Faraday's laws of electrolysis (4b).
1835	Auguste Comte declares that knowledge of the chemical composition of stars will forever be denied to man (9a).
1853	First observation of the spectrum of hydrogen (9b).
~1855	Geissler invents his mercury pump and his vacuum tube (1b, 4a).
~1855	Rühmkorff invents his version of the induction coil (1b, 4a).
1859	First measurements of frequencies of spectral lines (9b).
	Kirchhoff diagnoses sodium on the sun and discovers his blackbody radiation law (9b).
1860	Kirchhoff and Bunsen lay the foundations for systematic quantitative spectrum analysis (9b).
1864	Maxwell's *A dynamical theory of the electromagnetic field* (12a).
1867	Vortex model of the atom (9b).
	Death of Faraday (4c).
1869	Mendeléev's first version of the periodic table of elements (11a).
1871	The wavelengths of three lines in the hydrogen spectrum are found to have simple ratios (9b).
	Birth of Rutherford.
1872	Maxwell remarks that atoms remain in the precise condition in which they first began to exist (15a).
1874	First estimate of the fundamental charge *e* (4b).
1875	Maxwell notes that atoms have a structure far more complex than that of a rigid body (9b).
1876	Introduction of the term 'cathode rays' (4d).

1879	Birth of Einstein.
	Death of Maxwell.
1881	J. J. Thomson introduces electromagnetic mass (4f).
	Electricity is divided into definite elementary portions (Helmholtz, 4b).
1882	Rowland begins his work on spectral gratings (4c).
1885	Balmer's formula for the hydrogen spectrum (9e).
	Birth of Bohr.
1887	Birth of Schroedinger.
1892	First detection of fine structure in the spectral lines of hydrogen (10).
1893	July. First issue of the *Physical Review* (16b).
1895	Introduction of the Lorentz force in electrodynamics (4c).
	Cathode rays are negatively charged (4d).
	Rutherford arrives in Cambridge.
	C. T. R. Wilson starts developing the cloud chamber (4d).
	November 8. Roentgen discovers X-rays (2a).
1896	March 1. Becquerel discovers uranic rays, the first observation of radioactivity (2b).
1897	January 7. First statement in the literature that there may exist particles about a thousand times lighter than the hydrogen atom (Wiechert, 4d).
	April. Kaufmann and J. J. Thomson independently determine e/m for cathode rays (4d).
1898	Thorium is radioactive (3a).
	Radioactive energy is released from within atoms (6b).
	Discovery of polonium (3a).
	Discovery of radium (3a).
	Discovery of anomalies in the Zeeman effect (13b).
	Radioactive radiation has two distinct components: α- and β-rays (3b).
	Rutherford moves to McGill University (3b).
1899	J. J. Thomson measures e and thus completes his discovery of the electron. He also recognizes ionization to be a splitting of atoms (4d).
	The American Physical Society is founded (16b).
1900	Discovery of γ-rays (3b).
	FitzGerald asks if magnetism is due to rotation of electrons (13c).
	First determination of a half-life for radioactive decay (6e).
	December. Planck discovers the quantum theory (7).
	Birth of Pauli.
1901	Births of Heisenberg and Fermi.

1902	Transformation theory of radioactivity (6b).
	Birth of Dirac (13e).
1903	A hydrogen atom contains about a thousand electrons; plum pudding model of the atom (9c).
	First use of the term 'atomic energy' (6c).
1905	March. Einstein postulates the light-quantum (7).
	June. Einstein's first paper on special relativity (4f).
	September. His second paper on special relativity; $E = mc^2$ (4f).
1906	The number of electrons in a hydrogen atom does not differ much from unity (9c).
	Rutherford discovers α-particle scattering (8b).
	First experiments on β-spectra (8b).
	Death of Pierre Curie (3a).
1907	Potassium and rubidium are radioactive (6d).
	Rutherford moves to Manchester.
1908	Death of Becquerel (2b).
1909	Strong back scattering of α-particles observed (9d).
	Energy fluctuations in blackbody radiation; first statement about particle–wave duality (12a).
1910	First attempt to link atomic structure to Planck's constant (9e).
	Debye's derivation of Planck's law (15b).
1911	Rutherford proposes the nuclear model of the atom. (The term 'nucleus' is first used in 1912.) (9d)
	First formulation of the isotope concept (11c).
	First statement that Planck's law is related to the indistinguishability of light-quanta (13d).
	First Solvay conference.
1912	Discovery of cosmic radiation by manned balloon flights (17b).
	First attempt to link Planck's constant to angular momentum (9e).
	Paschen–Back effect (13b).
1912–13	Beginnings of nuclear spectroscopy (8g).
1913	Bohr's trilogy on the constitution of atoms and molecules (9e).
	First statement that β-decay is a nuclear process (Bohr, 11b).
	First recognition that A and Z are independent nuclear parameters (11d).
	Emergence of the proton–electron model of the nucleus (11e).
	Moseley's experiments lead to the definitive interpretation of the periodic table (11e).
1914	First detection of the continuous β-spectrum (8h).
1915	Introduction of the fine structure constant (10).
	Field equations of gravitation (Einstein, Hilbert, 12a).
1916	Bohr appointed professor of physics in Copenhagen (10).

1916–17	Einstein introduces the A- and B-coefficients for radiative emission and absorption (15a).
1918	First selection rules for atomic spectra (10).
	Noether's theorem.
1919	α-particle–hydrogen scattering shows strong deviations from the Rutherford formula (11h).
	Rutherford assumes the Directorship of the Cavendish Laboratory in Cambridge.
1921	New forces of great intensity exist inside the nucleus (11h).
	Landé introduces half-integer magnetic quantum numbers (13b).
	Compton conjectures that ferromagnetism is due to the rotation of an electron around its own axis (13c).
	Internal conversion; the first nuclear energy level schemes (14c).
1923	De Broglie introduces particle–wave duality for matter (12c).
	The core model of the anomalous Zeeman effect (13b).
1924	January. Bohr–Kramers–Slater theory of radiative processes (14d).
	February. Laporte discovers a new selection rule in the iron spectrum (20c).
	July. Bose introduces a new statistics for light-quanta (12c).
	July. Einstein applies Bose statistics to matter and deduces that matter should exhibit wave properties (12c).
	August. Pauli's theory of hyperfine structure (13c).
	October. Stoner's rule for the periodic system (13b).
	December. Pauli recognizes a two-valuedness of the quantum properties of valence electrons (13b).
	Ising proposes to accelerate charged particles by multiple traversal of a voltage difference (17b).
1925	January. Pauli exclusion principle (13b).
	June. First experimental demonstration of energy–momentum conservation in individual elementary processes (14d).
	July. Heisenberg's first paper on quantum mechanics (12c).
	August. Uhlenbeck and Goudsmit introduce half-integer quantum numbers for the hydrogen atom (10, 13c).
	September. Born and Jordan recognize that Heisenberg's theory is a matrix mechanics (12c) and that there is a need for a matrix electrodynamics (15b).
	October. Discovery of spin (13c).
	November. Quantum algebra (Dirac, 12c). Born, Heisenberg, and Jordan give a comprehensive treatment of matrix mechanics (12c) and recognize the concept of second quantization (15b).
1926	January. Derivation of the discrete hydrogen spectrum by matrix methods (12c).
	Schroedinger's first paper on wave mechanics (12c).

February. Fermi statistics (13d). Discovery of the Thomas factor (13c).
June. Born's first paper on the probability interpretation of quantum mechanics (12d).
August. Dirac relates symmetric (antisymmetric) wave functions to Bose–Einstein (Fermi–Dirac) statistics and derives Planck's law from first principles (13d).
October. G. N. Lewis introduces the name 'photon' for the light-quantum (12a).
November. Group theory enters quantum mechanics (13a).
December. Dirac's first paper on quantum electrodynamics (15c).
From 1926 to 1929 the proton–electron model of the nucleus leads to a series of paradoxes (14b).

1927 March. Uncertainty relations (12d). First detection of electron diffraction (12c).
May. Pauli matrices (13e).
August. Chadwick and Ellis' calorimeter experiment shows that the continuous character of β-spectra is of primary nature (14c).
September. Bohr states the notion of complementarity (12c).
October. Jordan–Klein matrices (15d).
December. Jordan and Pauli introduce covariant commutation relations for free electromagnetic fields (15e, g).

1928 January. The Dirac equation (13e). Jordan–Wigner matrices (15d).
February. Wigner introduces parity (20c). Death of Lorentz (10).
August. α-decay is interpreted as a barrier penetration (6f).
October. A system of N fermions satisfy Bose–Einstein (Fermi–Dirac) statistics if N is even (odd) (13d). The Klein–Nishina formula for Compton scattering (15f).
December. Klein paradox (15f).
Wideröe designs the first (linear) accelerator (17b).

1929 February. First observation of cosmic ray showers (17b).
March. Heisenberg and Pauli give the Lagrangian formulation of quantum field theory (15e). Weyl formulates gauge invariance and its relation to charge conservation (15e, 21e).
September. First systematic treatment of quantum electrodynamics in the Coulomb gauge and proof of its covariance (15e).
November. In a letter Heisenberg refers to Pauli's neutrino postulate (14d). First quantum electrodynamic self-energy calculation (16c).
December. Dirac introduces the notion of hole theory, identifying a hole with a proton (15f).
From 1929 to 1936 Bohr considers the possibility that energy is not conserved in β-decay (14d).

1931 April. First working cyclotron (4-inch) (17b).
May. Dirac proposes the positron (15f).
December. First publication of a positron cloud chamber picture (15f). Discovery of the deuteron (17b).

1932 February. Discovery of the neutron (17a).
April. First suggestion that the neutron may be as elementary as the proton (17c).
June. First nuclear process produced by an accelerator (17b). Heisenberg's first paper on nuclear forces and on the formal introduction of certain matrices which were to lead to isospin (17d).
November. Wigner introduces time reversal in quantum mechanics (20c).

1933 May. Discovery of the anomalous value of the proton magnetic moment (17c).
October. The positron theory implies scattering of light by light (16d).
December. Fermi theory of β-decay (17e).
1933–5: Calculations to leading order of the principal processes in the positron theory (16d).
1933–8: A period in which alternatives to standard electromagnetic theory are explored (16e).

1934 January. Discovery of β^+-radioactivity (17a).
De Broglie introduces the general term 'antiparticle' (17e).
Speculations about an electron–neutrino theory of nuclear forces (17g).
June. Heisenberg casts the positron theory in the Hamiltonian form used from then on. He recognizes the photon self-energy (16d). First calculations of electron self-energy in the positron theory (16d).
July. Pauli–Weisskopf theory of spin-0 particles (16d). First calculation of neutron β-decay (17e). Death of Marie Curie (3a).
October. Dirac calculates the polarization of the vacuum and introduces the first subtraction rules (16d).
November. Yukawa's first paper on mesons (17g).
1934–58. Evolution of the spin-statistics connection (20c).

1935 February. Wick ascribes the proton's anomalous magnetic moment to partial dissociation of the proton (17g).
April. Refined treatment of vacuum polarization (16d).
1935–40: It is believed that the shape of β-spectra is best fit by Konopinski–Uhlenbeck derivative coupling, until secondary distortions of the spectra are identified (17e).

1936 February. First calculation of a high energy neutrino scattering

cross section (17g). Serber introduces the term 'renormalization' (16d).
May. Proca introduces vector meson field equations (17g).
August. First good proton–proton scattering results lead to introduction of charge independence and the physical concept of isospin (17f).
October. Furry's theorem (16d). SU(4) symmetry in nuclear physics (17f).
November. First announcement of a meson detected in cosmic rays, the later muon (17g).
December. Theory of cascade showers (17g).

1937 June 1. First mention of the Yukawa meson in a Western publication (17g).
October 19. Death of Rutherford (17h).
October. Kramers insists that the particle mass introduced in a theory be from the outset the experimental mass (18a).
November. Kramers introduces the charge conjugation operation (16d). Majorana introduces self-conjugate neutrinos (16d).

1938 February. First introduction of pseudoscalar and pseudovector mesons (Kemmer, 17g).
April. Kemmer postulates a neutral meson to save charge independence (17g).
July. K-capture of electrons observed (17e).
First formulation of the conservation of heavy particles (protons and neutrons) (19d).

1939 Nuclear fission (8g).

1940 Strong coupling theory of nuclear forces (19f).

1941 Introduction of the term 'nucleon' (18a).

1942–3 Japanese theorists propose a second meson (18b).

1943 First papers on S-matrix theory (19f).

1944–5 Invention of phase stability (19a).

1946 November. The first synchrocyclotron produces 380 MeV α-particles (19a).
December. Negative cosmic ray mesons are anomalously absorbed in carbon (18b).
Introduction of the term 'lepton' (18a).
1946–7. Discovery of the first two 'V-particles' (20a).

1947 January. Birth of Brookhaven National Laboratory (19a).
May 24. Experimental discovery of a second cosmic ray meson, soon called pion, the Yukawa meson (18b).
June 2–4. Shelter Island conference. Lamb shift, hyperfine anomalies (18a), two meson hypothesis (18b), electron–neutron interaction (19c).

	The Berkeley proton linear accelerator produces 32 MeV free protons (21c).
	1947–9. First formulations of universality in the Fermi theory (20c).
1948	February. The first artificially-produced pions detected at Berkeley (19b).
	March 30–April 2. Pocono conference (18c).
	Beginnings of the systematic renormalization program (18c).
1949	April 11–14. Oldstone on the Hudson conference (18c).
	Model of the pion as a nucleon–antinucleon composite (19f).
1950	Discovery of the neutral pion (19b).
	December 16. First Rochester conference (18c).
1951–7	Evolution of the *CPT* theorem (20c).
1952	Discovery of the 33-resonance (Δ) (19d).
	Invention of strong focussing (19a).
	Associated production of the new *V*-particles (20b).
	The Cosmotron (Brookhaven) accelerates protons beyond 1 GeV (19a).
	1952–6. Formulation of *G*-parity (19d).
1953	First bubble chamber pictures produced (19e).
	Beginnings of nucleon resonance spectroscopy (19e).
	Conservation of leptons (20c).
	July. Bagnères de Bigorre Conference (20a). First reference to a hierarchy of interactions in which strength and symmetry are correlated; and to the need for enlarging isospin symmetry to a bigger group (20b).
	August. Strangeness scheme (20b).
	November. First experimental evidence for associated production (20b).
1954	June. Yang–Mills equations (21e).
	September 29. Birth of CERN (19a).
	November. K_0, $\bar{K}_0$ are particle mixtures (20b, 20d).
	Introduction of the term 'baryon' (20a).
	Death of Fermi (19c).
1955	Discovery of the antiproton (19d).
	Forward pion–nucleon dispersion relations (19f).
	Death of Einstein.
	1955–6. The theta-tau puzzle (20c).
1956	Discovery of the antineutron (19d).
	Direct detection of the β-decay neutrino (e-neutrino) (21c).
	First evidence for a long-lived neutral K-particle (20b).
	Conservation of parity (*P*) and of charge conjugation invariance (*C*) have never been checked in weak interactions (20c).

	First studies of colliders (21c).
1957	January. First experimental observations of P- and C-violation in β-decay, and in μ-decay (20c).
	CP-invariance; γ_5-invariance; two-component neutrino theory; V–A theory of weak interactions (20d).
	Pomeranchuk theorem (19f).
	1957–67: preludes to electroweak unification (21e).
1958	First observation of hyperon β-decay (21b).
	Conserved weak vector current (20d).
	Goldberger–Treiman relation (20d).
	First suggestions that weak interactions are mediated by charged vector bosons, later called W (21e).
	Death of Pauli.
1959	A single-ring collider suffices for producing collisions between particles of opposite charge (21c).
	Regge poles in potential scattering (19f).
	The '72-inch' hydrogen bubble chamber at Berkeley begins operations (19e).
	1959–60. The first strong focussing proton synchrotrons go into operation (CERN, Brookhaven), at energies of about 30 GeV (19a).
1960	Discovery of the first strange particle resonance (21b).
	Partially conserved weak axial current (20d).
1961	Beginnings of meson spectroscopy; discoveries of ρ, ω, and η (19e), and of K* (21b).
	Introduction of SU(3), later called flavor SU(3) (21b).
	Goldstone theorem (21e).
	Death of Schroedinger.
1962	Discovery of the second neutrino (μ-neutrino) (21c).
	Regge pole ideas are applied to high energy processes (19f).
	Introduction of the term 'hadron' (21a).
	Death of Bohr.
1963	Cabibbo theory of weak interactions (21b).
1964	Discovery of the Ω^- (21b).
	First observation of CP-violation (20e).
	First report of electron–positron interactions in a single-ring collider (21c).
	Higgs mechanism of spontaneous symmetry breaking (21e).
	Hypothesis that all hadrons known till then are composites of three species of quarks and antiquarks; current algebra; static SU(6) (21b).
	Introduction of a fourth quark species carrying 'charm' (21e).
1965	If the W exists its mass is larger than 2 GeV/c^2 (21a).
	Adler–Weisberger relation (21b).

	Introduction of 'color' and its symmetry $SU(3)_c$. First conjectures that all observed hadrons and the electromagnetic current are colorless, and that $SU(3)_c$ is associated with the dynamics of strong interactions (21b).
1966	SLAC (Stanford Linear Accelerator Center) goes in operation (21c).
1967	Electroweak unification in terms of a spontaneously broken $SU(2)\times U(1)$ symmetry; this involves a new current, 'the neutral current' (21e).
1969	Discovery that total inelastic electron–proton scattering is 'hard'; and that it obeys scaling laws (21d). Interpretation of scaling in terms of the parton model (21d).
1970	Charm serves to eliminate paradoxical properties of the neutral current ('GIM mechanism') (21e).
1971	The ISR (intersecting storage rings) at CERN, a proton–proton collider, begins operations. It produces the highest effective energies of the 1970s, 31 + 31 GeV (21c). Proofs of renormalizability of massless Yang–Mills fields and of massive Yang–Mills fields with spontaneously broken symmetry (21e).
1972	Fermilab begins operations at 200 GeV and reaches 400 GeV later in the year (19a). SPEAR (at SLAC), an electron–positron collider, goes into operation at $2\frac{1}{2}$ GeV per beam, later raised to 4 GeV (21e). First results on 'hard' hadron–hadron scattering at the ISR (21c). Theories with six or more quark species offer new options for incorporating *CP*-violation (21e).
1973	Discovery of neutral currents in neutrino reactions (21e). Discovery of a rise in the total proton–proton cross-section at high energies (21e). Beginnings of quantum chromodynamics: asymptotic freedom; conjecture of confinement (21e). 1973–4. Beginnings of grand unification (22).
1974	'The November revolution': discovery of the J/ψ and the ψ', 'hidden charm'; beginnings of charmonium spectroscopy (21e). Beginnings of lattice gauge theories (21e); and of supersymmetry in particle physics (22).
1975	June–July. Discovery of 'naked charm' (21e). August. Discovery of the τ-lepton (21e). October. First evidence (SLAC) for two-jet structures in high energy collision products; and for spin-1/2 of quarks (21e).
1976	At CERN the SPS (super proton synchrotron) goes into operation at 400 GeV (19a).

	Proposal for a proton–antiproton ($\bar{p}p$) collider (21c).
	Beginnings of supergravity (22).
	Death of Heisenberg (21e).
	1976–82. Studies of two-jet events at the ISR (21e).
1977	Discovery of several upsilon resonances, 'hidden bottom', bottom being the fifth quark; beginnings of upsilon spectroscopy (21e).
1978	Discovery of neutral-current effects in electron–deuteron scattering (21e).
	The PETRA collider at DESY (Hamburg) begins operation (21e).
1979	First evidence for three-jet events (PETRA) (21e).
1980	First evidence for 'naked bottom' (CESR, Cornell) (21e).
	The PEP collider at SLAC begins operations (21e).
1981	First $\bar{p}p$-collisions in the CERN $\bar{p}p$-collider (21c).
1983	Discovery of the W and the Z (21e).
1984	Fermilab reaches 800 GeV (19a).
	Death of Dirac.

Faraday, Michael (*cont.*)
induction principle 70; *see also (in subject index) entries beginning with* Faraday
Feather, Norman [1904–1978] 202, 398n, 438
Fechner, Gustav Theodor [1801–1887] 70
Feinberg, Gerald [1933–] 570
Feldman, David [1921–] 375
Fermi, Enrico [1901–54]; beta-decay theory 17, 18, 231, 318, 319, 380, 401, 402, 418–22, 580, 632; death 485, 585, 634; English spoken by 366; on meson theory 436, 484, 494–5, 519, 521; on neutrino 318; neutron bombardment studies 400–1; pion composite model 495, 519, 555; quantum field theory 354–5, 356; quantum statistics 255, 284, 285; on spinning electron 278n, 298–9; and Uhlenbeck 275, 366, 376; on V-particles 512n, 513, 517n; *see also (in subject index) entries beginning with* Fermi
Feynman, Richard Phillips [1918–]; clowning behavior 459; conversation (1961) with Dirac 23; Lamb shift calculations 460, 461; lecture notes 517n; papers (collected) 466; at Pocono conference 458, 459; at Shelter Island Conference 452; superposition principle used 523; on $V-A$ theory 542; on W-boson 581
Fierz, Markus Eduard [1912–] 91, 355, 387, 423, 428–9, 528
Finsen, Niels Ryberg [1860–1904] 96–7
Fischer, Emil Hermann [1852–1919] 63, 150, 151, 235
Fitch, Val Logsdon [1923–] 24, 25, 538, 542
FitzGerald, George Francis [1851–1901] 41, 80, 82n, 179, 279, 628
Fizeau, Armand Hippolyte Louis [1819–1896] 45
Fliess, Wilhelm [1858–1928] 139–40
Fock, Vladimir [1898–1974] 339, 379, 380
Fokker, Adriaan [1887–1968] 194, 208, 237
Foley, Henry Michael [1917–1982] 451
Fowler, Ralph Howard [1889–1944] 60, 283, 286, 366, 367
Fowler, William Alfred [1911–] 517n
Franck, James [1882–1964] 235–6, 260
Frankfurter, Felix [1882–1965]
Frankland, Edward [1825–1899] 72
Franklin, Benjamin [1706–1790] 70, 79
French, Bruce [1921–] 460
Frenkel, Yakov Ilyich [1894–1954] 303, 372
Fretter, William Bache [1916–] 542
Freud, Sigmund [1856–1939] 139–40
Freidrich, Walter [1883–1968] 42, 94
Frisch, Otto Robert [1904–1979] 158, 160, 319
Frobenius, Georg Ferdinand [1849–1917] 265
Fröhlich, Herbert [1905–] 435
Furry, Wendell Hinkle [1907–] 370, 379, 380, 381, 382, 383, 384n, 385, 633

Galbraith, John Kenneth [1908–] 5
Galison, Peter [1955–] 467, 603
Gamow, George [1904–1968]; on alpha-decay 124; alpha-scattering studies 296; on Fermi beta-decay theory 418; G.–Teller coupling 423, 567; inverse alpha-decay proposal 406; on neutrinos 417; textbook 297, 298, 299
Gardner, Eugene [1913–1950] 479
Garwin, Richard Lawrence [1928–] 542
Gassiot, John Peter [1797–1877] 68n
Geiger, Hans Wilhelm [1882–1945]; alpha-particle studies 61, 188, 189–90, 191–2, 226, 228n, 238; beta-particle spectra 154, 158, 159, 160; in World War 1 236
Geissler, Johann Heinrich Wilhelm [1815–1879] 10, 67–8, 69–70, 131, 167, 627; *see also (in subject index):* Geissler . . .
Geitel, Hans Friedrich Karl [1855–1923] 109–10
Gell-Mann, Murray [1929–]; on baryon names & properties 514 & n, 520; on current algebras 564n; on quarks 558, 564n, 591, 611; on Regge trajectories 504n; on S-matrix theory 500n, 505; strangeness scheme 519, 520, 545; on SU(3) symmetry 556, 557, 611; on SU(6) symmetry 559; $V-A$ theory 536, 542; on W-bosons 581; on weak interactions 520n, 542
Georgi, Howard Mason [1947–] 591, 592, 593–4, 622
Germer, Lester Halber [1896–1971] 255
Gibbon, Edward [1737–1794] 71
Gibbs, Josiah Willard [1839–1903] 5, 281
Giesel, Friedrich Otto [1852–1927] 99
Glaser, Donald Arthur [1926–] 491, 505
Glashow, Sheldon Lee [1932–] 591, 592, 601, 611
Glasser, Otto [1895–1964] 49, 101
Goethe, Johann Wolfgang von [1749–1832] 234
Goldberger, Marvin Leonard [1922–] 500 & n, 504, 538, 635
Goldhaber, Gerson [1924–] 605, 607, 611
Goldhaber, Gertrude Scharff- [1911–] 87
Goldhaber, Maurice [1911–] 87, 124, 189n, 412
Goldstein, Eugen [1850–1930] 79, 80, 81
Goldstone, Jeffrey [1933–] 593, 635
Goldwasser, Edwin Leo [1919–] 505
Good, Myron Lindsay [1923–] 523
Gordon, Walter [1893–1940] 292, 329, 345
Goudsmit, Samuel Abraham [1902–1978]; background 276; as detective 276; on electron spin 254; on hyperfine structure 279n, 317; on Landé factor 271–2; on Pauli 279n, 317; quantum numbers 214–15, 271, 273–4, 276, 277, 347, 630; in USA 280, 366; on Zeeman effect 269, 270; *other references* 99n, 417
Gouy, Louis-Georges [1854–1926] 108n
Green, Kenneth [1911–1977] 407
Greenberg, Oscar Wallace [1932–] 561, 562
Gross, David Jonathan [1941–] 578, 580, 590, 591
Grove, William Robert [1811–1896] 106
Groves, Leslie R. [1896–1970] 19, 117, 472
Guericke, Otto von [1602–1686] 68
Gupta, Suraj Narayan [1924–] 355
Gurney, Ronald Wilfrid [1898–1953] 104, 124, 296
Gürsey, Feza [1921–] 559

Haas, Arthur Erich [1884–1941] 197–8, 199n, 200
Haber, Fritz [1868–1934] 235
Hafstad, L. R. 416, 424
Hagenbach-Bischoff, Eduard [1833–1910] 172, 173n
Hahn, Otto [1879–1968] 147, 149–51, 153–5, 158, 236
Halpern, Otto [1899–1982] 385
Hammer, William Joseph [1858–1934] 35, 111
Han, Moo- Young [1934–] 562

Dalitz, Richard Henry [1925–] 524, 525, 542
Dalton, John [1766–1844] 221
Dam, W. J. H. 35, 39
Dancoff, Sidney Michael [1914–1951] 369, 455
Danysz, Jean [?–1914] 155, 304
Darrow, Karl [1891–1982] 450
Darwin, Charles Galton [1887–1962] 195, 230, 238, 240, 289, 292, 310–11
Darwin, Charles Robert [1809–1882] 177, 216
Daudel, Raymond [1920–] 124
Davisson, Clinton Joseph [1881–1958] 255
Davy, Sir Humphry [1778–1829] 70
de Broglie: *see* Broglie, ...
Debierne, André Louis [1874–1949] 123
Debye, Peter [1884–1966] 89, 134, 315, 332, 333, 353, 629
Delbrück, Max [1906–1981] 314, 386
Demarçay, Eugène [1852–1903] 118
Dennison, David Mathias [1900–1976] 285, 301, 365, 366
Descartes, René du Perron [1596–1650] 165
Devons, Samuel [1914–] 189n
Dewar, Sir James [1842–1923] 115, 137
Dickens, Charles [1812–1870] 195, 211
Dirac, Charles Adrien Ladislas 286
Dirac, Paul Adrien Maurice [1902–1984]; age when major work done 251, 362; alternative approaches 388, 389, 390–1; approximation methods 291–2; and Bohr 287, 288, 312–13, 374n, 378, 382; at Cambridge (1946) conference 7; at Cambridge University 286–7; Compton effect calculations 337–8, 353–4, 374; death 637; delta function development 339, 354; and Einstein 288; electron, relativistic wave equation (Dirac equation) 286, 290–1, 330, 340, 345, 631; on energy conservation 319, 361; family background 286; Feynman conversation (1961) 23; Hamiltonian methods used 335, 353; Heisenberg's rule, derivation 254; hole theory 349–50, 351, 360, 361, 378, 631; comments on 352, 360–1; many-time formalism 375; on mathematical complexity 15; negative energy photon concept 389; negative energy states theory 346, 347–8, 349; notation used 287; on old (pre-1925) quantum theory 218; on parity 25–6; on Pauli exclusion principle 273; on Pauli matrices 290; Ph.D. topic 251, 287; Planck's law derivation 284–5, 631; positron proposal 349, 351, 632; probability interpretation of wave mechanics 257; on proton–neutron–electron model 410; quantum electrodynamics papers 288, 334, 338, 464, 631; and quantum field theory 325, 329, 334, 361; quantum statistics 255, 280, 282, 284–5, 631; reminiscences 334, 347, 355; scalar wave equation criticized 289, 387, 388; scientific background 286–7; second-order perturbation methods 337, 373; second quantization 334; at Solvay Conferences 23, 340, 377, 378, 410; speech, manner of 286; spontaneous decay theory 9, 123; on vacuum polarization 382–3, 632; work habits 287; *see also (in subject index):* Dirac equation; Fermi–Dirac (FD) statistics
Dirac-Holten, Florence Hannah 286
Dodd, Walter James [1869–1916] 98–9
Dorfman, Yakov Gregorievitch [1904–] 303 & n
Dreitlein, J. 557
Drell, Sidney David [1926–] 465, 467
Dufay, Charles-François de Cisternai [1698–1739] 70
Dulong, Pierre Louis [1785–1838] 221, 627
Dyson, Freeman John [1923–] 462, 466

Earnshaw, Samuel [1805–1888] 181, 185
Eckart, Carl [1902–1973] 365
Edison, Thomas Alva [1847–1931] 35, 38, 75
Edlefsen, Niels Edlef [1893–1971] 408
Ehrenfest, Paul [1880–1933]; Bohr correspondence 99n, 279; Bose–Einstein statistics, objection 283; Einstein correspondence 255; and electron spin 278, 279; and Goudsmit 276; Lorentz correspondence 261, 277–8; and Oppenheimer 367, 368, 369; spinor, term introduced 292; and Uhlenbeck 215, 275, 276, 277–8; and Wigner's rule 285n; *other references* 99n, 216
Ehrlich, Paul [1854–1915] 63
Einstein, Albert [1879–1955]; blackbody radiation law (with Rayleigh & Jeans) 135n; on Bohr hydrogen atom model 208; Born influenced by 258–9; on causality 212; classical theory re-examined 389; death 634; and Dirac 288; on dual nature of light 248, 252; *E–m–v* relationship 87–8; on electron spin 278; on electron stability 371; energy fluctuation studies 248, 252–3, 259, 332, 353; on energy nonconservation 310, 311; general theory of relativity 245; ghost field ideas 259, 260, 261; gravitation equations 218, 629; light quantum introduced 135–6, 150, 193, 210–11, 245, 251, 252, 327, 629; and Lorentz force 77; mass-energy equivalence 104, 232, 629; on Maxwell electromagnetic theory 244; Nobel Prize 89; photo-electric effect relation 305; on Planck's radiation law 9, 123, 130, 134–5, 193, 324, 328, 332, 630; on quantum mechanics 255, 262; quantum statistics 252–53, 280, 630; on quantum theory 130, 134; on revolutionary nature of relativity theories 250; on Rutherford–Geiger value for electron charge 62; on Schroedinger's work 259–60; special theory of relativity 89, 136, 244, 327, 629; unification theory, attempts at 30; war, reactions to 235 & n; *other references* 149, 181, 216, 286, 366; *see also (in subject index):* Bose–Einstein statistics
Elgar, Sir Edward [1857–1934] 286
Ellis, Charles Drummond [1895–1980]; on beta-decay 303–8, 318, 631; at King's College, London 436; knighthood 308; Rutherford obituary 308; scientific background 304; textbook 64, 157, 233, 297, 398; wartime service 160, 236n, 309
Elster, Julius [1854–1920] 109–10
Empedokles of Akragas [*c*. 492-*c*. 432 BC] 3
Enz, Charles [1925–] 312n, 387
Euler, Hans [1909–1941] 386 & n
Evans, Robley Dunglison [1907–] 11, 100
Eve, Arthur Stewart [1862–1948] 59, 61, 63, 64, 202, 438

Fajans, Kasimir [1887–1975] 225n
Faraday, Michael [1791–1867]; as atomist 72; death 74, 627; electromagnetic nature of light 74–5; energy principle studies 106; electrical discharge studies 10, 68, 69; electrolysis laws 71, 627;

Index of names

Note: year of birth, and year of death (if known) are recorded after each name; footnotes are indicated by suffix 'n'.

Abraham, Max [1875–1922] 88, 132, 277
Adler, Hanna [1859–1947] 194
Adler, Stephen [1939–] 567, 577, 635
Akhiezer, Aleksandr Il'ich [1911–] 386
Alvarez, Luis Walter [1911–]; bubble chamber work 491, 492, 505; on financing 19; linear accelerator work 571–2; at New York (1946) meeting 8; *other references* 407, 412, 422
Ampère, André Marie [1775–1836] 70
Anderson, Carl David [1905–]; meson discovery 432, 433, 438, 513, 542; on Oppenheimer 363; positron discovery 17, 330, 351–2, 363, 375, 376, 403
Anderson, Herbert Lawrence [1914–] 488n, 505
Andrade, Edward Neville da Costa [1887–1921] 16 60, 62, 67, 68n, 193, 202, 236, 298, 437–8
Ångström, Anders Jonas [1814–1874] 166, 167, 171, 172
Appelquist, Thomas [1941–] 606
Aristotle [384–322 BC] 165
Arrhenius, Svante August [1859–1927] 39
Ascheim, Elizabeth Fleischman [1859–1905] 98
Aston, Frances William [1877–1945] 120, 233, 296
Auger, Pierre [1899–] 156, 304, 410

Bacher, Robert Fox [1905–] 299, 411, 412, 417, 423, 426
Back, Ernst [1881–1959] 270 & n, 274n, 629
Badash, Lawrence [1934–] 64, 112n
Baeyer, Otto von [1877–1946] 153–4, 155
Bainbridge, Kenneth Tompkins [1904–] 124, 233
Balmer, Johann Jakob [1825–1898] 12, 164, 167, 170–3, 255, 268, 273, 628
Barkla, Charles Glover [1887–1944] 42, 187n, 222, 226
Bartlett, James Holley [1904–] 416
Barthélémy, Toussaint [1850–1906] 43
Becker, Herbert 398
Becquerel, Alexandre Edmond [1820–1891] 44–5
Becquerel, Antoine César [1788–1878] 44
Becquerel, Antoine Henri [1852–1908]; age when radioactivity discovered 133; on atomic energy 115–16; beta-ray studies 87; character 36; death 49, 629; early work 45; further work on uranic rays 47–9, 108; Nobel Prize 49; phosphorence 108–9; physiological effects of radium 11, 99; as President of Académie des Sciences 49; pupils/students 183; radioactivity discovery 8, 42, 45–6, 628; on X-rays 43, 44, 137
Becquerel, Jean [1878–1953] 42, 45, 46
Bémont, Gustave [1857–1932] 55
Bertram, Franca [1901–], (Pauli's second wife) 314n
Bethe, Hans Albrecht [1906–]; on Lamb shift 456; on 1930s 401; nuclear coupling constant 482; on Oppenheimer 369; reviews of nuclear physics 417, 423, 426, 485; at Rochester Conferences 461n; *other references* 299, 376, 390, 422
Bhabba, Homi Jehaugir [1909–1966] 376, 427, 434, 452
Bieler, Etienne Samuel [1895–1929] 239–40
Biot, Jean-Baptiste [1774–1862] 44, 70
Birge, Raymond Thayer [1887–1980] 366
Biswas, S. 513
Bjorken, James Daniel [1934–] 467, 577, 578, 579
Blackett, Patrick Maynard Stuart [1897–1974] 318, 362–3, 375, 404, 511, 513
Blankenbecler, R. 504
Bleuler, Konrad [1912–] 355
Blewett, John Paul [1910–] 408n, 438
Bloch, Felix [1905–1983] 254, 376, 387, 412
Bludman, Sidney Arnold [1927–] 592
Bohr, Harald [1887–1951] 194
Bohr, Niels Henrik David [1885–1962] 193–202; age when quantum work done 252; Atoms for Peace Award 163; background 194; and Balmer formula 164 & n, 173; on beta-radioactivity 223–4, 629; on beta-spectra 309–10, 311–12, 413; BKS proposal 105n, 248n, 311, 319, 630; on Born's work 260; at Cambridge (1946) conference 7; at Cambridge University (1911–1912) 194–5; character 199, 200, 459; classical theory conflicted by his atomic model 211; complementarity principle 251, 255, 262, 631; on confusion and progress 4, 249; at Copenhagen 217, 218, 363, 629; correspondence principle used 247–8; death 585, 635; and Dirac 287, 288, 312–13, 349–50, 374n, 378, 382; on electron spin 278–9; energy conservation law questioned 17, 105, 309–13, 317, 318–19, 631; hydrogen atom model 13, 164, 195–96, 200–2, 251, 372; hydrogen atom model, reactions to 208–10; on hydrogen spectrum fine structure 213, 214; on Jordan–Pauli commutation relations 354; later life 459, 585; magneton (unit) 270, 299; marriage 196; on matrix mechanics 255; on Moseley 228–9; on neutrino hypothesis 320; on neutrons 412, 413–14; and Oppenheimer 367; and Pauli 268; Ph.D. thesis 194; quantum theory work 252, 313; reminiscences 164, 200, 211, 228–9, 249, 278–9; on Rutherford 163–4, 186, 437; Rydberg constant calculations 247, 347; on Schroedinger 261; on (J. J.) Thomson 195; on Thomson atomic model 185
Bohr-Nørlund, Margrethe [1890–1984] 196
Boltwood, Bertram Borden [1870–1927] 146, 154, 188, 231
Boltzmann, Ludwig [1844–1906]; counting procedures 281, 282, 285, 334; molecular dissociation studies 121, 174; statistics 135, 280, 282; suicide 151; time reversal invariance 526; in Vienna 150; on X-rays 41; *other references* 84, 275, 365; *see also (in subject index):* Boltzmann . . .

Born, Max [1882–1970]; age when major work done 252; Einstein correspondence 259–60, 261; fluctuation studies 332, 333, 334, 336, 338, 630; at Goettingen 218, 365; on Jordan–Wigner work 340; Lagrangian methods used 342; matrix mechanics 249, 254, 331, 352, 630; Nobel Prize 258; non-linearly modified Maxwell equations 390; on Oppenheimer 367; Ph.D. students/assistants 251, 268, 367; probability in quantum mechanics 251, 255, 256–8, 260–1, 631; Schroedinger's influence 259–60, 261; thinking processes described 258–60; on (J. J.) Thomson's work 187
Bose, Satendra Nath [1894–1974] 135, 251, 252, 280, 283, 630; *see also (in subject index):* Bose–Einstein (BE) . . .
Bothe, Walther Wilhelm [1891–1957] 398, 404
Bouchard, C. 99
Bouchez, R. 124
Bouguer, Pierre [1698–1758] 165n
Bowen, Ira Sprague [1898–1973] 170
Bragg, Sir William Henry [1862–1942]; at Adelaide University 145; on alpha particles 145–7, 151; Nobel Prize (with son) 148; Royal Society election 147–8; in World War 1 236
Bragg, Sir William Lawrence [1890–1971] 148
Brashier, J. A. 78
Breit, Gregory [1899–1981] 365 & n, 376, 405, 406, 424
Brickwedde, Ferdinand Graft [1903–] 402
Brink, David Maurice [1930–] 438
Brobeck, William Morrison [1908–] 407
Brode, Robert Bigham [1900–] 512
Broek, Antonius Johannes van den [1870–1926] 223, 227–8, 230
Broer, Lambertus Johannes Folkert [1916–] 355
Broglie, Louis de [1892–] 250, 252, 299, 418–19, 630, 632
Bromberg, Joan [1929–] 356, 438
Brown, Laurie Mark [1923–] 320, 414n, 438
Brown, Percy [1875–1950] 98, 101
Brueckner, Keith Allan [1924–] 486, 496, 525n
Bunsen, Robert [1811–1899]; burner 76, 168, 170; calorimeter 113; spectral analysis 167, 168, 169, 170, 177, 627
Burckhardt, Jakob Christoph [1818–1897] 447
Burgoyne, N. 528
Butler, Clifford Charles [1922–] 511, 512, 542

Cabibbo, Nicola [1935–] 563, 601, 609, 635
Callan, Curt [1942–] 578 & n
Campbell, Norman Robert [1880–1949] 119, 179–80
Carlson, Franklin [1899–1954] 369, 410, 427, 428
Carlyle, Thomas [1795–1881] 3, 129, 136, 137, 260
Carnot, Sadi [1796–1832] 106, 108n, 109, 111, 112
Caruso, Enrico [1873–1921] 286
Case, Kenneth Myron [1923–] 484, 523n, 535
Casimir, Hendrik [1909–] 261, 297, 354, 375, 376, 387
Cassen, Benedict [1902–] 424, 425
Chadwick, Sir James [1891–1974] ; alpha-hydrogen scattering 239–40; alpha-helium scattering 285; beta-spectra 11, 157, 158–60, 303, 304, 306, 631; Liverpool cyclotron 436; on neutron 409, 410, 411, 412, 438; neutron discovery 397, 398, 399; textbook 64, 157, 233, 297, 398; wartime internment 160, 236n
Chateaubriand, François René de [1768–1848] 234
Chew, Geoffrey Foucar [1924–] 497, 502, 503, 504 & n, 505
Christenson, James [1937–] 538
Christofilos, Nicholas [1917–1972] 476n
Christy, Robert Frederick [1916–] 369
Clausius, Rudolf Julius Emmanuel [1822–1888] 174, 177
Clerk Maxwell, James: *see* Maxwell, James Clerk
Cockcroft, Sir John Douglas [1897–1967] 17, 234, 405, 406, 478
Colding, Ludvig August [1815–1888] 106
Coleman, Sidney Richard [1937–] 590, 594
Compton, Arthur Holly [1892–1962]; electron spin and ferromagnetism 279, 630; gravitational effects on radioactivity 114n; photon–electron scattering 246; *see also (in subject index):* Compton effect . . .
Compton, Karl Taylor [1887–1954] 365
Comte, Auguste [1798–1857] 165–6, 627
Conant, James Bryant [1893–1978] 117
Condon, Edward Uhler [1902–1974] 104, 124, 296, 365, 411, 424, 425
Conversi, Marcello [1917–] 453
Cool, Rodney Lee [1920–] 490
Coriolis, Gaspard Gustave de [1792–1843] 107
Cornu, Alfred Marie [1841–1902] 268, 269
Coulomb, Charles Augustin de [1736–1806] 70; *see also entries beginning with* Coloumb *in subject index*
Courant, Ernest David [1920–] 476
Courant, Richard [1888–1972] 199
Cowan, Clyde Lorrain [1919–1974] 569
Cronin, James Watson [1921–] 534, 538, 539, 542
Crookes, Sir William [1832–1919]; on Becquerel and his experiments 45, 46, 47; corpuscular view 80, 81; Darwinian imagery 178; on radioactive energy source 109; X-ray scintillations discovered 110; in World War 1 236
Curie, Eve [1904–] 57, 58, 63
Curie, Irène [1897–1956] as a baby 53, 55, 56; *see also* Joliot-Curie, Irène
Curie-Sklodowska, Marie [1867–1934]; age when major work done 133n; alleged affair with Paul Langevin 57; background 52; beta-ray studies 87, 262; bibliography 63; death 58, 632; on energy source of radioactive substances 108–9, 111; experimental methods 53–4, 55; first scientific paper 53; League of Nations work 58; Lorentz tribute 262; as a mother 53, 55, 56; Nobel Prizes 56, 57; papers on radioactivity 8, 53, 54–5, 108–9; polonium discovery 55, 109; at Solvay Conferences 123, 223, 224; radioactive substances, search for 119; radioactivity as atomic property 8, 55, 144; radium discovery 55–6; Sorbonne professorship 57; on spark spectrum for radium 118; on transmutation 112; war (WW1) effort 58, 236
Curie, Paul-Jacques [1855–1941] 52–3
Curie, Pierre [1859–1906]; beta-ray studies 87; complete works 63; on criminal use of radium 116; death 57, 629; energy measurement for radium 113, 115; illness and fatal accident 56–7; Nobel Prize 56; his parents 53; on physiological effects of radioactivity 11, 35, 99–100; scientific background 52–3, 56; Solvay Conference discussions 123, 223; Sorbonne professorship 56

Hartree 361, 377, 380
Hasenöhrl, Friedrich [1874–1915] 236, 251
Heaviside, Oliver [1850–1925] 286
Heckius, Johannes [1557-*c.* 1620] 275
Heilbron, John [1934–] 164n, 187n, 202, 228n, 268n, 438
Heisenberg, Werner Carl [1901–1976]; age when major work done 251, 362; on antimatter 360; collected works 391; death 585, 587, 635; on Dirac equation 347, 348, 375, 377; dispersion formula 353; on electron self-energy 372, 632; on electron spin 278, 279, 280; energy fluctuation studies 332, 333n, 334, 336, 338, 630; on Fermi field theories 580, 585–6; on Fermi interaction 426 & n, 427, 428; on Laporte rule 525; matrix methods 249, 251, 253–4, 630; on neutrino 315, 631; nuclear forces theory 401, 410, 413, 414, 415 & n, 416, 423, 430; Pauli correspondence 17, 265, 318, 348, 361–2, 377, 380, 387, 388, 412, 426, 528, 585 & n, 631; Ph.D. topic 251; on photon self-energy 385, 632; on positron theory 380, 386, 497, 632; probability interpretation 260; quantum electrodynamics 329, 330, 341, 342–3, 346, 355, 361–2, 368, 369, & n, 631; quantum mechanics paper 251, 253–4, 630; comments on 255, 287; reminiscences 330, 347; *Rumpf* model 271 & n, 176, 277; *S*-matrix theory 7, 498; second quantization method 379, 418; subtraction techniques 379, 380, 381; on symmetrically coupled oscillators 271; uncertainty relations 249, 251, 255, 261–2; universal length concept 428, 498; vacuum polarization calculations 383; *other references* 215, 386, 459; *see also (in subject index):* Heisenberg forces; *S*-matrix theory; uncertainty relations
Heitler, Walter [1904–1981] 302, 356, 376 & n, 390, 426, 435, 486, 512n
Helmholtz, Hermann Ludwig Ferdinand von [1821–1894]; atom model 182–3; on applicability of thermodynamics 108n; electricity, components concept 70, 628; on energy conservation 106–7; on Faraday's laws of electrolysis 72, 73; Lagrangian methods used 342; vector potential, term used 343; on vortex motion 176; *other references* 75, 364
Henry, Charles [1859–1926] 42
Henry, Joseph [1797–1878] 183
Hermann, Armin [1933–] 202, 314n
Herschel, Sir John Frederick William [1792–1871] 170, 216
Hertz, Gustav Ludwig [1887–1975] 236
Hertz, Heinrich Rudolf [1857–1894] 37, 59, 69, 80, 81
Herzberg, Gerhard [1904–] 302
Herzfeld, Karl [1892–1978] 366
Hess, Victor Franz [1883–1964] 403
Hevesy, Georg Charles von [1885–1966] 208, 225n
Higgs, Peter Ware [1929–] 594, 635; *see also (in subject index) entries beginning with:* Higgs
Hirn, Gustave Adolfe [1815–1890] 106
Hittorf, Johann Wilhelm [1824–1914] 79, 168
Hoff, Jacobus Henricus van't [1852–1911] 275
Hofstadter, Robert [1915–] 502, 573
Holten, Florence Hannah (Dirac's mother) 286
Holton, Gerald [1922–] 400
Holtzmann, Karl Heinrich Alexander [1811–1865] 106
Houston, William [1900–1968] 365
Hoyer, Ulrich [1938–] 202
Huggins, William [1824–1910] 169, 172, 177
Hulthén, Lamek [1909–] 449
Hund, Friedrich [1896–] 285, 425

Ikeda, Mineo [1926–1983] 555
Iliopoulos, John [1940–] 601, 602
Inoue, Takesi [1921–] 453 & n
Ishiwara, Jun [1881–1947] 429
Ising, Gustav Adolf [1883–1960] 407, 571, 630
Itzykson, Claude [1938–] 467
Iwanenko, Dmitrij Dmitrievič [1904–] 303 & n, 409, 410, 411, 418, 426

Jaffe, George Cecil [1880–1965] 52 & n, 57, 84–5
Jammer, Max [1915–] 263
Jauch, Josef Maria [1914–1974] 467
Jeans, Sir James Hopwood [1877–1946] 121–2, 135n, 182, 209, 365
Johnstone Stoney: *see* Stoney . . .
Joliot-Curie, Irène [1897–1956] 398–9, 400, 416, 438
Joliot, Jean Frédéric [1900–1958] 125, 398, 399 & n, 400, 416, 438
Jordan, Pascual [1902–1980]; age when major work done 251; commutation relations papers 341, 354, 631; on electron self-energy 372; on electron spin 279; energy fluctuation studies 332, 333n, 334, 336, 338, 630; matrix mechanics 249, 254, 290, 331, 338, 339, 340, 352, 387, 630, 631; neutrino theory of light (poem) 419; Ph.D. topic 251; polarization treatment 337; second quantization method 329, 332, 338, 339, 340, 377, 379, 630; *see also (in subject index) entries beginning with:* Jordan
Jost, Res [1918–] 334, 355, 381, 387, 489–90, 529
Joule, James Prescott [1818–1889] 106

Kalckar, Jörgen [1935–] 260
Källen, Gunnar [1926–1968] 375
Kaluza, Theodor [1885–1954] 584, 624
Kamerlingh Onnes: *see* Onnes, Kamerlingh . . .
Karlik, Berta [1904–] 403
Kaufmann, Walter [1871–1947]; beta-ray studies 87, 88, 132, 153; cathode ray studies 10, 78, 83–4, 628
Kay, William [1879–1961] 216
Kayser, Heinrich Gustav Johannes [1835–1940] 165, 182, 202
Kekulé von Stradonitz, Friedrich August [1829–1896] 72
Kelvin, Lord, (William Thomson) [1824–1907]; atom model 174, 175, 176–7, 184, 185; on Becquerel rays 52, 111; death 216; on energy conservation principle 111; on radioactive decay 122, 180; on radium energy production 113; on radium chemical composition 118n; USA visit 365; Kemble, Edwin Crawford [1889–1984] 365
Kemble, Edwin Crawford [1889–1984] 365 & n
Kemmer, Nicholas [1911–] 387, 434, 435, 438, 486, 633
Kennard, Earle Hesse [1885–] 279
Kerst, Donald William [1911–] 574
Kevles, Daniel J. [1939–] 391

Killian, James [1904–] 163
Kimball, Arthur Lalanne [1856–1922] 179
Kinoshita, Toichiro [1925–] 465, 466, 467
Kirchhoff, Gustav Robert [1824–1887]; blackbody radiation law 167–8, 627; on Kelvin's vortex atom theory 177; sodium lines in solar spectrum 165, 167, 177, 627; spectral analysis studies 168, 169, 170, 177, 627
Kleeman, Richard Daniel [1875–] 146, 147, 151
Klein, Felix [1849–1925] 235
Klein, Martin [1924–?] 134n
Klein, Oskar [1894–1977]; Compton scattering, theory 348, 430, 631; at Copenhagen 217; on electron self-energy 372; on gauge theories 583n; matrix methods 338, 339, 387, 631; paradox 349, 631; Pauli correspondence 309, 316, 343, 348; scalar wave equation 289; unified field theory 584, 624; *see also (in subject index):* Jordan–Klein . . .; Kaluza–Klein . . .; Klein . . .
Kleist, Edward Georg von [*c.* 1700–1748] 70
Knipping, Paul [1883–1935] 42
Kobayashi, Makoto [1944–] 608
Kobayasi, Minoru [1908–] 435
Kockel, Bernhard 386
Kolhörster, Werner [1887–1946] 404
Köllicker, Rudolf Albert von [1817–1905] 39
Kohen, Heinrich Mathias [1874–1948] 173
Konen, Heinrich Mathias [1874–1948] 173
Konopinski, Emil John [1911–] 420, 421, 422, 438, 530, 570
Kovarik, Alois Francis [1880–1965] 154
Kowarski, Lew [1907–1979] 505
Kramers, Hendrik Anton [1894–1952]; book (on quantum mechanics) 247, 260; C-invariance introduced 381, 423, 633; death 449; degeneracy theorem 527; dispersion relation 353, 499 & n, 500; on electron spin 280; electron theory 448–9; energy nonconservation proposal (with Bohr & Slater) 105n, 248n, 259n, 311, 319, 630; renormalization ideas 391; at Shelter Island Conference 450, 451; *other references* 215, 217, 368n, 447
Kroll, Norman Myles [1922–] 460
Kronig, Ralph de Laer [1904–] 255, 280, 299–300, 302 & n, 387, 419n, 499, 500
Kuhn, Thomas [1922–] 130, 164n, 202, 286n
Kusaka, Shuichi [1916–1947] 369
Kusch, Polykarp [1911–] 451

Laborde, Albert [1878–1968] 113
Lamb, Horace [1894–1934] 145, 190
Lamb, Willis Eugene [1913–] 369, 447, 451, 460, 466
Lambert, Johann Heinrich [1728–1777] 165n
Landau, Lev Davidovich [1908–1968] 354, 533, 534, 541
Landé, Alfred [1888–1975] 269–70, 272, 274, 280, 366, 630
Langevin, Paul [1872–1946] 57, 216, 232
Laplace, Pierre-Simon, Marquis de [1749–1827] 107, 122
Laporte, Otto [1903–1971] 366, 525, 630
Larmor, Joseph [1857–1942] 181, 182, 185, 216, 271, 286, 342
Lattes, Cesare Mansueto Giulio [1924–] 454, 479, 505
Laue, Max Theodor Felix von [1879–1960] 42, 208
Lavoisier, Antoine Laurent [1743–1794] 107
Lawrence, Ernest Orlando [1901–1958] 8, 17, 407–9, 430, 436, 472, 474, 572
Lederman, Leon Max [1922–] 477, 608, 611
Lee, Benjamin W. [1935–1977] 602, 606, 611
Lee, Tsung Dao [1926–]; antinucleons paper 488; C, P, and T applications 527; CP-violation tested 25, 532, 542; CPT-theorem 529; G-parity 490; on neutral K-mesons 523; two-component neutrino theory 533; W-boson mass calculations 582n; weak interactions 534, 542
Lehmann, Harry [1924–] 500
Leibniz, Gottfried Wilhelm [1646–1716] 107
Lenard, Philipp [1862–1947]; age when doing major work; 133n; atom model 182; cathode ray studies 37, 40, 81 & 82n, 85; on spectra 197; in World War 1 235
Lenz, Wilhelm [1888–1957] 120, 233
Leprince-Ringuet, Louis [1901–] 8, 512, 542
Lewis, Gilbert Newton [1875–1946] 246, 365, 631
Liebig, Justus von [1803–1873] 106
Liénard, Alfred Marie [1869–?] 83
Livingood, John Jacob [1903–] 407, 478
Livingston, Milton Stanley [1905–] 407, 408 & n, 438, 476, 505
Llewellyn-Smith, Christopher Hubert [1942–] 580
Lockyer, Joseph Norman [1836–1920] 169, 178
Lodge, Sir Oliver Joseph [1851–1940] 41, 179, 236
Lofgren, Edward Joseph [1914–] 407 & n
Lohuizen, Tennis van [1883–1953] 270
Loomis, Wheeler [1889–1976] 365
Lorentz, Hendrik Antoon [1853–1928]; on atomistic interpretation of Maxwell equations 76–7; death and funeral 216, 631; E–m–v relationship 89; on electron spin 278; electron theory 76, 194, 233, 327, 343–4, 371 & n; Lagrangian methods used 342; motion equation 389, 390; Nobel Prizes 78, 89; on quantum mechanics 261; on Thomson's (1913) atomic model 209, 216; tribute by Marie Curie 262; on X-rays 41; on Zeeman effect 10, 76, 77, 132, 268, 269; *other references* 5, 75, 274, 365; *see also (in subject index) entries beginning:* Lorentz
Loschmidt, Johann Joseph [1821–1895] 74, 150, 174
Low, Francis Eugene [1921–] 497, 502
Ludwig, Guido [1907–] 387
Ludwig, Anna Bertha [1839–1919] (Roentgen's wife) 36, 38
Ludwig, Josephine Berta [1881–?] (Roentgen's niece) 36
Luttinger, Joaquin [1923–] 381

Ma, Shih-Tsun [1913–1962] 355, 390
Macaulay, Thomas Babington [1800–1859] 3
McClelland, James Alexander [1870–1920] 53
McCormmach, Russell [1933–] 200, 202
McGucken, William [1938–] 202
Mach, Ernst [1838–1916] 106
MacInnes, Duncan Arthur [1885–1965] 450
McMillan, Edwin Mattison [1907–] 8, 406n, 407, 409, 438, 474, 475, 505
McMillan, William George [1919–] 479
Madansky, Leon [1923–] 490
Mahmoud, Hormoz Massoud [1918–] 530, 570

Maiani, Luciano [1941–] 601, 602
Majorana, Ettore [1906–1938] 381, 402, 416, 423, 633
Marconi, Guglielmo, Marchese [1874–1937] 69
Margenau, Henry [1901–] 365
Marsden, Sir Ernest [1889–1970] 189, 191, 192, 226, 228n, 236, 238
Marshak, Robert Eugene [1916–]; books 466, 513, 518n, 542; pion calculations 485; and Rochester Conferences 461 & n, 466; two-meson hypothesis 452, 453; $V-A$ theory 536, 542; on W-bosons 581
Maskawa, Toshihide [1940–] 608
Matthews, Paul Taunton [1919–] 484
Maxwell, James Clerk [1831–1879]; on atomic model building 175, 176, 627; on atoms 72–3, 137, 175, 178, 326, 627; Avogadro's number, estimate 74; as Cavendish Professor (Cambridge) 58; death 628; on electrolysis 71–2, 73; electromagnetic theory 244, 343, 627; on experimental science 454; on Faraday's work on light 74–5; gas molecule size estimation 174, 175; on stellar spectroscopy 169; on theory and experiment 454, 455; *see also (in subject index):* Maxwell . . .
Mayer, Alfred Marshall [1836–1897] 183–4
Mayer, Julius Robert [1814–1878] 106
Mehra, Jagdish [1937–] 215n, 269n
Meitner, Lise [1878–1968]; beta-decay scheme 306 & n, 307, 308, 309; beta-spectra studies 147, 151, 154; Klein–Nishina formula verified 158, 348; scientific background 150–1; in World War 1 236
Mendeléev, Dmitri Ivanovich [1834–1907] 221 & n, 222, 227, 627
Merzbacher, Eugen [1921–] 555
Meyenn, Karl von [1937–] 268n, 314n
Meyer, Julius Lothar [1830–1898] 221n
Meyer, Stefan [1872–1949] 150
Michel, Louis [1923–] 490, 505, 530, 535
Michelson, Albert Abraham [1852–1931] 213, 236, 364
Mie, Gustav [1868–1957] 342, 344, 371 & n
Miehlnickel, Edwin 438
Miller, Arthur I. [1940–] 88n
Millikan, Robert Andrews [1868–1953] 20, 236, 351, 352, 363, 366, 404 & n
Mills, Robert Laurence [1927–] 585, 586, 587, 634
Mittag-Leffler, Magnus Gustaf [1846–1927] 188
Mohr, Carl Friedrich [1806–1879] 106
Møller, Christian [1904–1980] 375, 449, 450, 511
Mommsen, Theodor [1817–1903] 131
Monet, Claude [1840–1926] 286
Moore, Marianne [1887–1972] 33, 445
Morrison, Philip [1915–] 369, 471
Morse, Philip McCord [1903–1985] 365
Moseley, Henry Gwyn Jeffreys [1887–1915] 144, 223, 228–30, 237, 629
Mössbauer, Rudolf Ludwig [1929–] 124
Mott, Nevill Francis [1905–] 260–1, 311, 318
Moyer, Albert [1945–] 391, 414n
Mukherji, Visvapriya 438, 453n
Muller, Hermann Joseph [1890–1967] 95
Mulliken, Robert Sanderson [1896–] 365
Murphy, George Moseley [1903–1969] 402
Musschenbroek, Petrus van [1692–1761] 70
Mussolini, Benito [1883–1945] 317

Nafe, John Elliott [1914–] 451
Nagaoka, Hantaro [1865–1950] 183, 429
Nagel, Bengt [1927–] 110n
Nakano, Tadao [1926–] 519
Nambu, Voichiro [1921–] 562, 592
Natanson, Ladislas [1864–1937] 283
Neddermeyer, Seth Henry [1907–] 432, 433
Ne'eman, Yuval [1925–] 556, 611
Nelson, Edward Bryant [1916–] 451
Nernst, Hermann Walther [1864–1941] 235, 310, 365
Neumann, John von [1903–1957] 265, 266, 292 & n, 366, 422, 461
Newton, Sir Isaac [1642–1727] [Preface] v, 114, 163, 265, 216, 286
Nichols, Edward Leamington [1854–1937] 364
Nicholson, John William [1881–1955] 198–9, 200, 209, 211, 215
Nicolai, Georg Friedrich [1874–1964] 234, 235
Nielsen, Jeus Rud [1894–1979] 199
Nishina, Yoshio [1890–1951] 348, 429, 430, 432, 459, 631; *see also (in subject index):* Klein–Nishina formula
Nishijima, Kazuhiko [1926–] 505, 519, 520, 542
Nitske, W. Robert [1909–] 49
Nordsieck, A. 376, 426
Nørlund, Margrethe [1890–1984] (Bohr's wife) 196
Nuttall, John Mitchell [1890–1958] 124

Occhialini, Giuseppe Paolo Stanislao [1907–] 362–3, 375, 404, 454
Oehme, Reinhard [1928–] 529
Oersted, Hans Christian [1777–1851] 70
Ogawa, Shuzo [1924–] 555
Ohm, Georg Simon [1787–1854] 70
Ohnuki, Yoshio [1928–] 555
Okubo, Susumu [1930–] 557
Onnes, Heike Kamerlingh [1853–1926] 75, 115
Oppenheimer, J. Robert [1904–1967]; age when major work done 362; American physics influenced by 364, 365, 369–70; at Atoms for Peace Award ceremony (1957) 163; cosmic ray studies 410, 427, 428; emotional instability 367, 368; and Furry 379, 380, 382, 383, 385; on Heisenberg–Pauli papers 369 & n; on hole theory 351, 360; illness (1927–1928) 368; on mesons 433 & n, 496, 525; on neutrons 410–11; pair (electron–positron) formation 375; and Pauli 368; Ph.D. students 369, 410; Ph.D. topic 367; radiation damping quantum theory 390; renormalization hinted at 385; at Rochester Conferences 461n, 518n; scientific background 366, 367–8; self-energy work 369, 372–3, 384; on Shelter Island Conference 22, 450, 451; speech, manner of 369; and Wigner's rule 285n; *other references* 376, 407, 447, 461
Ornstein, Leonard Salomon [1880–1941] 299
Orthmann, Wilhelm [1901–1945] 308, 309
Ostwald, Friedrich Wilhelm [1853–1932] 235
Oudin, Paul [1851–1923] 43

Pais, Abraham [1918–]; Anderson interview 363; antinucleon papers 488, 489; at Atoms for Peace Award ceremony (1957) 163; baryon names and properties 514 & n, 520; and Bohr 4, 163–4, 194; in Copenhagen 449, 450; on *C*- and

Pais, Abraham (*cont.*)
P-invariance testing 532–3, 542; on *CP*-violation 538–9; and Dirac 25–6; on Einstein and photon 136; on electron charge 61–2; on electron self-energy 448; energy law, history 311n; and Fermi 485; and Hulthén 449; isospin group enlargement 519, 583; and Kramers 448, 449; and Lawrence 407; on Lee (& Yang) 532–3, 542; on leptons 450; and Møller 449, 450, 511; on neutral currents 602; and Oppenheimer 447; on particle physics development (1948–1968) 621; and Pauli 313–4; on positron 391; on quantum theory 248–9, 355, 356; and Rabi 19, 447; on Racah 555; on regeneration 522; and Rochester 512; at Rochester Conferences 461n, 485, 486; and Rubbia 2; selection rules 489–90, 518 & n; and Smyth 116–17; special relativity, history 14, 67; symmetry breaking papers 593–4; on tau-theta puzzle 525; and Uhlenbeck 215, 422, 423, 447, 463; and Watson 486; on weak interactions 542; and Wigner 266; on Yang (& Lee) 523–3, 542; and Yukawa 429
Pancini, Ettore [1915–1981] 453
Panofsky, Wolfgang Kurt Hermann [1919–] 407, 480, 571, 572, 573, 605, 611
Park, David Allen [1919–] 555
Paschen, Friedrich [1865–1947] 213, 214, 270 & n, 274n, 629
Pasternack, Simon [1914–1976] 451
Pauli, Wolfgang [1900–1958]; age when major work done 251, 362; assistants 387; on beta-spectra 309–310, 311–12, 315, 413; body movements 314, 360; on Born's work 260; commutation relations paper 341, 354, 387, 631; *CPT*-theorem 529; death 585, 635; on Dirac's work 360, 361, 375, 377, 387, 389, 412; divorce and remarriage 314 & n; on electron self-energy 372; on electron spin 278, 280; on electron stability models 371n; on energy conservation 311; exclusion principle 250, 253, 273–4, 284, 630; gauge field theories 583, 585; Heisenberg correspondence 265, 318, 348, 361–2, 377, 380, 387, 388, 412, 417, 426, 528, 585 & n; hyperfine structure paper 279, 630; on mass–energy equivalence 89, 104, 233; matrix methods 254, 255, 289–90, 291n, 631; neutrino hypothesis 17, 309, 315–6, 318, 351n, 410, 413, 631; on neutron decay 417; on nuclear spin 279 & n, 630; on Oppenheimer 368; personality 314; polarization treatment 337; at Princeton (1935–1936) seminar 361, 391; probability interpretation 258; quantum electrodynamics theory 329, 332–3, 341, 342–3, 346, 355, 368, 369 & n, 631; on quantum statistics 258, 285; review of quantum field theory 356; on Rydberg 173; scientific background 267–8; scientific correspondence 320; spin–statistics connection 528; spinless field theory 376, 387–8, 528, 632; as teacher 387; on vacuum polarization 383; on Zeeman effect 269, 270, 272; *other references* 7, 275, 525n; *see also (in subject index):* Pauli . . .
Pauli-Bertram, Franca [1901–] 314n
Pauling, Linus Carl [1901–] 365
Peierls, Rudolf Ernst [1907–] 354, 355, 382, 383, 387, 422
Perl, Martin Lewis [1927–] 608
Perrin, Francis [1901–] 42, 81, 183, 318 & n, 410, 418, 419
Perrin, Jean-Baptiste [1870–1942] 114
Petit, Alexis Thérèse [1791–1826] 221, 627
Peyrou, Charles [1918–] 466, 514, 542
Piccioni, Oreste [1915–] 453, 466, 490, 522
Planck, Max Karl Ernst Ludwig [1858–1947] 133–4; age when quantum theory formulated 133; blackbody radiation law 74, 133–4, 193, 210–11; on Carnot's estimate for mechanical equivalent of heat 106; electron charge studies 61–2, 74; on energy conservation 107–8; mass equivalent of molecular binding energy 232; *p–m–v* relationship 88; quantum theory 9, 130, 133–4, 137, 168, 251, 628; on time reversal invariance 526; in World War 1 235 & n, 236–7; *other references* 150, 151, 181, 262; *see also (in subject index):* blackbody radiation law . . . ; Planck . . .
Plessett, Milton Spinoza [1908–] 370, 375
Plücker, Julius [1801–1868] 68n, 79, 167, 168, 170
Poggendorf, Johann Christian [1796–1877] 68, 106n
Poincaré Henri [1854–1912]; on Becquerel rays 48, 49; electron structure theory 371n; on energy conservation violation 113–14; Lagrangian methods used 342; Nagaoka atom model quoted by 183; obituary by Wien 43 & n; on Roentgen rays 43, 48
Politzer, Hugh David [1949–] 590, 606
Pomeranchuk, Isaak Yakovlevich [1913–1966] 501, 635
Powell, Cecil Frank [1903–1969] 454, 479, 511, 512n, 516
Powell, Wilson March [1903–1974] 512
Present, Richard David [1913–] 424
Primack, Joel Robert [1945–] 602
Proca, Alexandre [1897–1955] 433, 633
Prout, William [1785–1850] 174, 221, 627
Pryce, Maurice Henry Lecorney [1913–] 419 & n
Pupin, Michael Idvorsky [1858–1935] 5

Rabi, Isidor [1898–]; on electron–neutron interaction 483; on Goudsmit 276; and hyperfine structure 451; on international consortia/centers 472–3; on military source of funding 19; on Oppenheimer 367, 368; on Schroedinger wave mechanics 255; *other references* 365 & n, 447, 458
Racah, Giulio [1909–1965] 267, 376n, 423, 555
Radicati di Brozolo, Luigi [1919–] 559
Raman, Chandrasekhara Venkata [1888–1970] 301 & n, 302
Ramsey, Norman Foster [1915–] 476, 477n, 505
Ramsey, Sir William [1852–1916] 60, 150, 169
Rasche, Günther [1934–] 438
Rasetti, Franco Rama Dino [1901–] 278n, 298, 302
Rasmussen, J. O. 124
Rayleigh, Lord, (John William Strutt) [1842–1919]; as director of Cavendish Laboratory 59; on Balmer's formula 164; blackbody radiation law (with Einstein & Jeans) 135n; on Bohr model of hydrogen atom, 208–9; on J. J. Thomson 85; *other references* 183, 216, 364
Re, Fernando 183
Rechenberg, Helmut [1937–] 215n, 269n
Regge, Tullio Eugene [1931–] 503, 504, 505, 566, 635

Regnault, Henri Victor [1810–1878] 183
Reid, Robert William [1933–] 63
Reines, Frederick [1918–] 569, 611
Reinganum, Maximilian [1876–1914] 236
Retherford, Robert Curtis [1912–1981] 447, 451
Richards, Theodore William [1868–1928] 35
Richardson, Owen Willans [1879–1959] 365
Richter, Burton [1931–] 604, 605, 611
Riesz, Marcel 390
Ritz, Walter [1878–1909] 164, 173–4
Robinson, Harold Roper [1889–1955] 155, 159
Rochester, George Dixon [1908–] 511, 512 & n, 542
Roentgen, Wilhelm Conrad [1845–1923]; age when X-rays discovered 133; character 36; demonstration before Emperor Wilhelm II 39; discovery of X-rays 1, 37–38, 131, 628; experimental approach 37; Geissler tube used by 69; illness and death 39; nickname 37; Nobel Prize banquet speech 39; photographs using X-rays 38, 95; public lecture (only one) on X-rays 39; Rühmkorff coil used by 69; scientific background 36–7; shyness 39; speculation on nature of X-rays 41, 130, 137; summary of paper on X-rays 40–1; his wife 36, 38; in World War 1 235
Roentgen-Ludwig, Anna Bertha [1839–1919] 36, 38
Rohrlich, Fritz [1921–] 467
Romer, Alfred [1906–] 38n, 49
Ron: *see* Sanchez Ron
Rose, Morris Erich [1911–1967] 383
Rosenfeld, Arthur [1926–] 195, 200, 202, 504n, 520n, 542
Rosenfield, Léon [1904–1974] 342n, 354, 355, 373n, 483
Rosner, J. L. 606
Rossi, Bruno [1905–] 405, 438, 452n, 466, 512n, 515–16
Rowland, Henry Augustus [1848–1901] 10, 75, 78, 131, 170, 364, 628
Royds, Thomas [1884–1955] 61
Rubbia, Carlo [1934–] 1, 2, 576, 610
Rubinowicz, Adalbert [1884–1979] 214
Rühmkorff, Heinrich Daniel [1803–1874] 10, 69, 131, 167, 168, 627: *see also (in subject index):* Rühmkorff . . .
Runge, Carl David Tolmé [1856–1927] 182
Russell, Alexander Smith [1888–1972] 225n
Russell, Henry Norris [1877–1957] 525
Rutherford, Eileen Mary [1901–1930] 60
Rutherford, Sir Ernest (Lord Rutherford of Nelson) [1871–1937]; on alpha-decay 180; on alpha-particles 61, 138; alpha-scattering 191, 223, 225, 237–9, 297, 311, 629; 'atomic energy' term used by 116; atomic model 12, 13, 190–1, 192, 296, 629; Armorial Bearings 217; background 59; Bakerian Lecture (1920) 397, 399; baronetcy 217; on Becquerel/uranic rays 59–60, 131–2; beta-radioactivity theory 401; on beta-ray absorption 153; on beta-spectra 155, 156–7, 303–4, 305; and Bohr 164, 195; on Bohr atomic model 209–10; Boltwood letters 154–5, 188, 231; books written 9, 64, 155, 156, 157, 233, 297, 398; on causality 212; as Cavendish Laboratory director 216–7, 218, 239, 365, 630; as Cavendish Laboratory research fellow 9, 59, 628; character 190; death and funeral 437, 633; delta-ray observations 148; on elements heavier than uranium 120; on energy nonconservation 311–2, 317; on energy source of radioactivity 111, 114 & n; and Fermi 401; on gamma-rays 62; Hahn letters 150, 153; half-life determinations 120–1, 125; hernia, fatal strangulation 436–7; on high-energy physics 405, 406, 436; on Kaufmann's work 132; knighthood 216; lifetime concept 120–2; lifetime concept 120–3; lifetime concept, implications not mentioned 129, 134, 137, 212; at McGill University 9, 60, 150, 628; at Manchester University 12, 63, 149, 188–93, 437–8, 629; 'metabolon' term used 119; on Moseley and his work 229, 237; mother of R. 62, 64; Nagaoka atom model quoted 183; nebulium mentioned 169; on neutron 397–8, 401, 410; Nobel Prize 63, 188; nuclear binding energy 233; nuclear charge 226; nuclear model 230–2, 237, 398; obituaries 308, 437–8; on polyelectron models 179; radioactivity conservation principle (with Soddy) 54, 115; on rapidity of advances in physics [Preface] v, 129, 217; on relativity theory 190; remembrances 437–8; scattering cross-section 191, 225; Silliman Lectures 9, 64, 129, 131–2, 134, 365; at Solvay Conferences 192, 193, 223, 224; as teacher 9, 437; and theoretical physics 12, 121, 190–3, 363; transformation theory (with Soddy) 112, 113 & n, 123, 190, 286, 297, 629; USA visits 364, 365; on van den Broek's rule 228; in World War 1 236; on X-rays 39; on Yale University 9
Rutherfurd, Lewis Morris [1816–1892] 169 & n
Rydberg, Johannes [1854–1919] 171, 173, 201, 202, 247; *see also (in subject index):* Rydberg constant

Sachs, Robert Green [1916–] 370
Sackur, Otto [1880–1914] 236
Sagnac, Georges [1869–1928] 114
Sakata, Shoichi [1911–1970] 422, 434, 435, 448, 453 & n, 555, 564n
Sakita, Bunji [1930–] 559
Sakurai, Jun John [1933–1982] 381, 527
Salam, Abdus [1926–] 462, 533, 534, 592, 594, 611
Sanchez Ron, J. M. 310n
Savart, Felix [1791–1841] 70
Schafroth, Max Robert [1923–1959] 387
Scharff-Goldhaber, Gertrude 87
Schiff, Lennart Isaac [1915–1971] 289, 291, 341n, 354, 370, 473
Schmidt, Gerhard Carl [1865–1949] 54 & n
Schmidt, Heinrich Willy [1876–1914] 148, 151
Schroedinger, Erwin [1887–1961]; age when major work done 251–2; death 635; English translation of papers 263; Ph.D. topic 251; on quantum probability 258; wave mechanics 249, 251–5, 256, 261, 288, 289, 345, 630; in World War 1 236; *see also (in subject index):* Schroedinger wave equations
Schur, Issai [1875–1941] 265
Schuster, Arthur [1851–1934] 41–2, 81, 83n, 164, 177, 188
Schwartz, Melvin [1932–] 570, 611
Schwarzschild, Karl [1873–1916] 236
Schweber, Silvan Samuel [1928–] 355, 380, 450n
Schweidler, Egon Ritter von [1873–1948] 123n
Schweigger, Johann Salomo Christoph [1779–1857] 70

Schwinger, Julian Seymour [1918–]; book and collected papers 355, 466; current algebra 568; electron anomalous magnetic moment calculations 456–7; Lamb shift calculations 458, 460, 461; and Oppenheimer 370; at Shelter Island Conference 452; on W-bosons 581n, 592
Scott, William Taussig [1916–] 181n, 245
Seebeck, Thomas [1770–1831] 70
Segré, Emilio [1905–] 49, 124, 400, 438
Séguin, Marc [1786–1875] 106
Serber, Robert [1909–]; and Lawrence 430, 479; on meson theory 433 & n, 434, 479, 523; on Oppenheimer 370, 391; reminiscences 438, 479; renormalization procedures 385, 633; vacuum polarization calculations 383, 385
Serwer, Daniel Paul [1945–] 94, 101
Shakespeare, William [1564–1616] 234, 483
Shaw, George Bernard [1856–1950] 234
Shaw, Ronald [1929–] 585
Sherrington, Charles Scott [1857–1952] 129
Silliman, Augustus Ely [1807–1884] 129
Sklodowska, Marie: *see* Curie-Sklodowska
Skobeltzyn, Dmitry [1982–] 404 & n, 438
Slater, John Clarke [1900–1976] 105n, 215n, 248n, 259n, 311, 319, 630
Smyth, Henry DeWolf [1898–] 116–7
Snell, Arthur Hawley [1909–] 407
Snyder, Hartland [1913–1962] 369, 476
Soddy, Frederick [1877–1956]; 'atomic energy' term used by 116; on Bragg's alpha-particle experiments 146; on Bohr atomic model 209; on (Marie) Curie's radioactivity theory 55; on decay lifetime concept 122; on electron absorption law 149, 152; on energy source of radioactivity 114; helium discovered ex-radium 60; isotopes discovered 224 & n, 225; 'metabolon' term used 119; radioactive substances listed (1904) 117–18; radioactivity conservation principle (with Rutherford) 54, 115; transformation theory (with Rutherford) 112, 113n, 123, 190, 286, 297, 629; on van den Broek's rule 228
Sommerfeld, Arnold [1868–1951]; age when quantum work done 252; atomic spectra structure theory 13, 89, 212–14; on Bohr atomic model 209; colloquium 134; on energy conservation 311; fine structure theory 311; fine structure theory 213–15, 276, 347; on half-integer quantum numbers 271; on nuclear constitution 232; and parastatistics 561; Ph.D. students 251, 268; on Planck's constant 199 & n; quantization conditions 284; spectroscopy book 166, 232, 273 & n; *other references* 215, 218, 365
Sopka, Katherine Russell [1921–] 391
Speiser, David [1926–] 556, 611
Stark, Johannes [1874–1957] 197, 398n
Stefan, Joseph [1835–1893] 150
Steinbeck, John [1902–1968] 286
Steinberger, Jack [1921–] 480
Steller, Jack Stanley [1921–] 480
Stern, Otto [1888–1969] 208, 278, 412
Stevenson, Edward Carl [1907–] 433
Stokes, Sir George Gabriel [1819–1903] 44, 47, 82n, 175, 176n, 302
Stoner, Edmund Clifton [1889–1973] 267, 273, 274, 630
Stoney, George Johnstone [1826–1911] 73, 74, 171, 174, 175, 176, 236
Strassman, Fritz [1902–1980] 158
Stratton, Julius Adams [1901–] 365
Street, Jabez Curry [1906–] 432
Strutt, John Williams: *see* Rayleigh, Lord
Strutt, Robert John (4th Baron Rayleigh) [1875–1947] 35–36, 179, 202, 208–9, 236
Stückelberg von Breidenbach zu Breidenstein und Melsbach, Ernest Carl Gerlach [1905–1984] 433, 458 & n, 459, 488
Stuewer, Roger [1934–] 231n, 320, 398n, 411n, 438
Sudarshan, Ennackel Chandy George [1931–] 536, 542, 581
Sutherland, William [1859–1911] 164
Symanzik, Kurt [1923–1983] 500, 578n

't Hooft, Gerardus [1946–] 587, 590n, 591, 597–8
Taine, Hippolyte Adolphe [1828–1893] 234
Tait, Peter Guthrie [1831–1901] 75, 91, 176
Taketani, Mituo [1911–] 435
Talbot, William Henry Fox [1800–1877] 170
Tamaki, Hidehiko [1909–] 429
Tamm, Igor Evgenievich [1895–1971] 349, 351, 426
Tanikawa, Yasutaka [1916–] 435, 453
Tarski, Jan [1934–] 556
Telegdi, Valentine Lories [1922–] 542
Teller, Edward [1908–] 423, 461, 479, 567
Thellung, Armin [1924–] 387
Thirring, Walter [1927–] 500n
Thomas, Lewis [1913–] 11, 93, 95
Thomas, Llewellyn Hilleth [1903–] 279, 366; *see also (in subject index):* Thomas factor
Thompson, Silvanus Phillips [1851–1916] 47, 48, 52, 202
Thomson, Elihu [1853–1937] 97
Thomson, Sir George Paget [1892–1975] 184–5, 202, 309, 312
Thomson, Sir Joseph John [1856–1940]; Adams Prize (1882) 177, 184; age when major work done 133, 184; atomic models 12; modified Coulomb law model 187–8, 209n, 216, 'plum pudding' model 166, 178–80; on Bohr atom model 209, 210; and Bragg 145; cathode ray studies 10, 81, 84–6; conduction of electricity through gases, studies 53, 59; death 216; delta-ray observations 148; as director of Cavendish Laboratory 59, 85, 184–5; dynamical models 122; electromagnetic mass 88, 628; electron charge/mass ratio measurements 83–4, 85–6, 131, 628; electron discovery 74, 628; electron model studies 12, 178–80; energy–mass equivalence 88, 628; on energy source of radioactivity 111; Geissler tubes used by 69; as Master of Trinity College (Cambridge) 85, 216, 304, 437; Nobel Prize 186; neon isotope separation 216; photon–electron scattering formula 187, 349, 370, 497; Rühmkorff coil used by 69; Silliman lectures 129, 179, 180, 184, 365; on spectra 197; in World War 1 236; X-ray scattering 187, 226; on X-rays 42, 59; *other references* 216, 365, 437; *see also (in subject index):* Thomson . . .
Thomson, William: *see* Kelvin, Lord
Thornton, Robert Lyster [1908–] 407
Ting, Samuel Chao Chung [1936–] 605, 611
Tolman, Richard Chase [1881–1948] 117, 365
Tomonaga, Sin-itiro [1906–1979] 429, 453, 455, 459–60, 466, 496n

Touschek, Bruno [1921–1978] 574
Treiman, Sam Bard [1925–] 489, 538, 559n, 602, 635
Trenn, Thaddeus [1937–] 192n
Turlay, René [1932–] 538
Tuve, Merle Antony [1901–1982] 405, 406, 416, 424
Tyndall, John [1820–1893] 183, 365

Uehling, Edwin Albrecht [1901–] 370, 383, 457
Uhlenbeck, George Eugene [1900–]; at American Physical Society meetings 376, 447; electron spin papers 254, 276, 277–8; family background 274, 275; and Fermi 275, 366, 376; and Goudsmit 275–8; half-integer quantum numbers 214–15, 271, 276, 277, 347, 630; on infinities in field theories 463; on neutrino theory 360; on Oppenheimer 367; on Pauli 360; on quantum-mechanical treatment of energy fluctuations 333; in USA 280, 360, 366; on wave mechanics 255
Urbain, Georges [1872–1938] 229
Urey, Harold [1983–1981] 280, 402

Van de Graaff, Robert Jemison [1901–1967] 406
Van der Meer, Simon [1925–] 2, 576, 610
van der Waals: *see* Waals . . .
Van Vleck, John Hasbrouck [1899–1980] 365
van't Hoff: *see* Hoff . . .
Varley, Cromwell Fleetwood [1828–1883] 79–80
Veksler, Vladimir Iosifovich [1907–1966] 474
Veltman, Martinus Justinus Godefridus [1931–] 587, 594, 611
Verrier, Urbain Jean Joseph Le [1811–1877] 177
Villard, Paul [1860–1934] 9, 62
Vogel, Hermann Carl [1841–1907] 172
Volkoff, George [1914–] 369
Volta, Alessandro, Conte [1745–1827] 70
von Guericke: *see* Guericke . . .
von Helmholtz: *see* Helmholtz . . .
von Laue: *see* Laue . . .
von Neumann: *see* Neumann . . .

Waals, Johannes Diderik van der [1837–1923] 174, 275
Waerden, Bartel Leendert van der [1903–] 262, 292
Waller, Ivar [1898–] 348, 349, 350, 373n
Walton, Ernest Thomas Sinton [1903–] 7, 234, 405, 406, 478
Warburg, Emil [1846–1931] 84
Warburg, Otto Heinrich [1883–1970] 63
Ward, John Clive [1924–] 497, 592
Watson, Kenneth Marshal [1921–] 486
Watson, William [1715–1787] 78–9
Weber, Wilhelm Eduard [1804–1890] 70
Weinberg, Erick James [1947–] 594
Weinberg, Steven [1933–] 355, 462, 489, 591, 594, 601, 602, 611
Weisberger, William [1937–] 567, 635
Weisskopf, Victor [1908–]; positron theory 384 & n, 385; renormalization hinted at 385; self-energy divergence 388; at Shelter Island Conference 450; spinless field quantum electrodynamics 387 & n, 388; vacuum polarization calculations 383; *other references* 355, 375, 460, 466; *see also (in subject index):* Pauli–Weisskopf . . .

Weizsäcker, Carl-Friedrich von [1912–] 427
Wells, W. H. 416, 424
Wentzel, Gregor [1898–1978]; books 356, 390, 433, 438; on point electron 389; reminiscences 355, 362; strong coupling theory 495, 591
Wess, Julius [1934–] 556
Weyl, Hermann [1885–1955]; on gauge invariance and charge conservation 329, 345–6, 631; group theory book 267; parity, Weyl's name for 525n; and Pauli 268; on positron theory 351, 488, 534; on Schroedinger 252; unified field theory 268, 344–5
Wheeler, John Archibald [1911–] 376, 467, 498n, 512n, 525
Whiddington, Richard [1885–1970] 226–7, 228
White, Milton Grandison [1910–1979] 407, 408, 416, 424
Whittaker, Edmund Taylor [1873–1956] 70
Wick, Gian Carlo [1909–] 422, 427, 483, 526, 632
Wideröe, Rolf [1902–] 407, 408 & n, 571, 631
Wiechert, Emil [1861–1928] 10, 78, 81n, 82, 83 & n, 133n, 628
Wiedemann, Gustav Heinrich [1826–1899] 81
Wien, Wilhelm [1864–1928]; blackbody radiation law 135; on cathode rays 81–2; nomination of Einstein and Lorentz for Nobel Prize 89; Poincaré obituary 43 & n; relativity discussion with Rutherford 190; Roentgen obituary 39–40; on spectra 197; in World War 1 235
Wightman, Arthur Strong [1922–] 355, 380, 505, 526, 529
Wigner, Eugene Paul [1902–]; on Einstein's ghost fields 259; on Fermi 485; group theory book 267; identical particle problem 265–6, 526; isotopic spin, term used 424; and matrix methods 249, 340, 377, 379, 631; N-particle system 285, 302 & n, 315n; on neutrino theory 315–16; on neutron–proton forces 416; nucleon conservation principle 488n; parity conservation 525–6, 631; on quantum theory 217–18; on symmetry 265, 425, 551, 559; time reversal invariance 527, 632; on wave mechanics 255, 259; *other references* 366, 402; *see also (in subject index):* Jordan–Wigner . . .
Wilczek, Frank Anthony [1951–] 588n, 590, 591
Wilson, Charles Thomas Rees [1869–1959] 86, 131, 628
Wilson, E. 124
Wilson, Kenneth Geddes [1936—] 591
Wilson, Robert Rathbun [1914–] 8, 407, 477, 505
Wilson, William [1887–1948] 152–3
Wolfe, Hugh Campbell [1905–] 422
Wood, Alexander [1879–1950] 119
Wooster, William Alfred [1903–] 303, 307–8, 309
Wu, Chien-Shiung [1912–] 523, 532, 533, 542

Yang, Chen Ning [1922–]; on antinucleons 488; C- and P-violation tested 25, 532, 542; composite model 495, 519, 555; CPT-theorem 529; on G-parity 490, 505; gauge theory 585, 586, 587, 587, 611, 634; on tau–theta puzzle 525; two-component neutrino theory 533; on weak interactions 534, 542; *other references* 375, 525; *see also (in subject index):* Fermi–Yang . . . ; Yang–Mills . . .
Young, Thomas [1773–1829] 107, 174

Yukawa, Hideki [1907–1981]; autobiography 429, 430 & n; capture process calculations 422, 483; lecturing style 429; meson proposed 20, 402, 430–2, 434, 435, 438, 452, 481, 580, 632, 633; nuclear forces theory 18, 430–1, 436, 481, 483; *see also (in subject index):* Yukawa . . .

Zeeman, Pieter [1865–1943]; age when spectral effects discovered 133n; electron charge/mass ratio, value 77; Nobel Prize 78; Rowland grating used by 10, 75, 76, 170; Rühmkorff coil used by 69; spectrum line-broadening effect 76–7; spectrum line-splitting effect 10, 77–8, 268–9; *see also (in subject index):* Zeeman effect
Zehnder, Ludwig [1854–1949] 37, 39, 49
Zeleny, John [1872–1951] 84
Zimmerman, Wolfhart [1928–] 500
Zola, Émile [1840–1902] 286
Zuber, Jean-Bernard [1946–] 467
Zwaardemaker, Hendrik [1857–1930] 99n
Zwanziger, Daniel [1935–] 562
Zweig, George [1937–] 558, 559, 564n, 611

Index of subjects

Note: footnotes are indicated by suffix 'n'.

A-coefficients, radiative emission/absorption 328, 630
A (atomic weight)-values 221, 222, 226, 629
AA (antiproton accumulator) 1–2, 576
Abelian group, definition of 584
Abraham model 88, 89, 132, 277
Académie des Sciences
Becquerel family as members 45, 49
Marie Curie's paper on Becquerel rays 53
perpetuum mobile not discussed 105
X-ray discussion 43
accelerator physics
first experimental result 234, 406
US developments 407, 408–9
accelerator-induced nuclear transmutations 234, 405
accelerators 1, 18–19, 27–8
early designs 406, 407, 408
early energies listed 26, 405, 407, 409
financing of 19, 407, 471, 472
highest energy machines (listed) 476, 477 & n
linear accelerators 27, 407, 473n, 571–3
synchrotrons 19, 407–9, 473n, 474–8
types (known in 1945) 473n
see also betatron; bevatron; colliders; linear . . . ; synchrocyclotrons; synchrotrons
actinium, nomenclature/half-life values 118
actinium C″, discovery 158
AdA (*Anelli di Accumulazione*] ring device 574
Adams Prize 177, 184
Adelaide University 145
Adler–Weisberger relation 567, 635
advanced renormalization theory 462
AEC (Atomic Energy Commission, USA) 472, 477
aether models 41
ages (of physicists) 132–3, 184, 211, 251–2, 362
all-embracing patterns of discovery, futility of search 130
allowed/forbidden transitions, nuclear beta-decay 421
α-decay
as nuclear process 193
quantum-mechanical tunneling explanation 124, 296, 631
Rutherford's view 180, 223
strong interactions responsible for 142–3
α-particles/rays
back-scattering effect 189, 629
beryllium reaction 398, 399
Bragg's studies 145–6
as bullets fired into wood 145
charge/mass ratio 61
discovery 9, 60, 628
electric charge 61
energy fine structure 143n
exact nature of 60
hard scattering behavior 12
hydrogen scattering studies 237–8, 239–40
models 240
monochromaticity 143, 146
range experiments 146
scattering 147, 189, 191–2, 223, 225, 237–9, 296, 297; *see also* Rutherford scattering
alternating gradient synchrotrons (AAGs) 476–8, 516
aluminium, alpha-irradiation reactions 400
America, *see also* USA
American Institute of Physics 5
American Physical Society 5, 8, 75, 317, 628
American Association (for the Advancement of Science) 317, 366
American Physical Society
first officers of 75, 364
formation of 5, 364
New York meeting (1946) 8, 447, 448, 450
New York Meeting (1983) 1
Pasadena (1931) symposium 317
positron symposium (1933) 376
American physics, quantum theory in 1930s 365
American X-ray Journal 98
Amsterdam conference (1971) 597
analyticity, *S*-matrix theory 501, 502, 503
Ångström units, definition 167n
angular momentum
introduced into atomic model 198–9
see also electron spin; nuclear spin; proton spin; spin, . . .
Ann Arbor summer schools 317, 360, 366, 410
lecturers listed 366
Annalen der Physik 39, 106n
annihilation processes 324
anomalous magnetic moments 427, 456, 481, 483, 632
anomalous Zeeman effect 14, 77, 269, 274, 628, 630
term first used 270n
anticommutation relations 340, 378, 462n, 565
anti-electron 351; *see also* positron
antimatter 15, 360
Heisenberg's comment 360
antineutrino 419, 420
antineutron 488, 634
antinucleons 488–9
antiparticles, first introduced 418–19, 632
antiproton, discovery 488, 634
antiproton accumulator 1–2, 576
antiproton–proton collider 1–2, 28, 552, 575, 576, 637
antisymmetrical eigenfunction solution 284, 285
antisymmetry 265, 284
approximations, 361
arc lamp, invention 70
Arithmetische Spielereien [playing games with arithmetic] 16
Armorial Bearings, Baron Rutherford of Nelson 217
Associated Press (of America) 406
associated production 518, 519, 634
Associated Universities Incorporated 472
astatine, discovery 229n
asymptotic freedom 28, 551, 580, 589, 590 & n, 636

Atombau und Spektrallinien 166, 232, 252, 273n
atomic energy 111–12
  discoverers of 109–11
  moral issues 115–16
  term first used 8, 116–17, 629
atomic models
  acoustic/harmonic model 171, 176
  Bohr's 13, 195–6
  Darwinian influence on 178
  early electron models 178–88
  electrons-in-motion models 181–2
  Maxwell's views 175, 627
  mechanical models 176 & n
  model-building criteria 181–2
  Nagaoka (Saturnian) model 183, 429
  pair models 182
  planetary models 182–3
  'plum pudding' model 166, 178–80, 185, 629
  polyelectron models 178–80, 629
  pre-electron models 174–7
  Rumpf model 271 & n, 276, 277
  Rutherford's 12, 13, 190–1, 192–3, 296, 629
  Thomson's modified-Coulomb-law model 187–8, 209n, 210
  Thomson's 'plum pudding' model 166, 178–80, 185
  vortex model 176–7, 627
atomic number (N) 222, 223
atomic stability, Thomson's view 180
atomic weight
  concept 221–2
  Dulong and Petit's values 221, 627
  Nuclear charge relationship 226
atomists
  Faraday and Maxwell as 72–3, 627
  Maxwell as 326
atoms
  Maxwell's views 326, 627
  nineteenth-century view 72–3, 137, 175, 178, 327
Atoms for Peace Award 163
Auger transitions 156, 304
Australasian Society for the Advancement of Science 145
Avogadro's number, early estimates 74
axial current, partial conservation of 538
axiomatic field theory 23, 326, 464, 494, 529

B-coefficients, radiative emission/absorption 328, 329, 334, 337, 630
B-quantum field theory 586–7
back-scattering effect, alpha particles/rays 189, 629
bacteriocidal properties, radium/X-rays 96, 99
bag models/quarks 567
Bagnères-de-Bigorre conference (1953) 514–15, 524, 634
Bakerian Lectures 169, 397, 399, 411
Balliol College (Oxford) 229
Balmer formula 12, 170–2, 628
  Bohr's awareness of 164 & n, 173, 199
  Bohr's correction factors 213
  Bohr's derivation 201
  international discussion 164
  matrix mechanical derivation 255
  Moseley's version 229
  principal quantum number in 268, 273
  Sommerfeld's generalization 214
Balmer series 167, 171
bare mass, definition 385
Bartlett forces 416
baryon, term first used 21, 514, 634
baryon spectroscopy 22, 490–1
Basel University 172
Batavia conference (1972) 602
Baton Rouge 52n
battery, invention 70
Becquerel rays
  compared with Roentgen (X-)rays 48
  Curies' work 52–8
  discovery 46
  Rutherford's work 59–60
  term used by Marie Curie 54
  see also radioactivity; uranic rays
BEIR-III Report 93
Berkeley 366, 367–8, 369
  accelerators 17, 407–9, 475, 476, 479, 480, 571–2, 634
  bubble chamber 491–2, 635
Berlin
  Fischer's Chemistry Institute 150, 151
  Physics Institute 83, 153
  Physikalisch Technische Reichsanstalt 36, 158
  University 84, 151, 168, 227, 235, 261
beryllium, α-particle bombardment/reaction 398, 399
β-decay
  allowed/forbidden transitions 421
  Bohr's work 223–4, 629
  chronology of discoveries/studies 138
  coupling constants equation 530
  as electron–neutrino emission 20
  Fermi's theory 17, 231, 318, 319, 380, 401, 402, 418–22, 580, 632
  fundamental processes 143
  Heisenberg's explanation 414
  interaction equation, 422–3, 527
  neutrino creation during 569
  Rutherford's view 223
  Thomson's view 180
  weak interactions responsible for 142, 143
  see also Fermi interaction
β-particles/rays
  absorption studies 148–51
  discovery 9, 60, 628
  electric charge determined 144–5
  exponential absorption law 151–2, 153, 154
  Kaufmann's first paper 88, 153
  monochromaticity theory 11, 151, 154, 157
  nature of 87
  velocity range 88
β^+ radioactivity 400, 422, 632
β-spectra
  continuous spectra 11, 16, 159, 297, 303, 309, 315, 413, 629, 631
    equation for 421
    missed 157–8
  discrete line spectra 11, 153, 154, 155, 303
    superimposed on continuous spectra 155–6, 303–4
  Ellis' studies 303–8
  Hahn and Meitner's work 147, 151
  monochromatic nature 143, 147, 306
  radium E 402
betatrons 19, 473n
bevatron 488, 492, 516
Bhabha (electron–positron) scattering 376

binding energy 232–4
  quark systems 589
Bjorken rule 579
blackbody radiation law
  Kirchhoff's version 167–8, 627
  Planck's version 74, 133–4, 137, 193, 210, 266, 629
    BE statistics 283, 284
    Boltzmann statistics 280
    Debye's derivation 332, 353
    Einstein's derivation/explanation 9, 123, 130, 134–5, 193, 328
  Rayleigh–Einstein–Jeans version 135n
Bloch–Nordsieck method 376
Bohr hydrogen atom model 13, 164, 195–6, 200–2, 328
  reactions to 208–211
Bohr magneton 270, 299
Bohr radius, definition 198
Bohr–Kramers–Slater (BKS) proposal 105n, 248n, 259n, 311, 319, 630
Bohr–Sommerfeld theory 213–15
Bohr's Institute 217
Boltzmann constant 74, 175, 281
Boltzmann statistics 135, 280, 282
Bonn conference (1973) 603
Bonn tubes 69
Bonn University 69, 79
bootstrap concept 503, 566
Born–Heisenberg–Jordan interpretation, energy fluctuation studies 332, 333n, 334, 336, 338
Bose–Einstein (BE) condensation 280, 284
Bose–Einstein (BE) statistics 135, 280, 282, 283–4, 285, 630, 631
  and second quantization 338
  and spin quantization 528
bosons, definition 340
bottom quark 608, 637
Bremsstrahlung [energy loss by radiation] 361, 376, 428, 480
brevity explained, modern times essay 550–2
Bristol University 286, 454, 479, 513
British Association (for the Advancement of Science) 73, 82n, 109, 188, 258, 436
Brookhaven National Laboratory (BNL)
  accelerators 476, 477, 635
  beginnings 19, 472, 633
  neutrino experiment 571
Brooklyn Medical Journal 96n
Brownian motion 108n, 328
bubble chambers 22, 491, 492, 634
Bureau of Standards 402

C-invariance 350, 381, 423, 527, 634
  violation of 25, 531, 532, 533–4
  see also charge conjugation . . . ; invariance principles
c-numbers 287
Cabibbo angle 563, 601, 609
Cabibbo currents 564, 588
Cabibbo hypotheses 563, 635
caesium, discovery 169
Caltech 351, 366, 368, 369
Cambridge (1946) physics conference 7
Cambridge University 58, 59, 85, 216, 226, 286, 304, 628
  as centre 218, 365
  Christ's College 366
  Lucasian Chair of Mathematics 286
  Trinity College 85, 216, 304, 437
  see also Cavendish Laboratory
cancer risks, recent reports 93
canonical transformations 254, 288
Canterbury College (New Zealand) 59, 437
Carnot's principle(s) 106
cascade phenomena, cosmic ray showers 427–8, 633
cascade resonance/particles 514n, 515, 553
Casimir effect 354
Casimir trick 375, 376
cathode rays
  charge 81, 82, 628
  charge/mass ratio 83–4, 85, 628
  discovery 131
  experimental details 40, 86
  kinetic energy 82
  nature of 10, 67, 80–1
  term first used 79, 627
  vacuum necessary for observation 79
causality
  Born's remarks 251, 255, 257
  first raised 9, 212
  S-matrix theory 499, 500, 501
Cavendish Laboratory
  accelerator 17
  cost of building 58
  established 58
  first directors 58, 59
  Jaffe's impressions 84
  positron detection studies 362
  Rutherford as director 216–17, 218, 239, 304, 630
  Rutherford as research fellow 9, 59
  study of conduction of electricity through gases 53, 59
  toast to electron 67
Cerenkov counters 478, 516
CERN (European center for nuclear research)
  beginnings 19, 473, 634
  intersecting storage rings 575, 576
  proton synchrotron 408, 476, 477
  proton–antiproton collider 28
  W-boson discovery 1–2
chalcite 55
chapter notations, this book, explained 4, 627
charge conjugation 25, 381, 633
  selection rule under 490
  see also C-invariance; C-violation
charge independence 424, 633
  and perturbation theory 435
charge renormalization 462, 463
charge–spin exchange 415, 416
charge symmetry 424 & n
charged K-mesons 515
charged mesons/pions 434, 435, 481
charged vector bosons: *see* W-bosons
charm (quantum number for quarks) 552, 601, 635
  discovery of 607, 636
charmed hadron spectroscopy 607, 636
charmed quarks 29, 552, 601, 602
charmonium 606, 607, 636
Chemical Society (London) 117
chemical valence theory 72
chemistry, beginnings as a science 174, 326
Chew–Low equation 497, 502
Chicago (1939) cosmic ray symposium 431, 438

chiral SU(3)×SU(3) symmetry 26–7, 565, 566
chirality, term first used 535n
chirality opeator 535
chronology (1815–1984) 627–37
circular proton accelerators 408–9, 473, 474–8
classical continuity equation, quantum version 352
classical physics
  and infinities 389
  quantum theory non-compatible 130, 133, 134, 210–11, 246
clerihew 137
cloud chamber
  counter-controlled 363, 404
  invention 131, 628
  techniques 10, 86, 145, 246, 311
    meson discovery 432
    positron discovery 351–2, 632
coarse-grained probability 281
cobalt (Co^{60}) beta-decay study 532, 533
cobalt–nickel anomaly 222
Cockcroft–Walton (voltage multiplier) device 406, 478
colliders 27–8, 552, 573–6
color (quantum number for quarks) 27, 551, 561–2, 636
color SU(3) symmetry 27, 551, 561–2, 588, 590, 591, 621, 622, 636
Columbia University 63, 97, 216, 364, 365, 371, 402, 429
commutation relations 341–2, 354, 554, 555–6, 631
  first introduced 254
  see also anticommutation . . . ; noncommutation
commutator, first introduced 254
Como conference 278
complementarity principle 248, 250, 251, 255, 262, 631
Comptes Rendus 43–4, 303
Compton effect 89, 136, 246, 248, 311
  α-beryllium irradiation experiment 399
  definitions 136, 246, 328
  Dirac's calculations 337–8, 353–4, 374
  and energy conservation law 319
  as possible origin of continuous β-spectrum 307
  theoretical predictions 325, 348, 349, 350, 361
  zero-energy/zero-frequency limit 187, 349, 497
  see also internal Compton effect; Klein–Nishina formula
computerized data-handling 22, 491
conferences, annual/biennial: *see* Rochester Conferences; Solvay Conferences
confinement (of quarks) 28, 591 & n, 622, 636
confusion (leading to progress) 5
conserved weak vector current 537, 635
consortium physics 2, 19, 472–8, 492, 610
constituent quarks 567
contents (of this book), explained 7
continuous spectra
  β-particles/rays 11, 16, 157–8, 159, 297, 303, 629, 631
  hydrogen atom spectrum derived 254
continuous stream clasical theory 391
Conversi–Pancini–Piccioni experiment 453
Copenhagen University 194, 217, 247, 268, 449, 450
  as centre 218, 365
coronium 170n, 198, 222
corpuscular approach 80, 81, 248 & n
correspondence principle 247, 498
cosmic ray particles 404
cosmic ray showers 404, 427, 631, 633
cosmic ray studies
  beginnings 19–20, 403–4, 629
  particle detection in 17, 351–2, 454, 511, 513–14
  positron discovered during 351–2
  research groups listed (1950s) 514n
cosmic rays
  absorptive components 432
  discovery of 19, 403, 629
  as fixed target devices 21
  literature (1930s) 405
  nature and properties of 404, 405
  term first used 20, 404
cosmology 623
Cosmotron 19, 476, 490, 491, 516, 519, 634
Coulomb barrier 192
Coulomb effects, nuclear forces 420, 421, 424
Coulomb electrostatic law 70, 187, 209n, 210, 237, 238, 245, 298
  modifications to 187, 209n, 240
Coulomb field 291
Coulomb gauge 329, 344, 346, 374, 631
Coulomb phase, nuclear physics 402
Coulomb potential 343, 430
Coulomb scattering 148
counter techniques, β-spectra 158, 159, 304
counter-controlled cloud-chamber 363, 404
counting procedures 280, 281, 282
coupling constants, β-decay/μ-capture/μ-decay 530
CP-invariance 534, 635
CP-violation 25, 538–41, 622, 635, 636
  sources of 541–2
CPT theorem 25, 528, 529, 532, 634
CPT-invariance 540, 542
creation processes 324
Crime and Punishment, Dirac's comment 286
cross-section ratio, e+e− collider beam 604, 605
crossing symmetry 500
Curie temperature 56
curie (unit), definition 99n
curiethérapie 100
current algebra 562–6, 635
  initial formulation of 564n
current quarks 567
cyclotrons 17, 18
  first working version 408, 632
cyclotron principle 408, 409, 474
  restricted validity of 409

D-lines, solar spectrum 167
Danish Academy of Sciences, Royal 194
Darwinian effects 177–8
de Broglie wavelength 299
deaths, radium poisoning/X-ray exposure 96–7, 100
deep inelastic scattering 573
deep tissue trauma, X-ray exposure 98
degenerate systems treatment, first introduced 254
delayed reaction, X-ray exposure 97
Delbrück scattering 386
Delft, Technische Hogeschool 275
delta-functions 339, 354
delta-rays (knock-on electrons) 148
delta-resonance (33-resonance) 487–8, 517, 553, 557, 634
DESY (Hamburg) 408, 476, 605, 637

detectors, accelerated particle reaction 21, 408–9, 478, 479, 491–2; *see also* bubble chambers; Cerenkov counters; cloud chambers; scintillation counters; photographic . . .
determinism: *see* causality
deuteron
  discovery 17, 233, 402, 632
  spin of 411
diffraction gratings, Rowland's version 10, 75, 76, 170, 628
Dirac equation 14, 286, 290–1, 330, 340, 345, 631
  background 286–90
  reactions to 347
  von Neumann covariants 292, 422
Dirac field quantized 377–81
discharge tubes 68–9, 78–9
discrete spectra
  β-particles/rays, 11, 153, 154, 155–6
  hydrogen atom spectrum derived 254; *see also* Balmer series
dispersion relations 494, 498–500, 634
displacement law (radiochemistry) 255
distinguishability 282
  abandonment of 334
diversity, meson physics 492–4
dividing engine, Rowland's design 75
double β-decay process 423
double SU(3) symmetry 561–2
Dover meeting (1899) 86
dual nature of light 248, 252
Dubna conference (1964) 559
Dubna Laboratory (USSR) 19, 475

E–m–v relations 87–8, 232
E–p relations 136, 245–6, 311, 342, 630
E–p–M relations 246, 327, 347
East Orange luminous dial painting plant 100
effective Lagrangian 386
elastic scattering, Born's studies 256–7
electric dipole moments, fundamental particles 541
electricity, positive and negative charges, terms first used 70, 628
electrochemical equivalent 71
electrolysis 70–1, 73, 627
  term first used 71
electromagnetic field theory 244–5
electromagnetic gauge group 342–6, 583
electromagnetic interactions, gamma-decay 142, 143
electromagnetic mass 88, 628
electromagnetic nature of light 74, 76
electromagnetism
  beginnings 70
  Lorentz theory 76, 194, 233
  nineteenth-century understanding of 70
electron
  anomalous magnetic moment explanation 456, 481
  charge estimated/measured 61–2, 73, 85–6, 131, 628
  charge/mass ratio measured 131
  discovery 9–10, 78–86, 131
  magnetic moment of 456–7, 465, 481
  passage through matter 148–9
  self-energy of 370, 372–3, 384, 389, 448, 632
  stability of, 370, 371
  term first used 74
  types identified 87
electron diffraction
  first detected 255, 631
  first predicted 252
electron models 12
  Abraham model 88, 89, 132, 277
  early models 178–88
  electronic configurations proposed 186
  marble model 370, 371
  non-relativistic quantum-mechanical model 371
  point model 372, 389
electron–neutrino (nuclear force) model 427, 632
electron–neutron interaction 483
electron–nucleon scattering studies 27, 502, 573
electron–positron collider 605, 609, 636
electron–positron scattering 376
electron radius
  equation 327, 370
  values quoted 239
electron rotation 279, 628, 630
electron spin 277–80, 628, 630
  discovery 215, 250, 254, 276, 277
  early ideas 279–80
electron synchrotrons 475, 476
electron volt, definition in electrostatic law 70
electroweak unification 591–600, 636
emigrés (to USA) 366 & n
emptiness of atom, first stated 182
Encyclopaedia Britannica 74, 75, 164
energy, term first used 107
energy conservation law 105, 106–7, 627
  and Becquerel rays 111
  Bohr's questioning of 17, 105, 309–13, 317, 318–19, 631
energy fluctuation studies 248, 252–3, 259, 332, 353
energy–mass equivalence 87–8, 88, 89, 104, 232, 628, 629
energy–momentum conservation laws 136, 246, 311, 342, 630
energy–momentum–mass relations 246, 327, 347
energy–momentum–velocity relations 87–8, 232
equipartition theorem 175, 280
essay, definition of 550
η-meson 491, 492
exclusion principle (Pauli) 250, 253, 267–74
  Fermi's application 284
  helium spectrum explained by 215
Exhibition of 1851 scholarships 59, 158
experiment, and theory 29, 454, 455
experimental physics, effect on quantum electrodynamics 363–4
experimentation, relation to theory 137, 139, 181
experiments, planning of modern 138
exposure limits, radium/X-rays 95, 100

faking of β-spectral lines 158, 159
false leads 4
families concept 518 & n
Faraday (constant), definition 72
Faraday dark space 79
Faraday lectures 72, 73, 311, 313
Faraday's laws of electrolysis 71, 627
Fermi constant 536, 537, 577, 580
Fermi–Dirac (FD) statistics 280, 282, 285, 631
  first introduced 255
  proton behavior 302
  and second quantization 339

Fermi interaction 420, 426 & n, 531
  breakdown of 582
generalized equation 422–3, 527
  high-energy behaviour 427, 428, 429
  nature of 551
  universality of 530, 535, 634
Fermi motion 479
Fermi–Yang (composite) model 495, 519, 555
Fermilab (Fermi National Accelerator Laboratory) 477, 576, 636, 637
Fermilab conference (1972) 575
fermions, definition 340
ferromagnetism 279, 630
field quantization 15, 329
fine structure constant 212–15, 629
  dimensionless number ($\alpha = e^2/\hbar c$) 16, 463, 482, 599
fine structure equation 214, 276, 347
fine-grained probability 281
first law of thermodynamics 105, 106
Fischer's Chemistry Institute (Berlin) 150, 151
fixed-target devices 27, 28, 573–4
flavor (quantum number for quarks) 588
flavor SU(3) symmetry 588, 635
floating magnets experiments 184, 186
fluctuation studies 248, 252–3, 259, 629
fluorescence 44 & n
footnotes, importance of 457
fourth-order effect, self-energy relations 385 & n
francium, discovery 229n
Frankfurter Zeitung 38
Franklin Institute, Thomson's lectures 210
Frascati 574, 604, 605
Fraunhofer lines 167
free-floating invariances, 485–87, 568
frequency-modulated cyclotrons 474–5
fundamental charge (*e*)
  early estimates of 73–4
see also cathode rays . . . ; electron, charge of
funding, accelerator-building 19, 407, 471–2, 477
Furry's theorem 381, 633

G-parity 489–90, 634
Galilean relativity 136
Galvani Conference (1937) 437, 448
galvanometer, invention 70
γ-decay, electromagnetic interactions responsible for 142, 143
γ rays
  compared with X-rays 62
  discovery 9, 62, 628
γ-spectroscopy 296, 304–5
Gamow–Teller couplings 423, 567
gases, electrical discharge through
  Faraday's work 68
  Thomson's work 53, 59
gauge, term first used 344
gauge fixing 343
gauge invariance
  beginnings 343–6, 631
  and charge conservation 329, 345–6, 583, 631
  definition 344
  local 584, 594
  photon self-energy zero value 385
  and subtraction techniques 383
gauge theories 28
  non-Abelian 494, 583–7
  Yang–Mills theory 585, 586, 587, 634
gauge transformations 344–5, 583, 594
Geiger–Nuttall relation 124
Geissler pump 10, 68, 627
Geissler tube 68–9, 627
  discoveries helped by 69, 167
Gell-Mann–Okubo mass formula 557
General Accounting Office (GAO), report on cancer risks 93
general relativity theory 14, 245, 246, 623
geophysics, Germany's first institute of 83
ghost field [*Gespensterfeld*] 259, 260, 261
Gibbs Lecture 266n
GIM (Glashow–Iliopoulos–Maiani) mechanism 601, 602, 636
Glasgow conference (1954) 520, 521
glueballs 591
gluons 28, 579, 588, 590, 591
  non-zero mass 591
Goettingen University 83, 88, 247, 251, 331–4
  as centre 218, 365
Goldberger–Treiman relation 538, 635
Goldstone bosons 593, 594
Goldstone theorem 593, 635
Goudsmit–Uhlenbeck (half-integer) quantum numbers 215, 271, 276, 277, 347, 630
Grand Canyon/Scotsman joke 397
grand unification theories 30, 622, 636
gravitation, possible effect on radioactivity 114n
gravitational field theory 245, 623, 629
Gross–Llewellyn Smith rule 580
ground state, Bohr atom model 199
group theory 15, 266, 267, 555, 631
GUT (grand unification theory) 30, 622–3, 624
Gymnasium, meaning of term 109n

H-particle–electron model 230, 231, 233; *see also* proton–electron model
Habilitationsschrift, term explained 172n
hadron spectroscopy 26, 490–1, 552–3
hadrons
  definition 26n
  term first used 26n, 550, 635
hafnium, discovery 229n
half-integer quantum numbers 215, 271, 273–4, 276, 277, 347, 630
half-life
  determination first made 120–1, 628
  discovery 9
  values quoted 118, 119
Halpern scattering 385
Hamburg University 268, 341
Hamiltonian methods 334–5, 337n, 341, 355, 375, 380
hard (electron–hadron) scattering 573, 575, 636
Harvard College 365, 366, 432
Harwell conference (1950) 513
heat production experiments, radioactive elements 113, 115, 308
heavy electron, term first used 433
heavy particle, term first used 431
Heisenberg forces 416
helicity 531–32

helium
phase transition 284
presence in solar spectrum 169
presence in uranium rocks 60
singly ionized spectrum, 21, 201–2, 277
spectrum 215, 267n
HERA (*Hadron Elektron Ring Anlage*) 624, 625n
hidden bottom 608, 637
hidden charm 606, 636
Higgs mechanism 594–8, 635
Higgs phenomenon 594, 598, 635
Higgses, definition of 594
high-energy physics
first experimental result 234, 406
Rutherford's views 405, 406, 435
higher symmetries 552–69
historian, Uhlenbeck as possible 275
history, subjective nature of 3
history of discovery, absence of general patterns 130–1
Hittorf vacuum tube 37
Hoboken, Stevens Institute 183
hole theory 349–51, 360, 361, 378, 631
homologous series 175
hydrodynamics, vortex model 177
hydrogen
α-scattering studies 237–8, 239–40
atomic model 13, 164, 195–6, 200–2, 251, 629
atomic spectrum 12, 166–7, 170–1, 213, 402, 628, 630
wavelength ratios 166–7, 627
nuclear structure 230, 276
hydrogen atom, *nS*-level displacement 383, 455, 456, 457
hydrogen bubble chambers 22, 491
hypercharge 554
hyperfine anomalies 451, 633
hyperfine structure 279, 301, 630
hyperfragments 515
hyperon β-decay 563, 635
hyperons 21
term first used 21, 514

ideal fluid, definition 176n
identical particle permutation symmetry 265–6, 526
importance of discoveries, lack of understanding 129, 134, 136
impurity parameter 541
induction coil, Rühmkorff's version 10, 69, 627
induction principle, Faraday's 70
inelastic scattering
electron–nucleon 573, 575
of photons 301
infinities 360, 361
alternative approaches 388–91
before positron discovery 370–4
quantum field theory 462–3
renormalization techniques, effect of, 388
information explosion 6, 533
information sources 49, 63–4, 202, 262–3, 320, 355–6, 391, 438, 466–7, 505, 542, 611
injectors, pre-accelerated particle 477–8
instrumentation, progress delayed by lack of 136, 140
interaction equation, nuclear 420, 422–3, 426 & n, 427: *see also* Fermi interaction
interactions, types listed 142
intermediate coupling 496n
internal Compton effect 307 & n
internal conversion electrons 156
internal conversion mechanisms 124, 156, 304, 305, 630
International Education Board 365
International Union of Pure and Applied Physics 461
invariance principles 25, 341, 381, 525–529: *see also* *C*- . . . ; *CP*- . . . ; *CPT*- . . . ; *P*- . . . ; *T*-invariance
inverse α-decay 406
ionium 146
ionization chamber meters 53, 132, 403
ions
as broken-up atoms 86, 94, 628
Faraday's definition 71n
Lorentz's use of term 76
iospin variance 480
isobaric spin 424n
isobars, nucleon 496
isospin 18, 423–5, 519, 632, 633
as (free-floating) invariance 485–7
isospin invariance 486–9
isospin symmetry 425 & n, 554
group enlargement 519, 583, 634
isotopes
developments in 1930s 403
'discovery' 13, 224–5, 296, 629
masses 120
studies 224–5
term first used 224
isotriplet
discovery of 480
prediction of 435, 537
ISR (intersecting storage rings) 575–6, 636

J/ψ resonance 605, 606, 636
Jahrbuch der Radioaktivität und Elektronik 63
jet events 609–10, 636, 637
John Simon Guggenheim Memorial Foundation 365
Johns Hopkins, first professor of physics 75
Joint Institute for Nuclear Research (USSR) 475; *see also* Dubna Laboratory
Jordan–Klein matrices 339, 631
Jordan–Klein relations 339, 387
Jordan–Wigner formalism (second quantization method) 340, 377, 379
Jordan–Wigner matrices 340, 631

K-mesons 21
charged 515
neutral 515, 521–3
particle mixing 521, 522, 523, 634
term first used 21, 514
two types 24, 515
K-resonance 553, 635
Kaiser Wilhelm Institute für Chemie (Berlin) 305
Kaluza–Klein (unified field) theory 584 & n, 624
Kelvin's law of conservation of (fluid) circulation 177
King's College (London) 436
Klein paradox 299, 312, 313, 319, 349, 631

Klein–Nishina formula 325, 349, 350, 361, 631
  deviations explained 414
  experimental verification 158, 348
  hole theory effect 374
klystron development 572
Knabenphysik [boy physics] 251
Königsberg University 82
Konopinski–Uhlenbeck (β-spectra) theory 422, 632
Kramers–Kronig relation 499
Kyoto conference (1953) 525

L-mesons, term first used 514
Lagrangian methods 341, 342, 386, 631
Lamb shift 451, 456, 457, 460, 633
λ-limiting process 389–90
λ-resonance 513, 517, 553
  decay 515, 517, 519
LAMPF (Los Alamos Meson Physics Facility) 572
Landé factor, 270, 271–2
Landé rule 270, 274
Langsworthy Professor of Physics 188
Laporte (selection) rule 525, 630
large-angle electron–nucleon scattering 573, 576–7
lattice gauge theories 591, 636
lattice (nuclear) model 413, 428
layout (of this book), explained 7
Le Matin 42
lead screening 98
League of Nations, International Committee on
    Intellectual Cooperation 58
L'Eclairage Electrique 43
Leeds University 148
Leiden conference (1953) 583
Leiden jar, invention 70
Leiden University 214, 227, 275, 276, 367
Leipzig University 347, 348, 459
Lenard window 40, 81
Leningrad 303 & n
LEP (large electron project) 624, 625n
lepton, term first used 450, 633
lepton conservation 20, 530–1, 634
lepton pairs, local action of 537
Liénard–Wiechert potentials 83
lifetime concept (of radioactive decay) 103, 120–3
  Rutherford's silence on implications 129, 134,
    212
light, dual nature of, 248
light elements, composition of 410
light-by-light scattering 386 & n, 632
light-quantum, discovery/introduction 135–6, 193,
    210–11, 245–6, 251, 252, 327, 629
L'Illustration 42
linear accelerators (linacs) 407, 473n, 571–2, 573,
    631
  versus cyclotrons 572
  see also SLAC
linearity 266
literature explosion 6, 533
lithium
  hyperfine structure measurements 302
  nuclear reaction 234, 405, 407, 408
Liverpool Science Congress 42
Liverpool University 436
locality condition 500
London
  Chemical Institute 150
  King's College 308, 436
London conference (1974) 591, 603, 604
Lorentz condition 344, 354, 355
Lorentz force 76, 628
Lorentz gauge 344, 346, 354
Lorentz transformation 346
Lorentz triplet 268
Los Alamos project 370
Louisiana State University 52n
Louvain, wartime bombing 235 & n
low level radiation exposure, definition 93–4
low-energy theorems 497
LSZ formalism 500–1
luminescence, definition 44n
luminosity, collider beam 574 & n
luminous dial painting 100
Lupus vulgaris (skin tuberculosis), treatment of 97,
    100

McGill University 9, 60, 150, 628
magnetic moments
  electron 456–7, 465, 481
  muon 466
  nuclear 299, 483
  ratio neutron/proton 560–1
magnetic separation techniques, β-spectra 152, 154,
    159, 304, 305
magnetism, electron spin as possible reason 279,
    628, 630
Majorana forces 416
managerial approach (to physics experiments) 22,
    491–2
managerial skills, project leader 22
Manchester University 149, 152, 189–93, 208n, 228
  cosmic ray studies 511, 513
  Rutherford 12, 63, 149, 188, 189–93, 216, 629
  submarine research 236
Manhattan Project 19, 472
Manifesto of the 93 234–5
many-time formalism 375, 459
Marburg University 150
Maryland University 183
mass renormalization 449, 455, 462, 463
mass–energy equivalence 87–8, 89, 104, 232, 629
  testing of 232, 233, 234
materialistic theory of radioactivity 112
mathematical complexity 15 [15a]
mathematical technology 255, 266, 267
matrix electrodynamics 331
matrix mechanics 331, 630
  introduction 249, 251, 254, 630
matter, as waves 250, 252, 630
matter, conservation law 107
Maxwell (1864) electrodynamic/electromagnetic
    theory 244, 430, 627
  Yukawa nuclear analogy 430–1, 483
Maxwell equations 331, 342, 354
  Lorentz' atomistic interpretation 76
  non-linear modifications of 385–7, 390
Maxwell phase, nuclear physics 402
Maxwell–Boltzmann distribution 282
Maxwell–Lorentz equations 194, 200, 371n
  Dirac's interpretation 390
Mayer–Joule principle 106
mechanical equivalent of heat 106
medical impact of physical discoveries 93–101
Medical News 95, 96
megalomorphs, term first used 518n

memoir, definition 18
Merchant Venturers' School (Bristol) 286
meson dynamics 484
meson field theory 325
  post–WW2 481–5
  pre-WW2 433–6
meson physics
  diversity in 492–4
  Fermi's views (1951) 494–5
meson resonances 22
  detection of 22, 490–1, 552–3, 635
meson theory
  problems in 1940s 23–4
  semiclassical nucleon in 495–7
meson–baryon interactions, as non-fundamental processes 29
mesons
  absorption data 453
  alternative names for 431
  cosmic ray mesons 17, 20, 432, 453, 633
  discovery of 17, 20, 325, 432–3
  lifetimes 452, 453
  scattering data 452–3
  term first used 431
  two-meson hypothesis 453, 454, 633
  types 448, 452
  Yukawa meson 20, 402, 430–2, 453, 632, 633
  see also η- . . . ; L- . . . ; M- . . . ; μ- . . . ; ω- . . . ; π- . . . ; ρ- . . . ; τ-mesons
metabolon, term explained 119n
metabolous matter 119
Michel parameter 530, 535 & n
Michigan University 280; *see also* Ann Arbor
microchemistry, term first used 167, 170
microorganisms: *see* bacteriocidal effects . . .
microtrons 473n
middle-aged men 133, 211
Midwestern Universities Research Association 473
military campaigns, X-ray techniques used in 96, 236
military funding 19, 471–2
MIT 365, 573, 576–7
molecular analogies, nuclear interactions 414, 415, 416, 430
molecular dissociation
  Boltzmann's work 121, 174
  effect on spectra 168
molecule, nineteenth-century meaning 72n, 171, 175
Möller scattering 375
moral issues, atomic energy 115–16
Mössbauer effect 124
Mount Wilson 170–513
μ-capture/decay, coupling constants equation 530
μ-meson, term first used 454, 633
μ-particle 231
Münich University 218, 247, 267, 365
muons 20, 454, 633
mutations, radiation-induced 95

N (atomic number) 222, 223
Nagaoka (atomic) model 183, 429
naked bottom 608, 637
naked charm 606
  discovery of 607, 636
National Academy of Science, documents on cancer risks 93
National Laboratories (USA) 407, 471
National Research Council (NRC) Fellowships 365, 367, 368, 369–70, 383, 406
Nature 39, 118n, 124, 209, 228, 418, 453
  founder of 169
Naturwissenschaften 302
nebulae, as gaseous clouds 169, 177
nebulium 169–70, 198, 199, 222
negative energy photons 389
negative energy states 346, 347–8, 349, 488
neon, inactive isotopes separated 216, 224n
neutral current 29, 552, 592, 600–1, 636
neutral current processes 550, 602–3, 636, 637
neutral K-mesons 515, 634
  particle mixture 521–3
neutral K-particle complex 24, 521–3, 534
  CP-violation discovery 539, 540, 541
neutral mesons 434, 435 & n, 633
neutral pions 480, 634
neutrino
  discovery of e-neutrino 569, 634
  discovery of μ-neutrino 571, 635
  mass of 420–1
  properties of 422
  scattering 17, 27
  term first used 318
  two-component theory 533, 534–5, 635
  zero mass 318, 419, 535
neutrino hypothesis 17, 18, 309, 313–20, 351n, 410, 413, 631
  chronology 316–20
neutrino theory of light 419
neutron
  β-decay 417–23
  decay life-times 422
  discovery of 17, 233, 310, 317, 397, 399, 405, 632
  Heisenberg's structural proposals 410, 413, 414
  mass of 412
  nature of 409–12
  search for 397, 398
  term first used 398 & n
neutron bombardment studies 400–1, 485
new physics, term explained 582
New York (1946) meeting 8, 447, 448, 450
New York (1983) meeting 1
New York Times 35, 38, 56, 317, 531, 607
New York University 385
New Yorker 276n
New Zealand Institute, *Transactions of the* 59
nitrogen
  induced nuclear transmutation of 216
  rotational band spectra 300–2
Nobel lectures 39, 56, 57, 63
Nobel Prize
  nominations
    Einstein 89
    Elster & Geitel 110
    Lorentz 89
    Rowland (self-proposal) 78
  recipients
    Becquerel 49
    Born 258
    the Braggs 148
    Chadwick 400

Nobel Prize (*cont.*)
  recipients (*cont.*)
    the Curies 56, 57
    effect of prize on 63
    the Joliot-Curies 400
    Lorentz 78
    Roentgen 39
    Rutherford 63
    (J. J.) Thomson 186
    Zeeman 78
noble gases, periodic table column 222
non-Abelian gauge theories 494, 583–7
noncommutation 287
nonconservation
  of energy 17, 105, 309–13, 317–20
    chronology 316–20
  of radioactivity 124
nonrelativistic quantum mechanics, electron model using 371
nonrelativistic time-dependent wave equation 356
Notre Dame, professor of physics 197
nuclear binding energy 231–4, 245
nuclear charge screening 192, 229
nuclear charge (Z) 223, 226
nuclear coupling constant 23, 482
nuclear electrons 87, 296, 303
  spin-statistics paradoxes 301–3, 411
nuclear energy, term first used 117
nuclear fission
  discovery 401, 438, 633
  term first used 158
nuclear force field, as game of 'catch' 430
nuclear forces
  early theory 413–17, 632
  Yukawa (meson) theory 430–2, 481–3, 632
nuclear isomerism, discovery 158
nuclear magnetic moments 299, 483
nuclear models
  electron–neutrino–proton 316
  first attempt 229–32
  lattice model 413, 428
  proton–electron model 230–2, 296, 298–303, 349, 372, 398, 410, 411, 629, 631
  proton–neutron model 401, 413
  Rutherford's version 13, 230–2, 237
nuclear physics
  beginnings 13, 223
  topics not discussed (in this book) 402
nuclear reactor, neutrino fluxes 18, 569–71
Nuclear Science, Annual Reviews of 63
nuclear size problems 298–9
nuclear spectroscopy, beginnings 11, 155–8, 629
nuclear spin
  Pauli's use of term 279n
  problems caused by, 299–301
see also proton spin
nuclear substructures 231
nuclear transmutations
  artificially induced 216, 240, 296
  first by accelerated particles 234, 405
nucleon conservation principle 488, 633
nucleon resonance spectroscopy 490–1, 634
nucleons 21
  magnetic moments of 483
  term first used 21, 450, 633
nucleus
  discovery by Rutherford 12, 13, 191–2
  term first used 192, 629

'old pros' 132–3
Oldstone-on-the-Hudson meeting (1949) 461, 634
Ω^- state 557, 635
ω-meson 490, 492, 502, 635
Origin of Species 177
Osaka Imperial University 429
Oxford University 229
Oxford University Press 297
oxygen, spectra types 168

p–m–v relationship 88
P-invariance 526, 527, 634
P-violation 25, 531, 532, 533–4, 598–9, 635
pair (atomic) models 182
pair formation, electron–positron 363
paradoxes 139, 298–320: *see also* Klein paradox
paraffin, proton ejection studies 399, 412
parallel-plate ionization chambers 53, 132
parastatistics 561, 562
Paris
  Academy of Science: *see* Académie des Sciences
  Conference (1900) 173, 202
  Conference (1932) 318
  Ecole Normale 62
  Museum of Natural History, physics professorship 44, 45
  School for Industrial Physics and Chemistry 53
parity
  not believed-in by Dirac 25–6
  term first used 525n, 631
  see also P-invariance; P-violation
parity conservation 525–6
parity nonconservation, K-decay 525
parity selection rule 526
particle mixtures 521, 522, 634
parton model 27, 579, 589, 636
partons 27, 579
Pasadena symposium (1931) 317
Paschen–Back effect 270, 274n, 629
Pauli exclusion principle 215, 250, 253, 267–74, 284
Pauli matrices 289–90, 291n, 631
Pauli–Weisskopf (spinless field) theory 376, 387–8, 528, 632
PCAC (partially conserved axial current) conditions 538, 565, 568, 578, 635
PEP (positron electron project) 609, 637
perihelion precession 214
periodic table 184, 186, 627
  cubic form 227
  interpretation of 221–3, 629
  and Mayer's magnets 184
perpetual machines 105
personalities, comments listed 4
perturbation theory 337, 374
  and charge independence 435
  first introduced 254
  see also second-order perturbation . . .
PETRA (positron electron tandem ring accelerator) 609, 637
Ph.D. thesis topics 118, 194, 251, 287, 367, 408
phase stability principle (of cyclotrons) 474, 633
phase transitions, experimental vs theory 137
phenomenological parameters 462
ϕ-resonance/particle 560
Philosophical Magazine 85, 190, 212, 365

phosphorescence
  Becquerel paper on 45–6
  definition 44n
  and uranic rays 46, 48, 108–9
  X-ray apparatus 47n
phosphoroscope, invention 44
photoelectric effect 86, 131
  Einstein relation 305
photographic detection techniques, β-spectra 154, 157, 159, 304, 305
photographic emulsion techniques 454, 513–14
photomeson production 485, 496
photon
  creation/annihilation of 324, 336–7
  formal introduction by Dirac 334–6
  inelastic scattering 301
  neutrino composition of 419
  prediction and discovery of 135–6
  scattering 329, 337–338; *see also* Compton effect
  self-energy of 385, 632
  term first used 246, 631
Physica 268
Physical Review 364, 365, 424, 628
  growth (1894–1980) 6
physics, growth of subject 5
Physics Today 621
physiological discoveries 93–101
π-meson, term first used 454
Pic-du-Midi 513
Pickwick Papers 195
piezo-electricity, laws discovered by Pierre Curie 56
pion composite model 495, 519, 555, 634
pion processes, second-order perturbation calculations 485
pion resonances 22, 487
pion–nucleon scattering experiments 22, 485
pions 20, 454, 634
  detection of 21–2
  discovery in accelerator experiments 479–81, 634
  discovery in cosmic rays 454, 479, 633
  mass 479–80
pitchblende 55
pitfalls of simplicity 130, 139, 140, 230, 309–10, 433
Planck length 623
Planck's constant 74
  and atomic structure 197–9, 629
  modified 287
Planck's radiation law: *see* blackbody radiation law
planetary (atomic) models 182–3
Platzwechsel [position exchange] 414, 416
'plum pudding' (atomic) model 166, 178–80, 185, 629
plutonium, medical casualties caused by 101
Pocono conference (1948) 458–9, 479, 512, 634
point model, for electron 372, 389
polarization 337
polonium
  α-irradiation from 398, 399
  discovery 55, 109, 628
  half-life 308
polyelectron models 178–80, 629
Pomeranchuk theorem 501, 635
positive (atomic) sphere concept 179, 185, 196
positron 15
  detection of 17, 319, 330, 351–2, 360, 403, 632
  prediction of 319, 349, 351
  term first used 352, 632
positron theory
  beginnings 377, 632
  light-by-light scattering calculations 386, 632
  and self-energies 384–5, 632
post-doctoral fellowships, USA 365
post-war years 18–31
potassium, radioactivity of 119, 629
Poynting vector 353
pre-electron models of atoms 174–7
precision, accelerator building 478
preprint distribution system (in Germany) 38n
preprint repositories, growth 6
Princeton
  European lecturers 365
  Institute for Advanced Study 163, 314, 429, 447, 518n, 555
privaatdocent/Privatdozent, term and conditions explained 75n, 172n
probability, introduction in quantum mechanics 250, 255, 631
promethium, discovery 229n
protactinium
  discovery 229n
  long-lived isotope, discovery 158
proton
  anomalous magnetic moment explanation 427, 483, 632
  disintegration 400
  ejection mechanism 399
  Heisenberg's structural proposals 414–16
  nature of 412
  spin 285, 301
  term first used 296
proton–antiproton collider 1–2, 28, 552, 575, 576, 637
proton–electron nuclear model 349, 372, 398, 410, 411
proton–electron (pe) model 230–2, 296, 297, 629
  problems with 298–303, 631
proton–neutron nuclear model 401, 413, 417
proton–proton collider 27, 552, 575, 636
Prussian Minister of Education 110
PS (proton synchrotron) 477, 575
pseudoscalar mesons 434, 436, 480, 481, 556, 633
pseudovector mesons 434, 633
ψ' resonance 605, 606, 636

q-numbers 287
QCD (quantum chromodynamics) 28, 551, 552, 588–91, 636
quantized spin-1/2 fields 417–18
quantum electrodynamics 15, 16
  advantages of 462, 463
  beginnings 326, 330, 332, 631
  Dirac's first paper on 288
  experimental physics, effect on 363–4
  leap forward in 455–66
  in 1930s 374–88
  publications on 466–7
  relativistic formulation 329, 341, 342–3, 346, 631
  on spin 299
quantum field theory
  beginnings 324, 329
  description 325
  first application 337
  infinities in 462–3
  'inner life' of 330, 331

quantum field theory (*cont.*)
  invariance principles incorporation, beginnings 341
  meson problems/uncertainties 492–505
  nuclear forces 426–7
  redux 580–611
quantum mechanics
  beginnings 14, 246–50, 630
  chronology 252–5
  early interpretational work 255–61, 631
  helium spectrum explained by 215
  interpretation 255–61
  introduction 251
  new theory 249
quantum numbers, notation 268, 273–4
quantum statistics 280–5
  beginnings 251
  connection with quantum fields 338–40
quantum theory
  in Bohr hydrogen atom model 199–200, 201
  as cause of crisis 217–18
  Einstein's work 123, 193
  first formulated 9, 133–4, 168, 193, 251
  new (post-1925) 14, 250–2
  not fitting into classical physics 130, 133, 134, 210
  old (pre-1925) 14, 246–50, 628
quarks 26, 28
  binding energy calculations 589
  current-algebra route 562–6
  discovery of 26, 551, 557–9
  masses of 567
  spin-1/2 604, 636
  types of 567
quaternions 290

R (cross-section ratio, electron–positron collider beam) 604, 605
R-matrix (of Dirac) 377
rad (unit), definition 95n
radiation damping, quantum theory of, 390
radiation dosage unit 94
radiation loss 181, 185, 186
radiation sickness 97
radiative corrections 455, 464–5
radio waves
  Marconi's work 69
  Rutherford's early work 59
radioactive constant (Rutherford) 121
  constancy of 125
radioactive decay
  exponential law of 125
  models 121–3
radioactive elements, general presence of 119–20
radioactive species, reason for 103
radioactive substances, listed (1904) by Soddy 117–18
radioactivity
  as atomic property 8, 55, 144, 628
  discovery of 45–7, 628
  energy source debated 103–4, 108–9, 110–11, 113–15, 327
  nature of 60, 628
  nonconservation of 124
  physical conditions, effects 110, 114 & n, 115
  popular books on 35
  Rutherford's (1912) views 193
  term first used 9, 54
radioactivity conservation law 54, 115
radiochemistry 8, 55, 218
  developments in 224n, 225
radiothorium, discovery 149–50
radium
  atomic weight (1903) estimate 118
  discovery of 56, 628
  elemental nature 117–18
  heat production experiments 113, 115
  helium liberated from 60
  maximum permissible body burden 100
  nomenclature 118
  physiological effects 99–100
  popular books on 35–6
  weight loss 232n
radium B 305
radium C 305
radium C″ 158
radium E 307–8, 318, 402
radium emanation 114, 118
radium therapy 100
rainbow effect 165
Raman spectrum 301n, 302
Ramsey Panel 476, 477n
rapidity of advances, Rutherford's comments [Preface] v, 129, 130, 217
rarefied gases, electrical discharge through 53, 59, 68–9, 78–9
Rayleigh–Einstein–Jeans law 135n
reactor experiments 569–70
recoil, radioactive transformation, discovery 158
reflection in space: *see* parity . . .
regeneration, K-meson 522–3
Regge analysis 504–5, 566
Regge poles 503–5, 553, 567n, 635
regularization 448
relativistic invariance 341–3
relativistic ion cyclotrons 473n
relativistic kinematics 87–9
relativistic wave equations 288, 290–1; *see also* Dirac equation
relativity, revolutionary nature of 250
relativity theories 14; *see also* general . . . ; special relativity
rem (unit), definition 95n
renormalizability 462, 484
  SU(2)×U(1) 597, 602
renormalizable theory, definition 462
renormalization 16, 22–3, 325, 388, 391, 633, 634
  fundamental types of 462
  introduction 385
  see also charge . . . ; mass . . . ; wave function renormalization
resistance law, invention 70
resonance, pion-nucleon scattering 22, 487
resonance acceleration technique 27, 407, 571–2
Reviews of Modern Physics 25–6, 365n
  Dirac's paper 25–6
  review of particle properties 517, 553
revolution, definition of 250
rhenium, discovery 229n
ρ-meson 490, 491, 502, 587, 635
ripeness of times, discovery of electron (1897) 131–2

Rochester Conferences 461
  1st (1950) 461 & n, 634
  2nd (1952) 461n, 485, 486, 496, 518n
  3rd (1952) 495
  4th (1954) 493, 516
  5th (1955) 516, 524
  6th (1956) 501, 524, 525, 532
  8th (1958) 531
  11th (1962) 557
Roentgen current 37
Roentgen rays 39
Roentgen Society, Rutherford's (1918) address to 217
roentgen (unit) 94
Rome, nuclear physics conference (1931) 299, 317
rotational band spectra 300, 301n
Rowland gratings 10, 75, 170, 628
  Zeeman's use of 75, 76
Royal Institution lectures 48, 85, 186, 225, 303
Royal Society (Edinburgh) 177
Royal Society (London)
  discussions 18, 193, 230, 296, 410
  Fellows 63, 147–8, 198
  Moseley's bequest 237
  papers submitted 150, 188
  Presidents 63, 148
  Royal Medal 41
  Rumford medal 63
rubidium
  discovery 169
  radioactivity of 119, 629
Ruhleben, Chadwick's internment 160, 304
Rühmkorff electromagnet 76
Rühmkorff induction coil 10, 37, 69, 627
  discoveries helped by 69, 167, 168
Rumford Medal 63
Rumpf model 271 & n, 276, 277
Russia: *see* USSR
Rutherford atomic model 13, 191
  luck in discovery 192
  reations to 192–3
Rutherford Memorial Lecture 163n
Rutherford scattering (of α-particles) 191, 629
  deviations 238–9, 296, 630
Rutherford scattering cross-section 191, 225
Rutherford–Soddy radioactivity conservation law 54, 115
Rutherford–Soddy transformation law 112, 113 & n, 123, 190, 286, 297, 629
Rydberg constant 173, 201, 214, 383
  Bohr's determination 247
Rydberg–Ritz constant 171, 173

s-coupling, meson theory 435
S-matrix theory 7, 498, 500–3, 633
St. Louis Exposition (1904) 35, 179, 405
St. Louis Post Dispatch, on radium 35
Saturnian (atomic) model 183, 429
Savannah River nuclear reactor 569
scalar field, meson theory 434, 448
scalar wave equation 288, 289, 345, 387, 431
scaling law 577–8, 636
Schroedinger wave equation 256, 288, 337n, 338, 345, 372
Schweidler fluctuations 123
Science 39, 351
Science News Letter 352
scintillation counters 478, 569
second law of thermodynamics 106
second-order perturbation calculations 329, 337, 373, 374–6, 485, 486
second-order subtraction 384
second quantization methods 329, 332, 333, 339, 340, 377, 379, 418, 630
selection rules 214–15, 518, 526, 630
self-conjugate neutrinos 423, 633
self-energy 384–5, 631
  of electron 370, 372–3, 384, 389, 448, 632
  of photon 385, 632
  of vacuum 384
semiclassical nucleon interaction 495–7
semiclassical theory (of radiation processes) 334 & n
separation devices, accelerated particles 478
Shelter Island Conference (1947) 22, 450–1, 452, 479, 483, 633
shielding, radiation 478
short-range interactions
  neutron–neutron 415–16
  neutron–proton 414–15
sigma-resonance 514, 515, 520, 553, 557
signature (early name for parity) 525n
Silliman Lectures
  establishment and lecturers 129, 365
  by Fermi 436
  by Rutherford 9, 64, 129, 134, 365
  by (J. J.) Thomson 129, 180, 184, 365
simplicity
  as necessary evil 138–40, 223
  an unnecessary evil 129–31
single ring colliders 574, 635
skiagraph 96
skin tuberculosis, treatment of 97, 100
SLAC (Stanford Linear Accelerator Center) 27, 552, 573, 635
  beginnings 573
  MIT electron scattering study 573, 576–7
  preprint library 6
  see also SLC, SPEAR
SLC (Stanford linear collider) 624, 625n
Smyth Report 116, 117
sodium (spectral) lines, solar spectrum 165, 167, 627
Solvay Conferences
  1st (1911) 192, 199n, 223, 310, 629
  2nd (1913) 123, 188, 193, 209, 224
  5th (1927) 262, 288, 340, 346, 347
  7th (1933) 318, 377, 378, 410, 417, 419, 436
  8th (1948) 483, 512n
  12th (1961) 23
Sorbonne 56, 57, 227
sources (of information) 49, 63–4, 202, 262–3, 320, 355–6, 391, 438, 466–7, 505, 542, 611
Soviet Academy of Sciences 303n
Soviet Union, *see* USSR
spark inductor 69
spatial reflection symmetry 25 [20c]
SPEAR (Stanford positron electron asymmetric ring) 605, 609, 636
special relativity theory 244, 246
  beginnings 14
  comments on 136
  and electron stability 370, 371
  increasing applicability 89
  kinematic rules 87
  radioactive energy release explained by 327

specilization, emergence (in 1950s) 493
specific gravity series 174, 627
specific heat anomalies 175, 193, 194, 280
spectral analysis
  beginnings 165, 627
  development 166–70
  elements discovered using 169
spectroscopist's 'bible' 166
spectroscopy
  experimental techniques 218
  reference books 166, 232
spin
  applied to helium spectrum 215
  discovery of 215, 250, 254, 267, 276–80, 630
  meson/pion 481
  proton–electron model affected by 298
spin–charge exchange 415, 416
spin–orbit coupling 278, 279, 290, 417
spin–statistics paradoxes, nuclear electrons 301–3, 411
spin–statistics theorem 15, 528, 632
spinless fields, quantum electrodynamics of 387–8
spinor, term first used 292
spontaneous emission, Dirac's treatment 9, 123, 334, 374
spontaneous symmetry breaking 592–8, 635
SPS (super proton synchrotron) 477, 576, 636
SSC (superconducting super collider) 625 & n
stable particles, definition 517
Standard (London Newspaper) 38
Stanford 27, 63: *see also* SLAC
static SU(6) symmetry 27, 559, 635
stellar energy, origin of 312, 313, 327
stellar spectroscopy 165, 169–70, 627
Stevens Institute, founder of 183
stochastic cooling technique, antiproton accumulator 576
Stokes (spectral) lines 302
Stoner's rule 273, 274, 630
storage rings 574; *see also* colliders
straggling 147
strange particle studies 523, 552–3, 635
strangeness (quantum number) 24, 520
strangeness scheme 519–21, 634
strong coupling theory 495, 591, 633
strong focusing (of synchrotrons) 27, 476, 634
strong interaction selection rules 518
strong interactions 23, 482, 517
  α-decay 142–3
  first hints 13, 237–40
  reason for term 23
  and weak interactions 536–8
strong-focusing synchrotrons 476–7, 635
structure-independent electron theory 449
Stuyvesant Estate 169n
SU(2) symmetry 425, 592 & n
SU(2)×U(1) symmetry 2, 28, 592–7, 600, 621, 622
SU(3) symmetry 26, 551, 553–7, 635
SU(3)-color symmetry 27, 551, 561–2, 588, 590, 591, 621, 622, 636
SU(3)-flavor symmetry 588, 635
SU(3)×SU(3) symmetry 26–7, 565
SU(4) symmetry 425, 551, 559, 633
SU(6) symmetry 27, 551, 559–61
subtraction procedures 379, 380, 383, 384, 448, 632
sudden-death (electron absorption) mechanism 148–9, 152
super-many-time formalism 459
superconductivity 137
supergravity 623, 637
superpartners 624
superposition principle 266, 523
superstring theory 624
superstrong interactions 562
supersymmetry 30, 623–4, 636
superweak mixing 541
Swedish Academy of Science, Royal 173
symmetrical meson theory 435
symmetrically coupled oscillators 265
symmetry, meson field theory modified by 485–90
symmetry groups
  chiral SU(3)×SU(3) 26–7, 565, 566
  color SU(3) 27, 551, 561–2, 588, 590, 591, 621, 622, 636
  flavor SU(3) 588, 635
  static SU(6) 27, 559, 635
  SU(2) 425, 592 & n
  SU(2)×U(1) 2, 28, 592–7, 600, 621, 622
  SU(3) 26, 551, 553–7, 635
  SU(4) 425, 551, 559, 633
  SU(6) 27, 551, 559–61
  U(1) 28, 584
synchrocyclotrons (SCs) 19, 474–5, 633
  principle of 474
  size limit for 475
synchrotrons 19, 473n, 407–9, 474–8
synopsis (of this book) 627–37

T-invariance 526–7
T-matrix calculations 500–1
T-violation 25, 540, 541
tau-theta puzzle 516, 524, 634
τ-decay 513
τ-lepton 608, 636
τ-mesons 512
technetium, discovery 229n
tellurium, atomic weight for 221
Tentativo (Fermi's paper on beta-decay) 418, 419–22
Tevatron 477
  principles of 478
theoretical physics
  faculty numbers (in 1900) 5
  Rutherford's view 12, 121, 190–3, 363
theory, and experiment 29, 454, 455
thermal disorder model, radioactive decay 123
thermodynamics
  applicability to living tissue 108n
  first law 105, 106
  second law 106
  third law 284n
thermoelectricity, beginnings 70
θ-decay 516, 523, 524
third law of thermodynamics 284n
Thomas factor 279, 290, 291, 347, 631
Thomson cross-section 187, 370
Thomson limit 187, 349, 497
Thomson scattering formula 370
thorium
  discovery 8, 54
  nomenclature 118
  radioactivity discovered 108, 628
thorium C″, discovery 158

thorium emanation
  discovery 9
  half-life determination 118, 120–1
thorium series, branching reactions 318
time reversal, invariance under 25, 526–7, 632
Times, *The* 118n, 209, 250, 438
Tisvilde (Bohr's country home) 210
Tokyo 145, 183, 429, 430, 432
top quark 608, 637
torr, definition 1n
towers (ground states of spectral series) 450
transformation groups, theory of 266
transformation (quantum mechanics) theory 288, 289
transformation (radioactivity) theory 112, 113 & n, 123, 190, 286, 297, 629
transition probability concept 257, 259, 336
transitions, excited-to-ground state 156
transmutation 112, 113n, 296; *see also* nuclear transmutations
transuranium elements 120
Trinity College (Cambridge) 85, 216, 304, 437
Tübingen 280, 315
Turkenstrasse (physics laboratory, Vienna) 150
TUT (totally unified theory) 624
two-component neutrino theory 533, 534–5, 635
two-meson hypothesis 453, 454, 633
two-valuedness [*Zweideutigkeit*] 272–3, 274

U(1) symmetry 28, 584
U-field 431, 432
UA1/UA2 (underground area) teams 1, 2, 610
Uetli, Schwur 208
uncertainty relations 249, 251, 255, 261–2, 631
  meaning of term 262
unified field theories 28, 552, 580, 584n
  SU(2)×U(1) 2, 28
  Weyl's work 268, 344, 345
unitarity, *S*-matrix theory 498, 501
unitary gauge 594n
unitary spin 556
universal energy principle 105, 106–7
universal Fermi interaction 530, 535, 634
universal length (Heisenberg's concept) 428
universal $V-A$ theory 535–6, 581
universality of weak interactions 21, 529–30
Universities Research Associates (URA) 477
unpublished material, debates about 280
unrenormalizable theory 462, 484
upsilon-resonance 608, 637
uranic rays
  discovery 45–7, 628
  exact nature 60, 628
  persistence 108
  see also Becquerel rays; radioactivity
uranium, nomenclature 118
uranium compounds, Becquerel's studies 44, 45
USA
  accelerator design (1948) 473
  military financing of research 19
  particle physicists 465
  see also American . . .
USSR, accelerators in 1948 473
Utrecht University 299, 301

$V-A$ theory 25, 535–6, 635
V-particles 21, 22, 512, 513, 516, 633
  decay mechanisms 24, 515, 517–8
  production mechanisms 24, 517, 634
  see also Λ-particles
vacuum expectation values 380–1
vacuum polarization 382–3, 632
  measurement of 447, 451
vacuum self-energy 384
vacuum symmetry 593
vacuums used 1, 2, 68, 168, 575
Valdespino (sherry) symbol 605
valence quarks 579
valencies, effect on atomic weight values 71–7, 221
Van de Graaff generator 406
van den Broek's rule 227–8
  reactions to 228
variety of discovery 136–8
VBA (very big accelerator) 625 & n
vector bosons 581; *see also* W-bosons
vector current, conservation of 537
vector potential, term first used 343
Vienna 150, 227, 236, 403
violation rules 489, 520
virtual annihilation/creation processes, neutron beta-decay 432
virtual electron–positron pair formation 350
virtual states 338
voltages, achievable in early twentieth century 405
von Neumann covariants, Dirac equation 292, 422
vortex (atomic) model 176–7, 627
  definitions 176n

W-bosons
  detection of 1, 2, 29, 552, 610–11, 637
  first clues to existence 18, 21, 25
  indicator equation for 420
  paper on discovery 1–2
  weak interactions mediated by 581, 635
war, effects of 96, 160, 234–7
Washington (DC), National Archives statue 625–6
wave function renormalization 464 & n
wave mechanics, introduction 251, 254, 255, 256, 288, 630
weak focusing 475
weak-focusing synchrotrons 19, 475–6, 552
weak interactions 21
  beta-decay 142, 143
  Cabibbo theory 563, 635
  definition of 21
  literature explosion for 533–8
  mediation by charged vector bosons 581
  and strong interactions 536–8
  universality of 21, 529–30, 580, 634
  $V-A$ theory 25, 535–6, 635
weak neutral current 29, 552, 600
weight loss, radioactive substances 232n
Weinberg angle 596
Weinberg–Salam theory 594, 598
Westminister Abbey 216, 437
Whiddington's law 227, 228
Wick mechanism 422, 427, 483, 526, 632
Wien law 135
Wiener Presse 38
Wigner forces 416
Wigner's rule 285, 302, 315n
Wolfenbüttel Gymnasium 109, 110

World War 1 (1914–1918) 160, 234–7, 471
World War 2 (1939–1945) 438, 471
Würzburg University 36

X^3-particle 231
X-ray crystallography, Bragg's work 148
X-ray diffraction/refraction experiments 42
X-ray equipment, wartime use of 96, 236
X-ray photographs
  as court evidence 96 & n
  first taken 38, 95
X-ray scattering studies 187, 226
X-ray spectra, Barkla's work 226
X-rays
  compared with gamma rays 62
  dangers of 97–9
  deaths from 96–7
  discovery 1, 37–8, 131, 628
  effect on existing concepts 130
  exposure limit 95
  as longitudinal propagations 41–2
  properties reported by Roentgen 40–1
  public reaction to discovery 38–9
  Roentgen's first paper 37–8, 40–1
  speculation on nature 41–2
  as ultra-ultraviolet light 42
Ξ-resonance 514 & n, 515, 520, 553, 557

Yale College
  first American Ph.D. in physics 364
  Rutherford offered professorship 9, 63
  Silliman lectures 129, 365
Yang–Mills theory 585, 586, 587, 634
  renormalizability of 587, 636
Yukawa forces 28, 29
Yukawa interactions 431, 551
Yukawa meson field theory 430–2, 481, 632, 633
  breakdown of 483
Yukawa potential 431

Z (nuclear charge) 223, 226, 629
Z-bosons
  detection of 2 & n, 29, 552, 610–11, 637
  indicator equation for 420
Zeeman effect 10, 76–7, 131, 180, 182, 268–269
  anomalous contrasted with normal 268, 269
  line-broadening effect 77
  line-splitting effect 77, 268–9
  Lorentz' interpretation 10, 77, 132, 268
  see also anomalous Zeeman effect
Zeitschrift für Physik 283n, 303
zero point infinities 360, 361
Zuiderzee reclamation scheme 216
Zürich, Institute of Technology 312, 317, 341, 368